COMPARATIVE VERTEBRATE NEUROANATOMY

Evolution and Adaptation

COMPARATIVE VERTEBRATE NEUROANATOMY

Evolution and Adaptation

ANN B. BUTLER

I. T. N. I.
Arlington, Virginia
and
Krasnow Institute for Advanced Study
George Mason University
Fairfax, Virginia

WILLIAM HODOS

Department of Psychology
University of Maryland
College Park, Maryland

WILEY-LISS

A JOHN WILEY & SONS, INC., PUBLICATION
New York • Chichester • Brisbane • Toronto • Singapore

Address All Inquiries to the Publisher
Wiley-Liss, Inc., 605 Third Avenue, New York, NY 10158-0012

© 1996 Wiley-Liss, Inc.

Printed in the United States of America

All rights reserved. This book is protected by copyright. No part of it, except brief excerpts for review, may be reproduced, stored in a retrieval system, or transmitted in any form or by any means, electronic, mechanical, photocopying, recording, or otherwise, without permission from the publisher.

Library of Congress Cataloging-in-Publication Data

Butler, Ann B.
    Comparative vertebrate neuroanatomy : evolution and adaptation/
Ann B. Butler, William Hodos.
        p.   cm.
    Includes bibliographical references and index.
    ISBN 0-471-88889-3 (alk. paper)
    1. Neuroanatomy.   2. Vertebrates—Anatomy.   3. Nervous system—
Evolution.   4. Anatomy, Comparative.   5. Nervous system—
Adaptation.   I. Hodos, William.   II. Title.
QM451.B895 1996
596'.048—dc20   95-49380
                                                                 CIP

The text of this book is printed on acid-free paper.
10 9 8 7 6 5 4 3 2 1

Dedication

This dedication is in three parts: to those special friends, mentors, and family members who are now deceased, to those special persons still living who have taught and guided us in our careers, and to our families.

In recognition of those special persons who are now deceased, Ann Butler dedicates her contribution to this work to the memory of Alexander and Ethel Benedict Gutman, Raymond C. Truex, and B. Raj Bhussry; William Hodos dedicates his contribution to the memory of his parents, Morris and Dorothy Hodos, and Walle J. H. Nauta.

In recognition of those special persons in our lives who have been teachers and mentors as well as friends and colleagues, we also dedicate this book to Warren F. Walker, Jr., Theodore J. Voneida, R. Glenn Northcutt, Ford F. Ebner, Sven O. E. Ebbesson, Boyd Campbell, and Harvey J. Karten.

Finally, we dedicate this book to our families, Thomas and Whitney Butler and Nira Hodos, who put up with much neglect during its writing but nevertheless gave us their loyal support, their patience, and their ceaseless encouragement.

Contents

Part Six

CONCLUSION

Preface

What is this book about?

This book is about the central nervous system of those animals that possess backbones (vertebrates) and how evolution has shaped and molded their bodies and their nervous systems, allowing them to thrive in their particular environments or to take advantage of new environmental opportunities. Thus, it is a book about the relationship between structure and function and about survival through effective design. It is a book about the past as well as the present and about the history of vertebrate nervous system evolution, to the extent that we can read that history from the present state of these animals.

Who is this book for?

This book has been written first and foremost for neuroscience students at the graduate or advanced undergraduate level. We have presumed that the reader will have taken one or more introductory undergraduate biology courses or otherwise be familiar with this material. In a more general sense, this book also is for anyone who is interested in the anatomy of the nervous system and how it is related to the way that an animal functions in its world, both internal and external.

This book is intended as an introductory work rather than as a handbook or reference work that scientists might refer to in their professional writing. We have modeled this book on several textbooks designed for advanced undergraduate to graduate levels: *Functional Anatomy of the Vertebrates: An Evolutionary Perspective* by W. F. Walker, Jr. and K. F. Liem (Saunders College Publishing, Fort Worth, TX, 1994); *Hyman's Comparative Vertebrate Anatomy* edited by M. H. Wake (The University of Chicago Press, 1979); *The Human Brain and Spinal Cord* by L. Heimer (Springer-Verlag, New York, 1983); *Core Text of Neuroanatomy* by M. B. Carpenter (Williams and Wilkins, Baltimore, 1991); *An Introduction to Molecular Neurobiology* by Z. W. Hall (Sinauer Associates, Sunderland, MA, 1992); and *Principles of Neuroscience* by E. R. Kandel, J. H. Schwartz, and T. M. Jessell (Appleton and Lange, Norwalk, CT, 1991). In keeping with the format of these texts, we have not cited references in the body of the text, but at the end of each chapter we have listed the references from which we drew material and additional papers that may interest the reader. Our aim has been to introduce the reader to the field and to synthesize information into a coherent overview, rather than to present an extensive catalog of individual data.

This is a book that we hope will be of interest to the general scientific reader and the nature enthusiast, as well as to advanced undergraduate or beginning graduate students in the neurosciences. Physicians and others with knowledge of the human central nervous system also should find much of interest here. To our colleagues who are specialists in the field of comparative neuroanatomy, we say that this is not the book that you might write; this is the book that students should read to give them the background to read your book and other scholarly publications in neuroanatomy and brain evolution.

What can be learned about the human brain from a book about the brains of many different vertebrates?

During the past 30 years, a great explosion of information about the anatomy of the brains of nonhuman and especially nonmammalian vertebrates has taken place. One of the lessons to emerge from this wealth of new data has been the reversal of the nineteenth century view that a dramatic change in brain evolution occurred with the evolution of mammals in general and humans in particular. In other words, once the powerful tools of modern neuroanatomy were applied to the brains of birds, fishes, reptiles, and amphibians, many of the same patterns of cell groups and their interconnections that were known to be present in mammals (including primates) were found to be present in nonmammalian vertebrates as well. Thus, comparative neuroanatomists came to recognize that the evolution of the vertebrate central nervous system had been far more conservative than earlier investigators had realized.

To be sure, great differences in specialization of the brain exist between animals that have become adapted to very different modes of existence. Indeed, those differences in form and function are what make the study of comparative neuroanatomy and brain evolution so fascinating—a fascination that we hope to share with you. In spite of these differences, however, all vertebrate central nervous systems share a common organizational scheme so that someone who is familiar with the brain of any vertebrate will also be on familiar ground when first encountering the brain of any other species. Someone who has read this book and retained the general principles of brain anatomy and organization that it presents will have little difficulty reading a medical school textbook of human neuroanatomy because much of it will be familiar both in overall conception and in many of the details.

What is new in this book and how does it differ from other texts?

This book incorporates several new approaches to the subject matter of comparative neuroanatomy and the orientation with which we study it. These new approaches include

- A recent reevaluation of the cranial nerves of vertebrates and their derivation and organization
- A new organizational approach to the various groups of vertebrates based on the degree of elaboration in their central nervous systems rather than the traditional, *scala naturae*-like ranking
- New insights into the organization and evolution of the dorsal thalamus and dorsal pallium in the forebrain
- A new and comprehensive overview of brain evolution in vertebrates that encompasses many of the evolutionary and developmental topics covered in the rest of the text

The first of these new approaches is based on the work of Northcutt, Baker, Noden, and others, on the organization of the cranial nerves. While constituting a radical departure from the established, traditional list of twelve cranial nerves with their functional components, we feel that this new approach is a marked improvement for two reasons: First, it takes into account additional cranial nerves, both long known and newly recognized ones, that are found in many vertebrates but are not included in the "traditional twelve." Second, it is based on embryological development and thus provides a coherent accounting of the segmentation of the head itself and of its component parts, including both the brain and other tissues (particularly the neural crest, epithelial placodes, and paraxial mesoderm). Chapter 9 on the embryology of the cranial nerves in relation to head segmentation covers this newly developed approach to cranial nerve organization and is crucial to understanding the subsequent chapters on the cranial nerves themselves.

The second new departure from tradition that we have taken is the order in which various groups of vertebrates are considered in the chapters on the various regions of the nervous system. This approach is based on the range of variation in brain structure within each of the major groups of vertebrates. It is intended to overcome the erroneous but culturally ingrained idea of a single, simple-to-complex, linear series of evolutionary stages leading from fish to frog to rat to cat to monkey to man, i.e., the myth of a *scala naturae*. Chapter 4 specifically addresses this issue.

A great diversity in brain organization has been achieved independently at least four separate times within four separate radiations of vertebrates; our approach is designed to highlight both the diversity itself and its multiple, independent development. Thus, in a number of the chapters on brain regions and systems, particularly in the midbrain and forebrain, we first consider and compare those species within each radiation in which the brain has relatively simple cellular organization, as for example, lampreys, dogfish sharks, gars, and frogs. We then consider and compare those species within each radiation in which the brain has relatively complex cellular organization, as for example, hagfishes, skates, teleost fishes, and amniotes (mammals, reptiles, and birds). We hope to convince the reader that the development of a more complex brain has been accomplished not just once for the "ascent of man," but multiple times. Moreover, we will show that mammals (including primates) do not always have the most sophisticated brain systems.

In line with this point, other chapters, including "Evolution and Variation," "Evolution and Adaptation of the Brain, Behavior, and Intelligence," and "Theories of Brain Evolution," seek to dispel further the myth of *scala naturae* and to deal with the actual range of variation in line with the known facts and processes involved. In this context, we hope that the reader will come to understand that while some vertebrates have simpler brains than others, all living vertebrates are equally successful in that they are alive and adapted to their environments.

A third new departure concerns the evolution of two major parts of the forebrain, the dorsal thalamus and the dorsal pallium, particularly in amniote vertebrates. Two fundamentally different divisions of the dorsal thalamus recently have been recognized in all jawed vertebrates: one that predominantly receives direct, lemniscal sensory and related inputs, called the lemnothalamus, and one that predominantly receives sensory inputs relayed to it via the roof of the midbrain, called the collothalamus. Correspondingly, two divisions of the dorsal pallium in amniote vertebrates that receive their respective inputs predominantly from the lemnothalamic and collothalamic divisions of the dorsal thalamus have been recognized as well. The way in which a number of the chapters on the forebrain have been organized and the material presented in them is based to some extent on these new concepts of forebrain evolution.

Finally, new insights into the evolution of the brain, not just in vertebrates but among some of the invertebrate chordates as well, are presented in the last chapter of this book. Recent findings on genetic patterning of central nervous system structure and on the anatomy of the brain and head region in the cephalochordate *Branchiostoma* allow for some of the features of the brain in the earliest vertebrates to be identified. Comparisons of landmarks in the brain and surrounding tissues in *Branchiostoma* and vertebrates suggest a greater degree of correspondence of both mesodermal segments and segments of the brain than ever before appreciated. A new survey of brain evolution is presented, beginning with a few but significant features that can be identified in the common ancestors of cephalochordates and vertebrates, identifying additional features that were present in the earliest vertebrates, including two novel tissues of the head (neural crest and placodes) and a number of cranial nerves associated with them, and then tracing the separate evolutionary histories of the brain in the major radiations of extant vertebrates.

Acknowledgments

A number of our colleagues have read portions of the book for us or have discussed a variety of the topics with us. They also sustained us with their enthusiasm for this project. We wish to express our gratitude to them for their advice and suggestions, and for intercepting various errors. We accept full responsibility for any errors that remain despite our best efforts. We wish to thank Catherine Carr, William Cruce, William Dingwall, Joseph Fetcho, Michael Fine, Katherine Fite, Jon Kaas, Harvey Karten, Wayne Kuenzel, Thurston Lacalli, Michael Lannoo, Rodolfo Llinás, Gloria Meredith, Donald Newman, R. Glenn Northcutt, Mary Ann Ottinger, Michael Pritz, Luis Puelles, Anton Reiner, Charles Sternheim, David Yager, and several anonymous reviewers.

We also owe a large debt of gratitude to Philip Zeigler, who served as editor of the entire volume and provided us with detailed and thoughtful commentary on all of the chapters. The book has been greatly improved as a result of his efforts.

A number of publishers granted us gratis permission for the use of material adapted from their publications. These include Academic Press Ltd., Annual Reviews Inc., Elsevier, W. H. Freeman and Company, Plenum Publishing Corporation, S. Karger AG, Springer-Verlag, The Johns Hopkins University Press, and The University of Chicago Press. We would like to thank them for their generosity and the support of scholarly endeavors that it demonstrates.

We offer our special thanks to several additional people, who are both friends and colleagues, and who had important influences on various aspects of the writing of this book. The first is Trev Leger, formerly of John Wiley & Sons, who played a major role in the inception of this project many years ago. Next, our friend and colleague, Boyd Campbell, also contributed to the inception of the book, advised us on many occasions, and offered numerous valuable suggestions about the overall conception and scope of the work. Arthur Popper, another friend and colleague, was instrumental in forming the partnership between Ann B. Butler and William Hodos for the task of writing the book. His seemingly modest proposal had major consequences for us. We also would like to thank Wally Welker, who provided us with encouragement for the project and supplied us with a number of excellent photomicrographs that we have included in the book. These photomicrographs are as valuable for their scientific content as they are asthetically pleasing to the eye. Finally, each of us also wishes to thank the other—for much intellectual stimulation, for mutual support, and, most important, for managing to remain friends!

Part One

EVOLUTION AND THE ORGANIZATION OF THE CENTRAL NERVOUS SYSTEM

1

Evolution and Variation

INTRODUCTION

One of the primary fascinations of the natural world is the vast diversity of living organisms within it. Diversity of organisms and their many body parts is a hallmark of biology. Biological diversity has been produced by the process of **natural selection,** part of which is a reflection of changing climates, geophysical phenomena, and habitats. The pressures of natural selection act on spontaneous variations of the phenotype that occur within a population and are the result of mutational changes within the genes.

Our current understanding of biological diversity began with the theory put forth by Charles Darwin (and independently by Alfred Russel Wallace) in 1858-1859. This theory states that a process of evolution by natural selection has produced the variation that is documented by the fossil record and among extant species. The source of the variation upon which natural selection acts was not identified until the science of genetics was established by the pioneering work of Gregor Mendel, who in 1865 formulated the principle of particulate inheritance by the means of transmitted units or genes, and the later work of Theodosius Dobzhansky and others in the first half of this century.

Modern evolutionary theory embodies the work of Darwin in the context of genetics, molecular biology, and other relevant sciences. This interrelationship of disciplines was termed an **evolutionary synthesis** by Julian Huxley in 1942; it refers to the recognition that both gradual evolutionary changes and larger evolutionary processes, such as speciation, are explainable in terms of genetic mechanisms. Since that time, a substantial body of work in genetics, systematic biology, developmental biology, paleontology, and related fields has yielded new and more complex insights on this subject. The title of a recent book by Niles Eldredge (1985), *Unfinished Synthesis,* reflects the continuing debate within the field.

Until Darwin's publication of his views, most Western scientists believed that animals living at that time had been unchanged since their creation by the Deity. Darwin's idea that living creatures had evolved over long periods of time was accepted fairly readily. Some resisted the idea of humans being animals in continuity with nature, that is, that humans are related to any other animals, such as apes and monkeys. That evolution is a directionless process employing natural selection is a concept that has been accepted by some but resisted with vigor by others, even today.

The idea that evolution is progressive in the sense that progress or continuous improvement occurs over time is seductive and comforting. It is seductive in its appeal to the egocentricity of our species and comforting as a source of moral principles. The Aristotelian concept of a ***scala naturae*** that living animals can be arranged in a continuous, hierarchical, ladder-like progression with humans at the pinnacle embodies this appeal, as does the Judeo–Christian tradition of creation that culminates with the human species. Julian Huxley promoted the idea that although evolution was without purpose, it was progressive. He believed that ethical principles and the meaning of human existence could be derived from the position of humans at its pinnacle.

The *scala naturae* concept of progressively ascending scales of life forms, such as the fish–frog–reptile–rat–cat–monkey–human sequence, is seen as intuitively correct. The pervasive flaw in all such rankings is that they are made from an anthropocentric point of view. The anthropocentric scale, however, is of no greater scientific value in an evolutionary context than one based, for example, on our assessment of the animals' beauty. The appeal of the idea of progress over evolution is based on the fact that progress itself, like beauty, is a human concept and value; it is not, however, a biological principle.

Inherent in the notion of a scale of nature is the idea that each animal has a natural rank on this scale. The more "advanced" animals (humans and the ones seemingly closest to humans) occupy ranks high on the scale, and those that seem to bear less resemblance to humans are relegated to the lower ranks. Thus, we have come to refer to some animals as "the higher vertebrates" and others as "the lower vertebrates," or "the submammalian vertebrates." Unfortunately, terms like "higher," "lower," and "submammalian" represent only homocentric value judgments; they are thus inappropriate ways of comparing animals and have no place in the vocabulary of evolutionary biology. Many extant species of vertebrates are only distantly related to humans and resemble them very little; nevertheless, these animals are just as successful and well adapted to their environments as humans and their closest relatives are to their own environments. The simple fact that animals are different does not confer any rank to them relative to each other.

EVOLUTIONARY MECHANISMS

Evolution can be simply defined as a *change over time.* In biological systems, random genetic variation occurring within a population allows for phenotype variation, which natural selection can then act upon. Darwin recognized that evolution occurs as a consequence of two separate processes. The first process, which we know today to encompass mutations and genetic recombinations as well as other factors, is a random process that produces variability. The second process, natural selection, is not random but rather opportunistic, and it acts on this variability. Natural selection acts on populations, affecting the frequency of particular genes within the population, rather than on individuals.

Mutations and changes in the frequency of certain gene alleles, that is, alternate forms of the gene (dominant vs. recessive), account for diversity within an interbreeding population. The individual members of the population are similar but not identical. Since a gene may exist in a large variety of allelic forms but an individual animal has only one pair of alleles for each gene, any given individual possesses only a small fraction of the total genetic variation that is stored in the population as a whole.

The relative frequencies of alternative alleles and genotypes reach an equilibrium and then tend to remain constant in a large, randomly mating population. In spite of this tendency, changes in the frequencies of different alleles do occur over succeeding generations. In addition to mutations, factors that affect the frequency of alleles in a species include **genetic recombination, gene flow,** and **isolating mechanisms.**

Gene recombination assembles an existing array of allelic forms of different genes into a variety of combinations. This does not increase the frequencies of these alleles but does increase variability. While mutation is the ultimate source of genetic variation, recombination generates numerous genotypic differences among individuals in a population. Consequently, recombination provides a large number of the variations acted upon by natural selection.

Gene flow is a change in the frequency of particular alleles caused by individuals of the same species migrating into and interbreeding within a given population. Gene flow is essential to maintaining various populations as members of a single species, since the most important aspect of the definition of a species is that it consists of a set of populations that actually or potentially interbreed in nature. Gene flow is responsible for genetic cohesion among the various populations that form the species. This process is a stabilizing influence on genetic variation and is responsible for the relatively slow rate of evolution that occurs in common, widespread species.

Biological mechanisms that isolate one population from another reproductively are in direct contrast to gene flow and define the limit of the species. Geographic isolation, such as islands separated from each other by the ocean or a peninsula being isolated as an island due to a rise in the level of the ocean, can result in changes in gene frequencies between the two populations. A small number of individuals that becomes isolated from the rest of the population will not necessarily have the same alleles in the same frequency distribution as the whole original population, resulting in a shift in allelic frequencies in the isolated population. Examples of such isolated populations are various species of birds on various islands in the Galapagos, Hawaiian, and other similar island groups. If the geographic isolation eventually ceases, reproductive isolation may nevertheless be maintained by newly established mechanical incompatibilities of the male and female or by behavioral isolation caused by differences in mating ritual or species recognition cues that exclude some formerly potential mates.

Natural selection acts on the variability and establishes certain variant types in new frequencies within a given population. Natural selection is the increase in frequency of particular alleles as a result of those alleles enhancing the population's ability to survive and produce offspring. The **fitness** of a variant is a measure of how strongly the variant will be selected for, that is, how adaptive it is. Thus, a novel variant that enables a population to capitalize on a vacant niche may rapidly establish itself. Alternatively, selective pressures that are too strong, such as a relatively sudden decrease in ambient temperatures during an ice age, may result in extinction of populations. In such a case, the variability within the population is simply not extensive enough to fortuitously have the number of variants that would allow for selection of adaptations to the cold.

If a mutant allele appears infrequently in a large population, the initial frequency of the allele will be low and will tend to remain low. If a mutant allele appears in a single individual and has no selective advantage or disadvantage, it can by chance alone readily become extinct. On the other hand, if it even slightly enhances the ability of its bearer to live and reproduce, it will increase in frequency. Selection is the most important means by which allele frequencies are changed. Natural selection can be considered to consist of the differential, nonrandom reproduction of particular alleles. As alleles are parts of whole genotypes, selection can also be thought of as the differential and nonrandom reproduction of particular genotypes.

Darwin considered natural selection to mean differential mortality. Contemporary evolutionists look upon it as differential reproduction. Natural selection does frequently take the form of differential mortality, but other strategies occur as well, such as increasing the number of offspring produced or improving the chance of successful mating by increasing efficiency in getting food or evading predators. Differential mortality can be regarded as one form of differential reproduction, if, for example, animals do not survive long enough to reproduce.

Gene alleles most often have multiple effects, that is, they are **pleiotropic.** Some of the characteristics determined by an allele may be advantageous to the individual, while others may be disadvantageous. An individual carrying an allele (A) might have a selective advantage over another individual carrying its matching allele (A'), but might be inferior to the phenotype of the second individual in some other character produced by allele A. If individuals carrying the allele A have a net superiority over individuals carrying the allele A', then allele A will increase in frequency despite its deleterious side effects.

The discussion thus far has been primarily concerned with selection of single alleles, but the same principles can be extended to encompass combinations of two or more genes. Many adaptations are based not on single genes, but on multiple genes or gene combinations and the resultant phenotype on which selection acts. Populations of organisms exist in a particular environment to which they must be fitted or adapted in order to live and reproduce successfully. If the environment remains stable and the population is highly adapted, selection operates primarily to eliminate peripheral variants and off-types that arise by mutation or recombination. If a change occurs in the environment, one or more of the peripheral variants may be better adapted to the new conditions than those with the more normal genotype. Selection now takes a different form, favoring the formerly peripheral variants and eliminating some of the standard genotypes.

Since natural selection acts on a population rather than any individual, traits such as "altruism" can be selected for. For example, in many species, an individual animal may do work or even sacrifice itself in the service of offspring that are related but not its own. By so doing, the animal is protecting the genes that it has in common with those offspring. Thus, rather than being truly altruistic, such acts are actually self-serving in that they are a mechanism to protect at least some genes that are the same as the animal's. The genetic basis for the potential to act in support and defense of related offspring is thus likely to be retained in the population.

MICROEVOLUTION: EVOLUTION AT THE SPECIES LEVEL

Darwin believed that the course of evolution resulted primarily from natural selection acting on variations within populations. In his view, this process produced gradual changes that could, over long periods, account for all the organic diversity that we observe today. The modern synthesis has incorporated the new knowledge of mechanisms provided by genetics, including that from the recent advances in molecular biology. Nevertheless, emphasis is still placed on the role of natural selection acting at the level of the population, as advanced by Darwin.

In 1972, Niles Eldredge and Stephen Jay Gould, after examining evidence from the fossil record of marine invertebrates, pointed out that in this group at least, few examples exist of species that undergo significant change gradually through time. In most cases, a particular morphology is retained for millions

of years and then changes abruptly over a short period of time. Eldredge and Gould used the term **punctuated equilibrium** to describe this pattern. Punctuated equilibrium is similar to the concept of saltatory evolution (the sudden origin of new taxa by abrupt evolutionary change, also referred to as macrogenesis) held by some of the nineteenth century biologists.

Evolutionary change is generally believed to have two components: **phyletic evolution** (also called **anagenesis**) and **speciation** (also called **cladogenesis**). Phyletic evolution is the process by which a single lineage, without branching into divergent lineages, undergoes change over time. The ancestral and descendant portions of the lineage can become sufficiently different that they are recognized as different species. Darwin thought that this process could also explain the origin of the so-called higher taxonomic categories of families, orders, and classes. The term speciation is used to describe a process in which a single species gives rise to a branch that becomes established as a new, sister species or splits into two new lineages that both become new species, that is, are reproductively isolated from each other.

Phyletic evolution can account for morphological change, but speciation accounts for most of the diversity of organisms. According to the proponents of the pattern of punctuated equilibrium, speciation is an important factor in producing morphologic change. Examples that indicate the occurrence of both gradual morphologic change and punctuated equilibrium have been found.

MACROEVOLUTION: EVOLUTION ABOVE THE SPECIES LEVEL

The evolution of the carnivore families of cats and dogs from their common ancestral stock can be understood conceptually without much difficulty. To envision the evolution of quite distinct groups with very different modes of life, for example, whales, primates, carnivores, and bats, from a common ancestral stock is more challenging. A number of workers have proposed that some special factors are involved in macroevolutionary events that are not found in species formation, but the identification of any such factors has proved to be elusive.

If one examines animals of different genera, and especially of different families and orders, the morphological differences between them may appear to be very great. After all, morphological differences are the major basis for creating these various higher categories. Were the ancestors of these groups also very different from each other and from their common ancestral stock near the point of their divergence? In the few cases of relatively large scale radiations that are well documented in the fossil record, the ancestors of what became distinct higher categories appear to be no more different from one another than are extant species or genera. The processes that lead to the formation of new species thus appear to be no different whether the daughter species go on to extinction or give rise to a new order.

One might think that some special factors might appear when examining transitions between major adaptive zones, for example, the transition from fish to amphibian or from amphibian to land vertebrate, but even here no special factors have been found. Only those parts of the skeleton that are specifically associated with the shift of habitats are found to be altered, and the transition may involve functional changes in body parts that remain structurally similar. Many of the characteristics that we now consider to be typical of members of the derived group evolved after the adaptive shift was completed. Evidence also exists that in some instances, major shifts in habitat can be initiated primarily by behavioral modifications without changes in anatomy or physiology.

DIVERSITY OVER TIME

Biological diversity is a result of natural selection acting on random variations within populations of organisms. The degree of biological diversity has increased over time in some ways. For example, twice as many species of marine animals (invertebrate and vertebrate) exist today as existed in the Paleozoic era. Biological diversity has dramatically decreased over time, however, in other ways. In the Paleozoic era, the number of groups of higher taxonomic rank was far greater than the number that exist today. Diversity in the range of basic body plans has decreased, while the diversity of species having any of the few, extant, basic body plans has increased. The greater number of extant species are grouped within the fewer number of higher categories.

One explanation for this more complex pattern of evolutionary change focuses on processes that tend to eliminate extremes in variation, such as competition under conditions of natural selection. Animals with the more successful body plans would ultimately survive, and the number of higher categories thus decrease over time. A more satisfactory explanation involves the random process of **extinction.** To consider this possibility, we need to examine evolutionary history in terms of the extreme, physical forces that shape it.

Extinctions of varying degree repeatedly occur and profoundly affect biological evolution. Some extinctions are of modest degree and limited extent, happening to isolated populations due to normal environmental fluctuations or accidental factors. Species most resistant to environmental fluctuations tend to be those with individuals that have larger bodies, longer lives, and a greater degree of social interaction related to breeding behaviors. Other extinctions are of greater consequence and related to **habitat fragmentation** caused by such factors as tectonic shifts, temperature changes, alteration in rainfall patterns, and changes in oceanic level. When habitat fragmentation occurs, species more resistant to extinction are those that are herbivorous versus carnivorous and, among carnivores, of smaller body size. Those species with more strictly defined habitat requirements—**habitat specialists**—are more prone to extinction than species that are **habitat generalists.** Species with populations of smaller size or lesser density are likewise more prone to extinction than those of greater size and/or density.

Of the greatest consequence are mass extinctions of hundreds or thousands of species, such as those that occurred at the end of the Permian and the Cretaceous periods. Not only are mass extinctions dramatic and of momentous impact on biological flora and fauna, but they appear to occur with a regular periodicity. Mass extinctions have recurred on a cycle of about 26 million years for at least the last 225 million years.

The Cretaceous extinction occurred 65 million years ago. To account for a cycle on such a long time scale, extraterrestrial causes, such as asteroid or comet impacts, have been considered, and evidence from the distribution of iridium, a relatively rare element of extraterrestrial (meteoritic) origin, in the earth's surface supports this possibility. A recurring disturbance of the Oort cloud—the cloud of comets that circle the sun—could release comets that could then impact the earth with catastrophic results.

In mass extinctions, some species and groups of species have better chances of survival that others. Categories of taxa that have a greater number of species (**species-rich clades**) tend to be composed of habitat specialist species and are thus more susceptible to extinction than the habitat generalist species. The latter tend to be in taxic categories with a fewer number of species (**species-poor clades**) and have wider geographic ranges.

In normal times, species-rich clades undergo a net increase in species number, off-setting losses due to limited environmental fluctuations, accidents, and habitat fragmentation. This speciation has a pattern of punctuated equilibrium. Species-poor clades, on the other hand, have a lower rate of speciation but are more resistant to environmental and habitat assaults. This balance permits both types of clades to flourish in the intervals between mass extinctions. In mass extinctions, the species-rich clades are more vulnerable, and thus over time, fewer higher categories survive. More species arise in at least some of the remaining higher categories due to new waves of speciation following each period of mass extinction. Biological evolution is thus the net result of multiple independent processes.

EVOLUTION OF THE VERTEBRATE CENTRAL NERVOUS SYSTEM

Extant vertebrates currently comprise diverse groups, each with diverse and numerous species. Nevertheless, in the subsequent chapters of this book, we will encounter many features of the central nervous system of vertebrates that are remarkably constant from one group to another. We will also encounter many that vary considerably. The features that are constant as well as those that are diverse have resulted from the pressures of natural selection, themselves derived from climatic and geophysical factors, acting on the randomly derived variation in the frequency of gene alleles in interbreeding populations.

The underlying processes that produce diversity in the central nervous system of vertebrates are relatively simple and few. Some of these changes occurred more than once independently in different lineages of vertebrates. Simple changes in the genome that result, for example, in the production of a new type of cells, in a lengthening of the time of proliferation of an area of the germinal epithelium, in migration of a set of neurons away from the germinal area, or in the radial alignment of the glial cells have had profound effects on the structure and complexity of the brain in the species and lineages in which they have occurred and been selected for. In a sense, this observation is the nub of the answer to the riddle of brain evolution: complexity is most often derived from simplicity, that is, great diversity and great complexity have arisen because they are merely the result of a few, simple random mutational events, the phenotypes of which were highly favored by natural selection. Comparative neuroscience addresses the questions of how brains can change as new species evolve and how much a given part of the brain can change over evolution. In order to assess the variation, we need first to be able to recognize the same structures in different brains and to then compare their similarities and differences. Defining what is meant by the word "same" in this context has been an important keystone of comparative neuroanatomical analysis.

HOMOLOGY

The concept of "same" is expressed in biology by the term homology. This term was first introduced by the influential British anatomist Richard Owen in 1843. Owen defined homologue (i.e., homologous structures) as "the same organ in different animals under every variety of form and function." Owen's definition preceded Darwin's theory of evolution, and modern concepts of homology have been affected by the revolutions in evolutionary biology and genetics that subsequently occurred.

Van Valen has defined homology as "correspondence caused by continuity of information," a definition that has been as criticized for its vagueness as it has been praised for its flexibility and utility. Another definition, proposed by Simpson, states that "homology is resemblance due to inheritance from common ancestry." These definitions both refer to similarity—"resemblance" or "correspondence"—but some structures that are present in related groups of animals and that have been inherited from a common ancestor may lack any vestige of resemblance. For example, the middle ear bones of mammals (the malleus and the incus) are very unlike the articular and quadrate jaw bones of other tetrapods and from which they are ancestrally derived. Only the data provided by the fossil record allows us to recognize the common derivation of these structures.

Other definitions stress phyletic continuity. E.O. Wiley proposed that "A character of two or more taxa is homologous if this character is found in the common ancestor of these taxa, or, two characters (or a linear sequence of characters) are homologous if one is directly (or sequentially) derived from the other(s)." A similar definition was proposed by Ghiselin: "Structures and other entities are homologous when it is true that they could, in principle, be traced back through a geneological series to a (stipulated) common ancestral precursor." The required stipulation is the basis of the homology. Without the stipulation, that is, specification, of the homology, any statement of homology is incomplete. Consider, for example, the following two statements, both of which are true:

- The wing of a bird is homologous to the wing of a bat.
- The wing of a bird is not homologous to the wing of a bat.

These two statements, while both true, are both incomplete, and hence are seemingly in conflict. The wing of a bird is homologous to the wing of a bat *as a derivative of the forelimb.* The common ancestors of birds and bats possessed forelimbs of a similar basic construction, from which the wings are derived in both cases. The wing of a bird, however, is not homologous to the wing of a bat *as a wing,* since the forelimbs of the common amniote ancestors of birds and bats did not have the form of wings. Saying that A is homologous to B is as incomplete a statement as saying that "Harriet is more than Jane." More what? More intelligent? More athletic? More sophisticated? In statements of homology, the specific characteristic being compared must be included in the statement for the statement to be meaningful.

Several different types of homologous relationships can be recognized. The most common type might be called a discrete homology. The wing of a bird being homologous to the wing of a bat as a forelimb derivative is an example of a discrete homology, in that a discrete structure in each of two or more taxa is being compared. Additional types of homology, as specified by Smith, are also applicable to neuroanatomical studies. A serial or iterative homology involves structures that are derived from the same ontogenetic division of two or more segments, such as the wing of a bird and the leg of a bird, as serially derived tetrapod limbs. A field homology involves structures that are derived from the same ontogenetic source in the same or different segments. An example of a field homology is the five digits of a human hand being homologous as a set of derivatives of a common embryonic field to the set of the lesser number of distal forelimb divisions in a variety of other mammals.

Since the features of any given structure may be altered in different lineages, fossil evidence has played a significant role in identifying homologous parts of the musculoskeletal system among vertebrates. The brain does not fossilize, however, so other criteria for proposing hypotheses of homology are needed for neuroanatomical work. These criteria are based on the degree of similarity and were proposed by Simpson in 1961. They include the minuteness of the resemblance and the multiplicity of the similarities. In neuroanatomical studies, the features that can be compared for a given group of neurons in two different extant taxa include:

- Topological similarity.
- Topographical similarity.
- Similarity of axonal connections.
- Similarity in their relationships to some consistent feature of the two species.
- Similarity of embryological derivation.
- Similarity in the morphological features of individual neurons that form the group.
- Similarity in the neurochemical attributes of the neurons.
- Similarity in the physiological properties of the neurons.
- Similarity in the behavioral outcomes of neuronal activity.

Not all of these criteria can be met in every case. Nevertheless, the more of these criteria that can be satisfied, the stronger the support for an hypothesis of homology, that is, phyletic continuity. In those cases in which structures that are homologous also meet most or all of the above criteria for similarity, the term homogeny, or its adjective homogenous, can be

Reminiscent of multi-causality / multifinal ability (handwritten margin note)

applied, although these terms are rarely encountered in the literature.

HOMOPLASY

Modern researchers recognize that structural similarity can occur in divergent lineages as a result of similar adaptive responses to similar environmental pressures rather than as an inheritance from ancestors. The term used to refer to structural similarity without phyletic continuity is **homoplasy,** and its adjective is **homoplastic.** The wing of a bird and the wing of a bat are not homologous as wings; they are homoplastic as wings. Note again that the relationship must be specified to make sense.

Two different types of mechanisms result in homoplasy: **convergence** and **parallelism.** These terms both refer to the same process of similar adaptive pressures that result in similar morphological (or behavioral) responses. Convergence refers to this process in remotely related animals and parallelism to it in closely related animals. One implication of parallelism is that the genetic material on which a given set of selective pressures acts is more likely to be the same in a case of parallelism than in a case of convergence. Homoplasy that results from parallelism is a serious problem because it can easily be mistaken for homology. In both homology and parallelism, similar structures are present in closely related animals with similar survival problems that have adapted in similar ways.

When such instances of similar structures being present in closely related species occur, distinguishing between parallelism and homology can sometimes be difficult. In these circumstances, assuming homology is regarded as the preferrable tactic. The German scientist Hennig codified this method in his **auxiliary principle:** "Never assume convergent or parallel evolution; always assume homology in the absence of contrary evidence." As we will discuss below, this method is based on the idea that it is simpler for a common ancestor to acquire a given structure once than for each of two descendent groups to acquire it independently.

ANALOGY

Structures with quite different morphology, phyletic origin, and embryological origin can have quite similar functions. The term "**analogy**" is used to describe structures with similar functions—irrespective of phyletic continuity. In other words, structures can be analogous if they serve the same function, whether they are homologous or homoplastic. The wing of a bird and the wing of a bat are homologous as forelimb derivatives, homoplastic as wings, and also analogous as wings. An elephant's trunk and a racoon's hand have nothing in common phyletically or embryologically, yet they are analogous as organs for manipulating objects in the external environment. As in the case of both homology and homoplasy, the analogy must be specified to make sense. Tables 1–1 and 1–2 offer some opportunities to see whether you understand the differences between homology, homoplasy, and analogy.

ANALYSIS OF VARIATION

Recognizing similar structures present in different taxa is the first part of the process of reconstructing evolutionary history, which, in the absence of a corroborating fossil record—as is the case with most features of the central nervous system—is essentially a guessing game. Although we may come up with sophisticated theories that seem to account for the data, there is no guarantee that these theories are correct. This is a problem that exists in many areas of science, such as astronomy, geology, psychology, and economics, to name a few. An approach to evolutionary reconstruction of the central nervous system that has been used with considerable success is a methodology called **cladistics.**

Cladistic Analysis

Cladistics is a formal method of analysis for classifying animals according to their inferred phyletic relations, based on sets of shared similar traits. This approach also can be applied to analyzing the variable occurrence of a given trait among different taxa. In the latter case, a highly corroborated hypothesis of phylogenetic relationships of the taxa under consideration is needed. This phylogenetic hypothesis should be one that is derived from sets of traits not related to the trait being analyzed and also is based on a large number of such traits. It is usually structured in the form of a **cladogram** or **dendrogram,** that is, a tree-like diagram of the species representing their genealogy, produced using cladistic methods of inference. For example, hypotheses of phylogeny, such as those presented in Chapter 4 and based predominantly on fossorial and osteological data, are appropriate to use in the analysis of the distribution of central nervous system traits. Traits that are **plesiomorphic,** that is, are similar to those present in a particular ancestral stock, need to be distinguished from traits that are **apomorphic,** that is, derived specializations within a particular taxon.

The observed distribution of the trait to be analyzed is plotted on the terminal branches of the cladogram, an example of which is shown in Figure 1–1. The pattern of distribution of the trait in various ancestral groups that can account for the observed distribution in the living (terminal branch) species via the *fewest* number of phylogenetic transformations is then inferred. This process thus generates an hypothesis about the evolutionary history of the trait based on its distribution in extant taxa. In our example (Fig. 1–1), we are considering four species: A, B, K, and Q. All are descended from a common ancestral stock, but A and B are more closely related to each other than to K or Q. Species A has a particular trait (X) but the related species B does not. Did the common ancestor of A and B have the trait, with the line leading to B subsequently losing it, or did the common ancestor lack the trait and the line leading to A alone gain it?

Cladistic analysis uses **out-group comparisons,** that is, comparisons of the trait in sister species or sister groups, which are taxa more closely related to the taxon being studied than to any other taxon. In our example, we thus examine species K and Q, the out-groups to species B and A, for the presence or absence of the trait and find that it is present in both. The scenario requiring the fewest number of transformations, that

TABLE 1-1. Comparisons with a Human Hand

Basis of the Relationship	Hand of a Monkey	Hand of a Raccoon	Forepaw of a Rat	Wing of a Bat	Wing of a Bird	Wing of a Moth
Homologous as a hand?	Yes—their common ancestor had hands	No—their common ancestor had forepaws, not hands	No—their common ancestor had forepaws, not hands	No—their common ancestor had forepaws, not hands	No—their common ancestor had forepaws, not hands	No—insect wings are not related to limbs at all
Homoplastic as a hand?	No—they are homologous	Yes—it looks roughly like a human hand	No—it does not resemble a human hand	No—it does not resemble a human hand	No—it does not resemble a human hand	No—it does not resemble a human hand
Homologous as a forepaw?	Yes	Yes	Yes	Yes	No—bird wings are forelimbs, not forepaws	No—insect wings are not related to limbs at all
Homoplastic as a forepaw?	No—they are homologous	No—they are homologous	No—does not resemble a human hand	No—bat wings do not look like paws	No—bird wings are forelimbs, not forepaws	No—insect wings are not related to limbs at all
Analogous?	Yes—used for manipulation	Yes—used for manipulation	Yes—used for manipulation	No—used for flight, not manipulation	No—used for flight, not manipulation	No—used for flight, not manipulation

TABLE 1-2. Comparisons with an Eagle's Wing

Basis of the Relationship	Wing of a Sparrow	Wing of a Crow	Wing of a Bat	Wing of a Moth
Homologous as a wing?	Yes—their common ancestor had wings	Yes—their common ancestor had wings	No—their common ancestor did not have wings	No—insect wings are not related to limbs at all
Homoplastic as a wing?	No—they are homologous	No—they are homologous	No—it does not resemble a bird wing	No—it does not resemble a bird wing
Homologous as a forelimb derivative?	Yes	Yes	Yes	No—insect wings are not related to limbs at all.
Homoplastic as a forelimb derivative?	No—they are homologous	No—they are homologous	No—it does not resemble a bird wing	No—insect wings are not related to limbs at all
Analogous?	Yes—used for flight	Yes—used for flight	Yes—used for flight	Yes—used for flight

is, two, is that X was gained at some point and was thus present in the common ancestral stock of A, B, K, and Q, and the trait was subsequently lost once in B. To hypothesize that Q, K, and A each gained the trait independently would require three transformations. Cladistics thus provides a rigorous method for inferring the likely nature of structures in common ancestors and, therefore, which structures are plesiomorphic (ancestral) and which are apomorphic (derived).

Parsimony

Generating an hypothesis based on the smallest number of phylogenetic transformations is in accordance with the **principle of parsimony.** In its simplist form, this principle states that if one is confronted by several competing theories or explanations, the simplist one (or the one with the fewest assumptions) is most likely to be correct. Please note that the principle states the simplest explanation is *most* likely to be correct. Parsimony is no guarantee of correctness. In biology, nonetheless, simple explanations seem to be supported by subsequent facts more often than more complex explanations. In the case of the distribution of trait X in Q, K, and A but not B, for

example, we would have no grounds to assume that if the common ancestor of A and B had the structure, then a subsequent ancestor of species B lost it, another subsequent ancestor of B regained it, and yet another subsequent ancestor lost it again. Unless there were specific evidence to support this having happened, most biologists would find such an elaborate hypothesis to be quite unconvincing. Hence, the value of parsimony: it tends to rule out overly elaborate and insupportable hypotheses.

The principle of parsimony is also in accord with current ideas about the mechanisms of evolution. Any given species does not have a good chance of success if it gains new structures that do not give it an advantage in maintaining itself or if it loses structures that were beneficial to survival. If a new structure allows for a new niche or adaptive advantage, the structure will be selected for. Changes in the genome of any population, however, must generally be parsimonious themselves. The simpler the alteration of the genome to produce a variant, on which natural selection can then act, the greater the probability of its occurring and becoming established in a population. This principle is the **principle of minimum increase in complexity,** as delineated by Saunders and Ho (1984).

Let us extend our example of using the principle of parsimony in an out-group analysis by analyzing the variation of trait "T" with some real taxa, as shown in Figure 1–2, in order to better demonstrate the correct method of analysis and the importance of being rigorous in applying it. Figure 1–2 shows a somewhat simplified cladogram of the major taxa in the vertebrate radiation. A plus sign is placed above each taxon where the trait is present and a minus sign above each where it is absent. We first note that the trait is present in the amniote vertebrates—mammals, diapsids (which include lizards, crocodiles, and birds), and turtles—but it is absent in amphibians. Was it present or absent in the common ancestor of amphibians and amniotes? The out-group to amphibians and amniotes, that is, tetrapods, is the crossopterygian *Latimeria,* and in our example, *Latimeria* lacks the trait, as do the lungfishes. We therefore hypothesize that the trait was absent in the common ancestor of lungfishes, *Latimeria,* and tetrapods. This would

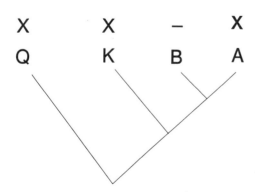

FIGURE 1-1 Cladogram showing the distribution of the trait X among the extant species Q, K, B, and A.

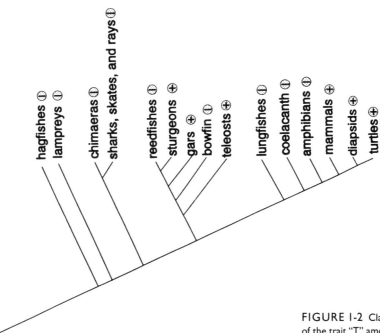

FIGURE 1-2 Cladogram showing the distribution of the trait "T" among extant groups of vertebrates.

mean that only one transformation occurred over evolution among these groups: the acquisition of the trait in the ancestral stock of amniotes. We can also conclude that the trait was not acquired in the common ancestor of tetrapods, because such a change would then have to be followed by another change (the loss of the trait in extant amphibians), and this scenario is not parsimonious. Two changes are more complicated than one, so this hypothesis must be discarded.

We now turn to the ray-finned fishes and find that a trait that appears similar to "T" is present in teleosts, absent in the bowfin, present in gars and sturgeons, and absent in reedfishes. Comparing only teleosts and the bowfin, we do not know whether the trait was present or absent in their common ancestor. Gars and sturgeons are the out-groups to the bowfin and teleosts, and they have the trait. Therefore, we hypothesize that the trait was present in the common ancestor of these four groups and that it has been lost once—in the bowfin. The trait is absent, however, in reedfishes. Was it present or absent in the ancestral stock of ray-finned fishes? The out-group to the ray-finned fishes is the cartilaginous fishes, and the trait is absent in both groups of cartilaginous fishes: chimaeras and the sharks, skates, and rays. Thus, we hypothesize that the trait was absent in the common ancestors of cartilaginous and ray-finned fishes. It was gained in the ancestral stock of sturgeons, gars, the bowfin, and teleosts and was subsequently lost in the bowfin.

Jawless vertebrates (hagfishes and lampreys) also lack the trait, so we can extend and summarize the above hypotheses to the following: the trait was absent in ancestral vertebrates, was gained once in the common ancestral stock of sturgeons, gars, the bowfin, and teleosts, was lost in the bowfin, and was independently gained a second time in amniotes. Two gains and one loss total three transformations. This is the minimum number of changes that can satisfy this distribution. If, on the other hand, we were to ignore parsimony and hypothesize that the trait was gained in the ancestors of ray-finned fishes and then lost in reedfishes, in the bowfin, in the coelacanth, in lungfishes, and in amphibians, we would have to ascribe to *six* transformations, twice the number that accounted for the pattern in the more parsimonious hypothesis.

Using the principle of parsimony has important implications for distinguishing homologous structures from homoplastic ones. Given the distribution of the trait "T" in our example, we must conclude that the trait in some of the ray-finned fishes and the trait in the amniotes are homoplastic and not homologous, even if they resemble each other closely. If two structures are homoplastic, we can examine to what degree they are similar and assess what constraints may be operating and how much potential exists for the development of new structures over evolution. A particularly good example of this type of analysis is that of the independent evolution of electroreceptive systems in two teleost taxa (gymnotids and mormyrids) that we will discuss in chapter 17.

A Word of Caution

We need to return to the example shown in Figure 1–2 to consider one last important point. A major pitfall can arise when using cladistic analysis that we need to be aware of. This pitfall involves ignoring the presence of out-groups to the assemblage being examined. If one of the more so-called "advanced" members of a taxon (i.e., those that are farther or more recently derived from the ancestral stock than others) differs from an extant species that is more closely related to the ancestral stock, one may be tempted to make the mistake of taking the condition in the latter species as the ancestral condition. Statements that reveal this line of thinking unfortunately appear with some frequency in scientific journals.

If we return to Figure 1–2 and consider the absence of the trait "T" in reedfishes, the potential pitfall in logic can be appreciated. We concluded, since the trait is absent in cartilaginous fishes and agnathans, that the trait was also absent in the common ancestor of ray-finned fishes and was not acquired in this line until after the divergence of reedfishes and the rest of the taxon. In this case, the absence of the trait in reedfishes does mirror the ancestral condition. Had the trait been present in cartilaginous fishes, lampreys, and hagfishes, however, we would have concluded that the trait was present in the common ancestor of all vertebrates and thus also present in the ancestral stock of all ray-finned fishes. In this case, its absence in reed-fishes would be a *specialization* within that group, as is the case in bowfins.

One of the most important concepts to bear in mind in comparative analyses is that while traits can be plesiomorphic or apomorphic, a particular species is neither. Any species is the sum of its traits—some of which are plesiomorphies and some of which are apomorphies. The position of any species on any type of phylogenetic tree does not have any correlation with the presence or absence of primitive or derived traits.

Reconstructing Evolution

The results of cladistic analyses can give us a picture of the various traits possessed by a particular ancestral group. In comparative neurobiology, we can thus reconstruct the neuroanatomical features in an ancestral stock and then compare them with the sets of features found in the various descendant, extant groups of that stock. Being able to identify the differences then allows us to formulate hypotheses about the mechanisms of how the evolutionary changes came about. For example, could the difference between one condition in an ancestor and a different condition in a descendant be accounted for by differences in the rate of cell proliferation during development? By a different migration pattern of neurons during development? By a lack of afferent axonal connections resulting in loss of a group of neurons during development?

In reconstructing the condition of the central nervous system in ancestral groups, we can thus gain a foothold on the challenge of finding out *how* brains have evolved. As many of the chapters in this book will demonstrate, the various parts of the brain have some very different evolutionary histories in different lineages of vertebrates. In all cases of extant vertebrates, the evolution of the brain has allowed for the particular species to be successful in its adaptation to its environment.

FOR FURTHER READING

Ayala, F. J. (1988) Can "progress" be defined as a biological concept? In M. H. Nitecki (ed.), *Evolutionary Progress*. Chicago: The University of Chicago Press, pp. 75–96.

Campbell, C. B. G. and Hodos, W. (1970) The concept of homology and the evolution of the nervous system. *Brain, Behavior and Evolution*, 3, 353–367.

Hall, B. K. (ed.) (1994) *Homology: The Hierarchical Basis of Comparative Biology*. San Diego, Academic Press.

Pollard, J. W. (ed.) (1984) *Evolutionary Theory: Paths Into the Future*. Chichester, England: Wiley.

ADDITIONAL REFERENCES

Bock, W. J. (1967) Evolution and phylogeny in morphologically uniform groups. *American Naturalist*, 97, 265–285.

Bock, W. J. (1969) Discussion: the concept of homology. In J. M. Petras and C. R. Noback (eds.), *Comparative and Evolutionary Aspects of the Vertebrate Central Nervous System, Annals of the New York Academy of Sciences*, 167, 71–73.

Bonner, J. T. (1988) *The Evolution of Complexity by Means of Natural Selection*. Princeton, NJ: Princeton University Press.

Brooks, D. R. and McLennan, D. A. (1991) *Phylogeny, Ecology, and Behavior: A Research Program in Comparative Biology*. Chicago: The University of Chicago Press.

Campbell, C. B. G. and Hodos, W. (1991) The *Scala Naturae* revisited: evolutionary scales and anagenesis in comparative psychology. *Journal of Comparative Psychology*, 105, 211–221.

Cracraft, J. (1967) Comments on homology and analogy. *Systematic Zoology*, 16, 356–359.

Darwin, C. (1836) *The Voyage of the Beagle*. New York: Mentor, 1988.

Darwin, C. (1859) *The Origin of Species*. New York: Random House, 1993.

Diamond, J. M. (1984) "Normal" extinctions of isolated populations. In M. H. Nitecki (ed.), *Extinctions*. Chicago: The University of Chicago Press, pp. 191–246.

Dobzhansky, T. (1951) *Genetics and the Origin of Species*, 3rd ed. New York: Columbia University Press.

Dobzhansky, T. (1970) *Genetics of the Evolutionary Process*. New York: Columbia University Press.

Eldredge, N. (1985) *Unfinished Synthesis: Biological Hierarchies and Modern Evolutionary Thought*. New York: Oxford University Press.

Eldredge, N. and J. Cracraft (1980) *Phylogenetic Patterns and the Evolutionary Process: Method and Theory in Comparative Biology*. New York: Columbia University Press.

Ghiselin, M. T. (1966) An application of the theory of definitions to systematic principles. *Systematic Zoology*, 15, 127–130.

Gould, S. J. (1977) *Ontogeny and Phylogeny*. Cambridge, MA: The Belknap Press of Harvard University Press.

Gould, S. J. (1985) *The Flamingo's Smile: Reflections in Natural History*. New York: W. W. Norton & Co.

Hodos, W. and Campbell, C. B. G. (1969) *Scala naturae*: why there is no theory in comparative psychology. *Psychological Review*, 76, 337–350.

Hull, D. L. (1988) Progress in ideas of progress. In M. H. Nitecki (ed.), *Evolutionary Progress*. Chicago: The University of Chicago Press, pp. 27–48.

Lauder, G. V. (1986) Homology, analogy, and the evolution of behavior. In M. H. Nitecki and J. A. Kitchell (eds.), *Evolution of Animal Behavior: Paleontological and Field Approaches*. New York: Oxford University Press, pp. 9–40.

Martin, P. S. (1984) Catastrophic extinctions and late Pleistocene Blitzkrieg: two radiocarbon tests. In M. H. Nitecki (ed.), *Extinctions*. Chicago: The University of Chicago Press, pp.153–189.

Mayr, E. (1963) *Animal Species and Evolution*, 3rd ed. Cambridge, MA: The Belknap Press of Harvard University Press.

Mayr, E. (1976) *Evolution and the Diversity of Life: Selected Essays*. Cambridge, MA: The Belknap Press of Harvard University Press.

Northcutt, R. G. (1984) Evolution of the vertebrate central nervous system: patterns and processes. *American Zoologist*, 24, 701–716.

Northcutt, R. G. (1985a) Brain phylogeny: speculations on pattern and cause. In M. J. Cohen and F. Strumwasser (eds.), *Comparative Neurobiology: Modes of Communication in the Nervous System*. New York: Wiley, pp. 351–378.

Northcutt, R. G. (1985b) Central nervous system phylogeny: evaluation of hypotheses. *Fortschritte der Zoologie*, 30, 497–505.

Northcutt, R. G. (1985c) The brain and sense organs of the earliest vertebrates: reconstruction of a morphotype. In R. E. Foreman, A Gorbman, J. M. Dodd, and R. Olsson (eds.), *Evolutionary Biology of Primitive Fishes*. New York: Plenum, pp. 81–112.

Northcutt, R. G. (1986) Strategies of comparison: how do we study brain evolution? *Verhandlungsbericht Deutsche Zoologische Gesellschaft,* 79, 91–103.

Northcutt, R. G. (1988) Sensory and other neural traits and the adaptionist program: mackerels of San Marco? In J. Atema, R. R. Fay, A. N. Popper, and W. N. Tavolga (eds.), *Sensory Biology of Aquatic Animals.* New York: Springer-Verlag, pp. 869–883.

Provine, W. B. (1988) Progress in evolution and meaning in life. In M. H. Nitecki (ed.), *Evolutionary Progress.* Chicago: The University of Chicago Press, pp. 49–74.

Roth, L. V. (1984) On homology. *Biological Journal of the Linnean Society,* 22, 13–29.

Saunders, P. T. and Ho, M.-W. (1981) On the increase in complexity in evolution. II. The relativity of complexity and the principle of minimum increase. *Journal of Theoretical Biology,* 90, 515–530.

Saunders, P. T. and Ho, M.-W. (1984) The complexity of organisms. In J. W. Pollard (ed.), *Evolutionary Theory: Paths into the Future.* New York: Wiley, pp. 121–139.

Scott-Ram, N. R. (1990) *Transformed Cladistics, Taxonomy, and Evolution.* Cambridge, England: Cambridge University Press.

Simpson, G. G. (1961) *Principles of Animal Taxonomy.* New York: Columbia University Press.

Smith, H. M. (1967) Biological similarities and homologies. *Systematic Zoology,* 16, 101–102.

Stanley, S. M. (1984) Marine mass extinctions: a dominant role for temperature. In M. H. Nitecki (ed.), *Extinctions.* Chicago: The University of Chicago Press, pp. 69–117.

Van Valen, L. (1982) Homology and causes. *Journal of Morphology,* 173, 305–312.

Wiley, E. O. (1981) *Phylogenetics: The Theory and Practice of Phylogenetic Systematics.* New York: Wiley-Liss.

2

Neurons and Sensory Receptors

INTRODUCTION

Virtually all animal cells react in some way to the physics and chemistry of the environments that they inhabit. Some multicellular animals, however, have evolved a special network of cells (neurons) that have the ability to communicate with specific groups of other neurons in a highly precise manner. This cellular communication network is the nervous system. Among the advantages of a nervous system are that it is able to take information about the surrounding environment and process it in some way before the animal reacts. This processing provides the animal with options such as to respond or not respond to a stimulus, or to respond one way or another way. In addition, a nervous system offers the ability to store information about the consequences of a particular response to a particular environmental stimulus; this information can then have an impact on the course of future action when a similar stimulus next occurs. Because of the wide range of chemical and physical events that are of importance to animals, certain neuron or neuron-like cells became specialized for the detection of these stimuli, such as light, pressure, chemical, and temperature detectors. These nervous system specializations, known as receptors, along with specializations of various body parts, permitted animals to enter and exploit new regions of the environment. To the extent that these explorations were successful, they led to further specialization and adaptation.

In this chapter we will examine some of the fundamentals of the anatomy of neurons and receptors as individual elements of the nervous system. In subsequent chapters we explore the organization of these elements into neuronal systems. Among these systems are

- Sensory systems that acquire information about the external and internal environments.

- Integrative systems that process the incoming information, evaluate this information, often in the context of past experience, and make decisions for action or inaction, depending on the circumstances.
- Motor systems that convert decisions into commands for action (or inaction) by effector organs (muscles and glands).
- Coordinating systems that organize the patterns of commands to the effector organs, especially muscle groups, to assure that the individual effector organs or groups of organs operate on the environment in a smooth, efficient, and orderly way.

THE NERVOUS SYSTEM

The nervous system, like all organs of the body, is made up of cells. Like many organs, the nervous system contains more than one specialized type of cell. Unlike other systems of the body, however, the nervous system has a great variety of cell types and sizes arranged in highly specific ways, which are fundamental to the operation of the nervous system. Indeed, these highly specific relationships between its cellular constituents are what give the nervous system its unique character, which permits us to have automatic central control over our internal organs, to sense the external and internal environments, to remember, to think, to communicate, and so on. These functions depend on precise interconnections between specific cell populations. In no other biological system, does the functioning of that system depend on such precise and rapid communication between one particular cell and another. Moreover, the sequences in which the cells communicate is fundamental to the way in which the nervous system functions. If, for example, these relationships are interfered with by injury, disease, or developmental malformation, important visceral and behavioral functions that the nervous system performs will be impaired.

Another major difference between the nervous system and other organs is the distance over which many of the cellular components communicate. In large animals such as humans, whales, elephants, and giraffes, the distances over which a single neuron communicates with other neurons can be a meter or more. In addition, in these large creatures, the lengths of the cells that carry information from the body surface to the nervous system and those that carry the nervous system's commands to the muscles can be many meters in length.

The cells of the central nervous system fall into two broad categories: **neurons** and **glia.** The neurons are the communication and information-processing elements of the nervous system. The glia are support elements; they protect and nurture the neurons and may play a subtle role in the processing of information. In addition to its cells, the nervous system contains a rich supply of blood vessels to bring oxygen and nutrients to the cells and to remove waste products.

Neurons communicate with each other by means of signals that are mostly chemical, sometimes electrical, and occasionally a mixture of the two. When the communication must be carried out at a distance, the transmission of the signal along the length of the neuron is carried out by means of an electrochemical process known as the nerve impulse or action potential. Because so much of the mechanism of transmission of the signal from cell to cell is by means of the rapid secretion of chemicals into the minute space between cells, neurons may have evolved from secretory cells that became specialized for secretion to one particular cell rather than to any cells that happen to be in its vicinity. As the nervous system grew in size and neurons became spatially separated, they developed the capability to maintain their specific-cell-to-specific-cell contacts over larger and larger distances.

Because the nervous system is the organ of behavior, it must acquire information about the external world and the condition of internal organs and systems. This information is acquired by means of specialized cells called **sensory receptors.** Many receptors are specialized neurons; others are neuron-like cells that have a number of properties in common with neurons and are innervated by neurons that relay the sensory signals to the central nervous system.

The following sections contain a brief description of some characteristics of neurons and receptors that will be especially useful for readers of this book. We assume that the reader already has a basic familiarity with the structure of neurons, their component parts, and the basic principles of axonal conduction and synaptic transmission. Readers who lack this background or wish to refamiliarize themselves with it will find a separate listing of introductory works on these subjects at the end of this chapter.

THE COMPONENTS OF NEURONS AND SENSORY RECEPTORS

The main components of the neuron are the cell body or **soma,** and its processes or outgrowths, the **axon** with its axon

terminals, and the **dendrites** with their dendritic branches and spines. Each of these components can be found in a seemingly endless variety of configurations. Figure 2–1 shows some examples of the variety of neuron types. In each of the examples, the arrows indicate the direction of flow of the nerve impulse. The neuron shown in Figure 2–1(A) shows the main components. The star-shaped, solid black region represents the soma, which consists of a cell membrane that contains cytoplasm and the **nucleus.**

The large extensions or processes that give the soma its star shape are the dendrites. The dendrites themselves may have further processes extending from them and are known as dendritic branches. These branches in turn often subdivide further into smaller and smaller branches until they take on the appearance of a leafless tree in winter. The soma and its dendritic tree are the most frequent points of contact between a neuron and those other neurons that are sending their communication signals to it. The size and shape of the soma can vary enormously among neuron types. The soma may be as small as a few micrometers (μm) to more than a millimeter (mm), as in the giant cell of Mauthner, which is discussed in Chapter 8. It may be star shaped, as in the example, or it may have many other forms, such as that of a pyramid (pyramidal cell), a pear (piriform cell), or a spindle (fusiform cell). Dendrites can vary in length from a fraction of a micrometer to many millimeters.

Also projecting from the soma is the axon, which is the component of the neuron that permits long-distance communication. Often axons are also referred to as "nerve fibers" or simply fibers. In this book, we will use the terms axons and fibers interchangeably. The axon leaves the soma from a gentle swelling called the axon hillock and travels over distances that can vary greatly among cell types from a few micrometers to

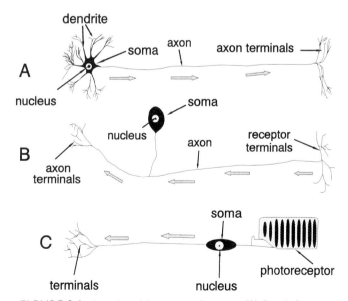

FIGURE 2-1. Examples of three types of neurons. (A) A typical motor neuron with a roughly star-shaped soma, dendrites, axon, and axon terminals. (B) A typical sensory neuron with a pear-shaped soma that is separated from the axon by a stem. This type of neuron does not have dendrites. (C) A receptor neuron, in this case a photoreceptor from the retina, that has neither an axon nor dendrites.

a meter or more. Most axons remain within the central nervous system and terminate on **interneurons,** which constitute the overwhelming majority of neurons in the central nervous system.[1] But some axons leave the nervous system and end on muscles or glands; these axons are called effectors or motor neurons. The neuron illustrated in Figure 2–1(A) is typical of motor neurons that control muscles of the skeleton. Its axon would leave the central nervous system and travel to a muscle where it would divide in a series of branches before terminating on the muscle fibers.

Not all neuron somata (plural of soma) have dendrites, however. Figure 2–1(B) illustrates a different neuron type that has no dendrites on its soma. Indeed, it has no dendrites at all, in the conventional sense. Its principal source of stimulation is not another neuron at all, but rather is some event outside of the central nervous system. It is a sensory receptor neuron and, like the effector neuron in Figure 2–1(A), it too has most of its extent in the environment outside of the central nervous system. At the right are its receptor branches, which might be under the surface of the skin or wrapped around a muscle fiber to detect muscle stretch. Other such sensory branches might be in contact with specialized receptor cells such as those that detect electric fields or one of the different types of hair cells that detect the movement of fluids in the auditory or vestibular systems. Mechanical deformation of the receptor branches or activation of the specialized receptor cell activates this sensory or receptor neuron to send a signal along the axon to the axon terminals that end on interneurons within the central nervous system. Other receptor types detect chemical changes in the external and internal environments.

The soma of this type of neuron typically remains outside of the central nervous system in a sensory **ganglion** and does not participate directly in the process of conduction of the axonal signal. It provides nutritional and metabolic support to the axon, but is not an active player in the flow of information into the central nervous system. Some sensory neurons, however, have their somata within the central nervous system.

Figure 2–1(C) represents a different type of sensory neuron, a photoreceptor, which is specialized for the detection of photons of light. These are the receptors of the eye that convert the energy of light into signals that can be conducted to the brain. The photoreceptor end is activated by photons of light and in some sense can be thought of as a highly specialized dendrite. But many sensory neurons or receptor cells belong to a class of neurons that have no axons. Activation of the photoreceptor eventually leads to the development of a signal that is conducted along the cell to a terminal where it contacts an interneuron that conducts the signal further into the central nervous system.

---

[1] The term "interneuron" is used in several ways in the contemporary literature. Some writers use the term as we do in this book, to indicate the broad class of neurons that are neither receptor nor effector neurons. Others use the term "interneuron" to refer to neurons that remain intrinsic to a neuronal population rather than interconnecting one population with another. In this text, this second concept of "interneuron" is referred to as a Golgi Type II neuron.

TRANSPORT WITHIN NEURONS

The soma produces many materials that are important for the maintenance of the internal and external workings of the neuron. These materials include enzymes and other substances that participate in the synthesis of neurotransmitters and neuromodulators; proteins for use in the formation of synaptic vesicles, ion channels, and membrane receptors; and proteins for the maintenance of the neuron's internal skeleton. Still other materials are necessary for maintenance of the cell membrane, which is the boundary between the inside and outside of the cell. Finally, used synaptic vesicles, depleted mitochondria, and other organelles must be returned to the soma for reuse or for digestion in the lisosomes and subsequent "recycling" into new membrane.

The often extreme separation between the soma and the axon terminals and between the soma and the tips of the dendritic branches is too great for simple diffusion to function effectively. Neurons therefore have a kind of intraneuronal circulatory system to move secretion products from the soma to the remote ends of the neuron. There are, in fact, three separate transport systems within the neuron: a fast anterograde (forward moving) transport system that carries materials from the soma towards the axon terminals and dendritic branches, a fast retrograde (backwards moving) transport system in the reverse direction, and a slow axoplasmic transport system.

The fast anterograde transport system, which can move as fast as a meter a day, makes use of one of the neuron's internal skeletal elements, the microtubules, which are slender tubes that run the length of the axon. Rather than flowing through these tubules, which have too narrow a diameter in any case, the organelles are transported along the surface of the tubules by a "motor molecule" called kinesin. The fast retrograde transport, which is involved in the return of used materials to the soma for recycling, moves at a slower speed than the anterograde fast transport. The returning materials are packaged in membranes and are transported along microtubules in a manner similar to that of the fast anterograde system except that a different motor molecule, in this case dynein, moves the membrane packages along the microtubules.

The slowest (1–10 mm/day or slower) of the three transport systems is the slow axoplasmic transport system, which consists of two components: a slow system and a very slow system. The slow system carries proteins that, among other things, coat the synaptic vesicles. The very slow system carries the proteins that maintain the filamentous internal skeleton of the neuron: the microtubules, neurofilaments, and microfilaments. The ability to chemically mark many of the substances being transported within neurons has served as one of the most powerful means of visualizing the connections between neuronal populations. The similarities and differences in the patterns of connections are among the major criteria for determining evolutionary trends within the nervous system.

CLASSIFICATION OF NEURONS

Somata

Neurons may be classified in a variety of ways. Figure 2–2 shows one type of classification. The figure shows a monopolar

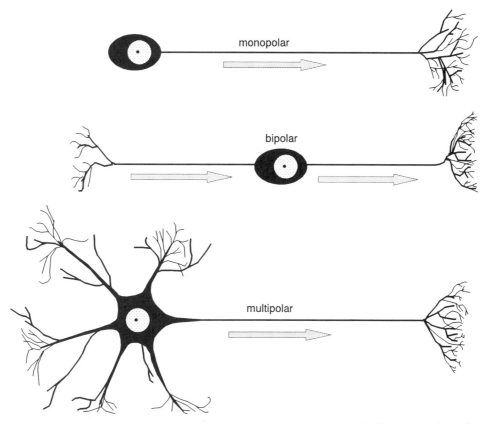

FIGURE 2-2. Three types of neurons classified according to the number of poles. The arrows indicate the direction of conduction of the nerve impulse.

cell without dendrites on the soma. These are typical of invertebrate nervous systems. The bipolar cell has fine dendritic branches at the left and its axon at the right. Such cells are found in the retina of the eye. The multipolar cell is the most common type found in the vertebrate central nervous system; it is shown in Figure 2–2 with many dendrites and finer dendritic branches on the left and an axon on the right. This cell type is typical of interneurons and motor neurons of the central nervous system.

Dendrites

Another system of neuronal classification is presented in Figure 2–3, which shows neurons with several different types of dendrites. The classification of dendritic types is based on the work of Enrique Ramon-Moliner. The dendrites of the neurons in Figure 2–3(A) are long and slender without many branches and are called **isodendrites.** The neurons in Figure 2–3(B and C) have branched dendrites and can achieve fairly high degrees of complexity and specialization. These are known as **allodendrites.** The most specialized are called **idiodendrites,** which are represented in Figure 2–3(D–G), typically are found in regions of the nervous system such as the olfactory bulb or cerebellar cortex.

Figure 2–4 depicts two neurons with their dendrites and a portion of their axons. The cell on the left is known as a pyramidal cell because its soma has the shape of a pyramid;

the other is called a piriform cell because its soma is roughly in the form of a pear. In each case, two sets of dendrites are shown: a long dendrite ascending from the peak or apex of the cell (hence called "apical" dendrites) and other dendrites protruding from the base of the cell (the "basal" dendrites). Each of these dendrites subdivides into dendritic branches that increase the dendritic surface area.

Dendrites offer an enormous surface area for axon terminals to end upon. The huge dendritic surface area provides termination sites for thousands of axon terminals. Some of these terminals are excitatory and contribute to the generation of the electrical signal; others are inhibitory and thus participate in suppression of the action potential. Rarely does a single axon terminal have sufficient influence to excite a neuron to produce an action potential or to inhibit it. Summation of the activity of many synapses therefore usually is required in order to influence a neuron's actions. Often the generation or suppression of an action potential is the result of a kind of algebraic summation of the excitatory and inhibitory influences on the cell. If the sum of the excitatory influences outweighs the sum of the inhibitory influences, and if the net excitatory influences are present in sufficient quantity, the action potential or nerve impulse will be generated in the axon hillock and conducted down the axon. Dendrites thus are the battleground on which the opposing forces of excitation and inhibition compete. Because of their varying thicknesses and varying distances from the axon hillock, dendrites serve not merely as the input end

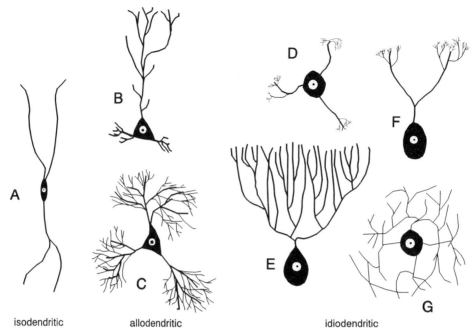

isodendritic   allodendritic   idiodendritic

FIGURE 2-3. A classification of neurons according to the type of dendritic tree. Their axons are not shown.

of the neuron, but as integrators of neuronal activity. They provide areas in which weak incoming signals can combine to form stronger influences on the ultimate action of the cell.

Axons

Neurons also may be classified according to their axons. At one extreme are the many small interneurons that have no axons at all. These axonless interneurons are involved only in local neuronal activity that is confined to a circumscribed cell population; since they do not exert an influence on distant cells, they need no axons. At the other extreme are neurons with extremely long axons, such as those that travel from the spinal cord down the length of the hind leg of a tetrapod and move its toes. Axons also vary in thickness, from the giant axon of the Mauthner cell to the tiny axons of the olfactory nerve. The effective thickness of an axon depends in part on the diameter of the axon itself, and in part on the diameter of its myelin sheath. The total diameter of the axon (axon plus sheath) is a major determiner of the velocity of conduction of the action potential. In general, larger diameter axons conduct more rapidly, although other factors play a role as well.

SYNAPSES

Chemical Synapses

The surface of the axon terminal where it contacts the neuron is known as the presynaptic membrane; the specialized surface of the neuron that receives the axon terminal is known as the postsynaptic membrane, below which is a specialized region of the dendrites's cytoplasm. The synapse itself is a small extracellular space that is about 20–30 nm across (1 nm is 1/

1000 of a μm). Within the axon terminal are small, membrane-bound packets or "vesicles," which contain chemical substances that are released into the synaptic space by the arrival of the action potential. These chemicals excite or inhibit the postsynaptic membrane and are one of the ways that one neuron can affect the activity of another neuron. Also contained within the axon terminal are one or more mitochondria. The mitochrondria are sources of energy for biological processes and are found in the soma, the synaptic vesicles, and wherever the neuron is highly active. Because the transmission of a signal across the synapse in this case is chemical, this synapse is known as a chemical synapse.

The events that occur in a chemical synapse are shown in Figure 2–5. Figure 2–5(A) shows a synaptic terminal at rest. The axon terminal filled with synaptic vesicles is shown just above the synaptic space. The terminal membrane that faces into the space is the presynaptic membrane. Below the synaptic space is the postsynaptic membrane with protein receptor sites. When the action potential arrives at the terminal, it causes calcium channels to open, which allows an influx of calcium ions. The arrival of the calcium influx in turn results in a forward movement of the vesicles with their contents of **neurotransmitter** substances towards the presynaptic membrane. As shown in Figure 2–5(B), the membranes of the vesicles fuse with the neuronal membrane allowing the contents of the vesicle to be released into the synaptic space where they can act upon protein receptors in the postsynaptic membrane. If the transmitter substance is excitatory (i.e., depolarizing), it generates an excitatory postsynaptic potential in the postsynaptic membrane. If the substance is inhibitory, its action on the postsynaptic membrane is as if it had increased the polarization of the cell; that is, increased the resting potential thereby making it more difficult to generate an action potential, a phenomenon called hyperpolarization. Figure 2–5(C) shows that excess neurotransmitter material that remains in the synaptic space may

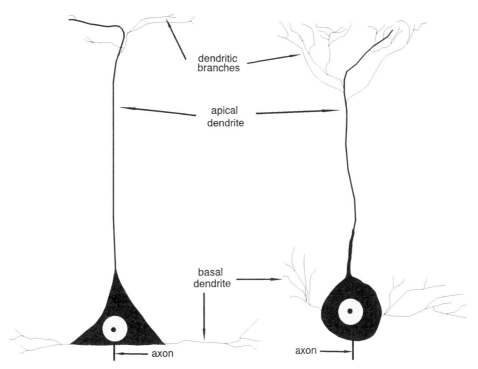

FIGURE 2-4. Examples of dendrites in neurons with different shaped somata. The initial segment of each cell's axon is shown at the bottom.

be taken back into the terminal by other channels to be reincorporated into vesicles for future use.

Neuro-Active Substances

Chemical synapses may be characterized by the chemical compounds present in them. Apart from their functions in synaptic transmission, these compounds are useful to comparative neuroanatomists because they can serve as markers of neuronal pathways and populations in different taxonomic groups. Thus they become another indicator of brain evolution and adaptation to the environment. Table 2–1 lists some of the major compounds just described. These fall into a variety of categories. The two major categories are neurotransmitters and **neuro-modulators.** Within each of these are several groups of chemicals. The neurotransmitter category contains relatively few compounds; the neuromodulator category has more than 30 compounds, only some of which are listed in the table. The cholinergic neurotransmitter is acetylcholine, which is excitatory or inhibitory. The biogenic-amine neurotransmitters, epinephrine (adrenalin), norepinephrine (noradrenaline), dopamine, serotonin, and histamine may likewise have excitatory or inhibitory actions depending on the type of receptor they encounter. The amino acid neurotransmitters consist of two excitatory transmitters, glutamate and aspartate, and two inhibitory transmitters, γ-aminobutyric acid (GABA) and glycine.

The neuromodulators generally do not directly affect the depolarization or hyperpolarization of neurons as do neurotransmitters (although some have been reported to have these properties). Instead, the neuromodulators influence the duration or intensity of the action of the neurotransmitters by affect-

ing: the reuptake of transmitters; the effectiveness of the enzymes present in the synapse; the rate of transmitter release; and a variety of other phenomena, which make the synapse very different from a simple on–off switch. A vast array of possibilities for subtle and sophisticated modifications of the transfer of information between neurons is made possible by these many substances.

In addition to modulating synaptic transmission, many of these versatile chemical compounds play other roles in the body; indeed many of their names may be familiar to you in other contexts. Some of these peptides are gastrointestinal peptides such as vasoactive intestinal polypeptide (VIP), substance P, cholecystokinin, and neurotensin. Others are hormones that are secreted by the posterior division of the pituitary (oxytocin and vasopressin) and are involved in the regulation of blood pressure and affect maternal functions. Others are releasing hormones that are secreted by the hypothalamus, such as thyrotropin releasing hormone (TRH), luteinizing hormone releasing hormone (LHRH), and growth hormone releasing hormone (GHRH). Still others are anterior pituitary hormones, such as adrenocorticotropic hormone (ACTH), growth hormone, and luteinizing hormone. Yet others are naturally occurring opioids, such as met-enkephalin, leu-enkephalin, and β-endorphin. Finally, some second messenger substances, such as cyclic GMP, also have been found to have neuromodulator properties.

Electrical Synapses

Not all synaptic junctions make use of chemical substances as the transmitter or modulator. At many synaptic junctions, the transmission is carried out by the passage of electrical cur-

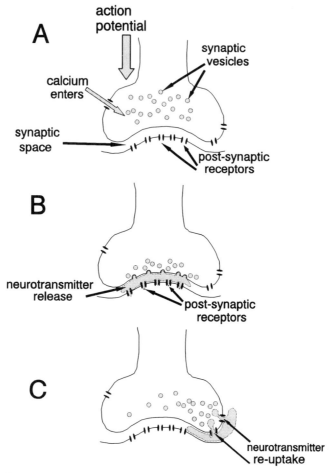

FIGURE 2-5. Transmission in a chemical synapse. (A) Neuroactive chemicals are stored in synaptic vesicles. Calcium enters through a calcium channel. (B) The vesicles move towards the synaptic membrane and fuse with it thereby releasing their chemical contents into the synaptic space where they are taken up by receptors in the postsynaptic membrane. (C) Reuptake of excess neurotransmitter material from the synaptic space and its reincorporation into vesicles for future use.

rent across the synapse. These are known as "electrotonic" or "electrical" junctions or simply "gap" junctions. The synaptic space of an electrotonic junction is only 2–4 nm across, which is only about one-tenth the width of a chemical synapse. Present on each side of the gap are matching pores or ion channels that can be opened and closed by means of a complex of proteins. When these proteins twist, the pore is opened, which allows the ionic current to flow across the gap. Electrical junctions provide for synaptic transmission with virtually no delay because no time is lost as vesicles move to the presynaptic membrane and discharge their chemical transmitters. Electrical junctions also are not subject to "fatigue" as are chemical junctions, which can deplete their supply of vesicles if stimulated too frequently. Most electrical junctions can transmit their electrical signals in either direction, whereas chemical junctions can only transmit from the presynaptic to the postsynaptic membrane. Finally, transmission at electrical synapses is much faster than at chemical synapses. In spite of these advantages, electrical junctions are less common than chemical junctions in vertebrate

central nervous systems than among the invertebrates because they tend to be rather stereotyped in their action and do not lend themselves well to the subtle and varied types of interactions and modulations that are possible in a chemical transmission system. Indeed, the huge variety of possible types of interactions that can occur in a chemical system is a major contributor to the vast array of complexity of behavior that is characteristic of vertebrates and that has played such an important role in their evolution.

NEURONAL POPULATIONS

Anatomists frequently refer to a discrete population of neurons, often situated within a well-circumscribed boundary, as a **nucleus.** The same term also is used to refer to the intracellular structure that contains the DNA of the cell. Fortunately, the potential for confusing these two uses of the term is minimal because the contexts in which they are used is so

TABLE 2-1. Some of the Major Neuroactive Substances in the Central Nervous System

Type	Chemical Group	Substance	Function
Neurotransmitters	Cholinergic	Acetylcholine	Excitatory or inhibitory depending on the type of receptor
	Biogenic amines	Norepinephrine	
		Epinephrine	
		Dopamine	
		Serotonin	
		Histamine	
	Amino acids	Glutamate	Excitatory
		Aspartate	
		GABA[a]	Inhibitory
		Glycine	
Neuromodulators	Peptides and hormones	VIP[b]	Modulation of synaptic transmission by affecting transmitter release or reuptake or by changing the sensitivity of the postsynaptic membrane for the transmitter. Some modulators have neurotransmitter-like activity
		Substance P	
		Met-enkephalin	
		Leu-enkephalin	
		Cholecystokinin	
		Somatostatin	
		Neurotensin	
		Bombesin	
		ß-Endorphin	
		Angiotensin II	
		Neuropeptide Y	
		Anterior pituitary hormones[c]	
		Posterior pituitary hormones[d]	
		Insulin	
	Second messenger	Cyclic GMP	

[a] Gamma-aminobutyric acid.
[b] Vasoactive intestinal polypeptide.
[c] These include: thyrotropin releasing hormone (TRH); gonadotropin releasing hormone (GnRH), also known as luteinizing hormone releasing hormone (LHRH) in mammals; growth-hormone releasing hormone (GHRH); adrenocorticotropic hormone (ACTH); growth hormone (GH); follicle stimulating hormone (FSH); thyroid stimulating hormone (TSH); and prolactin.
[d] Oxytocin and vasopressin.

different. Examples of such "population nuclei" are the dorsal division of the lateral geniculate nucleus of the visual system, the nucleus ovoidalis of the auditory system, and the motor nucleus of the trigeminal nerve, which are structures that will be described in detail in later chapters.

Golgi Type I and II Cells

Population nuclei often consist of more than one type of neuron. The nineteenth century Italian anatomist, Camillo Golgi, distinguished two types of cells within the boundary of a population nucleus. The Golgi Type I neuron tends to have a large soma and a long, thick, well-myelinated axon. This axon passes outside of the confines of the population nucleus and can travel considerable distances; those that pass from nuclei of the brain into the spinal cord of a large animal, such as a giraffe or a whale, can be more than a meter in length. Often

the axons of Golgi Type I cells have side branches, called collaterals, that permit the axon to contact other nuclear populations en route to its final destination. The various Golgi I axons from the same population nucleus typically travel together in bundles (often known as "tracts") as they make their journey to their target neurons.

In contrast, the Golgi Type II neurons have small cell bodies and short, often unmyelinated axons. These axons rarely pass outside the boundary limits of the population nucleus. Some Golgi II neurons have no axons at all. Thus their dendrites both receive input from other neurons (axo-dendritic terminations), but also make synaptic contact with the dendrites of other neurons (dendro-dendritic contacts). Similar points of contact can be found with the soma. The Golgi II neurons are critical for the functioning of the individual population components of the nervous system. They form local circuits within the population nucleus that perform whatever the function of the particular population nucleus might be, such as the processing of information or the patterning of rhythmic events. Figure 2–6 shows Golgi I and II neurons.

To understand the differences between the Golgi I and II cells, consider that the Golgi II cells are like a local telephone network that maintains communication within a factory and allows the workers to perform their tasks in an integrated and coordinated manner. The Golgi I cells are like a long-distance telephone network that permits factories that are located at considerable distances from one another to be in contact and to coordinate their activities.

Population Nuclei and Planes of Section

In order to study the anatomy of the central nervous system, anatomists often cut the neural tissue into very thin slices called "sections" (5–50 μm or thinner), so that they may be examined under a microscope. These sections pass through population nuclei and through the axonal bundles or tracts that pass from one population nucleus to another. While these sections give an accurate view of the cross-sectional extent of a population nucleus or axon bundle, they give no indication of how much of these structures may extend ahead of or behind the plane of the section. Many of the illustrations in this book consist of such sections through the brain or spinal cord, and the reader should understand that the ovals, ellipses, circles, and other shapes that appear in the section only represent a single slice through what may be a much larger and much more complex structure. This is diagrammed schematically in Figure 2–7. At the left, represented by the thicker cylinder, is a population nucleus with its large Golgi I neurons and smaller Golgi II neurons. The Golgi I axons leave the population nucleus and enter an axon bundle, represented by the smaller cylinder. The

striped rectangle represents a section through these two structures. To the right of the arrow is the section showing how the population nucleus and axon bundle would appear in this single plane. The Appendix describes the planes of section in detail.

Techniques for Tracing Connections between Population Nuclei

The second half of the twentieth century has been a period of unprecedented new knowledge about connections between neuronal populations in the central nervous system. This explosion of information was largely due to the development of a vast number of new techniques based on a variety of biological principles. Prior to this period, the most common method for labeling a specific set of axons was based on the degenerative processes that follow injury of the cell soma or separation of the axon from the soma.

Anatomists had been able to visualize neurons or parts of neurons for nearly a century using aniline dyes or metallic deposits. The Golgi technique, in which silver impregnation is used to visualize whole neuronal cell bodies and the full array of their dendritic processes, is an example of one of the metallic impregnation methods. The problem was how to identify a specific group of axons from among the vast number present in the microscopic image, and how to recognize them as the anatomist went from one section of brain tissue to the next. One of the first tracing methods consisted of injuring the neurons to be studied, and then to use the resulting degenerative process as a tag or label to trace the course of their axons. Degenerative changes in the soma separated from all or part of its axon (retrograde degeneration) could be observed with the traditional aniline dyes that had been used during the preceding

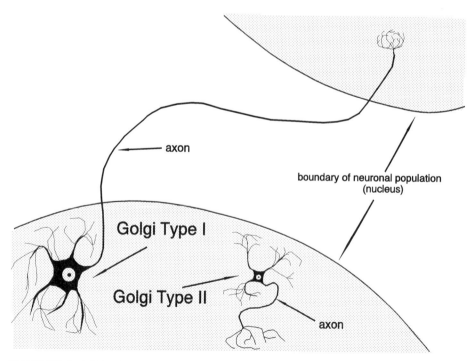

FIGURE 2-6. A Golgi Type II neuron, the axon of which remains within its neuronal population, and a Golgi Type I neuron, which sends its axon to a different neuronal population.

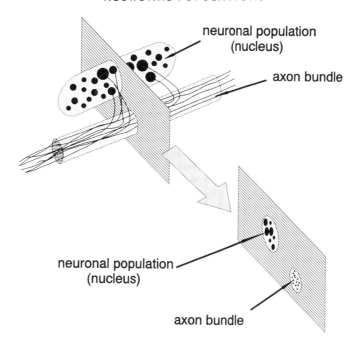

FIGURE 2-7. The appearance of a neuronal population and a nearby axon bundle in a three-dimensional view and in a transverse section through both.

century. The deposited metal then could be reduced so that it appeared black and thus visible under the microscope. The location of the degenerating axoplasm and the terminal endings of the axons could then be traced and charted in serial sections, and the axonal pathway from one point to another within the nervous system could be reconstructed. Degeneration of the separated axon (anterograde degeneration) could be studied by impregnating the degenerated axoplasm with silver or the degenerated myelin with osmium.

The initial phase of the technical revolution of the 1960s and 1970s depended on the newer methods of silver impregnation that were developed at that time (the Nauta–Gygax stain and its variants, and the Fink–Heimer stain); these methods allowed for much more precise tracing of connections by suppressing the appearance of silver in normal axons, which made the degenerating axons easier to see, and later by permitting visualization of the actual axon terminals rather than just the axons. These newer silver methods also were more effective in both mammalian and nonmammalian vertebrates alike than had been previous techniques; they opened the way for a greater exploration of connections in nonmammals than could previously have been attempted.

The techniques currently in use do not require lesions of nervous tissue but instead involve injecting one of a number of tracer substances into a particular site and allowing axonal transport to distribute the substance along the length of the axon to its terminals and its cell soma. Histological procedures are then carried out on serially sectioned material, as in the reduced silver methods, and the location of the labeled axons, axon terminals, and cell somas are charted under the microscope. A number of the labeling substances can be visualized at both the light and electron microscopic level. The tracing substances that are most commonly used include tritiated amino acids, horseradish peroxidase (HRP), HRP conjugated to wheat

germ agglutinin (WGA), and the plant lectin *Phaseolus vulgaris-*leukoagglutinin (PHA-L).

The large repertoire of axonal transport tracing techniques that has been developed has greatly enhanced our ability to visualize and trace connections. The tracing substances currently available vary widely in how they can be visualized. For example, some can be visualized by histological procedures in which the substance is reacted in various solutions to acquire a color or be tagged with a colored substance that is visible under the microscope. Other tracing substances fluoresce when illuminated with ultraviolet (UV) light. Still other tracing substances are radioactive and can be visualized by applying a thin layer of photographic emulsion over the sections, exposing the emulsion for a certain period of time, and then developing the emulsion just as one would develop a photographic print. The pattern of radioactive emissions then can be charted in serial sections to reveal the course and terminal site of the axonal pathway being studied.

This wide variety of tracing substances and visualization techniques has permitted much sophisticated and elegant experimental work. Double and even triple labeling of different components of a neuronal circuit has been developed to a fine art. For example, one might label one set of afferent axons of a nucleus with one fluorescing tracer, a second set of afferent axons with a different fluorescing tracer that fluoresces under a different wave-length, and the neuronal cell bodies and their dendrites in the nucleus with a retrogradely transported tracer that can be visualized with histological processing. The precise synaptic pattern of the two different sets of afferent fibers on the postsynaptic neurons within the nucleus can thus be worked out.

A number of additional tracing techniques exist that do not depend on axonal transport mechanisms. Some take advantage of the natural fluorescence of some of the neuronal transmitters, such as serotonin and dopamine. Widespread use is also currently made of immunohistochemical methods for antibody labeling of neuroactive substances, such as many of those listed in Table 2–1, which then can be visualized with histological techniques. In addition to axonal tracing methods, techniques allowing *in vivo* and *in vitro* study of particular parts of the brain, whole-brain scanning techniques, extra- and intracellular electrophysiological recording techniques, in situ hybridization techniques for localizing sites of gene expression, and an extensive array of additional such methods allow for a truly comprehensive approach to the study of the organization and function of the central nervous system.

Table 2–2 summarizes a number of the most widely used contemporary methods as well as some of the earlier methods. The table is not intended to be an exhaustive survey but rather gives an overview of many of the methods that you are likely to encounter if you read any of the works listed in the For Further Reading Sections at the ends of chapters in this book or elsewhere in the neuroanatomy literature.

RECEPTORS AND SENSES

The senses are our windows onto the world—both the world outside and the world within our bodies. We cannot know anything about either of these worlds except through our sensory systems. Of course we can know a lot about both worlds with the use of scientific instruments, but even this information must come to us via the sensory systems. The senses are our way of detecting energy or chemical substances in the inner and outer worlds. The detectors for the external world are known as **exteroceptors** and those of the inner world are known as **interoceptors.** Exteroceptors are the receptors for vision, taste, smell, touch, warmth, cold, and so on. Interoceptors provide the central nervous system with information about events within the body, such as the distension of the gastrointestinal system, pressure of the blood in certain blood vessels, and levels of various substances in the blood such as glucose, fat, and various hormones.

TABLE 2-2. Summary of Methods for Tracing Connections in the Central Nervous System

Method Type	Examples	Basis
Retrograde degeneration	Nissl stain	Aniline dye stain of a soma separated from part or all of its axon
Anterograde degeneration	Marchi stain	Impregnation with osmium of degenerating myelin of an axon separated from its soma
Reduced silver	Golgi stain Cajal stain Bielschowsky stain	Impregnation of neurons with reduced silver
	Nauta–Gygax stain	Especially effective on axons that are degenerating due to separation from their somata
	Fink–Heimer stain de Olmos stain	Especially effective on degenerating axon terminals
Anterograde axonal transport	Autoradiography	Tritiated amino acid transported from soma to terminals
	*Phaseolus vulgaris-*leuccoagglutinin (PHA-L)	Plant lectin transported from soma to terminals, visualized by fluorescence microscopy
Anterograde and/or retrograde axonal transport	Horseradish peroxidase (HRP)	Plant enzyme transported along axons
	Fluorescent dyes (Evans blue, fast blue, fluorogold, Di-I, Di-O)	Transported along axons, visualized with fluorescence microscopy
	Rhodamine beads	
	Cholera and other toxins	
Transneuronal transport	Wheat germ agglutinin-horseradish peroxidase (WGA-HRP), tritiated proline	Lectin + enzyme transported across synapses to postsynaptic neurons
Immunohistochemistry	Various neurotransmitters and neuropeptides	Antibodies to neurotransmitters and neuropeptides present within the neuron, visualized with fluorescence microscopy

The types of energy detected are:

- The electromagnetic spectrum, which includes visible light, UV, infrared (IR), and electricity and magnetism.
- Mechanical energy, such as is produced by bending, stretching, shearing, and compressing of the skin or other tissues.
- Chemical energy, which is the energy released by the reactions of chemical substances in the environment with the chemicals that make up the receptor.

How Many Senses?

How many senses are there? When this question is asked, we usually think of five: taste, touch, smell, hearing, and vision. Sometimes we refer to an elusive "sixth sense," by which we really mean "intuition" and not an energy detector. But in fact, there are many more than the proverbial five or even six senses. When we take into account the full range of senses available to vertebrates, we can count something in the neighborhood of 20 senses, depending on how fine one wishes to subdivide the different receptor types. These include the traditional five, plus the ability to detect electrical fields (lampreys, sharks, some ray-finned fishes, some amphibians, and monotremes), IR radiation in a manner similar to vision (some snakes), lateral-line sensations (nontetrapods, amphibian larvae, and some adult amphibians), and magnetic fields (birds and possibly other vertebrates).

Receptors and Awareness

Events that occur in the internal or external worlds that do not affect our interoceptors or exteroceptors are unknown to us. For example, many animals can detect light in the UV portion of the spectrum. No doubt they see it as some sort of color. We are unable to detect light energy in this portion of the spectrum, and so as far as our personal experience is concerned, this part of the spectrum does not exist. We can only know of its existence with the aid of specialized scientific instruments. Similarly, many mammals, including bats, many rodents, and dogs, can hear sounds that are much too highly pitched for us to hear. Even within our own bodies, receptors are at work, participating in the precise regulation of bodily functions such as blood pressure, the flow of materials through the digestive system, the secretion of hormones, all without any awareness on our part because the information from these interoceptors never reaches those brain regions that bring such information to our conscious awareness. For example, a complex series of neural circuits produces a precise regulation of the diameter of the pupil of the eye according to the intensity of light present in the external environment. We are totally unaware of these adjustments, which are being made each time the light level changes.

We can become aware of some of our senses if we produce certain types of disturbances of their mechanisms. For example, we normally are unaware of the operation of the vestibular organs in the middle ear, which provide us with information about the pull of gravity, the acceleration of our bodies, the position of our head, and so on, so that the necessary postural adjustments can be made to keep us in our erect posture and prevent us from falling over as we change position or try to walk on an uneven surface. These adjustments are, for the most part, totally transparent to us. If we spin ourselves round and round, however, as most of us did as children, we quickly become aware of a variety of sensations that result from the abnormal stimulation of this sensory system.

Sensory Experience as a Private Mental Event

To return to the detection of UV light, a feat that we are incapable of, we suggested that those animals that can detect UV light, probably experience it as some sort of color. What color is UV? We cannot imagine what UV color looks like any better than we could describe the colors of a sunset or autumn leaves to a person who has been born blind. Sensations are private mental events knowable only to the person or animal having the sensory experience. Indeed, you have no way of knowing for sure that what you experience as red when looking at an apple is what a bird or another mammal experiences when it sees the same apple. This may sound a bit philosophical, but it is of relevance to the fact that we cannot know, nor can scarcely imagine, what the subjective experiences are of animals that have senses very different from our own. For example, birds have more photopigments, which makes them more sensitive to certain regions of the light spectrum than humans, and therefore they almost certainly do not see objects as having the same colors as we do. What does a snake experience when it uses its IR pit organs to detect a mouse in total darkness? Undoubtedly it is something akin to vision, but we cannot know what it is like for the snake. Likewise, what does electroreception feel like to an animal that can detect minute changes in the electrical fields present in the surrounding environment? What is it like to have taste receptors all over one's body as do certain fishes? Some birds appear to have the capability to navigate by means of the earth's magnetic fields. What does a magnetic field feel like to them? There is no way for us to know the answers to any of these questions with any certainty.

Sensory Adaptation

A property of receptors is that they decrease their responsiveness to persistent stimuli. This phenomenon is referred to as **adaptation.** Adaptation is not a voluntary act like a human or an animal shifting its attention to and from specific stimuli at will; rather, it is more like a receptor fatigue phenomenon and can only be reversed by a period of absence of the stimulus or by changing the stimulus. You have most likely experienced adaptation of the olfactory system when you entered a room with a particularly strong odor, but after a few minutes in the room, you no longer notice the odor, even if you try. Some receptors become adapted to a particular level of stimulation and will only respond to a different level. This property sometimes results in strange effects. If you put one hand in very cold water and the other hand in very warm water, leave them for a few minutes and then put both hands in room temperature water, you will observe that the hand that was in the warm water feels cold and the hand that was in the cold water feels warm, yet both are receiving the same stimulus of room temperature water.

RECEPTOR TYPES

Table 2–3 presents a classification of receptors according to their functional classes, sensory modalities, and receptor types. All of these receptors are neurons or neuron-like cells. As is the case with all neurons, a resting potential is maintained across the cell membrane. When activated by the appropriate energy, the ionic events that maintain the resting potential are disturbed and a receptor potential results. The process by which mechanical, chemical, or electromagnetic energy is converted to neural events (i.e., the electrochemical process that produces receptor potentials) is known as **sensory transduction.**

Receptor potentials are produced by the receptor cells in response to stimulation by the appropriate stimulus. They are like the local potentials of neurons in that they are graded potentials; that is, the magnitude of the potential is proportional to the intensity of the stimulus that is being applied to it. If the receptor potential achieves a sufficient magnitude, an action potential develops in the axon of the sensory neuron. This action potential sweeps down the axon to the terminals located in the brain or spinal cord. These terminals release their neuroactive chemical and the process of sensation has begun.

Some receptor cells are true sensory neurons with either free sensory terminal endings or with specialized sensory terminals. Like other types of neurons, receptor cells with specialized endings have an axon that terminates within the central nervous system. This type of receptor cell is typical of the skin, muscle,

and joint senses of touch, pain, stretch, and so on. Some examples of these types of receptors are shown in Figure 2–7. Other receptor cells, however, are not modified neurons, but rather are specialized receptor cells; they lack axons and must be innervated by sensory neurons in order for their output to be transmitted to the central nervous system. Hair-cell mechanoreceptors, photoreceptors, and chemoreceptors are among those included in this category of receptor cells. These types of receptor cells are illustrated later in this chapter. Unlike other neurons, which have their cell bodies within the central nervous system, sensory neurons have their cell bodies outside of the central nervous system. The cell bodies are nearly always clustered together in a **ganglion,** so named because it resembles a "knot." Ganglia are located just outside of the sensory nerve's point of entry into the central nervous system. A sensory ganglion is illustrated in Figure 2–8.

Mechanoreceptors

Mechanoreceptors are among the most ubiquitous types of receptors. Although they exist in a variety of forms, their common feature is their response to mechanical deformation, such as bending, stretching, or twisting. Some mechanoreceptors are involved in the detection of touch or pressure on the body surface. Others respond to stretching such as those in joints, tendons and muscle spindles, and in the digestive system. Still other mechanoreceptors are located at the bases of surface

TABLE 2-3. A Classification of Receptors and Their Senses

Receptor Class	Sensory Modality		Receptor Type
Mechanoreceptor	Touch	Fast adapting	Meissner's corpuscles
		Slow adapting	Merkel's disks
	Tendon stretch		Tendon organs
	Joint position		Joint receptors
	Muscle contraction		Muscle spindles
	Pressure		Pacinian corpuscles
	Hearing		Hair cells
	Vestibular (gravity, acceleration, head position)		Hair cells
	Lateral line		Hair cells
Radiant-energy receptor	Light (including UV)		Photoreceptors
	Infrared radiation (pit organ)		Pit-organ receptors
	Infrared radiation (skin warmth)		Krause end bulbs
	Infrared radiation (skin cold)		Ruffini corpuscles
Chemoreceptor	Taste		Taste buds
	Smell		Olfactory receptors
Electroreceptor	Electric fields		Ampullae
			Tuberous receptors
Nociceptor	Pain		Free nerve endings
	General chemical sensitivity		Free nerve endings
Magnetoreceptor	Magnetic fields		???

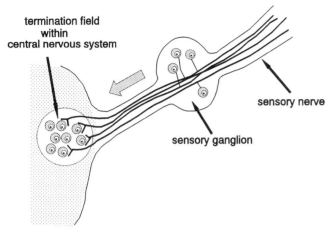

FIGURE 2-8. A sensory ganglion containing the somata of sensory neurons en route to their termination field in the central nervous system. The arrow shows the direction of conduction of the action potentials.

hairs, feathers, and scales and respond when the position of the surface structure is displaced. You can demonstrate this for yourself by lightly running your finger across the hairs on the back of your hand without touching the skin surface. The hairs that make up the facial whiskers or vibrissae of mammals are typical of such hairs. Mechanoreceptors are illustrated in Figure 2–9.

Mechanoreceptors for touch may be classified in several ways. One way is to categorize them according to the speed at which they adapt to continuous bending or deformation;

these are the **fast-adapting** and **slow-adapting** mechanoreceptors. Both types may be further subdivided according to the characteristics of their receptive fields: large with indistinct borders or small with distinct borders.

Lateral Line Organs. An important sense-organ system for aquatic anamniotes are the lateral line organs, which generally are found in the lateral line canals. These canals are located on the head and the body of fishes and larval amphibians. They contain a class of mechanoreceptors known as **neuromasts,** which are hair cell-like receptors. These neuromast receptors are situated in the lateral line organs that line the fluid-filled canals. Some canals are closed and their fluid is internally secreted; others canals have external openings that permit the surrounding water to enter. Neuromasts respond to low-frequency pressure changes in the surrounding water as might be produced by the swimming movements of other fish in a school, or by the approach of a predator.

The lateral line canals of a bony fish are shown in Figure 2–10. At least three canals are located on the head, one above the eye (supraorbital canal), one below the eye (suborbital canal), and one on the lower jaw or mandible (mandibular canal). Additional head canals are present in some fishes. The main or lateral canal runs the length of the body.

Hair Cells. One group of mechanoreceptors consists of hair cells, which are cells with one or more hairs or **stereocilia** protruding from them as well as a longer hair called a **kinocilium.** The stereocilia appear to be involved in the transduction of mechanical energy into a receptor potential, but the kinocilia do not seem to play a direct role in this process. Kinocilia are present in all vertebrates, but in mammals they degenerate during development, and hence are not found in adult mammals.

Figure 2–11 shows a typical hair cell. These cells generate receptor potentials in response to bending of their hairs. Hair cells typically are found in fluid-filled chambers, such as the auditory and vestibular organs of the ear and the lateral-line organs of the head and the body of aquatic anamniotes. This type of receptor is responsive to displacement waves in the surrounding fluid that have been transmitted from the external environment. The pressure waves that are detected by the lateral line system are of low frequency of the sort that would be

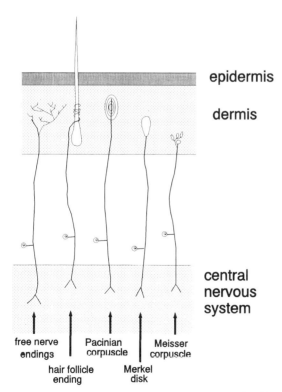

FIGURE 2-9. Examples of somatosensory receptors in the skin.

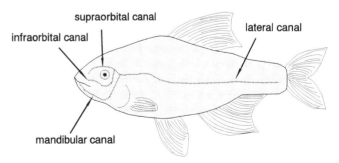

FIGURE 2-10. The lateral line canals of a fish. The main lateral canal runs the length of the body. Three or more canals are located on the head. The figure shows one above the eye (supraorbital), one below the eye (infraorbital), and one on the lower jaw (mandibular).

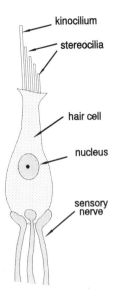

FIGURE 2-11. A hair-cell mechanoreceptor. Hair cells are found in the auditory, lateral line, and vestibular systems.

produced by the swimming movements of neighboring fish in a school or the bow wave of an approaching predator. Likewise, the hair cells of the vestibular system respond to the low-frequency waves that are set up in the fluids of the vestibular apparatus when the tilt of the head is changed or as the acceleration of the body changes. A third set of hair cells is tuned to respond to the higher frequencies of wave action that occur in the auditory organs when those pressure waves in the external environment that we call "sound" (i.e., pressure waves in the frequency range that we can hear) are transmitted to the fluid in the hearing organ. The external environment may be water, as in the case of fishes, or air, as in the case of terrestrial animals. Indeed, the transition from water to land resulted in dramatic changes in the ear and in hearing mechanisms as a result of the differences in the way sound waves are transmitted through water and air.

The auditory hair cells in tetrapods are located in the inner ear in a structure known as the **basilar papilla** or **basilar membrane.** In mammals, this structure (along with some related structures) is called the **organ of Corti.** The hair cells are arranged in rows in these structures. Some of the hairs are free and others are embedded in an overlying structure. In either case, the pressure of the in-coming sound waves ultimately causes these hairs to bend with resulting generator potentials. A more detailed discussion of the mechanics of hearing as well as the variety of hearing organs and their relationships to the animals' environments may be found in the specialized books and articles on this subject listed at the end of the chapter.

Radiant-Energy Receptors

Energy that is propagated in the form of electromagnetic waves (or streams of particles called photons) is known as **radiant energy.** A rather narrow portion of the radiant energy spectrum constitutes visible light. Please remember that when we use the term "visible," we mean visible to humans. Various nonhuman animals have no difficulty detecting portions of the

electromagnetic spectrum that are undetectable to us. Visible light is detected by specialized cells called **photoreceptors.** The two most common types of photoreceptors are the rods and cones of the retina of the eye. These are shown in Figure 2–12. Both types of photoreceptors consist of an inner and an outer segment. The outer segment consists of a series of stacked disks that capture light energy and transduce it into a receptor potential by means of a series of complicated chemical reactions in a group of compounds known as **photopigments.** Each photopigment reacts with light only in a specific range of the spectrum. Rods can be distinguished from cones by the appearance of the stacks of disks; in the rods, the disk diameter remains constant along the length of the outer segment, whereas the diameter of the cone disks become progressively smaller so that the cone outer segment tapers to a blunt point. The inner segments of the photoreceptors contain the nuclei and terminal branches of these cells. Unfortunately, the terms "inner" and "outer" segments are misleading. In fact, the tips of the outer segments of the photoreceptors are pointing towards the interior surface of the eyeball. Thus, the path of light in the retina is from the inner segments to the outer segments.

In nonmammalian vertebrates, rods and cones can be further differentiated by the presence of oil droplets located between the inner and outer segments of the cones. These droplets may be colorless or they may be yellow, greenish yellow, orange, or red. They appear to function as color filters that serve to restrict the spectral range of light that reaches the outer segment. Thus, a yellow droplet filters out short wavelength (blue) light and a colorless droplet filters out UV.

Figure 2–12 also shows some of the varieties of cones, which can include two types of paired cones: double cones and twin cones. These paired cones are characterized by the close proximity of the partners and the absence of pigment epithelium between their outer segments. The partners of these paired cones function in close coordination with each other because they are linked by fast acting, bidirectional electrical synapses. Moreover, they often are found in clusters. Their close proximity thus appears to be a device for increasing the local density of the photoreceptors, which increases the probability of capturing photons. Paired photoreceptors are not found in mammals.

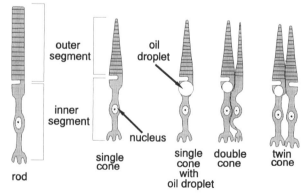

FIGURE 2-12. Photoreceptors of the retina. A rod and several types of cones are shown. The cones of many vertebrates contain colored oil droplets between the inner and outer segments. Vertebrate cones often are paired as may be seen in the double and twin cones.

The Retina. Figure 2–13 shows a diagram of the **retina.** Although we often discuss the retina as if it were a simple sensory surface. The retina is, in fact, a highly complex structure with very sophisticated processing capabilities. Indeed, the retina actually is a component of the central nervous system. A full treatment of the anatomy and function of the retina is beyond the scope of this introductory work, and we can only present some of the main features here.

The retina is the innermost layer of the eye. The outermost layer is the **sclera,** which is the tough coat of the eye that gives the eye its globular shape. The next layer is the **choroid,** which is highly vascular and provides the retina with access to the circulatory system. Deep to the choroid lies a layer of **pigment epithelium.** The tips of the photoreceptors are in contact with the pigment epithelium. As its name implies, the pigment epithelium contains dense pigment granules that exclude light. When the light level is high, the pigment epithelium extends down into the spaces between the outer segments to block excess light from reaching the photopigments. When the light level is low, the pigment epithelium retracts to permit the photopigments to have the maximum opportunity to capture photons.

The retina itself consists of several regions. The innermost region is the layer of photoreceptors. Unfortunately, this region is often referred to as the outer retina because it is the first stage of processing of light stimulation. As Figure 2–13 indicates, light must pass through all of the other retinal layers before it reaches the photoreceptor outer segments to begin the process of sensory transduction. The layer of photoreceptor nuclei is known as the **outer nuclear layer.** The next region is the middle retina, which consists of **bipolar cells, horizontal cells, amacrine cells,** and **interplexiform cells.** The retinal layer that contains the nuclei of these cells is known as the **inner nuclear layer.** The inner retina consists of the layer of ganglion cells, which are the source of efferent axons from the retina to the brain. These axons converge from all parts of the retina to a single location, known as the **optic nerve head,** to form the **optic nerve.** The bipolar cells connect the photoreceptors to the ganglion cells. The remaining cell types modulate the activity in this pathway from photoreceptors to bipolars to ganglions. The layer of the retina in which the photoreceptor terminals contact the dendrites of the bipolar cells is called the **outer plexiform layer** and the layer in which the terminals of the bipolar cells contact the dendrites of the ganglion cells is called the **inner plexiform layer.**

The horizontal cells and amacrine cells provide for lateral interactions within the outer and inner plexiform layers, respectively, while the interplexiform cells link activities between the two plexiform layers. The horizontal cells spread out within

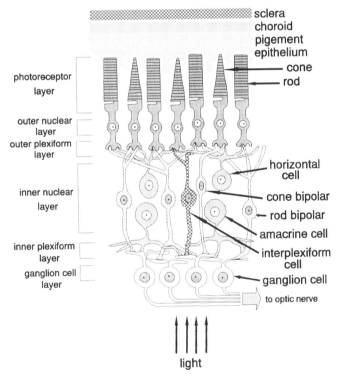

FIGURE 2-13. The retina. The photoreceptors are pointing towards the interior of the eye (sclera, choroid, and pigment epithelium). Their outer segments form the photoreceptor layer, their somata form the outer nuclear layer, and their zone of contact with other neurons of the retina forms the outer plexiform layer. The inner nuclear layer comprises the somata of the various types of integrative neurons of the retina: the horizontal cells, the bipolar cells, the amacrine cells, and the interplexiform cells. The efferent cells of the retina are the ganglion cells; their dendrites contact the terminals of the integrative cells in the inner plexiform layer. The horizontal cells modulate the activities of the outer plexiform layer, and the amacrine cells modulate the activities of the inner plexiform layer. The two plexiform layers are connected by the bipolar and interplexiform cells. Adapted from Dowling (1987) and Dowling and Boycott (1966).

the outer plexiform layer to provide a lateral interaction by coupling or uncoupling of the photoreceptors, which permits them to function as groups or as independent detectors of photons. The amacrine cells perform a similar function within the inner plexiform layer. The interplexiform cells provide feedback from the amacrine cells to the horizontal cells and to the bipolar cells.

The Optic Nerve. The optic nerve, which contains the efferent axons of the ganglion cells, leaves the globe of the eye and passes towards the brain. On route to its terminations in the brain, the optic nerve crosses the midline so that the axons from the right eye terminate in the left brain and vice versa. This point of crossing is known as the **optic chiasm.** Once the ganglion-cell axons have crossed the midline in the optic chiasm, they are thereafter referred to as the **optic tract.** Not all of the optic nerve axons decussate (cross the midline) in the optic chiasm. Some axons remain ipsilateral (on the same side) to the eye of origin. The proportion that remains ipsilateral varies considerably across vertebrate classes. In general, however, nonmammals tend to have the overwhelming majority of optic axons crossing to the contralateral (opposite) side. In mammals, the proportion of ipsilateral optic axons can be as many as 50%.

Centrifugal Axons. Not all of the axons in the optic nerve travel from the eye to the brain. In the optic nerve, as in virtually all cranial nerves, centrifugal axons are present that pass from the brain to the receptors. These centrifugal axons generally suppress the activity of the receptors and may play a role in attention mechanisms.

The Median Eye. The eyes that we have been discussing thus far may be considered lateral eyes in that they are located more or less at the sides of the skull. The optics of these eyes typically are adapted to forming detailed images of the external world on the retina. A considerable amount of this detail is represented in the neural signals that are transmitted to the brain. In addition to these lateral eyes, many vertebrates (but not mammals) have a single, unpaired median eye, located at the top of the skull. This median eye is known as the **pineal eye.** Because the pineal is also known as the **epiphysis,** we shall refer to this eye as the **epiphyseal eye.** Unlike lateral eyes, median eyes are not designed to form images, but merely to gather light. The median eye usually consists of a layer of photoreceptors that send their axons directly to the brain. Median eyes do not have the additional neuronal elements that are found in the retina, and therefore are greatly limited in their ability to transmit highly detailed information about the spatial properties of the visual world. In some vertebrates, a second median eye is present just below the first. The second eye is known as the **parapineal eye** or **paraphysis.** Frequently, the median eye is found below a translucent patch of skin that permits diffuse light to reach the photoreceptors. In some vertebrates, however, a lens-like element occurs that aids in the collection of light. The median eye synthesizes the hormone and neuroactive substance **melatonin** from serotonin in the absence of light. Melatonin is involved in daily biological rhythms and in seasonal rhythms that depend on the progressive lengthening or shortening of the day, metamorphosis, color change, and sexual development. Figure 2-14 shows such a median eye. The axons of median eyes terminate in the brain.

Pit Organs. Another type of radiant energy detector is the pit organ, which is found in a group of venomous snakes known as "pit vipers." Rattlesnakes and the other pit vipers have a sensory pit that is located on the snout between the lateral eyes and the nostrils. Certain nonvenomous snakes, such as boas and pythons, have rows of pit organs located on their upper and lower lips. Suspended between the inner and outer cavities of the sensory pit is a membrane that consists of a type of photoreceptor that is sensitive to the IR range of the electromagnetic spectrum. IR is invisible to the human eye, but it can be felt on our skin as heat. In contrast to the heat receptors of the skin, however, the pit membrane is extremely sensitive to subtle differences between the IR radiation coming from an object (such as a small mammal) and that coming from the background. The pit membrane is innervated by sensory axons of a nerve that terminates in the brain. A pit organ is illustrated in Figure 2-15.

Chemoreceptors

Chemoreceptors sense the chemical properties of the environment. A common feature among chemoreceptors is that they are capable of being stimulated by certain classes of water-soluble chemicals. The two chemosenses that humans are familiar with are gustation (taste) and olfaction (smell). In water, these senses act both as distance receptors (i.e, they can detect

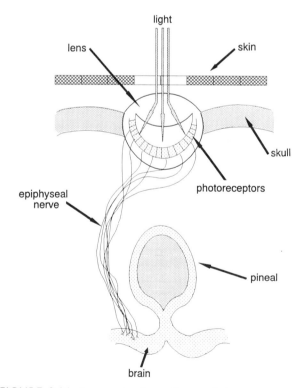

FIGURE 2-14. An example of a median eye. Light impinges on the photoreceptor layer, which is the origin of the epiphyseal nerve that terminates in the epithalamus near the pineal (epiphysis).

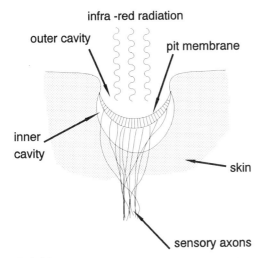

FIGURE 2-15. An IR pit receptor. An IR-sensitive membrane is suspended between the inner and outer cavities of the sensory pit. The membrane is innervated by sensory axons of the trigeminal nerve.

stimulus sources that are remote from the receptor, similar to hearing and vision) and as contact receptors (i.e., they detect stimuli that are in direct contact with or very close to a body surface as do touch, temperature, and pain receptors). This occurs because the chemical stimuli must be soluble in the water surrounding the animal. For animals that live in air, however, only the olfactory sense maintains its dual role as both a distance sense and a contact sense. Gustation, on the other hand, is only a contact sense in land animals because the taste stimuli are only soluble in water, not in air; these stimuli can only produce taste sensations when they dissolve in the epithelial coating of the taste buds within the oral cavity.

Gustation. The gustatory receptors are located within structures known as **taste buds.** In tetrapods, which have tongues, the taste buds are located mainly on the tongue, although some are in the throat as well. The muscular tongue evolved in those branches of the vertebrate lineage that left the water and adapted to life in air where there is no convenient column of sucked-in water to carry food from the mouth to a point where it could be swallowed into the digestive system. The tongue serves this purpose in land animals. In nontetrapods, taste buds not only are located in the mouth, but also in the throat, on the head and, sometimes, as in carp, all over the body surface. Some fishes, such as the catfishes, have evolved facial appendages (that give these animals their name) that are studded with taste buds. These appendages, known as **barbels,** are used to sample the gustatory qualities of the environment in the search for edible objects. A catfish typically has more than 100,000 taste buds. Other fishes have fewer but still quite substantial quantities. The number of taste buds on the tongue varies among different vertebrate classes. Fishes may have many thousands or even hundreds of thousands of taste buds. Among the land vertebrates, birds have the fewest taste buds and mammals have the greatest numbers. Even within classes the number varies considerably. Thus, among mammals, humans have approximately 10,000 taste buds (90% of which are on the tongue),

in contrast to rabbits and some ungulates (hoofed mammals), which can have two to three times that number.

Figure 2-16 shows a taste bud with its taste receptors. Taste receptors typically are located within a **gustatory papilla**, which can be a raised structure or a pit with a pore opening to the outside. The walls of these papillae contain small, barrel-shaped structures, which are the taste buds. Each taste bud contains dozens of taste receptor cells.

The primary taste qualities of humans are "sweet," "sour," "salt," and "bitter." Other more complex tastes are possible from combinations of these. Taste, however, must be distinguished from *flavor*, which is what most people mean when they refer to the "taste" of something. For example, This tastes like chocolate or This tastes exactly like the way my mother cooked it. These sentences really mean that something had the flavor of chocolate or had the same flavor as Mom's recipe. The reason that this distinction is necessary is because flavor is a different chemosense than taste; flavor is olfaction. We are all familiar with the loss of the flavor of food when we have a bad cold with a stuffy nose. We may complain that nothing seems to have any taste in such a circumstance, but in fact our sense of taste (sweet, sour, salty, and bitter) is perfectly intact. What has occurred is that olfactory stimuli from the mouth (or from the external world, for that matter) cannot reach our olfactory receptors. Gustatory neurons in nonhuman land vertebrates respond to the same or similar chemical compounds that produce the human taste qualities; among these are sugars, acids, salts, and alkaloids. Whether the subjective experiences of these animals to these chemicals is the same as ours, we cannot know, but the likelihood that they differ seems great. Gustatory neurons in aquatic vertebrates also respond to these same classes of chemical compounds, but in addition respond to a wide range of amino acids. We can only guess at what the subjective gustatory effect of mixing amino acids with the other taste chemicals in varying proportions might be.

Gustatory sensations are not uniformly distributed on the surface of the tongue. Figure 2-17 shows a schematic diagram of a human tongue with the zones that represent the locations

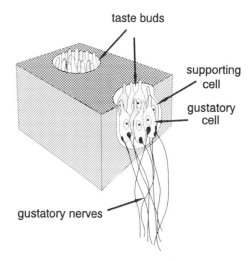

FIGURE 2-16. Taste buds. The cut-away view on the right reveals the structure of the taste bud.

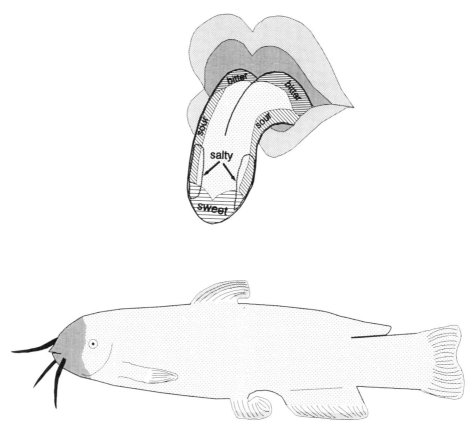

FIGURE 2-17. Top: The taste zones of a human tongue. Receptors that respond primarily to sweet are located at the tip of the tongue. On the sides, just behind the sweet zone, are the salty and sour zones. The bitter zone is located at the rear of the tongue. Note the overlap in the taste regions. Bottom: The relative distribution of taste buds on the body of a catfish. The taste buds are indicated by shading. Greater concentrations of taste receptors are shown by darker shading. Based on data from Atema (1980).

of the various taste sensations of humans: sweet (tip), salty (anterior sides), sour (posterior sides), and bitter (back of the tongue). Note the overlap of some of the taste zones. Figure 2-17 also shows the relative distributions of taste buds on the body surface of a catfish. The darker the shading, the greater the density of taste receptors. Note the presence of taste receptors on the fins and the very high concentrations of receptors on the barbels, the head, and the snout.

Olfaction. Olfactory stimuli are very complex. They emanate not only from food objects, but from a variety of sources. Because humans are **microsmatic**, which means that we have a relatively poor olfactory system, we cannot appreciate the rich and complex olfactory world of the **macrosmatic** animals, which have well-developed olfactory systems. The canidae, which include dogs, wolves, coyotes, and other dog-like animals, for example, are macrosmatic and their abilities to identify individuals by smell are well known.

Olfactory receptors are located in cavities within the nose. The openings of the nose (the nostrils or nares) allow the medium of the external environment (air or water) into the nasal cavity. Within the cavities is the olfactory epithelium that contains the olfactory receptors. These cavities, however, cannot be blind sacs, which would inhibit the continuous flow of

the medium from bringing a smooth and continuous sampling of the external environment to the olfactory epithelium. They therefore have both an anterior and a posterior opening. In fishes, the water flows into the anterior nares and out the posterior nares. Figure 2-18 shows the inflow and outflow of the water column through the anterior and posterior nares. Within the nasal cavities of fishes is a highly folded surface that contains the olfactory receptors. This folding greatly increases the receptor surface area permitting a greater number of receptors to be packed into a small space. In tetrapods, however, the air column is pulled into the nares during the inspiration phase of the respiratory cycle and drawn down into the lungs (Figure 2-19). Openings in the ethmoid bone, which forms the roof of the nasal cavity, lead to the olfactory epithelium. Special bones within the nasal cavity called **turbinate bones** insure a sufficient turbulence of the air column so that the volatile molecules can reach the olfactory epithelium. The olfactory epithelium is coated with mucus into which protrude the olfactory receptors. Olfactory stimuli (or odorants) are chemical substances that can dissolve in the mucus. Unlike the relatively small number of categories of taste stimuli, the number of odorant categories is vast.

The axons of the olfactory receptors terminate in the **olfactory bulb,** which is a specialized olfactory region of the brain

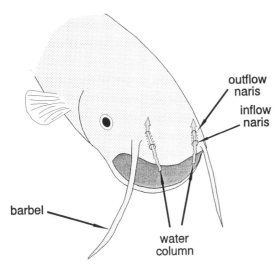

FIGURE 2-18. The nares (nostrils) of a catfish. A column of water enters the anterior naris, passes over the olfactory receptor sheet, and exits through the posterior naris. The barbels of the catfish are covered with taste buds as is much of the rest of its body surface.

that lies directly above the ethmoid bone. Macrosmatic animals have large, well-developed olfactory bulbs; microsmatic animals have small, less-well-developed olfactory bulbs. Figure 2-19 also shows how the olfactory receptors pass through the ethmoid bone to terminate in the overlying olfactory bulb.

Vomeronasal Organ. An important aspect of the olfactory world is the social aspect. One facet of this is the recognition of individuals that was mentioned above. Another facet of social

olfaction is chemical signaling or communication. Many animals signal danger, readiness to mate, or mark their territories by means of a class of chemical-communication substances called **pheromones.** A pheromone is a chemical substance secreted by one animal that produces a specific behavioral reaction in another animal. Alarm pheromones indicate some dangerous situation and sex attractant pheromones help males to locate sexually receptive females. Humans normally do not detect sex-attractant pheromones that are secreted by the bodies of other humans, although a few subtle exceptions have been noted. Instead we rely on a group of chemicals produced by human commercial manufacture known as perfumes.

In some amphibians, some diapsid reptiles, and many mammals, the chemoreceptors for pheromones are located in the nasal cavity. But in a number of vertebrate classes, these chemoreceptors are located in a separate cavity that is in the roof of the mouth and known as the **organ of Jacobson.** Because this cavity is located in a bone known as the vomer, the organ of Jacobson frequently is known as the **vomeronasal organ.** The vomeronasal organ is especially well-developed in snakes and some lizards, but absent in turtles; the rapid flicking in and out of the tongue serves to capture olfactory molecules and to bring them in contact with the vomeronasal organ on the roof of the mouth. Likewise, many ungulates curl their upper lips and inhale air. This behavior, known as the *flehmen* response, serves to draw air over the entrance to the vomeronasal organ. Figure 2-20 shows a vomeronasal organ in a diapsid reptile.

Primates have a vomeronasal organ during embryological development, but it becomes progressively reduced in size so that by birth it appears to have disappeared. Recent reports, however, indicate that a small vomeronasal organ is present in adult humans. Whether it has neuronal connections to the brain or whether chemical stimulation of it has any behavioral consequences are unknown at present.

Unlike the olfactory receptors of the nasal cavity, those of the vomeronasal organ do not have cilia extending into the

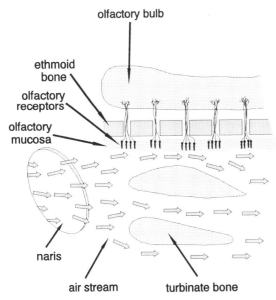

FIGURE 2-19. The olfactory organs of a tetrapod. A stream of air enters the nasal cavity from the outside. The air flows across the turbinate bones and comes in contact with the olfactory receptors located in the mucus layer below the ethmoid bone. The axons of the olfactory receptors pass through openings in the ethmoid bone and terminate in the overlying olfactory bulb.

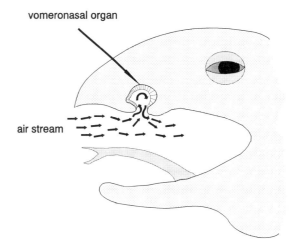

FIGURE 2-20. The organ of Jacobson or vomeronasal organ in a lizard. The flicking of the animal's tongue conducts the air stream into the vomeronasal organ, which is located in the vomer bone of the roof of the mouth. The vomeronasal organ detects pheromones, which are olfactory stimuli that can function as chemical-communication signals.

mucus that covers the receptor surface. A further difference between the nasal and vomeronasal systems is that they have different terminations within the brain. The nasal receptors terminate in a part of the brain known as the **olfactory bulb** (or sometimes as the **main olfactory bulb**), whereas the vomeronasal receptors terminate within an **accessory olfactory bulb** located near the main olfactory bulb.

Nervus Terminalis: An Unclassified Receptor

Still another type of receptor, the functions of which have not yet been established, is the **nervus terminalis** or terminal nerve, a microscopically small nerve that innervates the nasal septum. This nerve is present in all jawed vertebrates. Unlike the nasal and vomeronasal systems, the terminal nerve has no specialized receptor endings but rather ends in free nerve endings. Because many of the axons of the terminal nerve contain LHRH, which is involved in a number of reproductive neuroendocrine processes, the terminal nerve may function, among other things, in the detection of pheromones, especially in those vertebrates in which a vomeronasal organ is not present, although this has not been demonstrated experimentally. The terminal nerve also may detect temperature changes or other nonchemical, intranasal stimuli.

Electroreceptors

The detection of electric fields is widespread among vertebrates. Some aquatic predators can detect the electric fields produced by the contractions of the muscles of their prey as they swim. Even the electric fields produced by the contractions of the muscles of respiration may be sufficient for predators to detect at a short distance. Although predatory electroreception is typical of certain groups of cartilaginous fishes and ray-finned fishes, recent investigations report electroreception in lampreys, some amphibians, and in monotreme mammals as well; both platypuses and spiny anteaters have been reported to have electroreceptors in the skin of their snouts.

In addition to detecting the movements of prey, several groups of animals are capable of detecting the self-generated electric fields around their own bodies. These electric fields are much larger than those produced by routine muscle contraction. Rather they are produced by special organs known as **electric organs.** Many of you have heard of the electric eel that is capable of delivering a powerful electric shock to stun or kill prey or animals that threaten it. This powerful electric discharge is produced by electric organs. Other fishes are capable of powerful electric organ discharges such as the electric catfish, a ray-finned fish, and the torpedo, a cartilaginous fish from which the undersea weapon gets its name. These fishes, however, do not use their electric-organ discharges only for predation and defense; they also use them to detect objects in the environment around them. They do so by emitting weak electric discharges all the time and sensing the distortions of the electric fields surrounding them that result from the presence of nearby objects with high or low electrical conductivity. Objects that conduct electricity well, like another fish, distort the electric field differently than would a poor conductor like a rock. The detection of these weak, self-generated electric fields also is the specialty of several families of ray-finned fishes that are not capable of generating the strong discharges necessary for predation and defence. These animals use their electroreception of changes in the fields around them to locate food, but in a quite different manner than those animals that do not have electric organs; rather than locating prey by the electric fields that emanate from the prey, the weakly electric fishes detect the prey by the effects on the electric fields that they generate themselves. In addition to predation, these electric-organ fishes use their electroreception to detect others of their species, to maintain space around themselves, to identify individuals and potential mates, and for other aspects of social behavior. Some electric fishes also can detect the movement-generated electric fields around their prey. Figure 2-21(B) illustrates how stimuli of either high- or low-electrical conductivity affect the lines of the electrical fields actively generated around a weakly electric fish. Figure 2-21(C) shows a shark, which is using its predatory electroreception to detect the electric fields around a fish that it is about to capture.

The electroreceptors themselves fall into two broad categories: the **ampullae of Lorenzini** or **ampullary receptors** and **tuberous receptors.** Figure 2-22 shows examples of these two types of receptors. The electroreceptors of fishes are found in the lateral line canals and are innervated by branches of the lateral line nerves. The electroreceptors of tetrapods are located in the snout region and are innervated by the trigeminal nerve.

The ampullary and tuberous receptors differ in their sensitivity and function. The ampullary receptors are responsive to the sorts of low-frequency electric rhythms that would result from the muscular contractions of respiratory or swimming movements (0.1–50 cycles per second). Tuberous receptors, on the other hand, have a much higher frequency range (50–200 cycles per second), which is the frequency range of the electric-organ discharge. Further details about electrorecptors and electroreception may be found among the references at the end of this chapter.

Nociceptors

The term **nociception,** the detection of something unpleasant, is derived from the same roots as the word "noxious" and generally refers to pain "receptors" that usually are free nerve endings. These are activated when the tissue is cut, crushed, or otherwise damaged. Another type of noxious sensation arises from the stimulation of certain irritating chemicals such as strong acids, bases, and a variety of irritants and toxins produced by animals and plants as a defence against predation or as a form of predation, such as insect bites or jelly-fish stings. Many of the spicy foods activate general chemical sensitivity in the oral cavity. This sensitivity is not limited to the mouth region, however, as anyone who has been eating, spiced crabs with their fingers and then makes the mistake of rubbing their eyes with their spice-laden fingers can testify. This type of general chemical sensitivity is quite different from chemoreception because receptors are specialized for specific categories

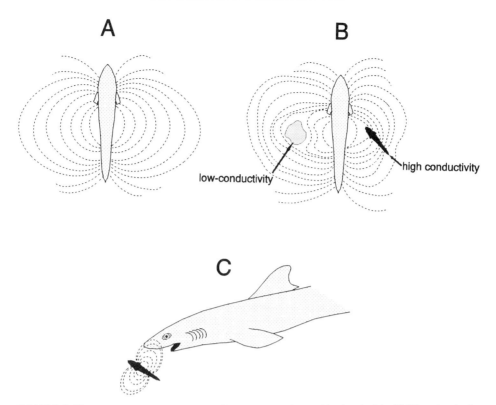

FIGURE 2-21. (A) Actively generated lines of current around a weakly electric fish. (B) Distortion in the current lines as a result of a nearby object of high conductivity (right) and another of lower conductivity (left). Based on Feng (1991). (C) A predatory electroreceptive shark detects the lines of current around a nearby prey fish. Based on Kalmijn (1988).

of chemical compounds and thus unique sensory experiences resulting from these different compounds do not occur. Like pain, general chemical sensitivity is mediated by free nerve endings located under the surface of the skin and in a number of internal organs.

Magnetoreceptors

Behavioral studies have reported that a variety of vertebrates appear to be able to detect various aspects of the earth's magnetic field that they use for orientation and/or navigation; that is, to go in a particular direction and/or to get from one place to another on the surface of the planet. For example, attaching small magnets to the heads of homing pigeons interferes with their ability to find their way home. Although the evidence for magnetoreception is convincing, where these receptors are and what they might be remains a mystery. One recent intriguing possibility that currently is being investigated is that the photopigment molecules of the retinal photoreceptors of birds become magnetized, which permits the bird to "see" the orientation of the magnetic field. Some investigators have reported that cartilaginous fishes also can use their electroreceptors to detect the electric fields that are produced in their bodies when they swim through the earth's magnetic field

the way that an electric current is produced when a coil of wire is moved around a magnet.

TOPOGRAPHIC ORGANIZATION

Another important feature of sensory systems is **topographic organization,** which means that the spatial organization of the receptor surface, such as the skin, the retina, or the basilar papilla, is preserved within the sensory parts of the central nervous system. In the case of the representation of the body in the central somatosensory system, those neurons that receive sensory information that originates on the body surface are organized in a sequential arrangement that roughly mirrors the sequence in which the individual parts are found on the body surface. Thus, head and forelimb representations are at one end of this area, while pathways that originate in the hind limb and tail areas are located at the other end. Within the body representation, the upper trunk is located near the head and forelimb representations and the lower trunk near the tail and hind limb representations.

Sometimes the topographic organization is crude so that only gross areas are represented; other topographic organizations have a very precise point-to-point mapping of the receptor surface within the brain. For example, in those mammals with

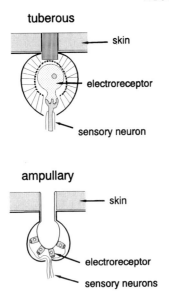

FIGURE 2-22. Two types of electroreceptors: tuberous (top) and ampullary (bottom). Adapted from Feng (1991).

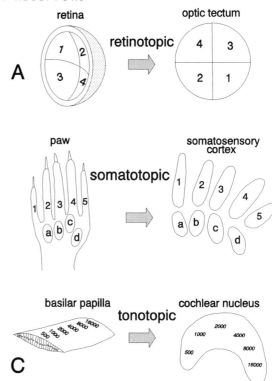

FIGURE 2-23. Topographical organization in the visual, auditory, and somatosensory systems. (A) Retinotopic organization: A topographical representation of the retina on the optic tectum in a reptile. (B) Somatotopic organization: A topographical representation of the paw on the somatosensory cortex in a mammal. (C) Tonotopic organization: A topographical representation of the basilar papilla on one of the cochlear nuclei of a bird.

facial vibrissae (whiskers), each whisker can have its own region of representation in the head area of the somatosensory brain. In general, the more an animal makes use of a sense to detect the fine details of the environment whether it be small objects in the visual world, or subtle acoustic changes in the auditory world, or microvariations in the objects that are touched, the more likely it is that the central representation of the receptor surface will be topographically organized. Moreover, the greater the degree to which a sense is used by an animal, the greater will be the size of the topographic map.

Special names are used to characterize topographic organization within the various senses: the central representation of the retinal map is known as a **retinotopic** organization, and the central representation of the body map is a **somatotopic** organization. In the auditory system, the spatial distribution of the basilar papilla or cochlea is a representation of the different frequencies of sound that the animal can detect with the low tones at one end and the high tones at the other. The preservation of this spatial mapping of tones is known as a **tonotopic** organization. Figure 2-23 shows examples of three receptor surfaces and their corresponding central representations. In the retinotopic organization, although the retinal map appears to have flipped over, the individual quadrants of the retinotopic map are in the same relationship to each other as they are on the retina. In highly visual animals, within each quadrant a point-to-point mapping of the retina to the retinotopic map exists. In the somatotopic map, the central somatotopic map of the digits and the palm are in the correct sequence and form a virtual map of the surface of the paw. Finally, the order of frequencies present in the auditory receptor surface is preserved within the brain. Topographic organizations may be found in any sensory system in which the ordering of either the position in space or specific sensory qualities is important to the animal's success in its environment. Thus, in animals in

which gustation and olfaction are highly developed, evidence for chemotopic maps has been reported. For example, each taste-bud studded barbel (or whisker) of the catfish has its own area of representation within the central gustatory system of this animal.

RECEPTIVE FIELDS

In those senses in which either the body surface or the surrounding environment are mapped in a topographic manner, the notion of the **receptive field** has proved quite useful. A receptive field is that area of a sensory surface (such as the retina or the skin) that must be stimulated in order to influence the activity of a particular sensory neuron in the central nervous system. In the case of vision, the receptive field is that area in visual space in which a stimulus must be presented in order for a neuron in the central visual system to respond. Similarly, a somatosensory receptive field is that region of the body that must be stimulated in order for a central somatosensory neuron to discharge. The existence of discrete receptive fields is the basis of retinotopic and somatotopic organization. The size of the receptive field represents the sum of the excitatory and inhibitory influences present in the sensory pathway that converge on the central sensory neuron. These excitatory and inhibitory influences may be from receptor cells, interneurons, or some combination of both.

THE SENSES AND EVOLUTION OF THE CENTRAL NERVOUS SYSTEM

The diversity of the sensory worlds of animals is matched only by the diversity of the environments that they inhabit. Those animals that achieved the capability to extend their exploration of the sensory world, either by sudden mutation or by more gradual changes in their structure, were able to invade and exploit new adaptive zones to the benefit of themselves and their offspring. These benefits led to further adaptive pressures for increased sensitivity or broader range of responsiveness or altered ranges. In the majority of instances, the development of new sensory adaptations was the wedge that opened the way to new sources of food, new systems for early warning of the approach of predators, more efficient methods for care of the young, better communication with conspecifics, easier detection of the location of potential mates and recognition of their sexual receptivity, and a host of other behavioral adaptations that have promoted both the survival of existing lineages and the development of new ones.

FOR FURTHER READING

Atema, J., Fay, R. R., Popper, A. N., and Tavolga, W. N. (1988) *Sensory Biology of Aquatic Animals.* New York: Springer-Verlag.

Brodal, P. (1992) *The Central Nervous System.* New York: Oxford University Press.

Bullock, T. H. and Heiligenberg, W. (eds.)(1988) *Electroreception.* New York: Wiley.

Bullock, T. H., Orkland, R., and Grinnell, A. (1977) *Introduction to Nervous Systems.* San Francisco: Freeman.

Dowling, J. E. (1987) *The Retina: An Approachable Part of the Brain.* Cambridge, MA: Harvard University Press.

Dowling, J. (1992) *Neurons and Networks.* Cambridge, MA: Belknap/Harvard University Press.

Dowling, J. E. and Boycott, B. B. (1966) Organization of the primate retina: electron microscopy. *Proceedings of the Royal Society (London)*, B, 166, 80–111.

Feng. A. S. (1991) Electric organs and electroreceptors. In C. L. Prosser (ed.), *Neural and Integrative Animal Physiology.* New York: Wiley, pp. 317–334.

Feng, A. S. and Hall, J. C. (1991) Mechanoreception and phonoreception. In C. L. Prosser (ed.), *Neural and Integrative Animal Physiology.* New York: Wiley, pp. 247–316.

Goldsmith, T. (1991) Photoreception and vision. In C. L. Prosser (ed.), *Neural and Integrative Animal Physiology.* New York: Wiley, pp. 171–245.

Kalmijn, A. (1988) Detection of weak electric fields. In J. Atema, R. R. Fay, A. N. Popper, and W. N. Tavolga (eds.) *Sensory Biology of Aquatic Animals.* New York: Springer-Verlag, pp. 152–186.

Kandel, E. R., Schwartz, J. H., and Jessell, T. M. (1991) *Principles of Neural Science.* Norwalk, CT: Appleton and Lange.

Nicholls, J. G., Martin, A. R., and Wallace, B. G. (1992) *From Neuron to Brain.* Sunderland, MA: Sinauer.

Shepherd, G. M. (1988) *Neurobiology.* New York: Oxford University Press.

Ulinski, P. S. (1984) Design features in vertebrate sensory systems. *American Zoologist*, 24, 717–731.

Walker, W. F. and Liem, K. F. (1994) *Functional Anatomy of the Vertebrates: An Evolutionary Perspective.* Second Edition. Fort Worth: Saunders College Publishing.

ADDITIONAL REFERENCES

Atema, J. (1980) Smelling and tasting underwater. *Oceanus*, 23, 4–18.

Collin, S. P. and Northcutt, R. G. (1993) The visual system of the Florida garfish, *Lepisoteus platyrhinchus* (Ginglymodii). III Retinal ganglion cells. *Brain, Behavior and Evolution*, 42, 281–352.

Crowe, A. (1992) Muscle spindles, tendon organs, and joint receptors. In C. Gans and P. S. Ulinksi (eds.), *Biology of the Reptilia, Volume 17, Neurology C, Sensorimotor Integration.* Chicago: University of Chicago Press, pp. 454–495.

Dowling, J. E. and Boycott, B. B. (1966) Organization of the primate retina: electron microscopy. *Proceedings of the Royal Society (London)*, B, 166, 80–111.

Dulka, J. G. (1993) Sex pheromone system in goldfish: comparisons to vomeronasal systems in tetrapods. *Brain, Behavior and Evolution*, 42, 265–280.

Gregory, J. E., Iggo, A. U., McIntyre, A. K., and Proske, U. (1989) Responses of electroreceptors in the snout of the echidna. *Journal of Physiology (London)*, 414, 521–538.

Heiligenberg, W. (1988) Neural mechanisms of perception and motor control in a weakly electric fish. In J. S. Rosenblatt, C. Beer, M-C. Busnel and P. J. B. Slater (eds.), *Advances in the Study of Behavior*, Vol. 18, New York: Academic, pp. 73–98.

Hundspeth, A. J. (1983) Mechanicoelectrical transduction by hair cells in the acoustico-lateralis system. *Annual Review of Neuroscience*, 6, 187–215.

Lohmann, K. J. and Lohmann, M. F. (1994) Detection of magnetic inclination by sea turtles: a possible mechanism for determining latitude. *Journal of Experimental Biology*, 194, 23–32.

Molenaar, G. J. (1991) Anatomy and physiology of infrared sensitivity in snakes. In C. Gans and P. S. Ulinski (eds.), *Biology of the Reptilia, Vol. 17, Neurology C, Sensorimotor Integration.* Chicago: University of Chicago Press, pp. 367–453.

Muske, L. E. (1993) Evolution of gonadotropin-releasing hormone (GnRH) neuronal systems. *Brain, Behavior and Evolution*, 42, 215–230.

Peterson, E. H. (1992) Retinal structure. In C. Gans and P. S. Ulinski (eds.), *Biology of the Reptilia, Vol. 17, Neurology C, Sensorimotor Integration.* Chicago: University of Chicago Press, pp. 1–135.

Ramon-Moliner, E. (1968) The morphology of dendrites. In G. H. Bourne (ed.), *The Structure and Function of Nervous Tissue. Volume 1. Structure.* New York: Academic.

Scheich, H., Langner, G., Tidemann, C., Coles, R. B., and Guppy, A. (1986) Electroreception and electrolocation in the platypus. *Nature (London)*, 319, 401–402.

Schroeder, D. M. and Loop, M. (1976) Trigeminal projections in snakes possessing infrared sensitivity. *Journal of Comparative Neurology*, 169, 1–14.

Taylor, R. (1994) Brave new nose: sniffing out human sexual chemistry. *Journal of NIH Research*, 6, 47–51.

Wagner, H-J. and Djamgoz, M. B. A. (1993) Spinules: a case for retinal synaptic plasticity. *Trends in Neuroscience*, 16, 201–206.

Wilczynski, W., Allison, J. D., and Marler, C. A. (1993) Sensory pathways linking social environmental cues to endocrine control regions of amphibian forebrains. *Brain, Behavior and Evolution*, 42, 252–264.

Wiltschko, W., Munro, U., Ford, H. and Wiltschko, R. (1993) Red light disrupts magnetic orientation of migratory birds. *Nature (London)*, 364, 525–526.

Zakon, H. H. (1986) The electroreceptive periphery. In T. H. Bullock and W. Heiligenberg (eds.), *Electroreception*. New York: Wiley, pp. 103–156.

3

The Vertebrate Central Nervous System

INTRODUCTION

The brain evolved in ancestral chordates in conjunction with the evolution of the head as a specialized body region. One of the principal evolutionary events that preceded the sequences of head and brain evolution was a novel embryological event: the development of the mouth as a distinct body opening separate from the gut opening, which eventually becomes the anus. Animals that possess these characteristics are known as **deuterostomes,** which means "second mouth." Deuterostomes comprise a number of groups of which the three major ones are echinoderms (starfishes and sea urchins), hemichordates (pterobranchs and acorn worms), and chordates. The latter include tunicates (sea squirts), cephalochordates (*Branchiostoma,* previously called *Amphioxus*), and vertebrates. Because some of the key evolutionary events in deuterostome and chordate evolution have been ones that occur relatively early in embryological development, we will begin this chapter with some of the early aspects of the embryological development of the brain in vertebrates.

Following the fertilization of a deuterostome egg, development proceeds with repeated cleavages of the cells. The pattern of cleavage is variable among the different groups of deuterostomes but in all groups results in the formation of a **blastula** [Fig. 3-1(A)], a sphere of cells around a central cavity, the **blastocoel.** Some differences also exist among deuterostomes in particular features of the course of development following the blastula stage, but a general pattern of development is consistent among all deuterostomes.

In vertebrates, the process of gastrulation [Fig. 3-1(B and C)] follows the blastula stage, with continued cell proliferation and the specification and arrangement of specific groups of cells relative to the parts of the body that they will eventually form. The **gastrula** initially consists of two cell layers: an outer layer of cells, the **ectoderm,** from which the skin and nervous system will form, and an inner layer, which consists initially of **endoderm.** The endoderm surrounds the cavity, called the **archenteron,** that will become the lumen of the gut. The archenteron expands during gastrulation, obliterating the blastocoel. The opening of the gut cavity is the **blastopore.**

As gastrulation proceeds, the roof of the archenteron, formed by the dorsal part of the inner layer of cells, differentiates to form mesodermal tissue, which in turn gives rise to a midline structure called the **notochord,** as well as to lateral sheets of tissue that will form muscle. The rest of the inner layer remains endoderm, the tissue of the gut. During the later stages of gastrulation, the ectoderm overlying the archenteron roof thickens to form the **neural plate.** A pair of folds develops along this plate and grows dorsally to meet and then fuses to form a tube, called the **neural tube** [Fig. 3-1(D)]. The lumen (inner space) of this tube forms the ventricular system, while the brain and spinal cord develop within the walls of the tube. The notochord induces the differentiation of a specific group of cells in the midline of the neural plate, called the **floor plate,** and these cells in turn regulate the further differentiation of neurons and the growth of axons within the neural tube. A population of cells that initially lie between the neural tube and the surface ectoderm, the **neural crest,** also contributes

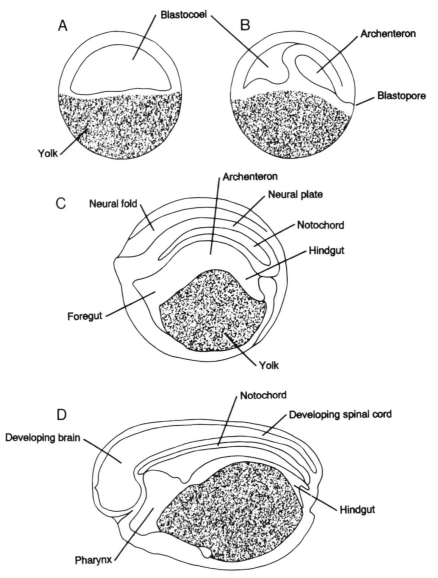

FIGURE 3-1. Drawings of sections through a gastrula (A) and stages of blastulation (B and C) to form a 3 mm embryo (D) in a frog.

to components of the developing nervous system, as do specific areas of neurogenic ectoderm in the head, called **placodes.**

THE BRAIN AND SPINAL CORD

In the head region of the earliest vertebrates, the neural crest and placodes were newly evolved tissues that resulted in the formation of paired sense organs and a new rostral (front) part of the head to house them. As the head developed special new functions, the central nervous system underwent a parallel development that provided the sensory innervation and motor control of the peripheral structures. A new rostral part of the brain, called the forebrain, was added in which sensory information could be analyzed, integrated, and remembered, allowing for sophisticated decision-making capabilities and for appropriate motor responses to a variety of stimuli.

The central nervous system consists of the brain and the spinal cord. Figure 3-2 illustrates the location of the brain and the rostral end of the spinal cord in the head of an adult vertebrate, a lizard in this example, and demonstrates some of the standard terms of orientation (also see the Appendix). The central nervous system traditionally has been divided into two great regions: the brain and the spinal cord. While this distinction is of some importance in medical education and neurology, it suggests a much greater dichotomy of function than actually exists. Indeed, this subdivision is often confusing to newcomers to the neurosciences because it emphasizes the relatively few differences and directs attention away from the many similarities between the two regions.

The boundary between the brain and spinal cord is not nearly as sharp as some textbooks suggest, nor does it correspond precisely to the junction of the skull and the vertebral column (spinal column or backbone), which surrounds and

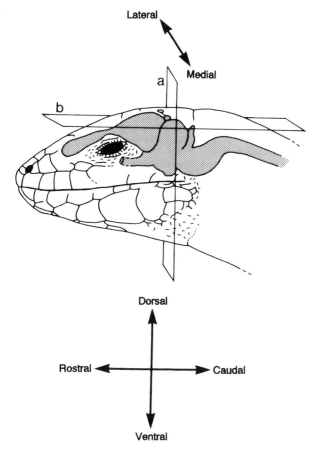

FIGURE 3-2. Lateral view of the head of a lizard (*Lacerta sicula*) with the position of the brain and the rostral part of the spinal cord (indicated by shading) shown *in situ*. Rectangle "a" represents the transverse plane, which runs from dorsal to ventral and medial to lateral. Rectangle "b" represents the horizontal plane, which is approximately parallel to the line of the mouth and runs from rostral to caudal and medial to lateral. The sagittal plane is parallel to the lateral view shown in the figure and runs from rostral to caudal and dorsal to ventral. Adapted from Senn (1979). Used with permission of Academic Press Ltd.

protects the spinal cord. Although a substantial portion of the brain consists of unique structures that have no counterpart in the spinal cord, much of the brain, especially its caudal region, is actually a continuation of the structures and general organization of the spinal cord.

The spinal cord is organized for the control of the body's limbs and trunk; similarly, the caudal brain is organized for the control of specialized structures in the head, such as the jaws, tongue, eye muscles and lids, and vocal organs. Moreover, the caudal brain and spinal cord share the task of control and regulation of the viscera, or internal organs, such as the heart, digestive system, and respiratory system.

CELLULAR ORGANIZATION OF THE CENTRAL NERVOUS SYSTEM

The cells in the brain and spinal cord are of two types: nonneural cells called glia, which fill in the spaces between neurons and wrap around parts of neurons, and the neurons themselves. The neurons can be categorized into three broad categories. The first category is composed of Golgi Type I cells (see Chapter 2) that are the **afferent neurons,** which are cells that bring information about the internal and external environments into the central nervous system. The second category is composed of Golgi Type I cells that are **efferent neurons,** which are motor or effector cells that bring instructions from the central nervous system to the body's effector organs, that is, the muscles and glands. The third category consists of both Golgi Type I and II cells that are **interneurons;** these neurons account for the greatest proportion of the mass of the central nervous system (Fig. 3-3).

We should note that the terms "afferent" and "efferent" can be used in conjunction with the primary sensory and the motor or effector neurons, respectively, or in reference to the direction of information relay in the central nervous system. These terms refer to the direction of conduction and are also used in talking about interneurons. Afferent means "coming into," and efferent means "going out of." A long-axon interneuron coming into and terminating within a specified local-circuit population, which we will call A, is afferent to A. Another long-axon interneuron going from A to a different local-circuit population, which we will call B, is efferent from A and afferent to B.

The interneurons form the neural chain between the primary sensory afferent neurons and the efferent motor or effector neurons. Rarely does a primary sensory neuron make direct contact with a motor neuron. Rather, a few or many thousands of interneurons typically receive the flow of information from sensory neurons into the central nervous system, process that information, and send the resultant outflow of instructions to muscles or glands via the motor neurons. The interneurons determine whether the instructions will be for action or for inaction, whether a response will be rapid or leisurely, vigorous or delicate, towards an object or away from it.

As described in Chapter 2, interneurons form discrete neuronal populations, called **nuclei** (singular: **nucleus**). In addition to **nuclei,** there are also areas of layered sets of interneurons in the brain. Some of these layered areas are called **cortex.** Cortices are present in some of the more dorsal, or roof, areas of the brain. A number of cortical areas are present in the rostral part of the brain in some vertebrates, and cortex is also present in the roof of the middle part of the brain (midbrain) and of the hindbrain.

This general description of central nervous system organization applies to both the spinal cord and brain. The rostral-most parts of the brain have fewer primary sensory, afferent components and fewer effector, efferent components than are present in the brainstem. The rostral brain consists mostly of interneurons arranged in a variety of specialized local-circuit populations that are interconnected by long-axon, Golgi I interneurons. Ascending projections from nuclei in the brainstem bring sensory information and also feedback information up to the rostral brain, and descending projections from the rostral brain carry its output. The heavy concentration of interneurons in the rostral brain has served to focus attention on this region as the possible "executive" component of the central nervous system, with major responsibilities for decision making, mem-

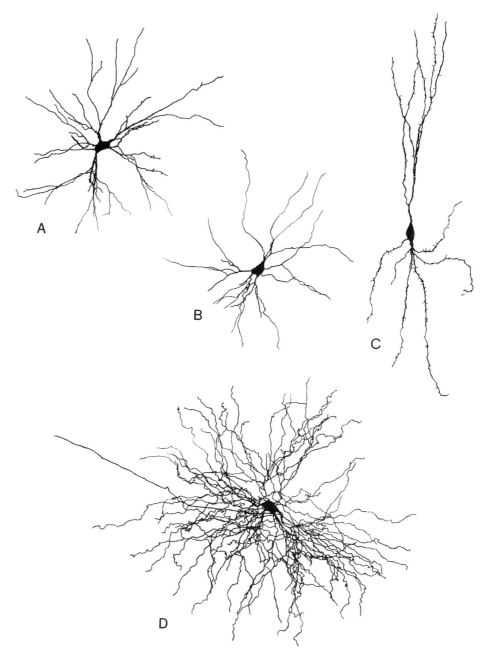

FIGURE 3-3. Examples of interneurons found in the forebrain in various mammals. (A and B) Both are from the dorsal thalamus of a cat, after Robson (1993); (C) is from the cerebral cortex of a monkey, after Lund and Lewis (1993); and (D) is from the dorsal thalamus of a monkey, after Havton and Ohara (1993).

ory, attention, communication, emotion, and other important, complex behavioral processes.

To summarize, the brain and spinal cord are organized to allow for the input of primary sensory information, the analysis and processing of the information, and the production and transmission of appropriate responses to the information. The axons of incoming, primary sensory, afferent neurons bring sensory information into the brain and spinal cord. Long-axon interneurons carry this information between multiple nuclei and cortices. Short-axon, local circuit interneurons within nuclei and cortices are involved with processing the information. Additional sets of long-axon interneurons carry the output for re-

sponses to the information to the outgoing, motor or effector neurons in the brain and spinal cord.

REGIONAL ORGANIZATION OF THE NERVOUS SYSTEM

The Spinal Cord

Both the brain and spinal cord can be subdivided into individual regions. The subdivisions of the spinal cord are named for regions of the spinal column through which the spinal nerves exit on their way to the various parts of the body. Thus, the region of the cord from which nerves exit through the cervical (neck) bones is called the **cervical cord.** This is the most rostral division of the cord and is continuous with the most caudal region of the brain.

Proceeding caudally within the spinal cord, the remaining divisions are the **thoracic (chest) cord,** the **lumbar** (abdominal) **cord,** and the **sacral** (pelvic) **cord.** In addition to receiving primary somatic, afferent fibers from the skin, muscles and joints of the neck, body, limbs (if any) and tail (if any), the spinal cord also sends out motor, efferent fibers (axons) that control the muscles of these body parts. The latter are known as somatic efferent fibers.

From the throacic and lumbar regions of the cord, a special group of efferent neurons sends its axons to innervate the smooth muscles of the digestive system and other internal organs and glands. These are known as **visceral efferent fibers.** The visceral efferents originating in the throacic and lumbar spinal cord are known as the **sympathetic nervous system.** They function to provide a rapid activation of various internal organ systems such as the cardiovascular and respiratory systems when severe demands are placed on these systems or in times of emergency when quick action is required. This system serves "fight or flight" and related responses.

The sympathetic nervous system is complemented by another visceral efferent system, the **parasympathetic nervous system,** which is composed of visceral efferent neurons from the sacral division of the spinal cord and from the caudal regions of the brain. The parasympathetic system controls the same organs as does the sympathetic system, but it does so under normal conditions, when no special stresses are present. This system functions in promoting the digestion of food, allowing urination to occur, and other, related, normal functions of the organs. Thus, when a high-demand situation arises, the sympathetic system takes control; when the high-demand situation ceases, control returns to the parasympathetic system. The sympathetic and parasympathetic systems together are known as the **autonomic nervous system.**

The Brain

Figure 3-4 is a drawing of the brain and rostral spinal cord of a vertebrate, in this case a ray-finned fish, which will serve as an example for the general organization of vertebrate brains. Each of the lobes and other swellings of the brain surface has a specific name, such as the optic lobe (tectum) or cerebellum. Groups of lobes or swellings are known by regional names, such as the mesencephalon (midbrain). Still more-encompassing regional names are forebrain and brainstem.

The nerves entering and leaving the brain are the **cranial nerves.** Like the spinal nerves, the cranial nerves carry primary sensory afferent neurons (in this case from the head rather than the body) into the central nervous system and motor or effector efferent neurons to control the muscles and glands of the head and neck. Although these nerves are given special names, such as trigeminal nerve and oculomotor nerve, the organization of at least some of them is essentially the same as that of the spinal nerves. Moreover, some of the interneuron networks of the more caudal cranial nerves are directly continuous with equivalent cell populations in the spinal cord.

Hindbrain. The caudal-most region of the brain is the **rhombencephalon** or **hindbrain,** which consists of the **medulla** and **pons.** The cranial nerves of this area include components that comprise most of the cranial division of the parasympathetic nervous system. However, a number of what have traditionally been called "special senses" are also represented by cranial nerves that enter the brain at this level. Such special senses are those that are unique to the head region, such as hearing and the sense of balance and acceleration. Other special senses, not present in all vertebrates, include the lateral line system for the detection of water displacement over the body surface and of electric fields in the aquatic environment and the infrared (IR) sense for the detection of body heat radiated by other animals. The cell bodies of the efferent neurons that control the muscles of the jaws and the superficial facial muscles, such as those of the lips and eyelids, are also located in this region of the brain.

In addition to the nuclei associated with some of the cranial nerves, the medulla and pons contain other nuclei and a number of fiber tracts. Many of the fiber tracts are long pathways between the spinal cord and the more rostral parts of the brain, while others are shorter and run between nuclei within this area. Some of the cell populations in the medulla and pons are more widely scattered and diffuse than most other nuclei in the brain. These scattered populations are collectively referred

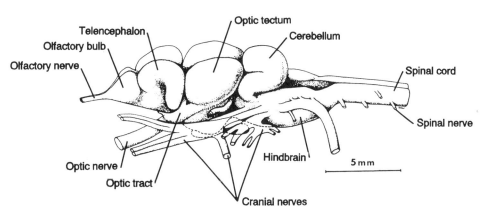

FIGURE 3-4. Drawing of a lateral view of the brain of a ray-finned fish, the longnose gar (*Lepisosteus osseus*). Adapted from Northcutt and Butler (1976).

to as the **reticular formation.** The nuclei of the reticular formation are involved in integrating inputs from a variety of sources, including the cranial nerve nuclei and more rostral parts of the brain. The reticular formation regulates and modulates activities elsewhere in the brain on the basis of the incoming information. A transverse section through the hindbrain in a fish (Fig. 3-5) illustrates the distribution of nerve cell bodies within the brain. Fiber tracts run in the areas in which there are few or no cell bodies. Closely related to the pons, both geographically and functionally, is the **cerebellum** (Fig. 3-4). The cerebellum is a cortical structure in the roof of the hindbrain. Among its other functions, the cerebellum is involved in balance, coordination, and the smooth execution of rapid movements.

Midbrain. The next major subdivision of the brain is the **mesencephalon** or **midbrain** (Figs. 3-4 and 3-5). The most prominent external feature of this area in most vertebrates is the **optic lobe,** or **optic tectum,** so named because of the large number of neurons of the optic nerve that terminate in it. The ventral portion of the mesencephalon, the region underneath the optic tectum, is known as the **tegmentum.** The mesencephalic tegmentum contains a number of nuclei and fiber tracts, including the nuclei of two of the cranial nerves that control the movements of the eyes. The operation of the intraocular eye muscles, which control the pupil and the focus of the lens, are controlled from this region as well.

Forebrain. The **prosencephalon** or **forebrain** is the most rostral division of the brain. It contains two major parts: the **diencephalon** and the **telencephalon.** The diencephalon (= interbrain) lies rostral to the midbrain, caudal-ventral to the telencephalon, and medial to the axons that form the optic tract, the central nervous system continuation of the optic nerve. The diencephalon is a large division composed of six principal areas. The caudal part of the diencephalon contains a dorsal area called the **pretectum** and a more ventral area called the **posterior tuberculum.** More rostrally, four areas are present. In dorsal to ventral sequence, these areas are the **epithalamus,** the **dorsal thalamus,** the **ventral thalamus,** and the **hypothalamus.** Each of these areas are composed of a number of nuclei, and there are also major fiber tracts that pass through this region.

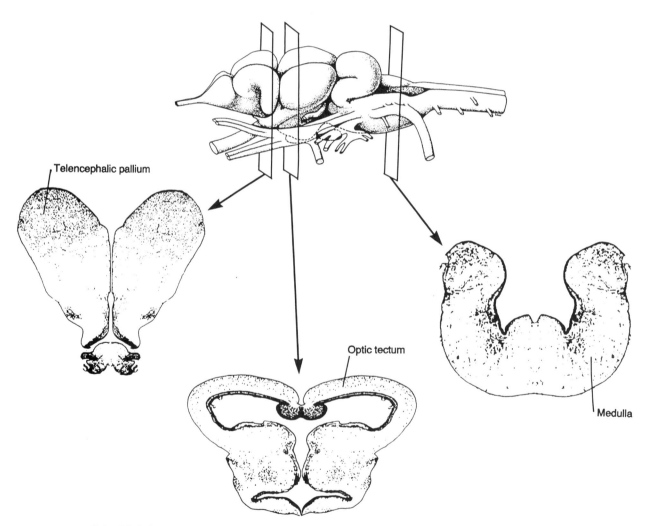

Telencephalic pallium

Optic tectum

Medulla

FIGURE 3-5. The brain of the longnose gar, as shown in Figure 3-4, with transverse sections through the telencephalon, midbrain, and hindbrain.

An additional neuroanatomical term frequently encountered is **brainstem.** This term sometimes refers to the more ventral parts of the brain except for any part of the telencephalon, that is, to the medulla and pons, the midbrain tegmentum, and the diencephalon. It is alternatively used to refer only to the medulla, pons, and midbrain tegmentum (and sometimes the midbrain tectum) without including the diencephalon. Context must be used to determine the specific sense in which this somewhat loose term is being used.

A number of the nuclei in the pretectum receive visual input from the retina and are involved in visuomotor behaviors. Pretectal nuclei and other visually related nuclei in the tegmentum influence eye movements in relation to prey and predator detection and to orientation of the body within space. The posterior tuberculum consists in part of a medially lying nucleus that contains neurons involved in regulating motor functions. In ray-finned fishes, more laterally lying nuclei of the posterior tuberculum are also present. These nuclei relay sensory inputs to the telencephalon. Similar nuclei may be present in cartilaginous fishes, but migrated posterior tubercular nuclei have not yet been identified in amphibians or land vertebrates.

The nuclei in the dorsal thalamus receive information via long-axon interneurons from the various sensory systems, which they transmit to various parts of the telencephalon. The nuclei in the dorsal thalamus constitute a gateway to the sensory areas of the telencephalon. A number of different parts within the telencephalon are involved in integrating the incoming sensory information for learning, memory, emotional responses, and motor responses.

Nuclei in the ventral thalamus are involved in modulating the activity of dorsal thalamic nuclei and also play a role, in concert with telencephalic structures, in motor control of the body and limbs. The epithalamus contains the **epiphysis** (pineal gland and related structures), which is located at the end of a stalk, the epiphyseal stalk. In some animals, such as reptiles, the pineal is a structure very similar to the eye. It contains light receptors and gives rise to primary sensory afferent neurons, which terminate in the epithalamus. In mammals and birds, the pineal is a glandular structure.

The hypothalamus is very much involved in the activities of the autonomic nervous system and the endocrine system. It controls the endocrine system's production of hormones by means of the **hypophysis** or **pituitary,** which is directly connected to the hypothalamus. The pituitary is located at the end of a stalk, the **infundibulum,** which is a direct outgrowth of the base of the hypothalamus. The hypothalamus thus is able to control and regulate behavior patterns that depend on the levels of hormones in the blood, such as sexual behavior, parental behavior, territoriality, migration, and hibernation, to name but a few. The hypothalamus also is concerned with feeding and drinking, aggression, temperature regulation, and a number of other important biological and behavioral functions.

The most rostral region of the brain, the **telencephalon** (Figs. 3-4 and 3-5), includes the **dorsal pallium,** or **cerebral hemisphere** (also called the **cerebrum**). At the rostral end of the cerebrum is the **olfactory bulb,** in which axons of olfactory receptor cells in the nasal mucosa terminate. In animals in which the sense of smell is highly developed, such as some sharks and bloodhounds, the olfactory bulb is rather impressive in size. In animals such as many birds and some mammals, in

which the sense of smell is not especially important for survival, the bulb is relatively small in comparison. The olfactory bulb projects to the **olfactory pallium** in the telencephalon via **olfactory tracts,** which can be short if the olfactory bulb lies adjacent to the telencephalon (as in the brain of the fish shown in Fig. 3-4) or elongated.

The cerebrum itself has a relatively smooth surface in most vertebrates. It is composed of nuclear areas in some vertebrate groups, such as cartilaginous and ray-finned fishes; in other vertebrate groups, particularly tetrapods, it is composed of both nuclei and cortex. In some animals, mostly among the mammals, although there are other instances as well, the surface of the cerebrum develops deep ridges and valleys known as convolutions. The ridges are called **gyri** (singular = **gyrus**) and the valleys are called **sulci** (singular = **sulcus**). The deepest of the sulci are known as **fissures.**

The telencephalon also includes several other major areas. The **limbic pallium** comprises a set of structures that include a cortical area known as the **hippocampus,** which is necessary for memory and is also involved in emotion. The hippocampus and some nuclei related to it, including the **septal nuclei** and **amygdala,** receive olfactory information from other telencephalic areas, which is integrated with other inputs. Several additional major areas lie in the more ventral part of the telencephalon and are involved with motor functions. These areas are collectively called the **striatum,** or **striatal nuclei.** They are interconnected with motor areas of the dorsal pallium and with nuclei in the more caudal parts of the brain.

The Meninges and the Ventricular System

The brain and spinal cord are covered by one or more layers of connective tissue, which are called the **meninges,** from the Greek word **meninx,** which means membrane. In fishes, only a single layer, the **primitive meninx,** is present. Amphibians and reptiles have two meningeal layers, an outer **dura mater** (meaning "hard mother") and an inner thin layer, the **secondary meninx.** In mammals and birds, three meningeal layers are present. The layer closest to the brain is a thin, vascular layer called the **pia mater** (meaning "tender mother"). The middle layer is a thin, avascular layer called the **arachnoid** due to its spider web–like appearance. The space between the pia mater and the arachnoid is the **subarachnoid space.** The outermost layer is the **dura mater** and is actually composed of two layers: an inner layer enclosing the central nervous system and an outer layer that lines the inside of the skull. The use of the word "mother" to describe these membranes comes from an ancient notion that they were the origin, or mother, of all membranes in the body.

Embryologically, the central nervous system develops from a hollow tube. The walls of the tube thicken to form the brain and spinal cord, and the hollow within the tube becomes the fluid-filled **ventricular system** of the adult. Instead of remaining a straight tube of uniform diameter, the ventricular system extends laterally into the variously expanded parts of the brain in different vertebrate groups, such as the olfactory bulbs, telencephalic hemispheres, the midbrain roof, and/or the cerebellum. This arrangement is shown in Figure 3-6.

In most groups of vertebrates, the ventricular system expands laterally within each of the telencephalic hemispheres,

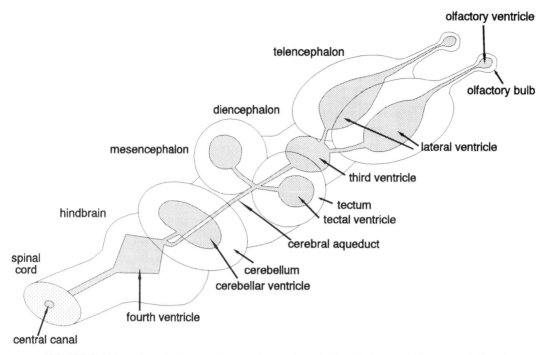

FIGURE 3-6. Dorsolateral schematic drawing of a vertebrate brain with the ventricular system, indicated by shading, projected onto it. Rostral is toward the upper right.

and this pair of laterally extending spaces are called the **lateral ventricles.** The lateral ventricles are in continuity, through paired openings called the **foramina of Monro,** with the unpaired, medial ventricular space of the diencephalon, called the **third ventricle.** The caudal continuation of the third ventricle is a narrow canal called the **cerebral aqueduct of Sylvius,** which in turn opens into the **fourth ventricle,** the unpaired, medial ventricular space of the hindbrain. The fourth ventricle is caudally continuous with the **central canal** of the spinal cord.

Parts of the ventricular walls consist of a thin ependymal epithelial layer and an inner, vascular meningeal layer, which together form the **tela choroidea.** A network formed by the blood vessels of the inner meningeal layer, called the **choroid plexus,** secretes **cerebrospinal fluid** into the ventricular spaces. The cerebrospinal fluid circulates in the ventricular system and also in the subarachnoid space; it reaches the latter by passing through three openings that connect the fourth ventricle with an enlarged part of the subarachnoid space, the **cisterna magna.** The three openings are a paired set, the **foramina of Luschka,** and an unpaired, medial foramen, the **foramen of Magendie.** After circulating, the cerebrospinal fluid passes out of the subarachnoid space into vascular sinuses through structures in the arachnoid called **arachnoid villi,** which act as one-way, pressure-sensitive valves. The cerebrospinal fluid provides support for the brain and cushions it from physical shocks by its buoyancy.

MAJOR SYSTEMS OF THE BRAIN

We now want to give you a very brief overview of some of the major systems of the brain. As we have discussed above, there are major sensory and motor systems in which information is relayed through various nuclei, being modified and sorted along the way. In later chapters, which cover various parts of the brain, a general idea of the basic organization of these systems will help you to make sense of the anatomy.

Sensory Systems

All of the receptors for touch, radiant-energy sense, pain, lateral line (for wave displacement and electrical field detection), hearing, vestibular sense, and gustatory sense have their initial points of termination within the central nervous system either in the spinal cord or in the medulla–pons region of the brain. The target cell populations of these primary pathways then project to more rostral levels of the brain. Most of the axons in the primary sensory pathways from the eyes terminate in part of the dorsal thalamus in the diencephalon and in the optic tectum. The olfactory pathways, in contrast to the more caudal sensory pathways, project directly into part of the telencephalon. Topographic organization is an important characteristic of all of these pathways; this feature provides for orderly maps of the sensory input within particular parts of the brain that correspond to the spatial map of the external world.

Each of the ascending sensory pathways to the telencephalon terminates in its own segregated region of the telencephalon. From here begins the complex and varied series of possible pathways that are the routes to other sensory, integrative, and motor systems. Functions including memory storage of sensory events and the consequences of reactions to them, decision making about subsequent reactions to such events, and the conscious awareness of these events reside within the telencephalon. In order to gain a general understanding of how ascending sensory systems are organized, we will briefly outline three of them here: the **auditory,** the **visual,** and the **somato-**

sensory, which includes touch and position sense from the body.

The Auditory Pathway. In the auditory pathway, axons arise from neurons in the inner ear and pass into the brain in the **eighth (octaval) cranial nerve.** These axons synapse on the dendrites of neurons in an auditory nucleus in the hindbrain [Fig. 3-7(A)]. The neurons in this nucleus give rise to axons that pass rostrally in a tract (called the **lateral lemniscus**) and synapse on the dendrites of neurons within a part of the midbrain roof called the **torus semicircularis** in most vertebrates and the **inferior colliculus** in mammals. The torus semicircularis lies ventral and/or caudal to the optic tectum. Neurons within the torus semicircularis give rise to axons that terminate on the dendrites of neurons in an auditory nucleus in the dorsal thalamus. Neurons in the dorsal thalamic auditory nucleus give rise to axons that pass into the telencephalon (via tracts called the **forebrain bundles**) and terminate on the dendrites of neurons within one or more auditory regions of the dorsal pallium.

The Visual Pathways. Neurons in the retina give rise to axons that enter the brain via the **optic nerve** [Fig. 3-7(B)] and its continuation, the **optic tract.** The retinal axons terminate on the dendrites of neurons located either in the **optic tectum** of the midbrain (called the **superior colliculus** in mammals) or in a visual nucleus in the dorsal thalamus. Visual neurons in the optic tectum also give rise to axons that terminate on the dendrites of neurons in a second visual nucleus in the dorsal thalamus. Neurons in each of the two visual dorsal thalamic nuclei give rise to axons that pass into the telencephalon via the forebrain bundles and terminate on the dendrites of neurons in two or more visual areas within the dorsal pallium.

The Somatosensory Pathways. Axons that carry somatosensory information [Fig. 3-7(C)] enter the spinal cord and pass rostrally to terminate on the dendrites of neurons within two cell groups called the **dorsal column nuclei.** These nuclei lie in the junctional area between the spinal cord and the brainstem. Neurons in the dorsal column nuclei give rise to axons that pass rostrally in a tract called the **medial lemniscus.** Some of these axons terminate on the dendrites of neurons in a somatosensory part of the midbrain tectum. Other somatosensory axons from the dorsal column nuclei bypass the midbrain and terminate on the dendrites of neurons in a somatosensory nucleus in the dorsal thalamus. Neurons in the somatosensory part of the midbrain also give rise to axons that pass rostrally and terminate on the dendrites of neurons in a second somatosensory nucleus in the dorsal thalamus. Neurons in each of the two dorsal thalamic somatosensory nuclei give rise to axons that pass into the telencephalon via the forebrain bundles and terminate on the dendrites of neurons in two or more somatosensory areas within the dorsal pallium.

In the telencephalon, sensory information is relayed through multiple sets of long- and short-axon interneurons—to secondary, tertiary, and further sensory and association cortical areas, into the limbic system for memory, into multisensory association cortices for integration, and so on. When the information has been processed and assimilated, appropriate motor responses follow.

Motor Systems

Complex control mechanisms for motor responses derive from neurons in the motor pallial areas of the telencephalon, and/or from neurons in other dorsally lying structures, such as the roof of the midbrain and the cerebellum. Motor responses are also regulated by a number of structures including the **striatum,** also known as the **basal ganglia,** in the ventrolateral part of the telencephalon [Fig. 3-7(D)] and nuclear areas within the hindbrain. Mammals have motor neurons in the dorsal pallium of the telencephalon that give rise to axons that terminate on the dendrites of neurons within the striatum. In turn, neurons within the striatum give rise to axons that pass caudally and terminate within a number of nuclei in the brainstem, including a nucleus in the diencephalon. Neurons in the latter nucleus give rise to axons that terminate on the dendrites of neurons within the midbrain tectum.

Other neurons within the motor part of the dorsal pallium (long-axon interneurons) pass caudally via the same major bundles in which sensory axons ascend: the forebrain bundles. These axons collectively form a tract that is rather like a major interstate highway. They pass caudally to synapse on the dendrites of neurons within nuclei in the midbrain and the brainstem and, in some cases, directly on the dendrites of neurons in the spinal cord. The midbrain tectum also gives rise to axons that pass caudally in a tract called the **tectospinal tract** and terminate on the dendrites of neurons in brainstem motor nuclei and in the spinal cord. Motor nuclei of the cerebellum similarly contain neurons that give rise to axons that pass ventrally to terminate on the dendrites of neurons within brainstem nuclei. Neurons in the various brainstem motor nuclei give rise to axons that pass caudally and terminate on the dendrites of neurons in the spinal cord. The reticular formation of the brainstem is composed of a number of different nuclear areas that also are involved in these descending motor pathways.

DEVELOPMENT OF THE BRAIN

During the embryological development of the brain from the neural tube, the rostral part of the brain flexes ventrally [Fig. 3-8(A)]. A transverse, ventral fold develops on the ventral surface of the tube in the position where the pituitary gland will form. A pair of lateral bulges in the rostral part of the forebrain forms the cerebral hemispheres of the telencephalon. More caudally, the eyes grow out of the diencephalon, and the pineal gland develops in the dorsal part of the diencephalon [Fig. 3-8(B)].

The midbrain lies caudal to the ventral fold in the neural tube. Paired bulges in the roof of the midbrain expand to form the optic tecta (plural of tectum), or roof of the midbrain, while the more ventral part of the midbrain becomes the tegmentum (the main body of the midbrain). The dorsal portions of both the caudal midbrain and the rostral hindbrain enlarge to form the cerebellum [Fig. 3-8(B)].

Within the neural tube, a layer of **neurogenic cells** gives rise to the neurons. The neurogenic cells encircle the ventricular lumen. As neuron cell bodies are produced, they migrate away from the region of the lumen. The degree of migration differs markedly among different vertebrates and in different regions

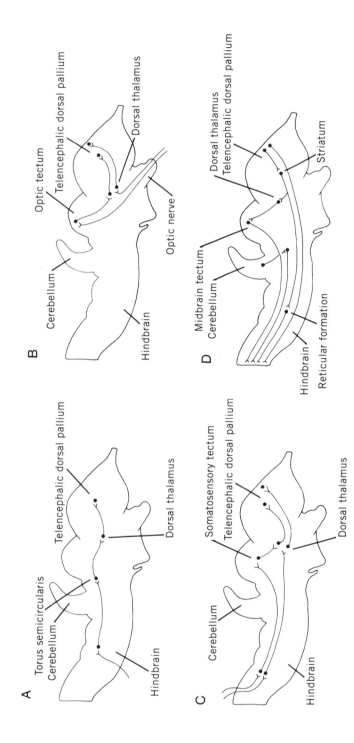

FIGURE 3-7. Schematic representation of the major sensory and motor pathways. Rostral is toward the right. Dots represent neuronal cell bodies, and lines represent the axons of the cell bodies with their terminal endings. Dendrites of cell bodies, on which the axons actually synapse, are not represented. (A) Ascending auditory pathways; (B) the two major ascending visual pathways; (C) two major ascending somatosensory pathways; and (D) some of the descending motor pathways.

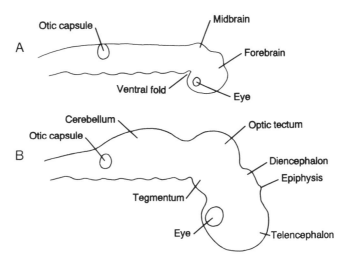

FIGURE 3-8. Diagram of the flexure of the neural tube during formation of the brain.

of the brain. The pattern of migration also varies, with nuclei and cortices being formed in different ways. In this section, we will examine some of the various ways in which the neurons are assembled into structures within different regions of the brain and among different groups of vertebrates.

The developmental processes involved in producing the individual neurons and their connections are complex and also subject to disruption by both local factors and more external influences. The timing of sequences in multiple groups of neurons must be properly phased for normal development to occur. The production and migration of populations of neurons to form nuclei and cortices also occurs in specifically timed sequences, since targets must be established for the growing, incoming axons to locate and reach.

Cortices and Nuclei

As we discussed above, two basic structural patterns for populations of neurons are **cortices,** which are laminated structures that tend to be located in the more dorsal parts of the brain, and **nuclei,** which are prevalent in the rest of the brain. Nuclei may consist of diffusely scattered, either loosely or more tightly packed cells or of cells aligned in a laminar pattern, reminiscent in some cases of cortex.

Cortical areas such as the cerebral cortex in mammals and the optic tectum in many vertebrates are formed by serial migrations of generated neurons. In the mammalian cerebral cortex, neurons that eventually lie in the outermost (layer I) and innermost (**subplate**) layer are produced first, but for the layers that form the bulk of the cortex, the sequence of production is altered. For most of the cortical layers, the neurons that lie nearer to the periventricular germinal zone (matrix) in the adult are produced earlier than those that lie nearer to the outermost layer of the cortex. The subsequently produced sets of neurons migrate through the layers of neurons formed earlier. This pattern of migration is called "**inside-out.**" Following this series of migrations, which result in particular sets of neurons lying in particular layers, growth of axons and dendrites occurs, and connections, both distant and local, are established.

A component of the guidance system for afferent cortical axons has been found within the cortex. The distal end of each in-growing afferent axon is a specialized structure called a **growth cone,** which is the site of elongation of the axon. The deepest layer of cortical neurons, produced first and called the subplate, normally disappears by cell death after the axons from cells in the dorsal thalamus reach their appropriate cortical areas. The axons arrive before their main target cell layer (layer IV) has been generated, but they wait for these target cells to be formed and do not grow beyond that area. However, if the subplate cells are selectively destroyed, the arriving dorsal thalamic axons do not halt in the proper position but continue growing beyond their normal area of termination and on into adjacent, unrelated areas of cortex. The subplate cells thus seem to provide a necessary sign post for the growth cones of the afferent dorsal thalamic axons.

There is a high degree of specificity in the connections of cortical structures, made possible by their geometric configuration. Different afferent neurons terminate on different, specified parts—superficial to deep—of the dendritic trees of the cortical neurons. The entire area of a given cortical region is also frequently in receipt of specifically ordered projections, so that the point-to-point mapping of sensory or motor space from the external world of the animal is mapped in order onto the cortex. In-growing axons may sometimes form connections in the wrong location with reference to the map, but these axons later retract and then form connections in the place that is congruent with the map of external space. These adjustments occur as sensory information comes over the afferent axons and into the cortex.

The nuclei in the brain, as well as some of the cortices, are formed by cellular migrations that proceed in an "**outside-in**" pattern, the opposite of that in the mammalian cerebral cortex. In this case, the more superficial, lateral groups of neurons are generated first. There are additional spatiotemporal gradients that have been identified in the dorsal thalamus. Caudal nuclei develop earlier than more rostral ones, and ventral nuclei earlier than more dorsal ones. Thus, different areas of the periventricular matrix give rise to different sets of nuclei at different times.

While nuclei form by different migration patterns than in some cortices, a number of nuclei have point-to-point maps that correspond to those found in cortex. The nuclei that have such maps are generally those that project to or have reciprocal connections with cortical, mapped areas. In mammals, the **dorsal lateral geniculate nucleus,** the dorsal thalamic nucleus that projects to primary visual cortex, has a point-to-point map of visual space in each of its multiple, separate layers of cells. A similar map is present in a midbrain nucleus of many vertebrates, **nucleus isthmi,** which has reciprocal, point-to-point projections with the optic tectum. Maps are also established and maintained in other ascending sensory systems, including the auditory and lateral line systems and the somatosensory system.

As nuclei are formed by migrations of cells, given regions of the periventricular matrix may give rise to one or more specific nuclei. Such a region is called a "**field.**" The number of distinct nuclei that arise from a given field may be different in different groups of vertebrates. Also, all the cells or cell groups produced by a field during development may not be retained in the adult if the proper connections are not made at the proper time. Such occurrences would affect the phenotype and may well play a role in evolutionary change. A subtle alteration in timing or in the presence or absence of local molecular cues, produced by a mutation established in the genome, could result in the lack of a particular cell group in the adult and/or the appearance of a "new" cell group with new connections.

Differing Patterns of Development

Within each of the major groups of vertebrates, one marked difference in brain development is present that will be discussed in detail in Chapter 4—that of wide variation in the degree of migration of neurons, both cortical and nuclear, away from the periventricular matrix. Those vertebrates in which relatively limited migration occurs, that we have designated as Group I, include some species from each of the four major vertebrate radiations: agnathans (jawless vertebrates), cartilaginous fishes, ray-finned fishes, and sarcopterygians (fleshy lobed-finned fishes and their tetrapod descendants). Other species in each of the same four radiations (Group II) have much more extensive migration of neurons and the formation of multiple, distinct layers in cortical structures as well as a greater number of discretely recognizable and larger nuclei. Two other major differences in the development of the brain among different verte-

brates will be considered in this chapter. These developmental differences result in marked differences in the organization of the telencephalon.

In the initial phase of development from the closed neural tube, the telencephalon develops by a process called **evagination** in most vertebrates, that is, agnathans, cartilaginous fishes, and amphibians and amniotes. The central lumen of the neural tube enlarges to form the telencephalic ventricles as the pallial (dorsal) part of the telencephalon bulges outward and expands, that is, evaginates. This process is shown in Figure 3-9. Following evagination, the part of the pallium that was originally in the most dorsal position (A) around the central lumen comes to lie in the most medial part of the telencephalon. This pallial area forms the hippocampal formation and limbic pallium (the functions of which include memory and emotion) in the adult. The originally intermediate pallial area (B) becomes the dorsal pallium, which forms the major sensory, integrative and motor pallial areas in the adult. The originally most ventral pallial area (C) comes to lie most laterally and gives rise to the olfactory pallium.

In contrast, a different process takes place in the development of the telencephalon in ray-finned fishes. This process is called **eversion.** The part of the roof of the neural tube over the central lumen thins and elongates, and the hemispheres bend outward (Fig. 3-10). Following eversion, the originally most dorsal part of the pallium (A) comes to lie in the most lateral position in the telencephalon. The originally intermediate part of the pallium (B) lies most dorsally, as in other vertebrates. The originally most ventral pallial area (C) comes to lie in the most medial position in the telencephalon of the adult. Thus, in comparison with other vertebrates, the mediolateral positions of the hippocampal (A) and olfactory (C) pallial areas are reversed in the ray-finned fishes. The areas in the basal part of the telencephalon are similarly aligned in all vertebrates, as this region does not undergo either evagination or eversion (Figs. 3-9 and 3-10).

A second major difference in telencephalic development occurs between the mammals on one hand and the nonmammalian amniotes on the other. The dorsal part of the evaginated pallium (B in Fig. 3-9) forms the regions of six-layered cortex called **neocortex,** or **isocortex,** in mammals (Fig. 3-11), which occupies the bulk of the cerebral hemispheres. In diapsid reptiles, birds, and turtles, part of the dorsal pallium also forms a superficial cortical area, but a larger part expands in a medial

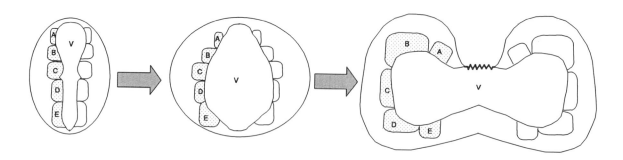

evagination

FIGURE 3-9. Diagram of the process of evagination in forebrain development, as occurs in most groups of vertebrates.

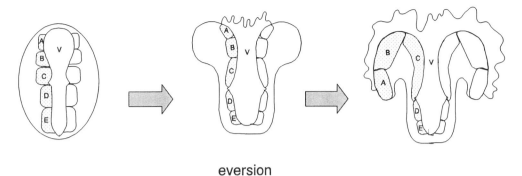

eversion

FIGURE 3-10. Diagram of the process of eversion in forebrain development, as occurs in ray-finned fishes.

direction to form a large, nuclear area called the **dorsal ventricular ridge** (Fig. 3-11). The dorsal ventricular ridge was long thought to represent a huge, basal telencephalic area for use in stereotyped motor behaviors, but many connectional and histochemical studies support the hypothesis that a large part of it is homologous as a collection of several embryologically derived fields to respective areas of part of mammalian isocortex.

The differences in the organization of both evaginated versus everted telencephalons and isocortex versus the dorsal ventricular ridge are differences of the topography of various groups of neurons. As discussed above, growth cones of axons require various local guidance cues in order to grow along the right path and reach their target. Developing dendrites require

the arrival of afferent axons within the proper time frame to continue developing and be maintained. Developmental differences can produce grossly different topographical relationships of areas in the telencephalon, but the other, multiple developmental factors necessary for the establishment of connections are obviously unaffected and sufficient despite the alterations of topography in these cases.

Ontogeny and Recapitulation

In closing this chapter, there is one final aspect of brain development to consider briefly. One of the most pervasive of "old wives' tales" is the notion that ontogeny recapitulates phylogeny: in other words, that the developmental sequence

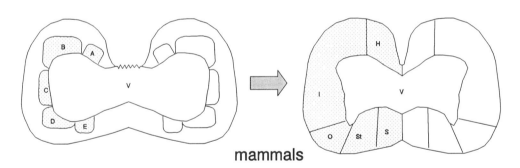

mammals

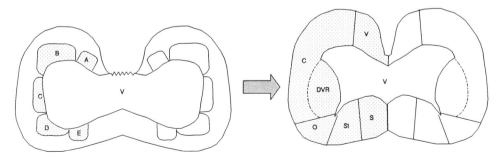

diapsid reptiles, birds, and turtles

FIGURE 3-11. Diagram of the differential development of the dorsal pallium (indicated by the letter B) in mammals versus diapsid reptiles, birds, and turtles. Abbreviations: C, cortex; DVR, dorsal ventricular ridge; H, hippocampal formation; I, isocortex; O, olfactory cortex; S, septum; St. striatum; V, ventricle.

is a kind of rerun of the species' evolutionary history. In the late 1800s, Haeckel promoted the idea that evolutionary change results from the addition of new ontogenetic stages to the terminal phase of the ancestor's development. Such **terminal additions** to existing sequences of development would be the basis of ontogeny recapitulating phylogeny. Terminal additions do, in fact, occur in phylogeny, but the sequence of development in vertebrate brains does not recreate adult ancestral stages of the species' evolutionary history.

From our analysis of differences in the development of the telencephalon, we can see that for ontogeny to recapitulate phylogeny, particularly in terms of the now discredited *scala naturae*, a highly improbable series of events would have to occur to produce the brain of an adult mammal. The telencephalic pallium would first have to evaginate, as in agnathans and cartilaginous fishes, then undo the evagination and evert as in ray-finned fishes, then undo the eversion and evaginate as in amphibians, and finally have all of the dorsal pallium develop into isocortex. Nevertheless, attempts persist to perceive a recapitulation of phylogeny in ontogenetic sequences, as with long projection neurons developing before locally projecting ones and motor neurons before sensory ones. Many of these attempts are based on erroneous *scala naturae* thinking and also fail to take into account the multitude of features that were gained quite suddenly with the evolution of neural crest and placodes (see Chapter 9) in ancestral chordates.

As in many incorrect ideas, there nontheless is a grain of truth in the ontogeny–phylogeny idea. In the early 1800s, Karl von Baer realized that developmental sequences do not recreate stages from "lower" to "higher" groups of animals. He did note, however, that resemblances occur among embryos within a group, and that the resemblances decrease as development proceeds. von Baer concluded that those features that are common to a group appear earlier in development than the more specialized features of individual taxa within the group. Thus, in the development of the brain of a seagull, for example, we would expect features common to chordates to develop first, followed by features common to vertebrates, and so on in sequence through features common to jawed vertebrates, sarcopterygians, tetrapods, amniotes, birds, and finally seagulls. von Baer thus believed that over evolution, there is conservation of a number of developmental stages. This concept is referred to as "**von Baerian recapitulation,**" and a growing body of data supports it.

The von Baerian view makes sense when we note that adult phenotypes are produced by a series of ontogenetic sequences. Changes over evolution in the phenotype are the direct result of changes in the ontogenetic sequences, which in turn are produced by mutations established in the genome. Features that are common to all vertebrates, such as the neural tube, arise before more specialized features present only in one radiation, such as the dorsal ventricular ridge in nonmammalian amniotes. Given the complexity and interactions of developmental events, modifications in early sequences would have much more profound, and in most cases disruptive, effects than modifications in later sequences. The reduction of particular systems and the maintenance of neotenic (embryonic) characteristics in the adults of some groups may in fact result from relatively early modifications in the rates of cell proliferation and differentiation.

Although later modifications constitute terminal or near terminal additions, the radiations of extant vertebrates have been separated long enough that such terminal additions reflect their independent histories rather than allowing for the construction of a single line of evolutionary history. Differences in the development of the telencephalon between the major vertebrate radiations, as well as marked differences in the degree of migration of cells from the periventricular matrix within all four major radiations, clearly demonstrate the separateness and independence of the radiations over a long period of time. On the other hand, those features that are common to most or all groups, in developmental structures and sequences and in the adults, can be used to reconstruct the condition of the brain in the common ancestral vertebrate stock.

FOR FURTHER READING

Jacobson, M. (1991) *Developmental Neurobiology*. New York: Plenum.

Nauta, W. J. H. and Karten, H. J. (1970) A general profile of the vertebrate brain, with sidelights on the ancestry of cerebral cortex. In F. D. Schmitt (ed.), *The Neurosciences, Second Study Program*. New York: Rockefeller University Press, pp. 6–27.

Northcutt, R. G. (1979) The comparative anatomy of the nervous system and the sense organs. In M. H. Wake (ed.), *Hyman's Comparative Vertebrate Anatomy*. Chicago: The University of Chicago Press, pp. 615–769.

ADDITIONAL REFERENCES

Altman, J. (1963) Autoradiographic investigation of cell proliferation in the brains of rats and cats. *Anatomical Record*, 145, 573–591.

Altman, J. and Bayer, S. A. (1979) Development of the diencephalon in the rat. IV. Quantitative study of the time of origin of neurons and the internuclear chronological gradients in the thalamus. *Journal of Comparative Neurology*, 188, 455–472.

Angevine, J. B., Jr. and Sidman, R. L. (1961) Autoradiographic study of cell migration during histogenesis of cerebral cortex in the mouse. *Nature (London)*, 192, 766–768.

Bolker, J. A. (1994) Comparison of gastrulation in frogs and fish. *American Zoologist*, 34, 313–322.

Campbell, R. M. and Peterson, A. C. (1993) Expression of a *lacZ* transgene reveals floor plate cell morphology and macromolecular transfer to commissural axons. *Development*, 119, 1217–1228.

Deacon, T. W. (1990) Rethinking mammalian brain evolution. *American Zoologist*, 30, 629–705.

Dodd, J. and Jessell, T. M. (1988) Axon guidance and the patterning of neuronal projections in vertebrates. *Science*, 242, 692–699.

Ghosh, A., Antonini, A., McConnell, S. K., and Shatz, C. J. (1990) Requirement of subplate neurons in the formation of thalamocortical connections. *Nature (London)*, 347, 179–181.

Gilbert, S. F. (1994) *Developmental Biology*, 4th ed. Sunderland, MA: Sinauer Associates, Inc.

Harris, W. A. and Holt, C. E. (1990) Early events in the embryogenesis of the vertebrate visual system: cellular determination and pathfinding. *Annual Review of Neuroscience*, 13, 155–169.

Havton, L. A. and Ohara, P. T. (1993) Quantitative analyses of intracellularly characterized and labeled thalamocortical projection neurons in the ventrobasal complex of primates. *Journal of Comparative Neurology*, 336, 135–150.

Karten, H. J. (1968) The ascending auditory pathway in the pigeon (*Columba livia*). II. Telencephalic projections of the nucleus ovoidalis thalami. *Brain Research,* 11, 134–153.

Karten, H. J. (1991) Homology and evolutionary origins of the "neocortex." *Brain, Behavior and Evolution,* 38, 264–272.

Karten, H. J. and Hodos, W. (1970) Telencephalic projections of the nucleus rotundus in the pigeon (*Columba livia*). *Journal of Comparative Neurology,* 140, 35–52.

Lund, J. S. and Lewis, D. A. (1993) Local circuit neurons of developing and mature macaque prefrontal cortex: Golgi and immunocytochemical characteristics. *Journal of Comparative Neurology,* 328, 282–312.

Noback, C. R., Strominger, N. L., and Demarest, R. J. (1991) *The Human Nervous System: Introduction and Review, Fourth Edition.* Philadelphia: Lea & Febiger.

Northcutt, R. G. (1981) Evolution of the telencephalon in nonmammals. *Annual Review of Neuroscience,* 4, 301–350.

Northcutt, R. G. (1990) Ontogeny and phylogeny: A reevaluation of conceptual relationships and some applications. *Brain, Behavior and Evolution,* 36, 116–140.

Northcutt, R. G. and Butler, A. B. (1976) Retinofugal pathways in the longnose gar *Lepisosteus osseus* (Linnaeus). *Journal of Comparative Neurology,* 166, 1–16.

Northcutt, R. G. and Butler, A. B. (1980) Projections of the optic tectum in the longnose gar, *Lepisosteus osseus. Brain Research,* 190, 333–346.

Northcutt, R. G. and Davis, R. E. (1983) Telencephalic organization in ray-finned fishes. In R. E. Davis and R. G. Northcutt (eds.), *Fish Neurobiology, Vol. 2: Higher Brain Areas and Functions.* Ann Arbor, Michigan: University of Michigan Press, pp. 203–236.

Placzek, M., Tessier-Lavigne, M., Yamada, T., Jessell, T., and Dodd, J. (1990) Mesodermal control of neural cell identity: floor plate induction by the notochord. *Science,* 250, 985–988.

Robson, J. A. (1993) Qualitative and quantitative analyses of the patterns of retinal input to neurons in the dorsal lateral geniculate nucleus of the cat. *Journal of Comparative Neurology,* 334, 324–336.

Roth, G., Nishikawa, K. C., Naujoks-Manteuffel, C., Schmidt, A., and Wake, D. B. (1993) Paedomorphosis and simplification in the nervous system of salamanders. *Brain, Behavior and Evolution,* 42, 137–170.

Senn, D. G. (1979) Embryonic development of the central nervous system. In C. Gans, R. G. Northcutt, and P. Ulinski (eds.), *Biology of the Reptilia, Vol. 9: Neurology A.* London: Academic Press, pp. 173–244.

Silver, J. and Rutishauser, U. (1984) Guidance of optic axons in vivo by a preformed adhesive pathway on neuroepithelial endfeet. *Developmental Biology,* 106, 485–499.

4

Vertebrate Phylogeny and Diversity in Brain Organization

INTRODUCTION

Animals in the phylum **Chordata** are characterized by the presence of a notochord, a dorsal nerve cord, and gill slits. Chordates (Fig. 4-1) comprise three groups: **urochordates** (tunicates, or sea squirts), **cephalochordates** (represented by *Branchiostoma,* previously known as *Amphioxus*), and **vertebrates** (or **craniates**). The four major, extant vertebrate radiations are hagfishes and lampreys, which are frequently grouped together as the **agnathans** (jawless vertebrates), **chondrichthyans** (ratfishes, sharks, skates and rays), **actinopterygians** (ray-finned fishes), and **sarcopterygians** (fleshy-finned fishes and their derivatives, the tetrapods). Most of the groups of vertebrates that form these four radiations are **anamniotes,** that is, their embryos lack an amniotic membrane. **Amniotes,** in contrast, have an amniotic membrane that encases the embryo in amniotic fluid during development, freeing amniote vertebrates from the need to lay their eggs in water for development; amniotes are the land vertebrate (i.e., nonamphibian tetrapod) descendants of ancestral sarcopterygians.

Within each of the four major vertebrate radiations, descendants of the ancestral forms have become diversified independently. Each species has some traits inherited from its ancestors, and most species also have specialized traits acquired through natural selection in changing environments. Specialized features can occur in any of the systems of the body, including the central nervous system.

Selective pressures may be many and diverse, but some general ones are active on most populations. Reproductive advantages may be gained by better avoidance of predators, better finding of prey, and caring for young to promote their survival. Within all four vertebrate radiations, such pressures have acted on the common genetic material inherited from the common ancestral stock. In some cases, the independently achieved results have some strikingly similar features. A notable example of this phenomenon is the evolution of the forebrain, the most rostral part of the brain. In all four vertebrate radiations, a set of some of the species have independently gained enlarged and more complexly organized forebrains, and these more elaborate forebrains are correlated with more complex behavioral repertoires.

In our overview of the vertebrate nervous system in Chapter 3, we covered the major regions of the brain and outlined some of the major systems that are common to many species. Within each of the four major vertebrate radiations, marked variation exists in the relative size of various regions involved in these systems and in the elaborateness of nuclei and cortical regions. In some species within each radiation, most neuronal cell bodies remain near the ventricular surface and form relatively few distinct, migrated, individual nuclei or cortical layers. In other species in each radiation, however, extensive migration of the neurons takes place, which results in the formation of complexly structured nuclear groups and elaborately layered cortices. In these latter species, the brains also tend to be larger, relative to body size, than the brains of species with less complexly structured brains. The selective pressures for such larger brains and for increase in behavioral complexity are discussed in detail in Chapter 5.

In this chapter we will first briefly review vertebrate phylogeny. Then we will identify the members of each of the various groups and provide a series of outline drawings of some of their brains for future reference. Next we will consider the variation in structural complexity within each major vertebrate

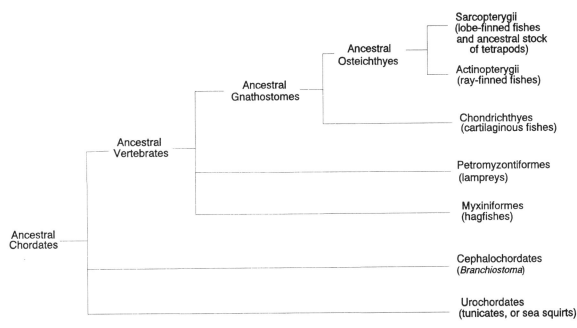

FIGURE 4–1. Diagram of the descent of extant groups of chordates from ancestral chordates.

group, recognizing that extant species can be divided into two major types of brain organization. We believe that approaching the study of comparative neuroanatomy by comparing these two major types of brain organization across the four major vertebrate groups can help to provide a clearer picture of the multiple, independent changes that have occurred in brains over the course of evolution.

VERTEBRATE PHYLOGENY

Jawless Fishes

One of the major vertebrate radiations, the **Agnatha** (a = without; gnathos = jaw), also referred to as **cyclostomes,** comprise two groups, lampreys and hagfishes, which are only distantly related. They are placed within Agnatha as they both lack the jaws common to the rest of vertebrates.

Hagfishes are marine, as are some lampreys, but all lampreys spawn in freshwater rivers. Lampreys undergo a larval (**ammocoete**) stage followed by a metamorphosis to the adult form. Lampreys are true parasites, which attach to the body surface of other fishes, while hagfishes are scavengers, which feed by entering the body cavities of dead or dying fishes.

Agnatha comprise five major groups: the extant hagfishes and lampreys and three extinct groups that constitute the majority of **ostracoderms** (Fig. 4-2). Ostracoderms were a diverse group, some of which had heavy, bony shields covering the rostral end of the body, with openings for two, dorsally placed eyes, a pineal eye, and a single, median nostril. They had a round, jawless mouth with up to 10 pairs of gill slits, and many were bottom dwelling, detritus feeders.

One group of ostracoderms, the anaspids, lacked heavy shields and were more active swimmers. The origin of jawed vertebrates may have been from a recently discovered group of thelodont agnathans, which may be related to the heterostracan ostracoderms. These animals had deep, compressed bodies and large eyes. They also had stomachs, which other agnathans do not, and appear to have been active predators.

Some phylogenetic studies indicate that hagfishes [Fig. 4-3(A)] are most closely related to the heterostracans, while lampreys [Fig. 4-3(B)] are derived from an ancestral stock in common with the anaspids. Another more recent analysis identifies the third major group of ostracoderms, the osteostracans, as the sister group of jawed vertebrates, with lampreys in a sister group to most ostracoderms and jawed vertebrates, and with hagfishes as the outgroup to all other vertebrates. An additional group of jawless vertebrates, the **conodonts,** which have been identified primarily by comb-like tooth elements, appear to have a phylogenetic position as an outgroup of anaspids, the rest of ostracoderms, and jawed vertebrates, with lampreys being a sister group of all of the latter and with hagfishes remaining the outgroup of all other vertebrates. The conodonts are thought to have had an actively predatory mode of life, which may thus also have characterized the earliest vertebrates.

Chondrichthyes

Cartilaginous fishes (Chondrichthyes) comprise two major groups: **Holocephali** and **Elasmobranchi** (Fig. 4-4). None of the cartilaginous fishes are known to have a larval stage. The Holocephali are the chimaeras [Fig. 4-5(A)], or ratfishes, and are distinctive in appearance with large eyes and tapering, thin tails. The elasmobranchs are the sharks, skates, and rays. Squalomorph sharks include hexanchids [cow sharks, Fig. 4-5(B)], squaliform sharks [such as the spiny dogfish shark, Fig. 4-5(C)], and pristiophoriform (saw) sharks. Squantinomorph sharks (angel sharks, or monkfish) have flattened bodies similar to rays but swim like other sharks by using their bodies and tails. Galeomorph sharks include bullhead, carpet, mackerel,

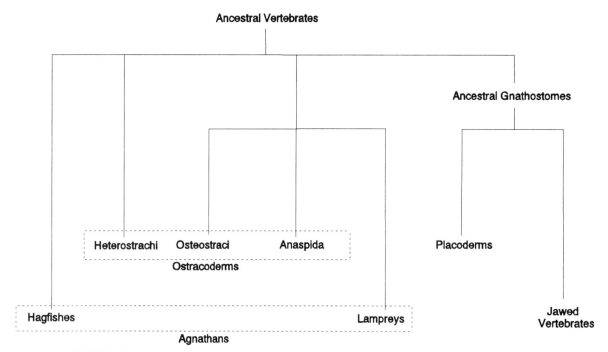

FIGURE 4–2. Diagram of the descent of extinct and extant groups of jawless and jawed vertebrates from ancestral vertebrates.

and requiem sharks. The mackerel sharks (Family Lamnidae) include the giant white shark (or man-eater), as well as mako and basking sharks. Most of the Batoidea are characterized by their flattened bodies. They include skates [Fig. 4-5(D)], sawfishes, guitar fishes, electric rays, stingrays , eagle rays, and manta rays.

Actinopterygii

The Actinopterygii comprise five radiations of **ray-finned fishes** (Fig. 4-6). Ray-finned fishes are one of the most widely distributed radiations of vertebrates. They inhabit Arctic to tropical waters, deep sea depths to surface waters to a partially air habitat, and both fresh and salt water. **Cladistia,** the reedfishes or bichirs (Fig. 4-7), are elongated in shape and live in shallow, tropical rivers and swamps. **Chondrostei** include sturgeons, some of which can reach a length of 4 m and weigh over 500 kg, and the smaller paddlefishes (or spoonbills). **Ginglymodi,** the gars [Fig. 4-8(A)], are distinguished by their elongated snouts and bodies. **Halecomorphi** have only one extant species, the bowfin *Amia calva* [Fig. 4-8(B)]. Reedfishes, sturgeons, paddlefishes, gars, and the bowfin all have skeletons that are largely cartilaginous. Highly vascularized air bladders can be used as auxiliary lungs for short periods of time, but no species in these groups is known to undergo a larval stage and metamorphosis.

The fifth group of ray-finned fishes, the **Teleostei,** differ from other ray-finned fishes in having bony skeletons and in their vast number and diversity of species. Larval stages followed by metamorphosis have been found in a number of taxa. They comprise four major groups (Fig. 4-9). The caution must be noted, however, that not all these groups, or some of the taxa within them, can be assumed to be monophyletic. Additional study is needed to clarify a number of relationships within

this extremely large and diverse radiation. Osteoglossomorphs include arawanas, arapaimas, mooneyes, and the freshwater butterfly fish, all of which have relatively large eyes and live in fresh water, notopterids (also known as featherbacks or knifefishes), which have exceptionally long anal fins, and mormyrids, which are weakly electric. Elopomorphs are saltwater fishes, which include tarpons, bonefishes, and eels. Clupeomorphs include herrings, anchovies, menhadens, shad, and aelwives. Many species of clupeomorph fishes form huge schools. Both elopomorphs and clupeomorphs possess metamorphosing species.

Euteleostei (Fig. 4-10) include the great majority of the more than 20,000 teleost species. Euteleosts comprise over 25 orders, 375 families, and 17,000 species. They include a number of diverse and familiar species, such as pikes, goldfishes, salmons, cods, and live bearers, in addition to their largest group, the percomorphs. Percomorphs include equally diverse species, including tunas, dories, stone fishes, porcupine fishes, flounders and soles, flying gunards, sea horses, perches, sunfishes, cichlids, snappers, and many others. Members of this diverse group occupy a great variety of niches. Some species have larval stages. Some are voracious predators. A number, particularly those that inhabit coral reefs, have complex and elaborate territorial displays and courtship rituals, pair-bonding, and biparental care of young. Those with the most complex repertoires also exhibit the most complex and enlarged brains.

Most studies on the central nervous system of fishes have been carried out on elasmobranchs and actinopterygians. A relatively few number of studies have been done on hagfishes and lampreys, and more are needed on these two important out-groups to jawed vertebrates. Relatively little information is available on holocephalians, and most studies on actinopterygians have focused on euteleosts.

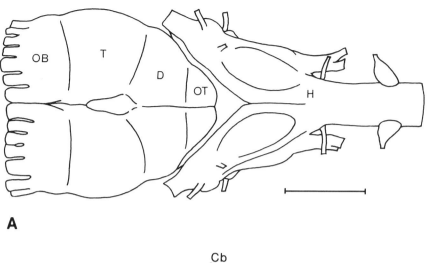

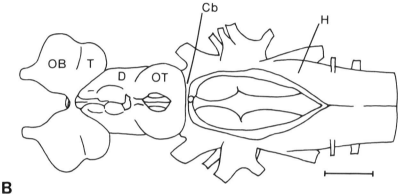

FIGURE 4–3. (A) Dorsal view of the brain of a Pacific hagfish (*Eptatretus stouti*). Scale = 2 mm. Adapted from Wicht and Northcutt (1992). Used with permission of S. Karger AG. (B) Dorsal view of the brain of a silver lamprey (*Ichthyomyzon unicuspis*). Scale = 1 mm. Adapted from Northcutt and Puzdrowski (1988). Used with permission of S. Karger AG.

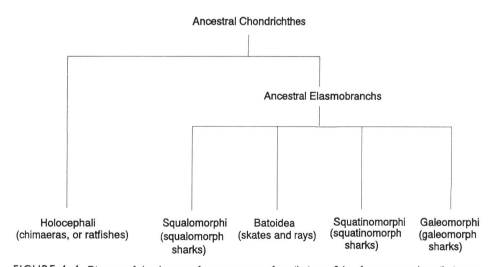

FIGURE 4–4. Diagram of the descent of extant groups of cartilaginous fishes from ancestral cartilaginous fishes.

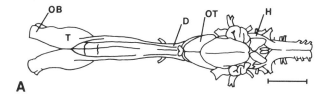

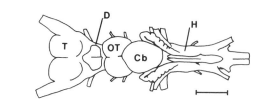

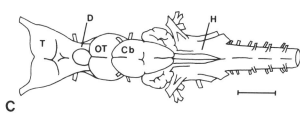

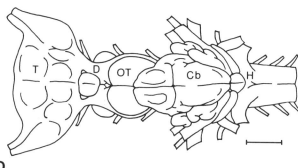

FIGURE 4–5. (A) Dorsal view of the brain of a chimaera (*Hydrolagus colliei*). Scale = 1 cm. Adapted from Northcutt (1978). (B) Dorsal view of the brain of a cow shark (*Notorynchus maculatus*). Scale = 1 cm. Adapted from Northcutt (1978). Note that for this and most other species illustrated in this chapter in which the olfactory bulbs are at the distal ends of elongated olfactory tracts, only the proximal stump of the olfactory tract is shown. (C) Dorsal view of the brain of a spiny dogfish shark (*Squalus acanthias*). Scale = 1 cm. Adapted from Northcutt (1979). Used with permission of The University of Chicago Press. (D) Dorsal view of the brain of the clearnose skate (*Raja eglanteria*). Scale = 5 mm. Adapted from Northcutt (1979). Used with permission of The University of Chicago Press.

Sarcopterygii

The sarcopterygian radiation of fleshy-finned fishes and tetrapod vertebrates (Fig. 4-11) is the large, sister radiation of the Actinopterygii. Three extant groups—**Dipnoi, Actinistia,** and **Tetrapoda**—are descended from the ancestral sarcopterygian stock. As is the case with ray-finned fishes, this radiation is widely distributed, with habitats ranging from the Arctic to the tropics and including fishes, amphibians, and land vertebrates. Actinistia are represented by only one extant species, the **coela-**

canth (or **crossopterygian**), *Latimeria chalumnae* [Fig. 4-12(A)]. Specimens of this fish were first taken from deep waters off the South African coast in the 1930s. They are regarded as "living fossils," as their skeletal form has not changed since the Devonian period (over 300 million years ago). Studies comparing the sequences of amino acids in hemoglobin chains indicate that *Latimeria* is more closely related to amphibians than to any other fish. Thus, coelacanths appear to be the sister group of tetrapods.

Dipnoi, the lungfishes [Fig. 4-12(B)], are also similar to species identified from the Devonian. Their air bladder is highly vascularized and enables these tropical, freshwater fishes to survive out of water if moisture is maintained on the skin. Neither coelacanths nor lungfishes undergo metamorphosis. Specimens of *Latimeria* are particularly rare, but some information on the organization of its brain is available. The brain is very similar in structure to that of some of the lungfish, on which more studies have been done. Analysis of neural characters supports the hypothesis that actinistians are the sister group of lungfishes.

Extant tetrapods comprise amphibians and amniotes, or land vertebrates. Amphibians arose in the Devonian period and include **gymnophionans (caecilians:** worm-like burrowers with minute eyes and persistent notochord), **urodeles** [newts and salamanders, Fig. 4-12(C)], and **anurans** [frogs and toads, Fig. 4-12(D)]. Whether amphibians are monophyletic or polyphyletic remains unresolved, however. In some amphibians, particularly salamanders, the skeleton is partially cartilaginous. Most amphibians have a larval stage and undergo metamorphosis, but some salamanders (*Necturus,* the mudpuppy) retain the larval stage throughout their life cycle. Many caecilians are viviparous.

A limited amount of data are available on the brain of lungfishes, and more research is needed on this important group. Very few studies have been done on *Latimeria* due to its extreme rarity. Most studies of the central nervous system in amphibians have been done on anurans; some data are available on urodeles, but data on the central nervous system of caecilians are very limited.

Extant amniotes (Fig. 4-11) comprise **mammals, diapsids,** which include diapsid reptiles and birds, and **turtles.** The term "reptiles" often is used to refer to extant **diapsid reptiles** (lizards, tuatara, snakes, and crocodiles) and turtles. Because these animals do not form a biologically unitary group, the blanket term "reptiles" will be used infrequently in this text.

The earliest group that diverged from the ancestral amniote stock was a group of synapsid amniotes that ultimately gave rise to mammals. The term **synapsid** refers to the presence of one, laterally placed temporal fenestra and one arch in the skull. Mammals comprise three major groups (Fig. 4-13). **Prototheria,** the **monotremes,** reproduce by laying eggs. They have pouches and mammary glands without nipples. Monotremes consist of the platypus and echidnas, or spiny anteaters (Fig. 4-14). **Metatheria,** the marsupials, are viviparous but lack a typical placenta. After birth, the embryos crawl into a skin pouch where they complete their development.

Eutherian mammals (Figs. 4-15 and 4-16) are viviparous and are those mammals in which embryological development occurs within a placenta formed by the apposition of the uterine lining and the extraembryonic membranes of the embryo. Eutheria are a diverse radiation, occupying niches in the ocean

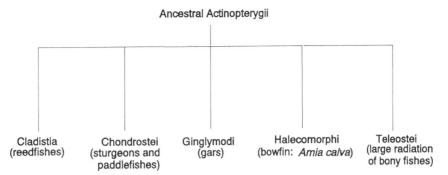

FIGURE 4–6. Diagram of the descent of extant groups of ray-finned fishes from ancestral ray-finned fishes.

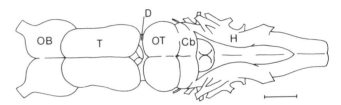

FIGURE 4–7. Dorsal view of the brain of a bichir (*Polypterus palmas*). Scale = 2 mm. Adapted from Reiner and Northcutt (1992).

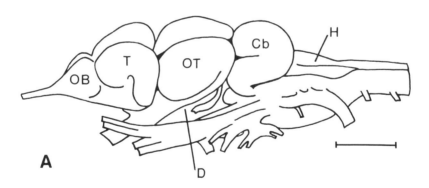

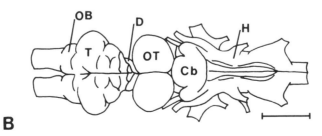

FIGURE 4–8. (A) Lateral view of the brain of a longnose gar (*Lepisosteus osseus*). Scale – 3 mm. Adapted from Northcutt and Butler (1976). (B) Dorsal view of the brain of a bowfin (*Amia calva*). Scale = 5 mm. Adapted from Butler and Northcutt (1992). Used with permission of S. Karger AG.

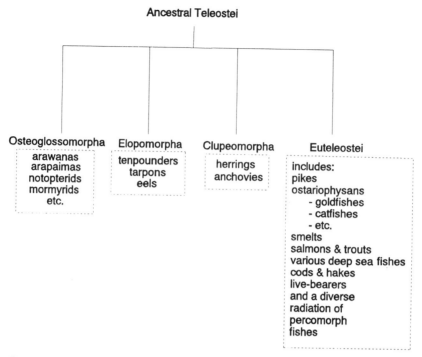

FIGURE 4–9. Diagram of the descent of extant groups of teleosts from ancestral teleosts.

(cetaceans) and partially in the air (bats) as well as on land. Specializations of the limbs and/or their distal portions, such as hooves and hoof-like structures in artiodactyls, perissodactyls, and other ungulates, opposing digits in primates, wing-like membranes in bats, and reduction in cetaceans and seals, are common. Specializations related to prey capture and consumption, such as the dentition in carnivores and rodents, elon-

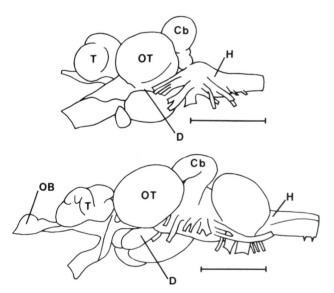

FIGURE 4–10. Lateral views of the brains of (top) a sunfish (*Lepomis cyanellus*) and (bottom) a goldfish (*Carassius auratus*). Scale = 5 mm. Adapted from Northcutt and Butler (1991) and Northcutt (1983), respectively, and used with permission of S. Karger AG and The University of Michigan Press.

gation of the snout in edentates, pangolins, and elephants, and tool making in primates are also diverse and pronounced.

As is the case with other vertebrate groups, the phylogenetic relationships of mammals (including those illustrated here) are still subject to revision as new data become available, particularly in genetic sequencing studies. For example, recent data indicate that whales are more closely related to artiodactyls than to perissodactyls, and rodents may be a sister group of the rest of the placental mammals. While at least some aspects of central nervous system organization have been studied in a number of mammalian species—including monotremes, marsupials, and, among placental mammals, insectivores—most mammalian studies have focused on the more common placental mammals, such as rodents, domestic cats, and macaque monkeys.

Extant diapsids comprise two groups (Fig. 4-11), one of which is the **Lepidosauria**, which includes **Rhynchocephalia**, the lizard-like tuatara native to New Zealand, and **Squamata**, the lizards and snakes. The second group is the **Thecodontia**, which comprises **Crocodilia** and **Aves**, as well as the extinct radiation of dinosaurs. The term diapsid refers to the presence of two temporal fenestrae and two arches in the skull. Extant anapsids are the turtles, which range in habitat from marine to fresh water to land. The term **anapsid** refers to a skull with a complete roof of bone in the temporal region, lacking any fenestrae or arches. The earliest, ancestral amniotes also had anapsid skulls.

Among the diapsids, birds [Fig. 4-17(A)] comprise a large radiation of 8700 species (compared to 6000 species of diapsid reptiles and turtles and 4500 species of mammals). Birds are uniquely characterized by feathers. Their bones are filled with hollow spaces for lightness, metacarpals are fused, and teeth

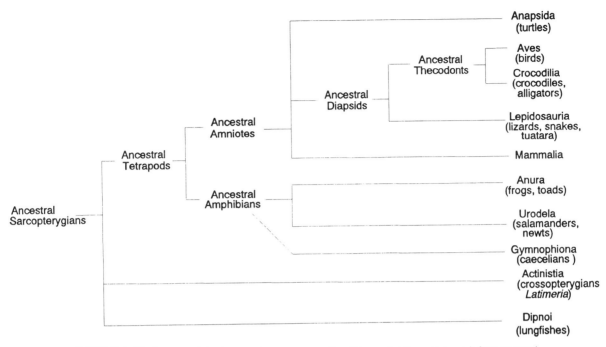

FIGURE 4–11. Diagram of the descent of extant groups of lungfishes, actinistia, and tetrapods from ancestral fleshy-finned fishes.

are absent. Numerous variations in size, habitat, structure of bill, and courtship and brooding behaviors exist. The largest group is the Passeriformes, which are perching birds, such as the swallows, and include over 5000 species. Other groups with between 250 and 400 species each are the Apodiformes (swifts and hummingbirds), Piciformes (including woodpeckers and toucans), Falconiformes (eagles, hawks, and vultures), Galliformes (gallinaceous, chicken-like birds), Charadriiformes (shorebirds and gulls), Columbiformes (pigeons and doves), and Psittaciformes (parrots). Smaller orders include penguins, rheas, ostrich (a single species), emus, kiwis, tinamous, grebes, albatrosses and petrels, pelicans and comorants, herons and flamingos, waterfowl, cranes, loons, cuckoos, owls, nightjars, mousebirds, trogons, and kingfishers.

The origin of turtles [Fig. 4-17(B)] is relatively obscure. They are derived from an ancestral stock in common with the diapsids, and they appeared after mammals had diverged from the common ancestral amniote stock. Turtles comprise four major groups: leatherback sea turtles, true sea turtles, side-necked turtles, and hidden-necked turtles.

Among diapsid reptiles and turtles, a fair sampling of species has been achieved in the study of central nervous system organization. Despite the great range of variation among extant species of birds, most studies of central nervous system organization in birds have been done on pigeons (Columbiformes) and owls (Strigiformes). A number of studies on songbirds have also been done that explore relationships among learning, reproductive functions, and the development of parts of the brain involved in song production. Over the past quarter century, our concepts of brain organization in these groups of non-mammalian amniotes, particularly relating to the evolution of sensory systems and forebrain organization, has been extensively reassessed vis à vis the organization of the brain in mammals.

TWO TYPES OF VERTEBRATE BRAIN ORGANIZATION

Within each of the four major radiations of vertebrates—agnathans, chondrichthians, actinopterygians, and sarcopterygians—two types of organization may be found in the brains of the various species. We will define the first type as those species in which the brains are characterized by the neuronal cell bodies being unmigrated or only partially migrated away from the embryonic, **periventricular matrix,** which is the zone from which neurons develop. Afferent projections to these neurons terminate on the more distal portions of their dendrites near the surface of the brain. This pattern of organization will be referred to as **laminar,** in reference to the periventricular lamina in which the majority of neuronal cell bodies are located. We will refer to the species, from all major radiations, with laminar brains as **Group I.** An example of the arrangement of the neuronal cell bodies within a representative slice of the brain in an animal with a laminar brain is shown in Figure 4-18.

Within each of the four major vertebrate radiations, other species have brains in which extensive migration of neuronal cell bodies away from the periventricular matrix has occurred. More individual nuclear groups are apparent, and the brains are generally relatively larger in size as compared with brains exhibiting laminar organization. The pattern of organization in brains with migration of the majority of neuronal cell bodies will be referred to as **elaborated.** Species with elaborated brains will be referred to as **Group II.** An example of the arrangement of the neuronal cell bodies in the brain of an animal with an elaborated brain is shown in Figure 4-19.

The differences in brain organization between Group I and Group II animals are not as notable in the spinal cord and the cranial nerves as in the brain itself. The organization of the spinal cord and cranial nerves is relatively very conservative among all vertebrate radiations, and variation in these structures is more related to habitat (aquatic vs. terrestrial vs. aerial) than to the type of organization of the brain itself. The types of

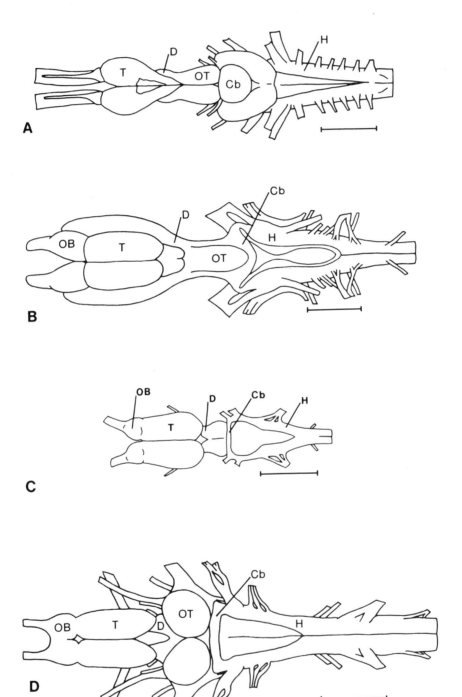

FIGURE 4–12. (A) Dorsal view of the brain of a coelacanth (*Latimeria chalumnae*). Scale = 1 cm. Adapted from Northcutt et al. (1978). (B) Dorsal view of the brain of an African lungfish (*Protopterus annectens*). Scale = 5 mm. Adapted from Northcutt (1977). (C) Dorsal view of the brain of a mudpuppy (*Necturus maculosus*). Scale represents approximately 3 mm. Adapted from Northcutt (1979). Used with permission of The University of Chicago Press. (D) Dorsal view of the brain of a bullfrog (*Rana catesbeiana*). Scale = 10 mm. Adapted from Northcutt and Royce (1975).

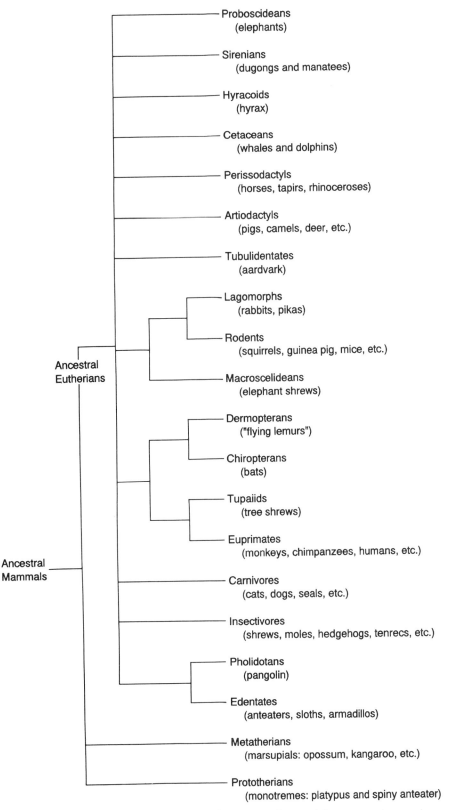

FIGURE 4–13. Diagram of the descent of extant groups of mammals from ancestral mammals.

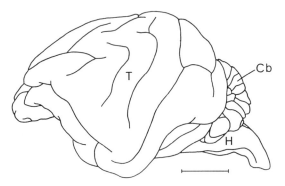

FIGURE 4–14. Lateral view of the brain of an echidna (*Tachyglossus aculeatus*). Scale = approximately 1 cm. Adapted from Rowe (1990). The olfactory bulb, diencephalon, and optic tectum are concealed by the large telencephalon. Used with permission of Plenum Publishing Corp.

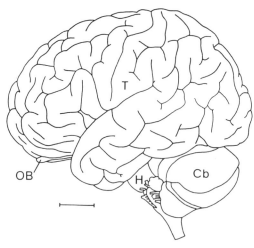

FIGURE 4–16. Lateral view of the brain of a primate (*Homo sapiens*). Scale = 2 cm. Adapted from Nieuwenhuys et al. (1978). The diencephalon and optic tectum are concealed by the large telencephalon. Used with permission of Springer-Verlag.

laminar and elaborated brain organization are clearly different among vertebrates in the hindbrain, in the midbrain, and, most prominently, in the forebrain, however. Thus, in a number of the sections on the various regions and systems in vertebrate brains, we will consider these two major types of organization separately.

The distinction between Group I and Group II vertebrates is useful for emphasizing the concept that diversity has been achieved independently in four separate radiations. It is not meant to imply, however, that all brains with either a laminar or an elaborated pattern necessarily will have particular specializations in common. Some features may be distributed in some but not all taxa with laminar brains and also be in some but not all taxa with elaborated brains. Other features may occur in only one or two taxa. Thus, at the end of many sections in the second half of the book (Chapters 15-31), we will summarize the distribution of the various features. Based on that distribution, we will interpret (1) whether given features are evolutionarily primitive (plesiomorphic) or evolutionarily derived (apomorphic) and (2) whether a feature present in more than one taxon is homologous (derived from the same feature present in a common ancestor) or homoplastic (derived independently, not having been present in a common ancestor).

Vertebrates with Laminar Brains

Table 4-1 lists those groups, as discussed below in the text, within each major radiation that have laminar brain organization. In lampreys, the vast majority of neuronal cell bodies are located within the periventricular zone throughout the brain. Parasitic lifestyles, such as lampreys have, are frequently associated with reduction of structures. Thus, whether the laminar organization of the brain in lampreys reflects the ancestral condition of vertebrate brains or, instead, a secondary process of reduction has yet to be established. The weight of evidence supports the theory that the earliest vertebrates were jawless, but the diversity of the ostracoderms precludes assuming that the brain in all of these early fishes closely resembled that of modern lampreys.

Cartilaginous fishes with laminar brains include cow sharks, sand sharks, spiny dogfishes, saw sharks, and angel sharks (or monkfish). They also include the chimaeras, or ratfishes. In these Group I, cartilaginous fishes, only a limited migration of cells away from the periventricular surface occurs.

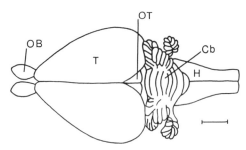

FIGURE 4–15. Dorsal view of the brain of a rabbit (*Lepus sp.*). Scale = approximately 5 mm. Adapted from Northcutt (1979). The diencephalon is concealed by the large telencephalon. Used with permission of The University of Chicago Press.

TABLE 4–1. Vertebrates with Laminar Brains (Group I)	
Agnatha	Lampreys
Chondrichthyes	Squalomorph sharks Squantinomorph sharks Holocephalians (chimaeras)
Actinopterygii	Cladistians (bichirs, or reedfishes) Chondrosteans (sturgeons and paddlefishes) Ginglymodians (gars) Halecomorphans (bowfin)
Sarcopterygii	Actinistians (coelacanth) Dipnoans (lungfish) Amphibians

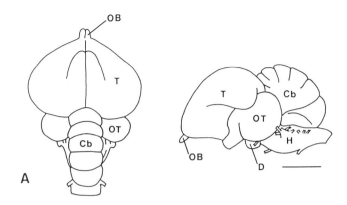

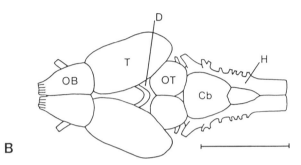

FIGURE 4–17. (A) Dorsal view (left) and lateral view (right) of the brain of a bird (*Columba livia*). Scale = 5 mm. Adapted from Northcutt (1979). (B) Dorsal view of the brain of a turtle (*Chrysemys picta*). Scale = approximately 5 mm. Adapted from Northcutt (1979). Both used with permission of The University of Chicago Press.

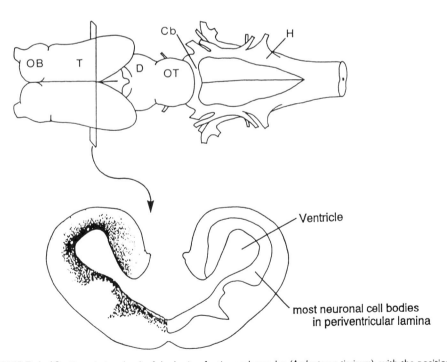

FIGURE 4–18. Dorsal view (top) of the brain of a tiger salamander (*Ambystoma tigrinum*), with the position of a transverse section through the telencephalon, shown below, indicated by the rectangle. The distribution of Nissl stained, neuron cell bodies in the transverse section is shown on its left side, and a mirror image drawing is shown on the right. Adapted from Northcutt and Kicliter (1980). Abbreviations (in rostral to caudal order) used in this and the following figure in this chapter: OB, olfactory bulb; T, telencephalon; D, diencephalon; OT, optic tectum; CB, cerebellum; H, hindbrain. Used with permission of Plenum Publishing Corp.

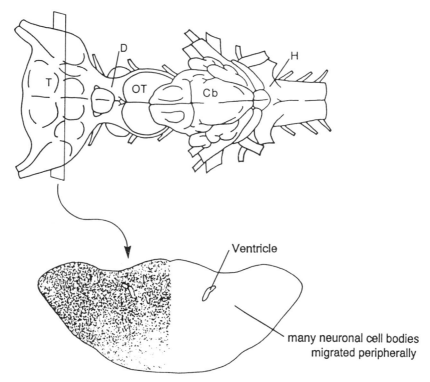

FIGURE 4–19. Dorsal view (top) of the brain of a clearnose skate (*Raja eglanteria*), with the position of a transverse section through the telencephalon, shown below, indicated by the rectangle. The distribution of Nissl stained, neuron cell bodies in the transverse section is shown on its left side, and a mirror image drawing is shown on the right. Adapted from Northcutt (1978).

Ratfishes are not vigorous swimmers. They live at varying depths and eat small invertebrates and fishes. Spiny dogfishes, in contrast, are voracious predators of a variety of other, fast moving fishes, such as cods and herrings. Thus, relatively simple and laminar brain organization does not necessarily preclude an active, predacious lifestyle.

Among ray-finned fishes, the brains of bichirs, sturgeons, gars, and the bowfin show only limited migration of cells away from the ventricular surface. The behavioral repertoires of these fishes do not vary greatly. Reedfishes, sturgeons, gars, and bowfins all inhabit relatively shallow and, for the most part, fresh water. Gars and bowfins are known to move very fast in order to attack and capture other fishes for prey.

Among sarcopterygii, the brains of the crossopterygian *Latimeria*, of lungfishes, and of urodele amphibians (Fig. 4-18) all show a pronounced degree of laminar organization. As in the case of lampreys, this morphology may be the result of a secondary process of reduction, in this case related to **neoteny**—the retention of embryonic characteristics in the adult. In anurans amphibians (frogs and toads) there is limited migration of neuronal cell bodies away from the periventricular matrix.

Vertebrates with Elaborated Brains

Table 4-2 lists those groups, discussed below in the text, within each major radiation that have an elaborated brain organization. The brain of hagfishes is strikingly complex, in contrast

to that of lampreys, with pronounced migration of cells and the presence of many large and distinct nuclear groups. Consistent with their scavenging lifestyle, hagfishes do not have complex behavioral repertoires, and the visual system is markedly reduced. They seem to use vision only for avoiding light, preferring to seek prey in the darkness. The functions and adaptive significance of their complex and enlarged brains are thus a puzzle. Modern jawed vertebrates currently are thought to be more closely related to lampreys than to hagfishes, although future DNA matching studies may be able to better clarify this phylogenetic relationship. Whichever is the case, the range of variation in brain organization among modern hagfishes and

TABLE 4–2. Vertebrates with Elaborated Brains (Group II)

Agnatha	Hagfishes
Chondrichthyes	Galeomorph sharks Skates Rays
Actinopterygii	Teleosts
Sarcopterygii	Mammals Diapsid reptiles Birds Turtles

lampreys suggests that a pronounced variation among the ancestral ostracoderms also existed.

Among cartilaginous fishes, an extensive migration of neuronal cell bodies occurs away from the periventricular surface in the brains of cat sharks, requiem sharks, and hammerheads, which we refer to as galeomorph sharks. Extensive migration and complex morphology also characterizes the brains of skates (Fig. 4-19) and rays. Galeomorph sharks are active predators, as are spiny dogfishes, but they also have modified jaws for deep biting and for successfully attacking larger sized prey. Skates and rays have a more sedate lifestyle, foraging for bottom dwelling invertebrates, but also have more complex modes of locomotion. They use undulating movements of the pectoral fins (skates and sting rays) or a flapping motion of the fins similar to the motion of a bird's wing (manta rays and eagle rays). As we will see below, the nuclei involved in the coordination of motor activities are all relatively enlarged and prominent in galeomorph sharks, skates, and rays. Likewise, forebrain areas involved in sensory processing and for additional, complex behaviors, such as social interactions, courtship, and mating, are all enlarged.

In teleosts, cells are more extensively migrated away from the ventricular surface than in other ray-finned fishes. Teleosts comprise a huge and diverse radiation, and there is considerable variation in various parts of the brain among the different groups that would fill another book or two. As we will discuss further below, teleosts include fishes that have very complex behavioral repertoires.

In amniotes, brain organization also is elaborated and complex. Among mammals, a large range of variation in brain structure and elaboration can be found, but, as in cartilaginous fishes with elaborated brains, teleosts, and other amniotes, the largest and most complex brains are in those mammals with the most complex behavioral repertoires. Among diapsid reptiles and turtles, crocodiles have the most complex behavioral repertoires known and also have the largest and most complex brains. Birds, which are more closely related to crocodiles than to other diapsid reptiles, have even larger and more elaborated brains. The areas of the forebrains related to sensory processing and integration are largest in those birds with complex behavioral repertoires and/or in which learning is of particular importance, such as songbirds, crows, and parrots.

As the preceding discussion shows, marked variation is found in brain structure and in behavioral repertoires within each of the four vertebrate radiations. Changes in the environment and the availability of new niches result in this range of variation by providing new evolutionary opportunities. The opportunistic nature of evolutionary change results in a great deal of variation both within and among the various taxa, rather than "trends." The approach that we have taken here, in first considering all Group I taxa with laminar brains and subsequently considering all Group II taxa with elaborated brains, will avoid the trap of *scala naturae* thinking and the fantasy of a single, directed, linear progression towards an anthropocentrically defined "goal," namely, human beings. Variation present *within* each vertebrate group is as important and meaningful as the variation *between* groups. The old idea of a fish-to-frog-to-rat-to-cat-to-monkey-to-human "progression" is neither historically accurate nor evolutionarily meaningful. Such notions should be filed on the reader's shelf in the appropriate place—next to Bullfinche's Mythology.

FOR FURTHER READING

Carroll, R. L. (1988) *Vertebrate Paleontology and Evolution*. New York: Freeman.

Hull, D. L. (1988) Progress in ideas of progress. In M. H. Nitecki (ed.), *Evolutionary Progress*. Chicago: The University of Chicago Press, pp. 27–48.

Roth, G., Nishikawa, K. C., Naujoks-Manteuffel, C., Schmidt, A., and Wake, D. B. (1993) Paedomorphosis and simplification in the nervous system of salamanders. *Brain, Behavior and Evolution*, 42, 137–170.

ADDITIONAL REFERENCES

Ayala, F. J. (1988) Can "progress" be defined as a biological concept? In M. H. Nitecki (ed.), *Evolutionary Progress*. Chicago: The University of Chicago Press, pp. 75–96.

Bernardi, G., D'Onofrio, G., Caccio, S., and Bernardi, G. (1993) Molecular phylogeny of bony fishes, based on amnio acid sequence of the growth hormone. *Journal of Molecular Evolution*, 37, 644–649.

Butler, A. B. and Northcutt, R. G. (1992) Retinal projections in the bowfin, *Amia calva*: cytoarchitectonic and experimental analysis. *Brain, Behavior and Evolution*, 39, 169–194.

Elzanowski, A. and Wellnhofer, P. (1992) A new link between theropods and birds from the Cretaceous of Mongolia. *Nature (London)*, 359, 821–823.

Evans, B. I. and Fernald, R. D. (1990) Metamorphosis and fish vision. *Journal of Neurobiology*, 21, 1037–1052.

Forey, P. and Janvier, P. (1994) Evolution of the early vertebrates. *American Scientist*, 82, 554–565.

Gabbott, S. E., Aldridge, R. J. and Theron, J. N. (1995) A giant conodont with preserved muscle tissue from the Upper Ordovician of South Africa. *Nature (London)*, 374, 800–803.

Gorr, T., Kleinschmidt, T., and Fricke, H. (1991) Close tetrapod relationship of the coelacanth *Latimeria* indicated by haemoglobin sequences. *Nature (London)*, 351, 394–397.

Gould, S. J. (1988) On replacing the idea of progress with an operational notion of directionality. In M. H. Nitecki (ed.), *Evolutionary Progress*. Chicago: The University of Chicago Press, pp. 319–338.

Graur, D. and Higgins, D. G. (1994) Molecular evidence for the inclusion of cetaceans within the order Artiodactyla. *Molecular Biology Evolution*, 11, 357–364.

Janke, A., Feldmaier-Fuchs, G., Thomas, W. K., von Haeseler, A. and Pääbo, S. (1994) The marsupial mitochondrial genome and the evolution of placental mammals. *Genetics*, 137, 243–256.

Janvier, P. (1995) Conodonts join the club. *Nature (London)*, 374, 761–762.

Jenkins, F. A., Jr. and Walsh, D. M. (1993) An early Jurassic caecilian with limbs. *Nature (London)*, 365, 246–250.

Jerison, H. J. (1973) *Evolution of the Brain and Intelligence*. New York: Academic.

Migdalski, E. C. and Fichter, G. S. (1976) *The Fresh and Salt Water Fishes of the World*. New York: Alfred A. Knopf.

Milinkovitch, M. C., Ortí, G., and Meyer, A. (1993) Revised phylogeny of whales suggested by mitochondrial ribosomal DNA sequences. *Nature (London)*, 361, 346–348.

Nieuwenhuys, R., Voogd, J., and van Huijzen, C. (1978) *The Human Nervous System: A Synopsis and Atlas.* New York: Springer-Verlag.

Northcutt, R. G. (1977) Retinofugal projections in the lepidosirenid lungfishes. *Journal of Comparative Neurology,* 174, 553–574.

Northcutt, R. G. (1978) Brain organization in the cartilaginous fishes. In E. S. Hodgson and R. F. Mathewson (eds.), *Sensory Biology of Sharks, Skates, and Rays.* Arlington, VA: Department of the Navy, pp. 117–193.

Northcutt, R. G. (1979) The comparative anatomy of the nervous system and the sense organs. In M. H.Wake (ed.), *Hyman's Comparative Vertebrate Anatomy.* Chicago: University of Chicago Press, pp. 615–769.

Northcutt, R. G. (1983) Evolution of the optic tectum in ray-finned fishes. In R. E. Davis and R. G. Northcutt (eds.), *Fish Neurobiology, Vol. 2: Higher Brain Areas and Functions.* Ann Arbor, MI: University of Michigan Press, pp. 1–42.

Northcutt, R. G. (1986) Lungfish neural characters and their bearing on sarcopterygian phylogeny. *Journal of Morphology Supplement,* 1, 277–297.

Northcutt, R. G. and Butler, A. B. (1976) Retinofugal pathways in the longnose gar *Lepisosteus osseus* (Linnaeus). *Journal of Comparative Neurology,* 166, 1–16.

Northcutt, R. G. and Butler, A. B. (1991) Retinofugal and retinopetal projections in the green sunfish, *Lepomis cyanellus. Brain, Behavior and Evolution,* 37, 333–354.

Northcutt, R. G. and Kicliter, E. (1980) Organization of the amphibian telencephalon. In S. O. E. Ebbesson (ed.), *Comparative Neurology of the Telencephalon.* New York: Plenum, pp. 203–255.

Northcutt, R. G., Neary, T. J., and Senn, D. G. (1978) Observations on the brain of the coelacanth, *Latimeria chalumnae:* external anatomy and quantitative analysis. *Journal of Morphology,* 155, 181–192.

Northcutt, R. G. and Puzdrowski, R. L. (1988) Projections of the olfactory bulbs and nervus terminalis in the silver lamprey. *Brain, Behavior and Evolution,* 32, 96–107.

Northcutt, R. G. and Royce, G. J. (1975) Olfactory bulb projections in the bullfrog *Rana catesbeiana Shaw. Journal of Morphology,* 145, 251–268.

Purnell, M. A. (1995) Microwear on conodont elements and macrophagy in the first vertebrates. *Nature (London),* 374, 798–800.

Radinsky, L. B. (1987) *The Evolution of Vertebrate Design.* Chicago: The University of Chicago Press.

Reiner, A. and Northcutt, R. G. (1992) An immunohistochemical study of the telencephalon of the Senegal bichir (*Polypterus senegalus*). *Journal of Comparative Neurology,* 319, 359–386.

Romer, A. S. (1962) *The Vertebrate Body, Shorter Version.* Philadelphia: Saunders.

Romer, A. S. (1966) *Vertebrate Paleontology.* Chicago: The University of Chicago Press.

Roth, G., Blanke, J., and Wake, D. B. (1994) Cell size predicts morphological complexity in the brains of frogs and salamanders. *Proceedings of the National Academy of Sciences USA,* 91, 4796–4800.

Rowe, M. (1990) Organization of the cerebral cortex in monotremes and marsupials. In E. G. Jones and A. Peters (eds.), *Cerebral Cortex, Vol. 8B: Comparative Structure and Evolution of Cerebral Cortex, Part II.* New York: Plenum, pp. 263–334.

Ruse, M. (1988) Molecules to men: evolutionary biology and thoughts of progress. In M. H. Nitecki (ed.), *Evolutionary Progress.* Chicago: The University of Chicago Press, pp. 97–126.

Thomson, K. S. (1991) *Living Fossil: The Story of the Coelacanth.* New York: W. W. Norton & Co.

Wake, M. H. (ed.) (1979) *Hyman's Comparative Vertebrate Anatomy, Third Edition.* Chicago: The University of Chicago Press.

Wake, M. H. and Hanken, J. (1982) Development of the skull of *Dermophis mexicanus* (Amphibia: Gymnophiona), with comments on skull kinesis and amphibian relationships. *Journal of Morphology,* 173, 203–223.

Walker, W. F., Jr. and Liem, K. F. (1994) *Functional Anatomy of the Vertebrates.* Fort Worth, TX: Saunders College Publishing.

Wicht, H. and Northcutt, R. G. (1992) The forebrain of the Pacific hagfish: a cladistic reconstruction of the ancestral craniate forebrain. *Brain, Behavior and Evolution,* 40, 25–64.

Wilson, M. V. H. and Caldwell, M. W. (1993) New Silurian and Devonian fork-tailed 'thelodonts' are jawless vertebrates with stomachs and deep bodies. *Nature (London),* 361, 442–444.

5

Evolution and Adaptation of the Brain, Behavior, and Intelligence

PHYLOGENY AND ADAPTATION

When we think about evolution, we usually have **phylogeny** in mind; that is, how different lineages are related to each other, how they have changed through time, and what their similarities and differences are. We typically do not consider the mechanisms by which such similarities and differences have come about except in a general sense of adjustment to environmental change. The general process by which a species adjusts to environmental change is known as **adaptation.** Adaptation to changing environmental conditions, however, is not the only situation that results in evolutionary change; sometimes a random change in a population, such as a shift in the direction of longer legs or more acute vision enables a species to begin to take advantage of environmental opportunities that were present all the time. Such "lucky accidents" then can serve as a stimulus to enhance the trend started by this chance occurrence to take still further advantage of the existing environment.

The usual approach to studying evolution of the nervous system, behavior, or any other characteristic that has a genetic basis, is to perform a comparative study. Although studies of fossils can be useful for some purposes, research on neural or behavioral evolution usually involves extant animals. Comparative studies involve research on a number of species that have something in common. When the basis for choosing a species for study is their lineage or degree of relatedness, the comparative study is being done in a phyletic context, and the investigators hope to learn something about the historical development of that lineage by studying its living representatives. This, however, is not the only reason to do comparative studies. One also could choose animals irrespective of their descent from a common lineage, because they vary in the degree to which they have some specific characteristic. For example, the animals could be chosen because they vary in the extent of development of their color vision or hearing. Or they might be chosen because they vary in the size of some brain structure. Such structure–function studies are vital for understanding how phyletic changes might have occurred. An important consideration in adaptation studies is that, unlike **phyletic studies,** the researchers need not necessarily concern themselves with the phyletic relatedness of the subjects. Indeed, similar environmental pressures have resulted in the independent development of remarkably similar adaptations in quite unrelated species. Thus, phyletic comparative studies are concerned with *what* has changed in a specific lineage; adaptation comparative studies are concerned with *how* such changes might have come about. Both types of comparative investigation are important for understanding the complete picture of evolutionary change and development.

Phyletic Studies

Phyletic comparative studies are those that attempt to reconstruct the evolutionary history of the development of some portion of the nervous system or perhaps of a particular behavior pattern in a given lineage. To do such studies, we must select animals that represent specific lines of descent in the evolution of a particular lineage. The most successful studies of this type often are limited to a fairly restricted lineage such as different species within a family or families within an order. In addition , the successful studies tend to compare large numbers of species. Those that study animals that are less closely related, such as those making comparisons at the class level (fishes vs. mammals vs. birds, etc.), often tend to be less successful because of the difficulty of finding two or three species that are truly representative of such a broad aggregation of characteristics as would be found in a class.

Why are historical reconstruction studies important? First, because of the human fascination with where we have come from—the same passion that guides all historical research. Can we say, for example, that a particular neural structure or process or even behavior pattern is uniquely characteristic of mammals or is solely possessed by carnivores or primates? Are there systematic differences between closely related families of rodents or birds? Are there general trends or levels of organization that are characteristic of certain lineages, such as the ray-finned fishes?

A second reason for understanding these historical trends is that they can yield important clues to the relationship between the specific behavior patterns that are used by an animal and the structure and organization of the central nervous systems that make this behavior possible. Thus, comparisons of the anatomy and physiology of related animals that vary in the extent to which they perform certain behaviors is a powerful research strategy that often can provide important clues to how the system works. In particular, much can be learned from the comparison of species that are highly specialized for a specific function with those that are less well adapted. Figure 5-1 illustrates how this approach can be used. This figure is based on the work of Leonard Radinsky, who studied the cerebral cortex of a number of species of otters. He ranked the species according to the extent to which they used their forepaws to manipulate food objects. He then observed how this ranking matched up with the somatosensory (e.g., touch and temperature) area of cortex devoted to the representation of the forepaws. The

figure shows representatives of the extreme ends of the rank order. The brain labeled A is from an otter species that makes extensive use of the forepaws in manipulation of food; the brain labeled B is from an otter that makes relatively little use of the forepaws for food manipulation. Also shown in the two diagrams is the sensory representation of the face. The amount of area devoted to the face is roughly the same in the two species, but a major difference can be seen in the amount of somatosensory cortex devoted to the forepaw. The animal shown in A has a much larger forepaw representation than does the animal shown in B. This relationship between the extent to which a body part is used and the amount of brain tissue associated with it will be discussed again later in this chapter.

Adaptation Studies

Although the two species shown in Figure 5-1 are closely related, commonality of lineage is not necessarily an important consideration in the adaptation type of research strategy. Because species often solve similar survival problems in similar ways whether they are related or not, all that matters in this approach is that each has evolved the desired level of development of the structure or process under investigation. An excellent illustration of the independent evolution of similar neural and behavioral processes may be seen in the comparative anatomy and comparative behavior of electroreception. This sensory system has independently evolved a number of times among remotely related groups of fishes as well as in other vertebrate classes, including mammals. Likewise, infrared (IR) detection has evolved independently in relatively unrelated groups of snakes.

An important benefit for biomedical science from adaptation studies is the awareness that the particular way that certain species have adapted *or failed to adapt* to the pressures of their environments makes them very useful as models for certain normal developmental processes as well as models for various pathological states, usually in humans. For example, about 20% of a strain of pigeons known as White Carneau have difficulty in discriminating between patterns that are mirror images of one another, such as < vs. > or [vs.], yet they have no difficulty discriminating patterns that are not mirror symmetrical, such as + vs. =. The failure of these birds to easily tell mirror-symmetrical patterns apart makes them useful as a model for the human reading disorder known as dyslexia, in which readers have difficulty telling apart the letters **p** from **q** and **b** from **d.** Such animal models can offer important clues to the morphological basis of this disorder and can provide a useful model system on which to test therapies.

Finally, another benefit to science of adaptation studies is that they frequently reveal unique experimental situations that permit experimenters to perform experiments that they otherwise could not. Three well-known examples are

- The squid giant axon, which allowed neurophysiologists to investigate the properties of the axonal membrane directly by providing a preparation with an axon diameter big enough to allow insertion of an electrode.
- Individually identifiable neurons, which allow experiment-

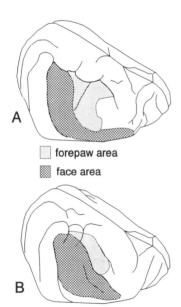

forepaw area

face area

FIGURE 5-1. The representations of the face and hand area in the somatosensory cortex of two species of otter. Species A uses its forepaws more extensively to manipulate food objects than species B. Although the representations of the face are about the same in both, the area of cortex devoted to the forepaws is greater in A than in B. (After Radinsky, 1968).

ers to investigate the same neuron in animal after animal after animal.

- Vocal-learning pathways in the brains of birds, which have permitted the elucidation of the complex central pathways from the auditory receptors in the medulla to the vocalization control regions of the telencephalon and back to the motor neurons of the medulla that control the actual production of sound.

The Phylogenetic Scale

One of the notions that has misguided comparative studies has been the notion of the **phylogenetic scale** or evolutionary scale, which is derived from the medieval notion of the *scala naturae* (scale of nature) with God at the top, angels next, humans somewhat lower, and then apes, monkeys, and assorted other animals that differ progressively in their resemblance to humans as the scale is descended. After the success of the Darwinian revolution, and the acceptance of the idea of evolutionary change in the scientific community, this notion served as the model for evolutionary change for many who had little knowledge of what is known about the history of animal evolution. One can readily see how this could come about: land animals came from the sea, reptiles gave rise to mammals, apes gave rise to humans, and so on. While this is correct in a general sense, one cannot take any living fish or reptile or nonhuman primate and use it to represent some stage in human evolution without first considering how or even whether it is related to the human lineage. Nevertheless, the scientific literature is filled with erroneous conclusions, sometimes by distinguished scientists, about evolutionary changes based on comparisons of a goldfish, a frog, a lizard, a rat, a cat, a monkey, and a human. Figure 5-2 shows the type of hierarchical arrangement that derives from this thinking. Each

species occupies a niche on this hierarchical ladder in which a higher position denotes both more recent evolution and greater evolutionary advancement or development. This model assumes that more ancient life forms are more primitive than more recent forms. While there can be no doubt that evolution has resulted in some dramatic improvements in the design of brains and bodies, one should not necessarily assume that ancient, ancestral forms were "primitive" in the sense of being crude and poorly adapted versions of modern animals. Animals generally are well adapted to their environments at whatever geological age they may have existed, and many of the adaptations of ancestral forms were quite sophisticated and efficient for the environments in which they lived. Indeed, we should only speak of "primitive" characteristics, not "primitive" species.

The Phylogenetic Tree

Figure 5-3 shows the way a systematic biologist would view the relationships among the animals shown in Figure 5-2. This figure is a family tree of animal species that shows how closely or remotely various animal groups are related and what the basis of the relationship is. In this model of evolution, the location of a group of animals on the tree only indicates the group's relative time of appearance in the fossil record; that is, the age of the oldest fossilized remains of the lineage thus far discovered. The vertical extent of the figure represents time with the present era represented at the top and more ancient eras represented below. The vertically oriented lines indicate lineages or branches on the family tree. Each of the animals indicated in Figure 5-1 is shown on its appropriate branch of the family tree. The complexity of the animals is not indicated in any way in this figure, nor are such judgmental evaluations as "advancement" or "progress" represented. One can see in

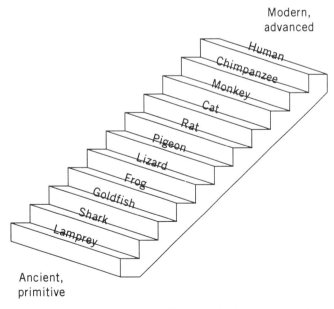

FIGURE 5-2. A phylogenetic scale of vertebrates. Those animals at the bottom of the scale are regarded as phylogenetically older and more primitive. Those at the top are regarded as being more modern and more advanced.

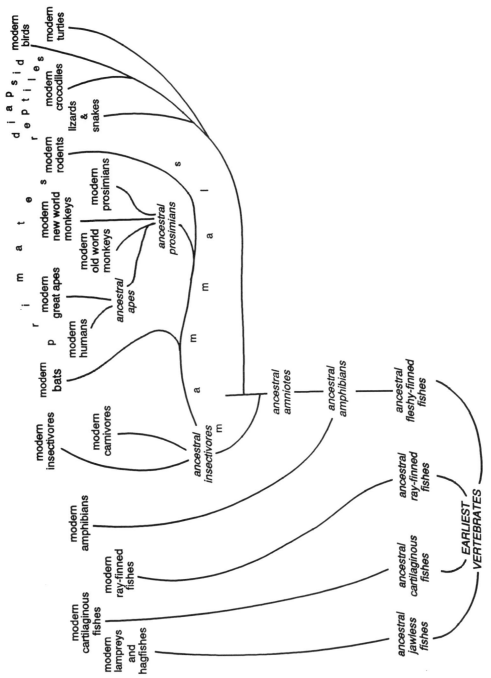

FIGURE 5-3. A phylogenetic tree of vertebrates. The animal groups represented at the bottom of the tree are phylogenetically older than those higher up on the tree. Evolutionary advancement or progress is not indicated on this diagram.

this figure the four great lineages of the vertebrates: the jawless fishes, the cartilaginous fishes, the ray-finned fishes, and the fleshy-finned fishes. Because these four great lineages evolved independently from one another (i.e., they developed in parallel), members of one group should not be used to represent a stage of evolution in another group. Thus no modern bony fish (a descendent of the ray-finned lineage) should be used to represent a stage in the development of any tetrapod (amphibians, mammals, diapsid reptiles, birds, or turtles). The same would apply to the modern sharks, which are members of the cartilaginous fish lineage. While these animals are descendants of very ancient lineages and can give some clues to what the earliest vertebrates were like, they are not members of the lineage that eventually produced land vertebrates.

Similar considerations apply within the land vertebrates, although the figure is not sufficiently detailed to show it. For example, one cannot simply study an extant diapsid reptile or turtle in order to gain understanding of what the reptilian ancestors of mammals were like. As discussed in Chapter 4, the amniote line leading to lizards evolved *after* mammals, so extant lizards could hardly represent mammalian ancestors. Likewise, crocodiles, alligators, and birds are derived from the same diapsid branch that gave rise to lizards. Finally, turtles appear to be even more recent evolutionary descendants of the ancestral amniote lineage than are the diapsids.

Table 5-1 compares the differences between the phylogenetic tree and the phylogenetic scale as models of animal evolution and development. As you can readily see, the two approaches to the systematic arrangement of living animals give very different sorts of information. The phylogenetic scale gives information that tends to be evaluative, whereas the phylogenetic tree gives information that is historical. Because the scale is a ranking, it is unilinear or unidimensional, whereas the tree model emphasizes the multilinearity of evolution and the radiation of lines of descent from common ancestral points.

COMPLEXITY AND EVOLUTION

One of the features of vertebrate evolution that has led to misconceptions is the observation that mammals in general and primates in particular seem to be the most complex; but this is probably only because they have been the most intensively studied. When detailed studies are made of nonprimate mammals or of nonmammals, levels of complexity as great, or sometimes greater than those of mammals emerge. A few examples: Olfaction is better in many nonprimate mammals than in primates. Many auditory capabilities of echo-locating bats and marine mammals are far superior to those of primates. Birds have a visual acuity that is far superior to that in most mammals, and the best of the birds, the birds of prey such as hawks and eagles, exceed human visual acuity by at least an order of magnitude. Bird color vision surpasses that of mammals, including primates, in terms of range of wavelength and ability to discriminate color. In certain species of fishes that generate electrical fields around themselves, which serve as social signals and are used to detect objects in the environment, the cerebellum is a far more complex organ than the cerebellum of any mammal. No mammal can come close to rivaling the taste system of a carp or a catfish, which have evolved large, elaborate lobes on the medulla to process information from more than 100,000 taste buds located not only in the mouth but all over the surface of the body. Many nonprimate animals posses senses that we lack totally: the detection of ultraviolet (UV), infrared (IR), and polarized light, the detection of electric fields, and the detection of taste and smell qualities that totally elude primates.

Another often misunderstood point is that although evolution frequently has produced an increase in complexity of brain organization and behavior, the earliest vertebrates were already quite complex animals in comparison to their invertebrate ancestors and were quite well adapted to their environments. By choosing a limited range of species, however, a false impression of a trend from simple to complex can emerge. An example of the effects of too narrow a selection can be seen in Figure 5-4. This figure shows the number of Purkinje cells (large neurons located in the cerebellum) per cubic micrometer of cerebellar cortex in five species of mammals based on a study by Winfried Lange. These particular mammals were selected for the figure from among 20 species of mammals reported by Lange because they represent a typical phylogenetic-scale sequence. If one were to assume that this sequence represents evolutionary history, which indeed it does not, this representation would suggest that there is a trend in the evolution towards humans of a decreased number of Purkinje cells in the cerebellum. Moreover, since many humans believe themselves to represent the pinnacle of evolutionary advancement, there must be something inherently better in having fewer Purkinje cells.

If one now looks at Figure 5-5, we see the same data represented, but with the addition of some further species from

TABLE 5–1. A Comparison of the Phylogenetic Tree and the Phylogenetic Scale as Models of Evolutionary Change and Development

Phylogenetic Tree	Phylogenetic scale
Multilinear	Unilinear
Based on scientific judgments	Based on scientific and value judgments
Nonhierarchical (animals not ranked)	Hierarchical (based on a ranking)
Presumed historical sequence is determined by historical data (the fossil record)	Presumed historical sequence based on increases in complexity
Complexity is independent of time; increased complexity and increased simplification can both be evolutionary trends	Complexity increases with time

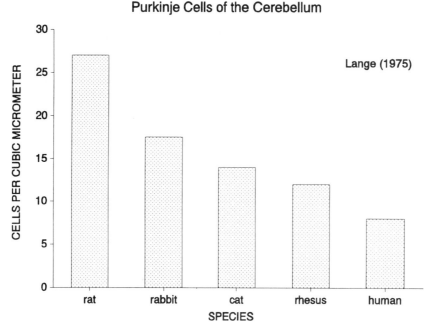

FIGURE 5-4. The relative number of Purkinje cells in the cerebellar cortex in five species of mammals. Data from Lange (1975).

Lange's report. We see that the squirrel monkey, which is a primate, has about the same relative number of Purkinje cells as does a cat or a fox, which are carnivores and supposedly lower animal forms. Moreover, a domestic bull has the same relative Purkinje cell number as does a rhesus monkey. Finally, the number of human Purkinje cells is nicely matched by the relative number of such cells in an elephant. How can we

interpret this? Lange's report tells us that there is an inverse relationship between relative Purkinje cell number and the absolute size of the cerebellum. Indeed, 80% of the variability in Purkinje cell number is due to the variability in cerebellum size. So rather than indicating a trend in the direction of human perfection, Lange's data mean that Purkinje cell number is dictated by cerebellum size. Similar data exist in birds, in which

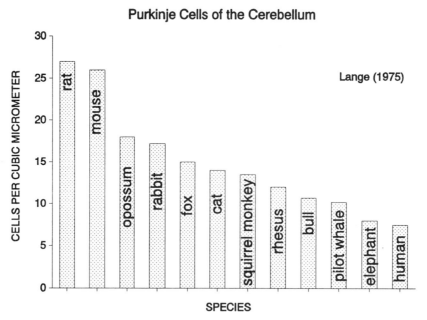

FIGURE 5-5. Seven additional species of mammals have been added to the data shown in Figure 5-4. Data from Lange (1975).

the cell counts in the ostrich cerebellum are comparable to those of the pilot whale, and those of the titmouse (a very small bird) exceed even those values for rats and mice.

Anagenesis

One of the ideas that has had great influence in comparative research is the notion that somehow primates in general and humans in particular must be superior to all other animals in virtually everything. This idea has been crystallized in the concept of "anagenesis," which was introduced into the biological literature in the late 1950s and refers to progressive advancement or improvement through time in a given lineage. Unfortunately, anagenesis has been associated too often with the notion that primates, and especially humans, are the peak of evolutionary perfection (after all we are studying the animals; they are not studying us). It also is derived from notions, prevalent since the fifteenth century at least, that the European body type, as well as European social systems, religion, and mores, represent the pinnacles of human development. All others are inferior, less perfectly formed, and so on. These highly anthropocentric, homocentric, and Eurocentric views have on more than one occasion lead to the massaging of data (or at least the interpretation of data) so that humans come out the best. Here is a quotation from Lange, the comparative neuroanatomist whose cerebellum data were shown in Figures 5-4 and 5-5. He offered the following lament in the discussion of an inconvenient result: "If the number of cells per unit volume alone is taken into account as the only parameter for the organization level of a brain, the elephant or some aquatic mammals would appear to show a higher degree of brain evolution than man."

This is not to say that humans are not very special and very unique creatures. Nevertheless, in spite of our unique and special characteristics that have permitted us to arrive where we are today (for good or for ill), we do not have a monopoly on special or unique characteristics, nor are we always to be found at the pinnacle of every hierarchical ranking. Figure 5-6 presents data that will help you to keep some perspective on human perfection. The figure, based on the work of Richard Passingham, shows the relationship between the area of the largest region of the cerebral cortex, known as the **isocortex** (often referred to as "neocortex") and the volume of the brain as a whole in primates. The figure shows that as brain volume increases, the volume of the isocortex increases proportionately. What about the isocortex of humans? Our large isocortical mass is often pointed out as a sign of our superior intellectual abilities compared to other primates. The figure, however, indicates that human isocortex falls close to the primate line. In other words, we have no more isocortex than would be expected for a primate of our size. To be sure, humans do have greater amounts of certain regions of the isocortex than would be expected from their brain size, and also certain structures in the depths of the brain are greater in size than would be expected from such curves. The purpose of this illustration is to indicate that humans will not always necessarily be at the top of any ranking of brain structures in terms of sheer volume or of organization.

Another problem with the anagenetic approach is that it is highly dependent on value judgments on the part of the observer as to what constitutes progress or advancement. The study of evolution has taught us, however, that the same adaptation to the environment in one circumstance can be the road to the development of a new species but in another circumstance leads inexorably towards extinction. There is nothing progressive about an adaptation in itself; it is the circumstances (i.e., the selective pressures) that make adaptation progressive.

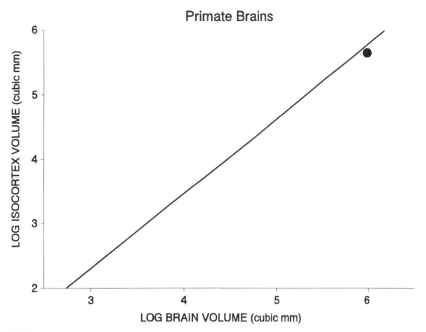

FIGURE 5-6. The diagonal line shows the relationship between the volume of isocortex (neocortex) to the total volume of the brain in primates. Larger brains have more isocortex. The dot indicates the cortical volume of human brains and indicates that humans have an amount of cortex that is appropriate to their brain size.

Anything that helps an animal to survive is "progress," whether it is based on increased complexity or increased simplification.

Grades of Evolutionary Advancement

The unit of anagenetic analysis is the "grade," which has been variously defined but is generally conceded to be a progressive improvement in a characteristic through a specified lineage. Figure 5-7 presents some comparisons from published data on various species of bats. The figure shows the relative size of the olfactory bulb (the volume of the olfactory bulb as a proportion of the volume of the cerebral hemisphere) in 40 different species of bats arranged according to their diet. The relative olfactory bulb sizes have been sorted according to whether the bats are insect eating, flesh eating, fruit eating, or fruit and insect eating. Since the majority of bats are insectivorous, these have the highest representation. The two groups of bats that are not fruit eaters have significantly smaller olfactory bulbs than do the two groups that have fruit as the exclusive or major part of their diet. One thus could consider that bat olfactory bulbs have two grades: an insect–flesh eating grade and a higher fruit eating grade, which is characterized by larger olfactory bulbs.

Unfortunately, the notion of evolutionary "progress" suggests an advancement from simple to complex, which in turn leads to a tendency to conclude that the more "advanced" level was the more recently evolved. Evolutionary history, however, is filled with nonlinearities in which a seemingly less complex form evolved from one that seemed more complex. One example is the evolution of the heart. Fishes have a two-chambered heart, amphibians and most diapsid reptiles and turtles have a three-chambered heart, and birds and mammals have a four-chambered heart. This progression would seem to suggest three grades of heart based on the number of chambers:

- Fish grade which consists of a two-chambered heart.
- Amphibian–diapsid reptile grade which consists of a three-chambered heart.
- Avian–mammalian grade which consists of a four-chambered heart.

On the face of it, this would appear to be a simple ranking of progressive advancement: two chambers, three chambers, and four chambers. But in fact, four chambers evolved directly from two chambers, and three chambers is an adaptation of four chambers rather than its predecessor. Modern evidence suggests that the four-chambered heart evolved in the immediate ancestors of land vertebrates (similar to modern lungfishes) as a consequence of the development of the lungs and the pulmonary circulation. Thus, the earliest land vertebrates had a four-chambered heart. The three-chambered heart of modern amphibians (who are not especially close relatives of the amphibian ancestors of mammals, diapsid reptiles, birds, and turtles) evolved as a consequence of the amphibian reliance on the skin as a supplementary respiratory organ in addition to lungs. Thus it would appear that mammals never went through a three-chambered heart stage in their evolution. On the other hand, for birds the history was two chambered to four-chambered to three-chambered to four-chambered. An important lesson about evolution may be learned from this example: simplification of a system can be just as adaptive (or progressive) as increasing complexity.

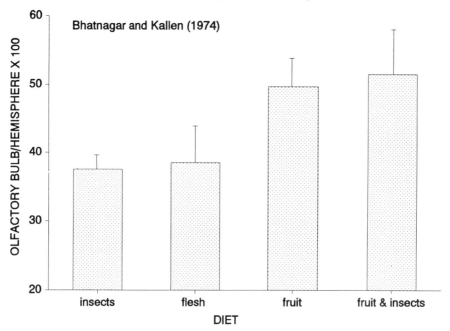

FIGURE 5-7. Relative size of the olfactory bulb in bats arranged according to their diet. Fruit eating bats have significantly larger olfactory bulbs than do insect or flesh eating bats.

A second example comes from grades of primate social behavior. Four grades have been noted:

- Solitary living.
- Harem system (one male and many females).
- Multiple males and females.
- Pair living.

Some might rank pair living as the highest grade since it is the one with which we identify most closely in Western societies. Any sequence, however, could be a plausible representation of the evolutionary history of primate social groupings and indeed all four of these grades can be found as widely distributed among the prosimians as among the simians and even among the hominoids, namely, the great apes and humans. If this were an evolutionary sequence, however, we would expect solitary living to be most widely distributed among the prosimians with the higher grades increasing in frequency in the groups most closely related to humans.

Our intention is not to leave you with the conclusion that grades have no utility in comparative research. Grades can be quite useful in analyzing structure function relationships. We must not fall into the trap, however, of assuming that the sequence of grades necessarily represents the historical development of the characteristic under study.

EVOLUTIONARY CHANGE

How do evolutionary changes come about and how are they reflected in the central nervous system? As we discussed in Chapter 1, evolutionary change often has its roots in pressures from the environment, either internal or external, although random mutations also can have great effects if they are not too quickly diluted in a large population. Environmental pressures can arise from decreased availability of food, increased presence of predators, increased competition for mates or nesting sites, and so on. Sometimes ample food supplies are available, but only to those who have the capability to gain access to them. Environmental pressures do not directly produce somatic or central nervous system changes; the changes have to occur independently, even if only to a limited extent, so that natural selection can operate on them. Some of these conditions exist as a byproduct of some other evolutionary trend. For example, the jaw bones that evolved into ear ossicles arrived at a location in the jaw where they could come under selective pressures that favored sound conduction not because the auditory system needed them but because of selective pressures related to feeding. Once they were in an appropriate position, a major evolutionary change occurred when their role in sound conduction became more advantageous to the animals' survival than did their role in feeding.

Let us return to the example of the use of forepaws that we discussed in connection with Figure 5-1. If more dexterous forepaws give access to ample food supplies, those creatures that already have slightly more dexterous paws than their fellows have a better chance of surviving long enough to reproduce and successfully rear their young who will carry the genes for those more dexterous forepaws. This process is shown diagrammatically in Figure 5-8. Of course, concurrent with the increased mechanical agility of these paws is the increased ability of the nervous system to control those paws with increased precision. As the figure indicates, if an advantageous tendency is already present in the gene pool, it will be selected for; if not, and if the environmental pressure is life threatening, the result will be extinction. Indeed extinction is a very common event in evolution. If these changes in the forepaws and their corresponding central nervous system adaptations can be put to good behavioral use to more effectively interact with the environment, the environmental pressure will be reduced and the probability of successful reproduction of successive generations will increase.

While subtle central nervous system changes could occur without being reflected in somatic changes (such as changes in dendritic or columnar organization in the cerebral cortex), the reverse (somatic changes without corresponding central nervous system changes) would be highly unlikely. To use a computer analogy, there is no point to adding more sophisticated peripheral devices to your computer if the computer lacks appropriate hardware and software to take advantage of the added sophistication of the peripheral.

BRAIN EVOLUTION AND BEHAVIORAL ADAPTATION

Most evolutionary biologists probably would agree that those animal species that are alive and flourishing today are uniquely adapted to the lives that they lead. Their brains are no less adapted than are other parts of their bodies. The brain is the executive organ of behavior, and often behavior plays a vital role in successful adaptation. Moreover, orderly relationships can be seen among differing levels of development, even in remotely related animals. For example, a vertebrate that uses its snout as an exploratory organ, whether it is a shark, an alligator, or a pig, will have highly developed sensory inner-

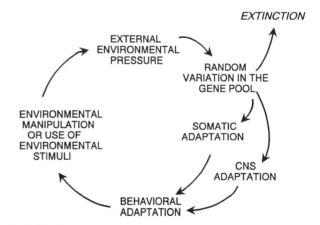

FIGURE 5-8. The cycle of environmental pressure and evolutionary change. Life threatening environmental pressures that do not act on appropriate variations already present in the gene pool of a species will lead to extinction. If the appropriate conditions are present, even to a small degree, they can be selected for to result ultimately in some reduction of the environmental pressure.

vation of the skin on the snout and a well-developed trigeminal cranial nerve. A fish, a lizard, or a mammal with a well-developed sense of smell will have a complex and highly organized olfactory system. A vertebrate with a powerful and adroitly used tail and/or hind limbs will have a large, well-developed lumber spinal cord.

Successful adaptation to the environment often means behavioral adaptation. Behavioral adaptation means brain adaptation. Overlaid upon a general vertebrate organization are the unique adaptations of the individual species. Animals (whether closely related or not) that have responded successfully to environmental challenges in similar behavioral ways frequently have accomplished their adaptations by having evolved remarkably similar neural substrates. Although much has been made of the rather impressive differences between the brains of different vertebrate groups, when seen in the context of a rather consistent overall organization with specialized adaptations, the differences seem less important, and the nervous system appears to have had a rather conservative evolution.

Brain Size and Intelligence

We have all heard the term "bird brain" used to suggest that someone's intellectual capabilities are rather low. In contrast, terms like "big brain" or "brainy" suggest a person who is well endowed with the capacity to absorb large amounts of information and to organize and utilize this information in creative and resourceful ways. Is this really the case? Does the size of a person's brain give an indication of their intellectual ability? Are animals with small brains less intelligent than animals with bigger brains? Has evolution been progressively shaping larger brains?

Many of our common sense ideas about relative brain size and its importance for behavior are derived from the period between the 1880s and the early 1920s. This was a period of intense activity in paleontology. A number of paleontologists of this period were interested in brain evolution as it was reflected in the size and shape of the cranium, which is the bony container of the brain. Among their observations was the finding that the cranial capacities of huge dinosaurs, such as *Brontosaurus* and *Stegosaurus,* were extraordinarily small in contrast to the great size of the animals' bodies. This disparity between brain size and body size led them to the conclusion that these animals must have been very stupid indeed. Were such conclusions justified? Later in this chapter, we will explore the nature of intelligence in humans and other animals and its relationship to brain size and structure.

Early Views of the Evolution of the Brain and Intelligence

One of the early pioneers in the field of paleoneurology (the study of ancient brains) was Othniel Charles Marsh. Marsh was an eminent paleoneurologist and discoverer of many important fossil finds in the American southwest. He was interested in the sizes of fossil brains and what they could reveal about brain evolution. In the material that he studied, he observed what appeared to him to be certain trends in the history of brain evolution. These observations led him to propose several "laws" of brain evolution. These laws had a major impact on the thinking of comparative neuroanatomists of his era, and a number of his ideas became incorporated into their own speculations about the history of brain evolution among vertebrates. Marsh intended his laws as a description of evolutionary trends within the amniotes, but his ideas appear to have been generalized by others to the anamniotes as well.

The most general of Marsh's laws states that as evolution progressed towards the modern era, brains tended to increase in size relative to the body. Another law stated that this increase was mainly reflected in the cerebrum. Still another law postulated that in some groups of mammals, evolution has produced a progressive increase in the complexity of cerebral convolutions.

The late Victorian era was also the period in which comparative psychology was developing the principles that would guide it, rightly or wrongly, for the next half century. One of these principles was that intelligence and the capacity for behavioral complexity increase as animals "ascend the phylogenetic scale." This principle seemed to fit in rather well with Marsh's laws. Thus, as evolution "advanced," nonhuman animals acquired relatively larger, more complex, more human-like brains, which made them progressively more intelligent, that is, more human-like in their activities. The seeming reasonableness of a parallel progression between brain development and intellectual development that advanced from simple to complex has kept these notions alive until very recent times.

Brain Allometry

Recent studies of relative brain size and comparative behavior in different lineages have led to some conclusions about brain size and intelligence that are rather different from the common sense model. In the light of present day investigations, Marsh's laws generally have come out rather well if their application is limited to the groups for which they were intended, namely, the amniotes. They do not, however, tell the complete story.

Figure 5-9 shows a survey of data about relative brain size from a number of vertebrate classes. The figure shows the average weight of the brain in grams plotted as a function of the average weight of the body in kilograms. Within these axes are groups of convex polygons. Each polygon is the outer boundary of a set of data points that represent the individual species of the vertebrate class that is contained within it.

Within the class of mammals, the shaded area represents the brains and bodies of the primates. A total of 198 species is plotted in the figure. The results of this analysis are quite revealing. First, note that the brains of bony fishes, which usually are assigned a rather lowly position on the "phylogenetic scale" of vertebrates, have brains that are comparable in relative size to those of diapsid reptiles and turtles, which are regarded as "higher" creatures because they are closer to humans on the scale. Moreover, the upper surface of the fish polygon is considerably higher than the upper surface of the diapsid reptile and turtle polygon. This indicates that the brains of the largest bony fishes, located along the top surface of the polygon, are heavier than are the brains of diapsid reptiles and turtles of equivalent body weight.

Higher up in the figure is the polygon representing the mammals. Notice that, as a group, the primates, which are at

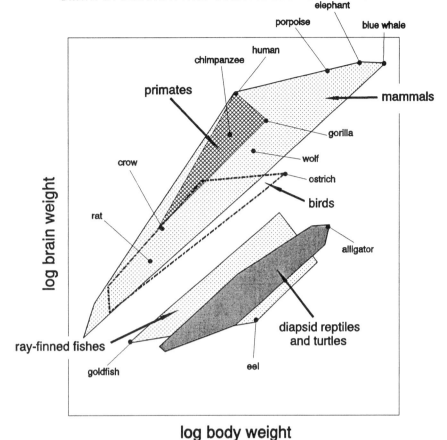

FIGURE 5-9. Log brain weight plotted as a function of body weight for 198 species of vertebrates. The geometrical figures are the minimum convex polygons that will enclose all of the data points. Data are shown for ray-finned fishes, diapsid reptiles and turtles, birds, and mammals. The location of primates within the mammals is shown separately. Individual data points for some species are shown. Data points that lie along the long axis of each polygon represent species that have a brain weight that would be expected from their body weight. Those at the top of the polygon have brain weights that are greater than would be expected from their body weights. Conversely, those at the bottom of the polygon have lower than expected brain weights. Thus, among the mammals, rats and wolves have brain weights that are typical for mammals, whereas primates tend to have larger brains than would be expected from their body weights alone. Data and figures adapted from Jerison (1973).

the top of the "phylogenetic scale," have brains that are, for the most part, only somewhat above average for mammals. Average-weight brains lie along a line that passes through the middle of the polygon from lower left to upper right. The bird polygon shows an impressive amount of overlap with the mammalian polygon, especially among the smallest birds and mammals. In this region, the polygons indicate that small birds and small mammals of comparable body weights have the same weight brains. Not only that, but the birds with the largest brains have brains that would be regarded as typical of a small primate of the same body weight.

Within Figure 5-9 some individual species have been identified within their respective polygons. The top surface of the mammalian polygon is occupied by the animals that are often regarded as having high intelligence: humans, porpoises, elephants, and whales. However, chimpanzees and gorillas, both close relatives of humans and considered to be quite intelligent, are only average among the mammals in relative brain size.

Before continuing further with diagrams of this sort, we should discuss some of the criticisms that have been leveled at this type of analysis. The method of comparing the relative development of different forms that has been illustrated in Figure 5-9 is **allometry.** Allometric analyses of brain and body development have been criticized on several accounts. First, the selective pressures that determine body weight are different for animals that have adapted to different types of environments. For example, vertebrates that fly, such as birds and bats, have a heavy selective pressure on them to reduce body weight in the interests of efficient utilization of energy. One of the ways birds have accomplished this has been by the development of a very light, porous bone structure. Similarly, vertebrates that live totally or largely in an aquatic environment have far less pressure on them to keep body weight to a minimum because of the buoyancy of their body masses in water. This is especially true of animals living in salt water. Thus, marine mammals, such as whales and porpoises, and some extinct marine reptiles, such as plesiosaurs, ichthyosaurs, and mossosaurs, were able to achieve enormous body weights.

On the other hand, selection pressures for brain development have not been uniform either. The brain is not a unitary organ; rather, it is a complex organization of separate but closely interacting organ systems. These systems have different functions and are subject to quite different selective pressures. For example, the autonomic nervous system, which is largely concerned with the regulation of the body's internal environment, is affected by a very different set of selective pressures than is the visual system or the motor system that controls fine movements of the digits in mammals. As we will see in later chapters, in some sharks and ray-finned fishes, the cerebellum has undergone an extraordinary development. In other fishes, in which taste has reached an extraordinary level of development, the region of the medulla into which the taste nerves enter has undergone a high degree of growth and specialization. To give one further example, in birds, many of which are highly specialized for vision, the optic tectum of the midbrain has reached a degree of size and complexity that is unrivaled in any other class of vertebrates. Thus, the variability in body-weight selective pressures seems to be matched by a comparable variability in brain-weight selective pressures. Nevertheless, even if these problems of variability may limit the mathematical power of allometric analysis, the type of plot that is shown in Figure 5-9 offers an opportunity to see the range of variation in brain size that can be found for a given body weight as well as the variation in body weight that can be found for a given brain weight.

Figure 5-10, which is similar to Figure 5-9, includes polygons representing the brains of cartilaginous fishes (sharks and rays) and jawless fishes. Notice that cartilaginous fishes, which are usually placed rather low on the "phylogenetic scale" and are commonly regarded (though perhaps unfairly) as rather stupid, actually have brains that are relatively large and are comparable in size to those of the medium weight mammal and bird brains from animals with equivalent body weights. In contrast, the jawless fishes or agnathans have brain weights that are outside of the range of any of the jawed vertebrates for which data have been described thus far.

Brain Size and Behavioral Adaptation

In general, the data seem to suggest a selective pressure for progressively larger brains, or at least for a progressive increase in certain brain components. Why should this be so? Many paleontologists believe that behavior plays an important role in an animal's adaptation to its environment. Brain components have evolved in ways that are appropriate to the animal's behavioral adaptations. Those organisms with more numerous and more complex neural networks in certain systems have increased opportunities to survive. These superior neural control and processing systems permit enhanced sensory capacity to better detect more subtle changes and differences in the environment. Improved motor control permits better locomotion through the environment for predation, escape, territorial defense, reproduction, and so on, as well as more sophisticated manipulation of objects. Expanded integrative systems permit more information to be filtered, encoded, stored, and retrieved, which permits the animals to make greater use of their previous experience. Moreover, in the presence of certain specific stimuli, entire behavioral sequences have emerged that have important social consequences for territorial establishment and defense, courtship and reproduction, rearing of the young, feeding, and social organization.

Thus, while animals with less complexly organized, relatively smaller brains continue to survive, animals with larger and more complex brains have also been selected for. Basically, larger and more complex brains allow for the occupation of new niches. The diversity of the range of brain sizes and degree of complexity is thus a direct function of the diversity of potential new niches. The evolution of species with larger brains is in fact a reflection of an increase in diversity, which is limited only by the availability of new niches. That an increase in diversity—rather than a unilinear increase in brain size per se—has occurred among vertebrates is demonstrated by the trends for reduction and simplification in the central nervous system of some vertebrate groups. Amphibians are a prime example. The niches that such species occupy are as compatible with a reduced, simplified brain organization as other niches are with an enlarged, complex brain organization.

At the neuronal level, the specific effects of adaptations for larger and more complex brains are increased numbers of neurons and a more extensive dendritic organization, thereby permitting greater numbers of terminals to contact an individual dendrite. These changes allow greater precision and greater elaboration of interneuronal relationships. Selective pressures have favored more efficient packing arrangements so that the more elaborate neuronal systems have not resulted in an excessively enlarged head. These more efficient packing and processing arrangements include:

- Gathering of scattered neurons into more dense aggregations.
- Segregation of myelinated axons of the Golgi Type I neurons from other axons and cell bodies so that the neuronal mass is encapsulated within the white-matter tracts formed by myelinated axons.
- Subdivision of the neuronal mass into specialized components with different sources of input or different output destinations.
- Specialized arrangements of local circuitry, such as **lamination,** which is the organization of the cell and fiber constituents of a neuronal mass into separate layers of cells and fibers.

Within a laminar organization, such as the cerebral cortex, **local-circuit** arrangements arise that are orthogonal to the plane of lamination, that is, they cut across the laminae at right angles to form a columnar organization of cells and fibers. Laminar organization appears to be the most sophisticated form of neuronal organization. It often is found as the neural substrate of the most complex behavioral processes.

Behavioral adaptation to the environment almost certainly involves selection for the chemical messengers and modulators that carry out many of the interneuronal communications. Selective pressures have acted on variations in the molecular structure of the chemicals and on their neuronal receptors as well. However, such adaptations would not be expected to add sub-

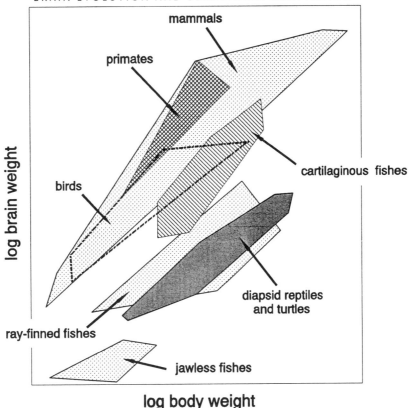

FIGURE 5-10. Minimum convex polygons for cartilaginous fishes and jawless fishes have been added to the data of Figure 5-9. Adapted from Northcutt (1985).

stantially to either the weight or volume of the brain or of its individual gross components.

Earlier in this chapter, we drew attention to the common-sense notion that brain size is indicative of intelligence. Some of the allometric plots shown in this chapter have been used to make the case that relative brain size indicates relative intelligence. Thus, if two species of the same body weight have different brain weights, the one with the heavier brain weight is presumed to be the more intelligent. Put the other way around, the argument sounds less convincing: If two species have the same brain weight but different body weights, the one with the lesser body weight is presumed to be the more intelligent.

Evolution of Intelligence

At the time that Marsh was making early efforts to reconstruct the history of amniote brain evolution, similar efforts were under way by psychologists interested in reconstructing intellectual evolution. In 1882, George J. Romanes published a book entitled *Animal Intelligence,* which was the first attempt at a scientific analysis of the subject. In Romanes' conception, intelligence was the capacity of an animal to adjust its behavior in accordance with changing conditions. For Romanes, the hallmarks of intelligent behavior were "conscious choice" and "intentional adaptation."

A contemporary of Romanes was Conway Lloyd Morgan, the author of an influential work, *Introduction to Comparative Psychology.* Lloyd Morgan's ideas were in the tradition of nine-

teenth century "faculty psychology," which held that the human mind was composed of various faculties or powers, among which were perception, memory, judgment, self-esteem, and so on. Humans shared the lower faculties with some other animals, but the highest were exclusively human. Lloyd Morgan, however, was greatly concerned about the then prevalent custom of "anthropomorphism," which was the attribution of the full range of human faculties to nonhuman animals. He summarized his views in his famous canon: "In no case may we interpret an action as an outcome of the exercise of a higher psychical faculty, if it can be interpreted as the outcome of one which stands lower in the psychological scale." Thus, we should not assume that nonhuman animal behavior is the result of complex, human-like processes if the behavior can just as well be explained by simpler mechanisms. In other words, you should not assume that your pet dog is showing love for you by licking the back of your hand; the dog may just be enjoying the salty taste of your perspiration. Lloyd Morgan's canon maintained an influence on our ideas about the evolution of animal intelligence and its relation to the evolution of the brain that still can be felt today. Its suggestion of a progressive scale of intellectual development was a natural companion to the progressive scale of brain development that Marsh's laws seemed to suggest.

What Is Intelligence?

What is intelligence and how can we determine its presence or absence in animals? Intelligence is a term that has been applied to human behavior in a wide variety of ways. A common feature of many definitions of intelligence is the notion of a purposeful adaptive behavioral response to the demands of the environment. Intelligence theorists argue at length about whether intelligence is a general characteristic of an organism that is applicable to many diverse situations, whether it is merely the sum of a number of rather specific abilities, or some combination of both. Whatever the definition of human intelligence, an important consideration in applying it to other animals is that intelligence is not a biological property of organisms, such as height, cranial volume, or cortical surface area. It is a value judgment on the part of the observer about the merits of the behavior observed. The intelligence tester decides which behaviors are important to measure. If the subject, whether human or other animal, performs those behaviors well, the tester concludes that the subject is very intelligent. Poor performance on the test results in a rating of low intelligence. In a different culture, a very different set of behaviors might be rated as "intelligent."

Contemporary testers of human intelligence have recognized that a person who does not value the abilities being tested or who may not have had an opportunity to develop those abilities may be incorrectly judged to be of low intelligence. Such persons may actually be using behavior to deal quite superbly with the changing demands of their environment, but the tests may not be assessing those abilities. Contemporary testers understand quite well that giving tests designed for urban school children and adults who have been educated in technically sophisticated societies to children or adults who were reared in the indigenous cultures of New Guinea, Greenland, or the Sudan, for example, would result in test scores that would indicate very low intelligence. They also know that the scores of the urban dwellers on tests based on behavioral characteristics required to survive in the nontechnological, rural cultures also would be extremely low. Hence, the search for intelligence tests that are "culture-fair" or "culture-free" continues.

The same cautions apply in the assessment of nonhuman animal intelligence. In their desire to compare nonhuman animal intelligence to human intelligence, comparative psychologists generally have sought to measure the performance of their subjects in behavioral situations that humans value. Moreover, they have tried to make the tests as similar to human tests as possible to facilitate the comparison. In many cases, the tests have built-in biases similar to the cultural biases in the human tests. For example, rats frequently do poorly on most tests intended to assess their intelligence. These tests, however, almost invariably involve visual stimuli. Vision is a very important distance sense for humans and other primates, but it is not very important for nocturnal rodents. The senses of smell and hearing are considerably more important distance senses for such animals. Indeed, when given a test of behavioral adaptability that uses olfaction rather than vision, rats perform as well as many primates do on visual tests. When attempting to characterize the intelligence of nonhuman animals, we should make every effort to test them in ways that are appropriate to their own mode of life.

We should not view nonhuman animal intelligence as a scaled-down version of human intelligence. To do so would fail to take into account the unique adaptations of nonhuman animals and would ignore the very special nature of human intelligence. Human intelligence is different from nonhuman animal intelligence in any situation in which our capacity for language can be applied. First, we can communicate with others in a rich and complex way. Second, we cannot only communicate with the present, but with the past and with the future by means of language. Very few nonhuman animals have the ability to pass on the benefits of their experiences to multiple generations. Third, the ability to formulate a problem and its solution in linguistic form provides us with reasoning capabilities that are well beyond the reach of any nonhuman animal. If you say to yourself "two lefts, one right, one left, and two rights" you have an enormous advantage over a nonhuman animal in attempting to learn the correct path through a maze.

In general, the greater the degree to which a psychological capacity is developed, the greater is the amount of neural tissue that is devoted to processing that capacity. The American psychologist and paleoneurologist, Harry Jerison, has referred to this relationship as the **principle of proper mass** and we will see many illustrations of this principle in the course of this book. It follows from this relationship that those parts of the brain that process intellectual information should be larger in animals that do a lot of such processing (i.e., are more intelligent) than in those that do less processing. Investigators of this problem are aware that large amounts of the brain are not devoted to intellectual activities but rather are involved in the "vegetative" or visceral aspects of being alive, as well as operating the musculoskeletal system. In order to separate intellectual from visceral and musculoskeletal functions, they have examined the ratio of brain weight to body weight or brain weight to spinal cord weight. By this reasoning, brains that are larger than average for a given body or spinal cord weight must be performing additional nonvisceral and nonmusculoskeletal processing. The assumption has been that this other neural processing represents the intellectual activities of the brain. This, in our view, is an oversimplification. Included in the nonmusculoskeletal and nonvisceral functions are arousal, wakefulness and attention, emotion, and sensory processes, as well as the work of major integrative and coordinating systems such as the cerebellum and the limbic system. Some investigators have attacked this aspect of the problem by studying the weight or volume of the isocortex (neocortex) of the brain in mammals as a proportion of total brain weight or volume. Because intellectual functions also are well represented in regions other than the isocortex, this approach is not completely satisfactory either.

One way to assess the merits of the notion that a big brain means big intelligence directly is to examine the human data on brain size and measures of intelligence. The literature in this field contains some well-known examples of eminent persons whose brains have either been heavier than normal (supporting the view that "brainy" and "intelligent" should indeed be synonymous) or surprisingly small, which has been used to argue for the contrary position. Neither of these observations makes a useful contribution because brain weight decreases considerably in a person's later years, and an elderly sage is likely to have a much lighter brain than one closer to middle age. Moreover, at any age, the longer a brain remains in the skull after death, the more it gains in weight due to the accumulation of body fluids (edema). These problems and others make many of the brain-weight values given for eminent persons of the past, such as

Anatole France and Oliver Cromwell, rather suspect as indicators of the relationship between brain size and intellect. Contemporary studies of intelligence and brain size have found that the relationship is not strong at all. In our judgment, the question of intelligence and brain size remains largely an open issue.

SUMMARY AND CONCLUSIONS

The comparative study of vertebrate nervous systems, whether from the perspective of anatomy, physiology, chemistry, or behavior, can be carried out in a phyletic context or in an adaptation context. Both approaches are equally valid. Moreover, data obtained from one approach often can aid in the interpretation of data from the other. Problems arise, however, when the goals of the two approaches are not kept clearly separated. The results of a study that is adaptational in character should not be taken to necessarily represent the historical course of evolution. Adaptation studies frequently organize results in a sequence of increasing complexity; this is not necessarily the course that was followed in the history of any lineage. Finally, concepts such as "progress" or "advancement" should be seen as highly dependent on a specific environment and not as absolutes. A sudden drastic change in environmental circumstances can quickly convert an adaptational success into an evolutionary disaster.

Intelligent behavior is one of the most sophisticated forms of adaptation to the environment. Behavioral adaptation involves many neural systems ranging from those that sense the environment and move the body to those that process and store information and make decisions. We must, however, look beyond relatively coarse indicators such as overall size or weight of the brain if we are to find anatomical correlates of intelligence. Moreover, we must not expect to find the same brain components necessarily to be correlates of intelligence in different taxonomic groups. The more closely related two species are or the more similar are their adaptations to the environment, the more likely are we to find a common substrate of intelligence. The points made in this chapter should be given consideration by the reader as we examine, in subsequent chapters, theories of how brains evolve, and as we describe the structure of the subdivisions and component systems of the brains of vertebrates and their roles in behavioral adaptation to the environment.

FOR FURTHER READING

Ayala, F. J. (1988) Can "progress" be defined as a biological concept? In M. H. Nitecki (ed.) *Evolutionary Progress*. Chicago: University of Chicago Press, pp. 75–96.

Campbell, C. B. G. and Hodos, W. (1991) The *scala naturae* revisited: evolutionary scales and anagenesis in comparative psychology. *Journal of Comparative Psychology*. 105, 211–221.

Gould, S. J. (1981) *The Mismeasure of Man*. New York: Norton.

Gould, S. J. (1988) On replacing the idea of progress with an operational notion of directionality. In M. H. Nitecki (ed). *Evolutionary Progress*. Chicago: University of Chicago Press, pp. 319–338.

Hodos, W. (1988) Comparative neuroanatomy and the evolution of intelligence. In H. J. Jerison and I. Jerison (eds.), *Intelligence and Evolutionary Biology*, Berlin: Springer-Verlag, pp. 93–107.

Hodos, W. and Campbell, C. B. G. (1969) *Scala naturae:* why there is no theory in comparative psychology. *Psychological Review*, 76, 337–350.

Jerison, H. J. and Jerison, I. (eds.), 1988, *Intelligence and Evolutionary Biology*. Berlin: Springer-Verlag.

ADDITIONAL REFERENCES

Bauchot, R. and Stephan, H. (1966) Données nouvelles sur l'encephalisation des insectivores et des prosimians. *Mammalia*, 30, 160–196.

Bhatnagar, K. P. and Kallen, F. C. (1974) Cribriform plate of ethmoid, olfactory bulb and olfactory acuity in forty species of bats. *Journal of Morphology*, 142, 71–90.

Brooks, D. R. and McLennan, D. A. (1991) *Phylogeny, Ecology, and Behavior: A Research Program in Comparative Biology*. Chicago: University of Chicago Press.

Carroll, R. L. (1988) *Vertebrate Paleontology and Evolution*. New York: Freeman.

Cronbach, L. J. (1960) *Essentials of Psychological Testing*. New York: Harper.

Griffin, D. R. (ed.), (1982) *Animal Mind—Human Mind*. Berlin: Springer-Verlag.

Griffin, D. R. (1992) *Animal Minds*. Chicago: University of Chicago Press.

Fasolo, A. and Malacarne, G. (1988) Comparing the structure of brains: implications for behavioral homologies. In H. J. Jerison and I. Jerison (eds.), *Intelligence and Evolutionary Biology*. Berlin: Springer-Verlag, pp. 119–141.

Gould, S. J. (1976) Grades and clades revisited. In R. B. Masterton, W. Hodos, and H. Jerison (eds.), *Evolution, Brain and Behavior: Persistent Problems*. Hillsdale, NJ: Earlbaum, pp. 115–122).

Harvey, P. H. (1988) Allometric analysis and brain size. In H. J. Jerison and I. Jerison (eds.), *Intelligence and Evolutionary Biology*. Berlin: Springer-Verlag, pp. 199–210.

Hoag, R. and Goldman, L. (eds.), 1986, *Animal Intelligence: Insights into the Animal Mind*. Washington, DC: Smithsonian Press.

Hodos, W. (1982) Some perspectives on the evolution of intelligence and the brain. In D. R. Griffin (ed.), *Animal Mind—Human Mind*. Berlin: Springer-Verlag, pp. 33–56.

Hodos, W. (1986) The evolution of the brain and the nature of animal intelligence. In R. Hoag and L. Goldman (eds.), *Animal Intelligence: Insights into the Animal Mind*. Washington, DC: Smithsonian Press, pp. 77–87.

Hodos, W. and Campbell, C. B. G. (1991) Evolutionary scales and comparative studies of animal cognition. In R. P. Kesner and D. S. Olton (eds.), *The Neurobiology of Comparative Cognition*. Hillsdale, NJ: Earlbaum, pp. 1–20.

Hofman, M. A. (1988) Brain, mind and reality: an evolutionary approach to biological intelligence. In H. J. Jerison and I. Jerison (eds.), *Intelligence and Evolutionary Biology*. Berlin: Springer-Verlag, pp. 437–446.

Horn, J. L. (1985) Remodeling old models of intelligence. In B. Wolman (ed.), *Handbook of Intelligence*. New York: Wiley.

Hull, D. L. (1988) Progress in ideas of progress. In M. H. Nitecki (ed.), *Evolutionary Progress*. Chicago: University of Chicago Press, pp. 27–48.

Humphreys, L. G. (1985) General intelligence. In B. Wolman (ed.), *Handbook of Intelligence*. New York: Wiley.

Jerison, H. J. (1955) Brain to body ratios and the evolution of intelligence. *Science*, 121, 447–449.

Jerison, H. J. (1973) *The Evolution of the Brain and Intelligence*. New York: Academic.

Jerison, H. J. (1982) The evolution of biological intelligence. In R. J. Sternberg (ed.), *Handbook of Human Intelligence*. Cambridge: Cambridge University Press.

Jerison, H. J. (1985) Animal intelligence as encephalization. *Philosophical Transactions of the Royal Society* (*London*), B308, 21–35.

Kruska, D. (1988) Mammalian domestication and its effect on brain structure and behavior. In H. J. Jerison and I. Jerison (eds.), *Intelligence and Evolutionary Biology*. Berlin: Springer-Verlag, pp. 211–250.

Lange, W. (1974) Regional differences in the distribution of Golgi cells in the cerebellar cortex of man and some other mammals. *Cell and Tissue Research*, 153, 219–226.

Lange, W. (1975) Cell number and cell density in the cerebellar cortex of man and some other mammals. *Cell and Tissue Research*, 157, 115–124.

Lloyd Morgan, C. (1894) *Introduction to Comparative Psychology*. London: Scott, p. 53.

Macphail, E. M. (1982) *Brain and Intelligence in Vertebrates*. Oxford: Clarendon Press.

Martin, R. D. (1974) The biological basis of human behavior. In W. R. Broughton (ed.), *The Biology of Brains*. New York: Wiley, pp. 215–250.

McCormick, C. A. (1982) The organization of the octavolateralis area in actinopterygian fishes: a new interpretation. *Journal of Morphology*, 171, 159–181.

Northcutt, R. G. (1977) Elasmobranch central nervous system organization and its possible evolutionary significance. *American Zoologist*, 17, 411–429.

Northcutt, R. G. (1985) Brain organization in the cartilaginous fishes. In E. S. Hodgson and R. F. Mathewson (eds.), *Sensory Biology of Sharks, Skates, and Rays*. Arlington, VA: Office of Naval Research, pp. 117–193.

Northcutt, R. G. (1985) Brain phylogeny: speculations on pattern and cause. In M. J. Cohen and F. Strumwasser (eds) Comparative Neurobiology: *Modes of Communication in the Nervous System*. New York: Wiley, pp. 351–378.

Passingham, R. E. (1979) Brain size and intelligence in man. *Brain, Behavior and Evolution*, 16, 253–270.

Passingham, R. E. (1982) *The Human Primate*. Oxford: Freeman.

Pickford, M. (1988) The evolution of intelligence: a palaeontological perspective. In H. J. Jerison and I. Jerison (eds.), *Intelligence and Evolutionary Biology*. Berlin: Springer-Verlag, pp. 175–198.

Radinsky, L. (1968) Evolution of somatic sensory specialization in otter brains. *Journal of Comparative Neurology*, 134, 495–505.

Ristau, C. A. (1991) (ed.) *Cognitive Ethology: The Minds of Other Animals*. Essays in Honor of Donald R. Griffin. Hillsdale, NJ: Earlbaum.

Romanes, G. J. (1882) *Animal Intelligence*. London: Kegan, Paul, Trench.

Roth, G., Nishikawa, K. C., Naujoks-Manteuffel, C., Schmidt, A., and Wake, D. B. (1993) Paedomorphosis and simplification in the nervous system of salamanders. *Brain, Behavior and Evolution*, 42, 137–170.

Sacher, G. A. (1970) Allometric and factorial analyses of brain structure in insectivores and primates. In C. R. Noback and W. Montagna (eds.), *The Primate Brain*. New York: Appleton.

Sternberg, R. J. (1985) Human intelligence: the model is the message. *Science*, 230, 1111–1118.

Walker, W. W. (1987) *Functional Anatomy of the Vertebrates: An Evolutionary Perspective*. Philadelphia: Saunders.

Weiss, S. R. and Hodos, W. (1988) Defective mirror-image discrimination in pigeons: A possible animal model of dyslexia. *Neuropsychiatry, Neuropsychology, and Behavioral Neurology*, 1, 161–170.

Wolman, B. (ed.), 1985, *Handbook of Intelligence*. New York: Wiley.

Yapp, W. B. (1965) *Vertebrates: Their Structure and Life*. New York: Oxford University Press.

6

Theories of Brain Evolution

INTRODUCTION

The Darwin–Wallace theory of biological evolution is one of a handfull of theories that can be said to have truly changed the way the world is viewed and the place of humans in it. Indeed, it is arguably the single most important theory in the history of biology. Unfortunately, not all biologists experience much formal education in evolutionary biology, and this lack often characterizes the training in the biomedical sciences, which include the neurosciences. Because of the importance of the theory, however, it is not at all unusual for a student of the brain to attempt to put the results of his/her research into some sort of "evolutionary" context or even to devise a theory of how the brain has evolved based on those findings. A number of such theories have been offered. Some have had no influence whatever, some have had influence in some form or other in the popular press, and some have been incorporated into the lore of neurobiology.

One of the most widely accepted of the earlier theories of brain evolution was that of C. Judson Herrick. Herrick's career began at the turn of the century and continued for more than 50 years, although he was most active in the 1920s and 1930s. His ideas and those of his student, Elizabeth Crosby, dominated comparative neurology well into the 1960s. So influential was Herrick that some of his ideas have attained the status of dogma in comparative neuroanatomy.

Herrick thought that the evolution of nervous tissue was similar to its embryology in that the brains of ancestral vertebrates were undifferentiated tissue. A stabilization of this tissue into permanent structural patterns occurred during phylogeny. He believed that with progressive increase in complexity of adjustment to the external and internal environments, a corresponding differentiation of structure occurred. Herrick further believed that the process of differentiation itself limits the future course of evolution within boundaries determined by the efficient working of the established organizational pattern. The result was an orthogenetic (i.e., proceeding in one direction), irreversible course of evolution of the nervous system. These ideas have been incorporated into most of the proposed theories of others:

- The ancestral vertebrate brain was diffuse and poorly differentiated.
- During evolutionary history brains become more and more complex and, therefore, more differentiated.
- The course of evolution is essentially unidirectional and not reversible.

SOME COMMON ASSUMPTIONS

In Herrick's defense, it should be stated that these ideas were not unique to him, but were a reflection of the views that were widespread in the biological and anthropological communities in the nineteenth and early twentieth centuries. As discussed in Chapter 1, these views were a result of an assumption of evolutionary progress, a product of combining Darwinism with the *scala naturae* theories of the mid-nineteenth century, and a pervasive anthropocentrism that continues to the present day. Since humans are supposedly the smartest creatures on the planet, the ideal brain must be like ours: large in size, complex, and highly differentiated. Conversely, the ancestral vertebrate brain must be the opposite of this: small, simple, and undifferentiated. Supposedly then, during the course of evolution, brains have gone from small to large, simple to complex, and undifferentiated to differentiated. This was, and still is for many, the dogma. Good evidence exists, however, that most, if not all, of these assumptions are wrong. As a result of starting with these false assumptions, many of the conclusions that have flowed from them have likewise been wrong.

The tradition has been to choose the brain of a single animal as the exemplar of the ancestral brain. In the 1950s and early 1960s, biomedical scientists usually used the brain of the laboratory rat to represent the brain of a "primitive" mammal. More recently, the hedgehog, a member of a group of so-called "basal insectivores," has become the exemplar, because it better fits the expectations cited above. Relative to the brains of other placental mammals, the brain of the hedgehog is small, seemingly simple, and relatively undifferentiated. Authors have frequently described it as **generalized.** Living insectivores probably do retain more ancestral features than most other living placental mammals, but they are not without their own specializations and derived characters. These animals and the other "basal insectivores" are mostly nocturnal and/or crepuscular. Consequently, their visual systems are probably secondarily reduced. Features of the visual system of these animals and other parts of the brain can only be identified as like that of ancestral mammals by out-group (sister group) comparisons, as discussed in Chapter 1, not by finding features that fit preconceived notions.

Along with the belief that the brains of some extant species, such as hedgehogs among placental mammals, resemble ancestral mammals, another assumption has been that adult structures in the brains of these species are the homologues of embryonic structures in other mammalian species, such as primates. In other words, at least some of the features of the brain of a primate would be the result of the brain of a hedgehog undergoing further embryonic development. Both developmental and comparative studies have amassed a great deal of evidence against this notion, but these recapitulationist assumptions are persistent in the literature and form components of many theories of brain evolution. We will note a different viewpoint on the recapitulation question at the end of this chapter, which makes sense of the grain of truth in this otherwise misbegotten idea.

THEORIES INVOLVING ADDITION OF STRUCTURES OR AREAS

MacLean

A number of theories assume that if organisms develop new adaptations and functions, then new structures must be added to already functioning brains. New structures may be added, of course, but old structures may be modified as well. In fact, modification of existing structures appears to be more common a process than addition in other organ systems. An example of an additive theory is Paul MacLean's **triune brain theory,** developed in the 1960s. MacLean believes that the human brain is composed of a hierarchy of three formations: the R-complex (or reptilian complex), which comprises the basal ganglia; the paleomammalian brain, which comprises the limbic system; and the neomammalian brain, which consists of the dorsal thalamus and the isocortex. This theory states that humans have inherited these three brains from their reptilian,

early mammal, and late mammal ancestors, respectively, and that with the addition of each superimposed brain, new cognitive abilities and behaviors have appeared.

Although MacLean's theory has had no significant impact on neurobiology, it has become popular in the lay press and among some psychological and educational therapists, partially due to the publication of *Dragons of Eden* by Carl Sagan in 1977. The extensive body of work in comparative neurobiology over the past three decades unequivocally contradicts this theory. First, homologues of the limbic cortical areas that MacLean considers to have been first present in early mammals have been found in nonmammalian vertebrates. Second, homologues of isocortical structures and of dorsal thalamic nuclei also have been found in nonmammals. Third, MacLean's observations on the behavioral differences between mammals and nonmammals are oversimplified and ignore the elaborate social and parental behaviors of nonmammalian vertebrates.

Flechsig and Campbell

Workers such as Paul Flechsig and Alfred W. Campbell (in the late nineteenth to early twentieth centuries) thought that an orderly addition of cortical areas accompanies a progressive elaboration of sensory-motor processes in mammals with the more complexly organized brains. This idea has found a place among the beliefs of many in neuroscience. For example, Broca's area for speech in the isocortex has been considered to be unique to humans and not to be found in the brains of monkeys. Recent work has shown that a homologue of Broca's area is present in the brains of monkeys, although it apparently plays no role in vocal communication in these animals. Similarly, Norman Geschwind elaborated a theory that the inferior parietal lobule of the isocortex, an "association area of association areas" believed to be involved in language processing, was unique to humans. Once again, recent work has shown it to be present in monkeys as well.

Sanides

In the 1970s, Friedrich Sanides promulgated a view of regional cortical evolution that was the opposite of that commonly held. Most workers believed that the primary sensory areas, that is, those areas of the isocortex receiving direct projections from dorsal thalamic sensory nuclei, appear first in phylogeny, and that association areas of isocortex appear later in phylogeny in mammals with more complex brains. Sanides thought that the association areas are older and the primary areas newer. He inferred a sequential evolution of cortical areas in progressive waves of differentiation due to a trend toward increasing architectural specialization in the arrangement of cortical neurons, and believed that the trend culminated in the appearance of specialized primary sensory areas. Specialized primary sensory areas, however, are found in all mammals and in a great variety of nonmammals as well. This now discredited theory is a good example of why trends perceived by humans and based on human value scales (such as a greater versus a lesser degree of architectural organization) are perilous at best.

THEORIES INVOLVING NEW FORMATION AND REORGANIZATION OF CIRCUITS

Herrick

Herrick put forth a specific theory on how the well-differentiated cerebral cortex of mammals has evolved. He held the view that in order to evolve well-differentiated cortex, well-defined tracts of projections from the dorsal thalamic sensory nuclei had to penetrate the telencephalon. These tracts would have to have different physiological properties, such as vision, audition, and somatic sensation. Herrick believed that fishes and amphibians lacked well-differentiated cortex because he thought that the entire forebrain was dominated by the olfactory sense, leaving no space for other modalities. Furthermore, no significant penetration by projection fibers of other modalities occurred in these groups of animals. Herrick thought that diapsid reptiles and turtles were the first groups of vertebrates to possess a forebrain with some areas free of olfactory influence. He thought that thalamic sensory projection fibers began to make an appearance and that a rudimentary cortical formation was present in these groups. He considered birds to have gone off in an abberant direction and to have developed no isocortical precursor, while in mammals most of the forebrain became a thalamic sensory projection target and only a small portion of it remained devoted to olfaction. The penetration of numerous thalamic projection fibers then supposedly allowed the development of six-layered isocortex. This scenario has been referred to by some as an **invasion hypothesis.**

Bishop

In 1959, George H. Bishop, a distinguished American neurophysiologist, described several generalizations bearing on the course of evolution of sensory systems in the brains and spinal cords of vertebrates. The framework upon which these generalizations rested was a general principle of nervous system evolution derived from Herrick's 1948 classic book, *The Brain of the Tiger Salamander,* and referred to as **cephalization** (not encephalization). This concept embodies the idea that during the course of vertebrate evolution, successively higher level structures were added to the rostral portion of the brain in relation to the evolution of more rostral special sensory levels. This concept implied a progressive shift towards the rostral portion of the brain in the processing of sensory information. For example, the observation that some nonprimates were rendered apparently blind by lesions of the optic midbrain, while primates were not so affected until lesions were made in the visual part of the isocortex suggested that ultimate control of vision had shifted from the midbrain to the forebrain. This result was considered to support the concept of cephalization.

Bishop proposed that during the course of vertebrate evolution, as cephalization progressed, the extension of each spinal pathway related to different sensory systems occurred not only by the addition of synaptic links above the old terminus, but also by adding more and successively larger caliber fibers in parallel with the nerve fibers already supplying lower level functions. Within each ascending afferent system, the supposedly newer, larger fibers were believed to bypass the more caudal and supposedly older terminal sites in favor of projecting to the newer, more rostral sites. The belief that smaller caliber fibers were developed earlier in evolution than larger caliber ones was a generalization of this concept.

Ariëns Kappers

Cornelius Ubbo Ariëns Kappers was an important Dutch comparative neurologist of the early years of the twentieth century. He formulated a theory of **neurobiotaxis** that described a phenomenon that he believed to be a force for change during vertebrate evolution. The theory states that if several centers of stimulation of neurons are present, the outgrowth of the chief dendrites of those neurons and eventually their cell bodies shift in the direction of greatest stimulation, resulting in lengthening of the axons. This response was believed to occur only if the source of stimulation was functionally related to the affected neurons. Good evidence confirming the existence of neurobiotaxis has not been found, however, since Ariëns Kappers put forth this notion.

Bowsher

David Bowsher, a British neuroanatomist, presented a "theory of the phylogenetic progression of central sensory systems" in 1973. Bowsher was greatly influenced by the theory of neurobiotaxis and used it to explain a number of observations on the topography of nuclear groups. He was also influenced by the work of **Ramon-Moliner** and **Nauta** in 1966 and Ramon-Moliner in 1968, in which the central reticular core of the mammalian neuraxis is described as being composed of neurons with a presumably primitive, simple dendritic morphology referred to as **isodendritic.** The morphology of these neurons was considered to be primitive because similar neurons were believed to make up almost the whole neuraxis of jawless vertebrates and cartilaginous fishes. These animals have been presumed by many neuroanatomists to be primitive types of living animals that can be used to represent ancestral vertebrates. Ramon-Moliner considered these isodendritic neurons to be a kind of cell from which many more elaborate types of neurons could be derived. Bowsher considered the primitive vertebrate nervous system to resemble a reticular nerve net in which each individual primitive cell would be synaptically connected with many other neurons. His opinion is at least partly based on a comment by Herrick that the neuraxis of amphibians is essentially a nerve net.

Bowsher believed phylogeny to be an "historical anterograde progression," and his theory was intended to answer the question of how specific, lemniscal (distantly projecting) systems evolved from a presumably primitive, reticular-like system. He believed that the earliest specific centers to appear were also composed of isodendritic neurons that were "captured" by a set of homogeneous afferent fibers carrying a single kind of sensory information. Several nuclei in the hindbrain of mammals—the nucleus of the solitary tract, which is involved in the taste system, and the vestibular nuclei, which are involved in balance and equilibrium—were given as examples of this earliest step in the march toward specificity. The next postulated stage was the induction of some morphological specialization in the central neurons receiving ascending branches of primary afferent sensory fibers. These newly evolved specific centers were believed to send projections to the isodendritic neurons of the reticular core but not to receive projections from them. Bowsher considered the principle (sensory) nucleus of the trigeminal nerve (that supplies the face) to be the first such nucleus to appear in phylogeny. To explain the existence of both isodendritic primitive neurons as well as specialized ones in many mammalian thalamic nuclei, Bowsher suggested that the efferent projections of newly evolved centers, such as the principal trigeminal nucleus and the dorsal column nuclei (that relay somatic sensation from the body), invade an originally isodendritic thalamus and trigger the differentiation of the complex neurons that become thalamocortical relay cells. When the process of specific "takeover" is incomplete, such nuclei still have reticular as well as specific sensory afferents, and thus both types of cells. Bowsher also indicated that the number of inhibitory Golgi type II interneurons increases "with the phylogenetic progression."

Diamond and Hall

In 1969, Irving Diamond and William Hall proposed a specific hypothesis of the course of evolution of the central visual system in the primate lineage as a part of an essay on the evolution of primate isocortex. They inferred the existence of four evolutionary stages. Three of these stages were represented by the central visual pathways of three living groups of animals, and the earliest stage by a hypothetical construct (a reptile-like mammal). The sequence of stages was reptile-like construct, tree shrew, prosimian primate, and anthropoid primate. In the first stage, the eye was believed to project only to the optic tectum of the midbrain, which in turn would project to an undifferentiated thalamic visual nucleus. This nucleus would then project to a general sensory cortex. The second stage was believed to be one in which the single thalamic nucleus has split into two, one receiving input from the eye directly and the other receiving visual input relayed through the optic tectum. These two thalamic nuclei then would project to a visual cortex differentiated into a visual "core" cortex and a surrounding belt of association cortex with both nuclei projecting to both cortices. In the third and fourth stages, progressive differentiation and separation of the two pathways to visual cortex and of the cortex itself were believed to have occurred. Serious doubt has been cast on this theory by more recent studies of visual pathways in many nonmammalian vertebrates, demonstrating that nonmammals also have multiple, separate visual pathways.

Ebbesson

Sven O. E. Ebbesson, an American neuroanatomist, published a **parcellation theory** of brain evolution in 1980 that rejected invasion hypotheses such as those of Herrick and Bowsher. Ebbesson proposed two main ideas. First, he proposed that ancestral vertebrate brains consisted of diffuse, relatively undifferentiated systems. These undifferentiated brains, never-

theless, had more extensive fiber connections with more overlap than is seen in living vertebrates. Second, he proposed that nervous systems have evolved from a simpler, more diffuse condition to a more complex condition by the parcellation of cell populations into subdivisions, brought about by the segregation of inputs, competition of inputs, and the loss of some connections. Ebbesson contended that abnormal connections seen in plasticity experiments represent ancient connections that were lost in modern organisms. The idea that early vertebrate brains were diffuse, undifferentiated, and containing only widely interconnected parts is in accord with the ideas of Herrick, Bowsher, and Diamond and Hall. Although parcellation can be seen in some neural systems, it does not seem to be as widespread a phenomenon as Ebbesson originally envisioned.

CRITIQUE OF THEORIES OF BRAIN EVOLUTION

Detailed critiques of the evidence for and against each of these theories are not appropriate here. Examining the underlying assumptions that form the conceptual framework for them may be worthwhile, however. First, most of these workers have an anthropocentric, *scala naturae* view of the animal kingdom. They refer to it as the **phylogenetic progression** or the **phylogenetic scale** or use other similar names. This fallacy is particularly apparent in the writings of Herrick, Bishop, and Bowsher. Second, most of these workers tend to view the course of evolution of whatever system is being examined as a progression of one kind of change from a simple to a complex condition. This progression is viewed as occurring in step with the ascent up the phylogenetic scale. Homoplasy, particularly parallelism and convergence, are not considered as factors in evolution. This unidimensional progression, seemingly under the direction of some imperative, is reminiscent of the now discredited, "predetermined path" theory of apparent steady lines of "progressive" evolution or a trend in one definite direction, referred to as **orthogenesis.**

Clearly, most of these workers have been greatly influenced by Herrick. If we remember the years during which this outstanding neurobiologist was most active and consider that notions of orthogenetic evolution, evolution in one direction, that is, from simple to complex states, and the *scala naturae* were common in biological thinking at that time, it is not difficult to understand how he came to these views. For example, Edward Drinker Cope, the American paleontologist, in 1896 inferred a **law of the unspecialized,** whereby phylogenetic sequences lead from generalized, primitive forms to more and more specialized forms. Since superficial observation often appears to confirm this notion, and indeed it is sometimes true by chance, this idea has been influential. G. H. Eimer introduced the concept of orthogenesis in 1888. These trends were thought to demonstrate a "directed" evolution.

An assult was made on the concept of orthogenesis by the leading neo-Darwinians, especially Glenn Jepsen and George Gaylord Simpson, in the 1940s and 1950s. A considerable amount of contrary evidence was gathered, and some of the most famous examples of orthogenesis were refuted.

Orthogenesis is mentioned in contemporary evolutionary works only as an historical curiosity. Nevertheless, the application of the concept still appears in the work of some comparative neuroanatomists.

Most of the theories described above imply a progression from a simple condition—usually an undifferentiated structure—to a structure divided into subdivisions or parcels. Surely such a sequence does often occur; however, to construct a theory in which only such a sequence is possible is to ignore a significant body of evidence. Evolution does not always produce a progression from simple to complex. Many examples exist where features evolve from complex states to simple ones.

Another problem with many of the theories is that ancestral states are often inferred by using inappropriate comparisons or models. Most often "primitive survivors" are used to infer what the ancestral state was—usually jawless vertebrates and elasmobranchs for vertebrates and hedgehogs for mammals. Neither jawless vertebrates nor elasmobranchs, within each group of which occurs a marked range of variation in brain structure and degree of complexity, can be considered to represent ancestors of extant amphibians or extant amniote vertebrates, however. Hedgehogs are equally useless as a model of mammalian ancestors.

CONTEMPORARY COMPARATIVE ANATOMICAL, CELLULAR, AND DEVELOPMENTAL APPROACHES

In the earlier part of this century, the predominant school of thought on forebrain evolution had held three major tenets. First, while the dorsal pallium of mammals (the cortical structure called neocortex or isocortex) had long been recognized to be the site of many of the so-called "higher" mental functions, only a small part of the more simply organized cortex in reptiles had been thought to be a "precursor" for the isocortex, and no potential homologues for it were thought to exist in anamniote vertebrates. Second, the telencephalons of the earliest vertebrates and of extant anamniotes were thought to be dominated by a massive olfactory input to the exclusion of other sensory system projections. Third, the *scala naturae* approach to evolution dictated a unilinear, progressive history of brain evolution. Beginning in the late 1960s, new findings on the organization of sensory system pathways and the telencephalon formed the basis for a new comparative anatomical approach to brain evolution that avoided the biases of the *scala naturae* and progressionism. Three series of experiments dramatically altered the traditional notions about the forebrain and its evolution.

Karten, Hodos, and their co-workers found that in birds, ascending sensory projections from the dorsal thalamus terminate in a large region within the dorsal pallium that is organized as a nucleus rather than as a cortex but, because of its connections and numerous other features subsequently found, is now widely recognized as a homologue of part of the mammalian isocortex. This large nuclear region, the dorsal ventricular ridge, also is present in diapsid reptiles and turtles, and subsequent studies by a number of workers showed that it, like the dorsal ventricular ridge in birds, also receives several ascending sensory pathways. Thus, a large dorsal pallial area, including the dorsal ventricular ridge and part of the cortex that is homologous as a set of neuronal cell populations to the isocortex, was found to be present in nonmammalian amniotes. Further, in 1971, Ebbesson and Schroeder published findings of similarly organized ascending sensory projections to a nuclear area in the telencephalic pallium in sharks, suggesting that the organization of sensory systems and their afferent areas within the telencephalon might be similar in all vertebrates.

The second series of experiments occurred during the same period and addressed the question of olfactory domination of the telencephalon in ancestral vertebrates. Scalia, Heimer, Ebbesson, and their co-workers found that olfactory projections in anamniotes are not widespread throughout the telencephalon. Rather, in contradiction to the traditional view, these projections were found to be limited to restricted areas in the lateral part of the pallium. Parenthetically, more recent work by Northcutt and his co-workers showing relatively extensive olfactory projections in hagfishes and lampreys has, however, reopened the question of the extent of olfactory projections in the earliest vertebrates. While olfactory projections were indeed restricted in the ancestral stock of all jawed vertebrates, they may not have been so in the earliest jawless vertebrates.

In the late 1960s, Northcutt had independently questioned the traditionally accepted theories of telencephalic evolution; the third series of experiments were carried out by Northcutt and his co-workers, extensively studying ascending sensory pathways in a wide variety of anamniote as well as amniote vertebrates. This work has led to the finding that enlargement and elaboration of the brain, and specifically of the forebrain, has occurred multiple times independently within different lineages of vertebrates—in jawless vertebrates, cartilaginous fishes, ray-finned fishes, and in amniotes—directly contradicting unilinear views of evolutionary change.

These findings and a host of additional experimental work that has augmented and extended them have led to a new perspective in evolutionary neurobiology. Multiple instances of independent evolution of neural structures and connections in different radiations have been documented. The importance of topology and the developmental preservation in the nervous system of all vertebrates of topological relationships, long recognized and studied by Nieuwenhuys, has recently been extended by studies of the genetic control of segmental brain development. The organization of multiple sensory pathways to the dorsal ventricular ridge in reptiles and birds has been explored in detail. The organization of the corresponding sensory pathways in mammals, first explored by Diamond, Harting, and their co-workers in the early 1970s, has also been studied extensively. More recently, Kaas and Allman proposed a theory of genetic replication to account for the evolution of the many sensory areas for a given modality present in the cortex in mammals, areas that appear to have evolved independently in various different mammalian radiations.

Karten presented an **equivalent cell hypothesis** to explain thalamic projections to parts of the dorsal pallium in amniotes, the nonlaminated dorsal ventricular ridge and the laminated isocortex. He proposed that connectional relationships between cell groups are specified by some factors specific to those cells, even if the cells are displaced or migrate. Thus, cell populations present in various layers of isocortex would also be present in the dorsal ventricular ridge in various locations within it forming nuclear groups. The basic circuitry of sensory projections terminating on a population of sensory afferent neurons in the dorsal pallium, which in turn project to a population of interneurons, which in their turn then project to a population of efferent neurons, is present in all amniotes, whether the cytoarchitecture consists of cortical layers or nuclear regions.

Additional theories relating to evolution from a developmental perspective have been proposed, such as the **displacement hypothesis** of Deacon. Deacon argued that where changes in connections occur over evolution—the loss of connections, aquisition of additional connections, or replacement of one class of connections by another—competitive axonal interactions are biased by contextual events during development. He believes that this mechanism can explain invasion-like and parcellation-like events. He considers a cause to be missing from both invasion and parcellation hypotheses, whereas he believes his displacement hypothesis presents a mechanism of change. The cause is in the form of regressive processes, for example, cell death or reduction of a peripheral sensory or motor system, or differential growth processes.

Continuing advances in cellular and developmental studies relevant to evolutionary neurobiology bode well for the future of this field. The methods of cladistic analysis, long recognized as valid and useful by zoologists, are also being applied to comparative neurobiology, and new insights into the relationship of ontogeny and phylogeny have clarified this long debated and confused subject and provide a new and fruitful approach to the data. As discussed in Chapter 1, cladistics is an objective method, based on the distribution of a character, for inferring which characters in the brains of a group of species are ancestral (plesiomorphic) and which are specialized (apomorphic) in each particular species. As this methodology is applied more widely in comparative neurobiology, the evolutionary history of many parts of the nervous system within each of the major vertebrate radiations will be better understood.

Cladistic analysis has one drawback in that it is based on characters expressed in the phenotype. Interpretive difficulties that arise from the possibility of genetic continuity of the genome for a particular character without phenotypic expression of the character in some taxa have yet to be fully addressed. Due to this possibility, some cases of homology may be mistakenly interpreted as cases of parallelism or convergence. Further advances in the field of genetics, particularly in mapping genomes, will help to alleviate this problem.

While ontogeny does not recapitulate phylogeny in the previously understood sense, it does recapitulate the features of an organism in terms of the organism's more general to more specific classification. As discussed in Chapter 3, four laws of development deduced by von Baer, a nineteenth-century German embryologist, can be applied to problems of nervous system evolution. These laws state that the more general features of an organism develop before the more specific features, so that, for example, a monkey embryo develops features common to all vertebrates before developing features specific to mammals and finally to monkeys. Thus, comparative studies of developing systems can reveal which features would been common to the *developing* ancestors of a specified group.

These new insights allow for exciting new perspectives in comparative neurobiology. The work of Karten, Northcutt, Ebbesson, and their co-workers was fundamental in establishing that differentiated, complex systems are widely distributed among vertebrates. A large body of recent work, including new findings on the role of neural crest in head development, the recognition that enlargement and elaboration of neural structures have occurred independently multiple times within different vertebrate lineages, and other related new findings allow for a reconstruction of many of the basic events that occurred in vertebrate brains over evolution. A recent application of cladistic methodology and the principles of von Baerian recapitulation to forebrain structures has also contributed to this reconstruction of evolutionary events, as have insights on developmental aspects of cellular organization in nervous systems, studies illuminating the role of homeobox genes during development in relation to the neuromeric organization of the brain, new perspectives on the evolution of specific cell populations as defined by their immunohistochemical profiles, and new perspectives on mammalian cortical organization and evolution. In the final chapter of this book (Chapter 31), we will be able to present a new synthesis of vertebrate brain evolution derived from these various lines of research.

FOR FURTHER READING

Karten, H. J. and Shimizu, T. (1989) The origins of neocortex: connections and lamination as distinct events in evolution. *Journal of Cognitive Neuroscience,* 1, 291–301.

Nieuwenhuys, R. (1994) Comparative neuroanatomy: place, principles, practice and programme. *European Journal of Morphology,* 32, 142–155.

Northcutt, R. G. (1981) Evolution of the telencephalon in nonmammals. *Annual Review of Neuroscience,* 4, 301–350.

ADDITIONAL REFERENCES

Allman, J. (1990) Evolution of neocortex. In E. G. Jones and A. Peters (eds.), *Cerebral Cortex, Vol. 8A: Comparative Structure and Evolution of Cerebral Cortex, Part I.* New York: Plenum, pp. 269–283.

Ariëns Kappers, C. U., Huber, G. C., and Crosby, E. C. (1967) *The Comparative Anatomy of the Nervous System of Vertebrates, Including Man.* New York: Hafner.

Baker, H. (1991) Evaluation of species-specific biochemical variation as a means for assessing homology in neuronal populations. *Brain, Behavior and Evolution,* 38, 255–263.

Balaban, C. D. and Ulinski, P. S. (1981) Organization of thalamic afferents to anterior dorsal ventricular ridge in turtles. I. Projections of thalamic nuclei. *Journal of Comparative Neurology,* 200, 95–129.

Bishop, G. H. (1959) The relation between nerve fiber size and sensory modality: phylogenetic implications of the afferent innervation of cortex. *Journal of Nervous and Mental Disease,* 128, 84–114.

Bowsher, D. (1973) Brain, behavior and evolution. *Brain, Behavior and Evolution,* 8, 386–396.

Brauth, S. E. (1990) Histochemical strategies in the study of neural evolution. *Brain, Behavior and Evolution,* 36, 100–115.

Bruce, L. L. and Butler, A. B. (1984a) Telencephalic connections in lizards. I. Projections to cortex. *Journal of Comparative Neurology,* 229, 585–601.

Bruce, L. L. and Butler, A. B. (1984b) Telencephalic connections in lizards. II. Projections to anterior dorsal ventricular ridge. *Journal of Comparative Neurology,* 229, 602–615.

Butler, A. B. (1994a) The evolution of the dorsal thalamus of jawed vertebrates, including mammals: cladistic analysis and a new hypothesis. *Brain Research Reviews,* 19, 29–65.

Butler, A. B. (1994b) The evolution of the dorsal pallium in the telencephalon of amniotes: cladistic analysis and a new hypothesis. *Brain Research Reviews,* 19, 66–101.

Campbell, A. (1905) *Histological Studies on the Localization of Cerebral Function.* Cambridge, England: Cambridge University Press.

Cope, E. D. (1896) *The Primary Factors of Organic Evolution.* Chicago: Open Court Publishing Co.

Deacon, T. W. (1990) Rethinking mammalian brain evolution. *American Zoologist,* 30, 629–705.

Diamond, I. (1973) The evolution of the tectal-pulvinar system in mammals: structural and behavioural studies of the visual system. *Symposium of the Zoological Society of London,* No. 33, 205–233.

Diamond, I. T. and Hall, W. C. (1969) Evolution of neocortex. *Science,* 164, 251–262.

Ebbesson, S. O. E. (1980) The parcellation theory and its relation to interspecific variability in brain organization, evolutionary and ontogenetic development, and neuronal plasticity. *Cell and Tissue Research,* 213, 179–212.

Ebbesson, S. O. E. (1984) Evolution and ontogeny of neural circuits. *Behavioral Brain Sciences,* 7, 321–331.

Ebbesson, S. O. E. and Heimer, L. (1970) Projections of the olfactory tract fibers in the nurse shark (*Ginglymostoma cirratum*). *Brain Research,* 17, 47–55.

Ebbesson, S. O. E. and Schroeder, D. M. (1971) Connections of the nurse shark's telencephalon. *Science,* 173, 254–256.

Flechsig, P. (1900) Über projektions und assoziations Zentren des menschlichen Gehirns, *Neurologie Zentralblatt,* 19.

Foster, R. E. and Hall, W. C. (1978) The organization of central auditory pathways in a reptile, *Iguana iguana. Journal of Comparative Neurology,* 178, 783–832.

Gans, C. and Northcutt, R. G. (1983) Neural crest and the origin of vertebrates: a new head. *Science,* 220, 268–274.

Goffinet, A. M. (1990) Cortical architectonic development: a comparative study in reptiles. In W. K. Schwerdtfeger and P. Germroth (eds.), *The Forebrain in Nonmammals: New Aspects of Structure and Development.* Berlin: Springer-Verlag, pp. 135–144.

Hall, W. C. and Ebner, F. F. (1970) Thalamotelencephalic projections in the turtle (*Pseudemys scripta*). *Journal of Comparative Neurology,* 140, 101–122.

Harting, J. K., Hall, W. C., and Diamond, I. T. (1972) Evolution of the pulvinar. *Brain, Behavior and Evolution,* 6, 424–452.

Harting, J. K., Glendenning, K. K., Diamond, I. T., and Hall, W. C. (1973) Evolution of the primate visual system: anterograde degeneration studies of the tecto-pulvinar system. *American Journal of Physical Anthropology,* 38, 383–392.

Heimer, L. (1969) The secondary olfactory connections in mammals, reptiles and shark. *Annals of the New York Academy of Sciences,* 167, 129–146.

Herrick, C. J. (1948) *The Brain of the Tiger Salamander.* Chicago: University of Chicago Press.

Jepsen, G. L. (1949) Selection, "orthogenesis," and the fossil record. *Proceedings of the American Philosophical Society,* 93, 479–500.

Kaas, J. H. (1982) The segregation of function in the nervous system: why do sensory systems have so many subdivisions? *Contributions to Sensory Physiology,* 7, 201–240.

Karten, H. J. (1968) The ascending auditory pathway in the pigeon (*Columba livia*). II. Telencephalic projections of the nucleus ovoidalis thalami. *Brain Research,* 11, 134–153.

Karten, H. J. (1969) The organization of the avian telencephalon and some speculations on the phylogeny of the amniote telencephalon. *Annals of the New York Academy of Sciences,* 167, 164–179.

Karten, H. J. (1991) Homology and evolutionary origins of the "neocortex." *Brain, Behavior and Evolution,* 38, 264–272.

Karten, H. J. and Hodos, W. (1970) Telencephalic projections of the nucleus rotundus in the pigeon (*Columba livia*). *Journal of Comparative Neurology,* 140, 35–52.

Karten, H. J., Hodos, W., Nauta, W. J. H., and Revzin, A. M. (1973) Neural connections of the "visual Wulst" of the avian telencephalon: experimental studies in the pigeon (*Columba livia*) and owl (*Speotyto cunicularia*). *Journal of Comparative Neurology,* 150, 253–278.

MacLean, P. D. (1990) *The Triune Brain in Evolution.* New York: Plenum.

Nauta, W. J. H. and Karten, H. J. (1970) A general profile of the vertebrate brain, with sidelights on the ancestry of cerebral cortex. In F. O. Schmidt (ed.), *The Neurosciences: Second Study Program.* New York: The Rockefeller University Press, pp. 7–26.

Nieuwenhuys, R. (1974) Topological analysis of the brainstem: a general introduction. *Journal of Comparative Neurology,* 156, 255–276.

Noden, D. M. (1991) Vertebrate craniofacial development: the relation between ontogenetic process and morphological outcome. *Brain, Behavior and Evolution,* 38, 190–225.

Northcutt, R. G. (1967) Architectonic studies of the telencephalon of *Iguana iguana. Journal of Comparative Neurology,* 130, 109–148.

Northcutt, R. G. (1970) The telencephalon of the Western painted turtle (*Chrysemys picta belli*). *Illinois Biological Monographs,* 43. Chicago: Univeristy of Illinois Press.

Northcutt, R. G. and Gans, C. (1983) The genesis of neural crest and epidermal placodes: a reinterpretation of vertebrate origins. *The Quarterly Review of Biology,* 58, 1–28.

Northcutt, R. G. and Puzdrowski, R. L. (1988) Projections of the olfactory bulb and nervus terminalis in the silver lamprey. *Brain, Behavior and Evolution,* 32, 96–107.

Northcutt, R. G., Reiner, A., and Karten, H. J. (1988) Immunohistochemical study of the telencephalon of the spiny dogfish, *Squalus acanthias. Journal of Comparative Neurology,* 277, 250–267.

Pritz, M. B. (1974) Ascending connections of a thalamic auditory area in a crocodile, *Caiman crocodilus. Journal of Comparative Neurology,* 153, 199–214.

Pritz, M. B. (1975) Anatomical identification of a telencephalic visual area in crocodiles: ascending connections of nucleus rotundus in *Caiman crocodilus. Journal of Comparative Neurology,* 164, 323–338.

Puelles, L. and Rubenstein, J. L. R. (1993) Expression patterns of homeobox and other putative regulatory genes in the embryonic mouse forebrain suggest a neuromeric organization. *Trends in Neurosciences,* 16, 472–479.

Rakic, P. (1988) Intrinsic and extrinsic determinants of neocortical parcellation: a radial unit model. In P. Rakic and W. Singer (eds.), *Neurobiology of Neocortex.* New York: Wiley, pp. 5–27.

Ramon-Moliner, E. (1969) The leptodendritic neuron: its distribution and significance. *Annals of the New York Academy of Sciences,* 167, 65–70.

Ramon-Moliner, E. and Nauta, W. J. H. (1966) The isodendritic core of the brain stem. *Journal of Comparative Neurology,* 126, 311–336.

Reiner, A. (1991) A comparison of neurotransmitter-specific and neuro-peptide-specific neuronal cell types present in the dorsal cortex in turtles with those present in the isocortex in mammals: implications for the evolution of isocortex. *Brain, Behavior and Evolution,* 38, 53–91.

Reiner, A. (1993) Neurotransmitter organization and connections of turtle cortex: implications for the evolution of mammalian isocortex. *Compartive Biochemistry and Physiology,* 104A, 735–748.

Reiner, A. and Northcutt, R. G. (1992) An immunohistochemical study of the telencephalon of the Senegal bichir (*Polypterus senegalus*). *Journal of Comparative Neurology,* 319, 359–386.

Sanides, F. (1970) Functional architecture of motor and sensory cortices in primates in the light of a new concept of neocortical evolution. In C. R. Noback and W. Montagna (eds.), *The Primate Brain: Advances in Primatology, Vol. 1.* New York: Appleton-Century-Crofts, pp. 137–208.

Scalia, F., Halpern, M., Knapp, H., and Riss, W. (1968) The efferent connections of the olfactory bulb in the frog: a study of degenerating unmyelinated fibers. *Journal of Anatomy,* 103: 245–262.

Simpson, G. G. (1953) *The Major Features of Evolution.* New York: Columbia University Press.

Stuesse, S. L. and Cruce, W. L. R. (1991) Immunohistochemical localization of serotoninergic, enkephalinergic, and catecholaminergic cells in the brainstem and diencephalon of a cartilaginous fish, *Hydrolagus colliei. Journal of Comparative Neurology,* 309, 535–548.

Wicht, H. and Northcutt, R. G. (1992) The forebrain of the Pacific hagfish: a cladistic reconstruction of the ancestral craniate forebrain. *Brain, Behavior and Evolution,* 40, 25–64.

Part Two

THE SPINAL CORD AND HINDBRAIN

7

Overview of Spinal Cord and Hindbrain

OVERVIEW OF THE SPINAL CORD

The spinal cord, which is located within the backbone (**vertebral column**), is the portion of the central nervous system that makes contact with the body. The spinal cord is continuous with the caudal end of the brainstem (**medulla** and **pons**), which is the portion of the central nervous system that makes contact with the specialized organs that make up the head (eyes, ears, jaws, tongue, salivary glands, etc.). By "make contact," we mean that the spinal cord receives information from the external and internal environments of the body and executes commands to the organs of the body, both internal organs and external body parts. In addition, the spinal cord contains neural networks that integrate the incoming sensory information from various body regions and organizes the specific commands to be given to the body parts. For example, a touch on one part of the body might, under one set of circumstances, provoke a movement of the animal to the right and, in another set of circumstances, a movement to the left. In addition to its own regulation of body function, the spinal cord supplies information about the body's internal and external environments to the brain and in turn is the route through which the brain controls the body. The integration of the actions of different body regions and communication with the brain are accomplished by long axons that run longitudinally within the cord.

Segmentation Within The Spinal Cord

During the embryonic development of the nervous system, the spinal cord develops in close accord with the vertebral

column. Thus a pair of right and left spinal nerves exit between each pair of vertebrae, which are the individual bones of the vertebral column. The region of the cord from which the spinal nerves emerge is known as a **spinal segment.** The term **spinal segmentation** is sometimes disconcerting to newcomers to spinal anatomy because they often expect to see something like the segmentation of a worm and this is not the case; the surface of the spinal cord is smooth with the exits of the spinal nerves giving the only clue to the location of the spinal segments. These segments, however, are not merely a device for teaching spinal anatomy; the spinal cord actually is organized according to segments as may be seen from the fact that spinal axons can be classified into two types: **intrasegmental** (those that remain within a segment) and **extrasegmental** (those that extend to one or more additional segments). Each spinal segment is named for its corresponding vertebra. Thus, spinal segment L2 is so named because its nerve (spinal nerve L2) exits at the second lumbar vertebra. Likewise, spinal segment T6 is so named because of its association with spinal nerve T6 and the sixth thoracic vertebra. This is shown schematically in Figure 7-1, which shows several spinal segments and their corresponding vertebrae. For clarity, the spaces between the vertebra have been exaggerated.

In most vertebrates, the spinal cord runs the entire length of the vertebral column. In some vertebrates, such as the angler fish (*Lophius piscatorius*) and humans, the spinal cord ends well anterior to the caudal end of the vertebral column. Consequently, a given spinal segment may no longer be located within the vertebra of the same name. This, however, does not prevent the spinal nerves from exiting between the appropriate vertebrae, even though this point of exit now is remote from

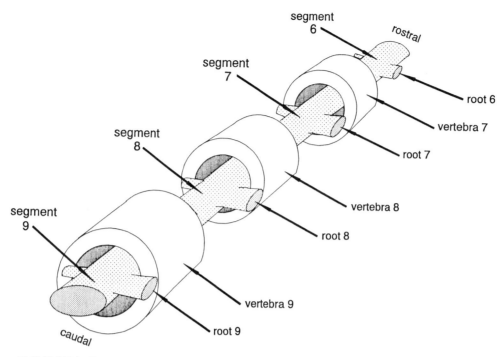

FIGURE 7-1. The relationship between segments of the spinal cord and their corresponding vertebrae. The spaces between the vertebrae have been exaggerated.

the originating spinal segment. The result of this anatomical arrangement is that many spinal nerves must pass well beyond the end of the cord before exiting. When early anatomists dissected the human spinal cord out of the vertebral column, they noted the resemblance of this collection of long spinal nerves that dangled well below the end of the spinal cord to the hairs of a horse's tail; they therefore named these nerves the **cauda equina,** which is horse's tail in Latin. Figure 7-2 shows the caudal end of the spinal cord. For clarity, the vertebrae that surround the cord have been omitted from the drawing. Only those vertebrae that exist caudal to the spinal cord are shown with the spinal nerves passing through them to form the cauda equina. Among nontetrapods, the cauda equina is often (but not exclusively) found in fishes with a large head and a relatively small body.

The segments of the spinal cord are named for their corresponding vertebrae of the vertebral column. These segments are grouped into regions according to the main body regions through which the vertebral column passes. The **cervical** vertebrae are the bones that support the neck. The **thoracic** vertebrae support the ribs that form the **thorax** (chest cavity). The **lumbar** vertebrae pass through the abdominal region and the **sacral** vertebrae are in the pelvic region. The spinal segment that sends its spinal nerve out at the fourth cervical segment is thus the fourth cervical segment or C4. Likewise, the segment that corresponds to the second thoracic vertebra is designated as T2 or the second thoracic segment. The numbers of vertebrae and their corresponding spinal segments vary according to the class of vertebrates; for example, virtually all mammals, irrespective of the lengths of their necks, have seven cervical vertebrae and seven cervical spinal segments—even giraffes.

Roots and Ganglia

Dorsal and Ventral Roots. The spinal nerves pass from the spinal cord to the periphery of the body (the limbs and internal organs such as the gut, the heart, etc.). When early anatomists traced these nerves back to the spinal cord, they observed that as these nerves neared the cord they seemed to divide into two **roots,** one dorsal and one ventral, which connected them to the spinal cord much as a tree's roots connect it to the earth. Later research revealed that as a general rule (with some exceptions) incoming sensory axons enter the spinal cord via the **dorsal root** and motor axons exit via the **ventral root.** Figure 7-3 is a transverse section through the spinal cord of a shark. A dorsal root with a single incoming sensory axon is shown at the top. Below, the axon of a motor neuron is seen exiting via a ventral root. For clarity, only one sensory axon and one motor axon are shown in the figure. The actual number of such axons, however, can be considerable.

In addition to the axons of motor neurons that control skeletal muscles, the ventral roots also contain two other types of axons: (1) axons of cells that regulate smooth muscles and glands via the autonomic system and (2) axons that terminate not on muscle fibers, but on the muscle spindles. These motor neurons are known as **gamma efferents.** Their function is to activate the muscle spindles that are nestled among the muscle fibers and that respond to the tension of the surrounding muscle fibers. The interplay between the receptors on the spindles and their state of contraction as regulated by the gamma efferents results in smooth changes in tension and elongation of the surrounding muscle fibers.

In agnathans the dorsal and ventral roots are not both present in the same spinal cord segment; the dorsal roots appear

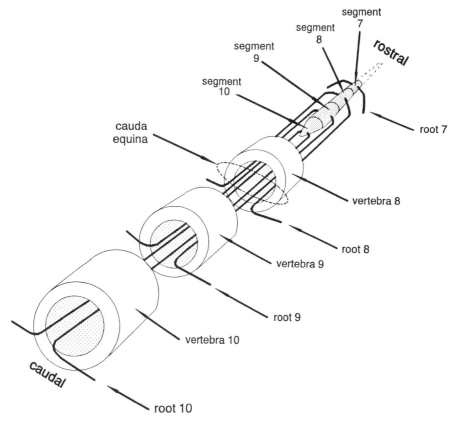

FIGURE 7-2. The cauda equina results from the shortening of the spinal cord so that it no longer entirely fills the length of the vertebral column. The spinal roots, however, continue to exit caudal to their corresponding vertebrae. For better visibility of the spinal cord, its surrounding vertebrae have been omitted.

in one segment and the ventral roots in the next. In lampreys, the dorsal and ventral roots continue as separate nerves along their entire course. In hagfishes and other nontetrapods, however, the two roots come together to form a single spinal nerve, although the dorsal and ventral roots exit from the cord at somewhat different levels. In tetrapods the dorsal and ventral roots exit at the same level of the spinal cord and shortly after exiting, combine to form a common spinal nerve.

Ganglia. A prominent feature of the dorsal root is a swelling known as the **dorsal root ganglion,** so named because it appeared to the early anatomists to resemble a knot. The dorsal root ganglion contains the cell bodies of the sensory axons that are entering the spinal cord from the pain, touch, temperature, and stretch receptors of the skin, muscles, tendons, and joints, as well as from the internal organs of the body. In agnathans, however, an additional ganglion cell type is located within the white matter of the spinal cord. These cells are known as **dorsal cells** and have been identified as mechanoreceptors for touch and pressure.

An additional ganglion occurs on the ventral root. This is known as the **sympathetic ganglion** and is an important part of the **sympathetic division of the autonomic nervous system.** The sympathetic division and its partner in the autonomic nervous system, the **parasympathethic division,** will be described in greater detail in Chapter 23. For now we simply

will note that the cell bodies of these sympathetic axons are located within the spinal cord and that their axons leave via the ventral root. A short distance from the cord, these autonomic axons, known as the **preganglionic sympathetic axons,** leave the ventral root and enter the sympathetic ganglion. The preganglionic axons synapse on the cell bodies that are located within the ganglion. These cell bodies give rise to axons that leave the ganglion and rejoin the ventral root and are known as the **postganglionic sympathetic axons.** The postganglionic axons travel to the target organs of the autonomic system where they regulate the contraction of smooth muscles and the secretion of glands. A sympathetic ganglion and its pre- and postganglionic axons may be seen in Figure 7-7. Postganglionic axons travel to the various internal organs of the body (collectively known as the viscera) and produce contractions of smooth muscles, such as those of the intestinal system, or stimulate glands to secrete.

The neurons of the parasympathetic division are located in the brainstem and in the sacral region of the spinal cord. The axons of the parasympathetic neurons that are located in the brainstem also travel to the viscera and constitute the visceral components of the cranial nerves. The actions of the sympathetic and parasympathetic divisions of the autonomic system complement each other. In general, the parasympathetic division functions under normal circumstances and the sympathetic division takes charge of the viscera during situations that require

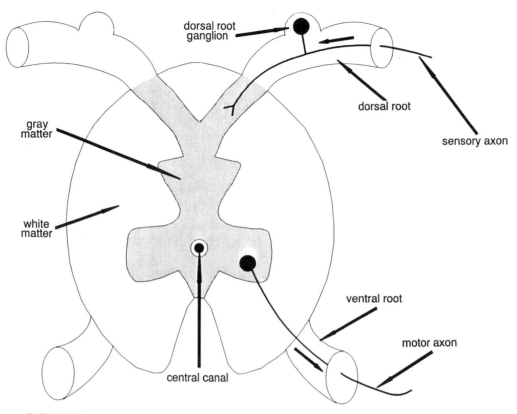

FIGURE 7-3. A transverse section of the spinal cord of a shark that illustrates the location of the dorsal and ventral roots, the dorsal root ganglion, sensory and motor axons, and the gray and white matter.

short-term levels of high activity, such as during periods of emergency or stress. Not all of the division of labor between the parasympathetic and sympathetic divisions, however, fits this generalization.

Columns of the Spinal Cord

We have used the term **vertebral column** rather than "spinal column" to refer to the column of bony rings that surrounds the spinal cord, so as to minimize confusion with the cell and axon columns that are internal to the spinal cord. The axon columns often are referred to as **funiculi** (singular = **funiculus**). The contemporary approach to the organization of the spinal cord is to consider the cell groups of the various segments as forming longitudinal columns that are more or less continuous.

The cell columns consist of motor neurons (either visceral or somatic), cells of origin of **intrasegmental** axons, which remain within their own segment, and cells of origin of **extrasegmental** axons, which pass to neighboring segments, remote segments, or even may leave the spinal cord altogether to terminate within the brain. This latter type of extrasegmental axon is called **suprasegmental.** The axon columns contain:

- Axons of incoming sensory neurons, the cell bodies of which are located in the dorsal root.
- Extrasegmental axons originating in the spinal cord, but remaining within the spinal cord.

- Suprasegmental axons originating in the spinal cord, but ascending to the brain.
- Suprasegmental axons that originate in the brain and descend to the spinal cord or brain.

The axon columns of the spinal cord constitute the so-called **white matter** of the cord; the cell columns make up the so-called **gray matter.** This gray matter is concentrated around the central canal of the cord, which is the remnant of the lumen of the embryonic neural tube and which contains cerebrospinal fluid. The gray matter tends to extend out dorsally and ventrally; the dorsal extensions are known as **dorsal horns** (because of their appearance in a transverse section); the ventral extensions likewise are known as **ventral horns.** Occasionally, **lateral horns** are present as well. These horns are shown in Figure 7-4. In general, the ventral horns contain the cell bodies of somatic motor neurons, and the lateral horns contain the cell bodies of the preganglionic sympathetic neurons. The dorsal horns and the remainder of the gray matter contain the cell bodies of interneurons that receive incoming sensory axons or axons from elsewhere in the spinal cord.

Another way of considering the columns of the cord derives from the results of experimental studies to trace the ascending and descending pathways in the spinal cord, particularly in tetrapods. These studies revealed that when the longitudinal cell columns are individually labeled, they take on the appearance of layers or **laminae** (singular = **lamina**). Figure 7-5 shows a hypothetical spinal cord with both axon

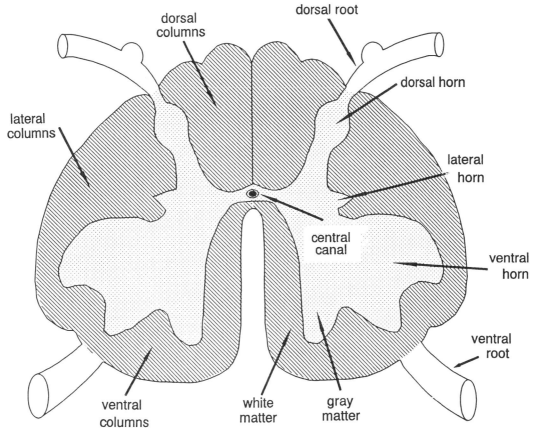

FIGURE 7-4. The white matter columns and the horns of the gray matter columns of a primate spinal cord.

and cell columns. For simplicity, only two axon columns and two cell columns are shown. The more medial of the two cell columns contains small cells and the more lateral contains larger cells. When a cross section of the cord is viewed, as in the shaded region at the left, the longitudinal organization of the columns of cell bodies cannot be easily recognized, but when appropriately labeled, the cells appear to be organized in a laminar arrangement. In other words, the "pie" is the same, it just has been sliced differently.

Pathways within the Spinal Cord

A number of axon pathways (known as **tracts**) arise in the spinal cord and pass to the brain. Likewise, pathways origi-

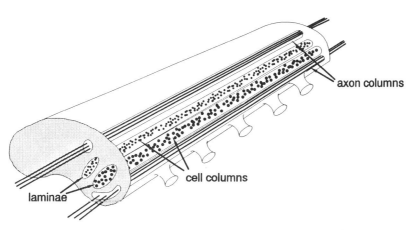

FIGURE 7-5. A hypothetical spinal cord showing axon and cell columns. When viewed in a transverse section (shaded region), the cell columns appear as laminae.

nate in the brain and descend to the spinal cord. The pathways, like most of the pathways or tracts in the central nervous system, have two-part names, such as the tectospinal tract or spinocerebellar tract. The convention for naming pathways in the central nervous system is for the first part of the name to indicate the source of the pathway (i.e., the location of its cell bodies) and for the second part to reflect the target of the pathway (i.e., the location of its axon terminals). Thus, the tectospinal tract originates in the tectum and terminates in the spinal cord. Likewise the spinocerebellar tract arises from cell bodies located in the gray matter of the spinal cord, and its axons terminate in the cerebellum.

Apart from those spinal pathways that remain within the cord, the longitudinal interneuronal pathways may be classified into three types:

- Ascending suprasegmental lemniscal and dorsal column pathways.
- Suprasegmental reticular pathways.
- Descending suprasegmental pathways.

Lemniscal Pathways. The **lemniscal** pathways that ascend from the spinal cord consist entirely of interneurons that terminate in brain regions such as the cerebellum, tectum of the midbrain, and thalamus. Examples of these are the spinocerebellar, spinotectal, and spinothalamic tracts, which travel in the lateral and ventral axon columns of the white matter. The spinocerebellar tract, which carries information about the state of contraction of the body's muscle groups and about the position of skeletal joints to the cerebellum, is present in all vertebrates. Spinotectal tracts, carrying somatosensory information to the tectum of the midbrain, are present in sharks, salamanders, and amniotes (mammals, diapsids, and turtles). Some evidence exists to suggest that these pathways evolved in ancestral jawed vertebrates, but subsequently were lost in ray-finned fishes and frogs. Spinothalamic tracts, which carry information about touch, pain, and temperature to the thalamus have been described in some sharks but typically are found in amniotes.

Dorsal Column Pathways. The dorsal column pathways contain interneuronal axons as well as the incoming axons of somatosensory neurons, the cell bodies of which are located in the dorsal root ganglia. These dorsal column axons terminate on cells located in the caudal medulla close to its junction with the spinal cord. The axons of the latter cells then join those of the lemniscal pathway as they ascend to higher levels of the brain. Dorsal column pathways have been reported in agnathans as well as some other anamniotes, but are a consistent and prominent feature of tetrapod spinal cords.

Reticular Pathways. The spinoreticular pathway arises primarily from interneurons in the spinal cord and terminates in the **reticular formation,** which is a collection of neuron groups that are located throughout the central nervous system from the spinal cord to the diencephalon. The principal cell groups that receive spinoreticular axons are located in the medulla, the pons, and the midbrain. The descending pathways of the reticular formation coordinate the activities of motor–neuron populations that control specific muscle groups, such

as those that move the eyes or jaws. More will be said about this important integrating system in Chapter 13. The spinoreticular tract appears to be common to all vertebrates, and hence probably is a heritage from the earliest vertebrates.

Descending Pathways. Descending pathways include the **tectospinal tract,** which carries motor commands from the midbrain tectum. The tectum contains "maps" of visual space, auditory space, somatosensory space, and, in those animals with electroreception, the electrical fields surrounding the animal. Another descending motor-command pathway is the **vestibulospinal** tract, which originates in the cell groups that receive sensory input from the vestibular organs. These organs, located in the inner ear, provide the animal with information about gravity and acceleration. Other movement-command pathways are the **rubrospinal tract** and the **reticulospinal tract,** which permit the cerebellum and striatum of the telencephalon to control spinal motor neurons. In mammals and birds, direct pathways from the cerebrum to the spinal cord end on interneurons as well as on motor neurons. These descending pathways are very modest in nontetrapods; they are complemented, however, by specialized, high-speed, giant-axon pathways that provide for rapid control of movements by the brain. These giant axons, which are relatively few compared to the number of axons present in typical descending pathways, are not present in sharks and lungfishes. In tetrapods, the descending pathways from the brain to the spinal cord are well developed, particularly in amniotes.

Reflexes

One of the primary functions of the spinal cord is the performance of reflexes. These spinal reflexes are important for providing an automatic reaction of muscles in response to a stimulus. Some of these stimuli are the result of stretching or contraction of muscles. Stimuli that are painful or unexpected result in the reflexive withdrawal of a limb or the entire body. Some reflexes are excitatory in that they result in the activation of a motor neuron that in turn causes a muscular contraction. Other reflexes, however, are inhibitory and, therefore, prevent the activation of the motor neuron. Inhibitory reflexes are crucial in order to produce a cooperative activity of muscles; for example, one would not get very far if the flexor and extensor muscles (those that cause limb or finger joints to bend or extend) both tried to contract at the same time. Inhibitory reflexes prevent these opposing sets of muscles from contracting at the same time. This is especially important in walking and running. Inhibitory reflexes, working in smooth synchrony with excitatory reflexes, produce this coordination. Other spinal reflexes are visceral; that is, rather than controlling skeletal muscles they involve the actions of the smooth muscles of internal organs as well as the action of glands, such as sweat glands.

The simplest skeletal–muscle reflex is the **monosynaptic reflex,** in which an incoming sensory axon from a muscle receptor enters through the dorsal root and terminates on a motor neuron in the ventral horn of the gray matter. As shown in Figure 7-6, the motor neuron's axon passes out through the ventral root and terminates on a muscle near the original receptor. Since the only synapse in this circuit is that between the sensory axon and the motor neuron, this reflex is said to

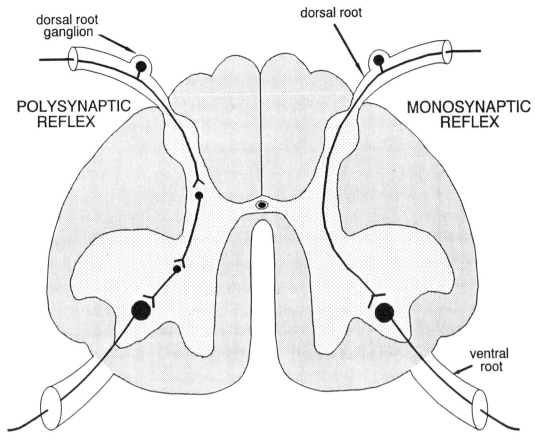

FIGURE 7-6. A transverse section through a primate spinal cord. The right side illustrates a monosynaptic reflex in which a sensory neuron enters via the dorsal root and passes directly to a motor neuron in the ventral horn. The left side illustrates a polysynaptic reflex in which two interneurons are interposed between the sensory and motor neurons.

be monosynaptic. The patellar reflex, in which the lower leg jerks forward in response to a tap just below the knee cap, is an example of such a monosynaptic reflex. Monosynaptic reflexes, however, are relatively rare. Most reflexes are **polysynaptic** and involve two, three, or more synapses between the incoming sensory axon and the motor neuron. Figure 7-6 also illustrates a polysynaptic reflex with two additional synapses between the sensory neuron and the motor neuron.

Figure 7-7 shows some of the pathways involved in a reflex of the sympathetic division of the autonomic nervous system. An incoming visceral afferent axon passes from the spinal nerve into the dorsal root and terminates in the gray matter on an interneuron. This interneuron, in turn, terminates on a preganglionic neuron located in the lateral horn of the gray matter. The axon of this preganglionic neuron leaves the spinal cord via the ventral root along with the somatic-motor axons. Unlike the somatic-motor axons, it does not leave the ventral root to enter the spinal nerve, but rather, almost immediately after leaving the spinal cord, it passes into the sympathetic ganglion where it terminates on the postganglionic, visceral-motor neuron. The visceral-efferent axon of the postganglionic neuron leaves the sympathetic ganglion and joins the somatic-efferent axons present in the ventral root as they leave the roots and enter the mixed, sensory and motor, spinal nerve. For clarity, the somatic-afferent and efferent axons are not shown.

Spinal Autonomy

One indicator of the degree of the brain's control over the spinal cord is the extent to which the longitudinal axons increase in number towards the rostral end of the cord. In nontetrapods, no substantial rostral increase in longitudinal axons is observed because most of the information remains within the cord. In contrast, in tetrapods, the existence of strong connections with the brain results in a build-up of longitudinal axons at the rostral end of the cord as each successive segment of the cord makes its axon contribution to the information flow to the brain and receives its axonal instructions from the brain.

The extent to which the spinal cord is an autonomous executive organ, more or less on a par with the brain, as opposed to being a dedicated subordinate that faithfully carries out the brain's instructions, depends on the type of environment that the animal inhabits and the evolutionary history of the species. In general, in nontetrapod vertebrates (agnathans, cartilaginous fishes, and ray-finned fishes), which inhabit an aquatic environment, the spinal cord has considerably more autonomous control over the musculoskeletal system than it does in tetrapods (amphibians, diapsids, turtles, and mammals). In other words, the nontetrapod spinal cord tends to carry out many of its postural and locomotor activities more or less independently of the brain. In particular, the spinal cord will coordi-

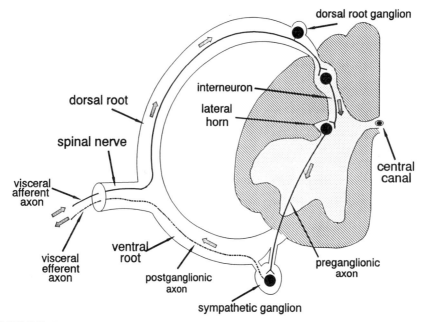

FIGURE 7-7. A sympathetic reflex arc. A visceral-afferent axon terminates in the gray matter of the spinal cord on an interneuron. The interneuron sends its axon to a preganglionic sympathetic neuron in the lateral horn, the axon of which passes out to the sympathetic ganglion where it ends on a postganglionic sympathetic neuron. The postganglionic visceral-efferent axon leaves the sympathetic ganglion and joins the axons in the ventral root as they enter the spinal nerve.

nate its activities with the senses that are located in the head (vestibular, visual, lateral line, etc.) by way of such brain structures as the tectum, the cerebellum, and the reticular formation, all of which will be discussed in subsequent chapters. On the other hand, in some situations, such as threat from a predator, a brain region, such as the midbrain tectum, may execute temporary, high-priority commands of the spinal cord and take complete charge of the body's movements during the duration of the emergency situation. In contrast, the tetrapod spinal cord generally is more under the control of the brain more of the time; this is especially the case in mammals, which appear to have the highest degree of brain control of the body's posture and movements.

Rhythmic Movements and Central Pattern Generators

The interneurons of the spinal cord form complex networks within an individual segment to coordinate the action of the muscles that are controlled by that segment. This is accomplished by networks of excitatory and inhibitory interneurons that are symmetrical with respect to the right and left sides of the body. Interneurons also coordinate the actions of large numbers of spinal segments to produce the coordinated activity necessary for propulsion of the body. An interneuronal network or circuit that coordinates and synchronizes the actions of motor neurons is known as a **central pattern generator.** At present, the central pattern generator is an explanatory concept rather than a specific, anatomically defined group of interneurons. Ample physiological evidence exists, however, to infer the existence of a central, interneuronal network that excites and

inhibits the interneurons in many segments of the spinal cord to produce the synchronized, rhythmic patterns of motor-neuron excitation and inhibition that result in smooth propulsion through the water or to locomote across the land or fly through the air. With a few exceptions (lampreys and frog embryos), the specific details of the location and neuronal circuitry of the central pattern generators have yet to be fully worked out either in nontetrapods or in tetrapods.

In fishes, central pattern generators cause oscillatory movements of the fins for steering, hovering, backward movement, and so on. These pattern generators are capable of producing rhythmic locomotor movements of the body (and limbs where present), even when the spinal cord has been surgically isolated from the brain. Respiration, although it involves muscles of the thorax and the diaphragm, is an autonomic function and is controlled by a central pattern generator located in the brainstem. In addition to the rhythmic movements of limbs and/or body for locomotion, tetrapods have certain other rhythmic body movements that are programmed by central pattern generators. These include scratching and chewing. Scratching is performed by spinal motor neurons. Chewing, being a specialized function of the head, is carried out by motor neurons located in the brain.

OVERVIEW OF THE HINDBRAIN

The hindbrain is one of the major constituents of the brain. It consists of the medulla and pons. Its distinguishing features include a large number of nerves, known as **cranial nerves,** and the cerebellum. The caudal end of the hindbrain is the transition zone between the spinal cord and the brain. The hindbrain is, for the most part, an extension of the spinal cord that has acquired certain specializations and elaborations. Like the spinal cord, it has the following:

- Afferent sensory nerves. These are the sensory cranial nerves. These nerves have ganglia similar to the dorsal root ganglia of the spinal cord. Like the spinal cord ganglia, they receive visceral afferents from internal organs of the body, as well as from internal organs of the head. The somatic afferent cranial nerves are essentially the same as the somatic afferent spinal nerves except that the somatic spinal nerves serve the body, and the somatic cranial nerves serve the receptors of the head.

- Sensory neuronal populations on which the afferent nerves terminate. These are the sensory nuclei of the cranial nerves, which are similar to (and sometimes a continuation of) the sensory neuronal columns or dorsal horns of the spinal cord. Like the dorsal horns, the sensory cranial nerve nuclei have both somatic-afferent components that innervate the skeletal muscles and joints, and visceral-afferent components that innervate the internal organs of the body. In general, however, the sensory nuclei of the hindbrain cranial nerves are specialized for head senses, such as hearing, taste, infrared detection, vestibular sense, lateral line mechanical sense, and electroreception, as well as pain, touch, and temperature sensations of the head, face, and oral cavity.

- Somatic-efferent motor neurons. These neurons, like their spinal counterparts, terminate on muscles of the skeleton, but in the case of the cranial nerves, the parts of the skeleton that they innervate are the head, neck, and jaws.

- Visceral-efferent neurons. Like the visceral-efferent neurons of the spinal cord, those of the hindbrain do not terminate directly on the target visceral organs. Instead they are preganglionic; that is, they terminate on neurons in an external ganglion. The axons of these ganglionic neurons (the postganglionic axons) are the ones that directly innervate the viscera. Although the visceral-efferent ganglia of the thoracic and lumbar regions of the spinal cord are known as sympathetic ganglia and are located close to the spinal cord, the visceral-efferent ganglia of the cranial nerves and those of the sacral region of the spinal cord are known as parasympathetic ganglia. Unlike the sympathetic ganglia, the parasympathetic ganglia are located close to their target organs. Further details on the similarities and differences between the sympathetic and parasympathetic ganglia may be found in Chapter 23. Although most of the visceral efferent axons of the cranial nerves remain in the head, one nerve, the **vagus,** follows a meandering course throughout the body as it innervates the internal organs of the thorax, abdomen, and pelvis.

- Integrating and coordinating systems. The spinal cord contains populations of neurons that are neither exclusively sensory nor exclusively motor. These coordinate the movements of muscle groups and provide central pattern generators for the production of rhythmic movements, such as walking, flight, and swimming. The hindbrain also has such integrating and coordinating systems, but they are

far more elaborate than those of the spinal cord. The major integrating and coordinating systems of the hindbrain are the **reticular formation** and the **cerebellum.**

In brief, the reticular formation coordinates the activities of related groups of motor nuclei in the hindbrain and midbrain so that they act together and are not at cross purposes. In many vertebrates, the efferent axons of the reticular formation that descend to the spinal cord are the principal pathway for the voluntary control of body movement by the brain. Other components of the reticular formation are involved in attention and wakefulness. Still others are the source of neuroactive substances that are transported to the forebrain.

The reticular formation consists of a large number of neuronal groups arranged in longitudinal columns of neurons (see Fig. 7-8). The medial-most column is known as the **raphe** nuclei. These are important sources of neuroactive substances that are transported elsewhere in the brain.

The cerebellum integrates sensory inputs from the somatosensory systems of the spinal cord and head with those of the vestibular system as well as other sensory systems of the head with the outflow from various components of the motor systems of the forebrain. The cerebellum has been implicated in a large number of sensory and motor functions, especially balance, coordination, and smoothing of motor activity. Injury to the cerebellum can result in a loss of the smoothness of voluntary movements. The cerebellum, like the reticular formation, is a very complex division of the brain with multiple functions. For example, in those fishes that make extensive use of electroreception, the cerebellum is highly developed and plays an important role in the processing of this sensory information.

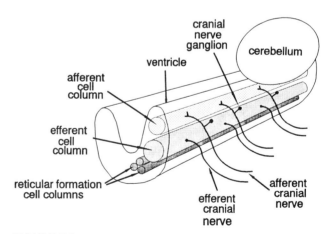

FIGURE 7-8. A schematic representation of the hindbrain. Caudal is to the left and rostral is to the right. Three cranial nerves, each represented by an afferent neuron and an efferent neuron, are shown. The afferent neuron, which could be somatic or visceral, is shown with its soma in a sensory ganglion and terminating in the afferent cell column, which corresponds to the dorsal horn of the spinal cord. The efferent neuron, which is shown with its soma in the efferent cell column, could be a somatic-efferent axon or a preganglionic visceral efferent axon. Also shown in the figure are two integrating and coordinating systems of the hindbrain: the reticular formation, which is organized in longitudinal columns that run parallel to the afferent and efferent columns, and the cerebellum, which is located at the rostral end of the hindbrain.

- Axons of passage. In addition to being the home of many specialized neuronal populations, the hindbrain is a major crossroads for axon systems ascending from the spinal cord to hindbrain, midbrain, and diencephalic sensory nuclei, as well as descending axon systems that originate in the telencephalon, diencephalon, mesencephalon, and hindbrain. These descending axon systems provide input to somatic motor neuron groups, preganglionic autonomic neurons, and integrating and coordinating systems in the hindbrain and spinal cord.

The Obex and the Fourth Ventricle

A major feature of the hindbrain is the fourth ventricle. This is the most caudal of the four great brain ventricles and is located in the medulla. The fourth ventricle is fully or partially covered by the cerebellum depending on the cerebellum's size and shape. The caudal end of the ventricle forms a small fold of tissue known as the **obex** (Latin for bolt or barrier). The obex marks the location of the dorsal column nuclei, the **nucleus gracilis** and the **nucleus cuneatus.** These nuclei are important structures in the ascending somatosensory pathways from the spinal cord.

The Pontine Nuclei

In mammals, the hindbrain contains a prominent bulge on its ventral surface in the region of the pons. Part of this bulge is caused by the large number of descending axons that pass through the ventral hindbrain in this region. These descending axons are the corticospinal tracts and corticopontine tracts, which originate in the cerebral cortex and terminate in the spinal cord and pons, respectively. Intermingled among these axon bundles are the **pontine nuclei.** These nuclei are the source of many axons that cross the midline to enter the contralateral cerebellum. These transverse axons that bridge the midline give the pons its name, which is the Latin word for "bridge."

Ganglia of the Cranial Nerves

The sensory neurons of the cranial nerves, like their spinal-cord counterparts, have their cell bodies located in ganglia on the roots of the nerves. Each ganglion has a unique name. Some of the names are descriptive, such as *semilunar* (half moon) or *spiral.* Sometimes they are named for their discoverer, such as the *ganglion of Scarpa* or *Gasserian ganglion.* Sometimes the ganglia are named for their cranial nerve, such as *trigeminal ganglion.* Finally, the same cranial nerve ganglion can have all three types of names, which are used interchangeably; thus, the ganglion of the trigeminal nerve is also known as the semilunar ganglion and the Gasserian ganglion. We will not present you with a table of the various names of the cranial nerve ganglia. Those of you who will have occasion to do intensive reading in the literature of a particular cranial nerve will soon become familiar with the interchangeable names of its ganglion.

Organization of the Cranial Nerves

Figure 7-8 shows a highly schematic organizational plan of the hindbrain showing the organization of the major cell columns. Dorsally and laterally is the afferent cell column, which consists of the somatosensory and visceral-sensory cranial nerve nuclei. Ventrally and more laterally is the efferent cell column that contains the somatic-motor neurons and the visceral preganglionic neurons. Medially and more ventrally are the cell columns that make up the reticular formation. At the caudal end is the junction of the hindbrain and spinal cord. At the rostral end is the cerebellum and the junction of the hindbrain with the midbrain.

Embryology of the Hindbrain and a New Classification of Cranial Nerves

In Chapter 9, we will discuss the embryological development of the head, the brain, and the cranial nerves in considerable detail. Embryological studies are important because they offer many clues to the evolutionary history of the nervous system as well as the functional relatedness of components that might otherwise appear to be quite unrelated in the adult animal. Recent data on the embryological origins of the cranial nerves suggest a functional relationship between certain groups of cranial nerves that was not suspected previously. These new relationships have led us to propose a new classification of the cranial nerves based on their embryological origins and functional organization. Students often have difficulty understanding the cranial nerves because some have only afferent components, some have only efferent components, and some have both. This new classification provides a framework for categorizing the afferent-only, efferent-only, and mixed nerves according to their embryological origin.

Neural Tube, Neural Crest, and Placodes. As discussed in Chapter 3, the central nervous system develops from the **neural tube,** the rostral end of which eventually develops into the brain. Lateral to the neural tube is the **neural crest,** which is the embryological origin of a variety of neuronal tissues and contributes to a number of the nerves and sensory ganglia of the trunk of the body as well as to a number of cranial nerves and their sensory ganglia. Another important embryological zone is formed by the **placodes.** Both the **dorsolateral placodes** and the **ventrolateral placodes** contribute to the cranial nerves and their sensory ganglia. Still a third group of cranial nerves develops from **evagination** of the forebrain. The specific embryological origin (placodal and/or neural crest) of the neurons in the various cranial nerves has yet to be fully established. Based on the present data, however, the cranial nerves can be categorized according to embryological origin as follows:

- Cranial nerves that appear to be predominantly derived from neural crest and/or neural tube. These are the visceral and somatic afferent and efferent nerves that serve the internal organs of the body and the muscles of the skeleton, muscular tongue (where present), and eyes. Their afferent components are mainly contact senses; that is, the stimulus source is at or near to the receptor, such as touch, temperature, pressure, or pain.

- Cranial nerves that are predominantly derived from placodes are only afferent and most serve distance senses such as the chemical senses of smell and taste (which can be a distance sense for aquatic animals), hearing, lateral-line sense, vestibular sense (which responds to the distant tugs of gravity and acceleration), and electroreception.

- The forebrain (neural-tube) derived cranial nerves. These too are purely sensory and serve the distance sense of light detection.

Three particular cranial nerves of the hindbrain are often troubling to students because they seem to mix together unrelated functions. These nerves are the facial, the glossopharyngeal, and the vagus nerve. The sensory components of these nerves are somatic and visceral, but they also serve the sense of taste. What does taste have to do with these other senses that it should be served by the same nerve? The answer is "nothing" and indeed these components *should not* be considered to be the same nerve because they have different embryological origins. The somatic and visceral afferent components of these three nerves are believed to be derived from neural tube and neural crest. The gustatory components, on the other hand, are believed to originate in placodes. They may travel together in the same peripheral nerve, but when they enter the central nervous system they immediately segregate themselves and terminate in quite distinct neuronal populations. We therefore have designated them as separate nerves.

Efferent Axons in Afferent Nerves. Some of the placodal and forebrain derived cranial nerves also contain efferent axons in their sensory roots. These efferents are not motor, but rather terminate on the receptors from which the afferent nerves receive their input. They appear to modulate the activity of the receptors and may play a role in attention.

EVOLUTIONARY PERSPECTIVES ON THE SPINAL CORD AND HINDBRAIN

The ancestors of land vertebrates were a group of fishes known as the **rhipidistians.** Rhipidistian fishes belong to the sarcopterygians, which are the fleshy-finned fishes, as distinguished from the overwhelming majority of fishes, the ray-finned fishes (actinopterygians). Among the modern sarcopterygian fishes are the lungfishes, which have the type of fleshy fins from which limbs evolved. These fleshy fins are quite different from the ray fins of teleost fishes, which typically lack sufficient mechanical structure to support the weight of the animal in very shallow water or on land where the aquatic buoyancy of the animal is absent.

The Transition to Land

One of the most far-reaching events in the evolution of the vertebrates was their invasion of the land. The entry of vertebrates into a vast and varying environment permitted the development of a large number of highly diverse species with an incredible variety of new adaptations for survival. More than anywhere else in the central nervous system, this major evolutionary change and its subsequent refinements and modifications are reflected in the spinal cord and hindbrain. Indeed,

since many of these adaptations involved behavior, they depended upon the central nervous system to supervise and perfect their execution. Many of the adaptations involved the head because feeding, hearing, and other behaviors are quite different in air and in water. These will be discussed in some detail in later chapters. For now, we will concern ourselves with the main adaptations of the body that are required for life outside of the water. These are the structures necessary to compensate for the body's loss of buoyancy when it leaves the water and locomotion across a solid surface.

Let us dispel a common misconception about the invasion of the land; it was not from the sea. The land was invaded from fresh water—not from sea water. This occurred because the invasion of the land was preceded by an earlier invasion; that is, from the sea to fresh water. Sea fishes became progressively adapted to the less saline waters of rivers and estuaries. As the invasion progressed farther upstream, some animals became trapped in ponds as flood waters receded. Those that were able to survive in the pond environment flourished there. One problem with ponds is that they often dry up. Those ancestral pond fishes that had stubby, fleshy fins, rather than the thin, delicate fins typical of most modern bony fishes, were able to use these stubby, fleshy fins to drag themselves across the land to a nearby pond that still had water. By so doing, they were able to survive and transmit this characteristic on to subsequent generations. Those that could not move across land remained trapped in their diminishing ponds. In other words, some of the fleshy-finned fishes left the water, not because life on land was better, but only as a means of getting to another body of water. The fleshy fins of these successful pond dwellers became the limbs of their tetrapod descendants.

Initially, the limbs could only permit these early tetrapods to drag their bodies across the ground. As anyone who has ever done anything similar will recognize, such dragging could cause injuries to the skin and abdominal muscles of the animals. Thus, animals with somewhat longer fins, or limbs as we will now call them, would have a lower risk of injury, and hence a better chance of survival. Thus began a trend to support the body off the ground that ultimately led to such specializations as the hoofed legs of horses and sheep, the wings of birds and bats, the hands of raccoons and primates, the paws of tigers and wolves, and many other variations including the upright, bipedal posture of birds and humans. The neural apparatus to operate these specialized structures in a coordinated, integrated way is present in the spinal cord and accounts for the differences between the nontetrapod and tetrapod spinal cords. With the development of long-axon descending pathways from the brain, spinal autonomy diminished considerably. Other specializations for terrestrial life included the development of the tongue and salivary glands to compensate for the loss of the water column that carries food through the mouth towards the digestive tract in nontetrapods. Modifications to the jaws permitted the capture of prey on land and the grinding and shearing processes necessary for the conversion of plant material into a digestible form.

Tetrapod Locomotor Patterns

The invasion of the land required a locomotor pattern quite different from that necessary for survival in an aquatic environment. Once out of the water, the body's buoyancy no longer counteracted the pull of gravity. On land, the tetrapod's limbs support its body's weight above the ground, and, in order to move forward, one or more limbs must be lifted off the ground. Once a limb is off the ground the body's weight must be carried by the remaining limbs. If the animal is to maintain its balance and not fall over, this weight must be so distributed that the center of gravity is now positioned appropriately over the limbs that remain on the ground. All of this shifting of weight must occur in smooth synchrony with the raising and lowering of the limbs in order to result in a smooth forward progression across the ground. To complicate matters further, a pattern of raising and lowering of limbs that is suitable for one speed of locomotion is less suitable for another. Tetrapods have evolved several patterns of locomotor progression to deal with these problems.

The simplest locomotor pattern is walking, in which one leg is raised and moved forward at a time. When one leg is off the ground, the body's weight is shifted over the tripod formed by the three legs that remain on the ground. When that leg is back on the ground, a leg on the opposite end of the body is raised. For example, if the right foreleg is raised first, then the next leg to be raised is the right hindleg. Perhaps the best way for a biped like you to understand tetrapod locomotor progression is to temporarily return to tetrapod status by getting down on your hands and knees and crawling forward like a human baby would. Notice that if you allow your hands to support your weight as well as your knees, when you lift one hand to move it to a forward position, your weight is still nicely supported by the other hand and the knees, although you will have to lean a bit toward the hand on the ground to keep from falling over. If you now try to move the other hand forward, without shifting your weight to your knees, you will fall on your face. The only way you can move forward is by lifting one of your knees; the most comfortable one to lift will be the one that is on the side opposite to the hand that was just moved forward. For faster forward progression tetrapods use a pattern in which a front and back limb are off the ground at the same time. When the two elevated limbs are on the opposite sides of the body (e.g., right forelimb and left hindlimb), this pattern is called trotting. When the two elevated limbs are on the same side of the body (such as right forelimb and right hindlimb), the pattern is called pacing. Readers who are familiar with horse racing will have come across these terms before. Finally there is running, which is actually a series of jumps in which the animal pushes off with its hind legs, passes through a brief period in which all four limbs are in the air, and then lands on its forelegs.

Birds, like humans and a few other mammals, are bipeds and, therefore, have different patterns for locomotor progression. Some birds walk like humans by lifting one of their two legs at a time while carrying the body's weight on the other. Others use a hopping progression, which is a series of short leaps in which both legs are in the air at the same time. The latter pattern of locomotion across the land also is used by kangaroos and anuran amphibians (frogs and toads).

Many tetrapods spend a portion of their time in the aquatic environment; examples of these are the majority of amphibians, some diapsid reptiles and turtles, a number of species of birds such as waterfowl, gulls and other shorebirds, and penguins as well as a number of mammals such as beavers, otters, seals, walruses, and some humans. Some mammals, particularly the whales and porpoises, are permanently aquatic. Many of these animals use the hindlimbs as the principal propulsion organs. These limbs often are used as paddles, alternating the right and left limbs. Frogs move forward through the water by forcing water from between their powerful hind legs by rapidly bringing them together. Those animals with powerful tails, such as alligators, propel themselves through the water by means of rhythmic side-to-side movements of their tails. Marine mammals use up-and-down rhythmic movements of the hind quarters of their bodies to move forward through the aquatic environment.

Another important mode of animal locomotion is flight. Birds and bats rhythmically move the large surface areas of their forelimbs to produce lift and forward movement through the air. Unlike other aquatic birds that paddle through the water with their hindlimbs, penguins locomote through the water with their forelimbs the way that other birds locomote through the air; that is, they "fly" though the water.

Each of these specializations for the tetrapod life has its corresponding specializations in the sensory, motor, and interneuron populations of the spinal cord and their continuations into the caudal hindbrain. Likewise these changes are reflected in various axon columns of the spinal cord and hindbrain. These changes will be discussed in detail in subsequent chapters.

FOR FURTHER READING

Cohen, A. H., Rossignol, S., and Grillner, S., (eds.) (1988) *Neural Control of Rhythmic Movements in Invertebrates.* New York: Wiley.

Dowling, J. E. (1992) *Neurons and Networks.* Cambridge, MA: Belknap/Harvard

Fritzsch, B. and Northcutt, R. G. (1993) Cranial and spinal nerve organization in Amphioxus and lampreys: evidence for an ancestral craniate pattern. *Acta Anatomica,* 148, 96–109.

Gans, C. and Northcutt, R. G. (1983) Neural crest and the origin of vertebrates: A new head. *Science,* 220, 268–274.

Heijdra, Y. F. and Nieuwenhuys, R. (1994) Topological analysis of the brainstem of the bowfin, *Amia calva., Journal of Comparative Neurology,* 339, 12–26.

Lumsden, A. (1990) The cellular basis of segmentation in the developing hindbrain. *Trends in Neurosciences,* 13, 329–335.

Northcutt, R. G. (1984) Evolution of the vertebrate central nervous system: patterns and processes. *American Zoologist,* 24, 701–716.

Northcutt, R. G. (1990) Ontogeny and phylogeny: a re-evaluation of conceptual relationships and some applications. *Brain, Behavior and Evolution,* 36, 116–140.

8

The Spinal Cord

THE SPINAL CORDS OF NONTETRAPODS

The principal form of locomotion of nontetrapods is based on rhythmic, undulatory movements of the trunk and tail muscles, often supplemented by movements of the fins. Much of the organization of their spinal cords is related to this locomotor pattern. In addition to normal locomotion to get from place to place, several sudden and very rapid forms of locomotion are required; these include locomotion to evade the approach of a predator and sudden strikes to capture prey.

Muscles, Interneurons, and Locomotion

The majority of nontetrapod animals locomote by means of rhythmic undulations of the body. The contractions of the corresponding muscles of the right and left sides must be coordinated to produce the smooth, rhythmic pattern of activity that propels the animal through the water. In those animals in which the fins assist in locomotion, the actions of the fins must be coordinated with each other and with the rhythmic movements of the body.

The muscles of nontetrapods differ from the familiar arrangement of muscles in tetrapods in which individual, specialized muscles are attached to the vertebrae, ribs, and limb bones. In contrast, the muscles of nontetrapods consist of segmented, sheet-like muscle masses known as **myomeres.** These myomeres are arranged in layers around the body cavity. The organization of muscle fibers into myomeres rather than into discrete muscles has resulted in a somewhat different organization and location of motor neurons within the spinal cord in nontetrapods.

Cell and Fiber Columns

The cell bodies of the motor neurons of the spinal cord in nontetrapods are located most often in the ventral horns of the cord. The periphery of the spinal cord is occupied by the **marginal dendritic plexus,** which is a dense network of dendrites of interneurons. The plexus serves as the main zone of interaction between neurons in the nontetrapod spinal cord. Incoming sensory axons terminate in the marginal dendritic plexus as do the axons of various interneurons. This plexus is shown in Figure 8-1.

The cell bodies of the interneurons of the cord are located in the gray matter, which is the area surrounding the central canal. Among these are a group of interneurons known as the **commissural cells,** the dendrites of which spread in all directions; an example of this is shown in Figure 8-1. The commissural cell axons cross the midline of the cord to the opposite side and then form a T-shaped **bifurcation,** with one branch of the axon running in the rostral direction and the other running caudally to one or more segments in each direction. This system of commissural interneurons is primarily an intrinsic spinal system that coordinates the actions of interneurons and motor neurons in one or more segments of the cord. Some commissural-cell axons, however, do ascend to the brain. Other ascending pathways from the cord to the brain include the spinocerebellar, spinoreticular, spinotectal, and spinothalamic tracts that were described above.

A number of axonal pathways descend from the brain to the spinal cord. These are present in varying amounts depending on the development of the particular systems within the brain, such as visual, auditory, gustatory, vestibular, and electrosensory. Prominent among these pathways are the vestibulospinal, tectospinal, and reticulospinal tracts. Other specialized descending pathways are trigeminospinal and gustatory-spinal.

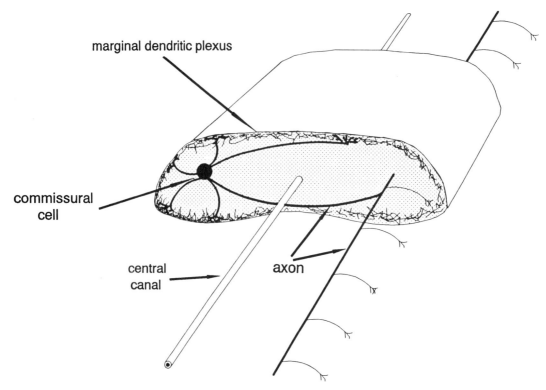

marginal dendritic plexus

commissural
cell

central
canal

axon

FIGURE 8-1. A commissural cell that sends its dendrites into the marginal dendritic plexus including a branch that passes over the central canal and enters the marginal dendritic plexus of the contralateral side. The cell's axon passes under the central canal to the contralateral side where it bifurcates into two long branches, one ascending and one descending. Each branch gives off many collaterals along its path.

These pathways carry commands from the brain to the spinal cord that are influenced by information acquired through the senses of the head (vision, audition, taste, vestibular sense, and electroreception). These pathways typically end on interneurons within each segment of the spinal cord and occasionally directly on motor neurons.

The motor neurons of the nontetrapod cord form a relatively uniform column in the ventral horn of the gray matter. In contrast, in tetrapods, the motor neurons of the ventral horn, while still forming a column, tend to be clustered into "pools" of neurons that correspond to specific muscles. Some motor-neuron pools exist in the nontetrapod cord as well, that is, those that control the muscles of moveable fins, such as the pectoral fins, which are used in locomotion and also in hovering in one location. Groups of motor neurons that innervate specific muscles of the pectoral fins are located at the rostral end of the spinal cord in the ventral horn. Another example of motor-neuron pools in nontetrapods are the motor neurons that control the generation of sounds in certain groups of fishes. These sounds, which play a role in courtship behavior, are generated by the drumming of specialized muscles on the walls of the swim bladder, which is a taut bladder filled with respiratory gases that these fishes otherwise use for regulating their buoyancy in the water. This motor-neuron pool is located at the junction of the rostral end of the spinal cord near its junction with the medulla. Figure 8-2 illustrates a lamprey motor neuron with a dendritic tree that extends into the marginal dendritic

plexus and an axon that leaves the soma and extends towards the periphery.

Two types of motor neurons have been distinguished in the ventral horns of the nontetrapod spinal cord. The **primary motor neurons** are large and are relatively few. The **secondary motor neurons** are smaller in size and are more numerous. These two motor-neuron types correspond roughly to two types of muscle fibers: the white fibers, which are fast in their contractions, and the red fibers, which are slower. Thus, in nontetrapods, the white muscles are used for fast swimming, such as pursuit of prey and escape, whereas the slower, red muscles are used for more leisurely swimming. In general, the primary motor neurons innervate the faster, white-muscle fibers, and the secondary motor neurons innervate the slower, red-muscle fibers, although in some nontetrapods, the secondary motor neurons also innervate some white fibers. The specific activities of these motor neurons are coordinated by the interneuron circuits known as central pattern generators.

GIANT AXONS AND ESCAPE

In addition to normal locomotion, escape from danger is an important function of the motor system. The detection of a potential threat, usually by the brain, breaks the normal locomotor pattern and results in a sudden, rapid change in direction and speed that is known as the escape pattern. In order to be

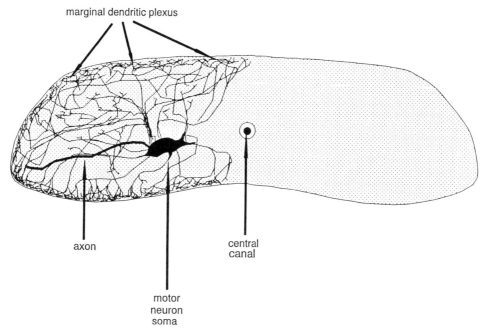

marginal dendritic plexus

axon

central canal

motor neuron soma

FIGURE 8-2. A lamprey motor neuron with numerous dendritic branches that enter the marginal dendritic plexus and an axon that is about to leave the spinal cord en route to muscles of the body.

effective, an escape pattern must be as rapid as possible since every wasted millisecond increases the likelihood of being someone's dinner. Since the velocity of a nerve impulse increases with the diameter of the axon, escape systems tend to be associated with axons of impressive diameter, which is to be expected since those fishes with smaller-diameter axons in their escape systems were less likely to escape than those with the larger-diameter axons. Two types of giant axons have been described in nontetrapods: the **Müller** and the **Mauthner axons.**

Müller Axons. The Müller axons are found in agnathans and at least some teleost fishes. They originate from a small number of neurons with very large somata that are located in the midbrain and in the medulla. The majority of these axons cross the midline and descend in the white matter of the spinal cord to the caudal region of the cord where they synapse on motor neurons. They control both normal swimming as well as escape and vestibular control of orientation of the body to acceleration and gravity.

Mauthner Axons. In agnathans, teleost fishes, and many amphibians, the escape response is orchestrated by one of the most remarkable interneurons to be found in any central nervous system: the Mauthner cell. And when we say "one," we mean precisely that; only one Mauthner cell exists on the right side of the nervous system and only one on the left. The Mauthner cell, which consists of a giant cell body located in the medulla, in the vicinity of the termination of the vestibular branch of nerve VIII, has been most thoroughly studied in teleost fishes and agnathans. These cells often are absent in bottom-dwelling species that do not move very much. An illustration of a Mauthner cell can be seen in Figure 8-3.

The Mauthner cell body consists of a banana-shaped soma with two main enormous dendrites, one pointing ventrally and one pointing laterally. The Mauthner soma often has a diameter as great as 100 μm (0.10 mm) and a total length of nearly 1.0 mm in goldfish. The main dendrites can extend as far as the outer margins of the brainstem. Each Mauthner cell can have tens of thousands of synaptic terminals on it. Smaller dendrites also are present on the soma, but they are not as spectacular as the two main dendrites. The total number of synaptic terminals can be several hundred thousand. Although most of these terminations are chemical synaptic junctions, a considerable number of junctions are electrical or mixed chemical/electrical, which result in a faster speed of transmission. Many of the axodendritic synaptic terminals form an unusual type of structure known as a "club ending" because of their shape. Other terminations on the Mauthner nerve are spiral in shape. These are axo-axonic and are wrapped around the axon hillock where the giant axon emerges. The spiral terminations are inhibitory and are ideally situated to suppress an action potential from being propagated down the giant axon.

The inputs to the Mauthner cell are mainly from the vestibular, auditory, and lateral line systems. The latter system contains two types of receptors: (1) mechanoreceptors that respond to such stimuli as vibrations, pressure waves, and water flow across the surface of the head and body, which are important in detection of predators, maintenance of schooling, and other behaviors; and (2) electroreceptors that indicate features of electrical fields in the surrounding water. These auditory, vestibular, and lateral line axons terminate mainly on the main lateral dendrite. Other sources of stimulation from the tectum, the cerebellum, and even the visual system have been described.

Shortly after its emergence from the axon hillock, the giant Mauthner axon sends a small collateral branch back

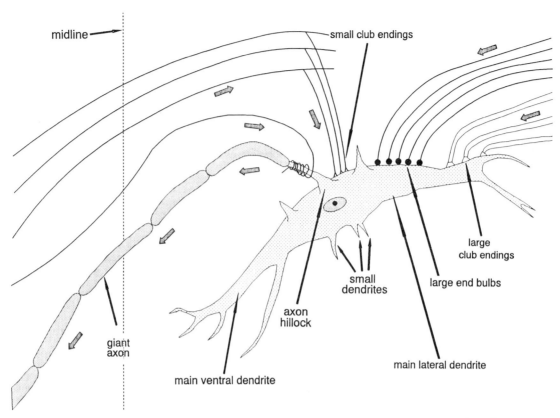

FIGURE 8-3. The giant Mauthner cell, which has a giant, heavily myelinated axon that crosses the midline and descends to the spinal cord on the contralateral side.

toward the axon hillock. This **recurrent** (flowing back) branch terminates on a small, inhibitory neuron, which in turn terminates on the axon hillock. This **recurrent inhibition** results in discrete pulses of muscle stimulation. Such a mechanism is known in engineering as "negative feedback"; it is essentially the same as the mechanism by which a thermostat shuts off the furnace when the room gets warmer than its set point.

The single, giant Mauthner cell body gives rise to a single giant, heavily myelinated axon that crosses the midline and descends in the white-matter column just ventral to the ventral horn. As it descends, the Mauthner axon gives off frequent collateral branches at each segment of the spinal cord. These axon collaterals have terminals directly on the primary motor neurons, which innervate the fast, white-muscle fibers. The Mauthner collaterals also terminate on descending interneurons that excite the primary motor neurons.

At the same time that the Mauthner cell is producing excitation of the primary motor neurons, it also produces inhibition of the corresponding primary motor neurons of the opposite side of the spinal cord by activating inhibitory commissural interneurons that cross the midline and inhibit these corresponding contralateral motor neurons. This contralateral inhibition is necessary to prevent the contraction of contralateral muscles from interfering with the sudden bending of the body into a C shape that quickly points the animal away from the direction of the threat. The relationship of

the Mauthner axon to interneurons and motor neurons is shown schematically in Figure 8-4. The Mauthner cell thus serves as the heart of a powerful integrating mechanism that insures a rapid pattern of contraction and relaxation of large numbers of muscles to result in a flexure of the body at a time when a sudden and speedy change in the direction of locomotion is required.

Until recently, amphibian Mauthner cells were thought to exist only in the animals' larval (tadpole) state and in those adult amphibian species that either remained in the water for their entire life cycle or that retained a tail in adulthood. Recent research, however, has revealed that these distinctions do not hold, and indeed Mauthner cells can be found in a wide range of amphibians irrespective of the presence of a tail or adaptation to an aquatic environment.

In tadpoles, Mauthner cells appear to play a role similar to that of their counterparts in lampreys and teleosts; that is, a sudden curvature of the body as the start of an escape reaction away from a source of threat. Animals such as frogs, however, cannot flex their bodies as a fish or tadpole can, and their locomotor pattern in water is very different due to the lack of a tail, the presence of legs, and the existence of specialized muscle groups rather than myomeres. In water, escape responses in such animals consist of a sudden contraction of the body followed by movements of the legs to propel the animal in a backward direction. On land, the response to threat is an escape jump.

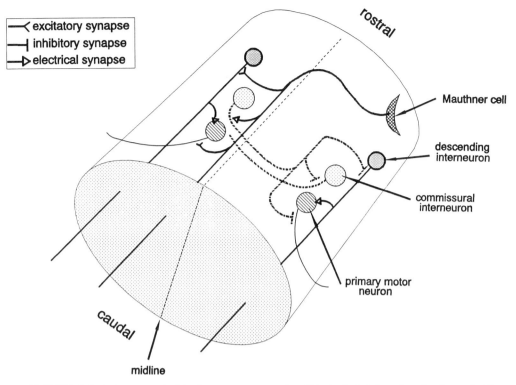

FIGURE 8-4. Pathways involved in the escape response in which the body is bent into the shape of a C. The Mauthner-cell axon crosses the midline and excites both the contralateral descending interneuron and the contralateral primary motor neuron via chemical synapses. It also excites the contralateral commissural neuron via an electrical synapse. The contralateral descending interneuron excites motor neurons along its path of descent towards the tail via an electrical synapse. Inhibition of the motor neurons of the side ipsilateral to the Mauthner soma is accomplished by the contralateral commissural interneuron, which sends its axon across the midline to the side of the Mauthner soma and inhibits the ipsilateral descending interneuron, the primary motor neuron and the corresponding commissural interneuron. Adapted from Fetcho (1991).

Electromotor Neurons

Although the term "motor neuron" typically is used to indicate efferent neurons that stimulate the contraction of muscles that move some body part, certain other neurons that are called "motor" do not actually move anything. For example, some of the visceromotor neurons of the autonomic system control the secretion of glands. Although strictly speaking they should be called "effector" neurons (indicating that they "do" something), they more often are called "motor" because of their cell type, location, or other reasons of convention. Among these effector neurons that do not move anything are the **electromotor neurons** that activate the electric organs of certain fishes. The most famous of these are the electric organs of fishes such as the electric eel, the electric catfish, and the torpedo. These electric organs, which are actually modifications of muscles, produce muscle action potentials that are so powerful they can stun prey or predators. Less well known are the weak electrical potentials, which are far too weak to stun another animal, that these animals and certain other groups of fishes produce. These weak potentials are important for social communication and for finding objects in murky waters or in dim illumination. The electrosensory system, by which these weak potentials are detected, and the ways in which they are used by these animals will be discussed in greater detail in Chapter 11. Here we will only point out that the electromotor neurons that control these electric organs are located in a separate column of the ventral horn.

The Curious Spinal Cords of Sharks

The observant reader might have noticed that our discussion of Mauthner and Müller axons did not include sharks and related animals. These animals have the same need to make rapid escape responses as do other aquatic nontetrapods; yet for reasons that we can only speculate about, they lack the high-speed, giant-axon escape systems of other nontetrapods. Their motor responses, both high-speed pursuit and escape as well as the initiation of more leisurely locomotion, are controlled by the reticulospinal, tectospinal, and vestibulospinal tracts. One might suppose that so effective a predator as a shark, which is high on the food chain, might not often need to make escape responses; this, however, is not always the case. There are often larger, more aggressive sharks to be avoided. Moreover, not all sharks are the high-speed predators that exist in the general public's image of a shark. Nevertheless, the relative freedom from predation that sharks enjoy may have played a role in the loss of the giant-axon systems that characterize the spinal cords of most nontetrapods.

THE ORGANIZATION OF THE TETRAPOD SPINAL CORD

Like its nontetrapod counterpart, the tetrapod spinal cord contains axons entering the cord from the body interior, exterior, and from the limbs. Likewise, axons leave the cord to innervate the muscles and glands of the body's interior, the body's exterior, and the limbs. The axons to and from the limbs, which are one of the principal features that distinguish the spinal cords of tetrapods from their nontetrapod counterparts, are present in great numbers so that they cause the cord to bulge in the regions of their entrance and exit. Thus, the axons to and from the forelimbs produce an enlargement of the cord in the cervical region (the **cervical enlargement**) and those to and from the hind limbs result in a **lumbar enlargement.** Animals with powerful tail musculature such as alligators, beavers, and dolphins also have large numbers of incoming axons as well as large numbers of motor neurons and their outgoing axons in the lumbar region. These too contribute to the lumbar enlargement. The combination of both powerful hind legs and a powerful tail, as in kangaroos and many varieties of dinosaurs (as may be seen in their fossilized skeletons), leads to an exceptionally large lumbar enlargement.

Locomotor Patterns and Spinal Cord Organization

Each of the locomotor patterns described in Chapter 7 has its representation in the local structure of the tetrapod spinal cord. In general, those segments of the cord that innervate the organs of locomotion are better developed by virtue of the presence of motor-neuron pools in the ventral horns to activate the locomotor muscles and increased numbers of interneurons in the dorsal horns to receive the larger numbers of sensory neurons from the locomotor organs as well as to receive commands from the central pattern generators. Like nontetrapods, tetrapods also have slow-acting red-muscle fibers and fast-acting white-muscle fibers.

An interesting and instructive comparison that reveals the relationship between mode of locomotion and spinal cord organization can be made among three types of reptiles: alligators, turtles, and snakes. Alligators are typical tetrapods in that they have well-developed limb muscles as well as well-developed trunk muscles and powerful tail muscles. The columns of motor neurons that supply these muscles are characteristic of most tetrapods; that is, they have a ventromedial column of motor neurons that extends along the length of the spinal cord, and lateral extensions of the ventral horns are present in the cervical and lumbar enlargements. Although they are legless, snakes nevertheless are tetrapods because they have descended from a lineage of tetrapod reptiles, and before that tetrapod amphibians. Snakes thus evolved from limbed reptiles into a body configuration that lacks external limbs, the muscles to move the limbs, and motor neurons with which to activate such muscles. Snakes therefore do not have ventrolateral columns of motor

neurons in the cervical and lumbar regions nor do they have cervical and lumbar enlargements. Instead they have well-developed ventromedial columns. In contrast, turtles have more or less conventional tetrapod limb muscles, but lack trunk muscles. The ventromedial cell columns are therefore very small in these animals although the ventrolateral columns are present in the cervical and lumbar regions. A similar situation exists in the spinal cords of birds. Finally, in mammals such as primates and raccoons, which have exceptional ability to manipulate objects with the digits of their forelimbs (hands), the ventrolateral columns of motor neurons are more elaborate than in other tetrapods.

Figure 8-5 shows a schematic representation of three types of spinal cords. At the left is a typical tetrapod spinal cord with cervical and lumbar enlargements and a thick thoracic region. The center diagram shows a snake spinal cord that lacks cervical and lumbar enlargements because of the lack of limbs, but which is thick throughout its length due to the powerful trunk musculature that is the basis of the animal's locomotion. At the right is the type of spinal cord that is characteristic of turtles and birds. It has lumbar and cervical enlargements due to the limbs of these animals, but the spinal cord is thin in the thoracic region. In the case of turtles, the thinness is due to the lack of trunk musculature in the thoracic and abdominal regions, which are contained within the armored shell of the animal. Moreover, since the shell is not innervated, the sensory input to these levels is mainly visceral, which also contributes to the thinness of the cord in these regions. The thin thoracic cord of birds derives from the lack of necessity to coordinate the actions of the hind limbs and forelimbs. When birds fly, their feet are neatly tucked flat against the body; when they walk, their wings are firmly positioned against the body. One can readily see the lack of coordination between the forelimbs and hind limbs of birds when observing a relatively flightless chicken running from some threat. Its wings flap and its legs run, but there seems to be little coordination between the two, which results in a rather clumsy escape. This contrasts sharply with the very smooth, precise, and rhythmic patterns that exist when birds walk or fly.

The Curious Spinal Cords of Birds

One of the more dramatic examples of how locomotor patterns affect spinal-cord morphology may be seen in the lumbar region of the avian spinal cord. Birds, like other tetrapods with well-developed hind legs, have a well-developed lumbar enlargement. Unlike other tetrapods, however, the dorsal white and gray columns have separated in the caudal lumbar and sacral regions to form a cavity known as the **lumbosacral sinus** or **rhomboid sinus.** This cavity appears to have been formed as a consequence of the extremely large avian sciatic nerve, which is formed by nerve roots from several segments in this region, and which is the main sensory and motor nerve of the leg. The cavity, however, is not a ventricle of the central nervous system as is the fourth ventricle of the brain, which contains cerebrospinal fluid. The central canal, which contains the cerebrospinal fluid, passes, like a water pipe suspended in air, through the lumbosacral sinus rather than opening into it as in the case of the fourth cerebral ventricle. The lumbosacral sinus is illustrated in Figure 8-6.

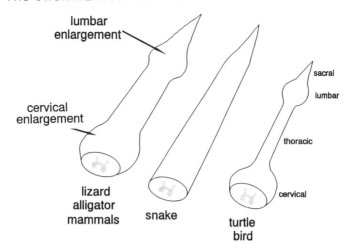

FIGURE 8-5. Schematic representations of three types of spinal cord. Left: Typical tetrapod spinal cord with cervical and lumbar enlargements and a thick thoracic region. Center: A snake spinal cord that is thick throughout its length and lacks cervical or lumbar enlargements. Right: A spinal cord characteristic of turtles and birds that has cervical and lumbar enlargements but is thin in the thoracic region.

The lumbosacral sinus is not empty, however; it contains a large gelatinous mass that surrounds the central canal. This mass, which is rich in glycogen, is known as the **glycogen body.** Glycogen is a stored form of glucose, typically found in the liver, which is vital for the nutrition of body cells. More modest glycogen stores have been found along the length of the central canal extending as far rostral as the medulla. The functions of the glycogen body have been the subject of much speculation. Perhaps the most plausible explanation at present is the suggestion that this storehouse of glycogen may play a role in the formation of myelin during development and may serve as an alternative source of energy for the central nervous system during periods of high metabolic stress. Glycogen bodies also occur in the spinal cords of the embryos of mammals and have been reported in adult newts. The presence of the glycogen body in the lumbosacral sinus may not have anything in particular to do with the lumbar region of the spinal cord, but rather

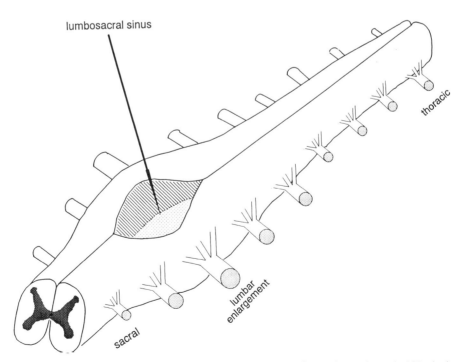

FIGURE 8-6. A pigeon lumbar cord showing a large cavity, the lumbosacral sinus, located within the lumbar enlargement.

may be a chance occurrence that took advantage of an unrelated evolutionary development.

Segmental Organization

In addition to entering sensory axons and motor-neuron cell bodies and their exiting axons, the tetrapod spinal cord contains a large number of interneurons. Many of these are intrasegmental interneurons, and their axons remain within their spinal segment. Others are extrasegmental and send their axons to one or more additional segments. Many of these extrasegmental interneurons send their axons beyond the spinal cord to targets located in the brain such as sensory nuclei of the medulla as well as the cerebellum, the thalamus, and the reticular formation. These brain structures are components of complex brain systems that control the activities of the spinal cord by means of descending axonal pathways from the brain to the spinal cord, such as the reticulospinal, tectospinal, vestibulospinal, rubrospinal, cerebellospinal, and hypothalamospinal pathways. These large numbers of ascending and descending axons result in a progressive increase in the diameter of the spinal cord from caudal to rostral as each successive segment makes its contribution to or receives its contribution from the traffic with the brain.

Lamination

When considered from a longitudinal perspective, the gray matter of the tetrapod spinal cord is organized into continuous columns of neurons. The spinal cord, however, also can be viewed in a transverse or segmental perspective as well. When we consider a transverse section through any spinal segment, the interneuron and motor-neuron cell columns can be viewed as forming laminae (layers) beginning in the dorsal horns and continuing ventrally into the ventral horns. This approach to the cytoarchitecture of the spinal gray matter was first proposed by the Swedish anatomist, Bror **Rexed,** who recognized nine laminae in the cat spinal cord. His system since has been shown to be applicable (with some minor variations) to the organization of the spinal cords of nearly all tetrapods. Figure 8-7 shows

some examples of this cytoarchitectonic organization in a lizard, a pigeon, and a human.

Each lamina consists of cells of a particular type. Many of the laminae are cross sections through neuronal cell columns; others consist of cell groups that exist only at specific levels of the cord, such as the cervical and lumbar enlargements; thus, not all laminae are present in each spinal segment. Although the laminae vary somewhat in form from region to region of the spinal cord, a more or less consistent pattern can be observed throughout. Laminae I–VI form the dorsal horns. The cells of these laminae are interneurons that are the sources of the ascending pathways to the brain as well as intrasegmental and extrasegmental spinal pathways.

Lamina VII forms an intermediate zone between the dorsal and ventral horns. In addition to being a principal contributor to the spinocerebellar pathways, lamina VII is also involved in the autonomic system. At its medial edge is a column of cells known as the **intermediomedial column;** these cells are the targets of the visceral-sensory (visceral-afferent) axons. At its lateral edge, in the lateral horn of the thoracic and lumbar regions, is the **intermediolateral column** that contains the preganglionic neurons of the sympathetic division of the autonomic nervous system.

Laminae VIII is a major coordinating zone of the gray matter. It contains interneurons that are the targets of many of the descending pathways from the brain and other levels of the spinal cord. These interneurons coordinate both reflex and voluntary commands from the brain with reflex actions of the spinal cord.

Lamina IX consists of separate groups of cells that form the motor-neuron columns that innervate the muscles of the skeleton. In the regions of the spinal enlargements, the groups of lamina IX cells are larger and better developed than elsewhere.

Intrinsic Spinal Interneurons

After passing through the dorsal white matter via a region known as the **zone of Lissauer,** incoming axons from pain and temperature receptors at the body surface as well as from

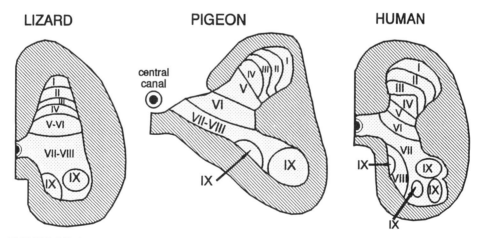

FIGURE 8-7. Lamination in the gray matter of three tetrapods: a lizard, a pigeon, and a human. Laminae I–VII constitute the dorsal horn. Laminae VIII and IX are the ventral horn. Many axons of descending pathways terminate in lamina VIII. Lamina IX neurons are the motor neurons.

receptors on the visceral surfaces of the body enter the gray matter of the cord. These axons terminate on interneurons in laminae I and II. Lamina II is also known as the **substantia gelatinosa** and takes its "gelatinous" appearance from the presence of large numbers of lightly myelinated and unmyelinated axons as well as many dendrites. The axons of the lamina II interneurons belong to the intrinsic spinal system; that is, they terminate within their own segment or travel to other spinal segments but they do not project to the brain. These interneurons may modulate or "gate" sensory information entering the spinal cord. Laminae III and IV interneurons also are mainly intrinsic to the spinal cord, although some lamina IV axons form part of the **spinothalamic** system. The interneurons of lamina V, which is present only in the cervical and lumbar enlargements, receive incoming axons from limb muscles as well as axons from descending spinal pathways. Lamina VIII interneurons coordinate reflex actions such as spinal reflexes and vestibulospinal reflexes, as well as cerebellospinal functions (via the rubrospinal tract) and voluntary movements via the reticulospinal route, and in some cases direct telencephalic pathways to the spinal cord.

Somatotopic Organization of the Ventral Horns

As illustrated in Figure 8-8 in a primate cervical spinal cord, the motor-neuron cell columns that we would encounter as we

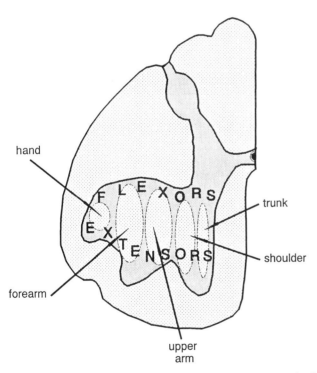

FIGURE 8-8. Somatotopic organization of the motor-neuron pools of the ventral horn that control the forelimb. The spatial distribution of motor neurons in the ventral horn corresponds to the spatial distribution of the body parts that they control. Thus, those motor neurons that control trunk muscles are located most medially in the horn, those that control the arm muscles are intermediate in the horn, and those that control the hand or paw are most lateral. Those motor neurons that control flexor muscles are located in the dorsal part of the horn; those that control extensor muscles are located ventrally.

progress from medial to lateral across the ventral horn are those that innervate the muscles of the trunk (most medial), then those that innervate the shoulder muscles, then the upper arm muscles, then the forearm muscles, and finally those motor-neuron columns that innervate the muscles of the hand (most lateral). This representation of the body parts in the nervous system in the same sequence as they exist in the body is known as a **somatotopic organization.** We will encounter somatotopic body "maps" elsewhere in the motor system as well as a number of places in the somatosensory system. A second type of organization also exists in the ventral horn; the more dorsal motor-neuron columns innervate the flexor muscles of the trunk, shoulder, upper arm, forearm, and hands while the more ventral motor-neuron columns innervate the corresponding extensor muscles.

The lateral horns of the gray matter, which are present only in the thoracic and lumbar regions, contain the intermediolateral column of visceral motor neurons that supply the preganglionic axons to the sympathetic ganglia located on the ventral roots of the spinal cord. Postganglionic axons travel to various internal organs of the body (collectively known as the viscera) and produce contractions of smooth muscles such as those of the intestinal system or stimulate the secretion of glands. The sympathetic division of the autonomic nervous system, along with the parasympathetic division, will be discussed in greater detail in Chapter 23.

Renshaw Cells

In our discussion of the Mauthner cell earlier in this chapter, we encountered the concept of the recurrent collateral. These small side branches that "flow back" towards the cell body are typical of motor neurons in the spinal cords of mammals. Recurrent collateral branches terminate on small, inhibitory neurons called **Renshaw cells.** Figure 8-9 illustrates a motor neuron with its recurrent collateral and associated Renshaw cell. The Renshaw cell, in turn, sends its axon with its inhibitory terminals to the cell body of the very axon from which the

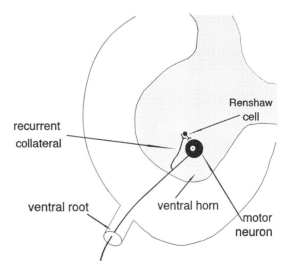

FIGURE 8-9. Recurrent inhibition via a Renshaw cell in the ventral horn of a mammal's spinal cord. The efferent axon of the motor neuron sends a recurrent collateral branch back towards the soma where it excites a Renshaw cell, which in turn inhibits the motor neuron.

recurrent collateral originated (recurrent inhibition). When the action potential sweeps down the axon towards the terminal in the muscle, it also travels in the recurrent collateral and excites the Renshaw cell. The Renshaw cell then inhibits the cell body. In other words, when the motor neuron is activated, it triggers a mechanism that results in its own inhibition. As in the case of the Mauthner cell's recurrent inhibition, the Renshaw cell inhibition results in the stimulation of the muscles being in the form of discrete pulses.

We have been unable to find any reports of Renshaw cells, recurrent collaterals, or recurrent inhibition (apart from the Mauthner cell) in any nonmammalian vertebrates. If indeed Renshaw cells are a uniquely mammalian phenomenon, they may be related specifically to direct telencephalic cortical control of motor neurons, which also seems to be a uniquely mammalian phenomenon.

Axon Columns and Cell Columns

Like the nontetrapod spinal cord, the tetrapod spinal cord consists of longitudinal columns of ascending and descending axons (the white matter) that surround a core of longitudinal columns of cells (the gray matter). Unlike nontetrapods, in which the gray matter shows considerable variation in shape, the spinal gray matter of many tetrapods shows a remarkable consistency in shape. The dorsal and ventral horns of the right and left side are connected by a small bridge of axons of gray matter that gives the gray matter the overall form of a letter H when viewed in transverse section. In the cervical and lumbar enlargements, however, the ventral horns extend laterally so that in these regions the gray matter takes on the shape of a butterfly. Also present in the thoracic and lumbar regions of mammals are small lateral extensions of the gray matter, just above the ventral horns; these are the lateral horns.

In general, the dorsal horns consist of the cell bodies of interneurons that are the sources of ascending (and some descending) axons of the white matter columns (extrasegmental axons) as well axons that remain within the spinal cord segment (intrasegmental axons). Among the dorsal column cells are those that are the sources of the long-axon pathways from the spinal cord to the brain, such as the spinocerebellar, spinotectal, and spinothalamic tracts. The ventral horns contain columns of motor neurons, the axons of which pass out of the spinal cord via the ventral roots. The most medial column consists of somatic motor neurons that terminate on the muscles of the body trunk; that is, the muscles of the back, the chest, the abdomen, and to some extent, the muscles of the neck. This medial cell column of the ventral horn (the ventromedial column) is present along the entire extent of the spinal cord. In the cervical and lumbar regions, the ventral horns expand laterally to accommodate the presence of the somatic motor neurons that send their axons to the muscles of the limbs. These lateral extensions are the ventrolateral columns.

Marginal Cells

Not all of the neurons of the spinal cord are located in the central cell columns. A small column of scattered cell bodies is located along the outer margin of the spinal cord in the lateral column of white matter. These cells, known as **marginal cells,** are found in mammals, diapsid reptiles, turtles, birds, and, among the nontetrapods, in lampreys. They form a continuous column along much of the spinal cord. These marginal cells have a number of characteristics in common with mechanoreceptors (cells that respond to mechanical bending or deformation). They are in contact with the **denticulate ligaments,** which are filaments of connective tissue that run lengthwise along the outside of the cord between the dorsal and ventral roots. These filaments are joined to the dura mater (the outermost and toughest of the series of membranes that envelop the central nervous system) by a series of tooth-like (and hence the name "denticulate") triangular attachments. The denticulate ligaments appear to provide mechanical support for the spinal cord. The anatomy and some physiological evidence suggest that the marginal cells are intraspinal mechanoreceptors that respond to flexing or bending of the spinal cord as the animal twists or bends its body. These cells may play a role in providing sensory feedback during rhythmic movements.

Ascending Spinal Pathways

In general, the dorsal horns contain the cells of origin of the ascending pathways that carry somatosensory information to the brain. For example, lamina I interneurons, in addition to some axons from laminae IV and V, are the sources of the **spinothalamic axons.** These axons carry information about light touch, pain, and temperature to the somatosensory region of the dorsal thalamus. These laminae also contain neurons that send their axons to the tectum of the midbrain, via the **spinotectal pathway.** The close association of this pathway with the spinothalamic pathway suggests that it may be carrying similar information to the tectum, which coordinates visual and auditory information with somatosensory information.

The interneurons of lamina VII, along with some cells of laminae V and VI, are the source of the spinocerebellar axons. Within lamina VII is a distinctive groups of cells, sometimes known as **Clarke's column,** that gives rise to the **dorsal spinocerebellar tract.** The axons of other lamina VII interneurons join axons originating in laminae V and VI to form the **ventral spinocerebellar tract.** Closely related to the spinocerebellar pathways are the **spino-olivary pathways** that terminate in the inferior olivary nuclei located at the junction of the medulla and pons. This complex of nuclear groups has reciprocal connections with the cerebellum as well as other components of the sensory and motor systems.

Lamina V interneurons are the spinal-cord component of the reticular formation. They participate in local reflexes and are the source of the spinoreticular pathway to the cell groups of the reticular formation in the medulla, pons, and midbrain tegmentum.

Not all ascending pathways originate from interneurons of the spinal cord. Some are made up of direct dorsal-root fibers. These axons are branches of incoming sensory neurons; one branch terminates near the segment of entry while the other enters the dorsal white-matter column and ascends to the caudal end of the medulla where it terminates in one of two groups of neurons, known as the **dorsal column nuclei.** In mammals the dorsal column is divided into two segments, a medial segment (**fasciculus gracilis**), which is present throughout the cord, and a lateral segment (**fasciculus cuneatus**), which is present only at the thoracic and cervical regions. The dorsal column nuclei are therefore sometimes known as the **nucleus gracilis** and the **nucleus cuneatus**. The cells of the dorsal column nuclei send their axons to the somatosensory region of the dorsal thalamus by a major ascending sensory pathway known as the **medial lemniscus**. The dorsal column–medial lemniscus system carries information about deep touch, pressure on the body surface, the ability to recognize the shape of an object that touches the body surface, the ability to discriminate the touch of one point from two, as well as kinesthesis (movement of body parts as indicated by muscle, tendon, and joint receptors).

Descending Spinal Pathways

The brain exerts its control over the interneurons and motor neurons of the spinal cord by means of a group of descending pathways. These pathways control voluntary and reflex movement and muscle tone as well as influencing visceral activity and the transmission of somatosensory information within the spinal cord.

Among the brain regions that control involuntary or reflex movements are the cerebellum and the vestibular nuclei. The vestibular nuclei receive information about gravity and acceleration from the semicircular canals of the middle ear. The cerebellum receives information from the vestibular system, from the spinal cord via the spinocerebellar pathway, and from the motor system of the cerebrum. Together with the cerebellum, the vestibular nuclei control muscle tone, balance of the body, and the coordination of the actions of the musculoskeletal system. The vestibular nuclei directly affect interneurons and somatic motor neurons of the spinal cord by means of the **vestibulospinal tract**. Vestibulospinal reflexes maintain body posture and balance as the body changes its rate of acceleration or suddenly encounters an uneven surface during locomotion. The cerebellum, which influences muscle tone and coordination of postural reflexes, has both direct and indirect links to the spinal cord. The direct input to the spinal cord is via the **cerebellospinal tract,** which originates in the **deep nuclei of the cerebellum.** An indirect route is via the vestibular nuclei. The major route of cerebellar influence on the spinal cord, however, is indirect via a specialized component of the midbrain reticular formation known as the **nucleus ruber** or the **red nucleus.** The efferents of the red nucleus are known as the **rubrospinal tract,** which is a pathway characteristic of animals that use their limbs for locomotion. Although they typically are found in land vertebrates and their descendants, a red nucleus and rubrospinal tract have been reported in a ray (*Raja clavata*) that uses its pectoral fins to "fly" through the water much as a bird flies through the air, as well as in squalomorph sharks, gars, lungfishes, and perhaps in ray-finned fishes as well.

Somatosensory cell groups of the brainstem also contribute to the descending spinal pathways. The dorsal column nuclei as well as those cell groups that receive somatosensory information from the head send their axons (**trigeminospinal**) to the spinal cord. These pathways permit body movements in response to touch stimuli on the head or body.

The control of voluntary movement is carried out by several descending pathways and the pattern of pathways varies among tetrapods. A common feature of these descending motor pathways is the **reticulospinal system.** The reticulospinal pathways are the main means by which the brain controls and coordinates voluntary movement of the body parts in most tetrapods. Pathways originating in motor regions of the telencephalon send their axons to the **lateral** and **medial reticular formation** of the caudal brainstem. The medial reticular formation and the lateral reticular formation are the sources of the reticulospinal tracts. These tracts terminate mainly on interneurons in laminae VII and VIII of the ventral horns and occasionally on motor neurons of lamina IX.

In addition to the reticulospinal pathways, mammals possess a direct pathway from the cerebral cortex of the telencephalon to the spinal cord, which is known as the **corticospinal tract.** This pathway leaves the telencephalon and passes through the brainstem to the caudal medulla where most of the axons decussate (cross the midline) and descend in the contralateral spinal cord; a smaller portion of these corticospinal axons remain ipsilateral. An exception to this rule may be seen in moles and hedgehogs, in which the corticospinal tract remains ipsilateral.

The axons of the corticospinal tract do not extend the full length of the spinal cord in all mammals. In marsupials, for example, the corticospinal axons extend no further than the cervical or anterior thoracic levels. However far they extend, these axons terminate mainly on interneurons of laminae VII and VIII of the ventral horn, although some terminate directly on motor neurons of lamina IX.

The decussation of the corticospinal tract often is regarded as the boundary between the caudal medulla and the rostral spinal cord. In the echidna or spiny anteater, which is a monotreme, and in some bats, the crossing occurs in the rostral medulla. In many mammals, such as primates, carnivores, and rabbits, the crossed corticospinal tract travels in the lateral columns of white matter and the uncrossed travels in the ventral white-matter columns. In other mammals, however, such as rodents, the corticospinal axons are found in the dorsal white-matter columns between the dorsal horns.

The corticospinal tract is relatively small in monotremes (the platypus and spiny anteater) and marsupials. It is somewhat better developed in placental mammals and is particularly well developed in those placentals that manipulate objects with the digits of their forepaws or hands such as rodents, raccoons, and primates.

The midbrain tectum, with its maps of visual and auditory space as well as a body map, is the structure responsible for the involuntary, reflex turning of the head in the direction of the source of a stimulus as happens when a sudden and unexpected visual or auditory stimulus occurs. Efferent axons of the tectum form the **tectospinal tract,** which ends on interneurons of the ventral horn in the cervical region. These interneurons in turn activate motor neurons that control the appropriate head and neck muscles to carry out this orientation response.

Just as the somatic-motor system is controlled by the brain, the visceral-motor or autonomic system is also under control of the brain. Descending axons from the hypothalamus (**hypothalamospinal tract**) and visceral cell groups (**solitariospinal tract**) in the medulla terminate on the preganglionic neurons of the sympathetic division of the autonomic nervous system that are located in the intermediolateral column or lateral horn of the spinal cord.

Tetrapod Central Pattern Generators

The execution of the complex patterns of contraction and relaxation of limb and trunk muscles that produce various locomotor patterns and other rhythmic movements of body parts are under the control of central pattern generators, which are interneuron networks located within the spinal cord, mainly in laminae VII and VIII, and in the brainstem. These networks coordinate and synchronize the actions of motor neurons on the left and right sides of the spinal cord to produce a smooth, locomotor progression of the body across the land, through the water, or through the air. Initiation of the actions of the central pattern generators of the spinal cord is controlled by interneuron networks located in the lateral reticular formation of the caudal brainstem and known as **command generators.** These command generators influence the central pattern generators of the spinal cord by means of reticulospinal pathways.

EVOLUTIONARY PERSPECTIVE

The spinal cord has remained relatively stable in its evolution. Its basic functions of gathering sensory information about touch, deformation of body surfaces, pain, and temperature from the external and internal environments, and executing reflex actions in response to this sensory information, have remained relatively unchanged throughout vertebrate history. What has varied in spinal cord evolution has been the shift from body undulation and tail propulsion (with some assistance from fins) as the means of locomotion in aquatic animals to rhythmic movements of limbs for locomotion in tetrapods. These changes are seen in the development of regional concentrations of sensory neurons, motor neurons, and interneurons to serve these organs and their functions, the development of the cell columns that give each spinal segment its laminar organization, and the development of somatotopic organization within these laminae. In concert with these changes, an increased communication between the spinal cord and the hindbrain and forebrain is reflected in the increasing development of long ascending and descending pathways not only for control of rhythmic movements of limbs, but also for the voluntary initiation of movement patterns. When flight developed in diapsid reptiles, mammals, and birds, hindlimb locomotion became separated from forelimb locomotion. The latter were used almost exclusively for flight, whereas the former were used only for walking, hanging, clinging, and limited manipulation of objects. This reduced coordination between forelimbs and hindlimbs is reflected in the overall form of the spinal cord. In mammals, with their ability for more adroit use of the digits of the limbs, manipulation of objects in the environment is an important aspect of their survival. The ability to dig, to pick up and displace objects, to manipulate food objects as they are eaten, and in other ways to modify their environment, is reflected in the greater development of long descending pathways from the cerebrum to the motor-neuron groups that control the limbs and digits. Similar pathways may be seen in birds, which also manipulate objects with their hindlimbs (although their

primary organ for manipulation is the bill), and to a lesser extent in reptiles.

FOR FURTHER READING

Ariëns Kappers, C. U., Huber G. C., and Crosby, E. C. (1960) *The Comparative Anatomy of the Nervous System of Vertebrates, Including Man.* New York: Hafner (reprint of 1936 edition).

Carroll, R. L. (1988) *Vertebrate Paleontology and Evolution.* New York: Freeman.

Cohen, A. H., Rossignol, S., and Grillner, S. (eds.), *Neural Control of Rhythmic Movements in Vertebrates.* New York: Wiley.

Dowling, J. E. (1992) *Neurons and Networks.* Cambridge, MA: Belknap/Harvard.

Faber, D. S. and Korn, H. (eds.) (1978) *Neurobiology of the Mauthner Cell.* New York: Raven.

Fetcho, J. R. (1987) A review of the organization and evolution of motoneurons innervating the axial musculature of vertebrates. *Brain Research Reviews,* 12, 243–280.

Kuypers, H. G. J. M. and Martin, G. F. (eds.) (1982) *Anatomy of Descending Pathways to the Spinal Cord. Progress in Brain Research,* Vol. 57. Amsterdam: Elsevier.

Northcutt, R. G. (1984) Evolution of the vertebrate central nervous system: patterns and processes. *American Zoologist,* 24, 701–716

Nudo, R. J. and Masterton, R. B. (1988) Descending pathways to the spinal cord: a comparative study of 22 mammals. *Journal of Comparative Neurology,* 277, 53–79.

Pearson, K. (1976) The control of walking. *Scientific American,* 235, 72–86.

Peterson, E. H. (1989) Motor pool organization of vertebrate axial muscles. *American Zoologist,* 29, 123–127.

Rexed, B. (1952) The cytoarchitectonic organization of the spinal cord in the cat. *Journal of Comparative Neurology,* 96, 415–496.

Wall, P. D. (1978) The gate control theory of pain mechanism: a reexamination and restatement. *Brain,* 101, 1–18.

ADDITIONAL REFERENCES

Armand, J. (1982) The origin, course and terminations of corticospinal fibers in various mammals. In H. G. J. M Kuypers and G. F. Martin (eds.), *Anatomy of Descending Pathways to the Spinal Cord. Progress in Brain Research,* Vol. 57. Amsterdam: Elsevier.

Bass, A. H. and Marchaterre, M. A. (1989) Sound-generating (sonic) motor system in teleost fish (*Porichthys notatus*): sexual polymorphisms and general synaptology of sonic motor nucleus. *Journal of Comparative Neurology,* 286, 154–169.

Butler, A. B. and Bruce, L. L. (1981) Nucleus laminaris of the torus semicircularis: projection to the spinal cord in reptiles. *Neuroscience Letters.* 25, 221–225.

Christianson, J. and Grillner, S. (1991) Primary afferents evoke excitatory amino acid receptor-mediated EPSPs that are modified by presynaptic GABA receptors in lamprey. *Journal of Neurophysiology,* 66, 2141–2149.

Cohen, A. H. (1988) Evolution of vertebrate central pattern generator for locomotion. In A. H. Cohen, S. Rossignol, and S. Grillner, (eds.), *Neural Control of Rhythmic Movements in Vertebrates.* New York: Wiley, pp. 129–166.

De Gennaro, L. D. and Benzo, C. A. (1991) Development of the glycogen body of the Japanese quail, *Coturnix japonica*: II. Observations of electron microscopy. *Journal of Morphology*, 207, 191–199.

Ebbesson, S. O. E. and Hodde, K. C. (1981) Ascending spinal systems in the nurse shark, *Ginglymostoma cirratum*. *Cell and Tissue Research*, 21, 313–331.

Faber, D. S., Korn, H., and Lin, J. W. (1991) Role of medullary networks and postsynaptic membrane properties in regulating Mauthner cell responsiveness to sensory excitation. *Brain, Behavior and Evolution*, 37, 286–297.

Fetcho, J. R. and Faber, D. S. (1988) Identification of motoneurons and interneurons in the spinal network for escapes initiated by the Mauthner cell in goldfish. *Journal of Neuroscience*, 8, 4192–4213.

Fetcho, J. R. Spinal network of the Mauthner cell (1991) *Brain, Behavior and Evolution*, 37, 298–316.

Gelfand, I. M., Orlovsky, G. N., and Shick, M. L. (1988) Locomotion and scratching in tetrapods. In A. H. Cohen, S. Rossignol and S. Grillner (eds.), *Neural Control of Rhythmic Movements in Invertebrates*. New York: Wiley, pp. 167–199.

Kasicki, S. and Grillner, S. (1986) Müller cells and other reticulospinal neurones are phasically active during fictive locomotion in the isolated nervous system of the lamprey. *Neuroscience Letters*, 69, 239–243.

Kremers, J.-W. P. M. and Nieuwenhuys, R. (1979) Topological analysis of the brain stem of the crossopterygian fish *Latimeria chalumnae*. *Journal of Comparative Neurology*, 187, 613–637.

Kuypers, H. G. J. M. (1982) A new look at the organization of the motor system. In H. G. J. M. Kuypers and G. F. Martin (eds.), *Anatomy of Descending Pathways to the Spinal Cord. Progress in Brain Research*, Vol. 57. Amsterdam: Elsevier, pp. 381–403.

Ladich, F. and Fine, M. L. (1992) Localization of pectoral fin motoneurons (sonic and hovering) in the croaking gourami *Trichopsis vittatus*. *Brain, Behavior and Evolution*, 39, 1–7.

Matsushita, M. and Hosoya, Y. (1978) The location of spinal projection neurons in the cerebellar nuclei (cerebellospinal tract neurons) of the cat. A study with the horseradish peroxidase technique. *Brain Research*, 142, 237–248.

Necker, R. (1989) Cells of origin of the spinothalamic, spinoreticular and spinocerebellar pathways as studied by the retrograde transport of horseradish peroxidase. *Journal für Hirnforschung*, 30, 33–43.

Nudo, R. J. and Masterton, R. B. (1989) Descending pathways to the spinal cord: II. Quantitative study of the tectospinal tract in 23 mammals. *Journal of Comparative Neurology*, 286, 96–119

Nudo, R. J. and Masterton, R. B. (1990) Descending pathways to the spinal cord, III. Sites of origin of the corticospinal tract. *Journal of Comparative Neurology*, 296, 559–583.

Pritz, M. B. and Stritzel, M. E. (1989) Reptilian somatosensory midbrain: identification based on input from the spinal cord and dorsal column nucleus. *Brain, Behavior and Evolution*, 33, 1–14.

Rhoades, R. W. (1981) Cortical and spinal somatosensory input to the superior colliculus in the golden hamster: an anatomical and electrophysiological study. *Journal of Comparative Neurology*, 153, 415–432.

Ronan, M. (1989) Origins of descending spinal projections in petromyzontid and myxinoid aganthans. *Journal of Comparative Neurology*, 281, 54–68.

Ronan, M. and Northcutt, R. G. (1990) Projections ascending from the spinal cord to the brain in petromyzontid and myxinoid aganthans. *Journal of Comparative Neurology*, 291, 491–508.

Rovainen, C. M. (1967) Physiological and anatomical studies of the large neurons of the central nervous system of the sea lamprey (*Petromyzon marinus*). I. Müller and Mauthner's Cells. *Journal of Neurophysiology*, 30, 1000–1023.

Schroeder, D. M. and Egar, M. W. (1990) Marginal neurons in the urodele spinal cord and the associated denticulate ligaments. *Journal of Comparative Neurology*, 301, 93–103.

Szabo, T., Libouban, S., and Denizot, J. P. (1990). A well defined spinocerebellar system in the weakly electric teleost fish *Gnathonemus petersii*. A tracing and immunohistochemical study. *Archives of Italian Biology*, 128, 229–247.

Takahashi, O., Satoda, T., Matsushima, R., Uemura-Sumi, M., and Mizuno, N. (1987) Distribution of cerebellar neurons projecting directly to the spinal cord: an HRP study in the Japanese monkey and the cat. *Journal für Hirnforschung*, (28) 105–113.

ten Donkelaar, H. J. (1976) Descending pathways from the brain stem to the spinal cord in some reptiles. *Journal of Comparative Neurology*, 167, 421–442.

ten Donkelaar, H. J. (1982) Organization of descending pathways to the spinal cord in amphibians and reptiles. In H. G. J. M. Kuypers and G. F. Martin (eds.), *Anatomy of Descending Pathways to the Spinal Cord. Progress in Brain Research*, Vol. 57. Amsterdam: Elsevier, pp. 28–67.

Trujillo-Cenoz, O., Echague, J. A., Bertolotto, C., and Lorenzo, D. (1986) Some aspects of the structural organization of the spinal cord of *Gymnotus carapo* (Teleostei, Gymnotoformes). I. The electromotor neurons. *Journal of Ultrastructure and Molecular Structural Research*, 97, 130–143.

Uehara, M. and Ueshima, T. (1986) Morphological studies of the spinal cord in tetraodontiformes fishes. *Journal of Morphology*, 190, 325–333.

Verburgh, C. A., Voogd, J., Kuypers, H. G., and Stevens, P. (1990) Propriospinal neurons with ascending collaterals to the dorsal medulla, the thalamus and the tectum: a retrograde fluorescent double-labelling study of the cervical cord in the rat. *Experimental Brain Research*, 80, 577–590.

Webster, D. M. and Steeves, J. D. (1991) Funicular organization of avian brainstem-spinal projections. *Journal of Comparative Neurology*, 312, 467–476.

Wilczynski, W. and Northcutt, R. G. (1977) Afferents to the optic tectum of the leopard frog: an HRP study. *Journal of Comparative Neurology*, 173, 219–230.

Will, U. (1991) Amphibian Mauthner cells. *Brain, Behavior and Evolution*, 37, 317–332.

Yasargil, G. M. and Sandri, C. (1987) Morphology of the Mauthner axon inhibitory system in tench (*Tinca tinca* L.) spinal cord. *Neuroscience Letters*, 81, 63–68.

9

Segmental Organization of the Head, Brain, and Cranial Nerves

"TWELVE" CRANIAL NERVES

Traditionally, 12 cranial nerves have been recognized, and the list of them, numbered with Roman numerals, has been taught to many generations of students. The nerves have been named according to various attributes, such as their appearance (vagus = wanderer), the region of innervation (facial), or their function (oculomotor = moving the eye). These nerves have been classified according to their function (sensory or motor) and their direction of conduction with respect to the brain (afferent, meaning towards, or efferent, meaning away from). The nerves were also classified according to the types of tissue that they were believed to innervate, either somatic or visceral. The branchiomeric muscles, innervated by some of the cranial nerves, were erroneously believed to be visceral in nature, as were some of the sense organs, such as those of taste and smell. Other senses were classified as "special somatic" since they are not as directly related to the gut as taste and smell appeared to be. The "special somatic" senses included vision (optic nerve), balance and hearing (vestibular and cochlear nerves), and mechanoreception and electroreception (lateral line nerves). Unfortunately, the embryological development of the brain and the cranial nerves was not correctly understood when the nerves were so classified, and hence the traditional classification schemes are to a greater or lesser degree misleading.

A list of the traditionally recognized 12 cranial nerves is shown in Table 9-1. This list is useful for the study of human neuroanatomy, for which is was devised, but it is in adequate for the comparative neuroanatomy of all the vertebrate classes. There are, in fact, many more than 12 cranial nerves. The termi-nology used in this text for all of the cranial nerves preserves the original Roman numeral designations of the "traditional" 12. These designations are so firmly established in the literature and so universally used that changing to a new set of numbers would be counterproductive.

SEGMENTAL DEVELOPMENT OF THE VERTEBRATE BRAIN

The central nervous system, in spite of the complexity of its structural components, is basically a hollow tube, the lumen of which is the central canal and the brain ventricles. During embryonic development, the neural tube is formed by the fold-ing of the neural plate, as discussed in Chapter 3. In the rostral part of the neural tube, which will eventually become the brain, a rostrocaudal series of bulges, called **neuromeres,** then devel-ops. The neuromeres, which are serial brain segments, were identified in the early 1900s, and formed the basis of a **neuro-meric theory.** This theory postulated that the neuromeres are specified by genetic fate determinants and constitute longitudi-nal and transverse segmental divisions, or compartments, of the neural tube that have independent and individual develop-mental fates.

In the hindbrain (rhombencephalon), the neuromeres are called **rhombomeres** (Fig. 9-1); at least seven rhombomeres are present, and they are delineated by a series of transverse grooves on the external and internal surfaces of the neural tube. In the midbrain (mesencephalon) and forebrain (prosencepha-lon), a series of transverse grooves is not readily apparent, but neuromeres are present there as well (Fig. 9-2). The neuromeres

TABLE 9-1. The Traditional Twelve Cranial Nerves			
Symbol	Name	Innervation
I	Olfactory	Olfactory epithelium
II	Optic	Retina
III	Oculomotor	Internal and external eye muscles
IV	Trochlear	External eye muscles
V	Trigeminal	Jaw muscles; touch to face, snout and oral cavity
VI	Abducens	External eye muscles
VII	Facial	Taste buds; facial muscles; salivary and tear glands
VIII	Vestibulocochlear	Cochlea, vestibular organs
IX	Glossopharyngeal	Taste buds; pharnyx; salivary glands
X	Vagus	Taste buds; viscera of thorax and abdomen; larynx; pharynx
XI	Spinal accessory	Neck and shoulder muscles
XII	Hypoglossal	Tongue

in the midbrain are called **mesomeres** and those in the forebrain **prosomeres.** To date, six prosomeres (p1–p6) and two mesomeres (m1 and m2) have been recognized.

The individual segmental specification of the rhombomeres has recently been correlated with the expression patterns of certain regulatory genes. These regulatory genes are involved in specifying anterior–posterior axial patterning in the developing embryo. Over the past decade, a new area of research on such regulatory genes has blossomed. One group of these genes are referred to as **homeobox-containing genes** (or **Hox** genes), because they contain a specific sequence of DNA nucleotide base pairs called the **homeobox.** Homeobox sequences are widely distributed in vertebrate genes, and are homologues of genes identified in the fruit fly *Drosophila* and various other invertebrates. Homeobox genes control segmental development by their role as DNA-binding transcriptional regulators, thus affecting the expression of a number of other genes involved in the developmental process. Similar regulatory genes with a variety of names, such as the "zinc-finger" gene (*Krox-20*) and **paired box** (**Pax**) genes, are also involved in the pattern specification of the brainstem.

A different and unique combination of homeobox-containing genes and other regulatory genes are expressed in each of the rhombomeres. In the upper part of Figure 9-1, the extent of each black bar indicates the part of the brainstem in which activity for the particular gene listed at the left has been detected. For example, the *Hox 2.8* gene is active in a long

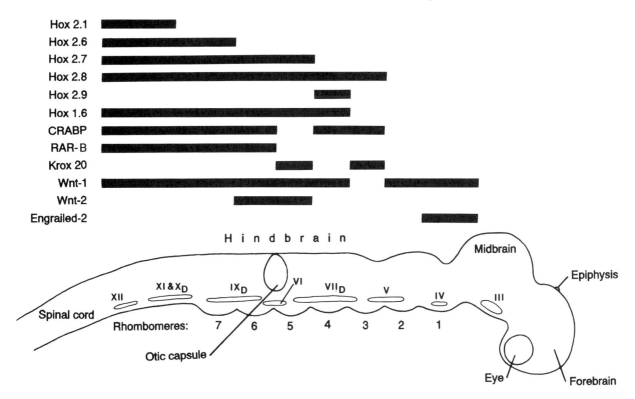

FIGURE 9-1. Schematic sagittal section through the developing chick brain, showing the expression patterns (indicated by the bars) of homeobox-containing and similar regulatory genes in the hindbrain and midbrain and the location of the populations of neurons for a number of the cranial nerve nuclei. The latter are indicated by Roman numerals, consistent with the terminology and classification of cranial nerves discussed in this chapter. Rhombomeres are indicated by Arabic numbers 1–7. Adapted from Noden (1991). Used with permission of S. Karger AG.

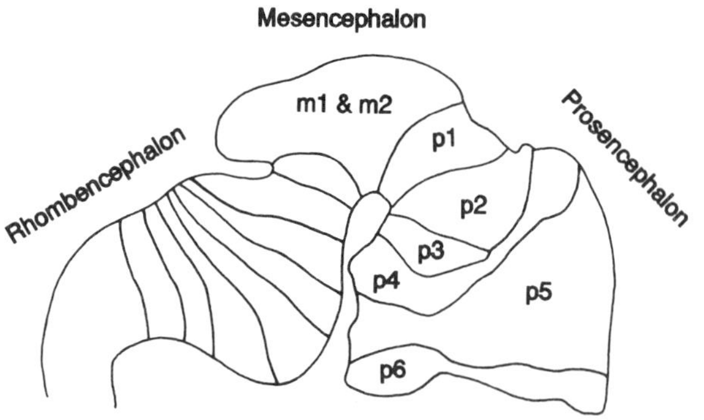

Mesencephalon

FIGURE 9-2. Drawing of a schematic lateral view of the developing brain of a chick embryo, with the segmental neuromeric divisions projected onto it. The prosomeres are identified as p1–p6 and the two mesomeres as m1 and m2. Adapted from Puelles and Rubenstein (1993). Used with permission of Elsevier.

area extending from the 2–3 rhombomere boundary caudally through the caudal-most part of the brainstem, whereas the *Krox-20* gene has two discrete areas of activity, one in rhombomere 3 and the other in rhombomere 5. The total pattern of gene activity thus forms a bar code or fingerprint-like profile that is unique for each individual rhombomere.

Whether the combined activity of these genes determines the specific identity of each rhombomere, the cranial nerve nuclei within the rhombomere, or both has not yet been established, however. During embryological development, the neuron cell bodies that form the motor nuclei of cranial nerves, which lie within the **basal plate** (the lower part of the tube) of adjacent pairs of neuromeres, migrate caudally within the neural tube to reach their final rhombomeric positions. Motor neurons of the seventh, ninth, and tenth cranial nerves and some of the neurons of the fifth and lateral line cranial nerves migrate over distances of one to three rhombomeres. How the activity of the regulatory genes specifies the various pools of motor nerve neurons in relation to the rhombomeric segments is thus unclear.

In the midbrain and forebrain, a number of similar, putative regulatory genes also have been identified recently that appear to specify the mesomeres and prosomeres. These include genes referred to as the *engrailed* homologue, *En-2,* as well as *Krox-20* in the midbrain (Fig. 9-2) and as *Wnt-1, Wnt-3, Wnt-3A, Wnt-4, Dix-1, Dix-2, Nkx-2.2,* and so on, in the forebrain. The same genes also have been discovered in a variety of invertebrates and vertebrates, and their names often reflect particular traits, such as the "zinc-finger" regulatory gene of the hindbrain mentioned above.

THE VERTEBRATE HEAD: SEGMENTAL ORGANIZATION

Understanding the organization of the cranial nerves depends on an appreciation of the segmental nature of the brain and head and of the embryonic derivation of the nerves, their ganglia, and the structures that they innervate. Questions that have been debated for a long time are whether or not the

brain and other tissues of the head are in fact developed in a segmental fashion and, if so, which structures are within each segment. Although neuromeres have yet to be identified in the developing spinal cord, the spinal cord in adult vertebrates clearly is a segmented structure. Each segment of the spinal cord has dorsal and ventral pairs of spinal nerves that innervate each successive segment of the body, but the segmental organization of the brain has not been as easily understood.

Recent advances from studies of the embryology of the head, using modern transplant and marking techniques, have led to a new understanding of the development of the head and brain in vertebrates through an understanding of the roles of two ectodermal (outer body layer) derivatives, called **neural crest** and **placodes,** and the role of mesoderm (middle body layer) in this process. In addition to the process of segmentation of neuromeres within the neural tube itself, at least some of these other tissues interact in a segmental fashion during the development of the brain and sensory organs. To approach this topic, first we need to review briefly some of the components of the skeleton of the head and then consider their relationship to the striated musculature, the neural crest, and placodes.

Gills and Arches

The skeleton of the head in vertebrates contains three components: the **chondrocranium,** which encloses the special sense organs and forms part of the braincase; the **splanchnocranium** (or **visceral arches**), which supports and moves the gills and, in jawed vertebrates, contributes to the jaws, and the **dermatocranium** (or **dermal bones**), which contributes to the braincase and the jaws. In vertebrates that have gills, the first step in the development of the gills is the formation of a paired series of pouches that arise as evaginations (out-growths) from the pharynx in the throat region. The visceral arches are a series of cartilaginous or bony arches that lie between the pharyngeal pouches. The cartilage of the visceral arches is mostly derived from neural crest tissue, as we will discuss below.

The first visceral arch, which is the **mandibular arch** in jawed vertebrates, lies rostral to the first pharyngeal pouch, and the second visceral arch, or **hyoid arch,** lies between the first and second pharyngeal pouches. Most jawed fishes have five additional visceral arches, which are called the first through the fifth **branchial arches.** The term branchial refers to the gills. In cartilaginous fishes, the mandibular arch consists of a dorsal palatoquadrate cartilage and a lower mandibular cartilage that form the upper and lower jaws, respectively. In bony fishes and tetrapods, the dermatocranium contributes substantially to the bony part of the jaws; remnants of the palatoquadrate and mandibular cartilages are retained in bony fishes, however, and the cartilaginous articulation of the jaws in non-mammalian tetrapods is derived from the mandibular arch. The dorsal part of the hyoid arch, called the hyomandibular cartilage, suspends the palatoquadrate cartilage in bony fishes, and its ventral part extends ventrally into the mouth and pharynx. In tetrapods, the hyomandibular forms one of the auditory ossicles, while most of the branchial arches are incorporated into the various cartilages of the throat (hyoid, thyroid, cricoid, etc.).

The Striated Musculature of the Head

One of the major breakthroughs in understanding the development of the head and central nervous system was the recent identification of the embryonic source of all of the striated muscles in the head. In the body, a part of the mesoderm called the **hypomere** (or **lateral plate mesoderm**) gives rise to the nonstriated, visceral muscles and connective tissues of the gut, while more dorsal mesoderm, called the **epimere** (or **paraxial mesoderm**) forms **somites** (segmental mesodermal divisions) that give rise to the striated muscles of the body (Fig. 9-3). The **branchiomeric muscles** (associated with the gills or throat cartilages) within the head were long thought to be derived from a rostral extension of the lateral plate mesoderm and therefore of visceral origin, and the cranial nerves supplying them were therefore classified as visceral motor nerves. Only the extrinsic muscles of the eye and the muscles of the tongue were thought to be somatic muscles, that is, derived from the embryonic paraxial mesoderm. The fact that the supposedly visceral, branchiomeric muscles are striated, like the eye muscles and the tongue, was an unresolved problem and ignored by most. These beliefs, which were the basis of the venerable classification of cranial nerves that is widely taught today, are, however, not supported by contemporary embryology.

Interspecies transplantation studies and cell lineage tracing studies have allowed a reexamination of the fate of the mesoderm within the head, revealing a rather different story. In the developing head, the paraxial mesoderm forms a longitudinal, segmented series of slightly elevated bulges called **somitomeres** (Fig. 9-4). The more ventrally lying, lateral plate mesoderm does not contribute to somitomere formation, so the somitomeres have no component that could be argued to be developmentally visceral. The somitomeres are delineated by a series of shallow depressions, rather than being completely divided into separate masses like the somites of the body. These somitomeres give rise to all of the striated muscles of the head (Fig. 9-4), just as the somites of the epimere in the body give rise to the striated muscles of the trunk and limbs. Thus, the voluntary, striated muscles in the head, which are innervated by cranial motor nerves, are all somatic muscles, and thus, their motor nerves are all somatic motor nerves. The lateral mesoderm of the head only contains precursors of cells that make a minor contribution to some of the throat cartilages and neighboring bone.

Neural Crest and Placodes

Studies of the movement and fate of neural crest cells and other populations of ectodermally derived cells have also revealed new findings about the development of the brain and head. Neural crest cells arise within the neural fold tissue lateral to the neural tube. These cells are segmentally specified at their point of origin by the pattern of expression of *Hox* regulatory genes, and they carry their segmental identity as they migrate ventrally between the paraxial mesoderm and the surface ectoderm (Fig. 9-5). The neural crest gives rise to many structures throughout the body, including nerves and sensory ganglia in the trunk, chromaffin cells, which secrete adrenalin, in the cortex of the adrenal gland, all pigment cells, and the Schwann cells that supply the myelin sheathing around peripheral nerves. In the head, neural crest cells migrate between the pharyngeal pouches. These cells form the cartilage of the branchial arches and the smooth muscle of the aortic arches. Within the rest of the head, neural crest gives rise to the anterior part of the neurocranium and to the meninges, the connective tissue layers that cover the brain. It also gives rise to a number of the cranial nerves and their sensory ganglia.

Parts of the surface ectoderm also contribute to the formation of cranial nerves and sensory ganglia. Patches of neurogenic tissue form within the ectoderm. These patches are called placodes, and they give rise to additional components of the nervous system. Neural crest cells migrate beneath the placodes, and the placodes then begin to produce neuroblasts (nerve cell precursors). A dorsolateral series, a ventrolateral series, and additional unclassified sets of placodes form the sensory receptors and ganglia of some of the cranial nerves. While neural crest occurs throughout the head and body, placodes are present only in the head.

Segmentation of the Head

The head in vertebrates differs from the rest of the body in two significant ways: lateral plate mesoderm is present but makes only modest contributions to cartilage and bone rather than giving rise to muscle, and placodes are present. The contributions of placodes and neural crest result in the formation of the paired sense organs and the anterior part of the neurocranium, that is, the rostral part of the head, which is thus new in vertebrates. This formation of a new part of the head due to the presence of placodes and neural crest and its implications for the origin of vertebrates was recognized by Carl Gans and R. Glenn Northcutt in 1985.

Like the body, the head is segmented, that is, composed of repeating units. These units may or may not be serially homologous to the segments of the spinal cord, but they exhibit organizational similarities. The segmental units include the expression patterns of regulatory genes, morphological neuroepithelial units (neuromeres), populations of motor neurons for cranial nerves, morphological paraxial mesodermal elements (somitomeres), and streams of segmentally specified neural

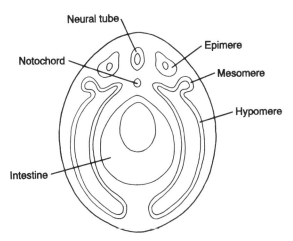

FIGURE 9-3. Schematic transverse section through a vertebrate embryo to show the organization of the epimeric and hypomeric parts of the mesoderm in the body.

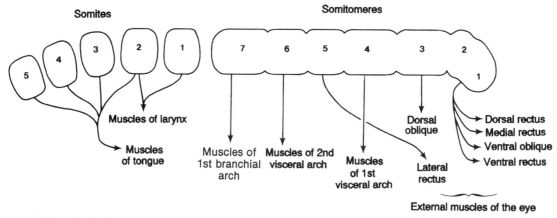

FIGURE 9-4. Schematic sagittal section through the epimeric mesoderm in the head and the muscles that each part of it gives rise to. Adapted from Noden (1991). Used with permission of S. Karger AG.

crest cells. Some of the cranial nerves are organized like segmented, spinal nerves. Other cranial nerves, including those derived from placodes, are unique to the head.

In creating a model of head segmentation, a simple, 1 : 1 correspondence among specific neuromeres, somitomeres, and visceral arches has not yet been defined across vertebrates. The series of rhombomeres that form the hindbrain is highly conserved among vertebrates, but variation occurs in the loca-

tion of some of the cranial nerve motor nuclei (Fig. 9-6). The series of somitomeres is also highly conserved, but a precise, topographic registration of somitomeres with neuromeres has yet to be confirmed.

Figure 9-7 shows a structural comparison among neurula stages of developing embryos of *Branchiostoma* (*Amphioxus*), a lamprey, a shark, and a chick. The neurula stage is reached with the first development of the nervous system following

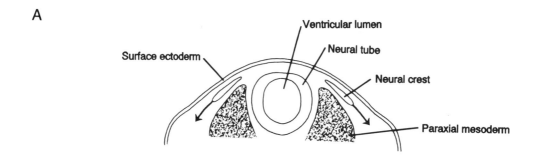

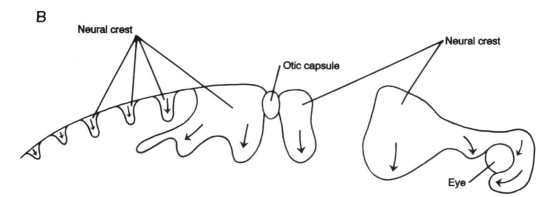

FIGURE 9-5. (A) Schematic drawing of a transverse section through the developing neural tube in the dorsal part of a vertebrate embryo, with arrows indicating the direction of migration of the neural crest. (B) A drawing to show the migration of the neural crest as seen in a parasagittal section. Adapted from Noden (1991) and used with permission of S. Karger AG.

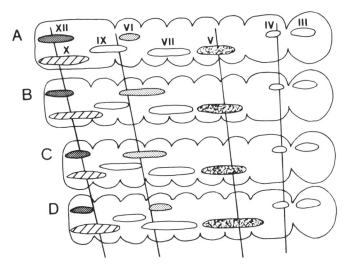

FIGURE 9-6. Schematic drawing to show the relationships between neuromeres and cranial nerve motor nuclei in a variety of vertebrate groups. Shadings and lines are used to facilitate comparisons of the relative positions and extent of the cranial nerve nuclei, which are indicated by their Roman numerals. (A) Elasmobranch, (B) teleost fish, (C) nonmammalian amniote, and (D) mammal. Adapted from Gilland and Baker (1992). Used with permission of S. Karger AG.

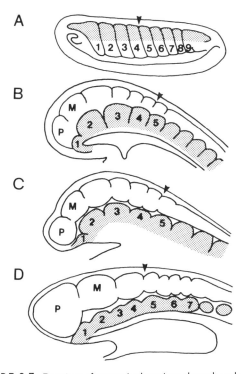

FIGURE 9-7. Drawings of parasagittal sections through early developmental stages of *Branchiostoma* (A) and three vertebrates—a lamprey (B), a shark (C), and a chick (D)—to show the comparative alignment of neuromeres, somitomeres (indicated by light shading), and the rostral limit of expression of the *Hox-3* (H3) regulatory gene (arrowheads). Comparable somitomeres are indicated by numbers, the prosencephalon by P and the midbrain by M. Rhombomeres are unnumbered. Adapted from Gilland and Baker (1992). Used with permission of S. Karger AG.

gastrulation. The dashed lines indicate homologous regions, showing that neuromeric units and somitomeric units do not have the same 1:1 relationship in all cases; nonetheless, a similar rostral to caudal sequence is maintained.

Two neurobiologists, E. Gilland and R. Baker, published a paper in 1992, which provides a new perspective for the evolution of the vertebrate head from this comparison. The rostral expression border of the *Hox-3* (H3) regulatory gene is indicated in each species shown in Figure 9-7. The position of H3 has been demonstrated experimentally in *Branchiostoma* and the chick but is theoretical for the lamprey and the shark. As Gilland and Baker discuss, the position of H3 in *Branchiostoma* implies that the first five somites of *Branchiostoma* are homologous as a field to the somitomeric region of jawed vertebrates. They argue that if this hypothesized homologous relationship is supported by further data, such as additional comparative information on homeobox gene expression patterns, then the cranial region of amniotes can be compared with the entire primary gastrula of *Branchiostoma*. In other words, the head of vertebrates may be the homologue of the primary gastrula of *Branchiostoma*.

THEORETICAL HEAD SEGMENTS

A head segment (or the concept of a head segment) encompasses elements of each of the segmentally organized tissues in the vertebrate head. Thus, a model head segment would contain a rostrocaudally defined portion of the neuromeres, at least some of the cranial nerve neurons, somitomeres, neural crest, visceral arches, and related tissues. Placodes may be influenced by segmentally specified neural crest but may not in themselves be segmentally organized. Since the rostrocaudal alignment of the various segmental tissues varies among different groups of vertebrates, the definitive head segment cannot yet be identified. Nonetheless, we can examine the organization of the cranial nerves with a theoretical model of head segments.

The overall organizational pattern that emerges reveals the presence of up to four separately derived nerves—**dorsal, ventral, dorsolateral,** and **ventrolateral**—for each side of each head segment. All four nerves are not present in all head segments, but we have illustrated all four schematically in Figure 9-8 to make the segmental organizational pattern clear. Figure 9-8 models a pair of somitomeres for each head segment. The more rostral somitomere of each pair gives rise to one or more of the muscles of the eye or to the muscles of the tongue. The more caudal somitomere of each pair, except the first, gives rise to branchiomeric muscles.

A ventral nerve for a given model segment is derived from cells in the ventral part of the neural tube. This nerve is the somatic motor nerve for the muscle(s) derived from the more rostral of the two somitomeres. This arrangement appears to correspond to that seen in the spinal cord, with the ventral nerve carrying somatic motor fibers.

In addition, a dorsal nerve for the segment is derived from the neural tube and is the somatic motor nerve for the branchio-

meric muscles(s) that are derived from the more caudal of the two somitomeres. In contrast to the ventral nerve, the dorsal nerve also has axons from cells derived from neural crest, which are sensory axons. This sensory component may correspond to that seen in the dorsal axons of the spinal cord. However, the presence of the somatic motor fibers in this dorsal nerve is unlike the condition of the dorsal spinal nerves, which are purely sensory.

The sensory ganglion cells of two additional nerves in the segment may be derived from placodes. Additional research is needed to clarify the origin from placodes and/or neural crest of a number of cranial nerve sensory ganglia. Further work is needed to correlate the embryological derivation of specific populations of sensory neurons with the specific identity of the sensory receptors that they innervate. The dorsolateral nerve of a model segment is known to be derived from the dorsolateral series of placodes. In the third head segment, multiple dorsolateral nerves are in fact present, and the placodes give rise to the receptors and sensory ganglion cells of the auditory, vestibular, and lateral line systems. A ventrolateral nerve for the segment can be modeled, subject to further research findings, as deriving in the hindbrain from the ventrolateral (or epibran-

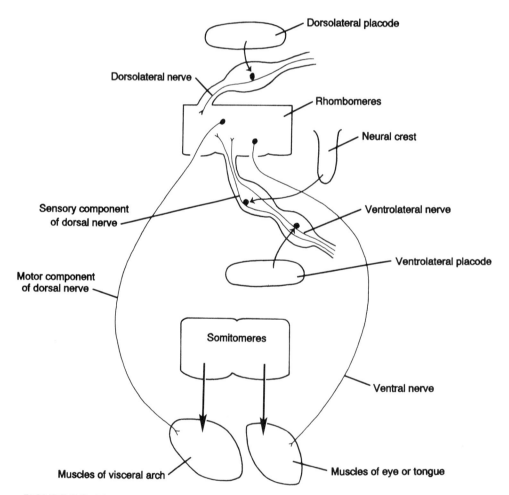

FIGURE 9-8. Schematic drawing of a model head segment to show the organizational plan of the cranial nerves. Rostral is toward the right. The embryological derivation of ganglion cells for a number of the cranial nerves has yet to be confirmed.

chial) series of placodes, which are thought to give rise to the sensory ganglion cells for the sense of taste, and in the telencephalon from more rostral placodes for other cranial nerves.

The most rostral cranial nerves are associated with the forebrain, which lies rostral to the first model head segment. Some forebrain cranial nerves are derived from placodes, which we have somewhat arbitrarily assigned to the ventrolateral category, but no dorsal, ventral, or dorsolateral nerves are present in the forebrain. Additionally, the forebrain has two unique nerves, derived from evaginations of the brain itself: the optic and the epiphyseal nerves.

From the above description, it should be apparent that there is a general scheme of organization of the cranial nerves. For a model head segment (Fig. 9-8), up to four separate cranial nerves are present: (1) a dorsal nerve, with a component derived from the neural tube that innervates branchiomeric muscles and a component derived from neural crest cells that is sensory; (2) a ventral nerve derived from the neural tube that is motor; (3) a sensory nerve derived from dorsolateral placodes; and (4) a sensory nerve derived from ventrolateral placodes and/or neural crest. As discussed above, the modeled embryological derivation of some of these components may be modified as further data become available. Variation occurs in the presence of some of the nerves among different model head segments, but the differential patterns of organization of the cranial nerves in each segment can be understood in terms of the basic theme. The model segments of the head thus resemble the acutal segments of the spinal cord in having dorsal and ventral nerves, but model head segments additionally have motor components of dorsal nerves and dorsolateral and ventrolateral nerves contributed by placodes and/or neural crest.

SEGMENTAL ORGANIZATION OF THE INDIVIDUAL CRANIAL NERVES

We will now examine the organization of cranial nerves in the forebrain and in a rostrocaudal series of head segments. These segments only represent a hypothetical, generalized vertebrate scheme, since a simple, 1:1 correspondence of neuromeres, somitomeres, and other segmental tissues of the head has not been defined across all vertebrate groups. With the overall scheme of a model head segment that was presented in Figure 9-8 in mind, the organization of the various, individual cranial nerves can be understood. Figure 9-9 is a comprehensive diagram that illustrates the cranial nerves of the forebrain and the successive head segments (or **holobranchs**) along with the neuromeres, somitomeres, and currently hypothesized contributions from neural crest and placodes. Some of the contributions from neural crest and placodes may be subject to revision as further data indicate. The alignment of the head segments with the neuromeres and somitomeres in Figure 9-9 is based only on our generalized, theoretical model of vertebrate head segmentation, as discussed above. Table 9-2 summarizes the segmental organization of the head and the cranial nerves.

The original terminology used for the cranial nerves has been modified only slightly for the new scheme. As we will discuss below, three of the traditionally recognized, 12 cranial

nerves actually comprise two separate nerves each. The facial, glossopharyngeal, and vagus nerves each contain the components of a dorsal, segmental, cranial nerve and a ventrolateral cranial nerve derived from the series of epibranchial placodes. In each of these cases, we refer to the former as a separate nerve from the latter. Thus, for example, we refer to the "dorsal facial (VII_D) nerve" and the "ventrolateral facial (VII_{VL}) nerve" rather than to just the "facial (VII) nerve." In cases where only one of the segmental nerves (i.e., dorsal, ventral, ventrolateral, or dorsolateral) is contained in a designated nerve or where a placodal component of a nerve remains undesignated, we do not use an adjective to modify the original name of the nerve. For example, although the trigeminal nerve is a dorsal cranial nerve that also has a placodal component, it is not subdivided here into dorsal and either ventrolateral or dorsolateral components. We refer to it simply as the trigeminal (V) nerve.

The model presented in Figure 9-9 is a neuromere–somitomere model of the development of the head and the cranial nerves. It is based on several recent publications, including those of Northcutt (1990), Noden (1991), and Walker and Liem (1994). This model may be subject to revision as new data continue to be obtained. For example, data and analyses published by Gilland and Baker (1992, 1993) and by Northcutt (1993) suggest a revised numbering system for the somitomeres such that somitomeres 1–3 as shown here would be the prechordal plate (mesoderm that is located rostral to the rostral end of the notochord), and seven somitomeres would then be recognized in the region presently considered to be somitomeres 4–7. Still other designations of the somitomeres have also been proposed. The rostrocaudal order of neuromere–somitomere model units remains relatively constant, however, and that order is the key point to keep in mind in analyzing the organization of the cranial nerves.

The Forebrain

The first several cranial nerves are associated with the forebrain, which lies rostral to the first model head segment. The most rostral cranial nerve is the **terminal nerve.** At least some of the ganglion cells of the terminal nerve are derived from a rostral, **olfactory placode** [Fig. 9-9(B)]. Some of the terminal nerve ganglion cells may derive from other sources, including neural crest and/or the forebrain itself. The dendrites of terminal nerve neurons are distributed over the nasal septum, and the axons project to cites in the forebrain; this nerve may play a role in reproductive behaviors.

Cranial nerve I, the **olfactory nerve,** also is derived from the olfactory placode. It carries olfactory information into the telencephalon. The **vomeronasal nerve** is another cranial nerve that is present in some but not all vertebrates and is associated with the olfactory nerve and likewise derived from the olfactory placode. This nerve also plays a role in chemical detection. It projects, however, to a different area of the telencephalon than the olfactory nerve. An additional placode present in this region (but not illustrated in Fig. 9-9) is the **adenohypophyseal placode.** The olfactory organ and adenohypophysis arise from a single placode in lampreys, and a single rostral placode may thus be the plesiomorphic condition for vertebrates.

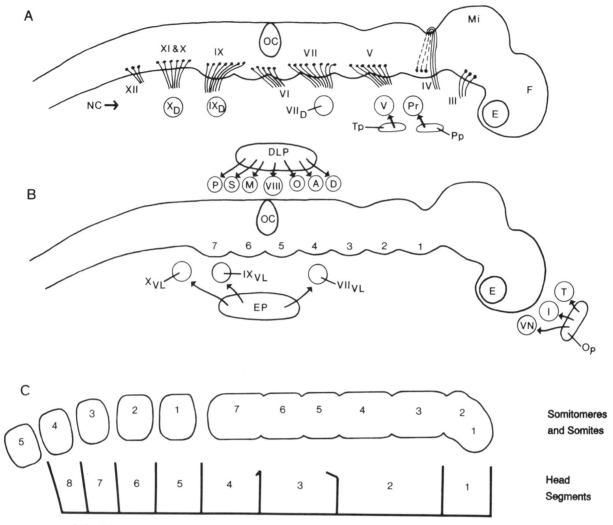

FIGURE 9-9. Schematic drawing of parasagittal sections through the neural tube (A and B) and the somitomeric mesoderm aligned with the model head segments (C). Rostral is toward the right and dorsal toward the top. The derivation of cranial nerve ganglia from neural crest (NC) and the trigeminal (T_P) and profundus (P_P) placodes and the location of their nuclei within the neural tube are shown in A. The derivation of cranial nerve ganglia and nerves from the dorsolateral (DLP) and ventrolateral, epibranchial (EP) series of placodes and the rostral olfactory (O_P) placode are shown in B. The rhombomeres are indicated by Arabic numerals 1–7 in B. The somitomeres (Arabic numerals 1–7) and rostral series of somites (Arabic numerals 1–5) along with model head segments (Arabic numerals 1–8) are shown in C in vertical alignment with the neural tube. Cranial nerve components are identified by their Roman numerals or by the abbreviations: A, anteroventral lateral line; D, anterodorsal lateral line; M, middle lateral line; O, otic lateral line; P, posterior lateral line; Pr, profundus; S, supratemporal lateral line; T, terminal; and VN, vomeronasal. Other abbreviations: OC, otic capsule; E, eye; F, forebrain; Mi, midbrain. Adapted from Noden (1991). Used with permission of S. Karger AG.

The **optic nerve,** cranial nerve II, arises from neurons in the retina, and the **epiphyseal nerve** (which in a number of groups of vertebrates has two divisions, a pineal nerve and a more rostral epiphyseal nerve) arises from neurons in the epiphysis. These cranial nerves should probably be considered as parts of the brain rather than as cranial nerves because, whereas other sensory nerves are derived from either neural crest and/or from placodes, the receptors and sensory ganglion cells of the retina in the eye and of the epiphyseal photoreceptor system are derived from evaginations of the neural tube. Only the lens of the eye, and probably the lens of the parietal eye

where present, are derived from placodal tissue. We thus have classified these nerves as evaginated sensory afferents.

The telencephalon thus has three nerves, the terminal, olfactory, and vomeronasal nerves, which are derived from placodes. We rather arbitrarily classify these nerves as ventrolateral nerves derived from a rostral placode; they are, however, different and distinct from the ventrolateral nerves that are derived from the more caudally lying series of epibranchial placodes and that are discussed below. The optic and epiphyseal nerves of the diencephalon are photosensory and may be nerves that are uniquely derived from the neural tube or may

TABLE 9-2. Head Segmentation and Cranial Nerve Organization

Head Segment	Origin	Visceral Arch	Branchial Arch	Neural Tube-Derived or Dorsal Cranial Nerve	Ventral Cranial Nerve	Dorsolateral Cranial Nerve	Ventrolateral Cranial Nerve
				II, Epiphyseal			T, VN, I
1	Somitomeres 1, 2			Profundus	III	LL_{AD}	
2	3, 4	1		V	IV	LL_{AV}	
3	5, 6	2		VII_D	VI	Otic, VIII	VII_{VL}
4	7	3	1	IX_D		LL_M	IX_{VL}
5	Somites 1	4	2	X_D	XII	LL_{ST}, LL_P	X_{VL}
6	2	5	3	X_D	Occipital		X_{VL}
7	3	6	4	X_D	Occipital		X_{VL}
8	4	7	5	X_D	Occipital		X_{VL}

represent the sensory components of dorsal nerves. Dorsolateral and ventral nerves, as present in the more caudal parts of the brain, are clearly absent in the forebrain.

The First Head Segment

The first head segment in our model contains the first two somitomeres [Fig. 9-9(C)]. The more rostral somitomere, and possibly the more caudal one, give rise to the extrinsic eye muscles (listed in Fig. 9-4) that are innervated by cranial nerve III, the **oculomotor nerve.** Thus, the oculomotor nerve [Fig. 9-9(A)] is the somatic motor nerve of the ventral part of the first head segment. The more caudal somitomere of this segment does not give rise to any branchiomeric muscle.

The **profundus placode** is currently not assigned to either the dorsolateral or ventrolateral series of placodes. It gives rise to most or all of the ganglion cells of the **profundus nerve** [Fig. 9-9(A)]; the possibility exists that the profundus ganglion also contains some cells derived from neural crest. The profundus is a sensory nerve, supplying a surface area on the snout of fishes. In the coelacanth *Latimeria,* the profundus nerve additionally supplies the mucosa of a series of rostral tubes that contain separately innervated electroreceptors. The profundus nerve has no motor component, as does the more caudal trigeminal nerve, but it can be considered to be the dorsal nerve of the first head segment. In mammals, in which the profundus nerve is called the **ophthalmic branch of the trigeminal nerve,** the ganglia of the profundus and trigeminal nerves merge. No ventrolateral nerve derived from an epibranchial placode is present in the first head segment. Dorsolateral nerves will be discussed in conjunction with the third head segment.

The Second Head Segment

Somitomeres 3 and 4 are the paraxial mesoderm of the second model head segment. Somitomere 3 gives rise to the dorsal oblique muscle of the eye. The dorsal oblique is innervated by cranial nerve IV, the **trochlear nerve,** which is the ventral nerve of the second (mandibular) model head segment [Fig. 9-9(A)]. Somitomere 4 gives rise to the branchiomeric

muscles of the jaw. The dorsal nerve of this segment is the **trigeminal nerve,** cranial nerve V, which has a somatic motor component innervating the muscles of the jaw and a somatic sensory component. The ganglion cells of the sensory component of this dorsal nerve lie in the proximal part of the trigeminal ganglion and may arise from both neural crest and/or a **trigeminal placode** [Fig. 9-9(A)]. Further studies are needed to clarify the embryological origin of all of the cells in both the profundus and trigeminal placodes. The second head segment, like the first, lacks ventrolateral nerves derived from placodes. Dorsolateral nerves will be discussed in conjuction with the third head segment.

The Third Head Segment

The paraxial mesoderm of the third (hyoid) model head segment is derived from somitomeres 5 and 6. Somitomere 5 gives rise to the lateral rectus muscle of the eye. The lateral rectus is innervated by the ventral, somatic motor nerve of the third head segment, cranial nerve VI, the **abducens nerve** [Fig. 9-9(A)], which courses rostrally to reach the muscle. Somitomere 6 gives rise to the branchiomeric muscles of the face. These muscles are innervated by the motor fibers in the dorsal nerve of the third head segment, the **dorsal facial nerve,** cranial nerve VII_D. Like the trigeminal nerve of the second head segment, the dorsal facial nerve also has a somatic sensory component. The ganglion cells for this sensory component are presumably derived from neural crest and lie in the proximal part of the ganglion of the nerve.

The third head segment is the central-most segment for nerves derived from **dorsolateral placodes** [Fig. 9-9(B)]. The dorsolateral series of placodes gives rise to the receptors and ganglion cells of a multiple set of nerves. Three **preotic placodes** (Table 9-3) give rise to the ganglion cells and receptors of three **lateral line nerves—anteroventral** (LL_{AV}), **anterodorsal** (LL_{AD}), and **otic** (O)—for electroreception and/or mechanoreception. An **octaval placode** gives rise to the receptors within the membranous labyrinth of the inner ear and to the ganglion cells of the vestibular and cochlear rami of the **octaval** (VIII) **nerve.** The **vestibular ramus** carries balance and posi-

TABLE 9-3. Lateral Line and Octaval Placodes and Nerves

Placodes	Nerves
Preotic	Anteroventral lateral line Anterodorsal lateral line Otic lateral line
Octaval	Cochlear ramus of eighth Vestibular ramus of eighth
Postotic	Middle lateral line Supratemporal lateral line Posterior lateral line

tion senses, and the **cochlear ramus** carries hearing. Three **postotic placodes** give rise to the ganglion cells and receptors of three additional lateral line nerves—**middle** (LL$_M$), **supratemporal** (LL$_{ST}$), and **posterior** (LL$_P$). We have illustrated the six lateral line placodes and the octaval (vestibulocochlear) placode as primarily associated with the third segment. Alternatively, as shown in Table 9-2, these dorsolateral placodes may be serially related to a number of head segments, such that the anterodorsal placode is the dorsolateral placode of the first head segment, the anteroventral placode is of the second, the otic and octaval placodes are of the thrid, the middle is of the fourth, and the supratemporal and posterior are of the fifth and more caudal segments. Further research is needed to clarify whether such registry of dorsolateral placodes and rhombomeres exists.

The third head segment also has a ventrolateral nerve, derived from a ventrolateral, **epibranchial placode,** which is thought to innervate taste buds. This nerve is the **ventrolateral facial nerve** [VII$_{VL}$ in Fig. 9-9(B)]. The neuron cell bodies of the axons of this nerve lie in the distal part of the ganglion of the facial nerve and are derived from placode.

The Fourth Head Segment

There is only one somitomere associated with the fourth model head segment. Within this segment, somitomere 7 gives rise to muscle associated with the first branchial arch (third visceral arch). These muscles are innervated by somatic motor fibers in the **dorsal glossopharyngeal nerve,** cranial nerve IX$_D$. The dorsal glossopharyngeal nerve also has a somatic sensory component, and neural crest cells give rise to the ganglion cells of these sensory fibers in the proximal ganglion of the nerve. The dorsal glossopharyngeal nerve is thus the dorsal nerve of the fourth head segment [Fig. 9-9(A)], and somitomere 7 resembles the more caudal of the two somitomeres present in the more rostral head segments. A more rostral somitomere apparently does not develop in the fourth head segment, and, correspondingly, no ventral, purely somatic motor nerve is present.

The dorsolateral nerve of the fourth head segment may be the middle lateral line nerve, as discussed above. The ventrolateral nerve of this segment, derived from an epibranchial placode, is the **ventrolateral glossopharyngeal nerve** (IX$_{VL}$). The ganglion cells for the sensory fibers that are thought to innervate the taste buds are derived from placode and lie in the distal ganglion of the nerve [Fig. 9-9(B)].

The Fifth Head Segment

A series of somites are present caudal to somitomere 7, which gives rise to muscles of the larynx and tongue, as well as to hypobranchial muscles (those below the gill arches in the neck), in the fifth model head segment. The ventral, somatic motor nerve, which comprises three separate nerves in most fishes, is derived from the neural tube of this segment and is cranial nerve XII, the **hypoglossal nerve** [Fig. 9-9(A)]. The hypoglossal nerve innervates the muscles of the tongue in tetrapods. The **dorsal vagus nerve,** cranial nerve X$_D$, is the dorsal nerve of the fifth head segment. The ganglion cells of the sensory component of the dorsal vagus nerve are derived from neural crest and lie in the proximal ganglion of the nerve. The **accessory nerve,** cranial nerve XI (not included in Table 9-2), is also associated with the fifth head segment but is not as clearly understood. It is present in a wide range of vertebrates rather than being present only in amniotes as previously believed. In birds and mammals, the accessory nerve is a composite nerve, with a spinal motor nucleus that is plesiomorphic for at least jawed vertebrates and a bulbar portion of the nucleus derived from the vagus.

The supratemporal and posterior lateral line placodes supply the dorsolateral nerve component for the fifth (and more caudal) head segment, as discussed above. The ventrolateral, epibranchial series of placodes gives rise to a ventrolateral nerve of the fifth head segment, the **ventrolateral vagus nerve** (X$_{VL}$). The ganglion cells of this nerve are thought to innervate taste buds for this head segment. They are derived from placode and lie in the distal ganglion of the nerve [Fig. 9-9(B)].

We have not included some cranial nerve components in this analysis for the sake of the clarity of the overall pattern. As in the spinal cord segmental nerves, there are also autonomic and visceral (i.e., innervating viscera of the body) sensory components of cranial nerves. In addition, there are efferent components in a number of the sensory cranial nerves that terminate on the sensory receptor cells. These additional components, however, do not negate the overall scheme presented above.

FOR FURTHER READING

Bulfone, A., Puelles, L., Porteus, M. H., Frohman, M. A., Martin, G. R., and Rubenstein, J. L. R. (1993) Spatially restricted expression of *Dlx-1, Dlx-2 (Tes-1), Gbx-2,* and *Wnt-3* in the embryonic day 12.5 mouse forebrain defines potential transverse and longitudinal segmental boundaries. *Journal of Neuroscience,* 13, 3155–3172.

Fritzsch, R. and Northcutt, R. G. (1993) Cranial and spinal nerve organization in amphioxus and lampreys: evidence for an ancestral craniate pattern. *Acta Anatomica,* 148, 96–109.

Gans, C. and Northcutt, R. G. (1983) Neural crest and the origin of vertebrates: a new head. *Science,* 220, 268–274.

Gilland, E. and Baker, R. (1993) Conservation of neuroepithelial and mesodermal segments in the embryonic vertebrate head. *Acta Anatomica,* 148, 110–123.

Holland, P. W., Holland, L. Z., Williams, N. A., and Holland, N. D. (1992) An amphioxus homeobox gene: sequence conservation, spatial expression during development and insights into vertebrate evolution. *Development,* 116, 653–661.

McGinnis, W. (1994) A century of homeosis, a decade of homeoboxes. *Genetics,* 137, 607–611.

Noden, D. M. (1991) Vertebrate craniofacial development: the relation between ontogenetic process and morphological outcome. *Brain, Behavior and Evolution,* 38, 190–225.

ADDITIONAL REFERENCES

Altaba, A. R. I., Prezioso, V. R., Darnell, J. E., and Jessell, T. M. (1993) Sequential expression of HNF-3β and HNF-3α by embryonic organizing centers: the dorsal lip/node, notochord and floor plate. *Mechanisms of Development,* 44: 91–108.

Barghusen, H. R. and Hopson, J. A. (1979) The endoskeleton: the comparative anatomy of the skull and the visceral skeleton. In M. H. Wake (ed.), *Hyman's Comparative Vertebrate Anatomy, Third Edition.* Chicago: The University of Chicago Press.

Boncinelli, E., Gulisano, M., and Pannese, M. (1993) Conserved homeobox genes in the developing brain. *Comptes Rendus d'Acadamie des Sciences Paris,* 316, 979–984.

Chalepakis, G., Stoykova, A., Wijnholds, J., Tremblay, P., and Gruss, P. (1993) Pax: gene regulators in the developing nervous system. *Journal of Neurobiology,* 24, 1367–1384.

Demski, L. S. (1993) Terminal nerve complex. *Acta Anatomica,* 148, 81–95.

Duboule, D. (ed.) (1994) *Guidebook to the Homebox Genes.* Oxford, Oxford University Press.

Finger, T. E. (1993) What's so special about special visceral? *Acta Anatomica,* 148, 132–138.

Gans, C. and Northcutt, R. G. (1985) Neural crest: The implications for comparative neuroanatomy. *Fortschritte der Zoologie,* 30, 507–514.

Gehring, W. J. (1993) Exploring the homeobox. *Gene,* 135, 215–222.

Gilland, E. and Baker, R. (1992) Longitudinal and tangential migration of cranial nerve efferent neurons in the developing hindbrain of *Squalus acanthias. Biological Bulletin,* 183, 356–358.

Hanneman, E., Trevarrow, B., Metcalfe, W. K., Kimmel, C. B., and Westerfield, M. (1988) Segmental pattern of development of the hindbrain and spinal cord of the zebrafish embryo. *Development,* 103,49-58.

Holland, P. (1992) Homeobox genes in vertebrate evolution. *BioEssays,* 14, 267–273.

Hunt, P. and Krumlauf, R. (1991) A distinct Hox code for the branchial region of the vertebrate head. *Nature (London),* 353, 861–864.

Keynes, R. and Krumlauf, R. (1994) *Hox* genes and regionalization of the nervous system. *Annual Review of Neuroscience,* 17, 109-132.

Kimmel, C. B. (1993) Patterning the brain of the zebrafish embryo. *Annual Review of Neuroscience,*16, 707–732.

Krauss, S., Maden, M., Holder, N., and Wilson, S. W. (1992) Zebrafish *pax* [b] is involved in the formation of the midbrain-hindbrain boundary. *Nature (London),* 360, 87–89.

Lumsden, A. (1990) The cellular basis of segmentation in the developing hindbrain. *Trends in Neurosciences,* 13, 329-335.

Lumsden, A., Sprawson, N., and Graham, A. (1991) Segmental origin and migration of neural crest cells in the hindbrain region of the chick embryo. *Development,* 113, 1281–1291.

Marsh, E., Uchino, K., and Baker, R. (1992) Cranial efferent neurons extend processes through the floor plate in the developing hindbrain. *Biological Bulletin,* 183, 354–356.

Meier, S. P. (1982) The development of segmentation in the cranial region of vertebrate embryos. *Scanning Electron Microscopy,* Pt. 3, 1269-1282.

Northcutt, R. G. (1990) Ontogeny and phylogeny: A reevaluation of conceptual relationships and some applications. *Brain, Behavior and Evolution,* 36, 116–140.

Northcutt, R. G. (1992) The phylogeny of octavolateralis ontogenesis: A reaffirmation of Garstang's phylogenetic hypothesis. In D. B. Webster, R. R. Fay, and A. N. Popper (eds.), *The Evolutionary Biology of Hearing.* New York: Springer-Verlag, pp. 21–47.

Northcutt, R. G. (1993) A reassessment of Goodrich's model of cranial nerve phylogeny. *Acta Anatomica,* 148, 71–80.

Northcutt, R. G. and Bemis, W. E. (1993) Cranial nerves of the coelacanth *Latimeria chalumnae* [Osteichthyes: Sarcopterygii: Actinistia] and comparisons with other craniata. *Brain, Behavior and Evolution,* 42, Suppl. 1, 1–76.

Northcutt, R. G. and Gans, C. (1983) The genesis of neural crest and epidermal placodes: a reinterpretation of vertebrate origins. *Quarterly Review of Biology,* 58, 1–28.

Placzek, M., Tessier-Lavigne, M., Yamada, T., Jessell, T., and Dodd, J. (1990) Mesodermal control of neural cell identity: floor plate induction by the notochord. *Science,* 250, 985–988.

Puelles, L. and Rubenstein, J. L. R. (1993) Expression patterns of homeobox and other putative regulatory genes in the embryonic mouse forebrain suggest a neuromeric organization. *Trends in Neurosciences,* 16, 472–479.

Romer, A. S. (1962) *The Vertebrate Body: Shorter Version, 3rd ed.* Philadelphia: Saunders.

Song, J. and Boord, R. L. (1993) Motor components of the trigeminal nerve and organization of the mandibular arch muscles in vertebrates: phylogenetically conservative patterns and their ontogenetic basis. *Acta Anatomica,* 148, 139-149.

Song, J. and Northcutt, R. G. (1991) Morphology, distribution and innervation of the lateral-line receptors of the Florida gar, *Lepisosteus platyrhincus. Brain, Behavior and Evolution,* 37, 10–37.

Sperry, D. G. and Boord, R. L. (1993) Organization of the vagus in elasmobranchs: its bearing on a primitive gnathostome condition. *Acta Anatomica,* 148, 150–159.

Szekely, G. and Matesz, C. (1993) The efferent system of cranial nerve nuclei: a comparative neuromorphological study. *Advances in Anatomy, Embryology and Cell Biology,* 128, 1–92.

Tam, P. P. L. and Trainor, P. A. (1994) Specification and segmentation of the paraxial mesoderm. *Anatomy and Embryology,* 189, 275–305.

Trevarrow, B., Marks, D. L., and Kimmel, C. B. (1990) Organization of hindbrain segments in the zebrafish embryo. *Neuron,* 4, 669-679.

Wahl, C. M., Noden, D. M., and Baker, R. (1994) Developmental relations between sixth nerve motor neurons and their targets in the chick embryo. *Developmental Dynamics,* 201, 191–202.

Wake, D. B. (1993) Brainstem organization and branchiomeric nerves. *Acta Anatomica,* 148, 124–131.

Wake, M. H. (1993) Evolutionary diversification of cranial and spinal nerves and their targets in the gymnophine amphibians. *Acta Anatomica,* 148, 160–168.

Walker, W. F., Jr. and Liem, K. F. (1994) *Functional Anatomy of the Vertebrates: An Evolutionary Perspective, Second Edition.* Fort Worth, TX: Saunders College Publishing.

Webb, J. F. and Noden, D. M. (1993) Ectodermal placodes: contributions to the development of the vertebrate head. *American Zoologist,* 33, 434–447.

Wilkinson, D. G., Bhatt, S., Cook, M., Boncinelli, E., and Krumlauf, R. (1989) Segmental expression of Hox-2 homeobox-containing genes in the developing mouse hindbrain. *Nature (London),* 341, 405–409.

Wilkinson, D. G. and Krumlauf, R. (1990) Molecular approaches to the segmentation of the hindbrain. *Trends in Neurosciences,* 13, 335–339.

10

Functional Organization of the Cranial Nerves

INTRODUCTION

The segmental organization of the cranial nerves, as discussed in Chapter 9, forms the basis for the classification and organization of the cranial nerves presented in this chapter. Table 10-1 is a list of all the cranial nerves in vertebrates. The reader should note that not all animals possess all of these nerves; many nerves and/or individual components of nerves are absent in various species depending on how they have become adapted to their particular environments. Table 10-2 presents a classification of the cranial nerves according to embryological origin and nerve type. This new classification is tentative, since the embryological origin of all the neurons in the various cranial nerve ganglia has yet to be confirmed experimentally. It represents, however, a radical departure from the traditional classification of cranial nerves, as we presented in Table 9-1 and as might be found in a medical school textbook of human neuroanatomy. This new approach to classification is necessary for a comparative and evolutionary approach to the cranial nerves because many nerves and/or their components are not found in humans. Moreover, this new classification is consistent with the recent findings about the embryological origins of these nerves.

An inspection of Table 10-2 reveals that a number of the nerves appear in more than one cell. The reason is that some cranial nerves, such as the trigeminal (V) and the dorsal vagus (X_D), which belong to the dorsal group of cranial nerves (see Chapter 9), contain both afferent and efferent fibers. Others, such as the oculomotor (III) and abducens (VI) nerves, which belong to the ventral group of cranial nerves, have only efferent fibers. Until fairly recently, anatomists believed that sensory nerves, such as the optic (II) and the octaval (VIII), contained only sensory afferent fibers. We now know that a number (and possibly all) of the sensory cranial nerves also contain some efferent fibers. Also in Table 10-2, afferents for taste in the epibranchial placodal series of nerves (VII_{VL}, IX_{VL}, and X_{VL}) are combined with the more rostral placodal afferents (T, VN, and I) in a single category called ventrolateral placodal afferents. The trigeminal (V) and profundal (P) cranial nerves are classified as somatic and of neural crest and/or placodal origin, a category that may be modified as additional data become available.

Introductory textbooks often equate afferent neurons in cranial nerves with sensory processes and efferent neurons with motor processes. These terms frequently are interchangeable. However, in a number of instances, efferent fibers are not motor in the strict sense. Some efferent fibers activate glands, and the adjective "effector" can be used for these fibers. Other efferent fibers produce changes in the sensitivity of sensory receptors; they travel from the brain to the receptors via the same nerve bundles through which the sensory neurons travel to the brain from these very same receptors. In order to distinguish the efferent fibers of the sensory nerves from the efferent motor (effector) fibers, we have designated the somatic motor fibers as general somatic efferents. The efferents in the sensory nerves have been listed separately as efferents to sensory receptors; they include the efferent fibers in the optic (II), octaval (VIII), lateral line (LL), and taste (VII_{VL}, IX_{VL}, and X_{VL}) nerves.

Table 10-3 is essentially the same as Table 10-2 except that the sensory and effector functions are substituted for the names of the cranial nerves. Although additional research is needed to verify the categories shown in Table 10-3, current information suggests that cranial nerves with similar functions have similar

TABLE 10-1. The Cranial Nerves of Vertebrates

Symbol	Name	Innervation
T	Terminalis	Nasal septum
I	Olfactory	Olfactory epithelium
VN	Vomeronasal	Vomeronasal organ
II	Optic	Retina
E	Epiphyseal	Pineal; parietal eye
III	Oculomotor	Internal and external eye muscles
P	Profundus	Skin of the snout
IV	Trochlear	External eye muscles
V	Trigeminal	Jaw muscles; touch to face, snout and oral cavity
VI	Abducens	External eye muscles
VII$_D$	Dorsal facial	Facial muscles; salivary and tear glands
VII$_{VL}$	Ventrolateral facial	Taste buds
VIII	Octaval or vestibulocochlear	Vestibular organs; cochlea; lagena
LL	Lateral line (6 nerves)	Lateral-line organs
IX$_D$	Dorsal glossopharyngeal	Pharynx; salivary glands
IX$_{VL}$	Ventrolateral glossopharyngeal	Taste buds
X$_D$	Dorsal vagus	Viscera of thorax and abdomen; larynx; pharynx
X$_{VL}$	Ventrolateral vagus	Taste buds
XI	Spinal accessory	Neck and shoulder muscles
XII	Hypoglossal	Tongue; syrynx

embryological origins. Thus, the neural crest and neural tube derived cranial nerves bring sensory input from the visceral and somatic surfaces of the body and head and supply the efferent control of these same body regions. The dorsolateral placodal derived cranial nerves innervate hair cells of various sorts in the auditory, vestibular, and lateral line systems, as well as electroreceptors. The forebrain derived cranial nerves are the cranial nerves that detect light. Finally, the efferent axons in the sensory nerves provide a central modulation of the input from these receptors.

THE CRANIAL NERVES AND THE SPINAL CORD

Some of the cranial nerves share a number of characteristics of the spinal nerves. The dorsal cranial nerves have sensory components with ganglia, similar to the dorsal roots of the spinal cord. The ventral cranial nerves have motor roots that arise from large motor-neuron cell bodies similar to the ventral roots of the cord. Like the spinal nerves, segmental organization of the cranial nerves exists in the pattern of innervation of peripheral structures. The notion that the lower brainstem is a continuation of the spinal cord with the same general organization is strengthened by the continuity of some of the sensory cell columns of the dorsal horn of the cord with sensory cell columns of the lower brainstem and by the continuity of the motor columns of the ventral horn of the cord with motor columns of the lower brain stem (Fig. 10-1).

THE ORGANIZATION OF SENSORY AND MOTOR COLUMNS OF THE CAUDAL BRAINSTEM

Sensory and motor components of cranial nerves are segmentally organized, and the points at which they enter or leave the brain reflects their developmental history. Separate motor nuclei are present for each of the various motor cranial nerves. In contrast, a large number of nerves exist by which sensory axons can enter the brain, but a unique, separate nucleus is not always present in the brain for every sensory component of each nerve. The zones of termination are in fact relatively few and are related to the particular system rather than to the individual components of individual nerves. For example, all somatic afferent neurons (i.e., those carrying information about pain, temperature, touch, pressure, muscle stretch, and the position of joints) entering the brainstem, no matter by which nerve they enter, whether by the trigeminal (V), the dorsal facial (VII$_D$), the dorsal glossopharyngeal (XI$_D$), or the dorsal vagus (X$_D$), terminate in a sensory cell column known as the somatic afferent column, which is continuous with the dorsal horn of the spinal cord [Fig. 10-1]. In a similar manner, the visceral afferent neurons (i.e., those carrying information about pain,

TABLE 10-2. A Classification of the Cranial Nerves of Vertebrates Based on Embryological Origins			
Predominant Embryological Origin	Nerve Type	Afferents[a]	Efferents to Muscles and Glands (Effectors)[a]
Neural tube and neural crest	Visceral	IX_D, X_D	III VII_D, IX_D, X_D
	Somatic	VII_D, IX_D, X_D	III, IV, VI VII_D, IX_D, X_D XI, XII
Neural crest and/or placodes	Somatic	V, P	
			Efferents to Receptors
Placodes	Dorsolateral	VIII, LL	VIII, LL
	Ventrolateral	T, VN, I VII_{VL}, IX_{VL}, X_{VL}	?
Forebrain	Evaginated	E, II	E, II

[a] See Table 10-1 for symbol identification.

temperature, pressure, and muscle stretch from the viscera in the body) that enter the brainstem, whether they enter through IX_D or X_D, terminate in the visceral afferent column [Fig. 10-1].

This continuity does not occur for all systems, however, because not all of the components of the body are also present in the head, and some tissues are unique to the head. For example, visceral afferent fibers are not present in the head. Placodes develop only in the head. Thus, placodal afferent fibers and the columns in which they terminate are unique to the head. All of the epibranchial placodal afferent neurons carrying taste information, whether they enter through VII_{VL}, IX_{VL}, or X_{VL}, terminate in the same column, as do the dorsolateral

placodal afferent neurons [Fig.10-2(B)]. Thus, while the individual nerves through which neurons enter the brain are related to the segmental development of the head and brain, the populations of interneurons within the brain where various components terminate are organized according to the particular functional system to which they belong.

Afferent Columns of the Brainstem

General Somatic Afferents. In Figure 10-2(A), somatic afferent fibers from the face and head that enter the brainstem through the trigeminal (V), profundus (P), dorsal facial (VII_D),

TABLE 10-3. A Classification of the Cranial Nerves of Vertebrates Based on Function			
Predominant Embryological Origin	Nerve Type	Afferents	Efferents to Muscles and Glands (Effectors)
Neural tube and neural crest and/or placodes	Visceral	From visceral surfaces of the head and body	Smooth muscles, cardiac muscles, and glands (head and body)
	Somatic	From somatic muscles and skin	Striated muscles of face, head, and neck
			Efferents to Receptors
Placodes	Dorsolateral	Hair cells and electroreceptors	
	Ventrolateral	Chemical senses (and others?)	Modulation of receptors
Forebrain	Evaginated	Light detection	

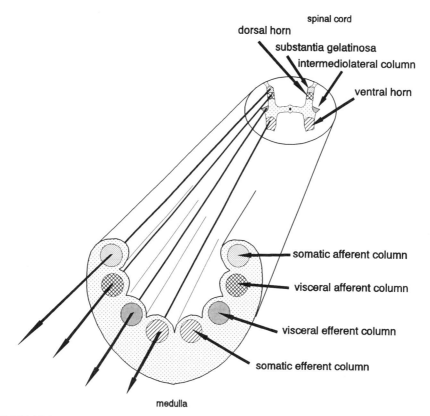

FIGURE 10-1. Schematic drawing to show the continuity of cell columns in the spinal cord with those in the medulla.

dorsal glossopharyngeal (IX$_D$), and dorsal vagus (X$_D$) nerves all are shown terminating among the cells of the **somatic afferent column.** This column is also known as the **descending nucleus of the trigeminal nerve,** or **descending nucleus of V.**

The profundus nerve is a distinct nerve only in fishes. It innervates the skin on the dorsal, rostral part of the head and on the snout. In mammals, the profundus nerve is the same as the most rostral (ophthalmic) branch of the trigeminal nerve. Further research is needed to fully clarify the trigeminoprofundal relationship; some research indicates that the trigeminal ganglion arises from neural crest, while the profundus ganglion cells are of placodal origin. Additional studies on a variety of vertebrate species are needed to answer this question. The central connections of the profundus nerve, particularly in non-mammalian vertebrates, are not well worked out; a comparative study of them would add to our understanding of the development of the head and the organization of the trigeminal sensory system as well.

Visceral Afferents. The cells of the visceral afferent column in the caudal-most part of the brainstem also have a special name: **nucleus solitarius.** Visceral afferent fibers from the viscera of the body that enter the brainstem through the dorsal glossopharyngeal (IX$_D$) and the dorsal vagus (X$_D$) terminate in the nucleus solitarius region of the **visceral afferent column.**

Ventrolateral Placodal Afferents. Ventrolateral, epibranchial placodal afferent fibers carrying taste information [Fig. 10-2(B)]

terminate in a cell column that lies rostral to the visceral afferent column. Because taste was previously regarded as a visceral sense, its afferent column was thought to be a rostral continuation of the visceral afferent column, and its position is consistent with this interpretation. The zone of termination of taste afferents, however, is now recognized as a separate nucleus, distinct from the more caudal, visceral afferent column. It is called the **gustatory nucleus** and receives the ventrolateral placodal afferents from the ventrolateral facial (VII$_{VL}$), ventrolateral glossopharyngeal (IX$_{VL}$), and ventrolateral vagus (X$_{VL}$) nerves.

Octaval Nerve. The vestibular and cochlear rami of the eighth cranial nerve are derived from the dorsolateral octaval placode, as is the otic capsule. Octaval nerve fibers terminate within a **dorsolateral octaval column** that lies near the somatic afferent column [Fig. 10-2(B)]. The dorsolateral octaval column contains a number of individual nuclei.

Vestibular receptors are located in parts of the inner ear called the utricle, saccule, and semicircular canals. They detect acceleration and position relative to gravity in order to maintain equilibrium and orientation in space. The vestibular nerve projects to nuclei within the dorsolateral placodal column called the **vestibular nuclei.** These nuclei lie within the pons. The cochlear ramus of the eighth cranial nerve carries auditory information from the cochlea in the inner ear. The auditory fibers terminate in nuclei in the dorsolateral placodal column of the pons that are called the **cochlear nuclei.** The fibers terminate in the nuclei in an ordered manner so that there are maps of

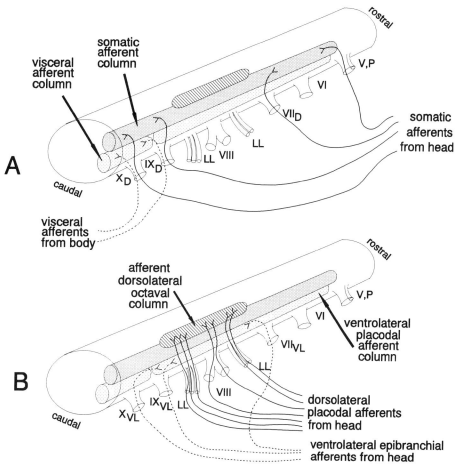

FIGURE 10-2. Schematic drawing of the caudal brainstem to show some of the major afferent cell columns in relation to the cranial nerves. Rostral is toward the right. Visceral and somatic afferent cell columns are shown in part A, and dorsolateral and ventrolateral, epibranchial placodal afferent cell columns are shown in part B.

lower to higher tones within the nuclei. This tonotopic organization is maintained in projections to auditory areas in the more rostral parts of the brain.

Lateral Line Nerves. The nerves of the caudal brainstem described thus far in this chapter are common to most vertebrates, including mammals. As discussed in Chapter 9, a set of as many as six nerves, collectively called the lateral line (LL) nerves, is present in aquatic anamniotes, including lampreys, some cartilaginous and ray-finned fishes, and some amphibians. The six lateral line nerves are the anterodorsal, anteroventral, otic, middle, supratemporal, and posterior, although not all six are present in all species that have a lateral line system. These nerves are derived from the dorsolateral series of placodes, as is the octaval nerve, and they terminate in nuclei of the **dorsolateral placodal column** [Fig. 10-2(B)]. Electroreception has recently been found to be present in monotremes and may be present in one or more placental mammals as well, but this sense in mammals is thought to have been evolved independently.

The lateral line nerves in aquatic anamniotes serve a series of organs known as the lateral line organs, which are arrayed in lines that generally lie in shallow grooves, or canals, several of which are on the surface of the head and one of which runs the length of the body. The lateral line canals contain receptors (see Chapter 2) that detect the displacement of water across the animal's head and body, as would be caused by another animal swimming nearby. These receptors are called **neuromasts** and are characterized by a hair-cell epithelium that is surrounded by nonneural support cells and mantle cells. They are said to be mechanoreceptive.

A second class of receptors, which are present in some species, are electroreceptive. These receptors are of two similar types, known as **ampullary organs** and **tuberous organs** (see Chapter 2). They are both formed by a receptor epithelium at the base of a small, encapsulated pouch that is connected to the surface of the epidermis by a small canal or pore. The electroreceptive, lateral line system is used to detect the weak electrical fields that are generated around living organisms. In those cartilaginous fishes, such as the torpedo, and in those ray-finned fishes, such as the electric eel, that generate more substantial electric fields around themselves, these organs serve to detect these fields. Many species of cartilaginous and ray-finned fishes that do not generate the stronger electric fields are, nevertheless, capable of detecting such fields. This electro-receptive sense is useful in the detection of prey in murky waters or just beneath the surface of sand or mud at the bottom of the water.

Efferent Columns of the Brainstem

Visceral Efferents. The **visceral efferent column** (Fig. 10-1) is a continuation of the visceral motor cell column of the spinal cord, which in tetrapods is called the **intermediolateral column.** This column of cells from spinal cord to caudal brainstem comprises the **autonomic nervous system.** The cranial nerve components (together with those of the sacral spinal cord) constitute the **parasympathetic division** and those of the throacic and lumbar levels of the spinal cord constitute the **sympathetic division.** The cranial nerve components of the autonomic nervous system control the internal eye muscles that dilate and contract the pupil of the eye and the muscle that controls accomodation (the focusing of the eye on close objects), the salivary glands, and tear glands in tetrapods, as well as the internal organs of the thorax, abdomen, and pelvis.

Somatic Efferents. The **somatic efferent column** (Fig. 10-3) is a continuation of the motor neuron column of the ventral horn of the spinal cord. Just as the motor neurons of the spinal cord control skeletal muscles of the body, so do their brainstem counterparts control the skeletal muscles of the head. The oculomotor (III), trochlear (IV), and abducens (VI) nerves together control the external eye muscles, which move the eye in its socket. In tetrapods, the hypoglossal nerve (XII) controls the muscles of the tongue. The motor neurons of this column that exit via the trigeminal (V) nerve control the muscles of the jaws, and the motor neurons of the dorsal facial (VII$_D$) nerve control the muscles of the face, such as lips, eyelids, cheeks, and nostrils, in those animals that have such structures and are capable of moving them. The motor neurons of the dorsal glossopharyngeal (IX$_D$), dorsal vagus (X$_D$), and accessory (XI) nerves control the muscles of the throat and neck.

FIVE CRANIAL NERVES ROSTRAL TO THE BRAINSTEM

Five nerves remain to be discussed in this introduction to the cranial nerves: the optic nerve (II), the epiphyseal nerve (E), and three placodal nerves: the olfactory nerve (I), vomeronasal nerve (VN), and terminal nerve (TN). All five of these nerves are in the forebrain. These nerves are exceptions to the general rules of cranial nerve termination, which apply to the cranial nerves discussed thus far in that their sites of termination are not at all related to the sensory columns of the brainstem.

Because the retina of the eye and the sensory receptors in the epiphysis are actually outgrowths of the brain, the optic nerve (II) and the epiphyseal nerve (E), which carry information from the receptors to the brain, are true pathways of the central nervous system rather than peripheral nerves, as are the cranial nerves of the caudal brainstem. The retina of the eye contains a number of receptors and neuron types arranged in a complex organization similar to that found in the central nervous system, so that a considerable amount of processing of the visual input goes on in the retina before the optic nerve conducts it to the brain. The optic nerve (II) terminates in a number of nuclei located in the **thalamus** and **hypothalamus,** which lie in the diencephalon, and in the **optic tectum,** which lies in the roof of the midbrain. The epiphyseal nerve comprises either one or two divisions in most vertebrate groups: one from the pineal

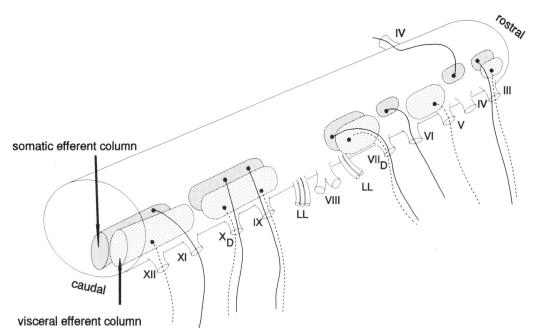

FIGURE 10-3. Schematic drawing illustrating the motor columns of the caudal brainstem. Rostral is toward the right.

part of the epiphysis and, sometimes, one from the more rostral part of the epiphysis (see Chapter 21). The one or two divisions of the epiphyseal nerve terminate in the **epithalamus,** which is the most dorsal part of the diencephalon.

The axons of the olfactory nerve (I) are, in some cases, among the shortest of the cranial nerve axons. The olfactory receptors are located in the mucous membranes that lie deep within the nose. The axons of the olfactory nerves travel through small openings in the roof of the nasal passage to the **olfactory bulb,** or **main olfactory bulb,** which lies just above. The main olfactory bulb is an extension of the cerebral hemisphere, which is located close to the olfactory receptors. It is connected to the main body of the cerebral hemisphere by a band of axons called the **olfactory tract** (Fig. 10-4). The length of the tract depends on the length of the animal's snout. In animals with a long snout, that is, those in which the nostrils are located relatively far rostral to the eyes, such as alligators, many hoofed mammals, and many fishes, the olfactory tract can be quite long. In short-snouted animals, such as many primates, the tract is rather short. In many birds, in which the nostrils are quite close to the eyes, the tract *per se* does not exist, and the main olfactory bulb is connected directly to the cerebral hemisphere. The olfactory nerve is derived from a rostral placode, as are the vomeronasal and terminal nerves.

The vomeronasal nerve (VN) is functionally similar to the olfactory nerve. It arises from a sensory area inside the nose called the **organ of Jacobson** and projects to a structure that lies next to the main olfactory bulb called the **accessory olfactory bulb.** The vomeronasal system is best developed in squamate reptiles.

Like the receptors for taste, the olfactory and vomeronasal receptors are sensitive to chemical substances present in the environment. In air-breathing animals, the chemicals to which the olfactory receptors are sensitive are volatile and disperse through the air, whereas the gustatory (taste) receptors are sensitive to chemicals that are water soluable. However, in aquatic animals, chemicals that are olfactory stimuli must also be water soluable.

The final cranial nerve on our list is the terminal nerve (TN). This nerve is present in most or all jawed vertebrates.

The terminal nerve was long thought to be absent in birds, but recent evidence suggests that it is present in this group as well as in other vertebrate groups. In aquatic animals, the terminal nerve dendrites are known to innervate the nasal septum, but the central connections of the terminal nerve are quite different from those of the olfactory and vomeronasal nerves. A unique characteristic of the terminal nerve is that many of its fibers contain a hormone called **luteinizing-hormone releasing hormone** (LHRH). This hormone is also present in some of the cells in the hypothalamus that project to the pituitary. This hormone causes the anterior pituitary to secrete luteinizing hormone, which controls the secretion of sex hormones from the testes and ovaries. A most peculiar property of the LHRH-positive terminal nerve fibers in fishes is that some of them turn from the terminal nerve into the optic nerve and terminate in the retina. The functional role of this olfactoretinal pathway in reproductive or other behaviors has not yet been elucidated.

GENERAL CONSIDERATIONS

A feature of all of the cranial nerves, one that is shared with structures in the central nervous system, is the relationship between the size and development of the nerve and the extent to which it is used. In animals with a well-developed olfactory sense, such as sharks and some mammals, the olfactory bulb is enormous in comparison to those animals with a more modest sense of smell. Similarly, animals with a prominent snout that is used for exploration of the environment with the sense of touch, such as pigs, platypuses, and anteaters, have very large sensory branches of their trigeminal nerves. Those with powerful jaws have large motor branches of their trigeminal nerves. Ray-finned fishes have an excellent gustatory system, but in certain species, especially goldfishes and catfishes, the gustatory system reaches an extraordinarily high degree of development. These fishes have many thousands of taste buds, not only in the mouth but on the head and all over the body as well. The gustatory nerves (VII_{VL}, IX_{VL}, and X_{VL}) are extremely large, and the central structures and cell groups that are associated with these nerves rival those of the visual system in size

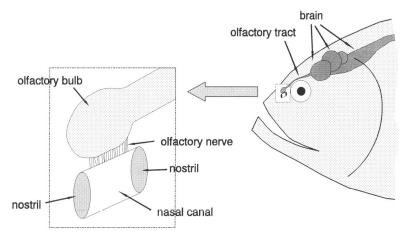

FIGURE 10-4. Schematic drawing of the olfactory afferent system in the head of a fish (right), with the area of the olfactory nerve and bulb enlarged (left).

and complexity of organization. In subsequent chapters, we will discuss each of the cranial nerves in greater detail and will point out how the structure and organization of these nerves and the cell populations that they interact with are intimately related to how animals survive in their environments.

FOR FURTHER READING

Finger, T. E. (1993) What's so special about special visceral? *Acta Anatomica,* 148, 132–138.

Northcutt, R. G. and Bemis, W. E. (1993) Cranial nerves of the coelacanth *Latimeria chalumnae* [Osteichthyes: Sarcopterygii: Actinistia] and comparison with other craniata. *Brain, Behavior and Evolution,* 42, S1, 1–76.

Szekely, G. and Matesz, C. (1993) The efferent system of cranial nerve nuclei: a comparative neuromophological study. *Advances in Anatomy, Embryology, and Cell Biology,* 128, 1–92.

ADDITIONAL REFERENCES

Fritzsch, R. and Northcutt, R. G. (1993) Cranial and spinal nerve organization in amphioxus and lampreys: evidence for an ancestral craniate pattern. *Acta Anatomica,* 148, 96–109.

Meyer, D. L., von Bartheld, C. S., and Lindörfer, H. W. (1987) Evidence for the existence of a terminal nerve in lampreys and in birds. In L. S. Demski and M. Schwanzel-Fukuda (eds.), *The Terminal Nerve (Nervus Terminalis): Structure, Function, and Evolution, Annals of the New York Academy of Sciences,* 519, 385–391.

Northcutt, R. G. (1990) Ontogeny and phylogeny: a reevaluation of conceptual relationships and some applications. *Brain, Behavior and Evolution,* 36, 116–140.

Northcutt, R. G. (1993) A reassessment of Goodrich's model of cranial nerve phylogeny. *Acta Anatomica,* 148, 71–80.

Sperry, D. G. and Boord, R. L. (1993) Organization of the vagus in elasmobranchs: its bearing on a primitive gnathostome condition. *Acta Anatomica,* 148, 150–159.

11

Sensory Cranial Nerves

INTRODUCTION

The sensory cranial nerves of the brainstem fall into three distinct categories: the dorsal cranial nerve sensory components of the trigeminal nerve that carry somatosensory sensation for the head, the ventrolateral placodal nerves that carry taste, and the dorsolateral placodal nerves that carry the lateral line and octaval senses. As discussed in Chapter 9, each of these three categories represents a developmentally distinct component of the neuromeric segments. The sensory cranial nerves of the forebrain are discussed in later chapters.

DORSAL CRANIAL NERVES: SENSORY COMPONENTS FOR GENERAL SOMATOSENSORY SENSATION

One of the most dramatic advances in the evolution of the head was the transformation of the first arch of the visceral skeleton into components of the jaws. Although the jaws continued to play a role in respiration by regulating the flow of oxygenated water to the gills, they also greatly altered the way that the animals fed and many aspects of their mode of living. Jawed fishes were able to become more active predators. The ability for predation had important consequences for subsequent evolutionary changes such as improved fins and tail propulsion for more rapid manuevering to approach prey or to avoid predators, changes in body form to reduce drag, and camouflage to deceive prey or to deceive predators. The central nervous system evolved concurrently with these changes by providing the sensory input and motor control that made these behaviors possible.

Somatosensory Innervation of the Head

The cranial nerve that is most closely associated with the development of the jaws is the trigeminal nerve (V). This nerve is derived from the nerve that supplies the first (mandibular) visceral arch (of the second head segment). Closely allied to the trigeminal nerve is the dorsal facial nerve (VII$_D$), which is derived from the nerve that supplies the second (hyoid) visceral arch (of the third head segment). These two nerves provide much of the sensory innervation from the skin and muscles of the head. These sensations include pain, temperature, touch, and proprioception.

The trigeminal nerve carries the sensory neurons from the jaws and elsewhere on the head as well as the motor neurons that control the deep muscles of the head, that is, the jaw muscles. The dorsal facial nerve carries sensory neurons from the skin of the head and face as well as from the muscles in anamniotes, but in amniotes, relatively few of these somatosensory fibers are found in VII$_D$. The majority of the somatosensory fibers from the head enter the brainstem via the sensory roots of V.

The trigeminal system is best developed in animals with a prominent snout, such as alligators, birds, and other animals, such as pigs, in which the snout is used for exploration and manipulation. The presence of vibrissae (whiskers) on the snout is also associated with an expansion of the trigeminal system. Vibrissae are important for tactile exploration and to provide an indication that the snout is about to bump into something. Many species of birds have a few rather untidy-looking feathers that stand out at the base of the bill. These feathers may be the avian equivalent of vibrissae and might possibly serve as air-speed indicators.

The trigeminal nerve has three great branches in mammals. The **ophthalmic branch** innervates the skin of the head region, the nonvisual parts of the eye including the muscles, and

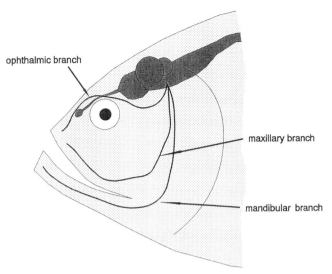

ophthalmic branch

maxillary branch

mandibular branch

FIGURE 11-1. Distribution of the branches of the trigeminal nerve in a bony fish.

the snout. The **maxillary branch** innervates the upper jaw (the maxilla) including the upper teeth, the roof of the mouth, and the upper lip. The **mandibular branch** innervates structures of the lower jaw (mandible) including the lower teeth, the tongue, the floor of the mouth, and the lower lip.

Similar branches of the trigeminal nerve are present in nonmammalian vertebrates (Fig. 11-1), but in many of these groups, a separate profundus nerve has been identified that corresponds to the ophthalmic branch of the trigeminal nerve in mammals. The profundus nerve may represent the dorsal nerve of the first head segment (see Chapter 9). In *Latimeria,* the profundus nerve innervates the mucosal walls of a series of rostrally located tubes that contain the lateral line receptors. The profundus nerve probably carries pain and temperature information from the walls of the tubes. The lateral line receptors themselves are innervated by the anterodorsal lateral line nerve, which we will discuss below. Some other special adaptations of the trigeminal nerve, such as infrared (IR) detection in snakes, which we will discuss below, also involve the ophthalmic, or profundus, part of the trigeminal nerve.

Central Terminations of the Trigeminal Nerve

The central terminations (Fig. 11-2) of the axons of the trigeminal nerve are rather consistent in vertebrates. The fibers of the maxillary, mandibular, and ophthalmic branches terminate in the **descending nucleus of the trigeminal nerve** (descending V) in the somatic afferent column, along with somatosensory axons entering from other cranial nerves. Descending V is continuous at its caudal end with the dorsal horn of the spinal cord and appears to serve an equivalent function for the head. The sensory trigeminal axons also terminate in the pons, in an expansion of the somatic afferent column called the **principal nucleus of the trigeminal** (principal V). In both the principal V and descending V, the terminations of the axons remain segregated according to their source. In other

words, a spatial map of the arrangement of structures in the head is maintained in the trigeminal nuclei.

The trigeminal nerve is not the only source of somatosensory input to the brainstem. In addition to the trigeminal nerve, general somatosensory fibers can arrive at the descending nucleus of V by way of three other nerves: the dorsal facial (VII_D), the dorsal glossopharyngeal (IX_D), and the dorsal vagus (X_D) nerves. In all cases, however, the destination of the somatosensory fibers in VII_D, IX_D, and X_D is the same, the descending nucleus of V. Their inputs are conveyed to the same dorsal thalamic and telencephalic sites, in somatotopic order, as the inputs of the trigeminal nerve itself.

The Mesencephalic Division of the Trigeminal System

An additional component of the trigeminal system remains to be described. This is the mesencephalic division, which is one of the most unusual components of any sensory system. It is a very consistent feature and can readily be identified in every class of vertebrates with the exception of agnathans. The unusual characteristic of this sensory nerve (the **mesencephalic root of V**) is the location of its cell bodies, which are known as the **mesencephalic nucleus of the trigeminal nerve,** or mesencephalic V (Fig. 11-3). All dorsal cranial nerves, with the exception of the mesencephalic root of V, have their cell bodies outside of the central nervous system in ganglia. The cell bodies in these ganglia are derived from neural crest. The cells that form mesencephalic V are likewise derived from neural crest, but unlike other neural crest cells, they do not migrate ventrolaterally. Rather than being in a ganglion or close to the other cells and fibers of the trigeminal system, these cells lie in the dorsomedial part of the midbrain, in or adjacent to the tectal commissure, which is a broad band of axons that connects the right and left tectal hemispheres. This arrangement may be a remnant of an earlier stage in the evolution of the brainstem when such internal ganglion cells may have been more common.

Similar cells are seen in the spinal cord and/or medulla of many aquatic vertebrates. The latter are called **Rohon-Beard cells** and were discovered in lampreys by Reissner in the mid-1800s. It was Sigmund Freud, however, who while still a medical student, identified these cells as unmigrated spinal ganglion cells. In lampreys, these cells are present in the spinal cord and also along the entire extent of the medulla; they may be serially homologous to the mesencephalic V cells of jawed vertebrates.

The mesencephalic root of V carries proprioceptive information (for position sense) from the jaw muscles and the connective tissue surrounding the teeth. In mammals and amphibians, mesencephalic V also carries proprioception for the extraocular eye muscles, that is, the muscles that move the eyes. In contrast, in fishes and birds, the proprioceptors in the extraocular muscles are innervated by trigeminal axons that terminate in descending V. In sharks, the fibers of mesencephalic V do not respond to stimulation of muscle receptors; rather, they respond to stimulation of the skin around the mouth and displacement movements of the teeth. The majority of the mesencephalic V fibers terminate in motor V and so would appear to play an important role in the regulation of jaw open-

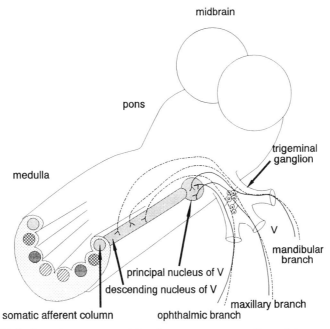

FIGURE 11-2. Central terminations of the afferent somatosensory fibers of the trigeminal nerve.

ing and closing and perhaps in the adjustment of jaw pressure.

Secondary Connections of the Trigeminal Nuclei

The majority of research on the secondary connections of principal V and descending V has been done in mammals and birds. The general organizational pattern of efferents from the trigeminal nuclei is that only the axons of the mesencephalic root of V terminate directly in the motor nucleus of V, which controls the contraction of the jaw muscles (see Chapter 12). In this case, a sensory neuron terminates directly on a motor neuron, forming a two-neuron (monosynaptic) reflex arc, which is the simplest type of neural network. In contrast, the efferents of principal V and descending V affect activity in motor V and motor VII$_D$ indirectly by way of their terminations in the reticular formation, which in turn sends axons to the motor nuclei. The reticular formation often serves to coordinate the activity in related motor nuclei, such as motor V, motor VII$_D$, and the three motor nuclei that control the eye muscles. Such coordination of jaw and eye muscles is important in feeding behaviors, especially in predatory animals.

Another secondary connection of the trigeminal system is a bundle of axons that ascends from descending V and principal V to a somatosensory nucleus in the dorsal thalamus. From the dorsal thalamus, axons pass up to the telencephalon and terminate in the somatosensory cortex or pallium. The somatotopic organization that exists in principal V and descending V persists throughout the entire pathway so that at the level of the telencephalon, the cells that are responsive to stimulation of various points on the head, snout, jaws, and mouth are arranged in separate groups.

An unusual variation on the general vertebrate pattern is found in birds, in which the target nucleus of principal V is not in the dorsal thalamus but in the telencephalon. This telencephalic cell group in birds is called **nucleus basalis.** It sends its efferents to a somatosensory telencephalic pallial area in a topographically organized fashion. Whether nucleus basalis is a homologue of the trigeminal, somatosensory, dorsal thalamic nucleus of mammals has not yet been resolved.

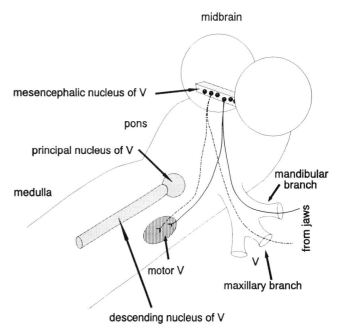

FIGURE 11-3. Connections of the mesencephalic nucleus of the trigeminal nerve.

Specialized Adaptations of the Trigeminal Nerve

A remarkable adaptation of the trigeminal nerve has occurred in two families of snakes: the boids (pythons and boa constrictors) and the crotalids or pit vipers (rattlesnakes, water mocassins, bushmasters, fer-de-lances, etc.). Both families have developed sensory pits on the head (Fig. 11-4). In the boids, the pits are principally on the lips; the pit vipers possess a prominent sensory pit below each eye. In both cases, the sensory pits are lined with photoreceptors that are sensitive not to the light that is visible to our eyes but to invisible IR radiation.

Infrared radiation is emitted by all objects that have warmth and especially by mammals and birds, which maintain a high body temperature. Such warm bodies are easily detected by IR sensors in the absence of all visible light. One study of the rattlesnake's IR detection sense reported that the snake could detect a temperature difference of 0.003°C from the background, which is roughly the amount of heat radiated by a human hand at a distance of one-half meter.

The IR detectors transmit their information to the brain via the maxillary branch of the trigeminal nerve. These neurons terminate in a special cell group: the **lateral trigeminal nucleus,** which is not found in non-IR detecting snakes. The nucleus is located lateral to descending V in the somatic afferent column. A similar, but independently evolved, lateral trigeminal nucleus has been reported in the vampire bat, which also detects IR radiation.

Recent physiological studies of the lateral trigeminal nucleus in rattlesnakes indicate that the neurons in this cell group possess many of the characteristics of cells in the visual system: small receptive fields and excitatory and inhibitory interactions that could serve to sharpen edges and provide a basis for the detection of motion and its direction. Of course, we have no idea what the rattlesnake's subjective experience of stimulation of its IR detectors is like, but the available evidence points to it being something very like vision.

In the crotalids, the lateral trigeminal nucleus sends its efferents to a special group of cells in the reticular formation: the **nucleus reticularis caloris** (i.e., the reticular nucleus of warmth). This reticular nucleus sends its efferents to the tectum, which is the region of the midbrain in which various maps of the surrounding environment, such as visual and auditory, are stored. The efferents of the boid lateral trigeminal nucleus project directly to the tectum. Tectal maps will be discussed in greater detail when we describe the tectum in a later chapter.

Another quite remarkable adaptation has recently been reported in the platypus. These monotreme (egg-laying) mammals have a prominent snout that resembles the bill of a duck. The "duckbill" of the platypus, rather than being keratinized as are bird bills, is leathery. Contained within the snout are rod-like structures that are mechanoreceptors and ampullary receptors similar to the electroreceptors of the lateral line system in aquatic anamniotes. These electroreceptors are capable of detecting the compound muscle-action potentials of the well-developed tail musculature of shrimp, on which platypuses feed. Similarly, mechano- and electroreceptive structures have been found on the beak of echidnas (spiny anteaters). Although the central termination of these receptors has not yet been determined, their location on the snout would seem to implicate the trigeminal system, and mechano- and electrosensory evoked potentials have recently been recorded in the somatosensory cortex in the platypus.

A third specialized adaptation of the trigeminal nerve may have occurred in birds and be related to magnetoreception. Some recent work suggests that some of the trigeminal neurons are responsive to magnetic stimuli; the neurons may be sensitive to the electromagnetic alignment of particles of ferromagnetic material (magnetite) that are present in the ethmoidal region of the skull. This system may thus contribute to migratory, homing, and orientation abilities.

VENTROLATERAL PLACODAL CRANIAL NERVES: TASTE

Some of the earliest vertebrates, derived from ancestral chordates, were adapted to life as filter feeders in an aquatic environment, probably very much the way that larval lampreys and some jawed fishes survive today. These animals filter food particles or small prey from the water column that is pulled in by the suction action of the branchial muscles. The filtration is done by sticky mucus or by the gill rakers acting as a sieve. The trapped particles or prey are washed into the digestive tract, and the excess water is expelled along with respiratory gasses. Some fishes are bottom feeders that suck up mouthfuls of bottom sediment, extract the edible morsels, and eject the remainder, and some predatory fishes suck in mouthfuls of water that carry the prey into their mouths. All of these methods of feeding require mechanisms for differentiating between the edible and the inedible, between objects that produce illness and those that do not, and between more favored and less favored foods. These mechanisms are the chemical senses of gustation (taste) and olfaction (smell). In this chapter, we will discuss gustation. Olfaction, which is used not only for feeding but also for social recognition, orientation, territorial marking,

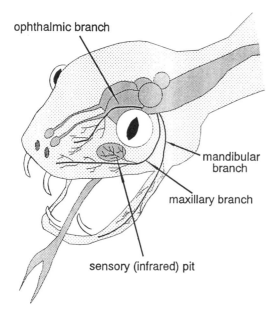

ophthalmic branch

mandibular branch

maxillary branch

sensory (infrared) pit

FIGURE 11-4. Distribution of the branches of the trigeminal nerve in a rattlesnake.

and courtship, is a more complex subject and will be discussed in Chapter 29.

The receptors for gustation are taste cells located in groups on small peg-like projections known as **taste buds.** An animal's sensitivity to taste stimuli depends on the number of taste buds. The greater the number of taste buds, the higher is the probability that a stimulus molecule will be detected.

In fishes, taste buds are located not only in the mouth and in the throat, but sometimes on the lips, on the surface of the head, and on the body skin. Many species of bony fishes have taste buds on the tips of their pectoral fins, which permits them to sample the taste qualities of the bottom sediment. The "whiskers" that give the catfish its name (technically known as **barbels**) are studded with as many as 20,000 taste buds and serve a similar, bottom-tasting function. In addition, the catfishes and the cyprinids, which includes the common goldfish, have in excess of 150,000 taste buds all over their body surface. Although we find it difficult to imagine what this skin-tasting sense is like, it probably gives the fish a taste map of the surrounding environment, just as the electrosense gives certain fishes an electrical image of nearby objects and organisms and the IR sense provides an IR picture to certain snakes.

In tetrapods, all of which have tongues, the taste buds are located on the tongue as well as in the mouth and throat. Amphibians, diapsid reptiles, and turtles have considerable numbers of taste buds, which suggests that their taste sensitivity is rather good. Birds, on the other hand, have relatively few taste buds compared to the other tetrapods and would appear to make less use of this sense.

The Gustatory System

The number of taste qualities that animals can detect is relatively few. The most commonly reported are sweet (sugars), salty (salts), sour (acids), and bitter (alkaloids). In addition, some taste neurons have been reported to be excited by certain amino acids and pure water. Because human taste sensitivity is limited to sweet, sour, salty, and bitter, we cannot imagine what amino acids or pure water taste like. Sensations in the mouth that are produced by chemical irritants such as pepper and various "hot" spices and seasonings are not transmitted either by the gustatory or by the olfactory nerves. These are somatic sensations just like pain and temperature and are carried by fibers of the maxillary and mandibular divisions of the trigeminal nerve (V) to the trigeminal nuclei in the somatic afferent column.

The Gustatory Nerves and the Nucleus Solitarius

The taste cells have no axons; instead, axons of gustatory fibers of the ventrolateral facial (VII_{VL}), ventrolateral glossopharyngeal (IX_{VL}), and ventrolateral vagus (X_{VL}) nerves detect changes in excited taste cells and transmit this information to the common target of taste axons no matter by which nerve they enter the brain: the gustatory nucleus (a rostral division of nucleus solitarius), which is at the rostral end of the visceral afferent column of the medulla (Fig. 11-5).

The nucleus solitarius consists of two divisions with their boundary being roughly at the level of the entrance of the vagus nerve. The rostral division of nucleus solitarius is the region that receives the gustatory fibers from VII_{VL}, IX_{VL}, and X_{VL} and is now recognized as a separate nucleus: the gustatory nucleus. This is a ventrolateral, epibranchial placodal afferent system rather than a visceral afferent system as previously believed. The caudal division of nucleus solitarius is the general visceral component of the visceral afferent column and receives fibers via the dorsal glossopharyngeal (IX_D) and the dorsal vagus (X_D) nerves from the viscera of the body, such as the digestive, circulatory, and respiratory systems. While not involved in the perception of taste, the visceral afferents of IX_D and X_D carry input that, along with taste afferents, affects feeding behavior.

A somatotopic organization exists in the gustatory nucleus such that the gustatory nerves from the different regions of the head (and the body in those animals with skin taste buds) enter the nucleus in approximately the same order as they are located in the body, that is, axons from more rostral regions of the mouth (or body) enter the more rostral portions of the gustatory nucleus. A similar, topographic arrangement has been found in the terminations of axons from the general viscera in the caudal division of nucleus solitarius. Axons from the various internal organs terminate in the caudal nucleus solitarius in the same order as the organs are arranged in the body.

Secondary Connections of the Gustatory Nucleus and Nucleus Solitarius

The pathways taken by efferent axons from the gustatory nucleus and the caudal division of nucleus solitarius have been studied in some detail in mammals, especially rodents and primates. In primates, the efferents of the gustatory nucleus follow the typical routes of the sensory systems of cranial nerves that terminate in the lower brainstem. From the target cell group of the primary axons of the cranial nerve, secondary axons ascend to the dorsal thalamus to end on tertiary cells whose axons in turn end in the telencephalon. The secondary gustatory fibers terminate on cells in the dorsal thalamus that are situated close to the cells that receive the secondary somatosensory fibers. The dorsal thalamic gustatory cells send their axons to one or more regions of the telencephalon—gustatory isocortex and areas in the ventral part of the telencephalon. As in the dorsal thalamus, the telencephalic gustatory cells are not far from the somatosensory cells.

An important additional nucleus in this system is the **parabrachial nucleus,** which is located in the dorsal pons. The medial part of this nucleus receives projections from the gustatory nucleus and is sometimes referred to as the **pontine taste area** (PTA). It contains a topographic representation of the gustatory receptors on the tongue. In rodents, the gustatory part of the parabrachial nucleus serves as an intermediary between the gustatory nucleus and the gustatory dorsal thalamus and also projects directly to gustatory isocortex. In primates, the parabrachial nucleus appears to be more involved with the caudal, visceral division of nucleus solitarius. In both primates and rodents, the parabrachial nucleus also sends many axons to ventral forebrain structures such as the hypothalamus and the amygdala, which lies in the basal part of the telencephalon. The gustatory pathways are shown in Figure 11-6.

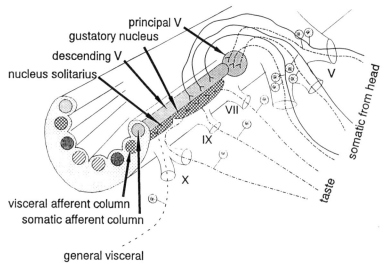

FIGURE 11-5. Diagram of the terminations of gustatory and tactile cranial nerves.

The cells of the caudal, visceral division of nucleus solitarius send some of their axons to the dorsal motor nucleus of the dorsal vagus (X_D) nerve. This nucleus, which is part of the visceral motor column, is the source of parasympathetic axons that distribute to the viscera of the thorax (chest) and abdomen and other brainstem motor nuclei that control salivation. Nucleus solitarius also projects to a more ventral nucleus in the brainstem: **nucleus ambiguus.** There is probably a stronger input to nucleus ambiguus from the gustatory nucleus than from the more caudal, viscerally related nucleus solitarius. Nucleus ambiguus gives rise to general somatic efferent axons in the dorsal glossopharyngeal, dorsal vagus, and accessory nerves that innervate the muscles of the pharynx (throat).

Cyprinid and Silurid Gustatory Specializations

Although taste is very well developed in fishes in general, two suborders of ray-finned fishes have become highly special-

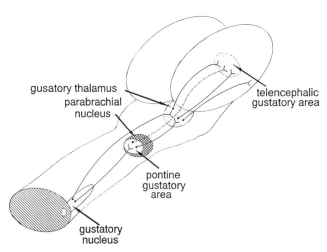

FIGURE 11-6. Ascending gustatory pathways in mammals.

ized for the use of the gustatory sense; these are the cyprinids, which include the carps, minnows, chubs, and goldfishes, and the silurids (or catfishes). As we discussed previously, these animals not only have mouth, throat, and skin taste buds, as do other species of fishes, but they have evolved a vast system of taste receptors over virtually the entire body surface. Along with this expansive gustatory surface, a system of elaborately organized central structures has developed that rivals the most complex neural organizations seen in any central nervous system.

Both groups of fishes are bottom feeders, and the mechanisms that they have evolved are highly sophisticated adaptations for the separation of food particles from inedible bottom sediment. In goldfishes, for example, two, opposing surfaces in the oropharynx (mouth and throat) manipulate bottom sediment in such a way as to separate the edible and tasty particles from those that are inedible or unpalatable. These surfaces are the **palatal organ,** which is a muscular structure attached to the roof of the mouth, and the surface of the gill arches. Both of these structures are studded with thousands of taste buds, which are innervated by branches of the ventrolateral vagus (X_{VL}) nerve. The gustatory branches of X_{VL} terminate in the gustatory nucleus, just as in other vertebrates, but what is so unusual in cyprinids and silurids is that the gustatory nuclei of the right and left sides have "ballooned" out to form rather prominent lobes on the caudal brainstem. They are known as the **vagal lobes** (or more properly the glossopharyngeal-vagal lobes because the axons of the ventrolateral glossopharyngeal (IX_{VL}) nerve in addition terminate in their rostral ends). The internal structure of the vagal lobes consists of a series of nine layers of neurons with a complex organization. In addition to a topographical organization of mouth structures in the vagal lobe, taste axons from each of the specialized structures terminate in specific layers. For example, axons innervating the taste buds of the palatal organ project to layer six; those innervating the taste buds on the gill arches project to layers two and four; layer nine receives both palatal and gill arch axons. The general visceral afferent axons from IX_D and X_D terminate in a separate, visceral afferent nucleus.

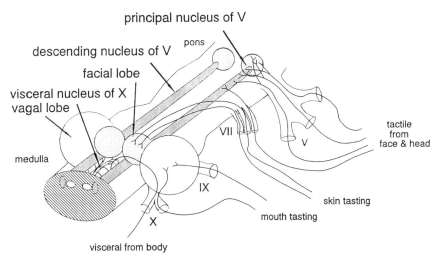

FIGURE 11-7. Afferent tactile and gustatory pathways in cyprinids and silurids.

In addition to the paired vagal lobes, these fishes have a second pair of lobes, called the **facial lobes,** which lie rostromedial to the vagal lobes. As the name implies, the facial lobe receives the gustatory axons of the ventrolateral facial (VII$_{VL}$) nerve. In catfishes, this nerve is quite elaborate with branches from the upper lip, the lower lip, the anterior palate, the pectoral fin, which also is used for bottom tasting, and a branch known as the recurrent branch because it "runs back" towards the tail. The recurrent branch serves the vast number of taste buds on the body surface. A topographical organization is found in the facial lobe with the more rostral structures being represented in the rostral portion of the lobe. The vagal and facial lobes and their relationship to the trigeminal system are shown in Figure 11-7.

Some of the efferents of the vagal and facial lobes (Fig. 11-8) are involved in the coordination of various feeding and postural reflexes. The vagal lobe projects to those general somatic efferent motor nuclei of the caudal brainstem that control the palatal organ and other muscles of the oropharynx. The facial lobe projects to a nucleus located near the border of the medulla and spinal cord, called the **spinotrigeminal funicular nucleus,** which provides a point of interaction between gustatory information and tactile information from the head and the body.

The major efferent projections of the facial and vagal lobes are ascending. Both lobes project to a nucleus located in the pons called the **superior secondary gustatory nucleus.** The superior secondary gustatory nucleus projects to an area in the caudal part of the diencephalon called the **preglomerular nuclear complex** (see Chapter 20) and to the hypothalamus. The preglomerular gustatory nucleus also receives direct projections from

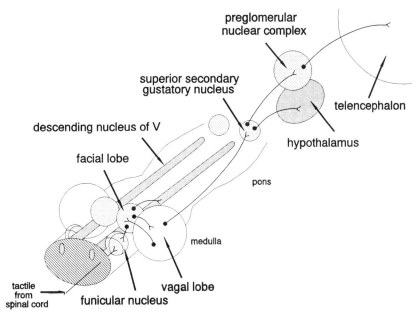

FIGURE 11-8. Ascending gustatory pathways in cyprinids and silurids.

the vagal and facial lobes; it relays gustatory information to an area in the telencephalon. The hypothalamic projection most likely provides the visceral and endocrine control cell groups in the hypothalamus with information about what has been tasted in the mouth. The superior secondary gustatory nucleus appears to be the homologue of the gustatory portion of the parabrachial nucleus, that is, the pontine taste area, of mammals.

In summary, we can see that the sense of taste has been developed to a superb degree in the cyprinids and silurids. At the peripheral end, the distribution of receptors on the body surface, fins, lips, and in the mouth provide the animal with a highly detailed taste map of the environment as well as the means for detecting food items and rejecting inedible or noxious objects. These processes are made possible by an extraordinary group of central structures: complex lobes with individual layers that receive very specific inputs and are organized topographically. The extreme development of the vagal lobe/facial lobe complex in these fishes is an excellent illustration of one of the ways that the nervous system responds to adaptive pressures for the increased use of a neural system.

A great deal of effort has been expended to work out the details of the vagal lobe/facial lobe system in cyprinids and silurids. We have good reason to believe that this research will greatly enhance our understanding of the way taste information (and other sensory information as well) is processed in the central nervous system of all vertebrates. One way to understand a biological system is to study it in an animal that is specialized for its use.

DORSOLATERAL CRANIAL NERVES: LATERAL LINE AND OCTAVAL SYSTEMS

In aquatic vertebrates, the dorsolateral series of placodes gives rise to multiple nerves that fall into as many as four main sensory categories:

- Mechanosensory lateral line.
- Electrosensory lateral line.
- Auditory.
- Vestibular sense.

In terrestrial vertebrates, the lateral line system is absent, and only the senses associated with the eighth nerve, that is, the auditory and vestibular senses, develop.

In the aquatic environment, the visual, auditory, and chemosensory systems aid in the detection of the presence of other animals. The lateral line system is an additional important detector of other animals. It functions over a greater range of environmental conditions than the other sensory systems—in calm, clear, illuminated water as well as in turbid, murky water and in darkness. The lateral line sensory system has two major components in many vertebrates: a mechanosensory system and an electrosensory system.

The mechanosensory lateral line allows an aquatic animal to perceive other moving animals at a distance. Current-like movements of the water caused by the motion of other animals (predators, prey, conspecific sexual partners, or other members in a school) are detected by the lateral line neuromasts and/or pit organs (see Chapter 2), the mechanosensory receptors. Behavioral studies have shown that such detections are responded to and are thus important to the animal. On the other hand, water displacements caused by the animal's own movements can be detected by the mechanoreceptors, but reactions to these stimuli are suppressed by other neural mechanisms. Motor control and related feedback mechanisms, rather than the lateral line, are used to regulate locomotory behavior.

The electrosensory lateral line is used for a wide range of functions. It is often used to detect and identify prey, whether the prey is another free-swimming animal or an animal concealed on the bottom under a layer of sand or mud. The electrical profile of such animals can be located and identified by the ampullary or tuberous receptors (see Chapter 2) of the lateral line electrosensory system. In the variety of fishes that can produce electrical signals—an electric organ discharge, or (**EOD**)—the electrosensory system plays an important role in social interactions, including the recognition of individual conspecifics within a social hierarchy.

The auditory and vestibular senses are common to aquatic and terrestrial vertebrates. Auditory receptors respond to pressure waves with accompanying particle motion, that is, to sound. The frequency of the waves and the location of their source are analyzed within the auditory system. Vestibular receptors are involved in maintaining equilibrium and respond to gravitational and other accelerational cues that allow for orientation of the body within the three spatial dimensions.

The Lateral Line System

The mechanosensory lateral line system is widely distributed in aquatic anamniotes. It was apparently present in the earliest vertebrates, as it has been identified in agnathans, cartilaginous fishes, bony fishes, lungfishes, the crossopterygian *Latimeria,* and aquatic amphibians. The mechanosensory lateral line system is thus a ubiquitous feature of anamniote vertebrates, and its evolutionary history is more conservative than that of electroreceptive lateral line systems.

An electroreceptive lateral line system has been evolved independently at least several times. Electroreception appears to have been evolved in the ancestral vertebrate stock of at least lampreys and jawed vertebrates, since morphologically similar electroreceptive systems are known to be present in lampreys, cartilaginous fishes, lungfishes, the crossopterygian *Latimeria,* and many amphibians. Among nonteleost bony fishes, a similar electroreceptive system is present in reedfishes and sturgeons. Electroreception appears to have been lost in the common ancestral stock of the Holostei (gars and the bowfin) and the Teleostei, however. It was then reevolved independently three or four times. These convergent electroreceptive systems are present in two different groups of ostariophysans, silurids and gymnotids, and in two different groups of osteoglossoforms, the mormyrids and notopterids.

The receptors for the mechanosensory lateral line usually lie within canals that are arrayed over the surface of the head and body. Figure 11-9 shows the distribution of the mechanosensory lateral line canals on the head of the coelacanth *Latimeria,* as well as the so-called rostral organ that contains electroreceptors in this species. The mechanoreceptors and electrore-

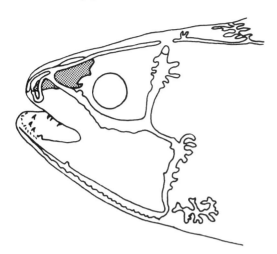

FIGURE 11-9. Drawing of the mechanoreceptive lateral line canals and the putative electroreceptive rostral organ (stippling) in the coelacanth *Latimeria*. Rostral is toward the left. Adapted from Northcutt (1986).

ceptors are innervated by up to six separate nerves, the **antero-dorsal, anteroventral, otic, middle, supratemporal, and posterior lateral line nerves.** Not all vertebrates with the lateral line system have all six nerves. The urodele amphibian *Ambystoma,* for example, has five lateral line nerves (Fig. 11-10), lacking an otic lateral line nerve.

In the medulla, the mechanoreceptive lateral line fibers terminate primarily in two structures (Fig. 11-11), the **eminentia granularis** of the cerebellum and part of the **lateralis column.** In bony fishes, the lateralis column consists of a large

nucleus that forms most of the column's rostrocaudal extent, called **nucleus medialis** or **nucleus intermedius.** The more caudal nucleus in the column, **nucleus caudalis,** also receives mechanosensory input. In cartilaginous fishes, the mechanoreceptive cell group is similarly called the medial (or intermediate) octavolateralis nucleus, and in amphibians, it is called nucleus intermedius.

The position of the column of nuclei that receive mechanoreceptive lateral line inputs is relative to one or two other cell columns present in the medulla. The mechanoreceptive nuclei generally lie dorsal and/or lateral to a cell column, the octaval column, which receives input from the eighth nerve [Fig. 11-12(A)]. In electroreceptive nonteleost bony fishes, the mechanoreceptive lateral line column additionally lies lateral or ventrolateral to the cell column that receives the electrosensory lateral line input [Fig. 11-12(B)], called **nucleus dorsalis,** while in electroreceptive teleosts, the mechanoreceptive lateral line column is medial or ventromedial to the electroreceptive area, called the **electrosensory lateral line lobe** (Fig. 11-13). In cartilaginous fishes, this cell column is called the **dorsal octavolateralis nucleus.** In electroreceptive amphibians, the mechanosensory nucleus intermedius lies ventral to the electrosensory column, which is called nucleus dorsalis.

The ascending projections that arise from both the mechanosensory and electrosensory lateral line nuclei travel with octaval fibers in the **lateral lemniscus.** This pathway is bilateral but projects predominantly to the contralateral side. The lateral lemniscus terminates in a part of the roof of the midbrain called the **torus semicircularis** in most anamniotes but is referred to as the **lateral mesencephalic nucleus** in some sharks and as a group of several nuclei—the **lateral mesencephalic complex**—in other sharks. Within the midbrain roof, the zones of

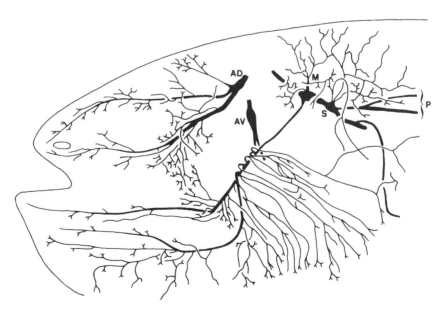

FIGURE 11-10. Drawing of the major branches of the five lateral line nerves present in the axolotl, *Ambystoma mexicanum,* which lacks an otic lateral line ramus. Rostral is toward the left. Abbreviations: AD, anterodorsal lateral line nerve; AV, anteroventral lateral line nerve; M, middle lateral line nerve; P, posterior lateral line nerve; S, supratemporal lateral line nerve. Adapted from Northcutt (1992).

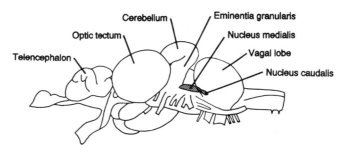

FIGURE 11-11. Lateral view of the brain of a goldfish (*Carassius auratus*) with the position of the lateralis column projected onto it. Adapted from Puzdrowski (1989). Used with permission of S. Karger AG.

termination of the mechanosensory, electrosensory, and octaval fibers remain separate (Fig. 11-14). Within each of the lateral line sensory systems, a spatial map of the input is also maintained. From the midbrain, lateral line input is relayed through the **preglomerular nuclear complex** in the caudal diencephalon in bony fishes (see Chapter 20), or its putative homologue in cartilaginous fishes, to the telencephalon.

The Octaval System

The octaval system consists of the auditory and vestibular rami of the eighth cranial nerve and their central structures and connections. These two rami are present in all vertebrate groups, but substantial evidence exists showing that at least part of the peripheral auditory receptor apparatus has been independently evolved in fishes and among tetrapods. Similarities in the central nervous parts of these systems allow for the possibility that at least some of the auditory nuclei and pathways were maintained through periods of change in the peripheral

receptor apparatus. Thus, some of the octaval nuclei may be homologous among the various groups of vertebrates, particularly at midbrain and more rostral levels, but some of the relationships of the various medullary octaval nuclei in different vertebrate groups remain open to question.

In most vertebrates, vestibular and auditory fibers terminate within various parts of the brainstem, including the cerebellum, reticular formation, and the **octaval column** in the medulla. The octaval column is the main auditory region involved in relaying the input to more rostral parts of the brain. Where a lateral line system is also present, the octaval column lies ventral to it, as referred to above.

In agnathans, the eighth nerve terminates within the octaval column but not in any other brainstem sites. In lampreys, the octaval column has three nuclei, two of which, the **ventral** and **octavomotor** nuclei, receive the bulk of the eighth nerve fibers. In hagfishes, the eighth cranial nerve projections are primarily confined to a single area, called the **ventral nucleus of the area acousticolateralis.**

In agnathans and jawed fishes, at least some of the octaval endorgans are thought to process both auditory and vestibular information. In fishes, the nerves from all of the octaval endorgans primarily terminate within the octaval column, which com-

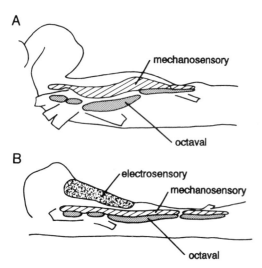

FIGURE 11-12. Projection onto a lateral view of the brainstem of nuclei receiving (A) mechanosensory (diagonal lines) and octaval (fine stippling) projections in the bowfin (*Amia*) and (B) electrosensory (random stippling), mechanosensory (diagonal lines), and octaval (fine stippling) projections in a sturgeon (*Scaphirhynchus*). Adapted from McCormick (1989). Used with permission of Springer-Verlag.

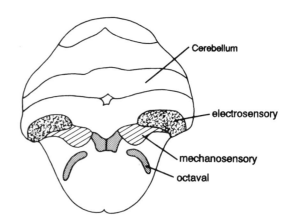

FIGURE 11-13. Drawing of a transverse section through the brainstem of the osteoglossomorph fish *Xenomystus* showing the relative positions of the electrosensory (random stippling), mechanosensory (diagonal lines), and octaval (fine stippling) nuclei. Adapted from McCormick (1989). Used with permission of Springer-Verlag.

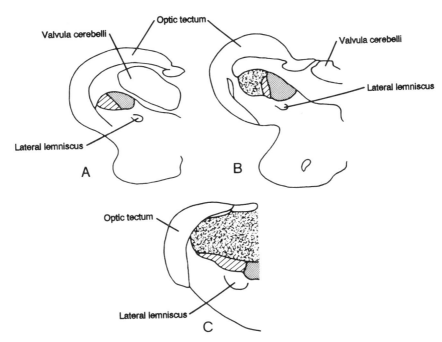

FIGURE 11-14. Drawings of transverse hemisections through the brainstem in three ostariophysan fishes: (A) *Cyprinus*, (B) *Ictalurus*, and (C) *Eigenmannia*. The areas of termination of ascending projections in the torus semicircularis are shown for electrosensory (random stippling, as in Figs. 11–12 and 11–13), mechanosensory (diagonal lines), and octaval (fine stippling) pathways. Adapted from McCormick (1989). Used with permission of Springer-Verlag.

prises at least four nuclei, the **anterior, magnocellular, descending,** and **posterior nuclei** (Fig. 11-12). Auditory input that is relayed through the octaval column to more rostral levels is primarily routed through the dorsal parts of the anterior and descending nuclei. Auditory input is relayed to the roof of the midbrain—the torus semicircularis in most anamniotes and the lateral mesencephalic nucleus or complex in cartilaginous fishes. The fibers ascend via the lateral lemniscus in a primarily contralateral pathway, as do the lateral line projections (Fig. 11-14). Just as a spatial map is maintained in the lateral line system, a tonotopic representation is maintained in the auditory pathway. In bony fishes, an auditory pathway from the midbrain to the dorsal thalamus, and thence to the telencephalon, has been traced. Auditory responses have been recorded in the telencephalon of sharks, although the anatomy of the ascending pathway remains to be studied.

In amphibians, two patterns of octaval nuclei in the medulla exist. One of these patterns is seen in nonanuran amphibians, in which an octaval column is present ventral to the lateral line sensory zone. This column can be divided into three nuclei that correspond to the anterior, magnocellular, and descending nuclei of fishes. The dorsal parts of these nuclei primarily receive auditory input, while the ventral parts primarily receive vestibular input.

The second pattern is present in anurans. In most adult anurans, the mechanosensory lateral line system is absent, and two columns in receipt of octaval input are present. The dorsal column contains one nucleus, the **dorsolateral nucleus,** which receives auditory input. The ventral column contains up to four nuclei: the **anterior, lateral octaval, medial vestibular,** and **caudal nuclei.** Auditory inputs are primarily confined to the dorsal parts of these nuclei and vestibular inputs to their ventral parts. The dorsolateral nucleus is the main nucleus for relay of auditory information to more rostral parts of the brain. The most prominent rostral projection is via the lateral lemniscus directly to the torus semicircularis and thence, via the dorsal thalamus, to the telencephalon. Pathways that relay auditory information to the torus semicircularis through secondary medullary nuclei, particularly a nucleus called the **superior olivary nucleus,** are also present.

In nonmammalian amniotes, the cochlear nerve terminates primarily in two medullary nuclei, a laterally lying **nucleus angularis** and a medially lying **nucleus magnocellularis.** Nucleus magnocellularis projects bilaterally to **nucleus laminaris** (Fig. 11-15), which is a third nucleus that lies between the two cochlear receptive nuclei. Nuclei angularis and laminaris give rise to bilateral pathways that project directly to the roof of the midbrain and also indirectly to it via relays through other brainstem nuclei, including the **superior olivary nucleus** and a nucleus embedded within the lateral lemniscal fibers, called the **nucleus of the lateral lemniscus.** In all amniotes, the auditory part of the roof of the midbrain projects, via the dorsal thalamus, to the telencephalon.

In the ancestral stock of mammals, bony elements of the jaw were incorporated into the middle ear, which allowed for the conduction of higher frequency sounds to the oval window and the auditory receptive neural apparatus, the cochlea. This change is correlated with the presence of additional, phylo-

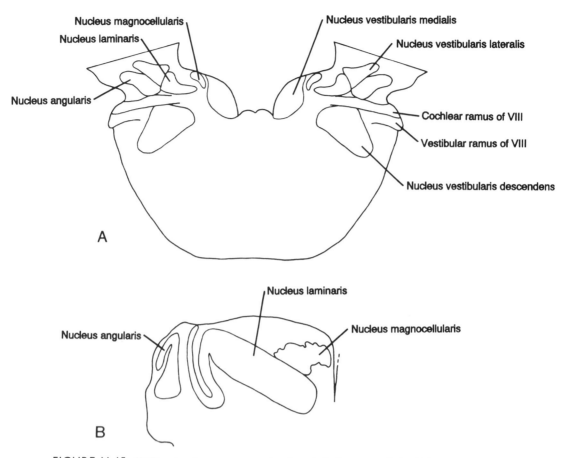

FIGURE 11-15. (A) Drawing of a transverse section through the brainstem of a bird (*Columba livia*) showing the position of octaval nuclei. Adapted from Karten and Hodos (1967). Used with permission of the Johns Hopkins University Press. (B) Drawing of a horizontal hemisection through the auditory nuclei in the brainstem of a barn owl (*Tyto alba*). Adapted from Carr and Boudreau (1991).

genetically new nuclei in the auditory brainstem of mammals. A number of the major nuclei present in mammals, however, are thought to be homologous to auditory nuclei in nonmammalian amniotes.

Of the cochlear nuclei present in mammals, the **anteroventral cochlear nucleus** appears to be homologous to the nucleus magnocellularis of nonmammalian amniotes, while the **posteroventral** and **dorsal cochlear nuclei** appear to be homologous as a field to nucleus angularis. The **medial superior olive** in mammals appears to be homologous to nucleus laminaris of nonmammalian amniotes, while the mammalian **lateral superior olive** is the homologue of the superior olivary nucleus of nonmammals. Ascending pathways, directly and indirectly to the midbrain roof, and thence via the dorsal thalamus to the telencephalon, similar to the ascending auditory pathways in nonmammalian amniotes, are present in mammals.

The vestibular nerve is relatively conservative among tetrapods, terminating in vestibular nuclei that lie in the same region of the medulla as the auditory nuclei. Two vestibular nuclei, a **superior** and a **lateral,** have been reported in anurans, with a possibility of two additional nuclei. In reptiles, four nuclei have been described: **dorsolateral, ventrolateral, ventromedial,** and **descending.** The most complex organization of

the vestibular nuclei has been found in birds, which have six: **descending, medial, rostral, tangential, dorsolateral,** and **ventrolateral.** In mammals, **superior, lateral, medial,** and **inferior vestibular nuclei** have been identified. The vestibular nuclei project to the cerebellum, some of the other motor cranial nerve nuclei, and the spinal cord. They also give rise to ascending projections to a region related to the somatosensory part of the dorsal thalamus.

FOR FURTHER READING

Bodznick, D. and Northcutt, R. G. (1981) Electroreception in lampreys: evidence that the earliest vertebrates were electroreceptive. *Science,* 212, 465–467.

Finger, T. E. (1988) Organization of chemosensory systems within the brains of bony fishes. In J. Atema, R. R. Fay, A. N. Popper, and W. N. Tavolga (eds.), *Sensory Biology of Aquatic Animals.* New York: Springer-Verlag, pp. 339–363.

Heiligenberg, W. (1988) Electrosensory maps form a substrate for the distributed and parallel control of behavioral responses in weakly electric fish. *Brain, Behavior and Evolution,* 31, 6–16.

Molenaar, G. J. (1992) Anatomy and physiology of infrared sensitivity of snakes. In C. Gans and P. S. Ulinski (eds.), *Biology of the*

Reptilia, Vol. 17: Neurology C, Sensorimotor Integration. Chicago: University of Chicago Press, pp. 367–453.

McCormick, C. A. (1989) Central lateral line mechanosensory pathways in bony fish. In S. Coombs, P. Görner, and H. Münz (eds.), *The Mechanosensory Lateral Line: Neurobiology and Evolution.* New York: Springer-Verlag, pp. 341–364.

McCormick, C. A. and Braford, M. R., Jr. (1993) The primary octaval nuclei and inner ear afferent projections in the otophysan *Ictalurus punctatus. Brain, Behavior and Evolution, 42,* 48–68.

ADDITIONAL REFERENCES

Amemiya, F., Kishida, R., Goris, R. C., Onishi, H., and Kusunoki, T. (1985) Primary vestibular projections in the hagfish, *Eptatretus burgeri. Brain Research, 337,* 73–79.

Arends, J. J. and Dubbeldam, J. L. (1984) The subnuclei and primary afferents of the descending trigeminal system in the mallard (*Anas platyrhynchos*). *Neuroscience, 13,* 781–795.

Arends, J. J. A., Zeigler, H. P., and Wild, J. M. (1988) Projections of the nucleus tractus solitarius in the pigeon. *Journal of Comparative Neurology, 278,* 405–429.

Atema, J. (1971) Structure and functions of the sense of taste in the catfish *Ictalurus natalis. Brain, Behavior and Evolution, 4,* 273–294.

Atema, J., Fay, R. R., Popper, A. N., and Tavolga, W. N. (eds.), 1988, *Sensory Biology of Aquatic Animals.* New York: Springer-Verlag.

Bangma, G. C. and ten Donkelaar, H. J. (1983) Some afferent and efferent connections of the vestibular nuclear complex in the red-eared turtle, *Pseudemys scripta elegans. Journal of Comparative Neurology, 220,* 453–464.

Beason, R. C., Dussourd, N. and Deutschlander, M. E. (1995) Behavioral evidence for the use of magnetic material in magnetoreception by a migratory bird. *Journal of Experimental Biology, 198,* 141–146.

Beason, R. C. and Semm, P. (1987) Magnetic responses of the trigeminal nerve system of the bobolink (*Dolichonyx oryzivorus*). *Neuroscience Letters, 80,* 229–234.

Beckstead, R. M., Morse, J. R., and Norgren, R. (1980) The nucleus of the solitary tract in the monkey: projections to the thalamus and brain stem nuclei. *Journal of Comparative Neurology, 190,* 259–282.

Beckstead, R. R. and Norgren, R. (1979) An autoradiographic examination of the central distribution of the trigeminal, facial, glossopharyngeal, and vagal nerves in the monkey. *Journal of Comparative Neurology, 184,* 455–472.

Bohringer, R. C. (1981) Cutaneous receptors in the bill of the platypus. *Australian Mammals, 4,* 93–105.

Boord, R. L. and Northcutt, R. G. (1988) Medullary and mesencephalic pathways and connections of lateral line neurons of the spiny dogfish *Squalus acanthias Brain, Behavior and Evolution, 32,* 76–88.

Bullock, T. H. and Heiligenberg, W. (eds.) (1986) *Electroreception.* New York: Wiley.

Bullock, T. H., Bodznick, D. A., and Northcutt, R. G. (1983) The phylogenetic distribution of electroreception: evidence for convergent evolution of a primitive vertebrate sense modality. *Brain Research Reviews, 6,* 25–46.

Bullock, T. H. and Fox, W. (1957) The anatomy of the infrared sense organ in the facial pit of pit vipers. *Quarterly Journal of Microscopic Science, 98,* 219–234.

Bullock, T. H., Northcutt, R. G., and Bodznick, D. A. (1982) Evolution of electroreception. *Trends In Neurosciences, 5,* 50–53.

Carr, C. E. and Boudreau, R. E. (1991) Central projections of auditory nerve fibers in the barn owl. *Journal of Comparative Neurology, 314,* 306–318.

Carr, C. E. and Boudreau, R. E. (1993) Organization of the nucleus magnocellularis and the nucleus laminaris in the barn owl: encoding and measuring interaural time differences. *Journal of Comparative Neurology, 334,* 337–355.

Coombs, S., Görner, P., and Münz, H. (eds.) (1989) *The Mechanosensory Lateral Line: Neurobiology and Evolution.* New York: Springer-Verlag.

Coombs, S. and Montgomery, J. (1992) Fibers innervating different parts of the lateral line system of an antarctic notothenioid, *Trematomus bernacchii,* have similar frequency responses, despite large variation in the peripheral morphology. *Brain, Behavior and Evolution, 40,* 217–233.

Ebbesson, S. O. E. (1981) Projections of the optic tectum and the mesencephalic nucleus of the trigeminal nerve in the tegu lizard (*Tupinambis nigropunctatus*). *Cell and Tissue Research, 216,* 151–165.

Eden, A. R. and Correia, M. J. (1982) Identification of multiple groups of efferent vestibular neurons in the adult pigeon using horseradish peroxidase and DAPI. *Brain Research, 248,* 201–208.

Finger, T. E. (1991) Gustatory nuclei and pathways in the central nervous system. In T. E. Finger and W. L. Silver (eds.), *Neurobiology of Taste and Smell.* Malabar, FL: Krieger Publishing Co., pp. 331–353.

Finger, T. E. (1993) What's so special about special visceral? *Acta Anatomica, 148,* 132–138.

Finger, T. E. and Morita, Y. (1985) Two gustatory systems: facial and vagal gustatory nuclei have different brainstem connections. *Science, 227,* 776–778.

Fritzsch, B. (1988) The lateral-line and inner-ear afferents in larval and adult urodeles. *Brain, Behavior and Evolution, 31,* 325–348.

Fritzsch, B. and Northcutt, R. G. (1993) Cranial and spinal nerve organization in amphioxus and lampreys: evidence for an ancestral craniate pattern. *Acta Anatomica, 148,* 96–109.

Fritzsch, B., Ryan, M. J., Wilczynski, W., Hetherington, T. E., and Walkowiak, W. (eds.) (1988) *The Evolution of the Amphibian Auditory System.* New York: Wiley.

Fuller, P. M. and Ebbesson, S. O. E. (1973) Central projections of the trigeminal nerve in the bull frog. *Journal of Comparative Neurology, 152,* 193–200.

Gonzalez, A. and Muñoz, M. (1988) Central distribution of the efferent cells and the primary afferent fibers of the trigeminal nerve in *Pleurodeles waltlii* (Amphibia, Urodela). *Journal of Comparative Neurology, 270,* 517–527.

Gordon, K. D. and Caprio, J. (1985) Taste responses to amino acids in the southern leopard frog, *Rana sphenocephala. Comparative Biochemistry and Physiology, Part A, 81,* 525–530.

Gregory, J. E., Iggo, A., McIntyre, A. K., and Proske, U. (1987) Electroreceptors in the platypus. *Nature (London), 326,* 387.

Gregory, J. E., Iggo, A., McIntyre, A. K., and Proske, U. (1989) Responses of electroreceptors in the snout of the echidna. *Journal of Physiology, 414,* 521–538.

Hartline, P. (1974) Thermoreception in snakes. In A. Fessard (ed.), *Electroreceptors and Other Specialized Receptors, Handbook of Sensory Physiology, Vol. III/3.* Berlin: Springer-Verlag, pp. 297–312.

Hiscock, J. and Straznicky, C. (1982) Peripheral and central terminations of axons of the mesencephalic trigeminal neurons in *Xenopus. Neuroscience Letters, 32,* 235–240.

Housley, G. D. and Montgomery, J. C. (1983) Central projections of vestibular afferents from the horizontal semicircular canal in the carpet shark *Cephaloscyllium isabella. Journal of Comparative Neurology,* 221, 154–162.

Karten, H. J. and Hodos, W. (1967) *A Stereotaxic Atlas of the Brain of the Pigeon (Columba livia).* Baltimore, MD: The Johns Hopkins Press.

Kishida, R., Goris, R. C., Nishizawa, H., Koyama, H., Kadota, T., and Amemiya, F. (1987) Primary neurons of the lateral line nerves and their central projections in hagfishes. *Journal of Comparative Neurology,* 264, 303–310.

Kishida, R., Goris, R. C., Terashima, S., and Dubbeldam, J. L. A. (1984) A suspected infrared-recipient nucleus in the brainstem of the vampire bat, *Desmodus rotundus. Brain Research,* 322, 351–355.

Kiyohara, S., Shiratani, T., and Yamashita, S. (1985) Peripheral and central distribution of major branches of the facial taste nerve in the carp. *Brain Research,* 325, 57–69.

Kojama, H., Kishida, R., Goris, R. C., and Kusunoki, T. (1989) Afferent and efferent projections of the VIIIth cranial nerve in the lamprey *Lampetra japonica. Journal of Comparative Neurology,* 280, 663–671.

Lamb, C. F. and Caprio, J. (1992) Convergence of oral and extraoral information in the superior secondary gustatory nucleus of the channel catfish. *Brain Research,* 588, 201–211.

Lannoo, M. J., Vischer, H. A., and Maler, L. (1990) Development of the electrosensory nervous system of *Eigenmannia* (Gymnotiformes): II. The electrosensory lateral line lobe, midbrain, and cerebellum. *Journal of Comparative Neurology,* 294, 37–58.

Luiten, P. G. M. (1979) Proprioceptive connections of the head musculature and the mesencephalic trigeminal nucleus in the carp. *Journal of Comparative Neurology,* 183, 903–912.

Manger, P. R. and Hughes, R. L. (1992) Ultrastructure and distribution of epidermal sensory receptors in the beak of the echidna, *Tachyglossus aculeatus. Brain, Behavior and Evolution,* 40, 287–296.

Manger, P. R., Pettigrew, J. D., and McLachlan, E. M. (1993) Platypus electroreception: behavior, physiology and anatomy. *Society for Neuroscience Abstracts,* 19, 374.

Marini, R. and Bortolami, R. (1982) A somatotopic and functional organization of the masticatory trigeminal nucleus of the frog. *Archives of Italian Biology,* 120, 385–396.

Meredith, G. E. and Butler, A. B. (1983) Organization of eighth nerve afferent projections from individual endorgans of the inner ear in the teleost, *Astronotus ocellatus. Journal of Comparative Neurology,* 220, 44–62.

Molenaar, G. J. (1978) The sensory trigeminal system of a snake in possession of infrared receptors. I. The sensory trigeminal nuclei. *Journal of Comparative Neurology,* 179, 123–136.

Molenaar, G. J. (1978) The sensory trigeminal system of a snake in possession of infrared receptors. II. The central projections of the trigeminal nerve. *Journal of Comparative Neurology,* 179, 137–152.

Molenaar, G. J., Fizaan-Oostveen, J. L. F. P., and van der Zalm, J. M. (1979) Infrared and tactile units in the sensory trigeminal system of *Python reticulatus. Brain Research,* 170, 372–376.

Montgomery, J. and Coombs, S. (1992) Physiological characterization of lateral line function in the antarctic fish *Trematomus bernacchii. Brain, Behavior and Evolution,* 40, 209–216.

Montgomery, N. M. (1988) Projections of the vestibular and cerebellar nuclei in *Rana pipiens. Brain, Behavior and Evolution,* 31, 82–95.

Morita, Y., Ito, H., and Masai, H. (1980) Central gustatory paths in the crucian carp, *Carassius carassius. Journal of Comparative Neurology,* 191, 119–132.

Morita, Y. and Finger, T. E. (1985) Reflex connections of the facial and vagal gustatory systems in the brainstem of the bullhead catfish,

Ictalurus nebulosus. Journal of Comparative Neurology, 231, 547–558.

New, J. G. and Northcutt, R. G. (1984) Primary projections of the trigeminal nerve in two species of sturgeon: *Acipenser oxyrhynchus* and *Schaphirhynchus platorhynchus. Journal of Morphology,* 182, 125–136.

New, J. G. and Singh, S. (1994) Central topography of anterior lateral line nerve projections in the channel catfish, *Ictalurus punctatus. Brain, Behavior and Evolution,* 43, 34–50.

Norgren, R. (1983) The gustatory system in mammals. *American Journal of Otolaryngology,* 4, 234–237.

Norgren, R. and Pfaffmann, C. (1975) The pontine taste area in the rat. *Brain Research,* 91, 99–117.

Northcutt, R. G. (1979) Central projections of the eighth cranial nerve in lampreys. *Brain Research,* 167, 163–167.

Northcutt, R. G. (1979) Experimental determination of the primary trigeminal projections in lampreys. *Brain Research,* 163, 323–327.

Northcutt, R. G. (1980) Anatomical evidence of electroreception in the coelacanth (*Latimeria chalumnae*). *Zentralblatt für Vetrinaermedizin. Reich C.; Anatomia, Histologia, Embryologia,* 9, 289–295.

Northcutt, R. G. (1986) Electroreception in nonteleost bony fishes. In T. H. Bullock and W. Heiligenberg (eds.), *Electroreception.* New York: Wiley, pp. 257–285.

Northcutt, R. G. (1992) Distribution and innervation of lateral line organs in the axolotl. *Journal of Comparative Neurology,* 325, 95–123.

Northcutt, R. G. (1994) Development of lateral line organs in the axolotl. *Journal of Comparative Neurology,* 340, 480–514.

Northcutt, R. G. and Bemis, W. E. (1993) Cranial nerves of the coelacanth *Latimeria chalumnae* [Osteichthyes: Sarcopterygii: Actinistia] and comparisons with other craniata. *Brain, Behavior and Evolution,* 42, S1.

Northcutt, R. G. and Brändel, K. (1995) Development of branchiomeric and lateral line nerves in the axolotl. *Journal of Comparative Neurology,* 355, 427–454.

Northcutt, R. G., Brändel, K. and Fritzsch, B. (1995) Electroreceptors and mechanosensory lateral line organs arise from single placodes in axolotls. *Developmental Biology,* 168, 358–373.

Northcutt, R. G., Catania, K. C. and Criley, B. B. (1994) Development of lateral line organs in the axolotl. *Journal of Comparative Neurology,* 340, 480–514.

Puzdrowski, R. L. (1989) Peripheral distribution and central projections of the lateral-line nerves in goldfish. *Brain, Behavior and Evolution,* 34, 110–131.

Puzdrowski, R. L. and Northcutt, R. G. (1993) The octavolateral systems in the stingray, *Dasyatis sabina.* I. Primary projections of the octaval and lateral line nerves. *Journal of Comparative Neurology,* 332, 21–37.

Roberts, B. L. and Witovsky, P. (1975) A functional analysis of the mesencephalic fifth nerve in the selachian brain. *Proceedings of the Royal Society, London (Biology),* 190, 473–495.

Ronan, M. (1988) The sensory trigeminal tract of Pacific hagfish. Primary afferent projections and neurons of the tract nucleus. *Brain, Behavior and Evolution,* 32, 169–180.

Roth, A. and Schlegal, P. (1988) Behavioral evidence and supporting electrophysiological observations for electroreception in the blind cave salamander, *Proteus anguinus* (Urodela). *Brain, Behavior and Evolution,* 32, 227–280.

Scheich, H., Langner, G., Tidemann, C., Coles, R. B., and Guppy, A. (1986) Electroreception and electrolocation in the platypus. *Nature (London),* 319, 401–402.

Schroeder, D. M. and Loop, M. S. (1976) Trigeminal projections in snakes possessing infrared sensitivity. *Journal of Comparative Neurology,* 169, 1–14.

Shingai, T. and Beidler, L. M. (1985) Response characteristics of three taste nerves in mice. *Brain Research,* 335, 245–249.

Song, J. and Northcutt, R. G. (1991) Morphology, distribution and innervation of the lateral-line receptors of the Florida gar, *Lepisosteus platyrhincus. Brain, Behavior and Evolution,* 37, 10–37.

Song, J. and Northcutt, R. G. (1991) The primary projections of the lateral-line nerves of the Florida gar, *Lepisosteus platyrhincus. Brain, Behavior and Evolution,* 37, 38–63.

Stanford, L. R. and Hartline, P. H. (1984) Spatial and temporal integration in primary trigeminal nucleus of rattlesnake infrared system. *Journal of Neurophysiology,* 51, 1077–1090.

Stanford, L. R., Schroeder, D. M., and Hartline, P. H. (1981) The ascending projection of the nucleus of the lateral descending trigeminal tract: a nucleus in the infrared system of the rattlesnake, *Crotalus viridis. Journal of Comparative Neurology,* 201, 161–173.

Torvik, A. (1957) The ascending fibers from the main trigeminal sensory nucleus. An experimental study in the cat. *American Journal of Anatomy,* 100, 1–15.

Vischer, H. A. (1989) The development of lateral-line receptors in *Eigenmannia* (Teleostei, Gymnotiformes). I. The mechanoreceptive lateral-line system. *Brain, Behavior and Evolution,* 33, 205–222.

Vischer, H. A. (1989) The development of lateral-line receptors in *Eigenmannia* (Teleostei, Gymnotiformes). II. The electroreceptive lateral-line system. *Brain, Behavior and Evolution,* 33, 223–236.

Walker, A. E. (1939) The origin, course and terminations of the secondary pathways of the trigeminal nerve in primates. *Journal of Comparative Neurology,* 71, 59–89.

Webb, J. F. (1988) Gross morphology and evolution of the mechanoreceptive lateral-line system in teleost fishes. *Brain, Behavior and Evolution,* 33, 34–53.

Webster, D. B., Fay, R. R., and Popper, A. N. (eds.) (1992) *The Evolutionary Biology of Hearing.* New York: Springer-Verlag.

Wild, J. M., Arends, J. J., and Zeigler, H. P. (1985) Telencephalic projections of the trigeminal system in the pigeon (*Columba livia*): a trigeminal sensorimotor circuit. *Journal of Comparative Neurology,* 234, 441–464.

Wild, J. M., Arends, J. J. A., and Zeigler, H. P. (1990) Projections of the parabrachial nucleus in the pigeon. *Journal of Comparative Neurology,* 293, 499–523.

Wold, J. E. (1975) The vestibular nuclei in the domestic hen (*Gallus domesticus*). II. Primary afferents. *Brain Research,* 95, 531–543.

12

Motor Cranial Nerves

INTRODUCTION

Like the sensory cranial nerves, the motor cranial nerves fall into three distinct categories: dorsal cranial nerves, ventral cranial nerves, and parasympathetic components that innervate organs such as the salivary glands. The head provides a site of attachment for many muscles, both external and internal. The dorsal cranial nerves innervate the external muscles of the head, while the ventral cranial nerves innervate the internal muscles.

Muscles that can be categorized as internal muscles of the head include those within the orbits and those that form the tongue. Muscles within the orbits (eye sockets) move the eyes and are innervated by three ventral cranial nerves, called the oculomotor nerves. Other internal muscles that form the tongue are innervated by the fourth ventral cranial nerve, the hypoglossal nerve. One of the most important roles of the tongue in terrestrial vertebrates is its use in feeding and swallowing, which we will consider in detail here. Over the course of evolution among tetrapods, additional uses of the tongue for a variety of other purposes were also selected for, including being a tasting organ for the substances that it manipulates (all tetrapods), a sticky instrument for the capture of prey (frogs, salamanders, and anteaters), an instrument for grooming the animal itself or grooming others (many mammals), a device for conducting pheromones (chemical signals) into the vomeronasal organ (many reptiles and mammals), and a tool for boring into trees in search of insect prey (woodpeckers). Another role for the tongue, one that is very likely derived from its ability to modulate the flow of a column of water on its way into the throat, is its ability to modulate the flow of a column of air on its way out of the throat. When such an air column is used for vocal communication as in vocalizing mammals, birds, and reptiles, the tongue also becomes an important part of this communication system.

The external muscles of the head operate the jaws, the eyelids, the nostrils, the external ears, the scalp, the lips, the cheeks, and the eyebrows in those animals that possess such organs. Other external muscles are attached at one end to the head and at the other end to the body skeleton and move the head itself in those animals with a moveable head. Finally, some muscles that are external muscles in most vertebrates are located within the middle ear in mammals and are involved in protecting the delicate inner ear mechanisms from intense sounds. The majority of the external muscles of the head are controlled by motor neurons whose axons exit from the brainstem by way of the trigeminal and dorsal facial nerves. In this chapter, we will first address the control of feeding and swallowing and the cranial nerves involved in these processes. We will then discuss another important function of the trigeminal and dorsal facial nerves, the acoustic reflex. Finally, we will consider the oculomotor muscles and their innervation.

FEEDING AND SWALLOWING

Animals that evolved in the water and have remained there throughout their history move food from the mouth into the throat and digestive tract by means of suction pressure. Animals that feed in air have to overcome the added difficulties of the friction of the food with the sides of the mouth, the food's lack of buoyancy, and the pull of gravity. In the transition from life in the water to life on land, the bones of the head, especially those of the jaws, underwent considerable adaptive modification in response to the new selective pressures. Along with these changes came alterations in the muscles of the head that open and close the jaws. The early amphibian tetrapods had to deal with these selective pressures in order to feed in the terrestrial environment. Those that could manipulate the muscles of the hyoid arch (a bony remnant of the second visceral arch, located in the floor of the mouth) to help move food along in the mouth and throat were able to feed successfully in air. Eventually, other muscles derived from somites in the fifth (second metaotic) head segment became well developed and evolved into a specialized organ: the tongue.

The problems of feeding and swallowing are very different in water and in air. For example, filter feeding, in which minute particles are sieved from the water being taken into the mouth, is only possible in an aquatic environment. Many aquatic feeders have evolved a negative-pressure mechanism that draws prey into the mouth. A number of terrestrial species have in fact readapted to aquatic feeding, such as turtles, crocodiles, penguins, and the various aquatic mammals (whales, porpoises, seals, otters, etc.), and some have reevolved suction feeding. In contrast, suction plays little role in the movement of food into and through the mouth and throat of air-feeding animals, except for drinking. Air-feeding animals instead depend on the tongue to move food through the mouth and into the throat for swallowing. In mammals, many of which chew their food to break it into smaller particles before swallowing, the tongue plays an important role in manipulating the food particles to and from the moving rows of teeth en route to the throat.

The Neural Control of Feeding and Swallowing

The fleshy tongue, which is present only in tetrapods, is controlled by a group of neurons located at the caudal end of the medulla, near its junction with the spinal cord. The axons of these neurons form one of the ventral cranial nerves, the **hypoglossal nerve (XII),** and their cell bodies are collectively called the **hypoglossal nucleus** (Fig. 12-1). The muscles of the tongue are complex, and their action is to change the shape of the tongue, as well as to raise and lower it and move it forward. Each muscle is controlled by a different group of neurons within the hypoglossal nucleus. The backward movement of the tongue is accomplished by muscles that elevate its hyoid skeleton. These muscles are innervated by axons of motor VII_D.

Neurons that form the nucleus of VII_D, the **motor nucleus of the dorsal facial nerve,** lie in the somatic efferent column of the brainstem (Fig. 12-1), rostral to the hypoglossal nucleus. In addition to innervating muscles of the hyoid apparatus, the efferent axons of **motor VII_D** innervate the superficial muscles of the head, including muscles of the cheeks, lips, nares (nostrils), and the forehead muscles. In mammals, in which the muscles of the head and face are very well developed, motor VII_D controls the movements of the pinna of the ear, which is the external, sound collecting portion of the ear, as well as the muscles of facial expression (lips, cheeks, eyebrows, etc.), which in some species are very important in social communications such as those of aggression and appeasement. The **motor nucleus of the trigeminal nerve** (motor V) lies in the somatic efferent column rostral to motor VII_D. The efferents of **motor V** exit the brainstem through the mandibular branch of V and terminate on the muscles of the jaws.

Two important influences over the actions of motor V and motor VII_D are the senses of taste and touch. The gustatory nucleus and the descending nucleus of V are the central cell populations that receive the incoming taste and somatic axons,

as discussed in Chapter 11. These cell groups, in turn, send some of their axons directly to motor neurons of the somatic efferent column that control muscles of the head, as well as to cell populations of the reticular formation. These reticular formation cell groups, scattered throughout the brainstem, serve to coordinate various reflexes and voluntary motor behavior. They also serve to integrate information from various senses. For example, a fish might detect something edible in the nearby water; not only would the reticular formation coordinate the activities of motor V and motor VII_D so that the muscles of the jaws and face would produce the necessary suction to ingest the food, but it would also stimulate or inhibit the activities of the oculomotor nerves in order to move the eyes in the direction of the food source so that both visual and gustatory information could be coordinated.

In air-feeding animals, motor V and motor VII_D control muscles of the jaws that press the food objects against the roof of the mouth and manipulate them in the direction of the throat. Of crucial importance to these motor responses are inputs from the mesencephalic nucleus of V, which provides feedback from the muscles of the jaws (see Fig. 11-3). Ascending sensory pathways to the telencephalon via the dorsal thalamus and descending motor pathways from the telencephalon to the reticular formation also play a role in the integration and coordination of these very precise and complex movements.

In ray-finned fishes, such as catfishes and carps, which have excellent gustatory systems, motor V receives projections from the facial lobe (see Chapter 11). Motor V and motor VII_D also receive projections from the descending nucleus of V, the cerebellum, and the midbrain tectum. These connections may serve to coordinate opening and closing the mouth in relation to other sensory systems such as somatosensory, vestibular, and lateral line. Motor V also receives axons from the reticular formation and the hypothalamus. In addition, reciprocal connections between motor V and motor VII_D serve to further coordinate their activities. Although motor V and motor VII_D control the ingestion of food into the mouth, swallowing is under the control of neurons located farther caudally, in a nucleus called **nucleus ambiguus.** This nucleus sends its efferent axons via the dorsal glossopharyngeal (IX_D), dorsal vagus (X_D), and accessory (XI) nerves to the muscles of the pharynx and palate. Nucleus ambiguus receives projections from the vagal (X_{VL}) lobe, which, together with the facial (VII_{VL}) lobe, is an expansion of the gustatory region of the caudal medulla. Thus, the facial gustatory system appears to be an ingestive system whereas the vagal gustatory system appears to be a swallowing system.

Motor V is well developed in birds, of which pigeons and ducks have been the most extensively studied. These studies have described the mechanisms for operation of the jaws during various motor acts related to feeding, such as opening the mouth in proportion to the size of the object to be taken in, pecking, grasping, manipulation of food in the mouth and throat, and swallowing. Motor V is also well developed in those mammals that chew their food.

In air-feeding animals, the lack of a water column to lubricate the food has been compensated for by the evolution of the salivary glands. These glands are present only in amniotes and are controlled by the parasympathetic nervous system. Their postganglionic fibers are innervated by axons from the

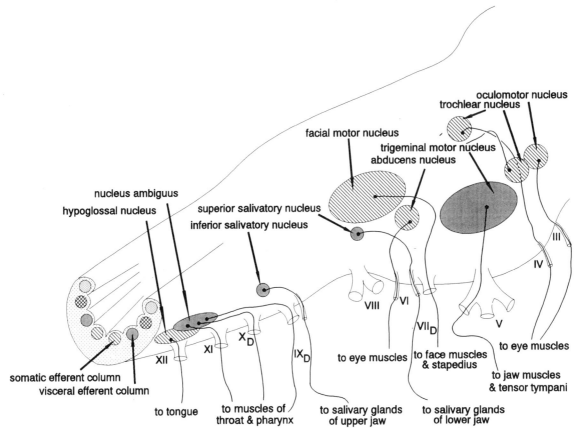

FIGURE 12-1. Schematic drawing of the motor nuclei in the brainstem that innervate the muscles of the head in a mammal. Both the dorsal cranial nerve nuclei discussed here and the ventral cranial nerve nuclei (III, IV, IV, and XII), which are discussed below, are included. Beginning at the rostral end of the brainstem, the nucleic of the oculomotor (III), trochlear (IV) and abducens (VI) nerves send axons to the muscles of the eyes. The motor nucleus of the trigeminal (V) nerve supplies axons to the muscles of the jaws and the tensor tympani muscle of the middle ear. The dorsal facial (VII$_D$) motor nucleus controls the muscles of the face, snout, external ear, and the stapedius muscle in the middle ear. The superior (VII$_D$) and inferior (IX$_D$) salivatory nuclei send parasympathetic axons to the salivatory glands. Nucleus ambiguus sends axons to the muscles of the pharnyx and throat via the dorsal glossopharyngeal (IX$_D$), dorsal vagus (X$_D$), and accessory (XI) nerves. The hypoglossal nucleus (XII) nucleus controls the muscles of the tongue.

parasympathetic components of motor VII$_D$ and motor IX$_D$ (Fig. 12-1). Salivary glands not only provide lubrication of food; they are also essential for the grooming behaviors in a number of groups of mammals, which involve the jaws and teeth as well as the tongue. In some snakes, the salivary glands produce neurotoxins and/or anticoagulants. The neurons of the parasympathetic division of motor VII$_D$ form the **superior salivatory nucleus;** this nucleus controls the submandibular and sublingual salivary glands in the lower jaw. The parasympathetic division of motor IX$_D$ consists of the neurons in the **inferior salivatory nucleus,** which innervates the parotid and infraorbital salivary glands in the upper jaw. The latter gland is absent in humans.

THE ACOUSTIC REFLEX

Among the more striking changes in the head in response to the transition from the aquatic to the terrestrial environment were those that involved the ear. Because air is a less dense medium for sound conduction than water, greater acoustic sensitivity was selected for in meeting the challenges of survival on land. Among the structural adaptations that resulted from this transition was the development of a chamber between the inner ear (in which the acoustic receptors are located) and the outer ear (which traps the sound waves from the surrounding environment). This chamber, the middle ear, is separated from the external ear by a membrane, the **tympanum** or ear drum. Initially, the middle ear contained a single bone, the **columella,** which was derived from the dorsal portion of the second gill arch. The columella connected the tympanum to the inner ear and served to conduct sound to the entrance of the inner ear.

In the ancestral stock of mammals, two jaw bones, the **quadrate** and the **articular,** and the muscles that were attached to them, migrated towards the ear canal under the influence of selective pressures related to feeding. Once they were close to the ear canal, however, they invariably became involved in the conduction of sound as are many of the bony structures in that region. At this point, their sound-conducting function became more important to the animal's survival than did their feeding function, and selection favored the bones being more intimately involved in the hearing process. Eventually, they became incorporated into the middle ear and took their associated nerves and muscles along with them. Extant mammals thus possess three bones in their middle ears, which are collectively known as the **ossicles.** In the mammalian middle ear, the bone derived from the columella is called the **stapes,** and the bones derived from the quadrate and articular are called the **maleus** and the **incus.**

Two muscles are attached to the ossicles: the **stapedius** and the **tensor tympani.** These muscles, together with the ossicles themselves, help to dampen intense sounds and thereby prevent damage to the acoustic receptors in the inner ear. This protective action is known as the **acoustic reflex.** Motor V innervates the tensor tympani, and motor VII$_D$ innervates the stapedius. If you want to recall which nerve innervates which muscle, remember that "tensor tympani" and "trigeminal" all begin with the letter "t," while "stapedius" and "seventh" begin with the letter "s."

In mammals, sound waves from the external world pass down the ear canal and vibrate the tympanum (ear drum) at the entrance to the middle ear. The vibrations are transmitted via the ossicles in the middle ear to the oval window, which is the entrance to the cochlea. The cochlea, a bony structure shaped like a snail shell, from which its name is derived, contains the receptors for hearing (Chapter 2). The receptors activate axons of the cochlear ramus of the eighth nerve which terminate in the medulla in the dorsal and ventral cochlear nuclei. As discussed in Chapter 11, axons from these cochlear nuclei pass to a nearby cell group called the superior olivary nucleus. This is an important coordinating nucleus for auditory reflexes. The superior olivary nucleus sends its axons to motor V and motor VII$_D$, which control the tensor tympani and stapedius muscles in the middle ear. The tensor tympani increases tension on the tympanum, which decreases the intensity of sound entering the middle ear; the stapedius pulls the stapes away from the oval window, thereby reducing the intensity of the vibrations entering the cochlea.

MOTOR CONTROL OF EYE MUSCLES

For the overwhelming majority of vertebrates, the eyes are an important source of information about distant objects. For many animals, such as birds, the eyes are the most important source of information about the surrounding environment. The eyes receive assistance in their information-gathering task from the eye muscles. These muscles are of two types: **extraocular muscles,** located on the outside of the eye and innervated by three ventral cranial nerves, and **intraocular muscles,** located within the globe of the eye and innervated by parasympathetic fibers (see Fig. 12-2).

The Extraocular Muscles in Jawed Vertebrates

Many animals have panoramic or near-panoramic vision, which means that they can see in all directions (or nearly so) without moving their eyes. These animals have their eyes situated on the side of the head, as many birds have, or near the top of the head, as do amphibians, diapsid reptiles, and turtles. Such animals still need to move their eyes in order to place objects into the zone of most acute vision on the retina. The extraocular muscles serve to move the eyes so that the center of gaze can be shifted from one position in space to another. The extraocular muscles also produce micromovements of the eyes so that images that fall on the retina do not remain constantly on the same receptors. Without these micromovements of the eye, the receptors would soon fatigue and the image would fade away. The major extraocular muscles comprise four recti muscles—the **inferior rectus, superior rectus, medial rectus,** and **posterior** (or lateral) **rectus**—and two oblique muscles—the **inferior oblique** and **superior oblique.**

In addition to the muscles that move the eyes, another extraocular muscle controls the eyelid. In those animals that have movable eyelids (amphibians and amniotes), the upper lid is raised by the **levator palpebrae muscle** or the lower lid is dropped by the **depressor palpebrae muscle.** Another function of the extraocular eye muscles is to control the **nictitating membrane** or so-called "third eyelid." The nictitating membrane is an internal eyelid that provides an additional layer of protection to the cornea of the eye as well as aiding in the distribution of the tear film. It is present in diapsid reptiles, birds, turtles, and some amphibians and mammals. Among mammals, the membrane is present in carnivores, rabbits, and hares, but not in primates. Some bottom-feeding ray-finned fishes and sharks have a similar protective membrane.

The action of this internal eyelid is quite different from that of the external eyelids. The latter are actively operated by muscles that open and close the lids. The nictitating membrane, however, is passively operated. The membrane slides in front of the cornea when the eyeball is retracted farther into the eyesocket by a special extraocular muscle, the **retractor bulbi.** In some reptiles, the retractor bulbi is assisted by one of two additional muscles: the **quadratus** (lizards) or the **pyramidalis** (crocodiles and turtles). In birds, the globe of the eye fits so tightly into the eyesocket that little room is left for retraction. Therefore, the retractor bulbi is not present and in its place are both the quadratus and the pyramidalis muscles. These muscles act to slide the nictitating membrane over the cornea without retraction of the eye.

The various extraocular muscles are innervated by the **oculomotor nerve (III),** the **trochlear nerve (IV),** or the **abducens nerve (VI),** which are exclusively motor in function. Axons of the oculomotor nerve also innervate the levator palpebrae muscle. The sensory innervation of the extraocular muscles is via the ophthalmic division of the trigeminal nerve (V). Table 12-1 lists the extraocular muscles that are most commonly found in jawed vertebrates and their cranial nerve motor innervation. The presence of four recti and two oblique muscles is plesiomorphic for at least jawed vertebrates.

Movement of the eyes is not the only use of the extraocular muscles. In an electric fish known as *Astroscopus* (the stargazer), these muscles have been independently modified to form electric organs, similar to the somatic electric organs of the electric eel and other electric fishes. The stargazer lies buried in the sand of the sea floor with only its eyes exposed. When prey approach closely, they are stunned by discharges from the electric organs. In spite of this dramatic change in function of

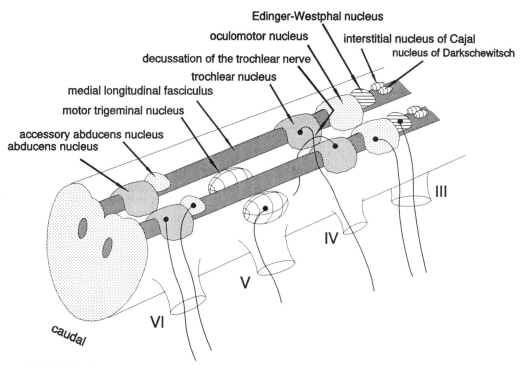

FIGURE 12-2. Schematic drawing illustrating the nuclei and nerves in the brainstem that control the muscles of the eye.

the extraocular muscles, their innervation remains the oculomotor nerve.

The Extraocular Muscles in Jawless Vertebrates

Hagfishes lack extraocular muscles altogether. Lampreys have extraocular muscles, but differences occur in the number and innervation pattern of the muscles as compared with the situation in jawed vertebrates. Whether the plesiomorphic condition for vertebrates corresponds to that in lampreys or to that in jawed vertebrates cannot be determined, since hagfishes, lacking these muscles, cannot be used for an out-group comparison.

In lampreys, two oblique extraocular muscles are present, a **rostral oblique** innervated by III and a **caudal oblique** innervated by IV, which appear to be homologous to the inferior and superior oblique muscles of jawed vertebrates, respectively. Three, rather than four recti are present in the main group of extraocular muscles. The **dorsal** and **rostral recti** of lampreys are innervated by III and appear to be homologous to the superior and inferior recti of jawed vertebrates, respectively. The **ventral rectus** of lampreys is innervated by VI and corresponds to the posterior rectus of jawed vertebrates. A medial (or nasal) rectus is absent in lampreys. Lampreys, like jawed vertebrates, have a muscle that retracts the globe of the eye and is innervated by an accessory abducens nucleus; in lampreys this muscle is called the **caudal rectus.**

The Intraocular Muscles

The intraocular eye muscles do not move the eye; rather, they control the action of structures within the eye. The iris is one such structure; it is opened and closed in response to the level of illumination of the retina by a pair of muscles: the **dilator pupillae,** which increases the diameter of the pupil, and the **sphincter pupillae,** which decreases the diameter. Another intraocular muscle, the **ciliary muscle,** controls the curvature of the lens to permit accommodation, which is the ability of the eye to focus on close objects.

The intraocular muscles are innervated by parasympathetic fibers. Neuron cell bodies of the parasympathetic component of the oculomotor nerve lie within the **Edinger-Westphal nucleus** (Fig. 12-2) within the oculomotor complex. Their axons course to the **ciliary ganglion** via the oculomotor nerve. Postganglionic parasympathetic fibers from the ciliary ganglion innervate the intraocular muscles.

Central Control of the Eye Muscles

The characteristic appearance and consistent position of the nuclei of the cranial nerves that control the eye muscles makes them easy to identify in all vertebrates (Fig. 12-2). These nuclei are located just below the floor of the third ventricle. The oculomotor nucleus (III) and the trochlear nucleus (IV) usually are located rather close to one another. The abducens nucleus (VI) is separated from III and IV in the caudal direction. In animals with a retractor bulbi, an accessory abducens nucleus is present, which controls this muscle.

TABLE 12-1. Innervation of Extraocular Muscles in Jawed Vertebrates

Muscle	Nerve	Function
Inferior oblique	III	Rotates eye upward and outward
Superior oblique	IV	Rotates eye downward and outward
Inferior rectus	III	Rotates eye downward and inward
Superior rectus	III	Rotates eye upward and inward
Medial rectus	III	Rotates eye medially
Posterior (lateral) rectus	VI	Rotates eye laterally
Levator palpebrae superioris	III	Raises upper eyelid
Depressor palpebrae inferioris[a]	V	Lowers lower eyelid
Retractor bulbi[b]	VI	Retracts glove of the eye; advances nictitating membrane
Pyramidalis[c]	III	Advances nictitating membrane
Quadratus[d]	III	Advances nictitating membrane

[a] Amphibians, reptiles, and some mammals.
[b] Birds, crocodilians, and turtles.
[c] Birds and lizards.
[d] Amphibians, birds, and reptiles.

In lampreys, the trochlear and abducens nuclei lie in a position that conflicts with the generally accepted interpretation of the somatic motor column. Trochlear motor neurons have a ventrorostral position early in development but then migrate to a dorsal position in the cerebellar plate, a position that is topologically a rostral continuum of the trigeminal motor nucleus and implies that these neurons are branchial rather than somatic. Abducens motor neurons also migrate, and their axons exit the brainstem with the trigeminal root.

The Oculomotor Complex

The oculomotor nucleus (III) is often considered along with several related nuclei as part of an **oculomotor complex.** In fishes such as the stargazer, in which the extraocular muscles have become specialized as electric organs, the oculomotor nucleus consists of an **oculomotor division** and an **electromotor division.** The oculomotor axons terminate in the extraocular muscles and control eye movements, and the electromotor axons terminate in the electric organs. Also included in this complex are the **nucleus of Darkschewitch,** the **interstitial**

nucleus of Cajal, and the **nucleus of the posterior commissure.** These nuclei are involved in the reflex control of those extraocular muscles that are controlled by the oculomotor nerve.

Coordination of Eye Muscle Action

As seen in Table 12-1, the effects of contraction of each of the individual eye muscles on eye movement is rather different. In order to produce the precise movements needed for the inspection and tracking of visual stimuli and to prevent fading of the retinal image, a very precise coordination of the activities of each of the eye muscle nuclei is required. This coordination is carried out by a system of neuronal groups: the reticular formation. Among the functions of the reticular formation is the coordination of the action of motor nuclei that innervate antagonistic muscles. The nuclei of Darkschewitch and Cajal are reticular formation nuclei that are specialized for coordinating the activities of the oculomotor nucleus with other nuclei of the brainstem, in particular, those of the vestibular system.

Coordination of the actions of eye muscles is also carried out by the reticular formation. An important pathway of the reticular formation is the **medial longitudinal fasciculus,** which consists of heavily myelinated axons and is a conspicuous feature of the dorsal brainstem. The reticular formation and the nuclei of III, IV, and VI are interconnected via this pathway.

EVOLUTIONARY PERSPECTIVE ON THE HINDBRAIN AND MIDBRAIN CRANIAL NERVES

The evolutionary development and organization of the cranial nerves of the hindbrain and midbrain have paralleled the evolutionary development of the head. The change from gill arches to jaws and hyoid arch as well as the later evolution of the muscular tongue in tetrapods were major adaptations. Other adaptations, such as the development of mechanisms for the lubrication and transport of food in the mouth and the ability to manipulate the air column in the mouth, throat, and body cavities for communication between individuals, also permitted greater and more efficient survival and success of future generations. Sill other adaptations involved the development and elaboration of the muscles of the eyes and the incorporation of jaw bones and muscles into the inner ear as devices to affect the conduction of sounds to the inner ear. All of these adaptations are reflected in the sensory and motor neuronal populations of the hindbrain and midbrain. Because nearly all of these structures operate in close coordination with one or more of the others, the midbrain–hindbrain region is a major zone for pathways involved in a large number of reflexes of the head and neck. In Chapter 13 on the reticular formation, which is another major component of the midbrain–hindbrain region, we shall see how these reflexes are coordinated and integrated with each other.

FOR FURTHER READING

Fritzsch, B., Sonntag, R., Dubuc, R., Ohta, Y., and Grillner, S. (1990) Organization of the six motor nuclei innervating the ocular muscles in lamprey. *Journal of Comparative Neurology*, 294, 491–506.

Gans, C. and Gorniak, G. C. (1982) How does the toad flip its tongue? Test of two hypotheses. *Science*, 216, 1335–1337.

Jacquin, M. F., Rhoades, R. W., Enfiejian, H. L., and Egger, M. D. (1983) Organization and morphology of masticatory neurons in the rat: a retrograde HRP study. *Journal of Comparative Neurology*, 218, 239–256.

Kishida, R., Onishi, H., Nishizana, H., Kadota, T., Goris, R. C., and Kusunoki, T. (1986) Organization of the trigeminal and facial motor nuclei in the hagfish (*Eptatretus burgeri*): a retrograde HRP study. *Brain Research*, 385, 263–272.

ADDITIONAL REFERENCES

Anderson, C. W. and Nishikawa, K. C. (1993) A prey-type dependent hypoglossal feedback system in the frog *Rana pipiens*. *Brain, Behavior and Evolution*, 42, 189–196.

Barbas-Henry, H. A. (1982) The motor nuclei and primary projections of the facial nerve in the monitor lizard (*Varanus exanthematicus*). *Journal of Comparative Neurology*, 207, 105–113.

Bennett, M. V. and Pappas, G. D. (1983) The electromotor system of the stargazer: a model for integrative actions at electrotonic synapses. *Journal of Neuroscience*, 3, 748–761.

Berkhoudt, H., Klien, B. G., and Zeigler, H. P. (1982) Afferents to the trigeminal and facial motor nuclei in pigeon (*Columba livia* L.): central connections of jaw motoneurons. *Journal of Comparative Neurology*, 209, 301–312.

Cabrera, B., Pasaro, R., and Delgado-Garcia, J. M. (1993) A morphological study of the principal and accessory abducens nuclei in the caspian terrapin (*Mauremys caspica*). *Brain, Behavior and Evolution*, 41, 6–13.

Carter, G. S. (1967) *Structure and Habit in Vertebrate Evolution*. Seattle: University of Washington Press.

Evinger, C., Graf, W. M., and Baker, R. (1987) Extra- and intracellular HRP analysis of the organization of extraocular motoneurons and internuclear neurons in the guinea pig and rabbit. *Journal of Comparative Neurology*, 252, 429–445.

Finger, T. E. and Rovainen, C. M. (1978) Retrograde HRP labeling of the oculomotoneurons in adult lampreys. *Brain Research*, 154, 123–127.

Fritzsch, B. and Northcutt, R. G. (1993) Cranial and spinal nerve organization in amphioxus and lampreys: evidence for an ancestral craniate pattern. *Acta Anatomica*, 148, 96–109.

Fritzsch, B. and Sonntag, R. (1988) The trochlear motoneurons of lampreys (*Lampetra fluviatilis*): location, morphology and numbers as revealed with horseradish peroxidase. *Cell and Tissue Research*, 252, 223–229.

Gonzalez, A. and Muñoz, M. (1988) Central distribution of the efferent cells and the primary afferent fibers of the trigeminal nerve in *Pleurodeles waltlii* (Amphibia, Urodela). *Journal of Comparative Neurology*, 270, 517–527.

Graf, W. and Brunken, W. J. (1984) Elasmobranch oculomotor organization: anatomical and theoretical aspects of the phylogenetic development of vestibulocochlear connectivity. *Journal of Comparative Neurology*, 227, 569–581.

Graf, W. and McGurk, J. F. (1985) Peripheral and central oculomotor organization in the goldfish *Carassius auratus. Journal of Comparative Neurology*, 239, 391–401.

Heaton, M. B. and Moody, S. A. (1980) Early development and migration of the trigeminal motor nucleus in the chick embryo. *Journal of Comparative Neurology*, 189, 61–99.

Heaton, M. B. and Wayne, D. (1983) Patterns of extraocular innervation by the oculomotor complex in the chick. *Journal of Comparative Neurology*, 216, 245–252.

Hildebrand, M., Bramble, D. M., Liem, K. F., and Wake, D. B. (eds.), 1985, *Functional Vertebrate Morphology*. Cambridge, MA: Belknap Press.

Kojama, H., Kishida, R., Goris, R. C., and Kusunoki, T. (1987) Organization of sensory and motor nuclei of the trigeminal nerve in lampreys. *Journal of Comparative Neurology*, 264, 437–448.

Leonard, R. B. and Willis, W. D. (1979) The organization of the electromotor nucleus and extra-ocular motor nuclei in the stargazer (*Astroscopus y-graecum*). *Journal of Comparative Neurology*, 183, 397–413.

Luiten, P. G. and van der Pers, J. N. (1977) The connections of the trigeminal and facial motor nuclei in the brain of the carp (*Cyprinus carpio* L.). *Journal of Comparative Neurology*, 174, 575–590.

Moody, S. A. and Meszler, H. A. (1980) Subnuclear organization of the ophielian trigeminal motor nucleus. I. Localization of neurons and synaptic bouton distribution. *Journal of Comparative Neurology*, 190, 463–486.

Moody, S. A. and Meszler, H. A. (1980) Subnuclear organization of the ophielian trigeminal motor nucleus. II. Ultrastructural measurements on motoneurons innervating antagonistic muscles. *Journal of Comparative Neurology*, 190, 487–500.

Morita, Y. and Finger, T. A. (1985) Reflex connections of the facial and vagal gustatory systems in the brainstem of the bullhead catfish *Ictalurus nebulosus. Journal of Comparative Neurology*, 231, 547–558.

Naujoks-Manteuffel, C., Manteuffel, G., and Himstedt, W. (1986) Localization of motoneurons innervating the extraocular muscles in *Salamandra salamandra* L (Amphibia, Urodela). *Journal of Comparative Neurology*, 254, 133–141.

Nauta, W. J. H. and Feirtag, M. (1986) *Fundamental Neuroanatomy*. New York: Freeman.

Nishikawa, K. C., Anderson, C. W., Deban, S., and O'Reilly, J. C. (1992) The evolution of neural circuits controlling feeding behavior in frogs. *Brain, Behavior and Evolution*, 40, 125–140.

Pombal, M. A., Rodicio, M. C., and Anadon, R. (1994) Development and organization of the ocular motor nuclei in the larval sea lamprey, *Petromyzon marinus* L.: an HRP study. *Journal of Comparative Neurology*, 341, 393–406.

Puzdrowski, R. L. and Leonard, R. B. (1994) Vestibulo-oculomotor connections in an elasmobranch fish, the Atlantic stingray, *Dasyatis sabina. Journal of Comparative Neurology*, 339, 587–597.

Regal, J. R. and Gans, C. (1976) Functional aspects of the evolution of frog tongues. *Evolution*, 30, 718–734.

Rosiles, J. R. and Leonard, R. B. (1980) The organization of the extraocular motor nuclei in the Atlantic stingray *Dasyatis sabina. Journal of Comparative Neurology*, 193, 677–687.

Sarkasian, V. A. and Fanardjian, V. V. (1985) Neuronal mechanisms of the interaction of Deiter's nucleus with some brain stem structures. *Neuroscience*, 16, 957–968.

Song, J. and Boord, R. L. (1993) Motor components of the trigeminal nerve and organization of the mandibular arch muscles in vertebrates: phylogenetically conservative patterns and their ontogenetic basis. *Acta Anatomica*, 148, 139–149.

Stuesse, S. L., Cruce, W. L. R., and Powell, K. S. (1983) Afferent and efferent components of the hypoglossal nerve in the grass frog, *Rana pipiens*. *Journal of Comparative Neurology*, 217, 432–439.

Szabo, T., Lázár, G., Libouban, S., Toth, P., and Ravaille, M. (1987) Oculomotor system of the weakly electric fish *Gnathonemus petersii*. *Journal of Comparative Neurology*, 264, 480–493.

Uemura-Sumi, M., Takakashi, O., Matsushina, R., Takata, M., Yasui, Y., and Mizuno, N. (1982) Localization of masticatory motoneurons in the trigeminal motor nucleus of the guinea pig. *Neuroscience Letters*, 29, 219–224.

Walls, G. L. (1942) *The Vertebrate Eye and its Adaptive Radiation*. Bloomfield Hills, MI: Cranbrook Institute of Science.

Wathey, J. C. (1988) Accomodation motor neurons in the foveate teleost *Paralabrax clathratus*: horseradish peroxidase labeling and axonal morphometry, with comparisons to other ciliary nerve components. *Brain, Behavior and Evolution*, 32, 1–16.

Wild, J. M., Arends, J. J., and Zeigler, H. P. (1985) Telencephalic connections of the trigeminal system in the pigeon (*Columba livia*): a trigeminal sensori-motor circuit. *Journal of Comparative Neurology*, 234, 441–464.

Zeigler, H. P., Levitt, P., and Levine, R. R. (1980) Eating in the pigeon: response topography, stereotypy and stimulus control. *Journal of Comparative and Physiological Psychology*, 94, 783–794.

13

The Reticular Formation

INTRODUCTION

Have you ever noticed that a person can speak at the same time as chewing a mouthful of food? Let us overlook the appallingly bad manners of such an act and consider instead how rarely such a person mistakenly bites his/her own tongue while happily chewing through the various other items that are in the mouth at the same moment. Consider what the tongue is doing during this process: not only is it manipulating the food to position it under the teeth for chewing and then transporting the chewed food back to the throat so that it can be swallowed, but it also is alternating these complex movements with the even more complex movements required for articulate speech (admittedly made less articulate by the presence of the food). What sort of neural system has taken notice of the location of the tongue, the position of the jaws, the states of contraction and relaxation of the flexor and extensor muscles that operate the jaws, and the muscles that control swallowing and respiration and other complex processes to permit these actions to occur as smoothly as they do? The answer is the **reticular formation,** a neural system that has input from sensory and motor pathways that control pattern generators for rhythmic motor patterns, such as swimming, walking, flying, the repetitive discharge of the electric organs in electric fishes, and chewing. In addition to providing these coordinating and organizing functions, the reticulobulbar and reticulospinal pathways are the principal routes of telencephalic control of voluntary movement of organs of the head and body in nearly all vertebrates, except mammals, where they work in close conjunction with the direct corticobulbar and corticospinal tracts. Regulation of muscle tone (the degree of tension present in a muscle) also is a function of the reticular formation. These motor activities are regarded as the functions of the "descending" reticular formation.

The "ascending" reticular formation consists of those reticular pathways that ascend to the diencephalon and telencephalon. The functions of the ascending pathways are quite different from those of the descending reticular formation. The ascending functions of the reticular formation include processes such as sleep, dreaming, and arousal, as well as attention, which is the ability to filter out extraneous stimuli so that the animal or person can concentrate on the relevant stimulus. Of particular importance in the functioning of the reticular formation are the ascending serotoninergic pathways and the noradrenergic pathways.

Much of the research on the ascending reticular formation has been done in mammals, partly for reasons of tradition, partly for convenience, and partly because the human scientists can more readily identify (and identify with) states of arousal, wakefulness, attention, and so on, in mammals than they can in diapsid reptiles, turtles, or fishes. Very good evidence does exist, however, for the presence of similar functions in nonmammalian amniotes and very likely in anamniotes as well.

In spite of its importance, the reticular formation is one of the more misunderstood and intimidating systems of the brain. To begin with it has a formidable nomenclature that is rivaled only by the diencephalon in the variety and complexity of the names of its components. To make matters worse, different anatomists have used different systems of nomenclature that make the correspondences between vertebrate classes or even between related species difficult for beginners to follow. Finally, the reticular formation is scattered throughout the central nervous system from the spinal cord to the forebrain, which tends to obscure its overall organization.

Much of the difficulty in understanding the reticular formation comes from attempting to deal with the individual components of the mammalian reticular formation before having a clear appreciation of either the evolution of the reticular formation or of the overall structure and organization of the system as a whole. Once the general organization is understood in the context of a uniform nomenclature and in the light of the evolutionary development of the system, the individual components become manageable details that can be dealt with as needed. In this chapter we will present the reticular formation in an evolutionary context and will show how its many component parts fit into a general pattern of organization, and we will attempt to make the nomenclature less intimidating and more consistent. We also will describe the relationship between the reticular formation and sensory, motor, and neurochemical systems. Finally, we will introduce you to some of the important functions of the reticular formation in the behavior of vertebrates.

THE ORGANIZATION OF THE RETICULAR FORMATION

The brainstem of vertebrates contains sensory nuclei of the cranial nerves (and their incoming sensory-root axons), the motor nuclei of the cranial nerves (and their outgoing motor-root axons), long ascending axons from the spinal cord, and long descending axons from the diencephalon and telencephalon. Interspersed among these are many clusters of neurons that are neither sensory nor motor. These clusters of neurons, which seem to fill in the spaces between the sensory and motor components of the brainstem, make up part of a coordinating system known as the reticular formation. The name is derived from the Latin word *reticulum*, which means a net or mesh. Like the other great coordinating system, the limbic system, the reticular formation has strong connections with sensory systems, somatic motor systems, and visceral motor systems. We have already encountered the reticular formation in our discussions of ascending and descending pathways in the spinal cord; we have seen that reticulospinal pathways are a major route through which the brain controls movements of the body in all vertebrates and are the exclusive route for the brain to control motor neurons in many of them, especially nonmammals. But the functions of the reticular formation are far more pervasive than that and include roles in:

- The coordination of movements of the head and body by facilitation and inhibition of both voluntary and reflex movements.
- Alteration of respiration and blood pressure.
- Serving as a "gate" to block out sensory inputs, including pain.
- Psychological processes such as arousal and attention.
- Sleep and dreaming.

Neurons of the Reticular Formation

Neurons of the reticular formation fall into three categories based on the size of the soma: small, large, and giant. At first glance they seem to be scattered throughout the tangle of dendrites and axons that make up the core of the medulla, pons,

and midbrain apart from the compact sensory and motor nuclei of the cranial nerves and several other well-demarcated nuclear groups. A more careful examination will reveal, however, that the cell groups of the reticular formation indeed have an organization even though they are not arranged in neat packages with clear boundaries such as the cranial nerve nuclei. The overall organizational arrangement of the reticular formation is that of columns of cells arranged longitudinally along the length of the medulla and pons. In general, the small cells comprise the lateral columns (**parvocellular** columns; "parvo-" = small), and the large and giant cells form more medial columns (**magnocellular** and **gigantocellular** columns; "magno-" = big and "giganto-" = giant). The gigantocellular cell columns also contain neurons of smaller size. The most medial column is the raphe column, which will be discussed later in this chapter. Within these columns are subdivisions that can be distinguished according to cell size and type as well as their connections with other parts of the nervous system. Figure 13-1 is a schematic representation of the columnar organization of the reticular formation within the caudal brainstem (medulla and pons); several subdivisions are shown within each column. Although this simplified organizational scheme is generally satisfactory, you should be aware that some exceptions can be found.

The small-celled, lateral column tends to be the afferent zone of the reticular formation; these neurons receive axons from sensory systems of the brainstem and spinal cord as well as from motor pathways. The more medial columns typically are the efferent divisions of the reticular formation; many of their axons descend to the motor neurons of the spinal cord and brainstem and ascend to higher levels of the nervous system.

Reticular formation neurons differ from neurons of cranial nerve sensory or motor nuclei in their cellular morphology. As shown in Figure 13-2, the dendrites of sensory or motor neurons of the cranial nerve nuclei tend to spread out like the branches of an oak or chestnut tree. The spread of these dendritic branches, however, remains within the sharply demarcated boundary of the nuclear group. In contrast, the dendrites of reticular neurons often are long and relatively straight. These dendrites tend to be of the isodendritic variety (see Chapter 2) and have a relatively consistent dorsal–ventral orientation within the brainstem. This dendritic pattern permits reticular neurons to be in contact with axons traversing the brainstem whether they rise dorsally from the ventral levels of the brainstem or descend ventrally from more dorsal portions.

The high frequency of isodendritic neurons in the reticular "core" of the mammalian brainstem led Ramon-Moliner and Nauta in the 1960s to speculate that the isodendritic pattern represented a primitive evolutionary condition from which more complex dendritic patterns evolved. Subsequent research on the reticular formation of lampreys, however, has revealed that while some divisions of the lamprey reticular formation possess the pattern of radially organized isodendrites, other divisions have a more dense dendritic organization than the mammalian pattern. These and other differences between lamprey reticular neurons and those of mammals suggest that the evolution of dendritic types may be a more complex story than a simple progression from an isodendritic organization to more complex and varied dendritic forms.

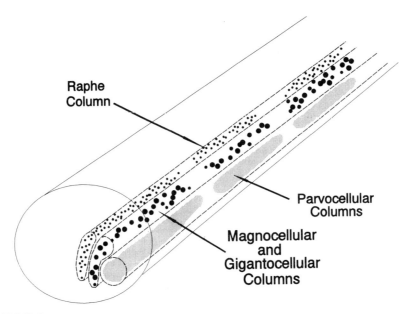

FIGURE 13-1. A schematic representation of the main cell columns of the reticular formation of the brainstem.

A characteristic of reticular neuron axons is that they are quite long and travel for considerable distances. In other words, they are Golgi Type I neurons. Short-axon, Golgi Type II neurons are not a feature of reticular formation anatomy. In most vertebrate classes, reticular formation axons give off numerous collateral branches as they travel the length of the brainstem. An exception appears to be lampreys in which the long reticular axons do not have the characteristic collateralization. In addition to the long axons, shorter axons connect the individual subdivisions of the reticular formation so that the actions of these subdivisions are highly coordinated. Figure 13-3 illustrates a reticular neuron with long dorsoventrally oriented dendrites.

An axon arises from the soma of this cell and bifurcates into two divisions that ascend and descend for long distances within the brainstem. As the axon travels its course, numerous collateral branches can be seen descending ventrally from it.

Giant Reticulospinal Neurons

Another characteristic of the reticular formation is the presence of giant neurons. In addition to the gigantocellular column, which contains neurons that are among the largest in the brainstem, the reticular formation of nontetrapods contains a relatively small number of neurons that have somata of enormous

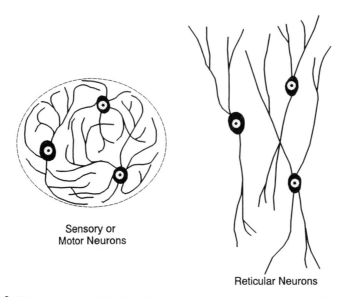

FIGURE 13-2. The arrangement of dendrites of reticular formation neurons is contrasted with the dendrites of sensory or motor neurons, which remain within the boundaries of their nuclear groups.

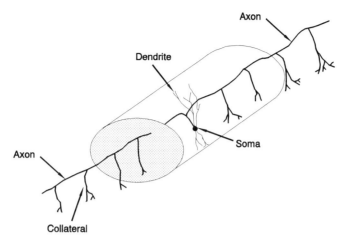

FIGURE 13-3. A reticular neuron with its long, bifurcated axon and many axon collaterals.

proportions. We have already encountered one of these in our discussion of the nontetrapod spinal cord, the Mauthner cell. Other giant cells, of similar proportions to the Mauthner cell may be scattered throughout the medulla, pons, and midbrain in various locations depending on the species of nontetrapod. These are the somata of the Müller cells. A common feature of these giant cells is that they send their axons to the spinal cord. The soma and dendrites of one such giant reticulospinal neuron from a larval lamprey are shown in Figure 13-4. The giant cells have their somata near the dorsal surface of the brainstem close to the floor of the ventricle. Their dendrites spread extensively in a ventral direction to the ventral margin of the brainstem where they intermingle with the passing axons of the ventral brainstem. Some dendrites occasionally cross the midline to the contralateral side.

These gigantic somata are associated with the giant axons of the spinal cord, which are involved in rapid evasion and escape responses. The principal advantage of a giant axon is its extremely rapid conduction time, because the speed of the action potential increases as the axon diameter increases. As they emerge from their somata, the initial segments of these giant axons are much narrower in diameter than the rest of the giant axon. Although this narrower diameter of the initial segment results in a somewhat increased conduction time in that region, it does have the advantage of partially insulating the axon hillock from the giant axon. Without this partial insulation, the local membrane potentials on the axon hillock would leak out along the giant axon and thus the interplay of excitatory and inhibitory processes that determine whether or not an action potential is initiated in the axon hillock would be lost.

Nomenclature of the Reticular Formation

One of the factors that complicates a comparative study of the reticular formation in various groups of vertebrates is the discrepancies in nomenclature used by different investigators. Early investigators of the reticular formation used variations on

FIGURE 13-4. A giant reticulospinal neuron of a larval lamprey.

the nomenclature used by J. B. Johnston and D. Tretjakoff in the first decade of the twentieth century to describe the reticular formation of nontetrapods; other anatomists preferred to apply the mammalian nomenclature used by Jerzy Olszewski (pronounced "Olshefsky") and Alf Brodal to name cell groups in nonmammalian reticular formations. The tendency in recent years seems to be in the direction of applying the Olszewski–Brodal nomenclature to nonmammals. The result has been a difficult situation for newcomers to the topic. Our approach to dealing with this problem has been to use the Johnston–Tretjakoff nomenclature for describing the agnathan reticular formation and the Olszweski–Brodal nomenclature for all other vertebrates. Table 13-1 shows these two nomenclatures. We recommend that you refer to the table frequently as you look at Figures 13-5–13-6 and 13-8–13-9, which are schematic representations of the reticular formation in several representative vertebrates. The table will help you to see the location of each reticular component in the brainstem and the correspondence between the two nomenclatures. Our treatment of the reticular formation has resulted in a simplification of certain details for the sake of clarity of presentation. Our goal here, as elsewhere in this book, however, is not an exhaustive presentation of the reticular formation, but rather to offer an introduction to the topic that gives the reader a framework onto which the specific details of any particular species can easily be attached.

The table shows the three regions of the brainstem that contain most of the reticular formation: the medulla, the pons, and the midbrain. Next, the components of the lamprey reticular formation in each brainstem region are presented in the Johnston–Tretjakoff nomenclature. Finally, the corresponding reticular formation components in the Olszewski–Brodal nomenclature are presented. Keep in mind that one should not expect all components to be present in all classes of vertebrates or even within all species of a given class. These reticular components have evolved to perform certain specific functions in the adaptation of the creatures to their environments. Each species of a particular taxonomic group has its unique survival needs and adaptations to the environment, which may be somewhat different from those of their sister groups in a taxonomic organization.

The Reticular Formation of the Medulla, Pons, and Midbrain

Figures 13-5–13-6 and 13-8–13-9 show schematic representations of the reticular formation of four representative vertebrates: a lamprey, shark, a lizard, and a rat. Bony fishes are not represented because of their similarity to sharks. Likewise, birds are not represented because of their similarity to diapsid reptiles. Amphibians, on the other hand, are not represented because of the unique condition of their brainstem, in which many of the neurons have remained in their unmigrated positions close to the ventricles. Amphibians do have a reticular formation that projects to the spinal cord, but the relatively undifferentiated state of their brainstems makes comparison with the other vertebrate classes extremely speculative at best.

The reticular formations of these animals have been chosen because they represent varying degrees of complexity and differentiation into specialized nuclei. We hasten to add that these reticular formations are not intended to represent an evolutionary sequence because lampreys and sharks evolved from separate lineages, and lizards evolved after mammals. Thus they represent an adaptive series rather than a phyletic series.

Figure 13-5 shows the brainstem of a lamprey with the components of the reticular formation in the Johnston-Tretjakoff nomenclature. Also shown is the location of the giant Mauthner cell body with its giant axon crossing the midline and descending to the spinal cord. The other giant cell bodies are not shown for the sake of simplicity of the diagram. One reticular cell group is shown in the medulla, the **nucleus reticularis inferior.** In the caudal pons, at the level of the vestibular nerve, is the **nucleus reticularis medius.** More rostrally in the pons, are two more reticular components that together make up the **nucleus reticularis superior.** One component, located near the roots of the trigeminal nerve is known as the **trigeminal group;** the other, located near the isthmus (the narrow neck of the brainstem) is called the **isthmus group.** Finally, in the midbrain, or mesencephalon, is the **nucleus reticularis mesencephali.**

Compare Figure 13-5 with the reticular formation of a shark as shown in Figure 13-6. Note that a simple correspondence of the components of the reticular formations of the two animals is not easily made because the shark reticular formation has more recognizable subdivisions. Another major difference is

TABLE 13-1. Nomenclatures of the Principal Components of the Reticular Formation of the Medulla, Pons, and Midbrain

Regions of Brainstem	Johnston-Tretjakoff Nomenclature[a]	Brodal-Olszewski Nomenclature[a]
Medulla	N. R. Inferior	Raphe Group
		N. R. Gigantocellularis
		N. R. Paragigantocellularis
		N. R. Magnocellularis
		N. R. Parvocellularis
		N. R. Dorsalis
		N. R. Ventralis
		N. R. Lateralis
Pons	N. R. Medius and N. R. Superior	Raphe Group
		N. R. Pontis caudalis
		N. R. Pontis oralis
		Locus coeruleus
		N. R. Subcoeruleus
		N. R. Tegmenti Pontis
		N. R. Pedunculopontinus
		Nucleus Cuneiformis
Midbrain	N. R. Mesencephali	Nucleus Subcuneiformis

[a] N. R. = Nucleus Reticularis.

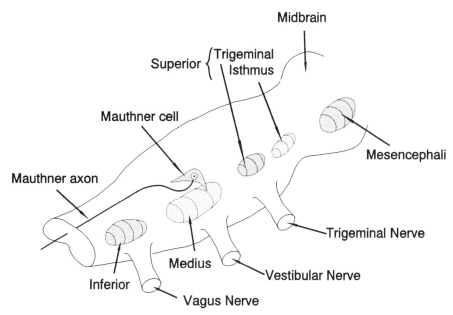

FIGURE 13-5. The main cell groups of the reticular formation of a lamprey.

that the shark reticular formation has a more obvious longitudinal orientation than that of the lamprey. This can be seen in the most medial column, which has been labeled as the **raphe group.** The word "raphe" is derived from the Greek work meaning to stitch or sew and means a "seam" in biological terminology. The raphe column of neurons lies along the median "seam" that divides the right and left halves of the brainstem. The raphe group, as we have called it, actually consists of a series of separate neuronal populations with different cell types and connections, with such names as nucleus raphe dorsalis, nucleus raphe magnus, and nucleus raphe pallidus. For purposes of this introductory presentation, we have treated them as a single entity, but the reader should be aware that further study of the reticular formation, and especially the

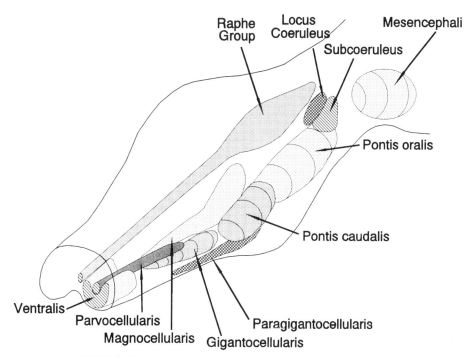

FIGURE 13-6. The main cell groups of the reticular formation of a shark.

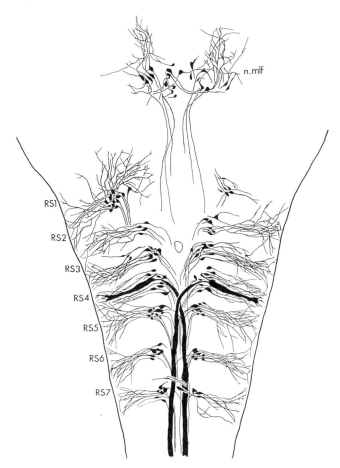

n.mlf

RS1

RS2

RS3

RS4

RS5

RS6

RS7

FIGURE 13-7. Drawing of a horizontal section through the hindbrain and midbrain of a goldfish showing the arrangement of reticulospinal neurons into longitudinal columns. Top is rostral. The numbers RS1–RS7 indicate rhombomeres. The large objects in RS4 are the lateral dendrites of the Mauthner cells. The giant Mauthner axons can be seen crossing the midline and descending towards the spinal cord. From Lee et al. (1993).

chemical pathways within it, will require more detailed knowledge of the individual components of the raphe group. In general, axons from the caudal raphe nuclei project to the spinal cord while those of the rostral raphe nuclei project to other regions of the brainstem, the diencephalon, and the telencephalon.

With the aid of Table 13-1, the correspondences can be seen between the shark reticular formation in Figure 13-6 and that of the lamprey in Figure 13-5. Moving from caudal to rostral, we can see that the caudal levels of the raphe group, and the **nuclei reticularis parvocellularis, magnocellularis, gigantocellularis,** and **paragigantocellularis** of the shark correspond to the nucleus reticularis inferior of the lamprey. These form the reticular formation of the medulla. Next, the middle and rostral levels of the raphe group as well as the **nuclei pontis caudalis** and **pontis oralis,** the **locus coeruleus** and **the nucleus subcoeruleus** of the pons of the shark correspond to the nuclei reticularis medius and reticularis superior of the lamprey. Finally, the nucleus reticularis mesencephali of the shark corresponds to the nucleus reticularis mesencephali of the lamprey.

Figure 13-7 is a drawing of a horizontal section through the hindbrain and midbrain of a goldfish and shows the arrangement of reticulospinal neurons in longitudinal columns. The rostral end is at the top of the figure. The cluster of neurons labeled "n.mlf" (nucleus of the medial longitudinal fasciculus) are in the midbrain. The numbers RS1–RS7 indicate rhombomeres. The large, prominent objects located in RS4 are the lateral dendrites of the Mauthner cell. The rest of the Mauthner cell is out of the plane of the section. The giant Mauthner axons decussate and descend towards the spinal cord.

Figure 13-8 represents the reticular formation of a reptile, the tegu lizard. The correspondence between this animal's reticular formation and that of the shark shown in Figure 13-6 is rather straightforward. The raphe group extends the length of the brainstem to the midbrain. The parvocellular, magnocellular, gigantocellular, and paragigantocellular groups are present in the medulla. Pontis caudalis, pontis oralis, the locus coeruleus/subcoeruleus complex, and the **nucleus reticularis pedunculopontinus** are present in the pons, and the nucleus reticularis mesencephali is present in the midbrain. In spite of the development of limbs in the tetrapod radiation, the reticular formation of nonmammalian amniotes does not differ significantly from that in fishes, although it is somewhat more complexly organized in birds than in diapsid reptiles or turtles. The reticular formation of alligators and crocodiles is more complex than that of other diapsid reptiles, which is consistent with their presumed closer evolutionary relationship to birds.

When we look at the mammalian reticular formation in Figure 13-9, which is represented by the reticular formation of a rat, we see an apparent explosion of additional reticular formation components. Serial transverse sections of the reticular formation of a rat are also shown in Figure 13-10 for comparison with Figure 13-9. More than 30 components of the mammalian reticular formation have been described thus far and the list is probably not yet complete. Part of this seeming increase in components may only be a reflection of the fact that the mammalian reticular formation has been far more intensively studied than that of any other class of vertebrates. Such detailed examination of a neural system invariably results in the recognition of new components and the subdivision of formerly recognized components into subcomponents, each with a separate name. Indeed, our simplified treatment of the reticular formation has glossed over a number of such subdivisions in nonmammals as well. One could suppose that after a more detailed examination of the reticular formation of nonmammals, especially with the use of the new, sophisticated tools now available, many more reticular components would emerge. On the other hand, mammals possess certain characteristics not frequently found in other tetrapods, particularly the ability to use the digits of the limbs as organs to manipulate objects in the environment. To be sure, some diapsid reptiles and turtles perform manipulatory movements with their forelimbs, such as digging and scratching, but more often than not, the snout is the principal organ for manipulation of objects. The same applies to birds. Many mammals, however, have developed the digits of the forelimbs into highly dexterous organs that can hold and turn objects for inspection and feeding or that can pull, lift, bend, or otherwise manipulate objects. Many primates and some birds have adapted their hindlimb digits as very sophisticated devices for grasping.

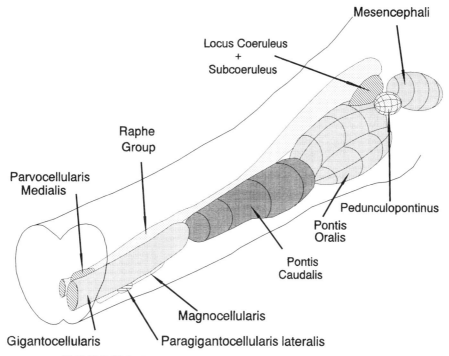

FIGURE 13-8. The main cell groups of the reticular formation of a lizard.

In spite of the greater complexity of the rat reticular formation, we can see that the basic pattern remains. The raphe group is present as the medial-most column, although the **nucleus reticularis paramedianus** has been added adjacent to it. The by-now-familiar parvocellular, magnocellular, gigantocellular, and paragigantocellular groups are present in the medulla, although paragigantocellularis now is seen as two divisions, a dorsal division adjacent to the raphe-group column and a lateral division that lies ventral to parvocellularis. Also present in the medullary reticular formation at its caudal end are the **nucleus reticularis ventralis,** nucleus **reticularis dorsalis,** and nucleus **reticularis lateralis.** Moving rostrally into the pons, we find nucleus reticularis pontis caudalis and nucleus reticularis pontis oralis in their familiar locations along with the locus coeruleus and the now, much enlarged nucleus subcoeruleus. At the level of the isthmus (the junction of the pons and the midbrain), we find the nucleus reticularis pedunculopontinus, which consists of two subdivisions, a compact part (**pars com-**

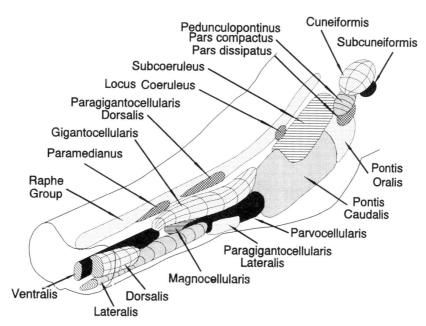

FIGURE 13-9. The main cell groups of the reticular formation of a rat.

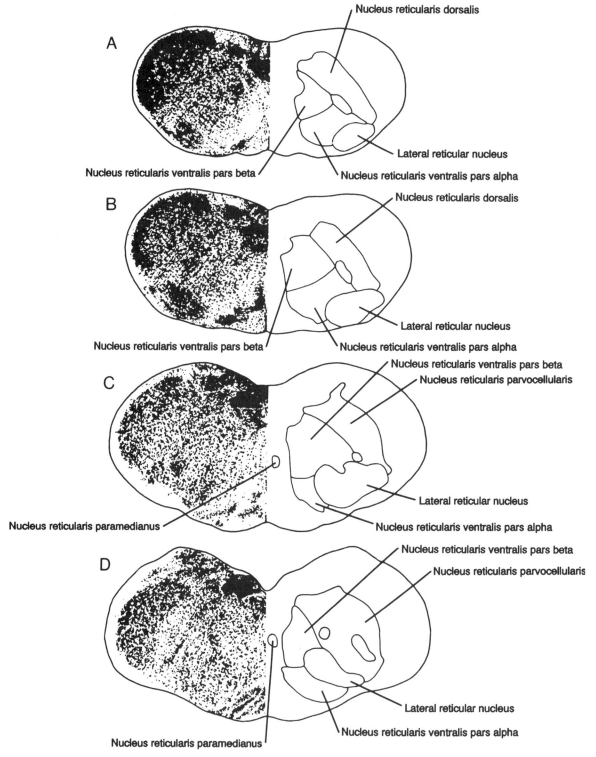

FIGURE 13-10. Transverse hemisections with mirror-image drawings through the reticular formation of a rat. Adapted from Newman (1985).

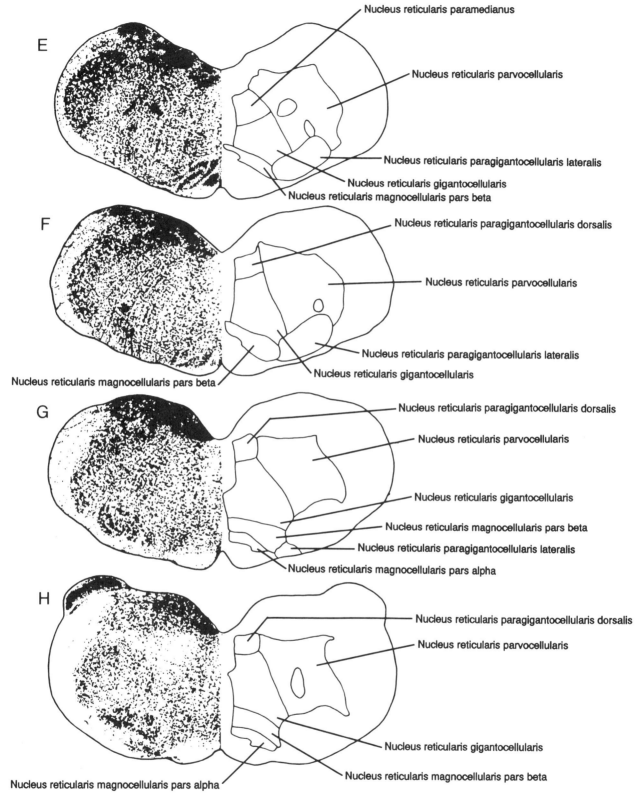

FIGURE 13-10. (Continued)

pacta) and a dispersed part (**pars disseminata**). At the rostral end, the mesencephalic reticular formation consists of the cuneiformis–subcuneiformis complex.

The Reticular Formation of the Diencephalon

The most rostral extension of the reticular formation is the **nucleus reticularis thalami.** In mammals, where it has been most extensively studied, this nucleus is a thin sheet of neurons that sits like a cloak over the dorsal thalamus at its rostral end. Its cells are characteristic of reticular neurons elsewhere in the brainstem; its axons project widely to other thalamic nuclei and

its dendrites receive collaterals of the axons that interconnect the dorsal thalamus and the cerebral cortex. The thalamic reticular nucleus also receives fibers from the striatum and basal forebrain of the telencephalon, from the substantia nigra in the mesencephalon, and from the raphe nuclei, pedunculopontine nuclei, and the locus coeruleus of the reticular formation. These axonal inputs to the nucleus also can be characterized by the neurotransmitters that are present in their terminals: acetylcholine, GABA, serotonin, and norepinephrine. Nuclei in the diencephalon with at least some of these properties have been described in all vertebrate classes, which suggests that the thalamic components of the reticular formation, like their medullary, pontine, and mesencephalic counterparts, are part of a phyloge-

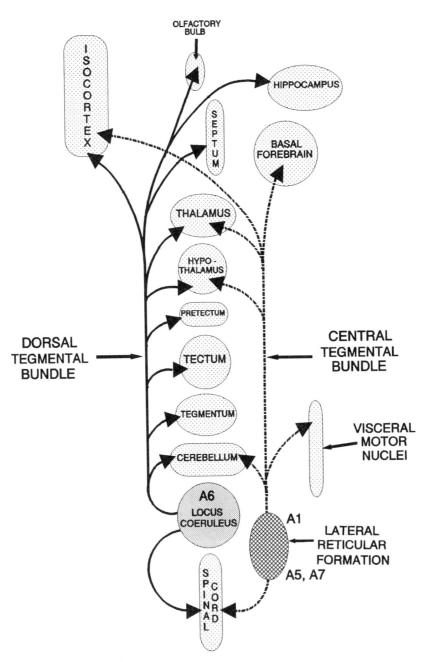

FIGURE 13-11. A summary of the main noradrenergic pathways.

netically old and relatively stable system. The thalamic reticular nucleus is thought to play a role in generating the slow oscillatory electrical rhythms observed in the thalamus and cerebrum during slow-wave (nondreaming) sleep.

PATHWAYS OF THE RETICULAR FORMATION

In all vertebrates, the descending pathways from the reticular formation are important for the brain's ability to control movements of the musculoskeletal system and the specialized organs of the head, such as the jaws and eyes, especially for voluntary movement. When limbs are present, these too are controlled by descending reticular formation pathways. The pathways from the reticular formation to the spinal cord, for control of limb and trunk musculature, are called reticulospinal pathways; those descending reticular formation pathways that end on cranial nerve motor nuclei in the brainstem are known as reticulobulbar pathways. In addition to reticulospinal and reticulobulbar control of their motor neurons, only mammals have corticospinal and corticobulbar pathways that work together with the descending reticular pathways to provide voluntary control over the musculoskeletal systems of the head and body. Indeed, in nonmammals, the reticulospinal and reticulo-

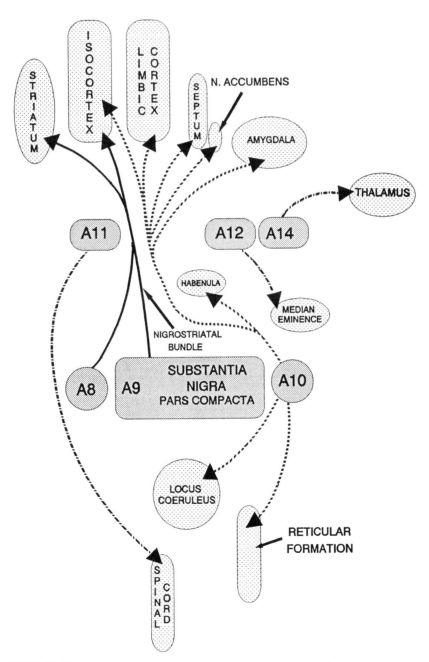

FIGURE 13-12. A summary of the main dopaminergic pathways of the midbrain and hypothalamus.

bulbar pathways constitute the bulk of all the descending pathways in the brainstem and spinal cord.

Contemporary studies of the reticular formation of non-mammals have revealed a pattern of descending pathways that is remarkably consistent among the vertebrate classes, including mammals. This suggests that once we make allowances for the reticular formation specializations associated with the development of limb locomotion and the development of highly dexterous digits on the limbs of mammals, the reticular formation emerges as one of the more conservative and stable of the neuronal systems.

The principal sources of descending pathways are from the reticular nuclei pontis caudalis, pontis oralis, parvocellularis,

magnocellularis, and gigantocellularis. The descending axons from these nuclear groups travel primarily in the **medial longitudinal fasciculus,** which is one of the main highways of the reticular formation. In addition, other descending pathways originate in the locus coeruleus–subcoeruleus complex and portions of the raphe group. These descending reticular projections typically travel in the lateral funiculus and the ventral funiculus of the spinal cord.

In addition to its connections with the motor-neuron pools of the brainstem and spinal cord, the reticular formation also has connections with the cerebellum, the diencephalon, and a number of structures within the telencephalon, including the cerebral cortex in mammals. These connections generally are

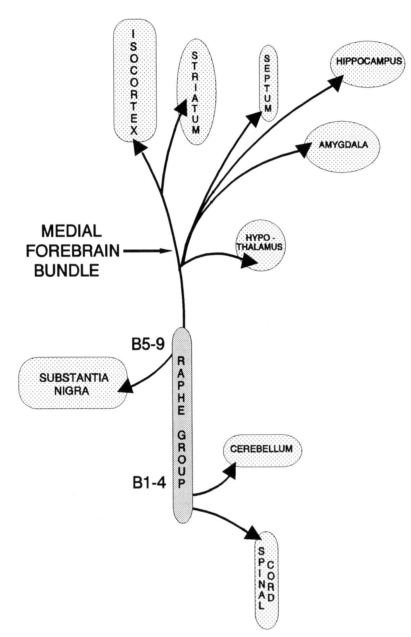

FIGURE 13-13. A summary of the main serotoninergic pathways.

via the medial longitudinal fasciculus, the central tegmental tract, and the dorsal tegmental and nigrostriatal bundles. These pathways will be discussed below and in subsequent chapters.

Chemical Pathways of the Reticular Formation

In addition to being classified according to the locations of their target neuronal populations, reticular formation pathways also can be classified according to the neurotransmitters that are produced in their neurons. These include norepinephrine, epinephrine, serotonin, acetylcholine, GABA, and several others, although the primary neurotransmitters of many reticular formation pathways are unknown at present. In addition to neurotransmitters, many of these neurons also contain neuropeptides and other neuroactive substances that modulate the action of the transmitters in a variety of ways (see Chapter 2). Many of the pathways are ascending, although some descend to the spinal cord as well. Figures 13-11–13-14 summarize these chemically defined pathways based largely on the mammalian literature, which has been the most intensively studied. Given the general stability of the reticular formation during vertebrate evolution, the figures tell a reasonably accurate story for vertebrates as a whole. Before studying these figures, the reader might want to quickly review the table of neuroactive substances in Chapter 2.

Early investigators of the chemically defined pathways originally referred to the cell groups of origin of these axons by letters of the alphabet: A and B. The A cell groups contain catecholamines and the B groups contain serotonin. Although virtually all of these groups have now been associated with specific, named aggregations of neurons in the brainstem, the A and B designations still appear frequently in the literature, and the serious student of this topic should become familiar with them. In Figures 13-11–13-13, we present these catecholaminergic and serotoninergic pathways with both the names and the A or B designations of their cells' origin. In these three figures, some details have been simplified for clarity in an introductory presentation. Thus, not all cell groups or all of their connections are shown.

Figure 13-11 summarizes the pathways that have been characterized by the presence of the neurotransmitter norepinephrine and/or epinephrine. Since another name for norepinephrine is noradrenaline these pathways often are referred to as "noradrenergic" pathways. The figure shows two noradrenergic pathways: one, on the left, originating in the locus coeruleus, and a second, on the right, that originates in catecholamine cell groups within the lateral reticular formation. Both pathways also have descending components to the spinal cord; the lateral reticular pathway also sends fibers to the visceral motor nuclei of the brainstem such as the motor nucleus of the vagus nerve. The locus coeruleus pathway ascends in the **dorsal tegmental bundle** and sends its axon terminations to a wide variety of target neurons including the cerebellum, the mesencephalon (the tegmentum, the tectum, and pretectum), the diencephalon (the hypothalamus and thalamus), and the telencephalon (the septum, the hippocampus, the olfactory bulb, and the cerebral cortex). The lateral reticular pathway's projections, by way of the **central tegmental bundle** (sometimes known as the **ven-**

tral noradrenergic bundle), are more limited and include the cerebellum, the thalamus and hypothalamus, and the basal forebrain and cerebral cortex (isocortex). The central tegmental bundle should not be confused with the **central tegmental tract,** which is more medially located and consists of ascending axons from the large-celled, medial reticular formation and terminates in the thalamus. The central tegmental tract is part of the ascending reticular formation. The neurotransmitters of this system are largely unknown, although one group, axons originating in the pars compacta of the nucleus reticularis pedunculopontinus, utilizes acetylcholine.

Figure 13-12 schematically illustrates two dopaminergic pathways: a midbrain-originating pathway from the A8, A9, and A10 cell groups, and a hypothalamus-originating pathway from A11, A12, and A14. The midbrain group consists of two main cell populations: (1) the A8 and A9 group, which project mainly to the corpus striatum in the depths of the telencephalon, and (2) the A10 group, which is the source of a major projection to the limbic system including the habenula, the septum, nucleus accumbens, limbic cortex, and amygdala. The A10 group (sometimes known as the **ventral tegmental area**) also has projections to the locus coeruleus and brainstem reticular formation. Because of its strong connections to the limbic system, the ascending dopaminergic pathway from the A10 group is known as the **mesolimbic pathway** (from mesencephalon to the limbic system). This pathway is thought to play a role in pleasure and reward mechanisms and may be related to drug addiction.

The dopaminergic cell groups of the tegmentum will be dealt with in detail in Chapter 17. The A8 cell group also is known as the **retrorubral nucleus** ("behind" the red nucleus). The A10 group consists of scattered cells in the ventral tegmentum. The A9 group is a subdivision of the **substantia nigra,** which is a prominent feature of the midbrain tegmentum of mammals and has been reported in various nonmammals as well. This particular subdivision of the substantia nigra is known as the **pars compacta** (compact part). The substantia nigra also contains a division known as the **pars reticulata,** which we will encounter shortly in connection with a different chemically defined pathway. The A8 and A9 groups are the source of a major ascending chemically defined pathway to the telencephalon known as the **nigrostriatal bundle** because one of its principal targets is the corpus striatum. This pathway also is known as the **mesostriatal dopaminergic pathway** (mesencephalon to striatum). Degeneration of the dopamine-producing neurons in the substantia nigra has been associated with the motor disorder known as Parkinson's disease.

The A11, A12, and A14 cell groups are the hypothalamic component of the **dopaminergic pathways.** These are scattered throughout the hypothalamus and around the third ventricle. The A11 group sends fibers to the spinal cord and the A14 group to the thalamus. The A12 group sends its axons to the **median eminence** of the hypothalamus, which is the transition zone between the hypothalamus and the posterior pituitary.

Figure 13-13 diagrams some of the B cell groups that make up sources of the **serotoninergic pathways.** Many of these B groups are components of the raphe-group cell column. For example, the B5, B6, B7, B8, and B9 groups are in the rostral end of the raphe-group column. These groups have widespread projections that include the substantia nigra, the cerebral cortex (isocortex), the striatum, and such limbic structures as the hypothalamus, amygdala, septum, and hippocampus. The B1, B2, B3, and B4 groups comprise the caudal end of the raphe-group column. They send their axons to the cerebellum and the spinal cord.

Finally, in Figure 13-14, we find pathways emanating from the pars reticulata of the substantia nigra. The neurons of the pars reticulata produce the inhibitory neurotransmitter, GABA. The targets of these GABA-ergic pathways are the parvocellular components of the reticular formation, which participate in control of muscles of the mouth and face region, the tectum of the midbrain, and various cell groups in the thalamus. The GABA-ergic neurons also are found in the nucleus reticularis magnocellularis; these cells project to the spinal cord and mediate the loss of muscle tone that occurs during the dreaming phase of sleep (REM sleep).

The chemical pathways of the reticular formation play a role in many neural processes. For example, the ascending serotoninergic and noradrenergic pathways play a role in sleep and wakefulness. The descending noradrenergic pathways function in the integration of the activities of the sympathetic nervous system, especially heart rate and blood pressure, via the preganglionic neurons in the intermediolateral column of the spinal cord, the nucleus solitarius, and the dorsal motor nucleus of the vagus nerve. The dopamine pathways are involved in the mechanisms of voluntary movement. In addition, many of the exogenous chemical compounds that have stimulating effects on behavior have been found to act on receptor sites in the dopamine system.

EVOLUTIONARY PERSPECTIVE ON THE RETICULAR FORMATION

The reticular formation has been rather conservative in its evolution. In virtually all jawed vertebrates, one can recognize without too much difficulty the main components of this system: the raphe group spanning the length of the hindbrain, the parvocellular, magnocellular and gigantocellular groups of the medulla, the pontis caudalis and pontis oralis group of the pons, the locus coeruleus, and the midbrain reticular formation. The development of limbs has added certain additional components in tetrapods, further structures have evolved in mammals and birds along with their more elaborate use of limbs and digits. The chemical pathways also have been similarly conservative in their evolution.

FOR FURTHER READING

Björklund, A. and Hökfelt, T. (eds.) (1984) *Handbook of Chemical Neuroanatomy. Vol. 2: Classical Neurotransmitters in the CNS. Pt 1*. Amsterdam: Elsevier.

Brodal, P. (1992) *The Central Nervous System*. New York: Oxford University Press.

Cruce, W. L. R. and Newman, D. B. (1984) Evolution of motor systems: the reticulospinal pathways. *American Zoologist,* 24, 733–753.

Kandel, E. R. (1991) Disorders of thought: schizophrenia. In E. R. Kandel, J. H. Schwartz and T. J. Jessell (eds.) *Principles of Neural Science*. Norwalk, CT: Appleton and Lange, pp. 853–868.

Kuypers, H. G. J. M. and Martin, G. F. (1982) *Anatomy of Descending Pathways to the Spinal Cord*. Amsterdam: Elsevier.

Newman, D. B. (1985) Distinguishing rat brainstem reticulospinal nuclei by their neuronal morphology. I. Medullary nuclei. *Journal für Hirnforschung,* 26, 187–226.

Newman, D. B. (1985) Distinguishing rat brainstem reticulospinal nuclei by their neuronal morphology. II. Pontine and mesencephalic nuclei. *Journal für Hirnforschung,* 26, 385–418.

Newman, D. B., Cruce, W. L. R., and Bruce, L. L. (1983) The sources of supraspinal afferents to the spinal cord in a variety of limbed reptiles. I. Reticulospinal systems. *Journal of Comparative Neurology,* 215, 17–32.

Parent, A., Poitras, D., and Dube, L. (1984) Comparative anatomy of central monoamingergic systems. In A. Björklund and T. Hökfelt (eds.), *Handbook of Chemical Neuroanatomy. Vol.2: Classical Neurotransmitters in the CNS. Pt I*. Amsterdam: Elsevier, pp. 409–439.

ten Donkelaar, H. J. (1982) Organization of descending pathways to the spinal cord in amphibians and reptiles. In H. G. J. M. Kuypers and G. F. Martin (eds.), *Anatomy of Descending Pathways to the Spinal Cord*. Amsterdam: Elsevier, pp. 25–67.

Webster, D. M. and Steeves, J. D. (1991) Funicular organization of the avian brainstem-spinal projections. *Journal of Comparative Neurology,* 312, 467–476.

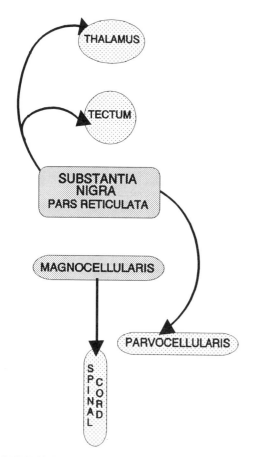

FIGURE 13-14. A summary of the main GABA-ergic pathways.

ADDITIONAL REFERENCES

Ayala-Guerrero, F. and Huitron-Resendiz, S. (1991) Sleep patterns in the lizard *Ctenosaura pectinata. Physiology and Behavior.* 49, 1305–1307.

Cabot, J. B., Reiner, A., and Bogan, N. (1982) Avian bulbospinal pathways: anterograde and retrograde studies of cells of origin, funicular trajectories and laminar terminations. In H. G. J. M. Kuypers and G. F. Martin (eds.), *Anatomy of Descending Pathways to the Spinal Cord.* Amsterdam: Elsevier, pp. 79–108.

Hlavacek, M. Tahar, M. Libouban, S., and Szabo, T. (1984) The mormyrid brainstem. I. Distribution of brainstem neurones projecting to the spinal cord in *Gnathonemus petersii.* An HRP study. *Journal für Hirnforschung,* 25, 603–615.

Johnston, J. B. (1902) The brain of petromyzon. *Journal of Comparative Neurology,* 16, 1–86.

Kimmel, C. B. (1982) Reticulospinal and vestibulospinal neurons in the young larva of a teleost fish, *Brachydanio rerio.* In H. G. J. M. Kuypers and G. F. Martin (eds.), *Anatomy of Descending Pathways to the Spinal Cord.* Amsterdam: Elsevier, pp. 1–23.

Lee, R. K. K. and Eaton, R. C. (1991) Identifiable reticulospinal neurons of the adult zebrafish, *Brachydanio rerio. Journal of Comparative Neurology,* 304, 34–52.

Lee, R. K. K., Eaton, R. C., and Zottoli, S. J. (1993) Segmental arrangement of reticulospinal neurons in the goldfish hindbrain. *Journal of Comparative Neurology,* 329, 539–556.

Livingston, C. A. and Leonard, R. B. (1990) Locomotion evoked by stimulation of the brainstem in the Atlantic stingray, *Dasyatus sabina. Journal of Neuroscience,* 10, 194–204.

Martin, R. J. (1979) A study of the morphology of the large reticulospinal neurons of the lamprey ammocoete by intracellular injection of procion yellow. *Brain, Behavior and Evolution,* 16, 1–18.

Newman, D. B., Hilleary, S. K. and Ginsberg, C. Y. (1989) Nuclear terminations of corticoreticular fiber systems in rats. *Brain, Behavior and Evolution,* 34, 223–264.

Newman, D. B. and Liu, R. P. C. (1987) Nuclear origins of brainstem reticulocortical systems in the rat. *American Journal of Anatomy,* 178, 279–299.

Ouimet, C. C., Patrick, R. L. and Ebner, F. F. (1985) The projection of three extrathalamic cell groups to the cerebral cortex of the turtle *Pseudemys. Journal of Comparative Neurology,* 237, 77–84.

Pritz, M. B. and Stritzel, M. E. A different type of thalamic organization. (1990) *Brain Research,* 525, 330–334.

Ramon-Moliner, E. and Nauta, W. J. H. (1966) The isodendritic core of the brain stem. *Journal of Comparative Neurology,* 126, 311–335.

Role, L. W. and Kelly, J. P. (1991) The brain stem: cranial nerve nuclei and the monoaminergic systems. In E. R. Kandel, J. H. Schwartz and T. J. Jessell (eds.), *Principles of Neural Science.* Norwalk, CT: Appleton and Lange, pp. 683–699.

Ronan, M. (1989) Origins of the descending spinal projections in petromyzontid and myxinoid agnathans. *Journal of Comparative Neurology,* 281, 54–68.

Smeets, W. J. A. J. (1981) Efferent tectal pathways in two chondrichthyans, the shark *Scyliorhinus canicula* and the ray *Raja clavata. Journal of Comparative Neurology,* 195, 13–23.

Stuesse, S. L., Cruce, W. L. R., and Northcutt, R. G. (1991) Localization of serotonin, tyrosine hydroxylase, and leu-enkephalin immunoreactive cells in the brainstem of the horn shark, *Heterodontus francisci. Journal of Comparative Neurology,* 308, 277–292.

Swain, G. P., Snedeker, J. A., Ayers, J., and Selzer, M. E. (1993) Cytoarchitecture of spinal-projecting neurons in the brain of the larval sea lamprey. *Journal of Comparative Neurology,* 336, 194–210.

ten Donkelaar, H. J. (1976) Descending pathways from the brain stem to the spinal cord in some reptiles. *Journal of Comparative Neurology,* 167, 421–442.

ten Donkelaar, H. J., Kusuma, A., and De Boer-Van Huizen, R. (1980) Cells of origin of pathways descending to the spinal cord in some quadrupedal reptiles. *Journal of Comparative Neurology,* 192, 827–851.

Tretjakoff, D. (1909) Das Nervensystem von Ammocoetes. II. Das Gehirn. *Archiv für Mikroskopische Anatomie,* 74, 636–679.

14

The Cerebellum

INTRODUCTION

An inspection of the gross anatomy of an avian or mammalian brain reveals two major masses of tissue on its dorsal aspect: a large mass, the cerebrum (the *great brain*) and a smaller, highly folded mass, the cerebellum (the *little brain*). A cerebellum with few folds or no folds at all and often of lesser relative size can be found in all vertebrates. In some fishes, the cerebellum assumes enormous proportions and, relative to the rest of the brain, dwarfs even the large cerebella of mammals and birds. What is this "little brain," and what are its functions in various vertebrate groups? How has evolution shaped its form and connections with other brain regions? We will try to answer these questions in this chapter.

In the nineteenth and early twentieth centuries, the cerebellum was the subject of a number of functional studies that observed that surgical removal of all or part of the cerebellum resulted in disorders of movement. With the advent of progressively more sophisticated techniques for recording electrical and chemical changes in individual neurons in the middle of the twentieth century, the cerebellum became the target of intense research that unraveled the excitatory and inhibitory relationships among its neuronal constituents. Because of its consistent pattern of internal organization, its relatively few cellular components, and only two routes in and one route out, the cerebellum (in particular, the mammalian cerebellum) soon became the subject of attempts at mathematical and computer modeling of the actions of this precise neuronal network. But in spite of its deceptive appearance of simplicity, our understanding of the functions of the cerebellum is far from complete and almost certainly is strongly biased by the heavy focus of attention on mammals as experimental subjects.

Among nonmammals, the cerebellum plays an important role in the analysis of information from the lateral line system in nontetrapods as well as some amphibians. In addition, in certain groups of fishes, the cerebellum is intimately involved in the detection of electrical fields. Finally, in all vertebrates, the cerebellum has a close relationship with the vestibular, somatosensory, visual, and auditory systems. The reader is urged to review Chapter 11, which covers the lateral line, electrosensory, and vestibular systems, before delving further into this chapter. A review of the ascending and descending pathways in the spinal cord (Chapter 8) would be useful as well.

OVERVIEW OF THE CEREBELLUM

The main divisions of the cerebellum that are common to most vertebrates are the **corpus cerebelli** (body of the cerebellum) and a **cerebellar auricle** (little ear). These are formed of an outer layer, the cerebellar cortex, and sometimes a distinct layer of white matter below. The cerebellar cortex consists of a **molecular layer** that contains relatively few cells, a **granule-cell layer** that consists of small, tightly packed cells, and a layer of **Purkinje cells,** which typically have their large, pear-shaped somata located in a layer that is only one cell thick and situated between the molecular and granule-cell layers.

In the typical vertebrate pattern, the input to the cerebellum comes from the primary or secondary nuclei of cranial nerves and from a specialized precerebellar nucleus, the **inferior olive.** Some cranial nerves, such as the vestibular and lateral-line nerves, terminate directly in the cerebellum. A variety of intrinsic, inhibitory cells modulate the cerebellar input and give it its unique network properties which will be described below. The output of the cerebellar cortex is mainly to one or more specialized **deep cerebellar nuclei** that often are located in the white matter below the cerebellum. The main targets of the cerebellar nucleus or nuclei are the reticular formation and the **nucleus ruber** (red nucleus), which are sources of important descending spinal pathways. Other consistent outputs of the cerebellum are to the vestibular nuclei, to motor nuclei of the brainstem, especially those that control the movements of the eyes, and to the region of the thalamus that is involved with the motor system.

Descriptions of the cerebella of mammals sometimes include terms such as **paleocerebellum, archicerebellum,** and **neocerebellum** to refer to the relative ages or stages of evolutionary development of the major components of these structures. These terms, however, do not reflect current knowledge of the comparative anatomy and evolution of the cerebellum and its relationship with other major brain systems, such as the cerebral cortex and the corticospinal pathways. In particular, the term "neocerebellum" was applied to large lateral extensions of the cerebellum in mammals. A number of investigators of the comparative anatomy of the cerebellum now regard the avian and mammalian cerebella as having a very similar organization; therefore, if the term "neocerebellum" has any utility at all, it should be applied both to birds and mammals.

Other descriptive terms found in the cerebellum literature are **spinocerebellum, visuocerebellum, vestibulocerebellum,** and so on. These terms were introduced to draw attention to the fact that input to the cerebellum from a particular source, such as the spinal cord, tends not to be to the entire cerebellum, but rather to specific regions. While these terms generally are useful, they give the impression that the inputs to the cerebellum are highly segregated according to their sources. This is not entirely correct and the reader should understand that even though such terms are used for the convenience of discussion, considerable overlap exists between these specialized regions.

A notable exception to this general plan are the ray-finned fishes, which have evolved certain other structures that are unique to this group of vertebrates; these include (1) a **caudal lobe** located just caudal to the corpus cerebelli; (2) a **valvula cerebelli** (little folding doors of the cerebellum); (3) several accessory "cerebelloid" structures, such as the **crista cerebellaris** (cerebellar crest), and the **torus longitudinalis,** which lies at the medial edge of the optic tectum in the midbrain; (4) an additional precerebellar nucleus, the **nucleus lateralis valvulae;** and (5) efferent axons of the cerebellar cortex that do not terminate in specialized deep cerebellar nuclei, but rather pass directly to the typical targets of the cerebellar nuclei. Other differences between the cerebella of ray-finned fishes and other vertebrates will be described below.

THE VARIOUS FORMS OF THE CEREBELLUM

The overall form of the cerebellum, its microscopic structure, and connections to other parts of the central nervous system vary considerably both within and between vertebrate classes. The overall form of the cerebellum can vary from a simple ridge or plate in agnathans, some fishes, and amphibians, to the large, highly folded structure that is typical of mammals and birds. The size of the cerebellum in relation to other brain structures varies greatly in the major radiations of vertebrates.

Corpus Cerebelli

Figure 14-1 shows a classification of the more common types of configuration of the corpus cerebelli. Each drawing represents a section through the corpus cerebelli in a parasagittal plane (except the hyperfolded, which is in a horizontal plane). At the upper left is an example of one of the simpler types of cerebellar shapes, a flat plate, which is typical of amphibians and turtles. Below that are two variations in which the flat plate has been extended, which increases the area of cerebellar cortex. The plates also have been curved so that the lengthened cerebellar plate will fit within the confines of the skull. These are characteristic of alligators and lizards. In the case of lizards, however, the curvature has been in the rostral direction, which results in the granule-cell layer appearing on the dorsal surface rather than the molecular layer. We have termed this a reverse curvature.

In the column of cerebellum types on the right of Figure 14-1 are other major types of variations on the flat plate cerebellum; these are various types of folding. A common form is the folded cerebellum that is characteristic of many ray-finned and fleshy-finned fishes. It consists of a single large fold, or **folium** (plural = **folia**) known as the corpus cerebelli (body of the cerebellum) and a rostral extension that consists of several smaller folds, the valvula. A double-folded structure often is found in cartilaginous fishes. Much more elaborate folding of the cerebellum as a consequence of a massive expansion in the total surface area of the cortex can be seen in sting rays, birds, and mammals. The drawing in Figure 14-1 labeled "multi-folded" shows the folding pattern of only two lobes of this type of multilobed, cerebellum in which each lobe contains several folds. Birds have 10 lobules or major folia in their cerebella. Mammals also have 10 lobules in the central region known as the **vermis** (worm) and many more folds in the lateral portions of the cerebellum.

Electroreception and the Cerebellum

The most elaborate form of the cerebellum is shown in Figure 14-1 as the "hyperfolded" type, which is found in the valvula cerebelli of mormyrid fishes. The cerebellum plays a major role in electroreception, and in these fishes it has evolved into a massive organ. The valvula, in particular, has developed into an enormous sheet of hyperfolded tissue. The mormyrid brain has carried the "packaging" technique of folding to an extreme by organizing the valvula into a long continuous ribbon that is folded over and over again along its length, like a wavy ribbon of toothpaste. These folds, shown in Figures 14-2 and 14-3, form thin, tight hyperfolds that provide a superb packaging arrangement for this vast sheet of cerebellar cortex. If the valvula of a mormyrid were to be unfolded and stretched out, it would be more than 10 times the length of the fish's body. Indeed, the weight of the entire cerebellum can account for nearly 80% of the total weight of the brain in these animals.

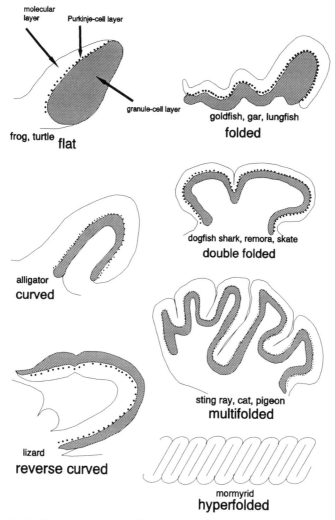

FIGURE 14-1. A classification of the form of the corpus cerebelli. Each drawing represents a parasagittal section through the corpus cerebelli except the hyperfolded type, which is a horizontal section. In each example, the rostral end is to the left and the caudal end is to the right. The shaded areas represent the granule-cell layers, the unshaded ares represent the molecular layers, and the dotted layers represent the Purkinje-cell layers.

A comparison of the cerebella of a mormyrid fish and a nonelectric, freshwater teleost fish, the goldfish, is shown in Figure 14-2. The goldfish cerebellum is of a size and shape that is typical of nonelectric ray-finned fishes, both teleost and nonteleost. Only the corpus cerebelli is visible in the intact brain; the valvula is tucked away within the convenient space provided by the large ventricle located in the depths of the lobe of the optic tectum.

In contrast, in the mormyrid brain shown in Figure 14-2, the cerebellum is hypertrophied (overgrown) and has expanded beyond the confines of the tectal ventricle; thus it covers the dorsal surfaces of the cerebrum and the tectum and overhangs the brain on the sides. Moreover, the valvula is so enormous that it protrudes caudally between the two massive lobes of the cerebellar hemispheres. Figure 14-3 shows a dorsal view of the brain of a mormyrid fish in which all of the brain except the medulla has disappeared beneath the massive cerebellum.

An important point worth noting here is that the valvula of weakly electric fishes is highly specialized for electrore-

ception and is a very different organ from the valvula of nonelectric fishes. The electroreception system has evolved a complex series of nuclei that have direct and indirect relations with the valvula; these are entirely lacking in nonelectric fishes.

Electroreception is not confined to fishes that generate electric fields around themselves. Agnathans, cartilaginous fishes, some species of ray-finned fishes, amphibians (except frogs and toads), and the monotreme mammals possess the specialized receptors necessary to detect the very weak electric fields that normally surround an animal or that might be produced by the action potentials in the muscles of prey as they move. In the electroreceptive nonmammals, a specialized region of the octavolateralis area known as the **electrosensory lateral line lobe,** which is closely related to the cerebellum, receives the electrosensory input. In the monotremes, the electroreceptors are a specialized division of the trigeminal nerve, the main sensory nucleus of which sends a projection to the cerebellum.

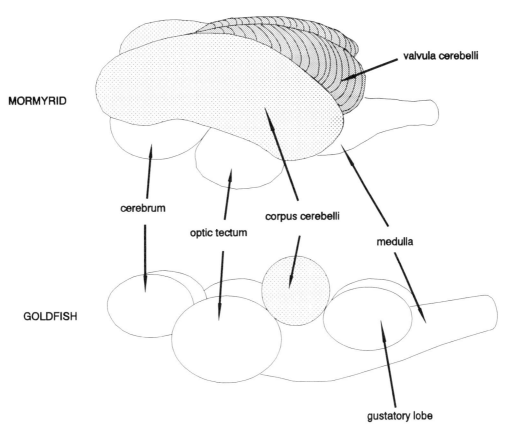

FIGURE 14-2. A comparison of the cerebella of an electroreceptive mormyrid fish (top) and a nonelectrore-ceptive goldfish (bottom).

The Cerebellar Auricle

In addition to the corpus and the valvula, the cerebellum contains a smaller, caudal structure, usually on the ventrolateral aspect of the cerebellum. This is the cerebellar auricle. The auricle often has a multifolded conformation, and in some animals, especially sharks and rays, it is subdivided into two lobes called the **upper leaf** and the **lower leaf.** The auricle is the region of the cerebellum that receives input from the vestibular system. The output from the auricular region is especially influential on those motor neurons that control the muscles that move the eyes. In tetrapods the auricle is known as the **flocculus.**

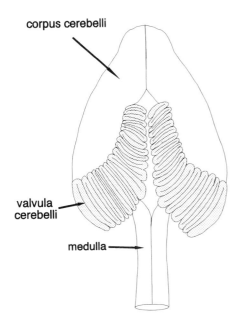

FIGURE 14-3. A dorsal view of a mormyrid brain. The massive cerebellar hemispheres completely overhang the cerebrum and optic tecta.

Phyletic Development of the Form of the Cerebellum

Figure 14-4 summarizes the main differences between the major vertebrate groups in the overall form of the cerebellum. Electrosensory ray-finned fishes, which do not fit the general vertebrate pattern, have been omitted from the diagram. The agnathan cerebellum, as represented by a lamprey, is shown at the left. In this class, the cerebellum is rudimentary, and its correspondence to the corpus cerebelli of jawed vertebrates is unresolved. Below is the cerebellum of a non-electrosensory ray-finned fish; note the corpus cerebelli and the auricle. In mammals, a foliated corpus cerebelli also is present as the midline **vermis.** In addition, the mammalian cerebellum shows a large, highly foliated, lateral expansion known as the neocerebellum. Similar patterns may be seen in the cerebella of cartilaginous fishes, amphibians, diapsid reptiles, and turtles. In birds, the corpus cerebelli becomes foliated into 10 folia. Whether this lateral expansion in mammals is in fact "new" or is merely a variation

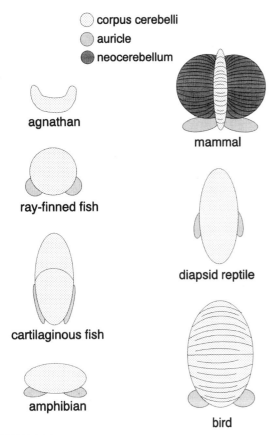

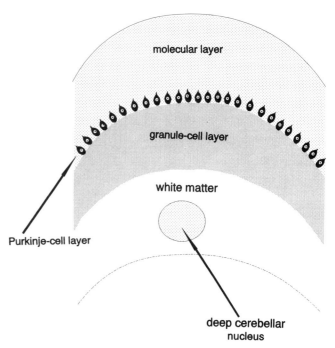

FIGURE 14-4. Schematic representations of the major components of the cerebellum in various groups of vertebrates.

FIGURE 14-5. The organization of the cerebellar cortex of tetrapods.

on the same foliated plan as seen in birds is a matter now being debated. Evidence does exist, however, to suggest that rather than being a new addition to the cerebellum, the neocerebellum is, in fact, merely a lateral expansion of the corpus cerebelli.

THE CEREBELLA OF TETRAPODS

n general, the cerebella of tetrapods are more consistent in their cellular organization and relationships to other cell populations in the brain and spinal cord than those of non-tetrapods, so we will describe those first. In tetrapods, the cerebellum comprises two principal zones: a three-layered cortex and an underlying white matter (Figure 14-5). The white matter consists of the incoming axons to the cortex and the outgoing axons from the cortex. The three layers of the cortex are a thick, superficial **molecular layer,** a thin, intermediate layer that typically is a sheet of large neurons only one cell thick, and a thick internal layer of very densely packed granule cells. The molecular layer consists mainly of dendrites, unmyelinated axons, and scattered cells of two distinctive types: the **stellate cells** and the **basket cells.** The intermediate layer of the cortex consists of a one-cell-thick sheet of neurons with large, pear-shaped somata and elaborate dendrites that extend into the molecular layer. These cells are the **Purkinje cells,** and were first described in the nineteenth century by the Czechoslovakian anatomist and physiologist Johannes Purkinje

(Purkyně in Czechoslovakian and pronounced "poor-kee-nyeh" although English speakers typically pronounce it "purr-kinn-jee"). The Purkinje cells tend to be spaced more or less uniformly with some distance separating each Purkinje cell from its neighbors. The axons of the Purkinje cells, which are the only route out of the cerebellar cortex, pass through the underlying granule-cell layer and enter the white matter below. Internal to the Purkinje-cell layer is the layer of **granule cells.** In contrast to the Purkinje cells, granule-cell somata are quite densely packed and indeed constitute the greatest density of neurons in the central nervous system. In addition to the granule cells, the granule-cell layer contains two other types of cell, **Golgi cells** and **Lugaro cells.** The granule-cell layer rests on the layer of white matter that contains the axons of the Purkinje cells, the axons that constitute the two afferent pathways to the cerebellar cortex (the **mossy fibers** and the **climbing fibers**), and one or more groups of neurons that constitute the deep cerebellar nuclei. The cells of the deep cerebellar nuclei are the targets of the Purkinje-cell axons. Figure 14-6 shows photomicrographs of the cerebella of a turtle and a bull frog. In nontetrapods, a white matter layer that is clearly distinguishable from the cortex is not present; the afferent and efferent axons of the cerebellum merely mingle with those of the underlying pons.

THE CEREBELLA OF NONTETRAPODS

Agnathans and Cartilaginous Fishes

In agnathans, the cerebellum consists of a simple bridge of gray matter between the right and left octavolateral areas of the rostral medulla; it is better developed in lampreys than in hagfishes. In lampreys, some Purkinje-like cells are present but

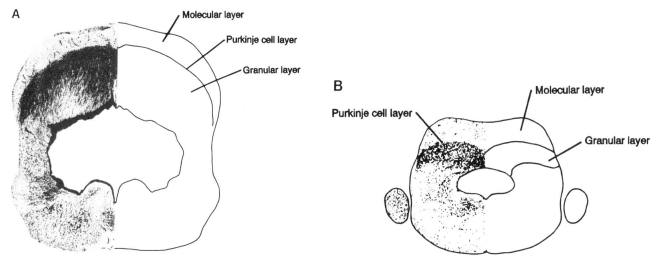

FIGURE 14-6. Transverse hemisections with mirror-image drawings through the cerebellum of (A) a turtle (*Pseudemys scripta*) and (B) a bullfrog (*Rana catesbeiana*). The latter was adapted from Wilczynski and North-cutt (1983).

are scattered throughout the molecular layer. In cartilaginous fishes, the granule cells, rather than appearing in a continuous sheet to form the granule-cell layer, have formed long, cylindrical columns known as the **eminenitae granulares (granular eminences).** Two such columns lie on the roof of the ventricle that is at the core of the corpus cerebelli, and two more are situated on the floor of the cerebellar ventricle. The latter two continue into the fourth ventricle of the medulla and pons. The eminentiae granulares are shown in Figure 14-7, which is a representation of the cerebellum and pons of the cartilaginous ratfish. The most ventral of the ventral cell columns is sometimes known as the **eminentia ventralis.** Also shown in the figure are the Purkinje cells, which congregate on the lateral walls of the corpus rather than being distributed uniformly across the cortex. The cerebellar auricles also appear in the diagram.

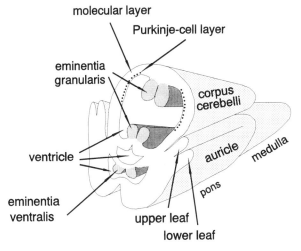

FIGURE 14-7. A drawing of the caudal half of the brain of a cartilaginous fish (a ratfish) that shows both the internal and external structure of the cerebellum.

Ray-Finned Fishes

The cerebellar layers show considerable variation in ray-finned fishes [Fig.14-8, (A and B)]. In some species, the Purkinje cells are organized in a single sheet everywhere as in tetrapods. In others, the Purkinje cells are in a sheet in the corpus, but are scattered in the molecular layer in the valvula. Still others have a sheet organization in some parts of the corpus and a scattered organization elsewhere in the corpus. Some species have the granule layers located in a column lateral to the molecular layer as well as ventral to it. These lateral columns are the eminentiae granulares. An example is shown in Figure 14-9, which is a drawing of the brain of a catfish.

THE CEREBELLAR CORTEX

The Purkinje-Cell Layer

Purkinje cells are easily recognized in all vertebrates except agnathans. In agnathans, the cerebellum contains large, scattered neurons with widely branching dendritic trees, rather than the flattened trees of Purkinje cells in jawed vertebrates. Moreover, they are not organized into the neat, monocellular layer that is characteristic of the Purkinje-cell layer of jawed vertebrates. Hence, these cells in agnathans often are referred to as Purkinje-like cells.

Dendrites. In order to understand the relationship between the dendrites of the Purkinje cells and other components of the cerebellar cortex, you must first understand the distinction between the short and long axes of a cerebellar folium. The short axis is the axis that cuts across the fold in contrast to the long axis, which runs along the length of the fold. These axes are diagrammed in Figure 14-10. Also shown in the figure are Purkinje cells and parallel fibers, which will be described below.

A

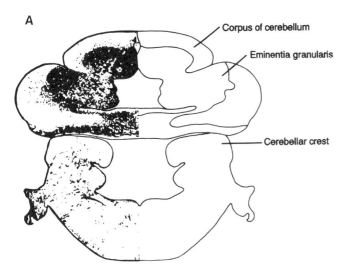

Corpus of cerebellum

Eminentia granularis

Cerebellar crest

B

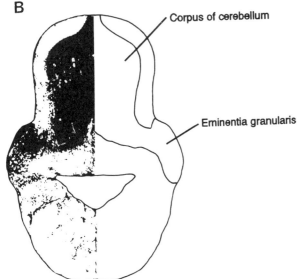

Corpus of cerebellum

Eminentia granularis

FIGURE 14-8. Transverse hemisections with mirror-image drawings through the cerebellum of (A) a gar (*Lepisosteus osseus*) and (B) a teleost (*Lepomis gibbosus*). Adapted from Parent and Northcutt (1982) and Parent et al. (1978).

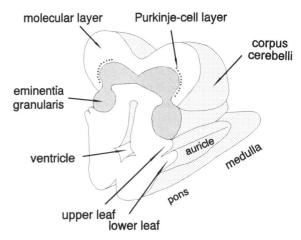

molecular layer   Purkinje-cell layer

corpus cerebelli

eminentia granularis

auricle

medulla

ventricle

pons

upper leaf
lower leaf

FIGURE 14-9. A drawing of the caudal half of the brain of a ray-finned fish (a catfish) that shows both the internal and external structure of the cerebellum.

Below is the dendritic domain of a Purkinje cell, which has the form of a thin slab, with the large surface being in the plane of the short axis and the narrow surface in the plane of the long axis. The bottom shows the same dendritic domain bisected through the Purkinje cell's soma to show the appearance of its dendritic tree in the plane of the long axis. The dendritic trees of the Purkinje cells of agnathans, however, do not follow the typical pattern; they are more like that of the Golgi cell shown at the top of the figure. Figure 14-12 shows examples of the varieties of dendritic trees that can be found in Purkinje cells. These range from the relatively sparse arborizations in catfishes, to the more candelabra-like formations of rays and mormyrids to the more tree-like arrangement in alligators and mice.

Somata. Except in agnathans in which they are scattered throughout the molecular layer, the pear-shaped somata of the Purkinje cells are located at the border between the molecular and granule-cell layers of the cerebellar cortex. They are spaced more or less uniformly apart with some separation between them. In general, at least within mammals where the most

The dendrites of the Purkinje cells extend up into the molecular layer and ramify (form branches) that are oriented along the short axis of the cerebellar folium. Unlike the dendritic trees of most neurons, however, which have a dendritic domain that is three dimensional, the Purkinje cell's dendritic domain is much closer to two dimensional because virtually the entire dendritic tree is flattened along the plane of the short axis. In other words, when seen in the plane of the short axis, the Purkinje cell appears to have a rich, highly elaborate dendritic tree, but when seen in the plane of the long axis, the same cell appears to have a thin, compressed dendritic tree. Figure 14-11 shows the dendritic domains of three cells. A Golgi cell, which is one of the cell types found in the cerebellar cortex of amniotes, is shown at the top of the figure. Its dendritic domain is roughly in the form of a cube and occupies approximately the same volume in both the long and short axes of the folium.

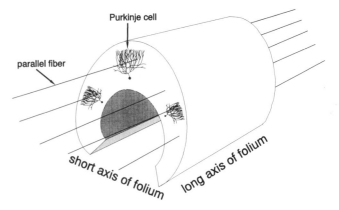

Purkinje cell

parallel fiber

short axis of folium   long axis of folium

FIGURE 14-10. A schematic representation of a single folium of a cerebellar cortex. The parallel fibers run parallel to the long axis of the folium, and the Purkinje-cell dendritic trees are flattened parallel to the short axis of the folium.

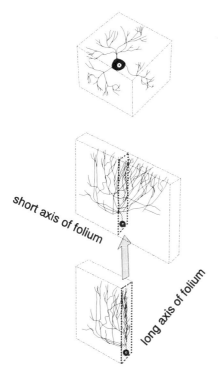

short axis of folium

long axis of folium

FIGURE 14-11. Dendritic domains of a Golgi and a Purkinje cell. Top: The dendrites of the Golgi cell occupy a space that is roughly a cube. Middle: The dendritic tree of the Purkinje cell, which spreads out across the short axis of the folium, occupies a space that is roughly a solid rectangle. Bottom: The solid rectangle of the Purkinje-cell domain has been cut in the plane of the soma to reveal the appearance of the Purkinje cell as it would been seen in the plane of the long axis of the folium.

detailed studies have been done, the larger the animal (and hence the larger the brain), the larger are the Purkinje-cell somata and the greater is the intercellular spacing. Although they appear morphologically identical within a single animal, individual Purkinje cells differ somewhat in size and can be classified according to the presence or absence of various neuroactive substances.

Axons. Like the giant axons of the Mauthner cells of the nontetrapod spinal cord, the initial segment of the Purkinje-cell axon is thinner than the rest of the axon. These axons descend through the granule-cell layer and leave the cerebellar cortex. As they do so, they give off recurrent collaterals that ascend to terminate on the somata and dendrites of Golgi cells and basket cells. In all jawed vertebrates except ray-finned fishes, the Purkinje-cell axons terminate in one or more specialized cerebellar nuclei and in nuclei associated with the vestibular system. A notable exception occurs in ray-finned fishes in which the Purkinje-cell axons are not the source of the critical efferents, but rather remain intrinsic to the cerebellar cortex, where they terminate on other Purkinje cells and on the deeper lying stellate cells. The task of carrying the results of cerebellar cortical processing to other regions of the brain falls to a group of specialized efferent cells. These cells, known as **eurydendroid** cells, also receive axon terminations of the Purkinje cells and differ from Purkinje cells in a number of ways. These

differences are in the spread of the dendritic tree and the lack of spines on their dendrites. Another major difference is that the extrinsic axons of the eurydendroid cells do not terminate in a deep cerebellar nucleus but rather pass directly to target cell populations in the brainstem and spinal cord.

The Granule-Cell Layer

Granule-Cell Somata. In contrast to the Purkinje cells, which are relatively large, relatively few, and can be separated by spaces at least as wide as a Purkinje-cell soma or wider, the granule cells are relatively small, highly numerous, and densely packed. Indeed, the cerebellar granule cells are probably the most densely packed cells in the central nervous system. The extent of the density of the granule cells, like that of the Purkinje cells, varies with the size of the animal and the weight of the brain. Thus, smaller animals tend to have more densely packed granule cells than larger ones. For example, the number of granule cells per cubic μm of cerebellar cortex is approximately 800–1100 in elephants and whales, 1200–1500 in sheep, bulls, and horses; 1600–2300 in humans, Old World monkeys, cats, foxes, and opossums; and 2500–3200 in rats, mice, and moles. Similar findings have been reported for birds in which cellular density in large birds, such as ostriches, is considerably lower than it is in small birds, such as a titmouse. With regard to the ratio of granule cells to Purkinje cells, in larger animals such as elephants, whales, porpoises, primates, sheep, bulls, and horses, the ratio varies from approximately 1500–3000 granule cells per Purkinje cell; in the smaller animals, however, such as mice, moles, hedgehogs, rats, and guinea pigs, the ratio varies from about 600–950 granule cells per Purkinje cell. Thus, even though both Purkinje-cell density and granule-cell density decrease with increasing weight of the cerebellum, at least in birds and mammals, the amount of change is greater in the granule cells.

Granule-Cell Dendrites. Unlike the Purkinje cells, which show a considerable degree of variation in the form of their dendritic arborization, the granule cells show a considerable degree of consistency across species as may be seen in Figure 14-13. Granule cells typically have dendrites that are relatively few (usually four or five per soma) and relatively short. The dendrites end in a characteristic claw-like formation or occasionally in a knob-like ending. These endings are the locus of termination of one of the two types of cerebellar input axons, the mossy fibers. The mossy fiber terminations are only axodendritic; axosomatic terminations do not occur in these cells. Because of the extremely dense packing and dense staining of the granule-cell somata, the regions of the granule-cell layer in which the dendrites are found appear in stained tissue as little islands of unstained tissue called **glomeruli.** In addition to the mossy-fiber input from outside of the cerebellar cortex, an intracerebellar axon also terminates on the granule-cell dendritic endings; these are the axons of the Golgi cells, which are also located in the granule-cell layer. Figure 14-14 shows a granule cell with one of its dendrites receiving axonal terminations on one of its dendritic endings to form a glomerulus. The reader should note that in this and subsequent figures that show individual granule cells, the size of the granule cell has been greatly exaggerated for the sake of visibility and understanding;

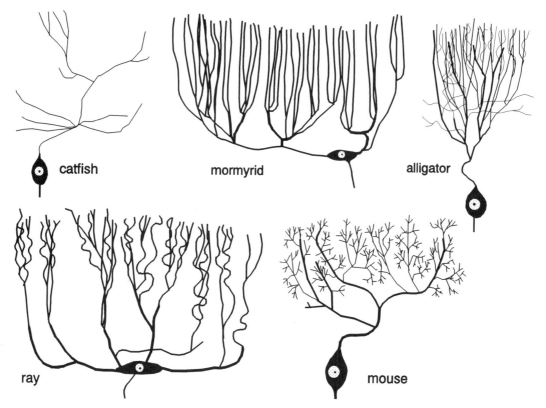

FIGURE 14-12. A sample of the differing forms of Purkinje cells among vertebrates.

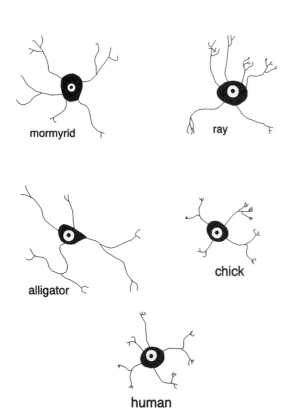

FIGURE 14-13. A sample of vertebrate granule cells. Note the similarity in dendritic organization.

the granule cells actually are much smaller in relation to the other cells of the cerebellar cortex. The figure shows an enlarged view of the glomerulus to reveal the claw-like ending of the granule-cell dendrite and the terminations of the mossy fibers and Golgi-cell axons.

Granule-Cell Axons. The granule-cell axons usually are unmyelinated. They ascend through the granule-cell and Purkinje-cell layers to the molecular layer where they bifurcate into branches that then travel at right angles to the main ascending axon. These branches run parallel to the long axis of the folium and are known as parallel fibers.

An illustration of a granule-cell axon is shown in Figure 14-15, in which the arrows show the direction of conduction of the granule-cell and Purkinje-cell axons. Because the Purkinje-cell dendritic tree is spread out across the plane of the short axis of the folium, the parallel fibers pass through Purkinje dendritic branches like telephone wires passing through the branches of a tree. This may be seen in Figure 14-15, which shows a Purkinje cell as it would appear in a section parallel to the long axis of the folium. In Figure 14-16, a slab of cerebellar cortex that has been cut in the plane of the short axis is shown. This view shows the Purkinje-cell dendritic tree in its wide extent. A segment of this slab has been removed to reveal the long-axis plane and to show the Purkinje-cell dendritic tree in its narrow aspect. Several granule cells are shown in the depths of the granule-cell layer. Their axons rise to the molecular layer and bifurcate to form the parallel fibers, which then run in the plane of the long axis through the branches of the

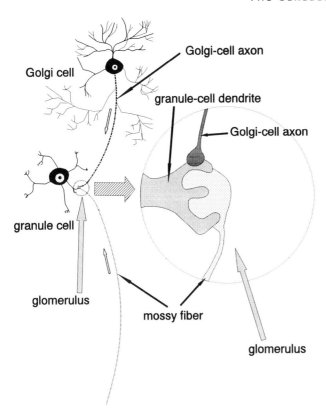

FIGURE 14-14. A cerebellar glomerulus formed by the conjunction of mossy-fiber and Golgi-cell terminations on the dendritic "claws" of a granule cell. The enlarged region shows the details of one such type of glomerulus. The small arrows show the direction of conduction of the axons.

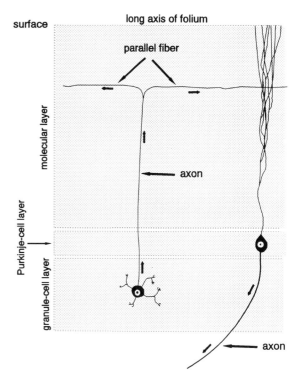

FIGURE 14-15. Granule-cell axons rise into the molecular layer and bifurcate to form the parallel fibers. The parallel fibers pass through the dendritic trees of the Purkinje cells. The arrows show the direction of conduction of the axons. The size of the granule-cell soma has been exaggerated for better visibility.

Purkinje-cell dendrites. The parallel fibers not only make synaptic contacts on the dendrites of the Purkinje cells, but they synapse as well on the dendrites of the other cellular constituents of the molecular layer: the basket and stellate cells. The parallel fibers of ray-finned fishes, however, tend not to bifurcate.

Golgi Cells. Also present in the granule-cell layer are the Golgi cells, which are characterized by their large somata (a little smaller than those of the Purkinje cells), which often have numerous indentations. A Golgi cell is shown in Figure 14-17. Two types of Golgi cells have been found: those in which the dendritic tree is confined to the granule-cell layer and those in which the dendritic tree extends upward into the molecular layer. The latter, which often is found in the vestibular regions of the cerebellum in birds and mammals, is receptive to input from the parallel fibers. The first type of Golgi cell mainly receives mossy fiber input on its dendrites as well as terminals from the vertical portion of the granule-cell axons as they ascend to the molecular layer. The dendrites of the second type of Golgi cell additionally receive terminals from the parallel fibers. The Golgi-cell axons terminate on the granule-cell dendrites in the glomeruli of the granule-cell layer. Golgi cells are found in cartilaginous fishes, ray-finned fishes, and tetrapods, although very few have been reported in amphibians.

Lugaro Cells. In addition to the Golgi cells, another constituent population of the granule-cell layer are the Lugaro cells,

which are located just below the Purkinje-cell layer. Their somata are elongated horizontally as are their dendritic trees. The latter receive axon terminals from Purkinje-cell recurrent collaterals and the parallel fibers. The targets of their axons are not yet known, nor is much known about their function. Lugaro cells thus far have only been reported in mammals.

The Molecular Layer

The molecular layer consists of the dendrites of Purkinje cells, the unmyelinated axons of the granule cells (the parallel fibers), the incoming extracerebellar afferents to the Purkinje-cell dendrites (the climbing fibers), and several cell types. The two principal cellular components of the molecular layer are the **stellate cells** and the **basket cells,** both of which are shown in Figure 14-17. The stellate cells are located in the more superficial regions of the molecular layer and the basket cells in the deeper regions. The basket cells sometimes are known as **inner-stellate** cells. The two cell types can be distinguished by their axonal terminations and their neurotransmitters.

Stellate Cells. The axons of the stellate cells extend in the long axis of the folium and run parallel to the parallel fibers. Along their course they send out a number of collaterals that end on the dendrites of the Purkinje cells. Stellate cells are found in all vertebrates. Their neurotransmitter is taurine.

Basket Cells. Like the more superficial stellate cells, the axons of the basket cells also extend in the long axis of the

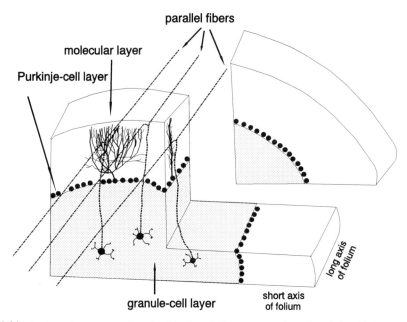

FIGURE 14-16. A schematic representation of a slab of cerebellar cortex to show the relationship between Purkinje and granule cells. A wedge of cortex has been removed to show the internal structure of the slab. The sizes of the granule-cell somata have been exaggerated.

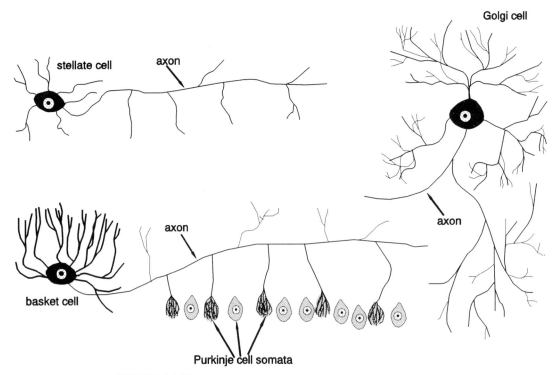

FIGURE 14-17. Illustrations of a stellate cell, a Golgi cell, and a basket cell.

folium and run parallel to the parallel fibers. Their axons send collaterals with densely branched terminals to neighboring Purkinje-cell somata where they form cage-like or basket-like formations (from which their name is derived) around the somata. A single mammalian basket cell typically contacts six or more nearby Purkinje-cell somata. Basket cells are found most commonly in birds and mammals and less frequently in reptiles and amphibians. They appear not to be present in nontetrapods. Their neurotransmitter is **GABA.**

Afferents and Efferents of the Cerebellar Cortex

Mossy Fibers. The mossy-fiber input originates mainly in the vestibular system, and contains both primary axons of the vestibular nerve as well as secondary axons from the vestibular nuclei, spinal cord, and various cell groups located in the pons, including the raphe group of the reticular formation. In addition, in aquatic animals with a lateral line system, the mechanosensory lateral line nerve axons have mossy-fiber terminations on the granule cells.

Each mossy fiber makes multiple terminations, each of which is on a granule-cell dendrite. These terminations are known as **rosettes.** Each morphological complex formed by the claw-like or club-like endings of the granule-cell dendrites, the mossy-fiber rosette, and the axon terminals of the Golgi-cell axons makes up a glomerulus. The mossy fibers excite the granule cells. Acetylcholine and several peptides have been reported as neuro-active substances in mossy-fiber rosettes, but the majority of such substances probably has yet to be determined.

Climbing Fibers. The climbing fibers are so named because they appear to intertwine themselves around the branches of the Purkinje-cell dendrites much as a climbing vine wraps its tendrils around the branches and twigs of a tree. The source of the climbing fibers is much more restricted than those of the mossy fibers; it is the inferior olive, a structure located in the rostral medulla, and which will be described below. In ray-finned fishes, the climbing fibers do not ascend as far as the molecular layer; instead, they terminate on that part of the dendritic tree that is closest to the soma.

Interconnections within the Cerebellar Cortex

The limited afferent and efferent pathways of the cerebellar cortex have held a particular fascination for neurophysiologists and those who do research on mathematical and/or computer models of neural networks. The only routes into the cerebellar cortex are the mossy fibers, which terminate on the granule cells and the climbing fibers, which end on the Purkinje-cell dendrites. The only route out of the cerebellar cortex is the axon of the Purkinje cell or, in the case of ray-finned fishes, the eurydendroid cell.

The basic circuitry of the cerebellar cortex, which is common to all jawed vertebrates, consists of

- The *mossy fiber → granule cell → parallel fiber → Purkinje-cell* input pathway
- The *climbing fiber → Purkinje-cell* input pathway

- The *Purkinje-cell axon* output pathway (eurydendroid cells in ray-finned fishes)
- The *stellate cell → Purkinje-cell dendrite* inhibitory pathway

As the cerebellum has evolved to satisfy the adaptive requirements of various species, these pathways have been augmented by a variety of additional pathways, among which are

- The *basket cell → Purkinje-cell soma and dendrite* inhibitory pathway
- The *Golgi cell → granule cell* inhibitory pathway

Regional Variations in the Cerebellar Cortex. Although the cerebellar cortex generally appears to be uniform from region to region, a regional organization can be recognized on the basis of its connections with other parts of the nervous system, the relative density of cellular components, and the presence of certain neuroactive chemical substances in specific regions. For example, in cartilaginous and ray-finned fishes, the corpus cerebelli mainly receives input from the spinal cord, while the valvula receives fibers from the lateral line system and the auricles from the vestibular system. Within the corpus, a topological representation of the body has been found in which the spinocerebellar axons from the caudal parts of the spinal cord terminate in the caudal levels of the corpus and those from the rostral levels of the cord terminate more rostrally in the corpus. The head of the animal is represented in the most rostral end of the corpus. This general pattern of topographic organization also occurs in the cerebella of tetrapods. Finally, the density of various cell types that make up the cerebellum also varies in different regions of the cerebellum.

Compartmentalization. With regard to locations of various neuroactive substances in the cerebellum, a rather dramatic parcellation or compartmentalization has been observed in a variety of vertebrates. Thus, acetylcholinesterase (AChE), an enzyme associated with the neurotransmitter acetylcholine, is present in the outer (molecular) layer of the cerebellum in distinct bands that are uniformly distributed. Likewise, another neuroactive enzyme, glutamic acid decarboxylase (GAD) and the peptide motilin also have been found located in bands or microcompartments that are separated by other bands or compartments that are free of these substances. Various other substances also have been found to exist in bands or compartments. For example, when the cerebellum is selectively stained to detect the presence of a class of peptides known as zebrins, the cerebellar cortex takes the stripped appearance of a zebra. What relationship the chemical compartmentalization has to topographic maps of the body is not at all clear at present. It does, however, suggest that we may be much farther from a comprehensive understanding of how the cerebellum works than we may have thought a few years ago.

Circuitry of the Cerebellar Cortex. A detailed description of the physiology of the cerebellar cortex is beyond the scope of this introductory anatomical description. In brief, however, the climbing fibers excite the Purkinje cells directly. The mossy fibers excite the granule cells, which in turn excite the Purkinje

cells via the parallel fibers. The parallel fibers also excite whichever inhibitory cells are present; these in turn inhibit the Purkinje and granule cells. An example from mammalian cerebellar
cortex is shown in Figure 14-18. In this figure, the Purkinje cell
is represented schematically with candelabra-like dendrites. A
climbing fiber is seen wrapped around one branch of the
Purkinje-cell dendritic tree; its terminals are excitatory. Beside
the climbing fiber is a mossy fiber that terminates in an excitatory
ending on a granule-cell dendrite. The granule-cell axon rises
to the molecular layer and gives off an excitatory terminal on a
Purkinje dendrite after which it bifurcates and forms the parallel
fibers. The parallel fibers have excitatory terminations on the
dendrites of the basket, the stellate, and the Golgi cell. The two
molecular-layer inhibitory cells send their axons to the Purkinje
cell (the stellate cell to the Purkinje-cell dendrites, and the
basket cell to the Purkinje-cell soma) where they terminate

in inhibitory endings. Similarly, the granule-cell-layer inhibitory cell, the Golgi cell, sends its axon to the granule-cell dendrite.

Although the spread of the parallel fibers along the long
axis of the cerebellar folium can be considerable (several millimeters in a large cerebellum) with synaptic endings on a large
number of Purkinje-cell dendrites, nevertheless a special relationship exists between a Purkinje cell and those granule cells
that are directly below it. This results because the vertical portion of the granule-cell axon makes synaptic contact with the
Purkinje-cell dendritic tree close to the soma (Fig. 14-18), which
makes its excitatory action much more effective than would be
the same stimulation farther out on the dendritic tree. The
vertical portion of the granule-cell axon, therefore, is much
more decisive in affecting the probability of an action potential
being generated in a Purkinje-cell axon than is the horizontal

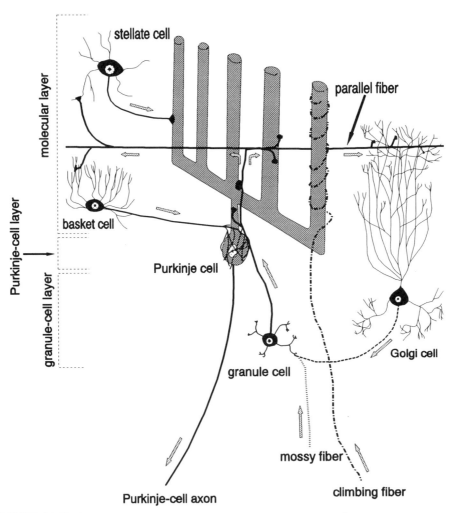

FIGURE 14-18. A schematic representation of the major cellular components of the cerebellar cortex and
their interrelations. The arrows indicate the direction of axonal conduction. Mossy fibers and climbing fibers
excite the granule cell and the Purkinje cell. The granule-cell axon excites the Purkinje cell as it ascends through
the molecular layer. The granule-cell axon then bifurcates to form parallel fibers that further excite the Purkinje
dendrites as well as the Golgi cell, stellate cell, and basket cell. The axon terminals of the latter three cell types
are inhibitory, and thus the stellate cell inhibits the Purkinje cell on its dendritic tree, the basket cell inhibits
the Purkinje cell on its soma, and the Golgi cell inhibits the granule cell. Arrows show the direction of
axonal conduction.

(parallel-fiber) portion of the same axon. A Purkinje cell and those granule cells that are directly below it in the upper levels of the granule-cell layer (which have a better chance of contacting Purkinje-cell dendrites before their axons bifurcate into parallel fibers) constitute a kind of functional unit. Thus, the vertical portion of the granule-cell axon seems to have the main responsibility for Purkinje-cell activation, whereas the parallel fibers seem to be more involved with the modulation of that excitation. One could say that the vertical granule-cell axon excites the Purkinje cell, and the horizontal (parallel-fiber) portion spreads the news of this excitation to other Purkinje cells and to those inhibitory cells that will modulate that excitation or suppress it entirely. This dual role of the granule-cell axon is an excellent example of how cellular architecture can be as effective in determining one neuron's influence on other neurons as are the types of neuroactive substances in its axon terminals. One should note that the region of termination of the upper granule cells on the lower region of the Purkinje-cell dendrite is the same region on which the climbing fibers terminate in ray-finned fishes.

Additional Input to the Cerebellar Cortex

In addition to the mossy-fiber and climbing-fiber excitatory pathways into the cerebellar cortex, an additional pathway, one that is inhibitory, has been described from the **locus coeruleus** (see Chapter 13) to the cerebellar cortex. Unlike the mossy and climbing-fiber inputs, which have very specific targets, the locus coeruleus input is widespread throughout the cortex, and its terminations occur on a variety of cortical components. The inhibition results from the action of norepinephrine, which ultimately results in a hyperpolarization (inhibition) of the Purkinje cell.

The Precerebellar Nuclei

Several neuronal populations send all or nearly all of their efferent axons to the cerebellum. These are collectively known as the precerebellar nuclei. Among these are the cell populations of the spinal cord that give rise to the spinocerebellar tracts, the funicular nuclei, the nucleus solitarius, the principal nucleus of the trigeminal nerve, the locus coeruleus, the inferior olive, and the pontine nuclei. In ray-finned fishes, an additional precerebellar nucleus is found; this is the **nucleus lateralis valvulae (lateral nucleus of the valvula).**

Inferior Olive. The inferior olive, which is located in the rostral medulla and is the source of the climbing fibers, is an unusual neuronal population because its neurons are capable of prolonged oscillatory activity. The inferior olive appears to be present in all vertebrate classes and is particularly well developed in species with a well-developed cerebellum. In birds and mammals, the inferior olive is subdivided into a **main inferior olivary nucleus** and one or more **accessory olivary nuclei,** which are together known as the **inferior olivary complex.** The inferior olivary complex receives afferents from very diverse sources, among which are the spinal cord, the reticular formation, the tectum, the red nucleus, the cerebrum, and various cranial nerve nuclei.

Nucleus lateralis valvulae. Unlike the inferior olive, which is located in the rostral medulla, the nucleus lateralis valvulae of ray-finned fishes is found in the tegmentum of the midbrain. Like the inferior olive, however, it receives axons from a number of neuronal populations and in turn supplies mossy fibers to the granule cells, in this case in both the corpus cerebelli and the valvula. The neurons of the nucleus lateralis valvulae are unusual in that they have no dendrites.

Pontine Nuclei. The pontine nuclei are the major source of mossy-fiber input to the cerebellar cortex in birds and mammals. In birds, the pontine nuclei consist of two small cell groups in the rostral medulla that receive their input mainly from the striatum of the telencephalon, the tectum, and the spinal cord. In mammals, six nuclear groups are recognized. They receive their afferents from the tectum, the spinal cord, and the isocortex of the telencephalon. Given the close affinity of crocodilians and birds, a detailed study of cerebellar afferents in alligators and crocodiles might reveal the presence of pontine nuclei in these animals also. A non-pontine relay between the telencephalon and the cerebellum has been reported in some ray-finned teleost fishes; this function is served by the **nucleus paracommissuralis,** which is located dorsal to the posterior commissure at the junction of the midbrain and the pretectum.

Cerebelloid Structures Associated with the Cerebellum in Ray-Finned Fishes

In addition to the valvula, the cerebella of ray-finned fishes have several other structures that have a cerebellum-like (cerebelloid) organization; these are the **crista cerebellaris** (cerebellar crest), **the torus longitudinalis,** and the **electroreceptive lateral line lobe** of electrosensory fishes. Some authorities have argued that despite their spatial separation from the main structures of the cerebellum, these cerebelloid structures should be considered as part of the cerebellum rather than as separate entities.

The Crista Cerebellaris. The crista cerebellaris is found just dorsal to the lateral line sensory areas in the rostral medulla and caudal pons. It consists of a layer of parallel fibers that have their granule-cell somata in the eminentiae granulares. Also present in the crista are Purkinje-like cells.

The Torus Longitudinalis. The torus longitudinalis is situated on the roof of the tectal ventricle in the midbrain, just dorsal (and sometimes rostral) to the valvula. It consists of a column of granule cells and receives some of its afferents from nucleus lateralis valvulae. The torus longitudinalis contains a population of neurons that projects to the marginal layer of the optic tectum (see also Chapter 18).

The Electrosensory Lateral Line Lobe. The electrosensory lateral line lobe of electrosensory fishes also bears a resemblance to the cerebellar cortex. In view of its close association with the electroreceptive valvula in these animals, this lobe should perhaps also be considered as a division of the cerebellum in these animals.

CEREBELLAR EFFERENTS AND THE DEEP CEREBELLAR NUCLEI

The sole pathway out of the cerebellar cortex is via the Purkinje-cell axons. The two main targets of these axons generally are one or more cellular populations located within the cerebellar white matter or in the body of the pons and nuclei of the vestibular system. As described above, ray-finned fishes lack deep cerebellar nuclei; the eurydendroid cells, which provide the efferents from the cerebellar cortex in these animals, pass directly to the vestibular nuclei, spinal cord, and other targets of the cerebellar cortex.

Deep Cerebellar Nuclei

The deep cerebellar nuclei appear to be unique to tetrapods. Amphibians have a single nucleus (with some suggestion of possible subdivisions). Reptiles have two nuclei (a **medial** and a **lateral**), and birds and mammals have three nuclei (a medial, a lateral, and an **interposed nucleus**). The medial nucleus of mammals is known as the **fastigial nucleus,** and the lateral nucleus is known as the **dentate nucleus.** The nucleus interpositus has an anterior division known as the **nucleus globosus** and a posterior division known as the **nucleus emboliformis.** The greater number of cerebellar nuclei in birds and mammals may be related to the greater lateral expansion and foliation of the cerebellum in these classes as well as the presence of a more elaborate system for telencephalic control of descending spinal and bulbar pathways.

In tetrapods, the Purkinje cells are organized into four longitudinal bands or zones according to the targets of their axons. In reptiles, the most medial band, band A, projects to the medial cerebellar nucleus. The next band, B, projects to the vestibular nuclei, followed by a band projecting to the lateral cerebellar nucleus, band C, and the fourth band, D, which projects to the vestibular nuclei. In mammals, band A projects to the medial (fastigial) cerebellar nucleus, band B to the lateral vestibular nucleus (Deiter's nucleus), band C to the two divisions of the nucleus interpositus, and band D to the lateral (dentate) cerebellar nucleus. Birds appear not to have a D band.

Efferents of the Deep Cerebellar Nuclei. The efferents of the cerebellar nuclei can be classified into two general groups: those that ultimately affect motor neurons of the spinal cord and those that ascend to those neurons of the thalamus that are involved in motor function. The descending pathways to the spinal cord can also be subdivided into two categories: **medial and lateral pathways.** The medial pathway efferents leave the cerebellum via a large fiber bundle known as the **brachium conjunctivum;** the lateral pathway exits the cerebellum via the **uncinate** (hook-like) **fasciculus.** As seen in Figure 14-19, the medial pathways ultimately terminate in the more medial cell columns of the ventral horn, where the motor neurons innervate the trunk musculature and those limb muscles that are closest to the trunk. These muscle groups are more involved in the maintenance of posture than in locomotion and manipulation of objects. The medial pathways include the **vestibulospinal, tectospinal, reticulospinal,** and **interstiti-**

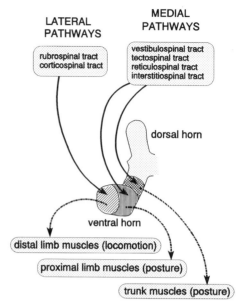

FIGURE 14-19. A summary of the lateral and medial descending pathways to the spinal cord. The medial pathways affect the activity of motor neurons in the more medial cell columns of the ventral horn of the spinal cord that control muscle groups that are involved in the maintenance of posture. The lateral pathways affect the activity of motor neurons in the more lateral portions of the ventral horn that control muscles used for locomotion.

ospinal pathways. The last of these pathways originates in the **interstitial nucleus of Cajal,** in the midbrain tegmentum. The lateral pathways include the **rubrospinal pathway,** which originates in the nucleus ruber (red nucleus) in the midbrain tegmentum in all tetrapods, and the corticospinal pathways in mammals. The lateral pathways terminate in the lateral "wings" of the ventral horn and influence those motor neurons that control the more distal limb muscles, which are more involved with locomotion and manipulation.

Figure 14-20 summarizes the differences in the efferents of the cerebellar nuclei in tetrapods. In amphibians, a single cerebellar nucleus gives rise to both a brachium conjunctivum route to the lateral pathways and an uncinate fasciculus route to the medial pathways. In reptiles and birds, the lateral cerebellar nucleus is the source of the brachium conjunctivum-lateral pathway, and the medial nucleus gives rise to the uncinate-medial pathway. In mammals, both the lateral (dentate) and interpositus nuclei supply axons to the brachium conjunctivum, lateral pathway, and the medial (fastigial) nucleus is the source of the uncinate-medial pathway. The brachium conjunctivum output of the cerebellum also is the source of fibers to the motor region of the thalamus.

THE EXCEPTIONAL CEREBELLA OF WEAKLY ELECTRIC FISHES

Among the aquatic animals that are capable of generating electric currents from a specialized electric organ are the well-known electric eel, among the ray-finned teleosts, and the tor-

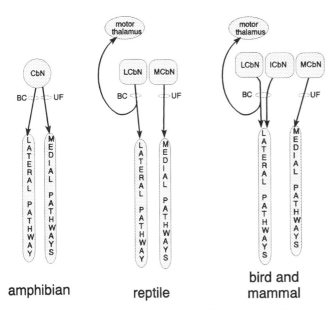

amphibian   reptile   **bird and mammal**

FIGURE 14-20. A summary of the major efferent pathways from the deep cerebellar nuclei of tetrapods. Abbreviations: BC, brachium conjunctivum; CbN, cerebellar nucleus; ICbN, interposed cerebellar nucleus; LCbN, lateral cerebellar nucleus; MCbN, medial cerebellar nucleus; UF, uncinate fasciculus.

pedo, among the cartilaginous fishes; these animals can generate quite powerful currents, which are capable of stunning or killing another animal. Less well known are two orders of freshwater ray-finned teleosts (the Mormyriformes and the Gymnotiformes) that generate electric fields much weaker than those of the electric eel and the torpedo and hence are known as the weakly electric fishes. Although many similarities exist between the electroreception systems and cerebella of these two orders, they appear to have evolved electroreception independently.

The cerebellum of the weakly electric fishes is the most elaborate and sophisticated of any of the fishes and, indeed, some might argue that it is one of the most specialized structures in any central nervous system. Like other fishes, the cerebellum consists of a corpus cerebelli and a valvula, as was described earlier in this chapter. The cerebella of the gymnotids are smaller and less developed than those of the mormyrids. In addition, the valvula does not show the highly sophisticated structure found in mormyrids. The corpus cerebelli of mormyrids consists of four lobes that have been designated as *C1, C2, C3,* and *C4.* The valvula is enormous and consists of a hyperfolded sheet that contains the same organization into granule cells, a molecular layer, and a Purkinje-cell layer as does the corpus cerebelli. Between the two divisions of the cerebellum is a smaller division, the **lobus transitorius.** But here the similarities to conventional cerebellar organization end. To begin with, although these animals have a precerebellar projection from the inferior olive to the cerebellum, the terminal branches of the precerebellar axons do not climb into the molecular layer. Moreover, the axons of the granule cells do not make the characteristic T-shaped bifurcation as they ascend into the molecular layer.

The electrosensory system of mormyrids arises from three types of electroreceptors: **mormyromasts, ampullae,** and **knollenorgan receptors.** The mormyromast and ampullary receptor axons terminate in the electrosensory lateral-line lobe, which has a cerebellar cortex-like structure; the knollenorgan pathway ends in the nucleus of the electrosensory lateral line lobe. The efferents of the nucleus of the electrosensory lateral line lobe, after synapsing in several intermediate nuclei, eventually end in the valvula. The efferents of the electrosensory lateral line lobe project to the lateral toral nucleus, which in turn sends its axons to the valvula.

The interrelations between the electrosensory system and the cerebellum in weakly electric fishes have been the subject of intensive research that has revealed a complex organization of nuclear groups that have rich interconnections. A fuller description of this system may be found in Chapter 11 and in the list of further readings at the end of this chapter.

The massive input from electroreceptors and other sensory systems, as well as other evidence, has led some workers to suggest that the cerebellum of mormyrids may have little to do with motor function, but rather it may have evolved into a highly sophisticated sensory-processing organ. A common feature of cerebellar processing, no matter what its role in the life of the animal, is the detection by the Purkinje cells of coincident activity of the parallel fibers. This common feature may have permitted the evolution of the mormyrid cerebellum to proceed in one direction and that of the cerebella of other fishes in quite another. We are quite certain that we have only begun to scratch the surface of the functions and behavioral capabilities of this amazing brain structure.

EVOLUTIONARY PERSPECTIVE ON THE CEREBELLUM

The cerebellum is a structure of the brain that has had a seemingly paradoxical evolutionary history. On the one hand, it has maintained a rather conservative pattern in its cell types and the relationships of these cells to each other and to other cell populations within the brain. On the other hand, its gross structure shows extreme variations both within and between the vertebrate classes. The cerebellum thus illustrates both phyletic continuity and adaptation to the demands of the environment.

With regard to the cellular organization of the cerebellum, once the Purkinje cell assumed its two-dimensional, flattened dendritic tree, the main phyletic trends appear to have been (1) an early change in the role of the climbing fibers by having them climb the Purkinje-cell dendrites rather than terminate on the soma, (2) a trend in the direction of refinement of inhibitory control over the granule-cell/Purkinje-cell circuitry by the addition of new inhibitory cell types in the tetrapods, and (3) the development of deep cerebellar nuclei as the targets of the Purkinje-cell axons in tetrapods.

In contrast to the relative conservatism of the cerebellar circuitry, major changes have occurred in the overall form of the cerebellum, such as the development of dramatic specializations in the valvula in mormyrids and other fishes with electroreception. Other changes were related to the transition from water to land, as may be seen in the cerebella of extant amphibians, diapsid reptiles, and turtles. As locomotor abilities advanced, along with the capacity to use the digits as organs of manipulation for digging, scratching, grasping, carrying, and so on, the cerebellum showed dramatic lateral expansion and a massive increase in foliation, especially in mammals.

FUNCTIONS OF THE CEREBELLUM

A review of the literature on the functions of the cerebellum yields a bewildering array of seemingly unrelated responsibilities for this division of the central nervous system. Because of its afferent connections with the vestibular system and the somatosensory system and its efferent connections with the motor system, a major role of the cerebellum has long been thought to be the maintenance of the balance of the body and the smoothness and coordination of muscular activity. In addition, a long series of studies also has suggested that the cerebellum plays a role in motor learning, and we have seen that in fishes with electroreception, the cerebellum is a major sensory organ as well.

To understand the variety of functions that the cerebellum performs, one must look for a common denominator of all of these functions. One such common denominator is the organization of the cerebellar cortex into a very regular network of parallel fibers with Purkinje cells placed at regular intervals. This spatial arrangement seems well suited to the task of rather precise timing of events. The presence of several groups of specialized inhibitory cells limits the actions of the Purkinje and granule cells to enhance and sharpen the precision of this timing. Temporal relations among events are important among all of the known functions of the cerebellum.

FOR FURTHER READING

Arends, J. J. and Zeigler, H. P. (1991) Organization of the cerebellum in the pigeon (Columba livia): II. Projections of the cerebellar nuclei. Journal of Comparative Neurology, 306, 245–272.

Bangma, G. C. and ten Donkelaar, H. J. (1982) Afferent connections of the cerebellum in various types of reptiles. Journal of Comparative Neurology, 207, 255–273.

Glickstein, M. , Yeo, C., and Stein, J. (1986) Cerebellum and Neuronal Plasticity. New York: Plenum.

Ito, M. (1984) The Cerebellum and Neural Control. New York: Raven.

Llinás, R. R. (1969) Neurobiology of Cerebellar Evolution and Development. Chicago: American Medical Association Institute for Biomedical Research.

Llinás, R. R. (1981) Electrophysiology of the cerebellar networks. In V. B. Brooks (ed.), Handbook of Physiology. The Nervous System. Section 1. Volume II. Motor Control. Bethesda, MD: American Physiological Society, pp. 831–876.

Llinás, R. R. and Sotello, C. (eds.) (1992) The Cerebellum Revisited. New York: Springer.

Meek, J. (1992a) Why run parallel fibers in parallel? Teleostean Purkinje cells as possible coincidence detectors in a timing device subserving spatial coding of temporal differences. Neuroscience, 48, 249–283.

Meek, J. (1992b) Comparative aspects of cerebellar organization. From mormyrids to mammals. European Journal of Morphology, 30, 37–51.

Nieuwenhuys, R. and Nicholson, C. (1969a) A survey of the general morphology, the fiber connections, and the possible functional significance of the gigantocerebellum of mormyrid fishes. In R. R. Llinás, (ed.), Neurobiology of Cerebellar Evolution and Development. Chicago: American Medical Association Institute for Biomedical Research, pp. 107–134.

Nieuwenhuys, R. and Nicholson, C. (1969b) Aspects of the histology of the cerebellum of mormyrid fishes. In R. R. Llinás, (ed.), Neurobiology of Cerebellar Evolution and Development. Chicago: American Medical Association Institute for Biomedical Research, pp. 135–169.

Paulin, M. G. (1993) The role of the cerebellum in motor control and perception. Brain, Behavior and Evolution, 41, 39–50.

ten Donkelaar, H. J. and Bangma, G. C. (1991). The cerebellum. In C. Gans (ed.), Biology of the Reptilia. Chicago: University of Chicago Press, pp. 496–586.

ADDITIONAL REFERENCES

Bell, C. C. and Szabo, T. (1986) Electroreception in mormyrid fish: central anatomy. In T. H. Bullock and W. Heiligenberg (eds.), Electroreception. New York: Wiley, pp. 375–421.

Bass, A. H. (1982) Evolution of the vestibulolateral lobe of the cerebellum in electroreceptive and nonelectroreceptive teleosts. Journal of Morphology, 174, 335–348.

Brochu, G., Maler, L. and Hawkes, R. (1990) Zebrin II. A polypeptide antigen expressed selectively by Purkinje cells reveals compartments in rat and fish cerebellum. Journal of Comparative Neurology, 291, 538–552.

Bullock, T. H. and Heiligenberg, W. (eds.), Electroreception. New York: Wiley.

Carr, C. E. (1993) Processing of temporal information in the brain. Annual Review of Neuroscience, 16, 223–243.

Finger, T. E., Bell, C. C., and Russell, C. J. (1981) Electrosensory pathways to the valvula cerebelli in mormyrid fish. Experimental Brain Research, 42, 23–33.

Finger, T. E., Bell, C. C., and Carr, C. E. (1986) Comparisons among electroreceptive teleosts. in T. H. Bullock and W. Heiligenberg (eds.) Electroreception. New York: Wiley, pp. 465–481.

Hawkes, R., Brochu, G, Doré, L. Gravel, C., and Leclerc, N. (1992) Zebrins: molecular markers of compartmentation in the cerebellum. In R. R. Llinás and C. Sotello (eds.), The Cerebellum Revisited. New York: Springer. pp. 22–55.

Ito, H. and Kishida, R. (1978) Afferent and efferent fiber connections of the carp torus longitudinalis. Journal of Comparative Neurology, 181, 465–476.

Lange, W. (1974) Regional differences in the distribution of Golgi cells in the cerebellar cortex of man and some other mammals. Cell and Tissue Research, 153, 219–226.

Lange, W. (1975) Cell number and cell density in the cerebellar cortex of man and some other mammals. Cell and Tissue Research, 157, 115–124.

Meek, J. and Nieuwenhuys, R. (1991) The palisade pattern of mormyrid Purkinje cells. A correlated light and electron microscopic study. Journal of Comparative Neurology, 306, 156–192.

Meredith, G. E. and Butler A. B. (1983) Organization of eighth nerve afferent projections from individual endorgans of the inner ear in the teleost, Astronotus ocellatus. Journal of Comparative Neurology, 220, 44–62.

Naujoks-Manteuffel, C. and Manteuffel, G. (1988) Origins of descending projections to the medulla oblongata and rostral medulla spinalis in the urodele *Salamandra salamandra* (Amphibia). *Journal of Comparative Neurology,* 273, 187–206.

Parent, A., Dube, L., Braford, M. R., Jr., and Northcutt, R. G. (1978) The organization of monoamine-containing neurons in the brain of the sunfish *(Lepomis gibbosus)* as revealed by fluorescence microscopy. *Journal of Comparative Neurology,* 182, 495–516.

Parent, A. and Northcutt, R. G. (1982) The monoamine-containing neurons in the brain of the garfish, *Lepisosteus osseus. Brain Research Bulletin,* 9, 189–204.

Ronan, M. and Northcutt, R. G. (1990) Projections ascending from the spinal cord to the brain in petromyzontid and myxinoid agnathans. *Journal of Comparative Neurology,* 291, 491–508.

Schnitzlein, H. N. and Faucette , J. R. (1969) General morphology of the fish cerebellum. In R. R. Llinás, (ed.), *Neurobiology of Cerebellar Evolution and Development.* Chicago: American Medical Association Institute for Biomedical Research, pp. 77–106.

Straka, H. and Dieringer, N. (1992) Chemical identification and morphological characterization of the inferior olive in the frog. *Neuroscience Letters,* 140, 67–70.

Wild, J. M. (1992) Direct and indirect "cortico"-rubral and rubro-cerebellar cortical projections in the pigeon. *Journal of Comparative Neurology,* 326, 623–636.

Wilczynski, W. and Northcutt, R. G. (1983) Connections of the bullfrog striatum: afferent organization. *Journal of Comparative Neurology,* 214, 321–332.

Wullimann, M. F. and Northcutt, R. G. (1989) Afferent connections of the valvula cerebelli in two teleosts, the common goldfish and the green sunfish. *Journal of Comparative Neurology,* 289, 554–567.

Yoram, Y. (1992) Electroneural hybridization: a novel approach to investigate rhythmogenesis in the inferior olivary nucleus. In R. R. Llinás and C. Sotello (eds.), *The Cerebellum Revisited.* New York: Springer. pp. 201–212.

Part Three

THE MIDBRAIN

15

Overview of the Midbrain

INTRODUCTION

In all vertebrates, a group of structures collectively known as the midbrain, or mesencephalon, links the sensory, motor, and integrative components of the hindbrain with those of the forebrain. The midbrain contains three major regions: the **tectum,** the **tegmentum,** and the **isthmus** (Fig. 15-1). One important feature of the midbrain is the presence of areas within the tectum, or roof, that receive topographically organized projections from the auditory, visual, and somatosensory systems, which form maps of the animal's sensory space. A second feature of the midbrain is the presence of nuclei intrinsic to the region and of ascending and descending fiber tracts that pass through it. Finally, like the hindbrain, the midbrain also contains several nuclei that control the distribution of certain critical neurotransmitters to other brain regions.

The tectum contains laminated structures—particularly the **optic tectum** and the **torus semicircularis.** The optic tectum receives its major input from the optic tract, which is the central continuation of the optic nerve. The optic tectum is involved in integrating the spatial aspects of visual and other sensory inputs and with spatially oriented motor responses. The torus semicircularis receives its major input from ascending auditory projections and, where present, from the lateral line system. The sensory maps in the midbrain roof are in register with each other; thus, visual and somatosensory stimuli, for example, that arise from the same part of the space around the animal result in neuronal activity in the same part of the optic tectum. Auditory stimuli are mapped onto parts of the torus semicircularis that are interconnected in a point-to-point manner with parts of the optic tectum that likewise correspond in spatial register. Both the optic tectum and the torus semicircularis project rostrally to nuclei in the dorsal thalamus.

The main part of the midbrain tegmentum lies ventral to the tectum and is separated from the hindbrain by a transitional area called the isthmus, or isthmal tegmentum. Parts of the tegmentum consist of the rostral continuations of areas present in the hindbrain: the reticular formation, the somatic motor and other columns of cranial nerve nuclei, and ascending and descending fiber systems. Additional nuclei, related to visuospatial functions, motor activities, ascending feedback for the integration and adjustment of motor activities, and relay to the forebrain of ascending sensory information, are also present. The midbrain tegmentum is thus a gateway (or clearinghouse) for incoming sensory information and outgoing motor responses to and from the forebrain. It assists the more rostral parts of the brain with analysis and modulation of the information.

In some species, especially in those with more laminar brains (Group I: see Chapter 4), not many experimental studies of connections have been done on tegmental nuclei. In these cases, we frequently do not yet know which populations of the cells that lie along the ventricular surface may project to a specific target and thus may correspond to particular nuclei in animals with more elaborated brains. Pinpointing these groups of cells must await future studies, but we need to note that a greater number of nuclei are present in these brains than can be identified for now. Specializations of nuclei and areas in the brainstem occur in correlation with the presence of specialized capabilities such as electroreception, motor adaptation for specialized forms of swimming, and motor adaptations for locomotion on land. In this chapter, we will present a generalized overview of the structures and systems of the isthmus, the tegmentum proper, and the tectum.

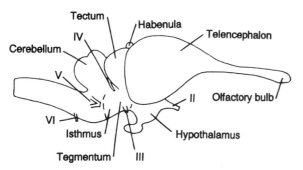

FIGURE 15-1. Lateral view of the brain of a crocodile, with the approximate boundaries of the isthmus, tegmentum, and tectum indicated by broken lines. Cranial nerves are indicated by Roman numerals. Data from ten Donkelaar and Nieuwenhuys (1979). Used with permission of Academic Press, Ltd.

THE ISTHMUS

The isthmus lies at the rostral end of the hindbrain. Within the isthmus are rostrally projecting cell groups, such as the locus coeruleus and raphe, cranial nerve nuclei including the trochlear, oculomotor, and mesencephalic nucleus of the trigeminal nerve, and the rostral parts of some of the structures of the hindbrain, such as the reticular formation. A major ascending tract from the cerebellum, the brachium conjunctivum, is present as are numerous other ascending and descending fiber tracts. Some of these structures are shown in a sagittal view of a rat brain for orientation in Figure 15-2 and in a generalized, schematic diagram in Figure 15-3(A), which is an amalgamation of structures present in this region in various vertebrates. The relative positions of some isthmal structures are shown in a lamprey and a crocodile in Figures 15-4 and 15-5. A partial list of structures in the isthmus is given in Table 15-1.

The rostral part of the **reticular formation** extends into the isthmus and lies in its central part. The reticular formation (as discussed in Chapter 13) has widespread connections with a number of structures. It gives rise to the **reticulospinal tract,** which courses through the ventral part of the tegmentum and is important in motor control.

The **nuclei of the raphe** lie near the midline. Cells in the raphe contain the neurotransmitter **serotonin** (5-hydroxytryptamine). These cells, along with serotonin-containing cells located more rostrally in the hypothalamus of the diencephalon, project to other brainstem nuclei, to motor cells in the spinal cord, to the optic tectum and other sensory processing structures, and to wide areas of the telencephalon. Serotonin has a variety of functions, including effecting the excitability of neurons in the forebrain. It plays a role in regulating mood and in perceptual integration.

The **locus coeruleus** is a nucleus that contains darkly pigmented cells bordering the central gray around the fourth ventricle. The cells are filled heavily enough with pigment to be visible to the unaided eye in fresh, unfixed tissue. Cells in the locus coeruleus contain the neurotransmitter **dopamine,** a catecholamine. Dopamine is present in some of the neuronal pathways that regulate movements and in others that participate in cognitive functions and affect. The locus coeruleus projects

to the cerebral cortex in the telencephalon and to the cerebellar cortex.

The **trochlear nucleus,** which gives rise to cranial nerve IV, lies in the dorsal part of the tegmentum. It is interconnected with the oculomotor and abducens nuclei and with vestibular nuclei by the **medial longitudinal fasciculus.**

Nucleus isthmi is present in the isthmus in some vertebrates. It is called the **parabigeminal nucleus** in mammals. Nucleus isthmi is reciprocally connected with the optic tectum in a point-to-point manner, and the map of visual space in the tectum is also topographically represented in this nucleus.

An **isthmo-optic nucleus** is present in the dorsomedial part of the isthmal region [rostral to the level shown in Fig. 15-3(A)] in some vertebrates. Unlike nucleus isthmi, this nucleus does not have tectal connections. Neurons of the isthmo-optic nucleus project centrifugally to the retina.

The **interpeduncular nucleus** lies in the ventral part of the isthmus. It receives descending projections from the habenula in the diencephalon, which receives input from olfactory and limbic related areas of the telencephalon. The tract between the habenula and the interpeduncular nucleus is the **fasciculus retroflexus.** The interpeduncular nucleus sends efferent projections to other areas within the dorsal part of the tegmentum.

The **brachium conjunctivum** is a tract that carries fibers from the contralateral cerebellum to the **red nucleus** in the tegmentum proper. The red nucleus projects to the spinal cord via the **rubrospinal tract,** which courses through the ventral part of the isthmus.

Ascending tracts carrying sensory information also course through the isthmus and tegmentum. The **spinal lemniscus,** a continuation of the lateral funiculus of the spinal cord, carries pain and temperature information. The **medial lemniscus** carries somatosensory information: touch, proprioception, and vibration senses. The **lateral lemniscus** carries ascending auditory projections in all animals and, in many, ascending electrosensory and/or mechanosensory lateral line projections.

We have included the **corticospinal** and **corticopontine tracts** here, as they are large and prominent in mammals. These tracts arise in the cerebral cortex and form large structures called the **cerebral peduncles** on the lateral edge of the tegmentum. The term "interpeduncular nucleus" is derived from the position of this nucleus in the **interpeduncular fossa** between the cerebral peduncles in mammals.

THE TEGMENTUM PROPER

The tegmentum lies ventral to the tectum in the midbrain. The tegmentum contains some nuclei involved in motor control, such as the cuneiform nucleus, substantia nigra, and ventral tegmental area, as well as the red nucleus, which receives ascending cerebellar projections. Many ascending and descending tracts pass through or originate in the tegmentum. Some tegmental structures are shown in the sagittal view of a rat brain in Figure 15-2 and in the generalized, schematic diagram of a transverse section in Figure 15-3(B). Figures 15-4 and 15-5 show the relative positions of some tegmental structures in a lamprey and a crocodile. A list of

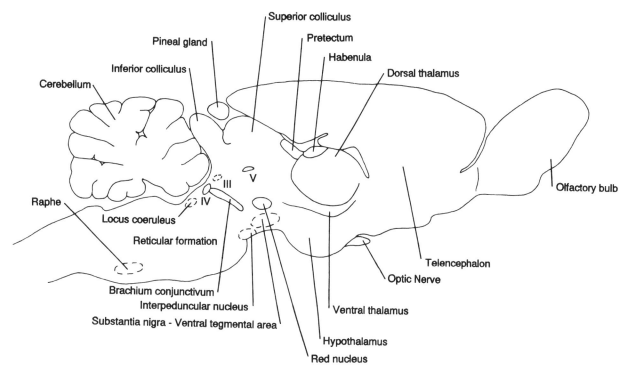

FIGURE 15–2. Drawing of a parasagittal section through the brain of a rat, showing the locations of some of the structures in the midbrain. Some nuclei, shown in broken lines, that are out of the plane of section have been projected onto this section to show their rostrocaudal and dorsoventral position relative to other structures. Cranial nerve nuclei are indicated by Roman numerals, with V indicating the mesencephalic nucleus of the trigeminal nerve. Data from Pelligrino et al. (1979). Used with permission of Plenum Publishing Corp. and the author.

a number of the structures in the tegmentum is provided in Table 15-2.

The **cuneiform nucleus** lies in the central part of the tegmentum. This nucleus projects to the reticular formation in the medulla, which in turn projects to the spinal cord. Stimulation of parts of the cuneiform nucleus and other nuclei in this area results in movements of the limbs. The area is called the **midbrain locomotor region** (MLR).

Large neurons of the mesencephalic nucleus of the trigeminal nerve (see Chapter 11) lie within the deep layers of the optic tectum. These neurons receive proprioceptive information from jaw muscles and are part of a monosynaptic reflex arc for regulation of the movements of the muscles.

The **oculomotor nucleus** lies in the dorsal part of the tegmentum in a medial position. It gives rise to the oculomotor nerve, which courses ventrally through the tegmentum to exit the brain. The **red nucleus** lies immediately lateral to the fibers of the oculomotor nerve. The nucleus is so named because of its pinkish color, due to the presence of abundant blood vessels, in fresh specimens of some animals. It receives cerebellar inputas well as descending projections from the telencephalon. Thus, the red nucleus can integrate and coordinate motor signals from both sources.The red nucleus projects to the spinal cord via the rubrospinal tract.

The **substantia nigra** and **ventral tegmental area** lie near the red nucleus. The cells of these nuclei contain the catecholamine dopamine. They project to a number of sites in the diencephalon and telencephalon and are involved in the initiation and control of movements. In humans, loss of cells in the substantia nigra results in the tremors, rigidity, and other symptoms of Parkinson's disease.

The **medial longitudinal fasciculus** lies next to the oculomotor nucleus and, as discussed above, interconnects the oculomotor, trochlear, and abducens nuclei with vestibular nuclei. The **fasciculus retroflexus** is also present at this level, carrying fibers from the habenula to the interpeduncular nucleus in the isthmus.

The **brachium conjunctivum** projects to the red nucleus, carrying fibers from the cerebellum. The tract lies ventromedial to the nucleus, as its fibers enter the nucleus to terminate there. Other ascending tracts pass through the tegmentum on the way to the tectum and to nuclei in the diencephalon. These include the spinal lemniscus and medial lemniscus, which carry pain and somatosensory information as discussed above.

Tracts that descend through the tegmentum include the corticospinal and corticopontine tracts, present in mammals, which form the cerebral peduncle on the lateral surface of the tegmentum. Other descending tracts include corticoreticular and corticobulbar projections. Fibers from a variety of sources cross to the contralateral side in the tegmentum via the **posterior commissure**. There is also a **tectal commissure** interconnecting the optic tecta.

The **torus lateralis** is a bulge on the lateral side of the tegmentum that is present in ray-finned fishes but has not been

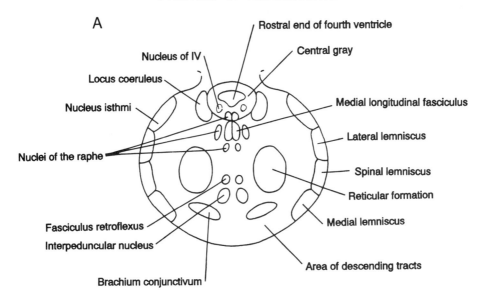

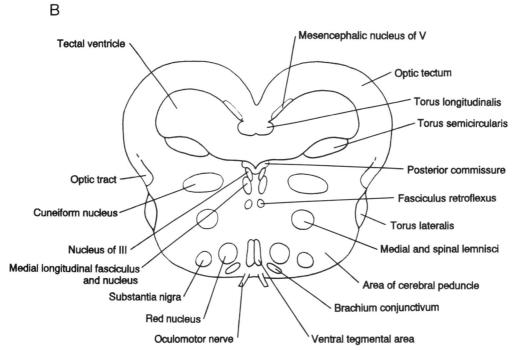

FIGURE 15–3. Schematic transverse sections through the isthmus (A) and the tegmentum and tectum (B) in a generalized vertebrate. The structures shown are not all present in any single species.

identified in other vertebrates. It has been found to project to the telencephalon, but its other connections and functions are unknown.

THE TECTUM

The two prominent structures within the tectum are the **optic tectum** and **torus semicircularis.** These two structures occur in all vertebrates and form rounded bulges over the tegmentum. In mammals, they are called the **superior collicu-**lus and **inferior colliculus,** respectively. Colliculus is Latin for "little hill," which each bulge resembles. A third structure—the **torus longitudinalis**—is present at the medial end of the optic tectum in ray-finned fishes. The optic tectum, torus semicircularis, and torus longitudinalis are shown in Figure 15-3(B).

The optic tectum is formed by a lobe on each side of the brain (Figs. 15-1 and 15-2). It is most properly called by its full name, to distinguish it from the entire midbrain roof, but is also referred to simply as the tectum in some contexts. In many nonmammalian vertebrates, the optic tectal lobes are relatively

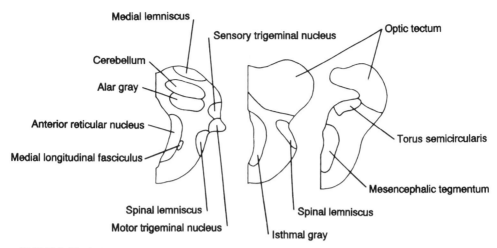

FIGURE 15–4. Drawings of transverse hemisections through the right midbrain of a lamprey, with the most caudal section on the left. Data for the most caudal section is from Northcutt (1979) and for the more rostral sections from Kennedy and Rubinson (1977). Note that in this figure and in figures in this and succeeding chapters on the midbrain, the ventricle is generally not labeled. Material from Northcutt (1979) used with permission of Elsevier.

expanded, and the third ventricle extends laterally into each tectal lobe, forming the tectal ventricle. The optic tectum consists of various layers of cells and fibers, with the dendrites of many cells oriented in a radial fashion and with afferent fibers coursing through it parallel to the surface. The cell and fiber layers can be divided into superficial, central, and periventricular zones. The optic tectum has multiple sources of input, among which are the visual and somatosensory systems. The optic tract terminates in the superficial and central zones, while somesthetic information terminates in the deeper part of the central zone. The inputs of both these modalities are mapped topographically and are in register with each other in regard to their spatial location of origin.

As a result of the expansion of the tectal lobes, the torus semicircularis lies ventral to the tectal ventricle in most cases. In mammals, the inferior colliculus lies caudal to the superior colliculus, rather than in a more ventral position, because the superior collicular lobes are not expanded laterally and the ventricle does not extend into them.

The torus semicircularis is the site of termination of auditory and lateral line fibers from the lateral lemniscus. The topography of these fibers is maintained so that there is a map of lateral line and/or auditory space in the torus semicircularis. The torus semicircularis is also reciprocally connected with the optic tectum, and the maps of visual and somatosensory space in the optic tectum and of auditory/lateral line space in the

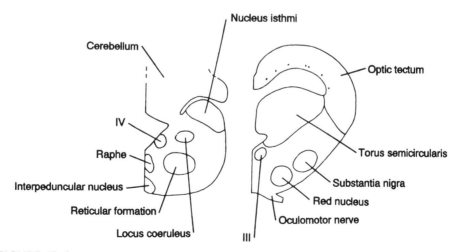

FIGURE 15–5. Drawings of transverse hemisections through the right midbrain of a crocodile, with the more caudal section on the right. Roman numerals indicate cranial nerve nuclei. The dots in the optic tectum represent cell bodies of the mesencephalic nucleus of the trigeminal nerve. Data from ten Donkelaar and Nieuwenhuys (1979). Used with permission of Academic Press Ltd.

TABLE 15-1. Some of the Nuclei and Tracts in the Isthmus

Reticular formation	Motor, autonomic, and activating system
Nuclei of the raphe	Serotonin-containing cells that have widespread connections
Locus coeruleus	Catecholamine-containing cells that project to cerebral and cerebellar cortices
Trochlear nucleus	Nucleus of cranial nerve IV
Nucleus isthmi	Connected with optic tectum
Isthmo-optic nucleus	Projects to the retina
Interpeduncular nucleus	Receives projections from the habenula in the diencephalon
Fasciculus retroflexus	Tract between habenula and interpeduncular nucleus
Brachium conjunctivum	Tract from cerebellum to red nucleus
Spinal lemniscus	Carries ascending pain and temperature information
Medial lemniscus	Carries ascending somatosensory information
Lateral lemniscus	Carries ascending auditory and lateral line information
Rubrospinal and reticulospinal tracts	Carry descending projections from red nucleus and reticular formation to spinal cord
Corticospinal, corticopontine, and other descending tracts	Carry descending projections from telencephalon to spinal cord and nuclei in the brainstem

TABLE 15-2. Some of the Nuclei and Tracts in the Tegmentum Proper

Cuneiform nucleus	Projects to reticular formation
Mesencephalic nucleus of the trigeminal nerve	Proprioception for jaw muscles; the cells lie in the deep part of the tectum
Oculomotor nucleus	Nucleus of cranial nerve III
Red nucleus	Receives cerebellar and descending projections and projects to spinal cord
Substantia nigra	Catecholamine-containing cells project to basal ganglia in the telencephalon
Ventral tegmental area	Catecholamine-containing cells similar to those in substantia nigra
Medial longitudinal fasciculus	Interconnects oculomotor and vestibular nuclei
Fasciculus retroflexus	Tract from habenula to the interpeduncular nucleus
Brachium conjunctivum	Tract from cerebellum to the red nucleus
Spinal lemniscus	Carries ascending pain and temperature information
Medial lemniscus	Carries ascending somatosensory information
Corticospinal and corticopontine tracts	Carry descending information from the telencephalon
Posterior and tectal commissures	Carry a variety of decussating fibers
Torus lateralis	Connections mostly unknown

torus semicircularis are maintained in register in both structures. Descending projections involved in motor movements to orient the head and/or body to novel stimuli in the environment arise from both the optic tectum and the torus semicircularis, and the mapping of space allows for rapid and precise orienting movements. This is equally important to the frog needing to catch a fly for lunch as to the human orienting to whatever goes bump in the night, or to motion in one's peripheral visual field. In addition to descending projections, the optic tectum and the torus semicircularis both project rostrally to nuclei in the diencephalon, which in turn project to the telencephalon.

The torus longitudinalis is a small structure that is unique to ray-finned fishes. It is involved in relaying motor inputs to the optic tectum, and thence to the reticular formation.

FOR FURTHER READING

Nieuwenhuys, R. and Pouwels, E. (1983) The brain stem of actinopterygian fishes. In R. G. Northcutt and R. E. Davis (eds.), *Fish Neurobiology, Vol. 1: Brain Stem and Sense Organs*. Ann Arbor, MI: University of Michigan Press, pp. 25–87.

ten Donkelaar, H. J. and Nieuwenhuys, R. (1979) The Brainstem. In C. Gans, R. G. Northcutt, and P. Ulinski (eds.), *Biology of the Reptilia, Vol. 10: Neurology B*. New York: Academic, pp. 133–200.

ADDITIONAL REFERENCES

Kennedy, M. C. and Rubinson, K. (1977) Retinal projections in larval, transforming and adult sea lamprey, *Petromyzon marinus*. *Journal of Comparative Neurology*, 171, 465–479.

Northcutt, R. G. (1979) Experimental determination of the primary trigeminal projections in lampreys. *Brain Research*, 163, 323–327.

Pelligrino, L. J., Pelligrino, A. S., and Cushman, A. J. (1979) *A Stereotaxic Atlas of the Rat Brain*. New York: Plenum.

Vanegas, H. (ed.) (1984) *Comparative Neurology of the Optic Tectum*. New York: Plenum.

16

Isthmus

INTRODUCTION

We have previously discussed a number of isthmic nuclei and tracts, such as the reticular formation, the trochlear and oculomotor nuclei, and the medial longitudinal fasciculus. Other tracts that pass through the tegmentum proper and the isthmus, including the medial and spinal lemnisci and a number of the descending tracts, are also covered elsewhere in conjuction with the specific systems to which they belong. This chapter focuses on the comparative anatomy of five isthmal structures: the **nuclei of the raphe,** the **locus coeruleus, nucleus isthmi,** the **isthmo-optic nucleus,** and the **interpeduncular nucleus.** Drawings of sections through the isthmus show a number of the nuclei and fiber tracts for a variety of animals with laminar brains (Group I)—a lamprey (Fig. 16-1), a nonteleost ray-finned fish (Fig. 16-2), a lungfish (Fig. 16-3), and an amphibian (Fig. 16-4)—and for some animals with elaborated brains (Group II)—a hagfish (Fig. 16-5), a teleost (Fig. 16-6) and, among amniotes, a mammal (Fig. 16-7) and a bird (Fig. 16-8).

NUCLEI OF THE RAPHE

Group I

The word raphe, derived from the Greek word for seam, refers to the midline of the brainstem. In most vertebrates, the nuclei of the raphe extend into the isthmus. In lampreys, the raphe has cells which contain serotonin but is confined to the hindbrain proper, even though this system is quite extensive within the hindbrain. It gives rise to widely distributed projections to the telencephalon, diencephalon, tectum, interpeduncular nucleus, and reticular formation. In squalo-

morph sharks and ratfishes, the raphe lies dorsal and caudal to the interpeduncular nucleus. The raphe nuclei in ratfishes are known to have serotonin-containing cells. These cells are also distributed through more lateral areas within the brainstem.

Among nonteleost ray-finned fishes, the serotonin system in the raphe is known to be extensive in gars. It extends caudally to upper spinal cord levels. In the isthmus (Fig. 16-2), the area of the serotonin-containing cells extends lateral to the raphe itself.

Differences occur in the degree of lateralization of this system in amphibians. In frogs, the serotonin-containing cells of the raphe extend from medullary to isthmic tegmental levels in the area of the midline and also extend into more lateral areas in the tegmentum. In urodele amphibians, however, the serotonin-containing cells are confined to the raphe nuclei in the area of the midline.

Group II

In hagfishes, the nucleus of the superior raphe extends rostrally to the level of the caudal isthmus (Fig. 16-5). Serotonin-containing cells are mostly confined to the area of the midline. In galeomorph sharks, skates, and rays, the rostral part of the raphe also extends into the isthmus, a part of it lying dorsal to the caudal part of the interpeduncular nucleus. Serotonin-containing cells are numerous within the raphe and also are distributed through a number of more laterally lying sites in the hindbrain.

The column of the raphe nuclei extends into the caudal part of the isthmus in teleosts (Fig. 16-6). The medially lying raphe nuclei have serotonin-containing cells, but this system is relatively limited in teleosts. The column of serotonin-containing cells in the raphe does not extend caudally through the medulla and into the spinal cord in most teleosts as it does in

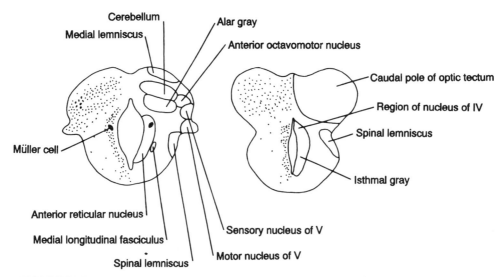

FIGURE 16-1. Drawings of transverse hemisections in lampreys (*Petromyzon marinus* and *Ichyomyzon unicuspis*), with mirror-image drawings, through isthmal levels, with the more caudal section on the left. Adapted from Kennedy and Rubinson (1977) and Northcutt (1979). The latter material used with permission of Elsevier. Please note that in this and most succeeding similar illustrations in the book, scales are not shown. The approximate scale can be determined in most cases by reference to Chapter 4, in which scales are shown on drawings of whole brains.

lampreys, cartilaginous fishes, and nonteleost ray-finned fishes such as gars. One exception to the limited extent of the system in teleosts is in mormyrid fishes where serotonin-containing cells extend throughout the brainstem. In most teleosts, the serotonin-containing cells are also limited in their lateral extent, being confined to the area of the midline. This lack of lateral migration of the cells contrasts with the more laterally migrated reticular formation and a number of other nuclei in teleosts.

In amniotes (Fig. 16-7), the serotonin-containing cells lie both within the nuclei of the raphe and more laterally in the tegmentum. This lateralization of the serotonin system is pronounced in mammals and most extensive in birds. In both mammals and birds, some serotonin-containing cell somata extend as far laterally as nuclei such as the locus coeruleus (discussed below), where they intermingle with the catecholamine-containing cells of that region. The number of such intermingling serotonin-containing cells is relatively few in mammals but quite high in birds. The functional consequences of having the cell somata intermingled have not yet been determined.

In mammals, widespread ascending serotoninergic projections of the raphe have been found to the cerebral cortex in the telencephalon, nuclei within the diencephalon, the superior colliculus in the roof of the midbrain, and the cerebellum. These projections have also been found to be topographically organized.

Evolutionary Perspective

The serotonin system is well developed and extends laterally from midline areas in most vertebrates. Hagfishes, teleosts, and urodele amphibians are the exceptions, having relatively reduced systems in which the cells are confined to the area of the midline. While we cannot determine whether or not lateralization of the serotonin system was present in the common ancestor of hagfishes and all other vertebrates, this condi-

tion does appear to have arisen very early in vertebrate evolution. Lateralization of the system has been maintained in most groups and, in some cases, augmented. The system has been secondarily reduced in teleosts and in urodele amphibians.

LOCUS COERULEUS

Group I

The presence of a locus coeruleus, which means "sky-blue place" in reference to the color of its catecholamine-containing cells, has not yet been investigated in lampreys, but this nucleus has been found in ratfishes along the border of the fourth ventricle. In nonteleost ray-finned fishes such as gars, the locus coeruleus is a small group of catecholamine-containing cells (indicated by the large dots in Fig. 16-2) in the isthmus, and within the telencephalon, there are only a few catecholamine-containing axons. The locus coeruleus is thus quite small, especially in contrast to the extensive serotonin system of the raphe. In amphibians, the catecholamine-containing cells of the locus coeruleus lie on the ventromedial wall of the fourth ventricle.

Group II

The presence of a locus coeruleus has not yet been investigated in hagfishes. Among cartilaginous fishes with elaborated brains, a circumscribed group of cells forms the locus coeruleus in galeomorph sharks, lying in the caudal part of the isthmus along the border of the fourth ventricle. This nucleus has not yet been identified in skates, but a locus coeruleus with scattered, catecholamine-containing cells has been found in a closely related animal, the guitarfish, caudal to the isthmus at the level of the trigeminal motor nucleus. The locus coeruleus projects to the spinal cord, cerebellum, and telencephalon. In teleosts,

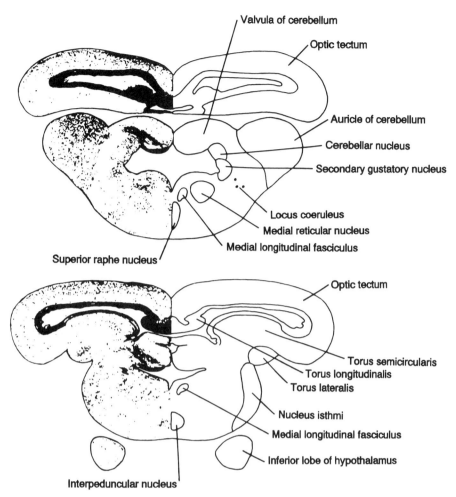

Valvula of cerebellum
Optic tectum
Auricle of cerebellum
Cerebellar nucleus
Secondary gustatory nucleus
Locus coeruleus
Medial reticular nucleus
Medial longitudinal fasciculus
Superior raphe nucleus

Optic tectum
Torus semicircularis
Torus longitudinalis
Torus lateralis
Nucleus isthmi
Medial longitudinal fasciculus
Inferior lobe of hypothalamus
Interpeduncular nucleus

FIGURE 16-2. Transverse hemisections in a longnose gar (*Lepisosteus osseus*), with mirror-image drawings, through isthmal levels, with the more caudal section at the top. Adapted from Northcutt and Butler (1980). Used with permission of Elsevier.

the locus coeruleus (Fig. 16-6), with numerous, large catecholamine-containing cells, lies dorsal to the reticular formation. In contrast with the serotonin system, the locus coeruleus is larger in teleosts than in nonteleost ray-finned fishes, and there is dense innervation of a number of areas within the midbrain

and forebrain, including part of the telencephalon, by catecholamine-containing fibers.

In amniotes, the locus coeruleus lies along the lateral edge of the rostral end of the fourth ventricle (Figs. 16-7 and 16-8). It is a relatively larger nucleus in birds than in mammals, diapsid reptiles, or turtles. In amniotes, the catecholamine-containing cells of the locus coeruleus project topographically to various parts of the forebrain, midbrain, hindbrain, and spinal cord. Like the serotonin-containing neurons of the raphe, the locus coeruleus plays a role in affecting the activity of sensory neurons and the general level of activity of the pallium in the telencephalon.

Evolutionary Perspective

A locus coeruleus is a common feature of the brains of jawed vertebrates. Whether or not it is also present in lampreys and hagfishes, we can conclude that this group of catecholamine-containing cells with widespread projections evolved early in vertebrate history.

In some cases, this set of cells seems to be developed in inverse proportion to the serotonin system of the raphe. In

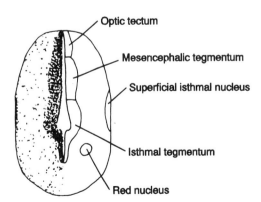

Optic tectum
Mesencephalic tegmentum
Superficial isthmal nucleus
Isthmal tegmentum
Red nucleus

FIGURE 16-3. Transverse hemisection in a lungfish (*Protopterus annectens*), with a mirror-image drawing through the isthmus. Adapted from Northcutt (1977) with additional data from Northcutt and Ronan (1985).

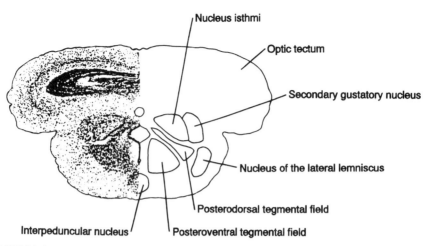

FIGURE 16-4. Transverse hemisection in a leopard frog (*Rana pipiens*), with a mirror-image drawing, through the isthmus. Adapted from Wilczynski and Northcutt (1977).

nonteleost ray-finned fishes, the locus coeruleus is relatively small while the serotonin-containing raphe cells are extensive. In teleost fishes, the reverse is true. However, in amniotes, serotonin-containing cells are numerous in the raphe and also extensively invade more lateral areas of the tegmentum, and the locus coeruleus is also well developed and has numerous and widespread projections.

NUCLEUS ISTHMI

Group I

A nucleus isthmi has not been identified in lampreys or in squalomorph sharks or ratfishes. In nonteleost ray-finned fishes, nucleus isthmi is present on the lateral edge of the isthmus (Fig. 16-2). In frogs, nucleus isthmi is a large and prominent nucleus that lies ventral to the caudal part of the optic tectum (Fig. 16-4). Nucleus isthmi is reciprocally and topographically connected with the optic tectum in both nonteleost ray-finned fishes and anuran amphibians. In the tegmentum of the coelacanth and lungfishes, a nucleus called the **superficial isthmal**

nucleus has been identified (Fig. 16-3). However, this nucleus, unlike the nucleus isthmi of ray-finned fishes and amphibians, receives retinal projections and may thus be a different, unrelated cell group. Its other connections are unknown.

Group II

A nucleus isthmi has not been found in hagfishes. A nucleus has been labeled nucleus isthmi in galeomorph sharks and skates. It corresponds in position to the nucleus named isthmi in squalomorph sharks but appears to lack reciprocal connections with the optic tectum. Nucleus isthmi is present and large in teleosts (Fig. 16-6), with a distinctive lamina of cells surrounding a relatively cell-free, central neuropil. It is reciprocally and topographically connected with the optic tectum.

Nucleus isthmi is called the **parabigeminal nucleus** in mammals and lies rostrally, at the level of the tegmentum proper. Parabigeminal roughly translates to "next to the two sets of twins," the two sets of twins being the two paired sets of the inferior and superior colliculi (= the torus semicircularis

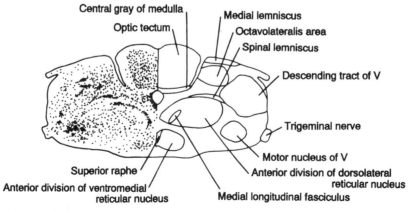

FIGURE 16-5. Drawing of a transverse hemisection in a hagfish (*Eptatretus stouti*), with a mirror-image drawing, through the isthmus. Adapted from Ronan and Northcutt (1990). Used with permission of S. Karger AG.

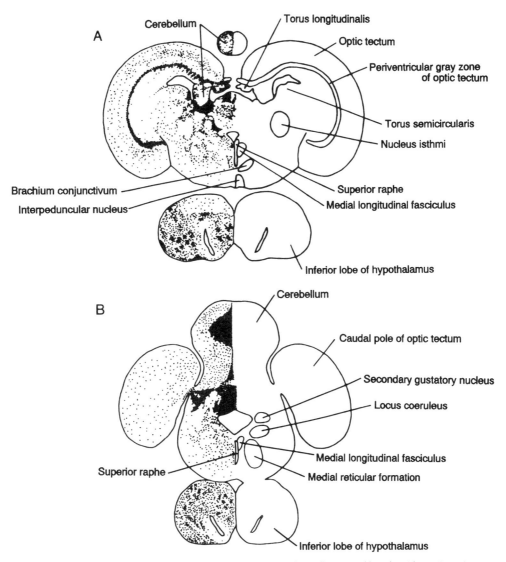

FIGURE 16-6. Drawing of transverse hemisections in a teleost (*Lepomis gibbosus*), with a mirror-image drawing, through isthmal levels, with the more caudal section at the bottom. Adapted from Parent et al. (1978).

and optic tectum, respectively). The nucleus is reciprocally and topographically connected with the optic tectum in amniotes, as it is in fishes and amphibians. In mammals, the parabigeminal nucleus has also been found to project to a major visual nucleus in the dorsal thalamus, the dorsal lateral geniculate nucleus. In the latter nucleus, parabigeminal projections overlap another set of afferent projections to it from the superior colliculus.

Electrophysiological studies in mammals have found that the responses of parabigeminal cells to light stimuli are very like the responses of tectal cells. The tectal and parabigeminal cells are most responsive to a point of light moving in a specific direction, that is, they are "direction selective." They do not selectively respond to the size or the speed of the light stimulus. The parabigeminal nucleus has been characterized as a "satellite" system for the optic tectum, used to monitor and modulate the responses of the tectal cells. Given the properties of the isthmal/parabigeminal cells, one can surmise that this nucleus helps the optic tectum help a frog to catch a fly by adjusting

for the directional component of the fly's movement. It would in like manner assist the optic tectum of fishes and amniotes in capturing moving prey.

In diapsid reptiles and turtles, nucleus isthmi lies in a dorsolateral position in the isthmal region. It has reciprocal connections with the optic tectum. Three nuclei present in the isthmal region in birds (Fig. 16-8), nucleus isthmi pars parvocellularis, nucleus isthmi pars magnocellularis, and nucleus semilunaris, receive tectal projections and project back to the optic tectum. These three nuclei may be homologous as a field to the nucleus isthmi of other vertebrates.

Evolutionary Perspective

A nucleus isthmi that is reciprocally and topographically connected with the optic tectum and that does not receive a direct retinal input is present in ray-finned fishes, amphibians, and amniote vertebrates. Such a nucleus has not been found

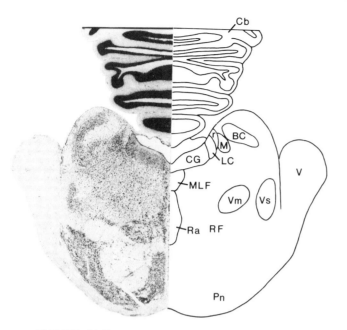

FIGURE 16-7. Transverse hemisection with mirror-image drawing through the isthmal region in a racoon (*Procyon lotor*). Abbreviations: BC, brachium conjunctivum; Cb, cerebellum; CG, central gray; LC, locus coeruleus; M, mesencephalic trigeminal nucleus; MLF, medial longitudinal fasciculus; Pn, pontine nuclei; r, mesencephalic root; Ra, raphe; RF, reticular formation; V, trigeminal nerve; Vm, motor trigeminal nucleus; Vs, principal sensory trigeminal nucleus. Photograph courtesy of Wally Welker.

in lampreys or hagfishes or in any cartilaginous fishes. Nucleus isthmi thus does not appear to have been present in the common ancestral stock of all vertebrates but to have evolved in the ancestral stock that gave rise to both ray-finned fishes and sarcopterygians.

If this nucleus is indeed absent in the coelacanth and in lungfishes, it was presumably secondarily lost as a result of the general process of reduction of the brain within these taxa. The lack of migration of most cell groups throughout the brain in the coelacanth, lungfishes, and urodele amphibians appears to be the result of a trend to secondary simplification related to neoteny (the retention of embryonic characteristics in the adult). These groups of vertebrates are in fact excellent examples of the principal that brain evolution is multidirectional; there is not a single trend from simple to complex, as complex can also give rise to simple. Phylogenetic analysis of such simplified and reduced structures is particularly difficult, however, as many features of the morphology of the ancestral adult phenotype have been lost.

ISTHMO-OPTIC NUCLEUS

Group I

An isthmo-optic nucleus, that is, a group of neurons located in the caudal mesencephalon that project to the retina, has been found in a fragmented distribution both among and within various vertebrate groups. Tegmental retinopetal projections have been found in lampreys and in two nonteleost ray-finned

fishes, bichirs and gars, but are absent in another nonteleost ray-finned fish, sturgeons. Similar projections also appear to be absent in Group I sharks as well as in amphibians.

Group II

Isthmo-optic neurons that project to the retina have been found in hagfishes in the medial, rostral part of the mesencephalic tegmentum. The nucleus in hagfishes that contains the majority of the retinopetal cells is called the **nucleus of the posterior commissure** (rostral to the level shown in Fig. 16-5; see Fig. 20-5); additional retinopetal cells lie scattered in an area lateral to this nucleus. An isthmo-optic nucleus has not been found in any Group II cartilaginous fish. Isthmo-optic neurons have been found in one teleost, a pike, but are absent in other teleosts studied.

Among amniotes, isthmo-optic neurons have been found in crocodiles and turtles. Only a few such neurons are present in lizards, and no retinopetally projecting neurons are present in the tegmentum of snakes. In birds, an isthmo-optic nucleus is present in the dorsal, medial part of the tegmentum; this nucleus is cytoarchitectonically well defined in some species, such as in the pigeon, as shown in Figure 16-8. A smaller number of retinopetal neurons also lie scattered in an area ventral to the isthmo-optic nucleus. The avian isthmo-optic nucleus is particularly well developed in birds that feed by pecking, such as pigeons, as opposed to those that feed on-the-wing, such as swifts and swallows. This nucleus in birds and its centripetal innervation of the retina may thus be involved in searching for and/or pecking at food on the ground.

Evolutionary Perspective

Retinopetally projecting isthmo-optic neurons are present in lampreys, hagfishes, some but not all nonteleost ray-finned fishes, at least one teleost, and, among amniotes, in some diapsid reptiles, and in birds and turtles. Such neurons appear to be absent in all cartilaginous fishes, at least one nonteleost ray-finned fish (sturgeons), most teleosts, amphibians, snakes, and mammals. From this distribution, the evolutionary history of isthmo-optic neurons cannot be clearly discerned, particularly among jawed vertebrates.

Due to their presence in both hagfishes and lampreys, isthmo-optic neurons were probably present in the earliest vertebrates and thus plesiomorphic for vertebrates. This population of neurons may have been lost in the common ancestral stock of jawed vertebrates, however, and subsequently regained independently within two groups of vertebrates—in some of the ray-finned fishes and in nonmammalian amniotes. Among the latter, since isthmo-optic neurons are present in lizards, crocodiles, and birds, their absence in snakes appears to be an apomorphy. Another possibility, that isthmo-optic neurons were retained in ancestral jawed vertebrates and subsequently lost multiple times cannot be entirely ruled out, however. Thus, whether the isthmo-optic neurons present in nonmammalian amniotes are homologous or homoplaseous to those in some ray-finned fishes cannot be determined from the present data.

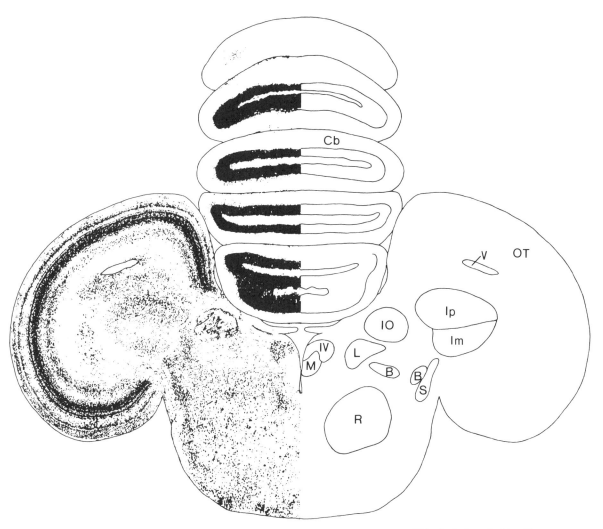

FIGURE 16-8. Transverse hemisection through the isthmal region in a pigeon (*Columba livia*) on the left with a mirror-image drawing on the right. Adapted from Karten and Hodos (1967). Abbreviations: B, brachium conjunctivum; Cb, cerebellum; Im, nucleus isthmi, pars magnocellularis; IO, isthmo-optic nucleus; Ip, nucleus isthmi, pars parvocellularis; L, locus coeruleus; M, medial longitudinal fasciculus; OT, optic tectum; R, nucleus reticularis pontis oralis; S, nucleus semilunaris; IV, nucleus of the trochlear nerve. Used with permission of the Johns Hopkins University Press.

INTERPEDUNCULAR NUCLEUS

Group I

An interpeduncular nucleus has recently been identified in the ventral part of the mesencephalon in lampreys, based on a major afferent input from the habenula via the fasciculus retroflexus. The interpeduncular nucleus in squalomorph sharks lies medial to the rostral part of the reticular formation and immediately rostral to the raphe and caudal to the ventral tegmental area. The fasciculus retroflexus traverses the medial tegmental region to project to the interpeduncular nucleus. In nonteleost ray-finned fishes, the interpeduncular nucleus (Fig. 16-2) and fasciculus retroflexus lie in similar positions. The interpeduncular nucleus is rather small and cell sparse in these fishes. In frogs, the interpeduncular nucleus is well developed and lies in the isthmus (Fig. 16-4).

Group II

In hagfishes, a large interpeduncular nucleus (and a large habenula) are present, but the interpeduncular nucleus lies farther rostrally than in most animals. In these fishes, it lies in a ventromedial position at levels through the diencephalon rather than in the isthmus or caudal tegmentum. The position of the fasciculus retroflexus has not been established.

The interpeduncular nucleus is larger and more distinct in cartilaginous fishes with elaborated brains than in squalomorph sharks. It lies in a medial position in the ventral tegmentum and receives projections via the fasciculus retroflexus. In teleosts (Fig. 16-6) and amniotes, the interpeduncular nucleus lies in the ventromedial part of the tegmentum, and the fasciculus retroflexus projects to it from the habenula. In mammals, the cerebral peduncle lies lateral to the nucleus.

Evolutionary Perspective

The fasciculus retroflexus, the habenula in the diencephalon from which it arises, and the interpeduncular nucleus to which it projects are common features of vertebrates and are assumed to have been present in the ancestral stock of all vertebrates. This system is the route by which input from olfactory and limbic related areas of the telencephalon reaches the tegmentum and can be integrated there with other incoming sensory information for coordination with the appropriate motor responses.

FOR FURTHER READING

Dube, L. and Parent, A. (1981) The monoamine-containing neurons in avian brain: I. A study of the brain stem of the chicken (*Gallus domesticus*) by means of fluroescence and acetylcholinesterase histochemistry. *Journal of Comparative Neurology,* 196, 695–708.

Ekström, P. and Van Veen, T. (1984) Distribution of 5-hydroxytryptamine (serotonin) in the brain of the teleost *Gasterosteus aculeatus* L. *Journal of Comparative Neurology,* 226, 307–320.

Fasolo, A., Franzoni, M. F., Gaudino, G., and Steinbusch, H. W. M. (1986) The organization of serotonin-immunoreactive neuronal systems in the brain of the crested newt, *Triturus cristatus carnifex* Laur. *Cell Tissue Research,* 243, 239–247.

Gruberg, E. R., Hughes, T. E., and Karten, H. J. (1994) Synaptic interrelationships between the optic tectum and the ipsilateral nucleus isthmi in *Rana pipiens. Journal of Comparative Neurology,* 339, 353–364.

Harting, J. K., van Lieshout, D. P., Hashikawa, T., and Weber, J. T. (1991) The parabigeminogeniculate projection: connectional studies in eight mammals. *Journal of Comparative Neurology,* 305, 559–581.

Malz, C. R. and Meyer, D. L. (1994) Interspecific variation of isthmo-optic projections in poikilothermic vertebrates. *Brain Research,* 661, 259–264.

Waterhouse, R. D., Border, B., Wahl, L., and Mihailoff, G. A. (1993) Topographic organization of rat locus coeruleus and dorsal raphe nuclei: distribution of cells projecting to visual system structures. *Journal of Comparative Neurology,* 336, 345–361.

ADDITIONAL REFERENCES

Baumgarten, H. G. (1972) Biogenic monoamines in the cyclostome and lower vertebrate brain. *Progress in Histochemistry and Cytochemistry,* 4, 1–90.

Bolliet, V. and Ali, M. A. (1992) Immunohistochemical study of the development of serotoninergic neurons in the brain of the brook trout *Salvelinus fontinalis. Brain, Behavior and Evolution,* 40, 234–249.

Dahlström, A. and Füxe, K. (1964) Evidence for the existence of monoamine-containing neurons in the central nervous system. 1. Demonstration of monoamines in the cell bodies of brainstem neurons. *Acta Physiologica Scandinavica,* 62 (Suppl. 232): 1–55.

Dube, L. and Parent, A. (1982) The organization of monoamine-containing neurons in the brain of the salamander, *Necturus maculosus. Journal of Comparative Neurology,* 211, 21–30.

Dubuc, R., Bongianni, F., Ohta, Y., and Grillner, S. (1993) Anatomical and physiological study of brainstem nuclei relaying dorsal column inputs in lampreys. *Journal of Comparative Neurology,* 327, 260–270.

Ekström, P., Honkanen, T., and Steinbusch, H. W. M. (1990) Distribution of dopamine-immunoreactive neuronal perikarya and fibers in the brain of a teleost, *Gasterosteus aculeatus* L. Comparison with tyrosine hydroxylase- and dopamine-β-hydroxylase-immunoreactive neurons. *Journal of Chemical Neuroanatomy,* 3: 233–260.

Feig, S. and Harting, J. K. (1992) Ultrastructural studies of the primate parabigeminal nucleus: electron microscopic autoradiographic analysis of the tectoparabigeminal projection in *Galago crassicaudatus. Brain Research,* 595, 334–338.

Feng, A. S. and Lin, W. (1991) Differential innervation patterns of three divisions of frog auditory midbrain (torus semicircularis). *Journal of Comparative Neurology,* 306, 631–630.

Feyerabend, B., Malz, C. R., and Meyer, D. L. (1994) Birds that feed-on-the-wing have few isthmo-optic neurons. *Neuroscience Letters,* 182, 66–68.

Frankenhuis-van den Heuvel, T. H. M. and Nieuwenhuys, R. (1984) Distribution of serotonin-immunoreactivity in the diencephalon and mesencephalon of the trout, *Salmo gairdneri. Anatomy and Embryology,* 169, 193–204.

Gruberg, E. R., Wallace, M. T., and Waldeck, R. F. (1989) Relationship between isthmotectal fibers and other tectopetal systems in the leopard frog. *Journal of Comparative Neurology,* 288, 39–50.

Gruberg, E. R., Wallace, M. T., Caine, H. S., and Mote, M. I. (1991) Behavioral and physiological consequences of unilateral ablation of the nucleus isthmi in the leopard frog. *Brain, Behavior and Evolution,* 37, 92–103.

Hunt, S. P. and Brecha, N. (1984) The avian optic tectum: a synthesis of morphology and biochemistry. In H. Vanegas (ed.), *Comparative Neurology of the Optic Tectum.* New York: Plenum, pp. 619–648.

Johnston, S. A., Maler, L., and Tinner, B. (1990) The distribution of serotonin in the brain of *Apteronotus leptorhynchus:* an immunohistochemical study. *Journal of Chemical Neuroanatomy,* 3, 429–465.

Kadota, T. (1991) Distribution of 5-HT (serotonin) immunoreactivity in the central nervous system of the inshore hagfish, *Eptatretus burgeri. Cell Tissue Research,* 266, 107–116.

Karten, H. J. and Hodos, W. (1967) *A Stereotaxic Atlas of the Brain of the Pigeon (Columba livia).* Baltimore, MD: The Johns Hopkins University Press.

Kennedy, M. C. and Rubinson, K. (1977) Retinal projections in larval, transforming and adult sea lamprey, *Petromyzon marinus. Journal of Comparative Neurology,* 171, 465–479.

Ma, P. M. (1994) Catecholaminergic systems in the zebrafish. I. Number, morphology, and histochemical characteristics of neurons in the locus coeruleus. *Journal of Comparative Neurology,* 344, 242–255.

Ma, P. M. (1994) Catecholaminergic systems in the zebrafish. II. Projection pathways and pattern of termination of the locus coeruleus. *Journal of Comparative Neurology,* 344, 256–269.

Meek, J. and Hoosten, H. W. J. (1989) Distribution of serotonin in the brain of the mormyrid teleost *Gnathonemus petersii. Journal of Comparative Neurology,* 281, 206–224.

Meyer, D. L., Gerwerzhagen, K., Fiebig, E., Ahlswede, F., and Ebbesson, S. O. E. (1983) An isthmo-optic system in a bony fish. *Cell Tissue Research,* 231, 129–133.

Nieuwenhuys, R. (1972) Topological analysis of the brain stem of the lamprey *Lampetra fluviatilis. Journal of Comparative Neurology,* 145, 165–178.

Nieuwenhuys, R. (1985) *Chemoarchitecture Of the Brain.* New York: Springer-Verlag.

Nieuwenhuys, R. and Pouwels, E. (1983) The brain stem of actinopterygian fishes. In R. G. Northcutt and R. E. Davis (eds.), *Fish Neurobi-*

ology, Vol. 1: Brain Stem and Sense Organs. Ann Arbor, MI: University of Michigan Press, pp. 25–87.

Northcutt, R. G. (1977) Retinofugal projections in the lepidosirenid lungfishes. *Journal of Comparative Neurology,* 174, 553–574.

Northcutt, R. G. (1978) Brain organization in the cartilaginous fishes. In E. S. Hodgson and R. F. Mathewson (eds.), *Sensory Biology of Sharks, Skates, and Rays.* Arlington, VA: Office of Naval Research, pp. 117–193.

Northcutt, R. G. (1979) Experimental determination of the primary trigeminal projections in lampreys. *Brain Research,* 163, 323–327.

Northcutt, R. G. (1980) Retinal projections in the Australian lungfish. *Brain Research,* 185, 85–90.

Northcutt, R. G. (1984) Anatomical organization of the optic tectum in reptiles. In H. Vanegas (ed.), *Comparative Neurology of the Optic Tectum.* New York: Plenum, pp. 547–600.

Northcutt, R. G. and Butler, A. B. (1980) Projections of the optic tectum in the longnose gar, *Lepisosteus osseus. Brain Research,* 190, 333–346.

Northcutt, R. G., Reiner, A., and Karten, H. J. (1988) Immunohistochemical study of the telencephalon of the spiny dogfish, *Squalus acanthias. Journal of Comparative Neurology,* 277, 250–267.

Northcutt, R. G. and Ronan, M. C. (1985) The origins of descending spinal projections in lepidosirenid lungfishes. *Journal of Comparative Neurology,* 241, 435–444.

Panneton, W. M. and Watson, B. J. (1991) Stereotaxic atlas of the brainstem of the muskrat, *Ondatra zibethicus. Brain Research Bulletin,* 26, 479–509.

Parent, A., Dube, L., Braford, M. R., Jr., and Northcutt, R. G. (1978) The organization of monoamine-containing neurons in the brain of the sunfish (*Lepomis gibbosus*) as revealed by fluorescence microscopy. *Journal of Comparative Neurology,* 182, 495–516.

Parent, A. and Northcutt, R. G. (1982) The monoamine containing neurons in the brain of the garfish, *Lepisosteus osseus. Brain Research Bulletin,* 9, 189–204.

Powers, A. S. and Reiner, A. (1993) The distribution of cholinergic neurons in the central nervous system of turtles. *Brain, Behavior and Evolution,* 41, 326–345.

Ronan, M. and Northcutt, R. G. (1990) Projections ascending from the spinal cord to the brain in petromyzontid and myxinoid agnathans. *Journal of Comparative Neurology,* 291, 491–508.

Sims, T. J. (1977) The development of monoamine-containing neurons in the brain and spinal cord of the salamander, *Ambystoma mexicanum. Journal of Comparative Neurology,* 173, 319–336.

Smeets, W. J. A. J., Nieuwenhuys, R., and Roberts, B. L. (1983) *The Central Nervous System of Cartilaginous Fishes.* Berlin: Springer-Verlag.

Smeets, W. J. A. J. and Steinbusch, H. W. M. (1988) Distribution of serotonin immunoreactivity in the forebrain and midbrain of the lizard *Gekko gecko. Journal of Comparative Neurology,* 271 419–434.

Stuesse, S. L. and Cruce, W. L. R. (1991) Immunohistochemical localization of serotoninergic, enkephalinergic, and catecholaminergic cells in the brainstem and diencephalon of a cartilaginous fish, *Hydrolagus colliei. Journal of Comparative Neurology,* 309, 535–548.

Stuesse, S. L., Cruce, W. L. R., and Northcutt, R. G. (1991) Localization of serotonin, tyrosine hydroxylase, and leu-enkephalin immunoreactive cells in the brainstem of the horn shark, *Heterodontus francisci. Journal of Comparative Neurology,* 308, 277–292.

Stuesse, S. L., Cruce, W. L. R., and Northcutt, R. G. (1991) Serotoninergic and enkephalinergic cell groups in the reticular formation of the bat ray and two skates. *Brain, Behavior and Evolution,* 38, 39–52.

ten Donkelaar, H. J. and Nieuwenhuys, R. (1979) The brainstem. In C. Gans, R. G. Northcutt, and P. Ulinski (eds.), *Biology of the Reptilia, Vol. 10.* New York: Academic, pp. 133–200.

Tóth, P., Lázár, G., Wang, S.-R., Li, T.-B., Xu, J., Pál, E., and Straznicky, C. (1994) The contralaterally projecting neurons of the isthmic nucleus in five anuran species: a retrograde tracing study with HRP and cobalt. *Journal of Comparative Neurology,* 346, 306–320.

Vesselkin, N. P., Ermakova, T. V., Repérant, J., Kosareva, A. A., and Kenigfest, N. B. (1980) The retinofugal and retinopetal systems in *Lampetra fluviatilis:* an experimental study using radioautographs and HRP methods. *Brain Research,* 195, 453–460.

Wicht, H. and Northcutt, R. G. (1990) Retinofugal and retinopetal projections in the Pacific hagfish, *Eptatretus stouti* (Myxinoidea). *Brain, Behavior and Evolution,* 36, 315–328.

Wilczynski, W. and Northcutt, R. G. (1977) Afferents to the optic tectum of the leopard frog: an HRP study. *Journal of Comparative Neurology,* 173, 219–230.

Wilson, M. A. and Molliver, M. E. (1991) The organization of serotonergic projections to cerebral cortex in primates: retrograde transport studies. *Neuroscience,* 44, 555–570.

Yañez, J. and Anadon, R. (1994) Afferent and efferent connections of the habenula in the larval sea lamprey (*Petromyzon marinus* L.): an experimental study. *Journal of Comparative Neurology,* 345, 148–160.

17

Tegmentum and Tori

INTRODUCTION

The tegmentum proper lies rostral to the isthmus and ventral to the tectum (or roof) of the midbrain. In addition to cranial nerve nuclei and ascending and descending tracts discussed earlier, the tegmentum contains some structures that are involved in the production and control of various motor functions: the **midbrain locomotor region (MLR),** the **mesencephalic nucleus of the trigeminal nerve,** the **red nucleus,** the **substantia nigra,** and the **ventral tegmental area.** Two additional midbrain structures are also discussed in this chapter: the **torus lateralis,** of which little is known, and the **torus semicircularis,** which receives ascending auditory and lateral line projections via the lateral lemniscus. The tori lateralis and semicircularis are part of the roof of the midbrain, lying in a region between the optic tectum and the tegmentum. A third torus, called the **torus longitudinalis,** will be discussed in conjunction with the optic tectum in Chapter 18. A number of the nuclei and fiber tracts in the tegmentum, with other structures labeled for orientation, are illustrated in some animals with laminar brains in Figures 17-1–17-6 and in other animals with elaborated brains in Figures 17-7–17-13.

MIDBRAIN LOCOMOTOR REGION

Group I

The midbrain locomotor region (MLR) appears to be a **central pattern generator,** or **command generator,** for sequences of movements. Stimulation of it produces locomotory movements, presumed to be mediated by projections to the reticular formation, and hence to the spinal cord. In lampreys, some cells in the mesencephalic tegmentum (Fig. 17-1) have been found to comprise an MLR that presumably sends descending projections to the reticular formation.

Group II

In stingrays, a midbrain locomotor region has been found that includes the interstitial nucleus of the medial longitudinal fasciculus and the medial part of the **cuneiform nucleus** (Fig. 17-8). Stimulation of this region evokes locomotory movements in the pectoral fin. The region projects bilaterally to the magnocellular/gigantocellular region of the medullary reticular formation, which in turn projects to the spinal cord. The MLR in cartilaginous fishes appears to correspond to the lateral subdivision of a similarly located MLR in mammals. Another nucleus in this region is the **intercollicular nucleus.** It lies near the cuneiform nucleus in a transition zone between the optic tectum and the torus semicircularis. In skates, this nucleus receives projections from the optic tectum and projects to the spinal cord, forming a tectal motor relay pathway complimentary to the tectoreticular pathway.

In teleost fishes, a midbrain locomotor region has been identified in the dorsocaudal tegmentum near the midline (Fig. 17-10), a position similar to that of the MLR in stingrays. Weak stimulation of the area produces movements of the pectoral fins, while stronger stimulation produces movement of both the pectoral and caudal fins.

Among amniotes, the midbrain locomotor region has been studied primarily in mammals. The MLR includes the caudal part of the cuneiform nucleus and also involves another, neighboring nucleus called the **pedunculopontine nucleus.** The MLR projects to the reticular formation via the medial longitudinal fasciculus, and produces walking movements when stimulated. In fact, this region takes its name from the observation (first made in cats) that stimulation of it results in rhythmic patterns of limb movements similar to those used by the animal in walking.

In birds, a locomotor region has been found in the region of a nucleus called the **lateral spiriform nucleus.** Stimulation in this region produces walking and wing flapping motions. The lateral spiriform nucleus lies farther rostrally, relative to

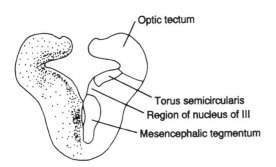

Optic tectum

Torus semicircularis
Region of nucleus of III
Mesencephalic tegmentum

FIGURE 17-1. Drawing of a transverse hemisection with mirror-image drawing through the tegmentum of a lamprey (*Petromyzon marinus*). Adapted from Kennedy and Rubinson (1977). Note that in this and in subsequent figures, ventricles are generally not labeled.

some other tegmental nuclei, than the cuneiform nucleus of other vertebrates, but it has the same set of neurotransmitters and presumably similar connections, so it is probably the homologue of the cuneiform nucleus.

Parts of the MLR are interconnected with other nuclei involved in motor control. These nuclei include the substantia nigra in the tegmentum and the striatum in the telencephalon. We will discuss these connections below, in conjunction with the motor pathways involving the substantia nigra.

Evolutionary Perspective

A midbrain locomotor region, which in jawed vertebrates produces coordinated, purposeful movements of the fins, limbs, or wings (i.e., the appendicular skeleton) when stimulated, has been identified in a wide variety of extant vertebrates. It has been found in both lampreys and hagfishes, cartilaginous fishes, teleost fishes, and in mammals and birds. An MLR was thus apparently present in the common ancestral stock of all vertebrates and represents a very old system for the production of locomotory motor behaviors.

MESENCEPHALIC NUCLEUS OF THE TRIGEMINAL NERVE

Group I

A mesencephalic nucleus of the trigeminal nerve (MesV) has not been found in lampreys. Both the trigeminal and facial nerves are known to be involved in producing movements of the sucking apparatus of the mouth, so it is possible that a MesV might be present in these fishes. In squalomorph sharks, as in other jawed vertebrates, MesV cells are present and receive proprioceptive feedback from the muscles of the jaws. The nucleus lies in the medial part of the periventricular gray layer of the optic tectum (Fig. 17-2) in these sharks, and it lies in a similar position in nonteleost ray-finned fishes and in amphibians.

Group II

In hagfishes, as in lampreys, a MesV has not been identified. In cartilaginous fishes with elaborated brains, cells of MesV

are present dorsally in the periventricular gray of the optic tectum (Fig. 17-8). They also occupy this position in teleost fishes. MesV is unusually large in some teleosts, such as in tarpons, which are elopomorphs. The cells of MesV lie in the periventricular gray of the tectum in mammals (Fig. 17-12), diapsid reptiles (Fig. 17-13), birds, and turtles. In many mammals, the cells are present throughout the rostrocaudal extent of the midbrain roof, extending caudally to the region of the caudal end of the inferior colliculus (Figs. 16-7 and 17-12).

Evolutionary Perspective

The mesencephalic nucleus of V is a very constant feature of the brain in all jawed vertebrates, and thus is thought to have been present in the common ancestral stock of this large group. Whether cells are present in lampreys and hagfishes that are homologous to the MesV cells in jawed vertebrates remains to be determined.

RED NUCLEUS AND RELATED NUCLEI

Group I

A red nucleus does not appear to be present in lampreys. In squalomorph sharks, a red nucleus is present in the ventral part of the tegmentum (Fig. 17-2) and receives projections from the cerebellar nuclei via the brachium conjunctivum. However, unlike the case in most other animals, the red nucleus does not appear to project to the spinal cord. A second nucleus related to the cerebellum, named **nucleus H,** is present at a more rostral level (Fig. 17-2).

Among nonteleost ray-finned fishes, a red nucleus has been identified in gars in the rostral tegmentum (Fig. 17-3) on the basis of projections to the spinal cord via the **rubrospinal tract.** In lungfishes, a red nucleus has similarly been identified on the basis of its giving rise to descending projections to the spinal cord (Fig. 17-5). A red nucleus is also present in amphibians. In frogs, it lies in the ventral part of the tegmentum (Fig. 17-6), projects to the contralateral spinal cord, and receives a sparse, descending projection from the striatum in the basal telencephalon. It has also been found to project to the cerebellar hemisphere and receive projections from the deep cerebellar nucleus.

Group II

Hagfishes are similar to lampreys in that they do not have a red nucleus. Cartilaginous fishes with elaborated brains have a relatively large red nucleus, bordered by the brachium conjunctivum. The nucleus lies in the ventral tegmentum (Figs. 17-8 and 17-9) at the level of the oculomotor nucleus, and it projects to the spinal cord. Scattered cells, which correspond to the nucleus H present in squalomorph sharks, lie in the lateral part of the tegmentum (Figs. 17-8 and 17-9). This nucleus is reciprocally connected with the cerebellum. More rostrally, an additional migrated nucleus, called **nucleus K** (Fig. 17-9) is present, and is also believed to be connected with the

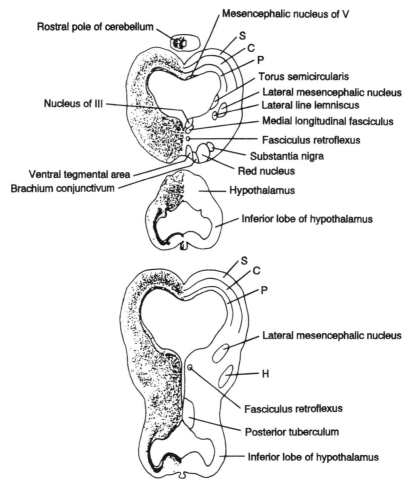

FIGURE 17-2. Transverse hemisections with mirror-image drawings through the tegmentum of a squalo-morph shark (*Squalus acanthias*). The more caudal section is at the top. Adapted from Northcutt (1978) with additional data from Boord and Northcutt (1988). In this and some subsequent figures, S, C, and P indicate the superficial, central, and periventricular zones of the optic tectum.

cerebellum. Like the red nucleus, nuclei H and K are devoid of catecholamine-containing cells.

In teleosts, the brachium conjunctivum (Fig. 17-10) runs rostroventrally through the tegmentum and decussates just dorsal to the interpeduncular nucleus. It projects to the red nucleus, a small, discrete group of cells (Fig. 17-10). The position of the teleost red nucleus is more dorsal and more medial than in cartilaginous fishes, and its cells appear larger and more tightly condensed.

In both mammals (Fig. 17-12) and birds, the red nucleus is a rather large and prominent tegmental nucleus. The connections of the red nucleus have been found to be very similar in these two groups of amniotes. In both groups, the afferent projections to the red nucleus arise in a somatosensory-related part of the pallium in the telencephalon and in cerebellar nuclei. The major efferent projection of the nucleus is to the contralateral spinal cord, via the rubrospinal tract.

In most diapsid reptiles and turtles, the red nucleus lies in the ventrolateral part of the rostral tegmentum (Fig. 17-13).

Descending projections from the striatum terminate just rostral to it in the **prerubral area.** The red nucleus gives rise to a rubrospinal tract that projects to the contralateral spinal cord. A rubrospinal tract is absent in pythons, but this absence cannot be correlated with the absence of limbs as the tract is present in colubrid snakes.

Evolutionary Perspective

A red nucleus is absent in agnathans, and its absence may be correlated with the limited development of the cerebellum in these animals. A red nucleus is present in cartilaginous fishes, ray-finned fishes, lungfishes, amphibians, and amniotes, and is thus plesiomorphic for jawed vertebrates. Descending projections to the contralateral spinal cord from this nucleus are also widely distributed among jawed vertebrates, but have not been found in a squalomorph shark and, among diapsid reptiles, are absent in pythons but present in colubrid snakes. The rubrospinal tract is thought to be primarily involved with move-

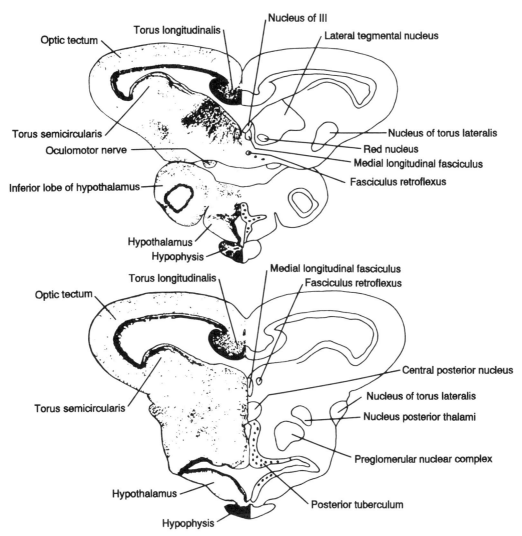

FIGURE 17-3. Transverse hemisections with mirror-image drawings through the tegmentum of a longnose gar (*Lepisosteus osseus*). The more caudal section is at the top. Adapted from Northcutt and Butler (1980) with additional data from Northcutt (1983). Used with permission of Elsevier.

ments of the limbs. This system is quite conservative, however; its organization is very similar in birds and mammals, for example, despite differences in the particular movements of the limbs. The rubrospinal system may have additional functions as well, as it is present in a least one snake and in both South American and African lungfishes, only the latter of which can use their fins for assistance in movements.

SUBSTANTIA NIGRA AND VENTRAL TEGMENTAL AREA

The substantia nigra and ventral tegmental area, where present, are characterized by the presence of catecholamine-containing cells; they project heavily upon the striatal region in the telencephalon and are part of a functional system for the initiation and regulation of movement. Other catecholamine-containing cell groups also occur in the brainstem, however,

and we need to clarify the nomenclature and anatomy of these groups before examining these structures in Group I and Group II vertebrates.

In 1964, A. Dahlström and K. Fuxe proposed a system of nomenclature for the catecholamine-containing cell groups in the brains of mammals that is now widely accepted and also applied to nonmammalian vertebrates where possible. They recognized 12 groups of such cells and labeled them **A1** to **A12** in their caudal-to-rostral sequence. At least 17 such cell groups are now recognized. In this system, some catecholamine-containing cells caudal to the red nucleus, in the **retrorubral area,** form the **A8** cell group. The substantia nigra is the **A9** cell group, and the ventral tegmental area is the **A10** cell group.

In many anamniote vertebrates, catecholamine-containing cells also occur in a nucleus in the caudal part of the diencephalon called the **posterior tuberculum** (see Chapter 20). The posterior tuberculum lies just rostral to the area of the tegmentum, which contains the ventral tegmental area and the substan-

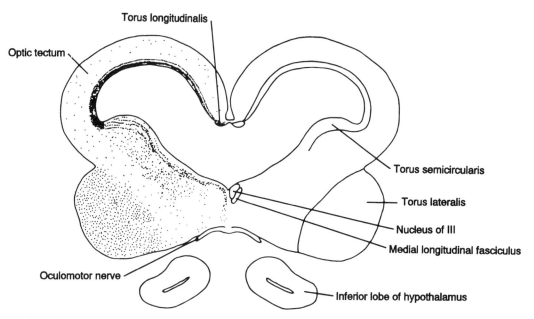

FIGURE 17-4. Drawing of a transverse hemisection with mirror-image drawing through the tegmentum of a bowfin (*Amia calva*). Adapted from Nieuwenhuys and Pouwels (1983).

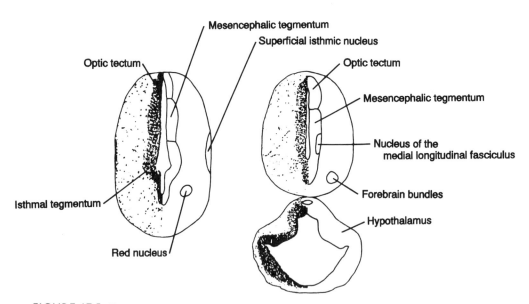

FIGURE 17-5. Drawing of a transverse hemisection with a mirror-image drawing through the tegmentum of a lungfish (*Protopterus annectens*). The more caudal section is on the left. Adapted from Northcutt (1977).

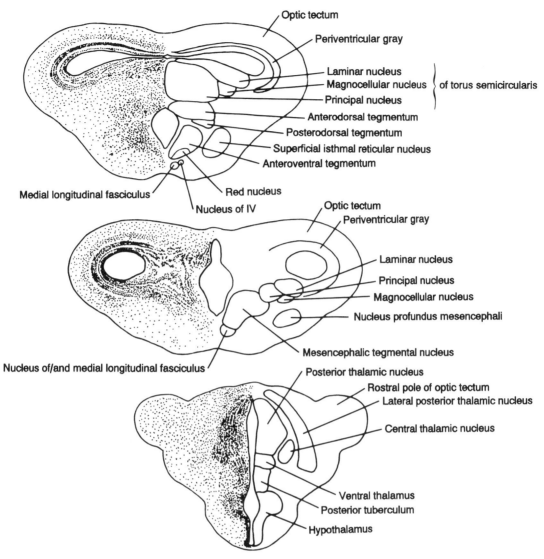

FIGURE 17-6. Drawings of transverse hemisections with mirror-image drawings through the tegmentum of a frog (*Rana catesbeiana*). The most caudal drawing is at the top. Adapted from Wilczynski and Northcutt (1983) and integrated with additional data from Feng and Lin (1991) on *Rana pipiens*.

tia nigra in some anamniote vertebrates. We will include some discussion of the posterior tuberculum here since it is located in close proximity to the catecholamine-containing cell groups of the mesencephalon, and the classification of its catecholamine-containing neurons is a topic of current debate.

Dahlström and Fuxe identified catecholamine-containing cells in the caudal diencephalon within the periventricular gray area of the hypothalamus, thalamus, and rostral midbrain as the **A11** cell group. In the diencephalon of anamniotes, the posterior tuberculum is located between the hypothalamus and the caudal part of the dorsal thalamus. Recent developmental studies indicate that the posterior tuberculum of anamniotes is most likely homologous to the A11 (caudal diencephalic) catecholamine neuron group and/or to the rostral part of the A8/A9 group in amniotes. Whether the posterior tuberculum is embryologically derived from the hypothalamus, a caudal nonhypothalamic part of the diencephalon, or the rostral part

of the mesencephalon—as have been variously proposed—is not yet resolved. The embryonic derivation of the substantia nigra (A9) and ventral tegmental area (A10) is likewise unclear, although some evidence suggests that these nuclei may be derived from multiple embryonic zones. Thus, whether the catecholamine-containing cells of the posterior tuberculum are homologous to the A11 group of amniotes or to the A8–A10 groups (or possibly to parts of both) is an open question.

Group I

Some cells within the mesencephalic tegmentum of lampreys (Fig. 17-1) contain catecholamine. These cells are located in the more rostral and ventral part of this region and may correspond to cells in the posterior tuberculum of other vertebrates. Among cartilaginous fishes, catecholamine-containing cells are present in the posterior tuberculum of ratfishes (chi-

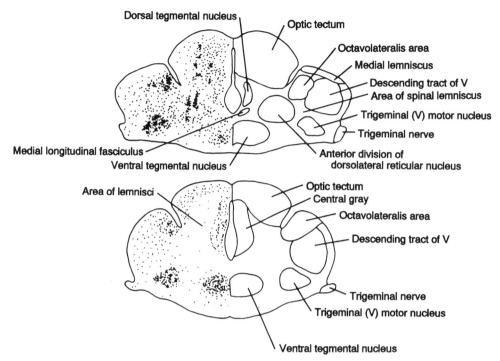

FIGURE 17-7. Drawings of transverse hemisections with mirror-image drawings through the tegmentum of a hagfish (*Eptatretus stouti*). The more caudal section is at the top. Adapted from Ronan and Northcutt (1990) and Wicht and Northcutt (1990). Material from the latter used with permission of S. Karger AG.

maeras) but not more caudally within the mesencephalon. In squalomorph sharks, however, there are catecholamine-containing cells in the posterior tuberculum and also more caudally in an area surrounding the red nucleus. The medial part of this latter area can thus be recognized as the ventral tegmental area because of its position rostral to the interpeduncular nucleus and medial to the red nucleus, and because it has catecholamine-containing cells (Fig. 17-2). The catecholamine-containing cells that lie around the lateral edge of the red nucleus appear to constitute the substantia nigra.

In nonteleost ray-finned fishes, separate nuclei forming a ventral tegmental area and substantia nigra are absent. Catecholamine-containing cells (represented by the large dots in Fig. 17-3) are for the most part confined to the posterior tuberculum of the diencephalon. A posterior tuberculum has been identified in both lungfishes and amphibians (Fig. 17-6), and, as in teleosts, the only catecholamine-containing cells in the rostral part of the brainstem in amphibians are a small number within the posterior tuberculum. In frogs, some of the cells in the posterior tuberculum have been found to project to the striatum.

Group II

In the mesencephalic tegmentum of hagfishes (Fig. 17-7), **dorsal** and **ventral tegmental nuclei** have been identified, which lie dorsal and ventral to the medial longitudinal fasciculus, respectively. The ventral tegmental nucleus is particularly large and extends rostrally. Unfortunately, we do not yet know whether catecholamine-containing cells are present in this region.

Both the ventral tegmental area and substantia nigra are present and relatively large in galeomorph sharks, skates, and rays. In Figure 17-9 we have shown four levels in between the two sections shown in Figure 17-8 for more detail of the nuclei within the ventral part of the tegmentum. The dots in Figure 17-9 represent catecholamine-containing cells. The most rostral level (D) in Figure 17-9 is through the posterior tuberculum, which also has numerous catecholamine-containing cells.

There are no nuclei corresponding to the ventral tegmental area or substantia nigra in teleosts. Catecholamine-containing cells in this region are confined to the posterior tuberculum (Fig. 17-10), as in nonteleost ray-finned fishes and in contrast to sharks, skates, and rays.

In amniotes, a great deal has been learned about the connections of the substantia nigra and the ventral tegmental area. The neurotransmitters of the various projections have also been identified in many cases. These nuclei and their connections, particularly those with the striatum, have received so much attention because of the involvement of the substantia nigra in diseases of humans that are characterized by abnormal, involuntary movements, such as Huntington's (chorea) and Parkinson's (paralysis agitans) diseases. The substantia nigra and ventral tegmental area are involved in the initiation and regulation of movements. We will discuss this system in more depth than many of the other systems because it can serve to some extent as a model of the complexity and interactions that occur in most systems but are worked out in detail in so few.

A region corresponding to the posterior tuberculum of anamniotes has not been found in amniotes, but catecholamine-containing cells are present in part of the substantia nigra and in the ventral tegmental area (Fig. 17-12 and 17-13). In mammals,

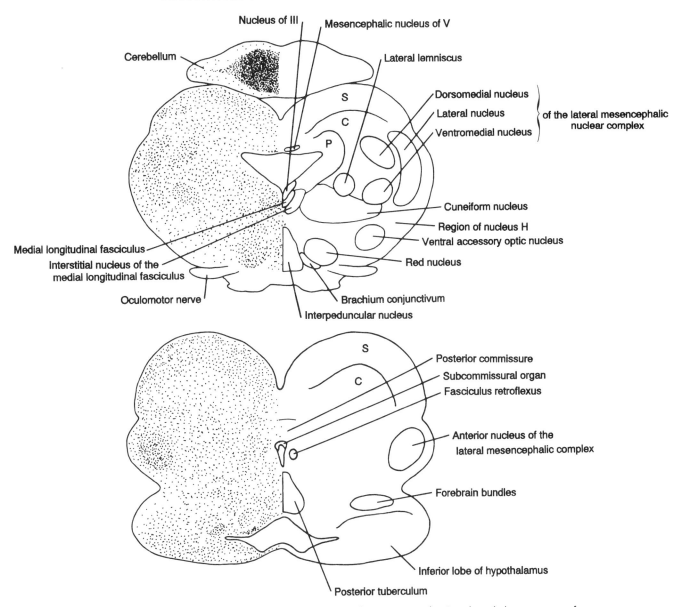

FIGURE 17-8. Drawings of transverse hemisections with mirror-image drawings through the tegmentum of a skate (*Raja eglanteria*). Adapted from Boord and Northcutt (1982).

in which the connections of these nuclei have been studied most extensively, the substantia nigra has two parts, the **pars compacta** and the **pars reticulata.** The pars reticulata lacks catecholamine-containing cells and is the area where a number of afferent projections terminate. The cells in the pars compacta of the substantia nigra and the ventral tegmental area contain catecholamine (i.e., dopamine) and project heavily to the striatum in the telencephalon. The loss of the cells that produce dopamine in the substantia nigra is the basis for Parkinson's disease.

The striatum, or **striatopallidal complexes** (see Chapters 19 and 24), in the ventrolateral part of the telencephalon, consists of a number of parts. We need to digress momentarily to list them here, since many of them are interconnected with the substantia nigra and ventral tegmental area. The dorsal striato-

pallidal complex contains the **dorsal striatum** and **dorsal pallidum**, and the ventral striatopallidal complex contains the **ventral striatum** and **ventral pallidum** (see Table 19-1). The dorsal striatum is composed of two structures called the caudate nucleus and the putamen, and the dorsal pallidum is a nucleus called the globus pallidus. The major components of the ventral striatum are nucleus accumbens and the olfactory tubercle, while the ventral pallidum consists of a population of scattered neurons in an area called the substantia innominata. The components of the striatopallidal complexes are related not only to the motor system but to the limbic parts of the telencephalon (that function in learning and emotion) as well. The connections of the striatopallidal complexes with the groups of catecholamine-containing neurons in the midbrain also encompass both the motor- and limbic-related forebrain systems.

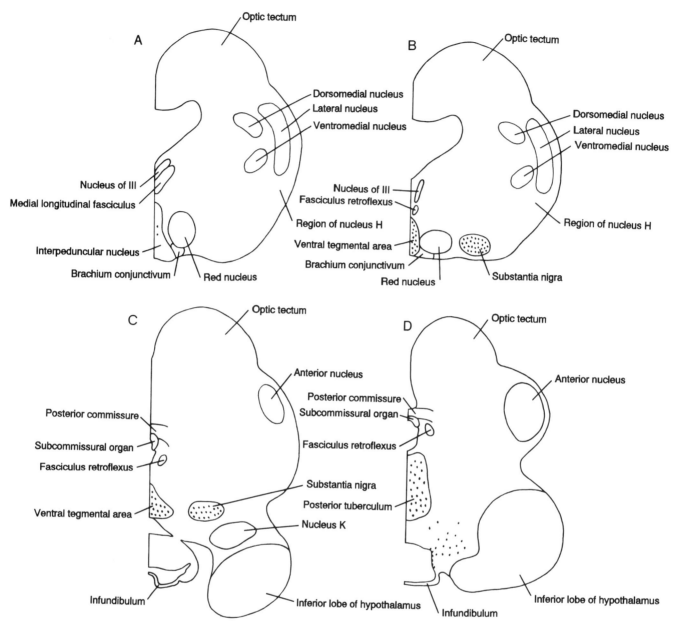

FIGURE 17-9. Drawings of caudal-to-rostral transverse hemisections through the tegmentum of a skate (*Raja radiata*) to show the positions of the substantia nigra, red nucleus, and ventral tegmental area. The levels shown in A and D are approximately the same as those shown in the caudal and rostral levels of *Raja eglanteria* in Figure 17-7, respectively, and the levels of B and C are intermediate. The dorsomedial, ventromedial, lateral, and anterior nuclei are components of the lateral mesencephalic nuclear complex and are shown for orientation. Dopamine immunoreactive cell bodies are indicated by dots. Adapted from Meredith and Smeets (1987).

Some of the connections of the substantia nigra are shown in Figure 17-14. In addition to projecting to the dorsal striatum, the substantia nigra projects to part of the dorsal thalamus, the optic tectum, and the pedunculopontine nucleus of the tegmentum. It receives reciprocal projections from the dorsal striatum and inputs from the globus pallidus, nucleus accumbens, the serotonin-containing cells of the raphe, a motor-related part of the diencephalon (the subthalamus), and the pedunculopontine nucleus.

The ventral tegmental area gives rise to ascending projections to the forebrain, some of which terminate in the dorsal and ventral striatum the ventral pallidum, the hippocampal formation, a telencephalic nucleus associated with the limbic system called the amygdala, and medial parts of isocortex. Within the brainstem, the ventral tegmental area projects to the raphe as well as to a number of other tegmental sites. Its sources of afferent projections include nucleus accumbens and the ventral pallidum, part of isocortex and limbic cortex, the amygdala, and, in the brainstem, the raphe and the interpeduncular nucleus.

The substantia nigra's ascending projection to the caudate-putamen is frequently referred to as the **nigrostriatal**

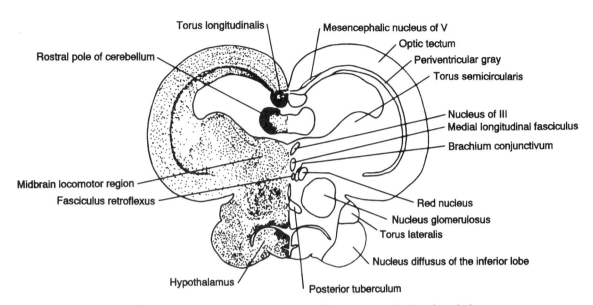

FIGURE 17-10. Drawing of a transverse hemisection with mirror-image drawing through the tegmentum of a teleost fish (*Lepomis gibbosus*). Adapted from Parent et al. (1978).

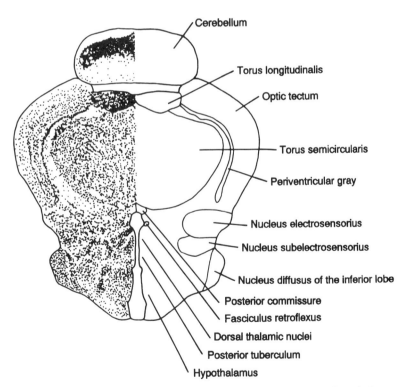

FIGURE 17-11. Drawing of a transverse hemisection with mirror-image drawing through the tegmentum of a weakly electric gymnotiform fish (*Apteronotus leptorhynchus*). Adapted from Maler et al. (1991).

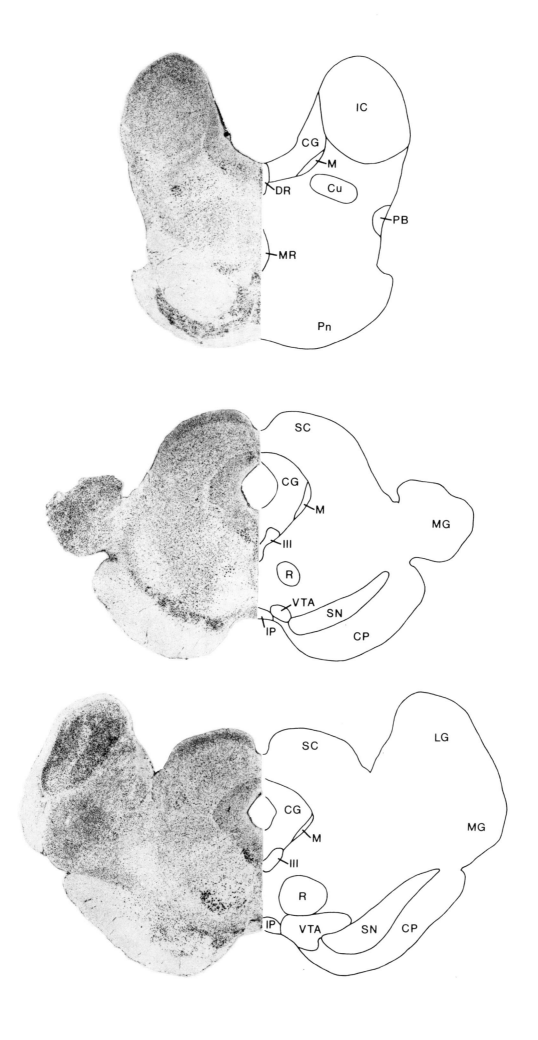

dopamine system. This system has historically been viewed as essentially separate and different from the set of ascending projections of the ventral tegmental area to more medial, limbic related structures in the telencephalon. A new concept of the midbrain dopamine projection systems in mammals has recently emerged, however, that recognizes a more integrated mixture of pathways. A dorsal–ventral division of the combined set of the A9 and A10 cell groups, rather than the medial (A10) versus lateral (A9) nuclear divisions, is more consistent with the pattern of connections. The ventral parts of A9 and A10 project to the dorsal striatum, while the dorsal parts project to both limbic and striatal strucutres. The total set of ascending dopaminergic midbrain projections is referred to as the **meso-telencephalic system,** which is composed of the **mesostriatal** and **mesocorticolimbic systems.**

The substantia nigra, ventral tegmental area, striatum, and nucleus accumbens have all been identified (although sometimes with differing terminology) in diapsid reptiles (Fig. 17-13), birds, and turtles on the basis of similar connections, and/or the identification of similar neurotransmitters. Unfortunately, the substantia nigra (i.e., A9) has been labeled as the pedunculopontine nucleus in some birds. Some differences in connections occur, however, which reflect the separate evolutionary courses of the amniote radiations. We now want to consider two examples of the variation that occurs in this system.

One of these differences occurs in the degree of interconnections between the substantia nigra–dorsal striatal projection system on one hand and the ventral tegmental area–ventral striatal projection system on the other (see Figure 17-15). In all amniotes, the substantia nigra is reciprocally connected with the striatum, and the ventral tegmental area is reciprocally connected with nucleus accumbens. In mammals, the number and magnitude of direct interconnections between the two systems is greater than in either diapsid reptiles or birds. These interconnections allow for interaction between the two telencephalic motor regions, the dorsal and ventral striatum, via relays through the tegmental nuclei.

A second and more complex example of variation has been discovered in the relative topographic arrangements of one of the input systems to the midbrain catecholamine-containing cell groups. In mammals, diapsid reptiles, birds, and turtles, some of the fibers that arise in the striatum and project to the substantia nigra contain a neurotransmitter that is a peptide called substance P. There are also substance P-containing afferent fibers to the ventral tegmental area, which, if they arise in the striatum, would constitute an additional, interconnecting pathway present in all groups of amniotes (not shown in Fig. 17-15). Differences occur in the degree to which the substance P-containing fibers overlap the area of the catecholamine-containing cells in both the substantia nigra and the ventral tegmental area. In lizards, turtles, and rats, there is very little overlap. In pythons, crocodiles, birds, and primates, there is extensive overlap (Fig. 17-16).

From this distribution, we cannot determine which trait (overlap or separation) of the substance P-containing fibers and the catecholamine-containing cells has been evolved multiple times. Outgroup comparisons suggest that a separation was the primitive condition for this system in ancestral vertebrates, as in both a lungfish and a cartilaginous fish, almost no overlap of the afferent fibers and the cells occurs. Overlap has therefore evolved at least three separate times independently: in pythons, in the thecodont ancestors of crocodiles and birds, and in primates. As the repertoires of movements in these groups are very diverse, it is hard to guess, at least for the moment, at the functional significance of greater overlap of substance P-containing fibers and catecholamine-containing cells in the substantia nigra and ventral tegmental area. In humans, the overlap is exceptionally extensive, and it is known that a loss of substance P occurs in cases of Huntington's and Parkinson's diseases.

One additional circuit needs to be discussed, as it is also related to the pathology of Parkinson's disease and is involved with the midbrain locomotor region that we discussed above. In mammals, the pedunculopontine nucleus lies next to the cuneiform nucleus (discussed above in relation to the midbrain locomotor region). The pedunculopontine nucleus has reciprocal connections with the pars compacta of the substantia nigra and receives descending projections from the cerebral cortex and striatum. It is also present in diapsid reptiles, birds, and turtles where it has been called the **intercollicular nucleus.** (The intercollicular nucleus of sharks, discussed above, may or may not be a homologue of this nucleus in amniotes; in sharks, the intercollicular nucleus is in a similar position, neighboring nucleus cuneiformis, but its known connections are different.) Chemicals that pathologically stimulate the neurons in the pedunculopontine nucleus in mammals have been injected into the nucleus, and the result has been a deleterious effect on the neurons in the target of the nucleus, the substantia nigra. The catecholamine-containing nigral cells are destroyed by this treatment, and it is suspected that Parkinson's disease may, at least in some cases, be the result of a similar mechanism.

Evolutionary Perspective

A posterior tuberculum is present in lampreys, all cartilaginous fishes, all ray-finned fishes, lungfishes, and amphibians. The posterior tuberculum is thus a plesiomorphic feature of vertebrate brains. In all species studied, cells in the posterior tuberculum have been found to contain catecholamine.

In contrast, nuclei identified as the ventral tegmental area and substantia nigra are present only in two groups: the sharks, skates, and rays (but not in ratfishes) and the amniotes. The

FIGURE 17-12. Transverse hemisections with mirror-image drawings through the tegmentum of a mammal (*Procyon lotor*). The upper section is most caudal. Abbreviations: CG, central gray; CP, cerebral peduncle; Cu, cuneiform nucleus; DR, dorsal raphe nucleus; IC, inferior colliculus; IP, interpeduncular nucleus; LG, dorsal lateral geniculate nucleus; M, mesencephalic trigeminal nucleus; MG, medial geniculate nucleus; MR, medial raphe nucleus; PB, parabigeminal nucleus; Pn, pontine nuclei; R, red nucleus; SC, superior colliculus; SN, substantia nigra; VTA, ventral tegmental area; III, oculomotor nucleus; Photograph courtesy of Wally Welker.

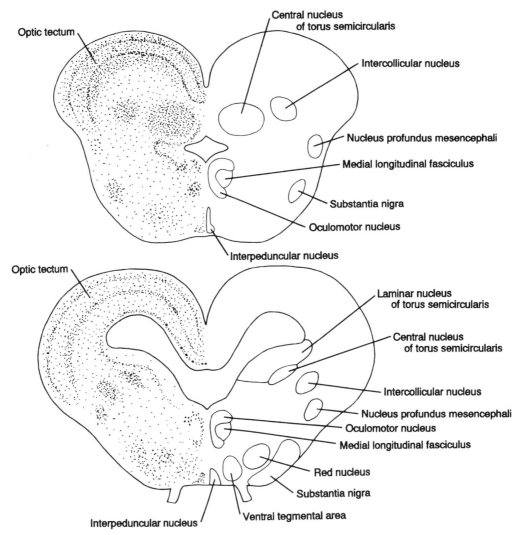

FIGURE 17-13. Drawings of transverse hemisections with mirror-image drawings through the tegmentum of a lizard (*Tupinambis nigropunctatus*). The more caudal section is at the top. Adapted from ten Donkelaar and Nieuwenhuys (1979) with additional data from Parent (1986). Cell bodies of the mesencephalic nucleus of the trigeminal nerve are indicated by the larger dots in the medial part of the optic tectum in the more rostral section. Used with permission of Academic Press Ltd.

absence of these structures in all ray-finned fishes, lungfishes, and amphibians implies that they have been independently evolved in cartilaginous fishes and in amniotes.

Is the posterior tuberculum derived from the hypothalamus and its catecholamine-containing cells homologous to the A11 cell group of amniotes, or could the posterior tuberculum comprise a field homology or a serial homology to the area or areas that give rise to some or all of the A8–A10 mesencephalic cell groups in both cartilaginous fishes and amniotes? In the adults of cartilaginous fishes, the cells in the posterior tuberculum are not in a rostrocaudal continuum with those in the ventral tegmental area and substantia nigra. In adult specimens of amniotes, no posterior tuberculum has been found, although it has recently been identified within the posterior parencephalon (or p2 prosomeric segment) in a segmental analysis of the brain in chick embryos, and further studies may allow for its identification in adult mammals and other nonavian amniotes. Recent evidence on the substantia

nigra in mammals indicates that it does not migrate during development from a more caudal region to its position in the adult, as previously believed, but may arise from the median proliferative zone within its own segmental sector in the mesencephalon, as does the ventral tegmental area. Whether the projection from the posterior tuberculum to the striatum in frogs is homologous (as a connection of serially homologous structures) or homoplaseous to that of the substantia nigra to the striatum in amniotes is unresolved. Few other data on posterior tubercular connections are currently available. Further embryological studies of the derivation of these nuclei in both cartilaginous fishes and amniotes and further studies of the connections of the posterior tuberculum will contribute to the resolution of this problem.

Variation in the pattern of projections to and from the substantia nigra and ventral tegmental area needs to be further examined in a number of species. At least six different neurotransmitters have been identified in this system of projections

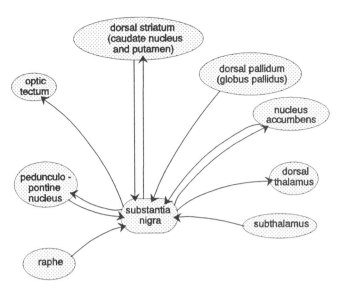

FIGURE 17-14. Some of the connections of the substantia nigra in mammals.

in mammals, and additional studies on the distribution of these neurotransmitters in nonmammalian vertebrates can be used to determine what specific changes and modifications this part of the motor control system has undergone in various radiations over evolution and how these changes are correlated with function.

TORUS LATERALIS

Group I

A cell group that might correspond to the torus lateralis has not been identified in lampreys or sharks. In nonteleost ray-finned fishes, the nucleus of the torus lateralis forms a bulge on the edge of the brain ventral to the optic tectum (Figs. 17-3). In reedfishes, sturgeons, and gars, the torus lateralis is relatively large in size, and it is enormous in the bowfin (Fig.17-4). In reedfishes, the torus lateralis projects to the telencephalon, but other connections are still unknown. A torus lateralis has not been identified in lungfishes or amphibians.

Group II

A torus lateralis has not been identified in hagfishes. The nucleus is present in most teleosts (Fig. 17-10) but is more modestly sized than in nonteleost ray-finned fishes. There is no known homologue of the torus lateralis in amniotes.

Evolutionary Perspective

The torus lateralis may be unique to ray-finned fishes. Its only known efferent connection is to the telencephalon in reedfishes, and its afferent connections are unknown. Additional studies of its connections may indicate that a homologue of this nucleus exists in other taxa, but for the moment, its function and origin remain an enigma.

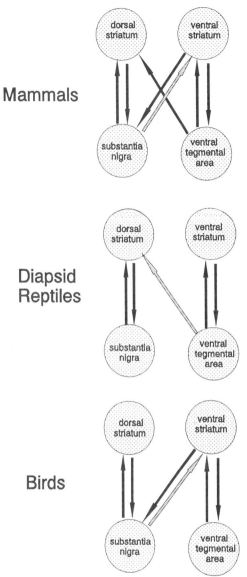

FIGURE 17-15. Interconnections between the substantia nigra–dorsal striatal projection system and the ventral tegmental area–ventral striatal projection system in mammals, diapsid reptiles, and birds. Gray arrows indicate a relatively sparse projection.

TORUS SEMICIRCULARIS

The torus semicircularis is the structure in the midbrain that receives ascending lateral line and auditory fibers. It is variable in both size and complexity and reflects the degree of development of the auditory system and both the mechanosensory and electrosensory components of the lateral line system. The torus semicircularis is an excellent example of how form is correlated with function.

Group I

In lampreys, the torus semicircularis (Fig. 17-1) lies ventral to the optic tectum along the ventral surface of the tectal ventricle. The cells form a lamina along the ventricle. The lateral line

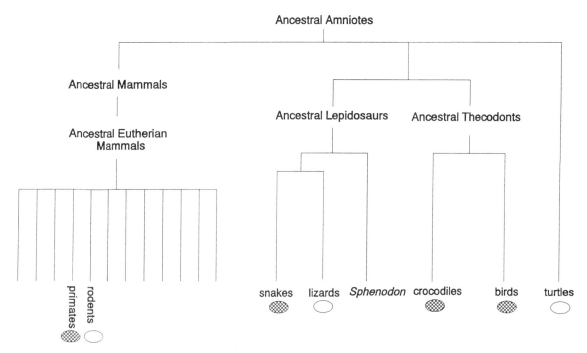

FIGURE 17-16. Distribution of substantial overlap (cross-hatching in ovals) versus little or no overlap (open ovals) of substance P-containing fibers and catecholamine-containing cells in the substantia nigra and ventral tegmental area among amniotes.

system in lampreys is both mechanoreceptive and electroreceptive. This is also the case in squalomorph sharks and ratfishes. In squalomorph sharks (Fig. 17-2), part of the torus semicircularis forms a slight bulge along the lateral surface of the ventricle. The major toral nucleus, the **lateral mesencephalic nucleus,** is partially migrated and lies in a more lateral position. Ascending projections from lateral line nuclei in the medulla project to the lateral mesencephalic nucleus via the lateral line lemniscus. The lateral mesencephalic nucleus is functionally subdivided in that electrosensory lateral line input is distributed to the more dorsolateral part of the nucleus, while mechanosensory lateral line input is distributed to its more ventromedial part. The most ventral part of the lateral mesencephalic nucleus may receive ascending auditory projections via the same lemniscal pathway.

The torus semicircularis forms a prominent bulge on the ventricular surface in nonteleost ray-finned fishes (Fig. 17-3). Ascending octavolateral fibers form the lateral lemniscus and terminate within subdivisions of the torus semicircularis. While reedfishes and sturgeons are electroreceptive, gars and bowfins are not. The electroreceptive system was apparently lost in the common ancestor of gars, bowfins, and the large teleost radiation.

Within the sarcopterygian radiation, the crossopterygian fish *Latimeria* and lungfishes have both mechanoreceptive and electroreceptive lateral line systems. All amphibians have a lateral line system in their aquatic larval stages, and both mechanoreceptive and electroreceptive components are present in urodele and larval apodan (gymnophionan) amphibians. Only a mechanoreceptive lateral line system is present in larval anurans. During metamorphosis, however, the lateral line systems

are lost in most anuran (including the frog *Rana*) and apodan amphibians. The lateral line organs are not functional in air. A few species, including the clawed toad *Xenopus*, retain the lateral line system, and this retention is correlated with an aquatic lifestyle and/or aquatic breeding.

In amphibians with the lateral line system, the information (electroreceptive and/or mechanoreceptive) is relayed to the torus semicircularis along with auditory information via the lateral lemniscus. The torus semicircularis in frogs is quite large and is composed of several subdivisions: the **principal, laminar,** and **magnocellular nuclei** (Fig. 17-6). In the frog *Xenopus*, which retains the mechanoreceptive lateral line system, the large principal nucleus receives ascending auditory projections, while the magnocellular nucleus receives lateral line projections. A small area is also present between these nuclei, which receives ascending somatosensory information for the integration of touch stimuli. In the frog *Rana*, which has no lateral line system in the adult, the auditory input is also to the principal nucleus, but the ascending somatosensory projection extends throughout the magnocellular nucleus. In all frogs, the laminar nucleus receives projections from the magnocellular and principal nuclei, and cells in the laminar nucleus also have dendrites that extend into these two nuclei. The laminar nucleus is the main source of projections from the torus semicircularis to the optic tectum and to the auditory area of the dorsal thalamus, the central thalamic nucleus.

With the emergence of vertebrates onto land, the lateral line system could no longer be used to detect predators or prey or to communicate. An increased use was made of both vision and sound. The only amphibians that use sound for communication are frogs, and the development of the auditory parts of the brain in frogs reflects the importance of the system.

Sound communication has been extensively studied in frogs. This system is useful as a model for the processing of sound and its content by the central nervous system, as there are a set of rather constant and simple signals used for specific situations, most relating to territory and reproduction. Males produce an "advertisement" call when ready to breed, and these calls are answered with either "phonotaxic" calls from receptive females or "antiphonal" calls from other males. Males produce "courtship" calls when approached by females. "Release" calls are made by females improperly clasped during amplexus (mating) or by clasped males, while "aggressive" calls are used by males confronting each other. "Distress" calls are made when a frog is seized by a predator. Each call can be characterized in terms of wave form and pulse rate, and each set of calls is species specific. Studies of central processing of the stimuli in the brainstem nuclei and midbrain, as well as information on individual mating selection, speciation (behavioral mechanisms of isolation), and variation of behavior with habitat have all been done on this system of sound communication. The releasing response, for example, has been found to be mediated by cells in the region of the anterodorsal tegmental nucleus, which lies immediately ventral to the torus semicircularis.

Group II

A torus semicircularis has not been identified in hagfishes. The lateral line system in hagfishes is solely mechanoreceptive and confined to the head. Studies of ascending auditory and lateral line projections which could reveal the location of the torus semicircularis have not yet been done.

While one nucleus, the lateral mesencephalic nucleus, is present lateral to the ventricular bulge of the torus itself in squalomorph sharks, the galeomorph sharks, skates, and rays have several distinct nuclei in this region (Fig. 17-8). These nuclei are collectively called the **lateral mesencephalic nuclear complex (LMN).** Ascending fibers from the electrosensory lateral line nuclei project via the lateral lemniscus to the **lateral nucleus** of the LMN. Mechanosensory lateral line projections terminate in the **dorsomedial nucleus,** and auditory projections terminate in the **ventromedial nucleus** of the complex. A more rostral nucleus in the complex, the **anterior nucleus,** does not receive direct, ascending octavolateral projections but instead receives an electrosensory input from the lateral nucleus.

In most teleost fishes, the torus semicircularis (Fig. 17-10) is well developed, but the nuclear cell masses that receive ascending octavolateral input are not migrated laterally away from the bulge in the ventricular surface as they are in cartilaginous fishes. The torus semicircularis projects to a nucleus in the dorsal thalamus, the central posterior nucleus, which, in turn, projects to the telencephalon. The central posterior nucleus also projects back to the torus semicircularis. The torus semicircularis has reciprocal connections with several other structures (the medullary reticular formation, the contralateral torus semicircularis, and the anterior tuberal nucleus in the hypothalamus), and it receives a direct projection from an area in the caudal telencephalon. Additionally, the torus semicircularis is reciprocally connected with the optic tectum, and these projections are topographically mapped so that the visual map

of space from the tectum is in register with the auditory and lateral line maps of space in the torus semicircularis and vice versa (see Chapter 18).

While electroreception was lost in the ancestral stock of gars, bowfins, and teleosts, an electroreceptive lateral line system was apparently subsequently reevolved independently in several groups of teleosts. This has occurred separately in two groups of osteoglossomorphs (Chapter 4)—notopterids and mormyrids—and in two groups of ostariophysans (Euteleostei)—gymnotids and silurids.

In some of these teleosts that are electroreceptive, the torus semicircularis is massive. A drawing through this region in a weakly electric, gymnotid fish (*Apteronotus*) is shown in Figure 17-11. In these fishes, the dorsal part of the torus semicircularis receives electrosensory input, and the ventral part receives auditory and mechanoreceptive lateral line inputs. While the torus semicircularis in *Apteronotus* has a number of connections similar to those in other teleosts, it also projects to a large electrosensory nucleus in the caudal diencephalon called **nucleus electrosensorius.** Nucleus electrosensorius, in turn, projects to another nucleus called the **prepacemaker nucleus,** which modulates the discharges of the electric organ for signals associated with courtship, aggression, and other behaviors. The location of nucleus electrosensorius is roughly similar to the location of the electrosensory-receptive, anterior nucleus of the lateral mesencephalic nuclear complex (LMN) in cartilaginous fishes with elaborated brains; thus, these structures may possibly be an example of convergence.

The lateral line system is absent in all amniote vertebrates except monotremes, having been lost in ancestral amniotes with the transition from water to land. Since the torus semicircularis is solely auditory in most amniotes, it is somewhat reduced in size and complexity relative to its structure in some anamniotes. In mammals, it is called the **inferior colliculus** (Fig. 17-12). In diapsid reptiles (Fig. 17-13) turtles, and birds it is called the torus semicircularis. In birds, two parts, the **dorsal part of the lateral mesencephalic nucleus** and the **intercollicular nucleus,** have long been recognized. Three subdivisions for it have more recently been proposed, based on neurochemical findings: an **intercollicular area, toral nucleus, and preisthmic superficial area.**

In all amniotes, the torus semicircularis (inferior colliculus) receives ascending auditory projections via the lateral lemniscus. Its major efferent projection is to an auditory nucleus in the dorsal thalamus, which, in turn, relays the information to the telencephalon (auditory cortex in mammals and a part of the dorsal ventricular ridge in diapsid reptiles, birds, and turtles).

Evolutionary Perspective

A torus semicircularis has been identified in all radiations of vertebrates with the exception of hagfishes, where its presence has not yet been investigated. This structure is clearly one that was evolved very early in the vertebrate radiation. Among anamniotes, the size and complexity of the torus semicircularis are directly correlated with the degree of elaboration of the lateral line system, particularly its electrosensory component. The electrosensory lateral line system is thought to have been evolved multiple times independently, and thus, the correlated augmentation of the torus semicircularis in these cases is an example of convergence.

FOR FURTHER READING

Bernau, N. A., Puzdrowski, R. L., and Leonard, R. B. (1991) Identification of the midbrain locomotor region and its relation to descending locomotor pathways in the Atlantic stingray, *Dasyatis sabina*. *Brain Research*, 557, 83–94.

Echteler, S. M. (1984) Connections of the auditory midbrain in a teleost fish, *Cyprinus carpio*. *Journal of Comparative Neurology*, 230, 536–551.

Feng, A. S. and Lin, W. (1991) Differential innervation patterns of three divisions of frog auditory midbrain (torus semicircularis). *Journal of Comparative Neurology*, 306, 613–630.

Meredith, G. E. and Smeets, W. J. A. J. (1987) Immunocytochemical analysis of the dopamine system in the forebrain and midbrain of *Raja radiata*: evidence for a substantia nigra and ventral tegmental area in cartilaginous fish. *Journal of Comparative Neurology*, 265, 530–548.

Smeets, W. J. A. J. and Reiner, A. (eds.) (1994) *Phylogeny and Development of Catecholamine Systems in the CNS of Vertebrates*. Cambridge, England: Cambridge University Press.

ADDITIONAL REFERENCES

Baumgarten, H. G. (1972) Biogenic monoamines in the cyclostome and lower vertebrate brain. *Progress in Histochemistry and Cytochemistry*, 4, 1–90.

Boord, R. L. and Northcutt, R. G. (1982) Ascending lateral line pathways to the midbrain of the clearnose skate, *Raja eglanteria*. *Journal of Comparative Neurology*, 207, 274–282.

Boord, R. L. and Northcutt, R. G. (1988) Medullary and mesencephalic pathways and connections of lateral line neurons of the spiny dogfish *Squalus acanthias*. *Brain, Behavior and Evolution*, 32, 76–88.

Brauth, S. E., Reiner, A., Kitt, C. A., and Karten, H. J. (1983) The substance P-containing striatotegmental path in reptiles: an immunohistochemical study. *Journal of Comparative Neurology*, 219, 305–327.

Bullock, T. H. and Heiligenberg, W. (eds.) (1986) *Electroreception*. New York: Wiley.

Capranica, R. R. (1965) *The Evoked Vocal Response of the Bullfrog. A Study of Communication By Sound*. Cambridge, MA: MIT Press.

Carr, J. A., Norris, D. O., and Samora, A. (1991) Organization of tyrosine hydroxylase-immunoreactive neurons in the di- and mesencephalon of the American bullfrog (*Rana catesbeiana*) during metamorphosis. *Cell Tissue Research*, 263, 155–163.

Corio, M. and Doerr-Schott, J. (1988) The monoaminergic system in the diencephalon of the newt tadpole, *Triturus alpestris* (Mert). A histofluorescence study. *Journal für Hirnforschung*, 29, 377–384.

Corwin, J. T. and Northcutt, R. G. (1982) Auditory centers in the elasmobranch brain stem: Deoxyglucose autoradiography and evoked potential recording. *Brain Research*, 236, 261–273.

Dahlström, A. and Fuxe, K. (1964) Evidence for the existence of monoamine-containing neurons in the central nervous system. I. Demonstration of monoamines in the cell bodies of brain stem neurons. *Acta Physiologica Scandinavica*, 62, Suppl. 232, 1–155.

Domesick, V. B. (1988) Neuroanatomical organization of dopamine neurons in the ventral tegmental area. In P. W. Kalivas and C. B. Nemeroff (eds.), *The Mesocorticolimbic Dopamine System. Annals of the New York Academy of Sciences*, 537, 10–26.

Ebbesson, S. O. E. and Campbell, C. B. G. (1973) On the organization of cerebellar efferent pathways in the nurse shark (*Ginglymostoma cirratum*). *Journal of Comparative Neurology*, 152, 233–254.

Ekström, P., Honkanen, T., and Steinbusch, H. W. M. (1990) Distribution of dopamine-immunoreactive neuronal perikarya and fibers in the brain of a teleost, *Gasterosteus aculeatus* L. Comparison with tyrosine hydroxylase- and dopamine-β-hydroxylase-immunoreactive neurons. *Journal of Chemical Neuroanatomy*, 3, 233–260.

Fallon, J. H. (1988) Topographic organization of ascending dopaminergic projections. In P. W. Kalivas and C. B. Nemeroff (eds.), *The Mesocorticolimbic Dopamine System, Annals of the New York Academy of Sciences*, 537, 1–9.

Frankenhuis-van den Heuvel, T. H. M. and Nieuwenhuys, R. (1984) Distribution of serotonin-immunoreactivity in the diencephalon and mesencephalon of the trout, *Salmo gairdneri*. *Anatomy and Embryology*, 169, 193–204.

Fritzsch, B. (1988) Diversity and regression in the amphibian lateral line and electrosensory system. In S. Coombs, P. Görner, and H. Münz (eds.), *The Mechanosensory Lateral Line*. New York: Springer–Verlag, pp. 99–114.

Fritzsch, B., Sonntag, R., Dubuc, R., Ohta, Y., and Grillner, S. (1990) Organization of the six motor nuclei innervating the ocular muscles in lamprey. *Journal of Comparative Neurology*, 294, 491–506.

González, A., Marín, O., Tuinhof, R., and Smeets, W. J. A. J. (1994) Ontogeny of catecholamine systems in the central nervous system of anuran amphibians: an immunohistochemical study with antibodies against tyrosine hydroxylase and dopamine. *Journal of Comparative Neurology*, 346, 63–79.

González, A., Russchen, F. T., and Lohman, A. H. M. (1990) Afferent connections of the striatum and the nucleus accumbens in the lizard *Gekko gecko*. *Brain, Behavior and Evolution*, 36, 39–58.

Graybiel, A. M. (1978) A satellite system of the superior colliculus: the parabigeminal nucleus and its projections to the superficial collicular layers. *Brain Research*, 145, 365–374.

Hoogland, P. V. (1977) Efferent connections of the striatum in *Tupinambis nigropunctatus*. *Journal of Morphology*, 152, 229–246.

Hornby, P. J. and Piekut, D. T. (1988) Immunoreactive dopamine β-hydroxylase in neuronal groups in the goldfish brain. *Brain, Behavior and Evolution*, 32, 252–256.

Hornby, P. J. and Piekut, D. T. (1990) Distribution of catecholamine-synthesizing enzymes in goldfish brains: presumptive dopamine and norepinephrine neuronal organization. *Brain, Behavior and Evolution*, 35, 49–64.

Kalivas, P. W., Churchill, L., and Klitenick, M. (1993) GABA and enkephalin projection from the nucleus accumbens and ventral pallidum to the ventral tegmental area. *Neuroscience*, 57, 1047–1060.

Karten, H. J. and Hodos, W. (1967) *A Stereotaxic Atlas Of the Brain Of the Pigeon (Columba livia)*. Baltimore, MD: The Johns Hopkins Press.

Kennedy, M. C. and Rubinson, K. (1977) Retinal projections in larval, transforming and adult sea lamprey, *Petromyzon marinus*. *Journal of Comparative Neurology*, 171, 465–479.

Kitt, C. A. and Brauth, S. E. (1981) Projections of the paleostriatum upon the midbrain tegmentum in the pigeon. *Neuroscience*, 6, 1551–1566.

Kitt, C. A. and Brauth, S. E. (1986) Telencephalic projections from midbrain and isthmal cell groups in the pigeon. II. The nigral complex. *Journal of Comparative Neurology*, 247, 92–110.

Kokoros, J. J. and Northcutt, R. G. (1977) Telencephalic efferents of the tiger salamander *Ambystoma tigrinum tigrinum* (Green). *Journal of Comparative Neurology*, 173, 613–628.

Larson-Prior, L. J. and Cruce, W. L. R. (1992) The red nucleus and mesencephalic tegmentum in a ranid amphibian: a cytoarchitectonic and HRP connectional study. *Brain, Behavior and Evolution*, 40, 273–287.

Maler, L., Sas, E., Johnston, S., and Ellis, W. (1991) An atlas of the brain of the electric fish *Apteronotus leptorhynchus*. *Journal of Chemical Neuroanatomy*, 4, 1–38.

McCormick, C. A. (1988) Evolution of auditory pathways in the amphibia. In B. Fritzsch, M. J. Ryan, W. Wilczynski, T. E. Hetherington, and W. Walkowiak (eds.), *The Evolution Of the Amphibian Auditory System*. New York: Wiley, pp. 587–612.

Medina, L., Puelles, L. and Smeets, W. J. A. J. (1994). Development of catecholamine systems in the brain of the lizard *Gallotia galloti*. *Journal of Comparative Neurology*, 350, 41–62.

Meek, J., Joosten, H. W. J., and Steinbusch, H. W. M. (1989) Distribution of dopamine immunoreactivity in the brain of the mormyrid teleost *Gnathonemus petersii*. *Journal of Comparative Neurology*, 281, 362–383.

Montgomery, N. M. (1989) Somatomotor connectivity in the midbrain of *Rana pipiens*. *Brain, Behavior and Evolution*, 34, 96–109.

Narins, P. M. and Capranica, R. R. (1980) Neural adaptations for processing the two-note call of the Puerto Rican treefrog, *Eleutherodactylus coqui*. *Brain, Behavior and Evolution*, 17, 48–66.

Nieuwenhuys, R. (1972) Topological analysis of the brain stem of the lamprey *Lampetra fluviatilis*. *Journal of Comparative Neurology*, 145, 165–178.

Nieuwenhuys, R. and Pouwels, E. (1983) The brain stem of actinopterygian fishes. In R. G. Northcutt and R. E. Davis (eds.), *Fish Neurobiology, Vol. 1: Brain Stem and Sense Organs*. Ann Arbor, MI: University of Michigan Press, pp. 25–87.

Northcutt, R. G. (1977) Retinofugal projections in the lepidosirenid lungfishes. *Journal of Comparative Neurology*, 174, 553–574.

Northcutt, R. G. (1978) Brain organization in the cartilaginous fishes. In E. S. Hodgson and R. F. Mathewson (eds.), *Sensory Biology of Sharks, Skates, and Rays*. Arlington, VA: Office of Naval Research, pp. 117–193.

Northcutt, R. G. (1979) Experimental determination of the primary trigeminal projections in lampreys. *Brain Research*, 163, 323–327.

Northcutt, R. G. (1980) Retinal projections in the Australian lungfish. *Brain Research*, 185, 85–90.

Northcutt, R. G. (1983) Brain stem neurons that project to the spinal cord in garpike (Holostei). *Anatomical Record*, 205, 144A.

Northcutt, R. G. and Butler, A. B. (1980) Projections of the optic tectum in the longnose gar, *Lepisosteus osseus*. *Brain Research*, 190, 333–346.

Northcutt, R. G., Reiner, A., and Karten, H. J. (1988) Immunohistochemical study of the telencephalon of the spiny dogfish, *Squalus acanthias*. *Journal of Comparative Neurology*, 277, 250–267.

Panneton, W. M. and Watson, B. J. (1991) Stereotaxic atlas of the brainstem of the muskrat, *Ondatra zibethicus*. *Brain Research Bulletin*, 26, 479–509.

Parent, A. (1986) *Comparative Neurobiology of the Basal Ganglia*. New York: Wiley.

Parent, A., Dube, L., Braford, M. R., Jr., and Northcutt, R. G. (1978) The organization of monoamine-containing neurons in the brain of the sunfish (*Lepomis gibbosus*) as revealed by fluorescence microscopy. *Journal of Comparative Neurology*, 182, 495–516.

Parent, A. and Northcutt, R. G. (1982) The monoamine-containing neurons in the brain of the garfish, *Lepisosteus osseus*. *Brain Research Bulletin*, 9, 189–204.

Puelles, L. and Medina, L., Development of neurons expressing tyrosine hydroxylase and dopamine in the chicken brain: a comparative segmental analysis. In W. J. A. J. Smeets and A. Reiner (eds.), *Phylogeny and Development of Catecholamine Systems in the CNS of Vertebrates*. Cambridge: Cambridge University Press, pp. 381–404.

Puelles, L., Robles, C., Martínez-de-la-Torre, M. and Martínez, S. (1994) New subdivision schema for the avian torus semicircularis: neurochemical maps in the chick. *Journal of Comparative Neurology*, 340, 98–125.

Rand, A. S. (1988) An overview of anuran acoustic communication. In B. Fritzsch, M. J. Ryan, W. Wilczynski, T. E. Hetherington, and W. Walkowiak (eds.), *The Evolution Of the Amphibian Auditory System*. New York: Wiley, pp. 415–431.

Reiner, A. and Northcutt, R. G. (1987) An immunohistochemical study of the telencephalon of the African lungfish, *Protopterus annectens*. *Journal of Comparative Neurology*, 256, 463–481.

Roberts, B. L., Meredith, G. E., and Maslam, S. (1989) Immunocytochemical analysis of the dopamine system in the brain and spinal cord of the European eel, *Anguilla anguilla*. *Anatomy and Embryology*, 180, 401–412.

Ronan, M. (1989) Origins of descending spinal projections in petromyzontid and myxinoid agnathans. *Journal of Comparative Neurology*, 281, 54–68.

Ronan, M. and Northcutt, R. G. (1985) The origins of descending spinal projections in lepidosirenid lungfishes. *Journal of Comparative Neurology*, 241, 435–444.

Ronan, M. and Northcutt, R. G. (1990) Projections ascending from the spinal cord to the brain in petromyzontid and myxinoid agnathans. *Journal of Comparative Neurology*, 291, 491–508.

Schmidt, R. S. (1988) Mating call phonotaxis in female American toads: lesions of central auditory system. *Brain, Behavior and Evolution*, 32, 119–128.

Schmidt, R. S. (1990) Releasing (unclasping) in male American toads: a neural substrate in the lateral subtoral tegmentum. *Brain, Behavior and Evolution*, 36, 307–314.

Smeets, W. J. A. J. (1991) Comparative aspects of the distribution of substance P and dopamine immunoreactivity in the substantia nigra of amniotes. *Brain, Behavior and Evolution*, 37, 179–188.

Smeets, W. J. A. J., Nieuwenhuys, R., and Roberts, B. L. (1983) *The Central Nervous System of Cartilaginous Fishes*. Berlin: Springer-Verlag.

Smeets, W. J. A. J. and Reiner, A., Catecholamines in the CNS of vertebrates: current concepts of evolution and functional significance. In W. J. A. J. Smeets and A. Reiner (eds.), *Phylogeny and Development of Catecholamine Systems in the CNS of Vertebrates*. Cambridge, England: Cambridge University Press, pp. 463–481.

Stuesse, S. L. and Cruce, W. L. R. (1991) Immunohistochemical localization of serotoninergic, enkephalinergic, and catecholaminergic cells in the brainstem and diencephalon of a cartilaginous fish, *Hydrolagus colliei*. *Journal of Comparative Neurology*, 309, 535–548.

Stuesse, S. L., Cruce, W. L. R., and Northcutt, R. G. (1991) Localization of serotonin, tyrosine hydroxylase, and leu-enkephalin immunoreactive cells in the brainstem of the horn shark, *Heterodontus francisci*. *Journal of Comparative Neurology*, 308, 277–292.

ten Donkelaar, H. J. (1976) Descending pathways from the brain stem to the spinal cord in some reptiles. I. Origin. *Journal of Comparative Neurology*, 167, 421–442.

ten Donkelaar, H. J. and de Boer-van Huizen, R. (1981) Basal ganglia projections to the brain stem in the lizard *Varanus exanthematicus* as demonstrated by retrograde transport of horseradish peroxidase. *Neuroscience*, 6: 1567–1590.

ten Donkelaar, H. J., de Boer-van Huizen, R., Schouten, F. T. M., and Eggen, S. J. H. (1981) Cells of origin of descending pathways to the spinal cord in the clawed toad (*Xenopus laevis*). *Neuroscience*, 6: 2297–2312.

ten Donkelaar, H. J., Kusuma, A., and de Boer-van Huizen, R. (1980) Cells of origin of pathways descending to the spinal cord in some quadrupedal reptiles. *Journal of Comparative Neurology*, 192, 827–851.

ten Donkelaar, H. J. and Nieuwenhuys, R. (1979) The brainstem. In C. Gans, R. G. Northcutt, and P. Ulinski (eds.), *Biology of the Reptilia, Vol. 10*. New York: Academic, pp. 133–200.

Volman, S. F. and Konishi, M. (1990) Comparative physiology of sound localization in four species of owls. *Brain, Behavior and Evolution*, 36, 196–215.

Wenstrup, J. J., Larue, D. T., and Winer, J. A. (1994) Projections of physiologically defined subdivisions of the inferior colliculus in the mustached bat: targets in the medial geniculate body and extrathalamic nuclei. *Journal of Comparative Neurology*, 346, 207–236.

Wicht, H. and Northcutt, R. G. (1990) Retinofugal and retinopetal projections in the Pacific hagfish, *Eptatretus stouti* (Myxinoidea). *Brain, Behavior and Evolution*, 36, 315–328.

Wilczynski, W. and Northcutt, R. G. (1977) Afferents to the optic tectum of the leopard frog: an HRP study. *Journal of Comparative Neurology*, 173, 219–230.

Wilczynski, W. and Northcutt, R. G. (1983) Connections of the bullfrog striatum: afferent projections. *Journal of Comparative Neurology*, 214, 321–332.

Wilczynski, W. and Northcutt, R. G. (1983) Connections of the bullfrog striatum: efferent projections. *Journal of Comparative Neurology*, 214, 333–343.

Wild, J. M., Cabot, J. B., Cohen, D. H., and Karten, H. J. (1979) Origin, course and terminations of the rubrospinal tract in the pigeon (*Columba livia*). *Journal of Comparative Neurology*, 187, 639–654.

Will, U. (1988) Organization and projections of the area octavolateralis in amphibians. In B. Fritzsch, M. J. Ryan, W. Wilczynski, T. E. Hetherington, and W. Walkowiak (eds.), *The Evolution of the Amphibian Auditory System*. New York: John Wiley & Sons, pp. 185–208.

Will, U. (1989) Central mechanosensory lateral line system in amphibians. In S. Coombs, P. Görner, and H. Münz (eds.), *The Mechanosensory Lateral Line*. New York: Springer-Verlag, pp. 365–386.

Wullimann, M. F. and Northcutt, R. G. (1988) Connections of the corpus cerebelli in the green sunfish and the common goldfish: a comparison of perciform and cyprinid teleosts. *Brain, Behavior and Evolution*, 32, 293–316.

Wurtz, R. H. and Albano, J. E. (1980) Visual-motor function of the primate superior colliculus. *Annual Review of Neuroscience*, 3, 189–226.

Zook, J. M., Winer, J. A., Pollak, G. D., and Bodenhamer, R. D. (1985) Topology of the central nucleus of the mustache bat's inferior colliculus: correlation of single unit properties and neuronal architecture. *Journal of Comparative Neurology*, 231, 530–546.

18

Optic Tectum

INTRODUCTION

The roof of the midbrain is known as the tectum. In most vertebrates, particularly those with a well-developed visual system, the tectum is dominated by a retinorecipient region called the **optic tectum** (Fig. 18-1) or, as it is often called in mammals, the **superior colliculus** (Fig. 18-2). Although its major input is from the retina, the optic tectum is more than just a visual processing structure. The three major hallmarks of its organization are (1) it is multisensory (visual, auditory, somatosensory, and where present, infrared-sensory and electrosensory), (2) the inputs from different systems are organized as "maps" of the external world or the body, and (3) these maps are in register with each other, for example, a given location on the map of external visual space corresponds to the same location in external auditory space. One of the major functions of the optic tectum is to localize a stimulus in space and to cause the animal to orient to the stimulus by moving its neck and/or its eyes.

OVERVIEW OF TECTAL ORGANIZATION

The cell bodies of neurons within the optic tectum are organized in a layered pattern from the outer surface to the deep, periventricular area. Although the terminology used varies according to taxonomic groups and authors, the optic tectum can, in most cases, be divided into three major zones, or groups of layers: **superficial, central,** and **periventricular.** Figure 18-3 illustrates these three zones in a ray-finned fish. A fourth zone, called the **deep white zone,** which is composed primarily of fibers, can be distinguished in some cases, either immediately superficial or deep to the periventricular gray zone.

The morphology of tectal neurons has been studied in a number of different animals using the Golgi technique. In this histological process, for an unknown but fortuitous reason, only a few of the many neurons become impregnated with silver, so that the entire morphology of the neuron and its processes can be seen and is unobscured by the surrounding neurons. This method is particularly well suited for the study of laminated structures, such as the optic tectum.

Golgi studies have revealed a population of tectal neurons that have radially (i.e., centrifugally) oriented dendrites. These neurons are called **piriform cells** in reference to their pear-shaped cell bodies. The dendrites of the piriform cells often extend through most of the tectal layers. Each population of incoming axons tends to be confined to a particular, tangentially oriented band, or sublayer, within the optic tectum. This physical arrangement allows for incoming information from different systems to be segregated on the dendritic shafts of individual neurons (Fig. 18-4). The very limited degree of branching of the dendrites of the radially oriented, piriform neurons also allows for a precise, point-to-point pattern of termination for each incoming system.

These morphological features provide for a high degree of precision of spatial mapping in the optic tectum, as well as permitting the different sensory systems to be spatially in register. The mapping of sensory information in the optic tectum is in topologic, if not topographic order, due to transformations of the information along the visual pathway through the layers of the retina and the optic nerves and tracts. The term topographic is that which is used in the literature in sensory system analyses, however, and will likewise be used here. An example of this topography is shown in Figure 18-5; successive points along the edge of a door in your visual field, which fall on the retina, will be kept in topographic order through the visual pathway to your optic tectum. A similar illustration of visual field mapping onto the optic tectum of a fish is shown in Figure 18-6.

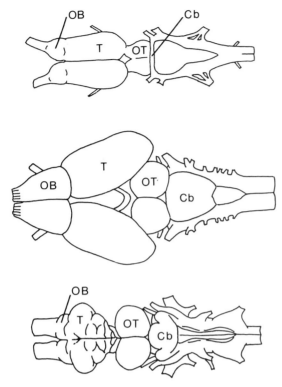

FIGURE 18-1. Dorsal views of the brains of a mudpuppy, *Necturus maculosus* (top), a turtle, *Chrysemys picta* (middle), and a ray-finned fish, *Amia calva* (bottom), showing the position of the optic tectum (OT) in relation to the olfactory bulb (OB) and telencephalon (T) rostrally and the cerebellum (Cb) caudally. Mudpuppy and turtle drawings adapted from Northcutt (1979b) and *Amia* drawing adapted from Butler and Northcutt (1992). Used with permission of The University of Chicago Press and S. Karger AG, respectively.

Visual information, carried by retinal cell axons, terminates primarily in the superficial zone of the optic tectum. Somatosensory information, relayed by nuclei in the hindbrain, terminates in the deeper half of the central tectal zone. Given points in space are mapped in register in the optic tectum for the two systems. For example, touch from the face is mapped in register with the superior and nasal visual fields, corresponding to what would be seen in looking up and in looking towards the nose; touch from the body is similarly mapped in register with what would be seen in looking to the side and down. In those species of snakes that have an infrared (IR)-detecting system (see Chapters 2 and 11), an IR map of space is also present in the optic tectum. This map is correlated with the visual map, particularly in the most rostral part of the optic tectum, which maps the area directly in front of the snake. Since this frontal field is important for striking prey, a greater area of the optic tectum is devoted to this part of the field for IR input than for visual input.

OVERVIEW OF TECTAL CONNECTIONS

The major source of afferents to the optic tectum is the retina. The optic tectum is also the target of descending projections from the forebrain and a variety of ascending projections,

such as those originating in the diencephalon, especially from nuclei in the thalamus and pretectum. In some vertebrates, direct projections from the telencephalon reach the optic tectum, but in others, telencephalic input is relayed to the optic tectum through nuclei in the diencephalon. Reciprocal commissural input to the optic tectum arises in the contralateral optic tectum. Nucleus isthmi (see Chapter 16) also has reciprocal connections with the optic tectum. Much of the tectal input and output is topographically organized. Ascending projections from the spinal cord and a number of brainstem regions provide topographic somatosensory and motor feedback information to the optic tectum.

The optic tectum sends ascending fibers to nuclei in the diencephalon. This projection is bilateral, with the greater number of fibers projecting to the ipsilateral diencephalon. As will be discussed in detail in Chapter 22, the major ascending pathway is to a nucleus in the dorsal thalamus that does not receive retinal input but does relay projections to a part of the telencephalon. The optic tectum also projects more sparsely to a dorsal thalamic nucleus that receives a substantial retinal projection and, in turn, projects to another part of the telencephalon. Thus, two major ascending pathways for visual information to the telencephalon are present.

Descending projections of the optic tectum are organized so that motor responses to sensory stimuli are appropriately oriented in space. Most descending tectal projections are to ipsilateral sites, although a minority of fibers cross to the contralateral brainstem. The three major tectal projection sites are regions of the midbrain tegmentum that are associated with motor functions, the reticular formation, and the rostral spinal cord. The optic tectum initiates motor commands that control sequences of movements directed towards a specific location in space. These commands reach the spinal cord through tegmental and reticular pathways, as well as via direct tectospinal pathways.

THE OPTIC TECTUM IN GROUP I VERTEBRATES

Lampreys

In lampreys, as in other vertebrates with laminar brains, the majority of the cell bodies of tectal neurons lie in the periventricular gray zone (Fig. 18-7). A deep white zone (not labeled) lies just superficial to the periventricular gray zone, and some neuron cell bodies are scattered within the central and superficial zones.

Piriform cells with radially oriented dendrites [Fig. 18-8a and b] are present in the optic tectum, as are neurons with more horizontally oriented dendrites [Fig. 18-8c and d]. Cells with more complex patterns of dendritic branching are also present [see Figure 18-8e].

The retina projects to the optic tectum via the optic tract, which enters the optic tectum from a superficial position, and the individual retinal ganglion cell axons terminate in the superficial and central tectal zones (Fig. 18-9). The retinal terminations are heavier in the superficial zone than in the central zone. Piriform cells in the periventricular gray zone, such as that in Figure 18-8b, receive a retinal input on the more distal portion of their dendrites.

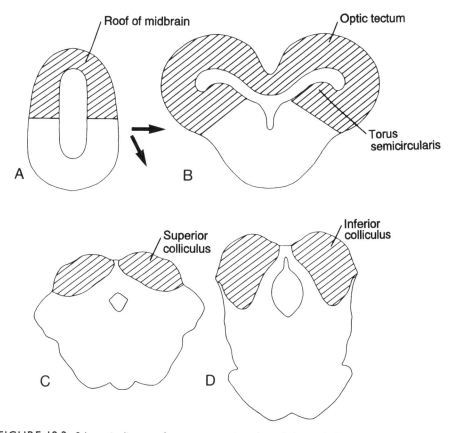

FIGURE 18-2. Schematic diagram of transverse sections through the roof of the midbrain in a generalized vertebrate embryo (A) that gives rise in nonmammalian vertebrates (B) to the optic tectum dorsally and the torus semicircularis ventrally and that gives rise in mammals to the superior colliculus rostrally (C) and the inferior colliculus caudally (D).

A sparse projection from the spinal cord also reaches the optic tectum. These fibers terminate along the ventral border of the optic tectum (Fig. 18-9). Presumably, the optic tectum in lampreys gives rise to ascending and descending projections, but these pathways remain to be studied.

Squalomorph Sharks and Ratfishes

In the optic tectum of squalomorph sharks, such as *Squalus*, neuronal cell bodies are located predominantly in the periventricular gray zone (Fig. 18-10), although limited migration into the central zone occurs. In squalomorph sharks, most of the periventricular cells are piriform in shape and have radially oriented dendrites that extend into the central and superficial zones. In ratfishes, approximately 90% of the cell bodies of tectal neurons are confined to the periventricular gray zone. The central zone consists almost exclusively of heavily myelinated fibers, and the superficial zone consists of a superficial marginal layer, retinal fibers, and neuropil.

The optic tectum receives projections from the retina that terminate in the superficial zone. Most of the retinal fibers course along the deep border of the superficial zone, rather than running over the surface of the optic tectum, to reach their points of termination. Other connections of the optic tectum have been studied only in sharks with elaborated brains and in skates. It is likely that cartilaginous fishes with laminar brains

have generally similar inputs to the optic tectum from the telencephalon, contralateral tectum, and spinal cord, as well as similar efferent projections to the diencephalon and to the more caudal parts of the brainstem. All of these pathways need to be investigated in squalomorph sharks and in ratfishes.

Nonteleost Ray-Finned Fishes

In all ray-finned fishes, the optic tectum can be divided into superficial, central, and periventricular gray zones, as in other vertebrates. A deep white zone of fibers is also present just superficial to the periventricular gray zone. In nonteleost ray-finned fishes, the largest collection of neuronal cell bodies is in the periventricular gray zone, and the cells form multiple, distinct layers (Fig. 18-11). Small, piriform cells with radially directed dendrites are abundant in the periventricular gray zone [Fig. 18-12e–i]. Cells with dendrites oriented horizontally are also present in the optic tectum [Fig. 18-12a and d], as are a variety of cell types with dendrites radiating in multiple directions [Fig. 18-12b,c,j and k].

Retinal fibers, as shown on the left in Figure 18-13, terminate primarily in the superficial tectal zone, a few also entering the more superficial part of the central zone. Other afferent projections to the optic tectum originate from a diverse array of nuclei in the telencephalon, diencephalon, and brainstem. The sources of tectal projections in the brainstem include nu-

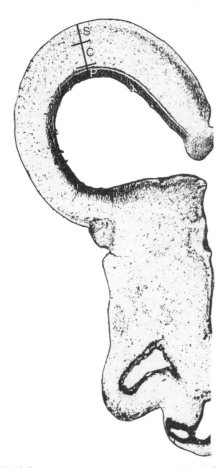

FIGURE 18-3. Transverse hemisection through the left side of the midbrain in the ray-finned fish *Amia calva,* with the superficial (S), central (C), and periventricular (P) zones of the optic tectum indicated.

cleus isthmi, the locus coeruleus, the reticular formation, and the cerebellum.

The efferent projections of the optic tectum are also diverse. Ascending tectal projections terminate in a number of different nuclei in the diencephalon. Of critical importance for visual perception are two nuclei in the dorsal thalamus that receive ascending tectal projections. The optic tectum projects sparsely to a nucleus in the dorsal thalamus, nucleus anterior, that is also in receipt of retinal projections and that, in turn, projects to the telencephalon. The optic tectum projects densely to another dorsal thalamic nucleus, the dorsal posterior nucleus, that does not receive retinal projections but does, in turn, also project to the telencephalon. The presence of two, separate, dorsal thalamic nuclei that receive tectal projections, one that also receives retinal projections and the other that does not, with both projecting to the telencephalon, is a widespread feature of visual system organization in vertebrates (Fig. 18-14) and will be discussed further below.

Many of the descending projections of the optic tectum are reciprocal in nature, such as those to nucleus isthmi and the reticular formation. The optic tectum also projects to the nucleus lateralis of the valvula cerebelli. The latter nucleus is a precerebellar nucleus that projects to the cerebellum. Recipro-

cal projections between the two tectal lobes cross the midline via the tectal commissure.

Ray-finned fishes have a structure associated with the optic tectum that is not present in other vertebrates. This structure is called the **torus longitudinalis** and consists of a rostrocaudally extending bulge of the tectal layers along the midline (Fig. 18-11). In teleosts, this structure has been found to receive inputs relayed to it from the telecephalon and to project in turn to the optic tectum, as we will discuss below. The torus longitudinalis is part of the circuitry that affects descending motor output through the optic tectum.

Amphibians

In the optic tecta of lungfishes and salamanders, neuronal cell bodies are almost totally confined to the periventricular gray zone (Fig. 18-15). In frogs, some of the neuron cell bodies are migrated more peripherally into the central and superficial zones (Fig. 18-16). Small piriform neurons [Fig. 18-17a–c] with radially oriented dendrites are numerous in the tectal layers in frogs. Among other cell types, larger, multipolar cells with far-reaching dendrites, called **ganglionic cells** [Fig. 18-17d and e] are also present.

In amphibians, the retinal projections to the optic tectum terminate in the superficial zone and are topographic. Most studies of tectal afferents have been done in frogs, where projections from nuclei in the diencephalon, the contralateral optic tectum, nucleus isthmi, several areas within the tegmentum, and the dorsal part of the cervical spinal cord have been identified. These nuclei, with the exception of nucleus isthmi (discussed below), project to the central zone of the optic tectum.

In frogs, input to the optic tectum from the telencephalon is via a multisynaptic relay in the diencephalon. In salamanders, a direct projection has been found from the telencephalon to the optic tectum. It terminates in the central zone, deep to the retinal input in the superficial zone.

As in ray-finned fishes, projections to the optic tectum from nucleus isthmi are topographic, and the optic tectum projects reciprocally and topographically to nucleus isthmi. The tectal cells that project to nucleus isthmi are primarily piriform neurons in the periventricular gray zone [Fig.18-17c]. Some tectal ganglionic cells project to nucleus isthmi as well. All of the neurons in nucleus isthmi project exclusively to the optic tectum. Their axons terminate in the superficial tectal zone, overlapping the zone of termination of retinal cell axons. The isthmal input to the optic tectum, which is cholinergic, modulates tectal responses to visual stimuli. Nucleus isthmi is believed to play a role in allowing the animal to avoid large, threatening stimuli and to attack smaller, prey-sized stimuli. These behavioral sequences have been studied extensively and will be discussed further below.

Efferent projections of the optic tectum are widespread. Ascending projections to the dorsal thalamus and pretectum arise from small piriform neurons in the central zone (such as in Fig. 18-17b). These projections terminate in a number of different nuclei, including the anterior lateral nucleus in the dorsal thalamus, which does not receive a direct retinal projection. It projects to the telencephalon. A second, dorsal thalamic nucleus, nucleus anterior, receives a minor tectal input relayed

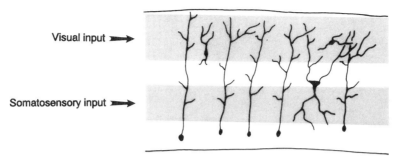

FIGURE 18-4. Segregation of afferent input on different parts of the dendritic shafts of individual tectal neurons. The retinal (visual) and somatosensory areas of termination are indicated by shading.

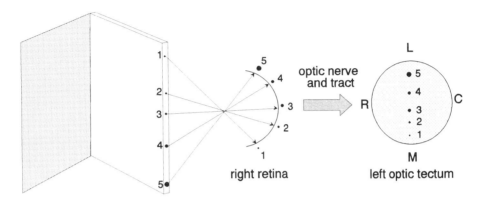

FIGURE 18-5. Maintenance of topographic order in the visual pathway from, in this example, the edge of a door to the retina of the eye, in the axons of the optic nerve and in their respective points of termination in the optic tectum.

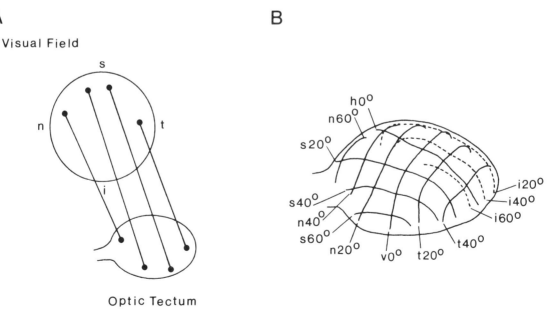

FIGURE 18-6. (A) Schematic diagram of the mapping of points in the visual field onto a dorsal view of the optic tectum. (B) Projection of the left visual field onto the dorsal surface of the right optic tectum. The optic tectum in both A and B is oriented so that lateral is toward the top, medial toward the bottom, rostral toward the left, and caudal toward the right. Abbreviations: s, superior; i, inferior; n, nasal (medial); t, temporal (lateral); h, horizontal; v, vertical. Adapted from Vanegas (1983).

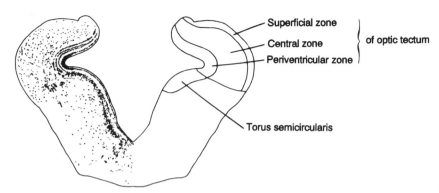

FIGURE 18-7. Transverse hemisection through the left midbrain of a lamprey (*Petromyzon marinus*) with mirror-image line drawing. Adapted from Kennedy and Rubinson (1977).

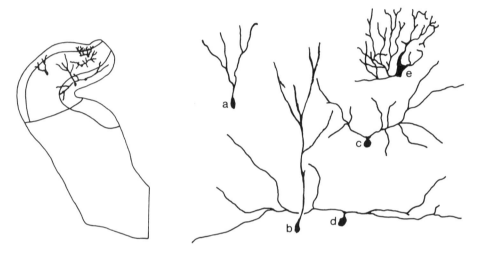

FIGURE 18-8. Neurons in the optic tectum of a lamprey as revealed with the Golgi method. A drawing of a left transverse hemisection is shown on the left for the orientation and location of the neurons shown on the right in an enlarged format. Data from Ariëns Kappers, Huber, and Crosby (1967).

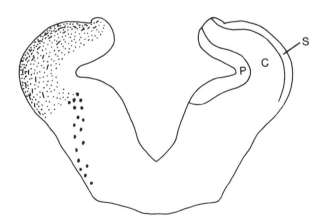

FIGURE 18-9. Retinal (small stippling) and spinal inputs (larger dots) to the optic tectum in a lamprey shown on the left side in a drawing of a transverse section. Abbreviations used in this and some following figures for zones within the optic tectum: S, superficial; C, central; P, periventricular. Data from Kennedy and Rubinson (1977) and Ronan and Northcutt (1990).

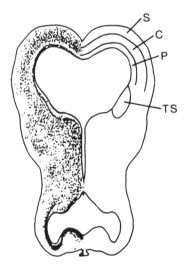

FIGURE 18-10. Transverse hemisection through the left midbrain of a squalomorph shark (*Squalus acanthias*) with mirror-image line drawing. TS: torus semicircularis. Adapted from Northcutt (1978).

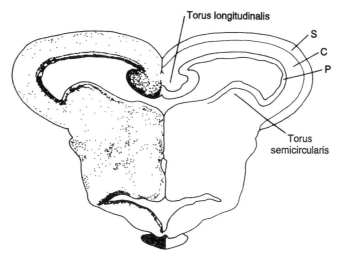

FIGURE 18-11. Transverse hemisection through the left midbrain of a longnose gar (*Lepisosteus osseus*) with mirror-image line drawing. Adapted from Northcutt and Butler (1976).

through another nucleus in the posterior thalamus and receives a substantial retinal projection as well. It also projects to the telencephalon.

The commissural and descending projections of the optic tectum in frogs are similar to those in other vertebrates and serve motor output functions. These projections arise from piriform and large ganglionic neurons [Fig. 18-17c and d] in the periventricular gray zone and the deeper part of the central zone.

The optic tectum projects to the contralateral tectum, nucleus isthmi (as discussed above), the magnocellular nucleus of the torus semicircularis, and multiple sites in the tegmentum. The latter include nucleus profundus mesencephali [see Fig. 17-6(B)], the anterodorsal tegmentum [Fig. 17-6(A)], the tegmental central gray, and the nuclei of the extraocular muscles and the medial longitudinal fasciculus. More caudally extending tectal projections reach the reticular

formation and other sites in the hindbrain as well as the rostral part of the spinal cord. Thus, there are multiple pathways by which tectal neurons influence neurons in the spinal cord—directly and by relays through the tegmentum and the reticular formation.

A number of behavioral studies designed to define the function of various afferent and efferent tectal connections have been carried out in frogs. In fact, the optic tectum of frogs is now regarded by many as a model for tectal function in anamniotes. For example, when a prey-sized black disk, small enough to resemble a fly or other similarly appetizing insect, is presented to a normal frog in a particular part of the visual field, the frog spatially orients towards the stimulus (employing the topographically mapped retinal input to the optic tectum) and then attempts to capture and eat it (employing spatially mapped motor patterns initiated by the optic tectum). When a predator-sized black disk, one that is large enough to resemble

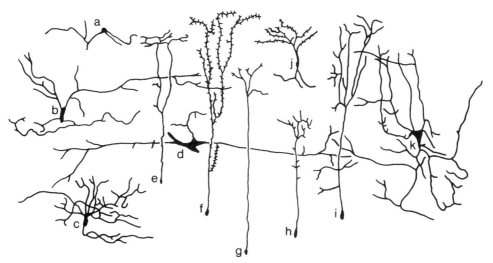

FIGURE 18-12. Neurons in the optic tectum of the bowfin (*Amia calva*) as revealed with the Golgi method. Data from Northcutt (1983).

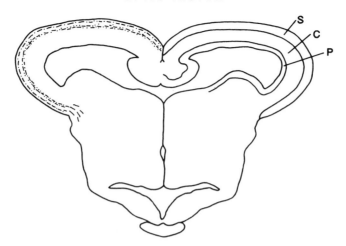

FIGURE 18-13. Retinal (small stippling) input to the left optic tectum in a longnose gar shown in a drawing of a transverse section. Data from Northcutt and Butler (1976).

a threatening stimulus, is placed in a particular part of the visual field, the frog orients away from this potential threat and leaps away, using similar tectal mechanisms. These responses do not involve learning; they are determined by the neuronal circuitry of the optic tectum and its related structures and pathways. Such natural behaviors allow one to ask experimental questions about both the sensory processing (size, location, movement, etc. of the disk) and the motor output (reaction, orientation, attack or escape, etc.) functions of the optic tectum.

In response to prey-sized stimuli, for example, the optic tectum initiates a program of behavioral events—orienting, fixating, snapping, gulping, and wiping—that are regulated both in space and, sequentially, in time. The program can be modified after it has been initiated, if, for example, a barrier to the stimulus is encountered or the position of the stimulus changes. The various parts of the program are carried out by the tectotegmental, tectoreticular, and tectospinal pathways. They are continually adjusted in response to tectal inputs from the retina,

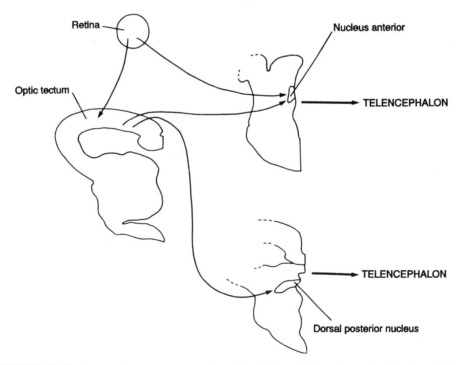

FIGURE 18-14. Diagram showing two major pathways by which visual information reaches the telencephalon in ray-finned fishes and some other groups of anamniote vertebrates. The retina projects directly to both the optic tectum and to a dorsal thalamic nucleus, nucleus anterior. The optic tectum projects to nucleus anterior and to a second dorsal thalamic nucleus, the dorsal posterior nucleus. Nucleus anterior and the dorsal posterior nucleus project to separate regions within the telencephalon.

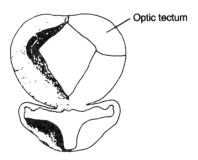

Optic tectum

FIGURE 18-15. Transverse hemisection through the left midbrain of a salamander (*Salamandra salamandra*) with mirror-image line drawing. Adapted from Fritzsch (1980). Used with permission of Springer-Verlag.

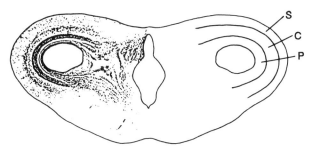

S
C
P

FIGURE 18-16. Transverse hemisection through the left midbrain of a bullfrog (*Rana catesbeiana*) with mirror-image drawing. Adapted from Wilczynski and Northcutt (1983).

forebrain, nucleus isthmi, contralateral optic tectum, brainstem feedback systems, and other sources.

Responses of avoidance to threatening stimuli are initiated by nuclei in the posterior thalamus and pretectum that project to the optic tectum. If these nuclei are surgically destroyed, the frog is "disinhibited" and readily attacks large stimuli as if they were prey. The normal tectal programs for avoidance postures and leaping are not triggered. Thus, the particular use of the descending tectal motor pathways, for attack or for escape, is determined by both tectal and forebrain processing of stimuli. The spatial mapping of topographically organized projections allows for correct orientation towards prey-sized stimuli and away from threatening stimuli.

THE OPTIC TECTUM IN GROUP II VERTEBRATES

Hagfishes

In the optic tectum of hagfishes, a number of neuron cell bodies are migrated away from the ventricular surface and are scattered throughout most of the thickness of the optic tectum (Fig. 18-18). There is an outer, relatively cell free layer, called a molecular layer, and an inner cell layer, although an extension of the third ventricle into the optic tectum is not present.

The optic tract in hagfishes runs through the brain in a deeper position than in other vertebrates and correspondingly

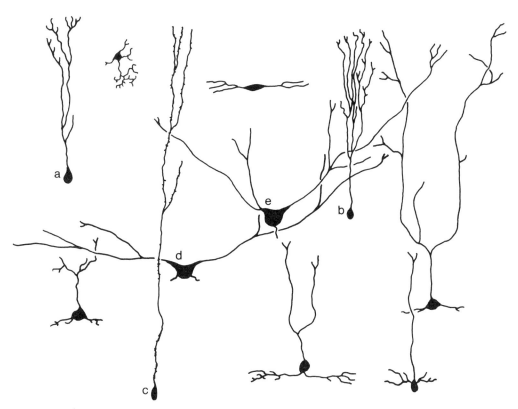

FIGURE 18-17. Neurons in the optic tectum of a frog as revealed with the Golgi method. Data from Székely (1971).

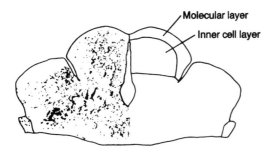

FIGURE 18-18. Transverse hemisection through the left midbrain of a hagfish (*Eptatretus stouti*) with mirror-image drawing. Adapted from Ronan and Northcutt (1990).

enters the optic tectum from its ventral aspect. Nonetheless, the retinal axons terminate primarily within the more superficial part of the optic tectum. The terminal zone lies in a transition area between the molecular and cell layers. The molecular layer and the area of retinal terminations may thus correspond to the superficial zone of the optic tectum in other vertebrates.

Other afferent connections of the optic tectum need to be investigated. The optic tectum may receive a spinal projection, since the spinal lemniscus lies just ventral to it (see Fig. 16-5). The spinal lemniscus is in a similar position in lampreys (Fig. 16-1) relative to the optic tectum and terminates sparsely in the ventral part of the optic tectum. Efferent projections of the optic tectum in hagfishes remain to be studied.

Galeomorph Sharks, Skates, and Rays

The optic tectum in cartilaginous fishes with elaborated brains is characterized by cell migration away from the periventricular zone into both the superficial and central zones. The amount of cellular migration is extensive, and the highest density of neuronal cell bodies is in the superficial zone (Fig. 18-19).

The neuronal cell types in the optic tectum consist of a predominance of large, multipolar neurons, which have den-

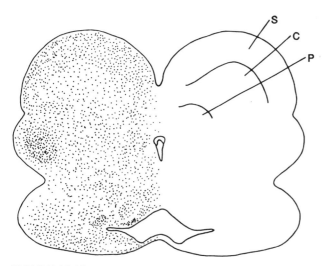

FIGURE 18-19. Drawing of a transverse hemisection through the left midbrain of a skate (*Raja eglanteria*) with mirror-image drawing. Adapted from Northcutt (1978).

drites radiating in all directions from the cell body (unlabeled, Fig. 18-20). Some neurons with horizontally oriented dendrites are present [Fig. 18-20a–c], but the small, piriform neurons with radially oriented dendrites, so characteristic of the optic tectum of other vertebrates, including squalomorph sharks, are conspicuously absent. Pyramidal cells—neurons that have an apical, radially oriented dendrite and two shorter dendrites off the basal part of the cell body—are also absent. Thus, the anatomical basis for precise, point-to-point, topographic input of visual and other information would seem to be missing.

In skates, recordings of electrical activity of neurons in the optic tectum in response to visual stimulation, called evoked potentials, are, however, similar to visually evoked potentials in the optic tecta of other vertebrates, such as frogs and birds, where the small piriform cells are present in abundance. In all these cases, as the recording electrode is lowered through the tectal layers, a reversal of polarity of the signal occurs. The small piriform cells are responsible for this "sink/source" (reversal) phenomenon in other vertebrates. In skates, a set of more migrated cells, which have relatively small cell bodies and dendrites extending both towards and away from the surface of the optic tectum [similar to that in Fig. 18-20d], have been identified as the population primarily responsible for the "sink/source" distribution of the evoked potentials. One of these cells is shown in Figure 18-21a, in comparison with similarly scaled piriform cells from the optic tecta of a frog [Fig. 18-21b] and a teleost fish [Fig. 18-21c]. This type of tectal cell in the skate appears to be a migrated version of a piriform cell, with some additional branching of the apical dendrite and the addition of a basal dendrite.

Evolutionary changes in the optic tecta of cartilaginous fishes thus appear to consist of (1) increased migration, accompanied by an increase in neuronal cell density and (2) elaboration of the configuration of the dendritic profiles. Further study of the cells with radially oriented dendrites in skates, in comparison with the piriform tectal cells of other vertebrates, could provide a window onto the evolution of neuronal cell structure.

Afferent projections to the optic tectum arise from the retina and from numerous sites within the brain. Retinal fibers enter the optic tectum deep to its surface and terminate primarily within the superficial zone. The optic tectum receives projections to its central zone from neurons lying in the caudal part of the telencephalon, in a number of nuclei in the diencephalon, and in the contralateral optic tectum. It receives ascending projections from lateral and ventrolateral tegmental nuclei and from the red nucleus. More caudal sites projecting to the optic tectum include vestibular, reticular, trigeminal, and octavolateral nuclei. A cerebellar nucleus and the dorsal part of the cervical spinal cord also project to the optic tectum and terminate in the central zone. Clearly, the optic tectum receives multiple inputs of a feedback nature from structures involved in motor functions in addition to sensory information from multiple systems.

The optic tectum in cartilaginous fishes has three major sets of efferent projections: descending, commissural, and ascending. Descending projections terminate in the reticular formation but do not extend to reach the spinal cord. No connections are present between the optic tectum and a nucleus that has been called nucleus isthmi but whose identity is thus in doubt. Commissural projections terminate in the contralateral

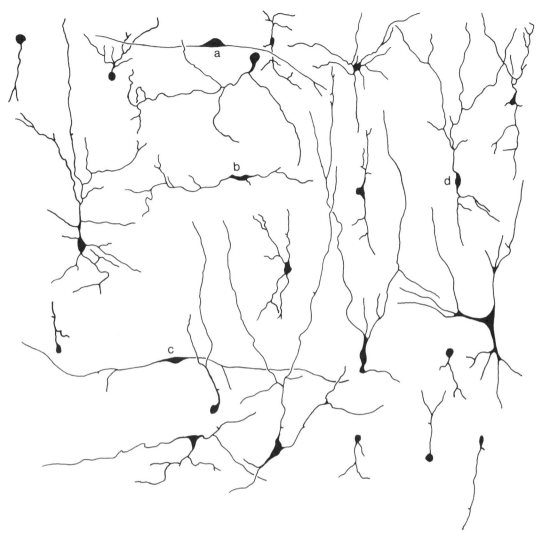

FIGURE 18-20. Neurons in the optic tectum of a dogfish shark (*Scyliorhinus canicula*) as revealed with the Golgi method. Adapted from Manso and Anadón (1991).

optic tectum and also in the intercollicular nucleus, which, like the reticular formation, projects to the spinal cord.

Ascending tectal projections terminate in a number of diencephalic nuclei. Two of these nuclei, which are in the dorsal thalamus, are nucleus anterior and the dorsal posterior nucleus. As in a variety of other vertebrates, nucleus anterior also receives a substantial retinal projection and projects to the telencephalon. The dorsal posterior nucleus does not receive a retinal projection but, like nucleus anterior, projects to the telencephalon. Thus, in cartilaginous fishes, there are two major visual pathways (see Fig. 18-14) by which tectal information is relayed to the telencephalon, one of which also has retinal input and one of which does not.

Teleosts

The periventricular gray, central, and superficial zones in the optic tectum of teleosts are each subdivided into multiple layers of cells and fibers (Fig.18-22). The layers are clearly demarcated, unlike the more diffuse array of cells seen in the optic tecta of cartilaginous fishes with elaborated brains. Small piriform cells [Fig. 18-23a and b] with radially oriented dendrites are present in abundance. An unusual population of pyramidal cells [Fig. 18-23c] is also present. These cells have an extensively branched and spiny apical dendrite and a basal dendrite that branches extensively in the horizontal plane and have not been found in nonteleost vertebrates. Cells with horizontally oriented dendrites and multipolar cells are present, as in the optic tecta of other vertebrates.

A presumably nonneural cell, called a **tanyocyte** or ependymoglial cell, present in the optic tectum of teleost fishes, should also be mentioned here. These cells are notable for their distinctive morphology, as revealed by the Golgi method and shown in Figure 18-24. The cell body lies in the ependymal layer, and the cell has a long apical dendrite that extends to the tectal surface and is studded with a multitude of short processes.

Afferent fibers from the retina terminate in two layers within the superficial zone of the optic tectum in teleosts. A

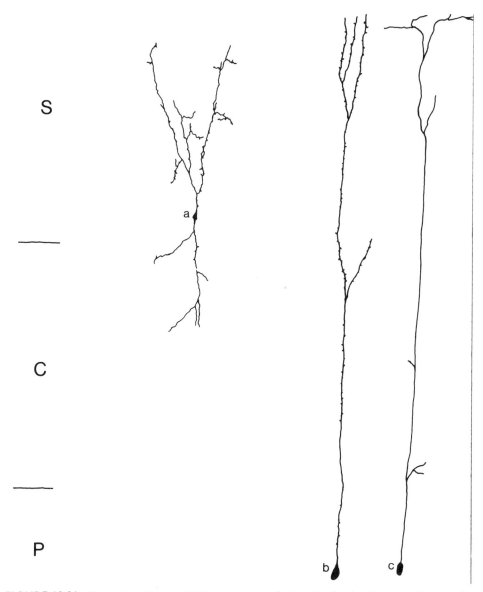

S

a

C

P

b   c

FIGURE 18-21. Comparison of small cell (a) in optic tectum of a skate that has dendrites extending towards both the surface and the periventricular zone with piriform neurons in the optic tecta of a frog (b) and a teleost (c). Data from Witkovsky et al. (1980), Székely (1971), and Vanegas et al. (1974), respectively.

few fibers also enter and terminate in the deeper part of the central tectal zone. The retinal projection is topographically mapped, as in other vertebrates. Other tectal afferents arise from nuclei in the telencephalon and diencephalon. In the mesencephalon, cells projecting to the optic tectum lie in the contralateral optic tectum, torus semicircularis, nucleus isthmi, and dorsolateral tegmentum and in the ipsilateral torus longitudinalis. Tectal afferents have been reported originating from multiple sites in the more caudal part of the brainstem in teleosts, including the reticular formation, inferior raphe, locus coeruleus, medial octavolateral nucleus, and rostral spinal cord.

The efferent projections of the optic tectum in teleosts are widespread and, in some cases, reciprocal to sites that project to the optic tectum. In the mesencephalon, cells in the tectal hemisphere project to the contralateral optic tectum, dorsolateral tegmentum, and nucleus isthmi. Descending projections have been traced to the reticular formation. Axons from the optic tectum that project rostrally to the diencephalon terminate in a number of nuclei, particularly in the pretectum and dorsal thalamus. In the pretectum, a nucleus known as the magnocellular (i.e., "large celled") superficial pretectal nucleus receives a tectal projection and, in turn, projects to nucleus isthmi, forming an additional relay path for tectoisthmal communication.

In the dorsal thalamus, two nuclei receive tectal projections: nucleus anterior and the dorsal posterior nucleus. Nucleus anterior also receives a substantial retinal projection, while the dorsal posterior nucleus does not. Both of these dorsal thalamic

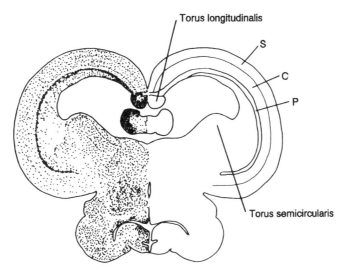

FIGURE 18-22. Transverse hemisection through the left midbrain of a teleost fish (*Lepomis gibbosus*) with mirror-image drawing. Adapted from Parent et al. (1978).

nuclei project to the telencephalon. Thus, two pathways from the optic tectum through the dorsal thalamus to the telencephalon are present in teleost fishes. This condition is like that in nonteleost ray-finned fishes with laminar brains and in other groups of vertebrates as well.

The tectal connections with nucleus isthmi are reciprocal and topographically organized, as discussed above. The tectal cells that project to nucleus isthmi are a portion of the small piriform cells with radially oriented dendrites that lie in the periventricular gray zone. Topographic mapping of the tecto-isthmal and isthmotectal fibers is a common feature in this

system in teleosts and tetrapods. In tetrapods, as discussed above for frogs, neurons in nucleus isthmi have relatively small receptive fields, as would be expected from the point-to-point organization of the tectal input. In teleosts, however, neurons in nucleus isthmi respond to small stimuli throughout the visual field. In other words, the receptive fields of individual isthmal neurons are very large.

These large receptive fields of isthmal neurons in teleosts may allow for a role in "alerting" the optic tectum to novel, and therefore important, stimuli that enter any part of the visual field. That such large receptive fields in nucleus isthmi are

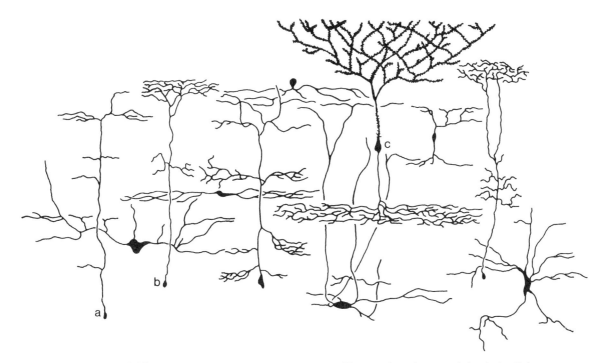

FIGURE 18-23. Neurons in the optic tectum of a teleost (*Eugerres plumieri*) as revealed with the Golgi method. Adapted from Vanegas et al. (1974).

FIGURE 18-24. A tanyocyte in the optic tectum of a teleost (*Eugerres plumieri*) as revealed with the Golgi method. Adapted from Vanegas et al. (1984). Used with permission of Plenum Publishing Corp.

present is an unexpected finding, however, in light of the topographic tectal input from the small piriform cells. Possible explanations for this phenomenon include branching and overlapping terminal areas of tectal axons onto isthmal neurons, a possibly nontopographic input from the magnocellular superficial pretectal nucleus, or electrical coupling among isthmal neurons (achieved through very close junctions, called tight junctions, between cell bodies).

The tectum also receives an input from the torus longitudinalis (Fig. 18-22), a structure unique to ray-finned fishes. This structure receives projections from several sources, including the lateral nucleus of the valvula of the cerebellum, oculomotor neurons, a dorsal tegmental nucleus, and a lateral part of the posterior tuberculum—the dorsal preglomerular nucleus—in the diencephalon. The latter two nuclei relay descending inputs from the telencephalon to the torus longitudinalis.

The torus longitudinalis gives rise to axons that run in the marginal layer on the outer surface of the optic tectum and that synapse, in topographic order, on the distal dendrites of pyramidal tectal neurons, such as that shown in Figure 18-23c. These pyramidal neurons also receive retinal projections, and the retinal and toral synapses on the pyramidal neurons are both excitatory. The toral fibers synapse on a more distal part of the dendritic tree of the pyramidal cell than do the retinal fibers, so the toral input can acitvate (depolarize) the dendritic tree, making it more receptive to visual input.

The torus longitudinalis thus serves as a relay for telencephalic inputs to the optic tectum. A tectal projection to the lateral nucleus of the valvula allows for tectal inputs to be relayed back to the torus longitudinalis as well. The torus longi-

tudinalis is thought to play a role, along with the optic tectum itself, in visuomotor, spatially oriented functions, such as movements in response to specific stimuli from various points in visual space. The torus longitudinalis can participate in the descending motor system by influencing the output of the optic tectum to the reticular formation.

Mammals, Diapsid Reptiles, Birds, and Turtles

In mammals, the optic tectum, which is often called the superior colliculus, is generally less impressive in size, relative to other brain structures, than in most other vertebrates. Among mammals, this structure is best developed in those with arboreal or partly arboreal habitats, such as squirrels, tree shrews, and some primates. The superior colliculus is laminated in mammals and can be divided into superficial, intermediate (central), and periventricular gray zones.

These three zones are subdivided so that six major layers are generally recognized. As shown in Figure 18-25, the superficial zone is divided into the **stratum zonale** (SZ), the **stratum griseum superficiale** (SGS), and the **stratum opticum** (SO). The bulk of the retinal fibers projecting to the superior colliculus pass through the latter. The central zone is divided into the **stratum griseum intermediale** (SGI) and the **stratum album intermediale** (SAI), while the deep zone comprises only the **stratum griseum profundum** (SGP). Within the various tectal layers, piriform neurons with radially oriented dendrites and a variety of horizontally oriented and larger, multipolar cells are present (Fig. 18-26).

Afferent fibers from the retina terminate topographically in the superficial zone of the superior colliculus. In mammals, three physiologically different types of retinal ganglion cells have been identified and are referred to as X, Y, and W cells (see Chapter 26). While all three types project to visual targets in the diencephalon, only the Y and W retinal ganglion cells project to the superior colliculus. Other afferent projections arise in parts of the visual cortex in the dorsal pallium of the

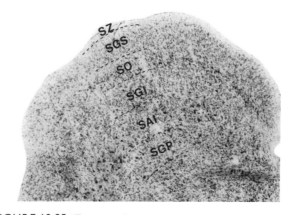

FIGURE 18-25. Transverse hemisection through the left superior colliculus of a rat. Abbreviations: SZ, stratum zonale; SGS, stratum griseum superficiale; SO, stratum opticum; SGI, stratum griseum intermediale; SAI, stratum album intermediale; SGP, straum griseum profundum. From Lane et al. (1993).

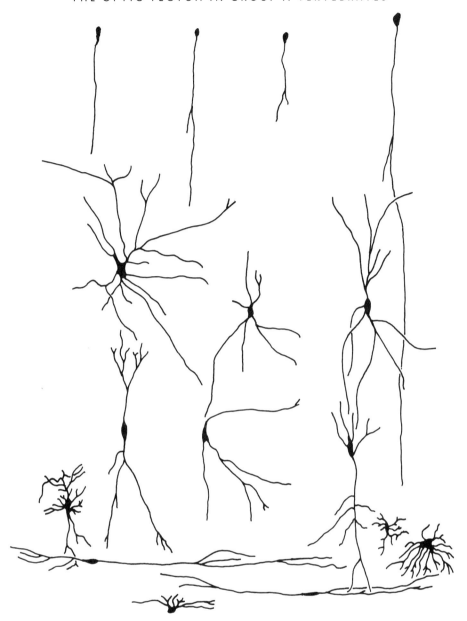

FIGURE 18-26. Neurons in the optic tectum of a mammal as revealed with the Golgi method. Data from Ariëns Kappers et al. (1967) as redrawn from Ramón y Cajal.

telencephalon and terminate in the superficial zone. Likewise, some diencephalic projections and those from the mammalian homologue of nucleus isthmi, the parabigeminal nucleus, terminate in the superficial zone.

Telencephalic cortical areas, including somatosensory and visual cortices as well as a more rostral part of cortex involved in eye movements, also project to the central and periventricular zones of the superior colliculus, as do fibers from the nucleus of the posterior commissure and substantia nigra that relay telencephalic striatopallidal input to the superior colliculus. The spinal (descending) trigeminal nucleus projects to the deeper zones of the superior colliculus, providing topographic representation of somatosensory input from the face. Other sources

of afferent input include nuclei in the hypothalamus and thalamus, the contralateral superior colliculus, the reticular formation, the spinal cord, and a variety of other brainstem sites.

Efferent projections can be divided into ascending, commissural, and descending pathways, as in other vertebrates. The major ascending pathway from the superior colliculus to the dorsal thalamus terminates in nuclei that are collectively called the **lateral posterior–pulvinar** (or **LP/pulvinar**) **complex** and that in turn project to visual pallial areas in the telencephalon. A minor tectal projection is present to the **dorsal lateral geniculate nucleus,** which also receives direct and substantial retinal projections and projects to visual pallium. These two tectofugal pathways, one via a nucleus that receives retinal input and one via a nuclear area that does not receive retinal input, are thus present in mammals as they are in other vertebrates.

The connections of the superior colliculus with the contralateral superior colliculus and with the parabigeminal nucleus (the homologue of nucleus isthmi) are topographic and reciprocal. Descending tectal fibers project to numerous sites, including the reticular formation, motor relay nuclei in the pons, and the spinal cord.

The superior colliculus plays an important role in visuomotor behavior in mammals, as it does in other vertebrates. The superior colliculus is involved in orienting to visual stimuli by head movements and/or **saccades,** which are rapid, directed eye movements. Tectal neurons respond to features of stimuli that include location, movement, brightness, and pattern. Deficits in the ability to notice and to then pay attention to objects in the visual field occur when the superior colliculus is damaged.

In cats and other mammals that use both head and body movements as well as saccadic eye movements to orient to a stimulus, the superior colliculus functions in all these behaviors, as it does in nonmammalian vertebrates. In primates, the primary response to an object of interest in the peripheral part of the visual field is to move the eyes, in preference to or before moving the head or body. The saccade is made in order to bring the object into the central visual field, where acuity is greatest. One of the primary roles of the superior colliculus in primates is in initiating and controlling the motor sequence for saccades.

The superior colliculus in primates also plays a key role in visual attention. A clinical case study in a human with damage to the major dorsal thalamic tectal target, the pulvinar, revealed neglect of stimuli in the peripheral part of the visual field. The degree of neglect was dependent on how far peripherally the stimulus was presented, the size and brightness of the stimulus, and the length of presentation of the stimulus. In the trials in which saccadic eye movements were made to the stimulus, the latency (interval between stimulus presentation and response) was abnormally long, and spontaneous eye movements were observed to be fewer than normal.

The superior colliculus additionally plays a significant role in predation. The degree to which the tectospinal tract is developed in mammals is directly correlated with the degree of predation in given species. This differential development of the tract has occurred independently at least four times within mammals.

The optic tectum of diapsid reptiles and turtles is characterized by many, clearly demarcated, cellular laminae (Figs. 18-27 and 18-28) in the superficial, central, and periventricular gray zones. As shown in a lizard in Figure 18-29, these zones can be subdivided into layers similar to the layers identified in mammals. However, note that the stratum opticum (SO), in which the afferent retinal fibers lie, occupies the most superficial part of the optic tectum, rather than the deeper part of the superficial zone as in mammals. Also shown in Figure 18-28, the layers (or strata) can in turn be subdivided into a total of 14 layers. An abundance of small piriform cells, with long, radially oriented dendrites [Fig. 18-29a–h], small cells with more complexly branched dendritic trees [Fig. 18-29i–k], small horizontally oriented cells [Fig. 18-29l–p], and large multipolar cells [Fig. 18-29q–s] are present.

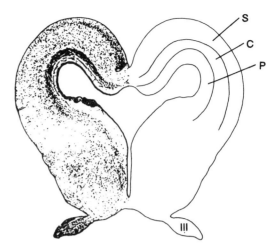

FIGURE 18-27. Transverse hemisection through the left optic tectum of a turtle (*Chrysemys picta*) with mirror-image drawing. III: oculomotor nerve. Adapted from Bass and Northcutt (1981).

Retinal fibers terminate topographically in the superficial zone. Somatosensory input from the dorsal spinal cord (and dorsal column nuclei) terminates in the central tectal zone in spatially mapped register with the visual map. In snakes that have IR-detecting pit organs, IR input is relayed through a brainstem trigeminal nucleus to a nucleus, nucleus reticularis caloris, which lies in the medulla, and thence to the optic tectum. The map of IR input is partly distorted, with the area of the central, frontal field magnified on the optic tectum, relative to the visual map. Several regions in the tegmentum proper,

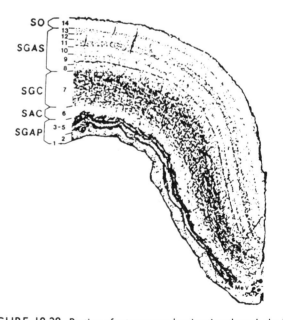

FIGURE 18-28. Portion of a transverse hemisection through the left optic tectum of a lizard (*Iguana iguana*), with medial toward the right. Abbreviations: SO, stratum opticum; SGAS, stratum griseum et album superficiale; SGC, stratum griseum centrale; SAC, stratum album centrale; SGAP, stratum griseum et album periventriculare. Numbers indicate layers 1–14. From Foster and Hall (1975).

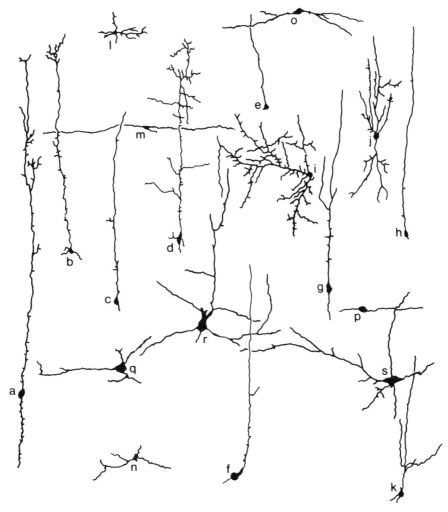

FIGURE 18-29. Neurons in the optic tectum of a lizard (*Tupinambis nigropunctatus*) as revealed with the Golgi method. Adapted from Butler and Ebbesson (1975).

nucleus isthmi, and a number of nuclei in the diencephalon and pretectum also project to the optic tectum.

There are several pathways from the telencephalon to the optic tectum in diapsid reptiles and turtles. The visual pallium projects directly to the optic tectum. Additionally, parts of the striatum in the telencephalon project to the optic tectum via two relays: a nucleus in the pretectum, called the **dorsal nucleus of the posterior commissure,** and the **substantia nigra** in the midbrain tegmentum. The striatum (also known as the basal ganglia in amniotes) is involved in the control of movements. These relayed inputs to the optic tectum thus modulate the orienting and other movements directed by the optic tectum in response to localized sensory stimuli.

Efferent projections of the optic tectum in diapsid reptiles and turtles include those descending to terminate in nucleus isthmi, parts of the tegmentum, the reticular formation, and the rostral spinal cord. These projections allow for the motor programs of orienting to stimuli, capturing prey, and escaping predators. The optic tectum also projects to the contralateral optic tectum and to nuclei in the pretectum and dorsal thalamus.

In the dorsal thalamus, the nucleus that receives most of the tectal projections is **nucleus rotundus,** a large, round nucleus that, in turn, projects to the telencephalon. This pathway is one of the two major visual pathways to the telencephalon. The other pathway consists of both a minor tectal projection and a substantial retinal projection to a nucleus called the **dorsal lateral optic nucleus** that, in turn, also projects to the telencephalon. The primary sites of termination in the telencephalon for both of these nuclei are in the pallium—not in the striatum as is the case in amphibians.

The optic tectum in birds is layered even more elaborately than it is in diapsid reptiles and turtles. Horizontal cells, large, multipolar cells, a plethora of small piriform cells, and other elements are similar to those in diapsid reptiles and turtles. Retinal afferents terminate in the superficial zone, and somatosensory and other inputs terminate in the central zone. These afferent projections include input relayed from the striatum in the telencephalon through two nuclei: **nucleus spiriformis lateralis** (the homologue of the dorsal nucleus of the posterior commissure of diapsid reptiles and turtles) and the substantia nigra. Part of the visual pallium in the telencephalon also projects directly to the optic tectum. Additional tectal afferents arise in tegmental and other caudal areas, three isthmal nuclei (nuclei isthmi, pars parvocellularis and pars magnocellularis, and nucleus semilunaris), and thalamic and pretectal nuclei, as they do in diapsid reptiles and turtles.

Tectal efferent projections in birds are similar to those in diapsid reptiles and turtles. Ascending projections terminate in a number of dorsal thalamic and pretectal nuclei, including nucleus rotundus. This nucleus projects to part of the pallium in the telencephalon, forming one of the two major ascending visual pathways. Retinal information is relayed to the telencephalon through a separate dorsal thalamic nucleus, homologous to the dorsal lateral optic nucleus of diapsid reptiles and turtles. No tectal projections to the latter nucleus have been found, however. The optic tectum and the ascending tectofugal system in birds have been found to play an important role in visual discrimination tasks—distinguishing one pattern from another or the relative degree of brightness or colors—as well as in localizing stimuli in visual or somatosensory space.

Other tectal connections include commissural projections to the contralateral optic tectum, reciprocal projections to the parvocellular, magnocellular, and semilunar isthmal nuclei, and a projection to the neighboring isthmo-optic nucleus that, in turn, projects to the retina. Descending projections of the optic tectum do not extend to the spinal cord directly, as they do in many other vertebrates, but terminate in areas in the brainstem tegmentum and the reticular formation, which, in turn, project to the spinal cord. These pathways serve motor functions for orienting to localized stimuli with rapid head movements.

EVOLUTIONARY PERSPECTIVE

The optic tectum is one of the most conservative structures in the brains of vertebrates. Although the optic tectum varies in the extent of its development in different vertebrate classes, its layered structure, its cell types, and its afferent and efferent connections are quite similar in all vertebrates. Many of its connections are topographic; this feature is basic for tectal function in localizing stimuli and responding to them in spatially appropriate ways. In spite of this conservative pattern, some significant differences have occurred in the structure and connections of the optic tectum in certain vertebrate radiations.

The migration of neurons into the central and superficial zones of the optic tectum has occurred independently in all four of the major radiations of vertebrates. Other specializations have evolved independently as well, in the types of tectal neurons present, in connections of the optic tectum, and in the presence of some structures associated with the optic tectum.

The presence of small piriform neurons with radially oriented dendrites is a common feature of tectal organization except in Group II cartilaginous fishes, the galeomorph sharks, skates, and rays. In these animals, small piriform cells are conspicuously absent, and instead, one finds a population of cells with a more elaborate dendritic profile that may be a migrated, elaborated form of piriform cell.

In teleosts (the Group II ray-finned fishes) piriform, horizontal, and multipolar tectal neurons are present as in other vertebrates. Additionally, a population of neurons with exten-

sively branched and spiny apical dendrites and extensively branched, horizontally extending basal dendrites is present. This population of branched, spiny neurons appears to be unique to teleosts.

The presence of such unique neurons in teleosts and the presence of the migrated variant of piriform cells in Group II cartilaginous fishes suggest specializations of tectal function in these two groups. These specializations have been evolved independently of each other, as neither of these cell types are present in any Group I vertebrates, including the nonteleost ray-finned fishes and the squalomorph sharks.

Variation also occurs in connections and in structures associated with the optic tectum. For example, a nucleus isthmi, which is reciprocally connected with the optic tectum, has been found in both Groups I and II of the ray-finned fishes and in all tetrapods as well: the Group I amphibians and the Group II amniotes. Such a nucleus has not been found in any of the cartilaginous fishes or in lampreys or hagfishes. This structure may thus be a specialization (apomorphy) unique to ray-finned fishes and tetrapods.

A second structure associated with the optic tectum, which is present in both Groups I and II of ray-finned fishes, is the torus longitudinalis, which receives relayed information from the telencephalon and projects to the optic tectum. A torus longitudinalis is not present in any tetrapod or in the out-groups to ray-finned fishes, the cartilaginous fishes and agnathans. Thus, it is an apomorphy in ray-finned fishes that was probably absent in the common ancestral group of the ray-finned fishes and the tetrapods.

Variation related to afferent tectal input from the telencephalon is present. Further study to clarify the situation is needed as a number of the studies in vertebrates other than amniotes have involved large telencephalic lesions or large applications of axon tracers and have not revealed the specific location(s) of origin of tectally projecting neurons within the telencephalon.

In amniotes, the striatum projects via two relay nuclei to the optic tectum, thus providing information on motor control systems. One of these nuclei is the nucleus of the posterior commissure, and the other is the substantia nigra. Of these two pathways, the one through the nucleus of the posterior commissure is developed better in nonmammalian amniotes, while the one through the substantia nigra is developed better in mammals. Additionally, in mammals, several parts of the cortex project directly to the optic tectum, providing feedback information, and in diapsid reptiles, birds, and turtles, the visual pallium in the telencephalon likewise projects to the optic tectum.

Salamanders have a direct telencephalotectal pathway, whereas in frogs, only a telencephalic projection to the optic tectum via a thalamic relay nucleus has been identified. Both Group I and Group II teleosts have a direct telencephalotectal pathway, as do Group II cartilaginous fishes. Information is not available for Group I cartilaginous fishes or for any agnathans.

If direct telencephalotectal projections arise in the visual pallium rather than the striatum in all cases, and if such projections are present in Group I cartilaginous fishes, then such projections would be present in all jawed vertebrate groups with the exception of frogs. Such projections would thus be a generalized (plesiomorphic) feature of jawed vertebrates and would have been lost in the ancestral stock of frogs.

With a similarly sparse amount of data, we can tentatively argue that tectal input relayed from the striatum *may* be an apomorphy (specialization) of land vertebrates, present in frogs (if not salamanders) and in amniotes. Such connections might be related to motor behaviors related to quadrepedal locomotion on land.

Ascending tectal projections to the dorsal thalamus are better understood, although more information would be welcome here as well. Tectal projections to two separate thalamic nuclei are present in all vertebrate groups studied to date: Group II cartilaginous fishes as well as the ray-finned fishes and tetrapods in both Groups I and II. This distribution suggests that such projections were present in the ancestral group of jawed vertebrates. It is probable, in fact, that they were present even earlier in the first vertebrates.

The optic tectum has a minor projection to one dorsal thalamic nucleus that also receives a substantial and direct retinal input, variously called nucleus anterior, the dorsal lateral optic (or geniculate) nucleus, or other terms. In addition, the optic tectum has a major projection to a second dorsal thalamic nucleus that does not receive direct retinal input, and is variously called the dorsal posterior nucleus, nucleus rotundus, or the lateral posterior/pulvinar complex. These two visual pathways to the telencephalon were established early in vertebrate evolution.

FOR FURTHER READING

Ebbesson, S. O. E. (ed.) (1984) *Comparative Neurology of the Optic Tectum*. New York: Plenum.

Gaither, N. S. and Stein, B. E. (1979) Reptiles and mammals use similar sensory organization in the midbrain. *Science*, 205, 595–597.

Hartline, P. H., Kass, L., and Loop, M. S. (1978) Merging of modalities in the optic tectum: infrared and visual integration in rattlesnakes. *Science*, 199, 1225–1228.

Northcutt, R. G. (1983) Evolution of the optic tectum in ray-finned fishes. In R. E. Davis and R. G. Northcutt (eds.), *Fish Neurobiology, Vol. 2: Higher Brain Areas and Functions*. Ann Arbor, MI: University of Michigan Press, pp. 1–42.

Reiner, A. (1994) Laminar distribution of the cells of origin of ascending and descending tectofugal pathways in turtles: implications for the evolution of tectal lamination. *Brain, Behavior and Evolution*, 43, 254–292.

ADDITIONAL REFERENCES

Abplanalp, P. (1970) Some subcortical connections of the visual system in tree shrews and squirrels. *Brain, Behavior and Evolution*, 3, 155–168.

Ariëns Kappers, C. U., Huber, G. C., and Crosby, E. C. (1967) *The Comparative Anatomy Of the Nervous System Of Vertebrates, Including Man, Vol. II*. New York: Hafner Publishing Co.

Barton, R. A. and Dean, P. (1993) Comparative evidence indicating neural specialization for predatory behaviour in mammals. *Proceedings of the Royal Society of London, B*, 254, 63–68.

Bass, A. H. and Northcutt, R. G. (1981) Retinal recipient nuclei in the painted turtle, *Chrysemys picta*: an autoradiographic and HRP study. *Journal of Comparative Neurology*, 199, 97–112.

Benevento, L. A. and Fallon, J. H. (1975) The ascending projections of the superior colliculus in the rhesus monkey (*Macaca mulatta*). *Journal of Comparative Neurology*, 160, 339–362.

Boord, R. L. and Northcutt, R. G. (1982) Ascending lateral line pathways to the midbrain of the clearnose skate, *Raja eglanteria*. *Journal of Comparative Neurology*, 207, 274–282.

Butler, A. B. and Ebbesson, S. O. E. (1975) A Golgi study of the optic tectum of the tegu lizard, *Tupinambis nigropunctatus*. *Journal of Morphology*, 146, 215–228.

Butler, A. B. and Northcutt, R. G. (1971) Ascending tectal efferent projections in the lizard *Iguana iguana*. *Brain Research*, 35, 597–601.

Butler, A. B. and Northcutt, R. G. (1992) Retinal projections in the bowfin, *Amia calva*: cytoarchitectonic and experimental analysis. *Brain, Behavior and Evolution*, 39, 169–194.

Campbell, C. B. G. and Hayhow, W. R. (1972) Primary optic pathways in the duckbill platypus *Ornithorynchus anatinus*: an experimental degeneration study. *Journal of Comparative Neurology*, 145, 195–208.

Corvaja, N. and d'Ascanio, P. (1981) Spinal projections from the mesencephalon in the toad. *Brain, Behavior and Evolution*, 19, 205–213.

Ebbesson, S. O. E. (1984) Structure and connections of the optic tectum in elasmobranchs. In H. Vanegas (ed.), *Comparative Neurology of the Optic Tectum*. New York: Plenum, pp. 33–46.

Ebbesson, S. O. E. and Goodman, D. C. (1981) Organization of ascending spinal projections in *Caiman crocodilus*. *Cell and Tissue Research*, 215, 383–395.

Ebbesson, S. O. E. and Ramsey, J. S. (1968) The optic tracts of two species of sharks (*Galeocerdo cuvier* and *Ginglymostoma cirratum*). *Brain Research*, 8, 36–53.

Ebbesson, S. O. E. and Schroeder, D. M. (1971) Connections of the nurse shark's telencephalon. *Science*, 173, 254–256.

Ebbesson, S. O. E. and Vanegas, H. (1976) Projections of the optic tectum in two teleost species. *Journal of Comparative Neurology*, 165, 161–180.

Edwards, S. B. (1977) The commissural projection of the superior colliculus in the cat. *Journal of Comparative Neurology*, 173, 23–40.

Ewert, J.-P. (1984) Tectal mechanisms that underlie prey-catching and avoidance behaviors in toads. In H. Vanegas (ed.), *Comparative Neurology of the Optic Tectum*. New York: Plenum, pp. 247–416.

Feig, S., van Lieshout, D. P., and Harting, J. K. (1992) Ultrastructural studies of retinal, visual cortical (area 17), and parabigeminal terminals within the superior colliculus of *Galago crassicaudatus*. *Journal of Comparative Neurology*, 319, 85–99.

Fiebig, E., Ebbesson, S. O. E., and Meyer, D. L. (1983) Afferent connections of the optic tectum in the piranha (*Serrasalmus nattereri*). *Cell Tissue Research*, 231, 55–72.

Finger, T. E. (1978) Cerebellar afferents in teleost catfish (Ictaluridae). *Journal of Comparative Neurology*, 181, 173–182.

Foster, R. E. and Hall, W. C. (1975) The connections and laminar organization of the optic tectum in a reptile (*Iguana iguana*). *Journal of Comparative Neurology*, 163, 397–426.

Fritzsch, B. (1980) Retinal projections in European Salamandridae. *Cell and Tissue Research*, 213, 325–341.

Frost, B. J., Wise, L. Z., Morgan, B., and Bird, D. (1990) Retinotopic representation of the bifoveate eye of the kestrel (*Falco sparverius*) on the optic tectum. *Visual Neuroscience*, 5, 231–239.

Goldman, P. S. and Nauta, W. J. H. (1976) Autoradiographic demonstration of a projection from prefrontal association cortex to the superior colliculus in the rhesus monkey. *Brain Research*, 116, 145–149.

Graeber, R. C. and Ebbesson, S. O. E. (1972) Retinal projections in the lemon shark (*Negaprion brevirostris*). *Brain, Behavior and Evolution*, 5, 461–477.

Graham, J. (1977) An autoradiographic study of the efferent connections of the superior colliculus in the cat. *Journal of Comparative Neurology*, 173, 629–654.

Graham, J. and Casagrande, V. A. (1980) A light microscopic and electron microscopic study of the superficial layers of the superior colliculus of the tree shrew (*Tupaia glis*). *Journal of Comparative Neurology*, 191, 133–151.

Graybiel, A. M. (1972) Some extrageniculate visual pathways in the cat. *Investigative Ophthalmology*, 11, 322–332.

Graybiel, A. M. (1972) Some fiber pathways related to the posterior thalamic region in the cat. *Brain, Behavior and Evolution*, 6, 363–393.

Graybiel, A. M. (1978) Organization of the nigrotectal connection: an experimental tracer study in the cat. *Brain Research*, 143, 339–348.

Gruberg, E. R., Hughes, T. E., and Karten, H. J. (1994) Synaptic interrelationships between the optic tectum and the ipsilateral nucleus isthmi in *Rana pipiens*. *Journal of Comparative Neurology*, 339, 353–364.

Gruberg, E. R., Kicliter, E., Newman, E. A., Kass, L., and Hartline, P. H. (1979) Connections of the tectum of the rattlesnake *Crotalus viridis*: an HRP study. *Journal of Comparative Neurology*, 188, 31–42.

Gruberg, E. R. and Lettvin, J. Y. (1980) Anatomy and physiology of a binocular system in the frog *Rana pipiens*. *Brain Research*, 192, 313–325.

Gruberg, E. R. and Udin, S. B. (1978) Topographic projections between the nucleus isthmi and the tectum of the frog *Rana pipiens*. *Journal of Comparative Neurology*, 179, 487–500.

Gruberg, E. R., Wallace, M. T., and Waldeck, R. F. (1989) Relationship between isthmotectal fibers and other tectopetal systems in the leopard frog. *Journal of Comparative Neurology*, 288, 39–50.

Harting, J. K., Glendenning, K. K., Diamond, I. T., and Hall, W. C. (1973) Evolution of the primate visual system: anterograde degeneration studies of the tecto-pulvinar system. *American Journal of Physical Anthropology*, 38, 383–392.

Harting, J. K., Hall, W. C., Diamond, I. T., and Martin, G. F. (1973) Anterograde degeneration study of the superior colliculus in *Tupaia glis*: evidence for a subdivision between superficial and deep layers. *Journal of Comparative Neurology*, 148, 361–386.

Harting, J. K., Huerta, M. F., Hashikawa, T., and van Lieshout, D. P. (1991) Projection of the mammalian superior colliculus upon the dorsal lateral geniculate nucleus: organization of tectogeniculate pathways in nineteen species. *Journal of Comparative Neurology*, 304, 275–306.

Harting, J. K., Updyke, B. V., and van Lieshout, D. P. (1992) Corticotectal projections in the cat: anterograde transport studies of twenty-five cortical areas. *Journal of Comparative Neurology*, 324, 379–414.

Hayle, T. H. (1973) A comparative study of spinal projections to the brain (except cerebellum) in three classes of poikilothermic vertebrates. *Journal of Comparative Neurology*, 149, 463–476.

Huerta, M. F. and Harting, J. K. (1984) The mammalian superior colliculus: studies of its morphology and connections. In H. Vanegas (ed.), *Comparative Neurology of the Optic Tectum*. New York: Plenum, pp. 687–773.

Hunt, S. P. and Brecha, N. (1984) The avian optic tectum: a synthesis of morphology and biochemistry. In H. Vanegas (ed.), *Comparative Neurology of the Optic Tectum*. New York: Plenum, pp. 619–648.

Hunt, S. P. and Künzle, H. (1976) Observations on the projections and intrinsic organization of the pigeon optic tectum: an autoradiographic study based on anterograde and retrograde, axonal and dendritic flow. *Journal of Comparative Neurology*, 170, 153–172.

Ingle, D. (1973) Two visual systems in the frog. *Science*, 181, 1053–1055.

Ito, H. and Kishida, R. (1978) Afferent and efferent fiber connections of the carp torus longitudinalis. *Journal of Comparative Neurology*, 181, 465–476.

Ito, H., Tanaka, H., Sakamoto, N., and Morita, Y. (1981) Isthmic afferent neurons identified by the retrograde HRP method in a teleost, *Navodon modestus*. *Brain Research*, 207, 163–169.

Jakway, J. S. and Riss, W. (1972) Retinal projections in the tiger salamander, *Ambystoma tigrinum*. *Brain, Behavior and Evolution*, 5, 401–442.

Karten, H. J. and Revzin, A. M. (1966) The afferent connections of the nucleus rotundus in the pigeon. *Brain Research*, 2, 368–377.

Kennedy, M. C. and Rubinson, K. (1977) Retinal projections in larval, transforming and adult sea lamprey, *Petromyzon marinus*. *Journal of Comparative Neurology*, 171, 465–479.

Kennedy, M. C. and Rubinson, K. (1984) Development and structure of the lamprey optic tectum. In H. Vanegas (ed.), *Comparative Neurology of the Optic Tectum*. New York: Plenum, pp. 1–13.

Kobayashi, S., Kishida, R., Goris, R. C., Yoshimoto, M., and Ito, H. (1992) Visual and infrared input to the same dendrite in the tectum opticum of the python, *Python regius*: electron-microscopic evidence. *Brain Research*, 597, 350–352.

Kokoros, J. J. and Northcutt, R. G. (1977) Telencephalic efferents of the tiger salamander *Ambystoma tigrinum tigrinum* (Green). *Journal of Comparative Neurology*, 173, 613–628.

Künzle, H. and Woodson, W. (1982) Mesodiencephalic and other target regions of ascending spinal projections in the turtle, *Pseudemys scripta elegans*. *Journal of Comparative Neurology*, 212, 349–364.

Lane, R. D., Bennett-Clarke, C. A., Allan, D. M., and Mooney, R. D. (1993) Immunochemical heterogeneity in the tecto-LP pathway of the rat. *Journal of Comparative Neurology*, 333, 210–222.

Lázár, Gy. (1984) Structure and connections of the frog optic tectum. In H. Vanegas (ed.), *Comparative Neurology of the Optic Tectum*. New York: Plenum, pp. 185–210.

Lázár, Gy., Tóth, P., Csank, Gy., and Kicliter, E. (1983) Morphology and location of tectal projection neurons in frogs: a study with HRP and cobalt-filling. *Journal of Comparative Neurology*, 215, 108–120.

Manso, M. J. and Anadón, R. (1991) The optic tectum of the dogfish *Scyliorhinus canicula* L.: a Golgi study. *Journal of Comparative Neurology*, 307, 335–349.

Masino, T. (1992) Brainstem control of orienting movements: intrinsic coordinate systems and underlying circuitry. *Brain, Behavior and Evolution*, 40, 98–111.

Masino, T. and Grobstein, P. (1990) Tectal connectivity in the frog, *Rana pipiens*: tectotegmental projections and a general analysis of topographic organization. *Journal of Comparative Neurology*, 291, 103–127.

Masino, T. and Knudsen, E. I. (1992) Anatomical pathways from the optic tectum to the spinal cord subserving orienting movements in the barn owl. *Experimental Brain Research*, 92, 194–208.

Masino, T. and Knudsen, E. I. (1993) Orienting head movements resulting from electrical microstimulation of the brainstem tegmentum in the barn owl. *Journal of Neuroscience*, 13, 351–370.

Medina, L. and Smeets, W. J. A. J. (1991) Comparative aspects of the basal ganglia-tectal pathways in reptiles. *Journal of Comparative Neurology*, 308, 614–629.

Meek, J. (1983) Functional anatomy of the tectum mesencephali of the goldfish. An explorative analysis of the functional implications of the laminar structural organization of the tectum. *Brain Research Reviews*, 6, 247–297.

Meredith, M. A., Wallace, M. T., and Stein, B. E. (1992) Visual, auditory and somatosensory convergence in output neurons of the cat superior colliculus: multisensory properties of the tecto-reticulo-spinal projection. *Experimental Brain Research*, 88, 181–186.

Montgomery, N. and Fite, K. V. (1989) Retinotopic organization of central optic projections in *Rana pipiens*. *Journal of Comparative Neurology*, 283, 526–540.

Montgomery, N. and Fite, K. V. (1991) Organization of ascending projections from the optic tectum and mesencephalic central gray in *Rana pipiens*. *Visual Neuroscience*, 7, 459–478.

Naujoks-Manteuffel, C. and Manteuffel, G. (1988) Origins of descending projections to the medulla oblongata and rostral medulla spinalis in the urodele *Salamandra salamandra* (Amphibia). *Journal of Comparative Neurology*, 273, 187–206.

Newman, E. A. and Hartline, P. H. (1981) Integration of visual and infrared information in bimodal neurons of the rattlesnake optic tectum. *Science*, 213, 789–791.

Niida, A. and Ohono, T. (1984) An extensive projection of fish dorsolateral-tegmental cells to the optic tectum revealed by intra-axonal dye marking. *Neuroscience Letters*, 48, 261–266.

Northcutt, R. G. (1977) Retinofugal projections in the lepidosirenid lungfishes. *Journal of Comparative Neurology*, 174, 553–574.

Northcutt, R. G. (1978) Brain organization in the cartilaginous fishes. In E. S. Hodgson and R. F. Mathewson (eds.), *Sensory Biology of Sharks, Skates, and Rays*. Arlington, Virginia: Office of Naval Research, pp. 117–193.

Northcutt, R. G. (1979a) Retinofugal pathways in fetal and adult spiny dogfish, *Squalus acanthias*. *Brain Research*, 162, 219–230.

Northcutt, R. G. (1979b) The comparative anatomy of the nervous system and the sense organs. In M. H. Wake (ed.), *Hyman's Comparative Vertebrate Anatomy*. Chicago: The University of Chicago Press, pp. 615–769.

Northcutt, R. G. (1980) Retinal projections in the Australian lungfish. *Brain Research*, 185, 85–90.

Northcutt, R. G. (1981) Evolution of the telencephalon in non-mammals. *Annual Review of Neuroscience*, 4, 301–350.

Northcutt, R. G. (1982) Localization of neurons afferent to the optic tectum in longnose gars. *Journal of Comparative Neurology*, 204, 325–335.

Northcutt, R. G. (1983) Evolution of the optic tectum in ray-finned fishes. In R. E. Davis and R. G. Northcutt (eds.), *Fish Neurobiology, Vol 2: Higher Brain Areas and Functions*. Ann Arbor, MI: The University of Michigan Press, pp. 1–42.

Northcutt, R. G. (1984) Anatomical organization of the optic tectum in reptiles. In H. Vanegas (ed.), *Comparative Neurology of the Optic Tectum*. New York: Plenum, pp. 547–600.

Northcutt, R. G. (1991) Visual pathways in elasmobranchs: organization and phylogenetic implications. *Journal of Experimental Zoology*, Suppl. 5, 97–107.

Northcutt, R. G. and Butler, A. B. (1974) Evolution of reptilian visual systems: retinal projections in a nocturnal lizard, *Gekko gecko* (Linnaeus). *Journal of Comparative Neurology*, 157, 453–466.

Northcutt, R. G. and Butler, A. B. (1976) Retinofugal pathways in the longnose gar *Lepisosteus osseus* (Linnaeus). *Journal of Comparative Neurology*, 166, 1–16.

Northcutt, R. G. and Butler, A. B. (1980) Projections of the optic tectum in the longnose gar, *Lepisosteus osseus*. *Brain Research*, 190, 333–346.

Northcutt, R. G. and Butler, A. B. (1991) Retinofugal and retinopetal projections in the green sunfish, *Lepomis cyanellus*. *Brain, Behavior and Evolution*, 37, 333–354.

Northmore, D. P. M. (1982) Visuotopic mapping and corollary discharge in the torus longitudinalis of goldfish. *Investigative Ophthalmology and Visual Science* (Suppl.), 22, 237.

Northmore, D. P. M. (1991) Visual responses of nucleus isthmi in a teleost fish (*Lepomis macrochirus*). *Vision Research*, 31, 525–535.

Northmore, D. P. M., Williams, B., and Vanegas, H. (1983) The teleostean torus longitudinalis: responses related to eye movements, visuotopic mapping, and functional relations with the optic tectum. *Journal of Comparative Physiology*, 150, 39–50.

Nudo, R. J., Sutherland, D. P., and Masterton, R. B. (1993) Inter- and intra-laminar distribution of tectospinal neurons in 23 mammals. *Brain, Behavior and Evolution*, 42, 1–23.

Parent, A., Dube, L., Braford, M. R., Jr., and Northcutt, R. G. (1978) The organization of monoamine-containing neurons in the brain of the sunfish (*Lepomis gibbosus*) as revealed by fluorescence microscopy. *Journal of Comparative Neurology*, 182, 495–516.

Pentney, R. P. and Cotter, J. R. (1978) Structural and functional aspects of the superior colliculus in primates. In C. R. Noback (ed.), *Sensory Systems of Primates*. New York: Plenum, pp. 109–134.

Reiner, A. (1994) Laminar distribution of the cells of origin of ascending and descending tectofugal pathways in turtles: implications for the evolution of tectal lamination. *Brain, Behavior and Evolution*, 43, 189–292.

Reiner, A., Brauth, S. E., and Karten, H. J. (1984) Evolution of the amniote basal ganglia. *Trends in Neurosciences*, 7, 320–325.

Reiner, A., Brauth, S. E., Kitt, C. A., and Karten, H. J. (1980) Basal ganglionic pathways to the tectum: studies in reptiles. *Journal of Comparative Neurology*, 193, 565–589.

Repérant, J., Miceli, D., Rio, J.-P., Peyrichoux, J., Pierre, J., and Kirpitchnikova, E. (1986) The anatomical organization of retinal projections in the shark *Scyliorhinus canicula* with special reference to the evolution of the selachian primary visual system. *Brain Research Reviews*, 11, 227–248.

Repérant, J., Rio, J.-P., Miceli, D., Amouzou, M., and Peyrichoux, J. (1981) The retinofugal pathways in the primitive African bony fish *Polypterus senegalus* (Cuvier, 1829). *Brain Research*, 217, 225–243.

Repérant, J., Vesselkin, N. P., Ermakova, T. V., Rustamov, E. K., Rio, J.-P., Palatnikov, G. K., Peyrichoux, J., and Kasimov, R. V. (1982) The retinofugal pathways in a primitive actinopterygian, the chondrostean *Acipenser güldenstädti*. An experimental study using degeneration, radioautographic and HRP methods. *Brain Research*, 251, 1–23.

Robards, M. J., Watkins, D. W. III, and Masterton, R. B. (1976) An anatomical study of somesomesthetic afferents to the intercollicular terminal zone of the midbrain opossum. *Journal of Comparative Neurology*, 170, 499–524.

Ronan, M. and Northcutt, R. G. (1990) Projections ascending from the spinal cord to the brain in petromyzontid and myxinoid agnathans. *Journal of Comparative Neurology*, 291, 491–508.

Rubinson, K. (1968) Projections of the tectum opticum of the frog. *Brain, Behavior and Evolution*, 1, 529–561.

Sakamoto, N., Ito, H., and Ueda, S. (1981) Topographic projections between the nucleus isthmi and the optic tectum in a teleost, *Navodon modestus*. *Brain Research*, 224, 225–234.

Schlussman, S. D., Kobylack, M. A., Dunn-Meynell, A. A., and Sharma, S. C. (1990) Afferent connections of the optic tectum in channel catfish *Ictalurus punctatus*. *Cell and Tissue Research*, 262, 531–541.

Schneider, G. E. (1969) Two visual systems. *Science*, 163, 895–902.

Schroeder, D. M. and Ebbesson, S. O. E. (1975) Cytoarchitecture of the optic tectum in the nurse shark. *Journal of Comparative Neurology*, 160, 443–462.

Schroeder, D. M. and Vanegas, H. (1977) Cytoarchitecture of the tectum mesencephali in two types of siluroid teleosts. *Journal of Comparative Neurology*, 175, 287–300.

Schroeder, D. M., Vanegas, H., and Ebbesson, S. O. E. (1980) Cytoarchitecture of the optic tectum of the squirrelfish, *Holocentrus. Journal of Comparative Neurology*, 191, 337–351.

Sligar, C. M. and Voneida, T. J. (1976) Tectal efferents in the blind cave fish *Astyanax hubbsi. Journal of Comparative Neurology*, 165, 107–124.

Smeets, W. J. A. J. (1981) Efferent tectal pathways in two chondrichthyans, the shark *Scyliorhinus canicula* and the ray *Raja clavata. Journal of Comparative Neurology*, 195, 13–23.

Smeets, W. J. A. J. (1981) Retinofugal pathways in two chondrichthyans, the shark *Scyliorhinus canicula* and the ray *Raja clavata. Journal of Comparative Neurology*, 195, 1–11.

Smeets, W. J. A. J. (1982) The afferent connections of the tectum mesencephali in two chondrichthyans, the shark *Scyliorhinus canicula* and the ray *Raja clavata. Journal of Comparative Neurology*, 205, 139–152.

Smeets, W. J. A. J., Nieuwenhuys, R., and Roberts, B. L. (1983) *The Central Nervous System of Cartilaginous Fishes*. New York: Springer-Verlag.

Sprague, J. M. (1972) The superior colliculus and pretectum in visual behavior. *Investigative Ophthalmology*, 11, 473–482.

Stein, B. E. (1984) Multimodal representation in the superior colliculus and optic tectum. In H. Vanegas (ed.), *Comparative Neurology of the Optic Tectum*. New York: Plenum, pp. 819–841.

Stein, B. E. and Edwards, S. B. (1979) Corticotectal and other corticofugal projections in neonatal cat. *Brain Research*, 161, 399–409.

Stone, J. and Freeman, J. A. (1971) Synaptic organization of the pigeon's optic tectum: a Golgi and current source-density analysis. *Brain Research*, 27, 203–221.

Striedter, G. F. (1990) The diencephalon of the channel catfish, *Ictalurus punctatus*. II. Retinal, tectal, cerebellar and telencephalic connections. *Brain, Behavior and Evolution*, 36, 355–377.

Striedter, G. F. and Northcutt, R. G. (1989) Two distinct visual pathways through the superficial pretectum in a percomorph teleost. *Journal of Comparative Neurology*, 283, 342–354.

Székely, G. (1971) The mesencephalic and diencephalic optic centers in the frog. *Vision Research*, Suppl. 3: 269–279.

Ulinski, P. S. (1977) Tectal efferents in the banded water snake, *Natrix sipedon. Journal of Comparative Neurology*, 173, 251–274.

Ulinski, P. S., Dacey, D. M., and Sereno, M. I. (1992) Optic tectum. In C. Gans and P. S. Ulinski (eds.), *Biology of the Reptilia, Vol. 17: Neurology C, Sensorimotor Integration*. Chicago: University of Chicago Press, pp. 241–366.

Vanegas, H. (1983) Organization and physiology of the teleostean optic tectum. In R. E. Davis and R. G. Northcutt (eds.), *Fish Neurobiology, Vol. 2: Higher Brain Areas and Functions*. Ann Arbor, MI: University of Michigan Press, pp. 43–90.

Vanegas, H., Ebbesson, S. O. E., and Laufer, M. (1984) Morphological aspects of the teleostean optic tectum. In H. Vanegas (ed.), *Comparative Neurology of the Optic Tectum*. New York: Plenum, pp. 93–120.

Vanegas, H., Williams, B., and Freeman, J. A. (1979) Responses to stimulation of marginal fibers in the teleostean optic tectum. *Experimental Brain Research*, 34, 335–349.

Vanegas, H. and Ito, H. (1983) Morphological aspects of the teleostean visual system: a review. *Brain Research Reviews*, 6, 117–137.

Vanegas, H., Laufer, M., and Amat, J. (1974) The optic tectum of a perciform teleost. I. General configuration and cytoarchitecture. *Journal of Comparative Neurology*, 154, 43–60.

Vesselkin, N. P., Ermakova, T. V., Repérant, J., Kosareva, A. A., and Kenigfest, N. B. (1980) The retinofugal and retinopetal systems in *Lampetra fluviatilis*. An experimental study using radioautographic and HRP methods. *Brain Research*, 195, 453–460.

Wang, S.-J., Yan, K., and Wang, Y.-T. (1981) Visual field topography in the frog's nucleus isthmi. *Neuroscience Letters*, 23: 37–41.

Wang, S. R. and Matsumoto, N. (1990) Postsynaptic potentials and morphology of tectal cells responding to electrical stimulation of the bullfrog nucleus isthmi. *Visual Neuroscience*, 5, 479–488.

Wicht, H. and Northcutt, R. G. (1990) Retinofugal and retinopetal projections in the Pacific hagfish, *Eptatretus stouti* (Myxinoidea). *Brain, Behavior and Evolution*, 36, 315–328.

Wilczynski, W. and Northcutt, R. G. (1977) Afferents to the optic tectum of the leopard frog: an HRP study. *Journal of Comparative Neurology*, 173, 219–230.

Wilczynski, W. and Northcutt, R. G. (1983) Connections of the bullfrog striatum: afferent organization. *Journal of Comparative Neurology*, 214, 321–332.

Williams, B., Hernández, N., and Vanegas, H. (1983) Electrophysiological analysis of the teleostean nucleus isthmi and its relationship with the optic tectum. *Journal of Comparative Physiology*, 152, 545–554.

Witkovsky, P., Powell, C. C., and Brunken, W. J. (1980) Some aspects of the organization of the optic tectum of the skate *Raja. Neuroscience*, 5, 1989–2002.

Wullimann, M. F. (1994) The teleostean torus longitudinalis: a short review on its structure, histochemistry, connectivity, possible function and phylogeny. *European Journal of Morphology*, 32, 235–242.

Wurtz, R. H. and Albano, J. E. (1980) Visual-motor function of the primate superior colliculus. *Annual Review of Neuroscience*, 3, 189–226.

Zihl, J. and von Cramon, D. (1979) The contribution of the 'second' visual system to directed visual attention in man. *Brain*, 102, 835–856.

Part Four

THE FOREBRAIN: DIENCEPHALON

19

Overview of the Forebrain

INTRODUCTION

The forebrain consists of two parts—the **diencephalon** and the **telencephalon,** which are extensively interconnected. This chapter is partly an overview of the forebrain, as the title proclaims, but it is more extensive than the previous overview chapters. Some parts of the forebrain, particularly the telencephalic pallium, some of the nuclei in the ventral thalamus, and the striatum, need to be introduced in greater detail here than a simple overview would encompass in order to adequately deal with the content of some of the following chapters.

Four major divisions of the diencephalon are present in all vertebrates: the **epithalamus, dorsal thalamus, ventral thalamus,** and **hypothalamus** (Fig. 19-1). These divisions were recognized by Charles Judson Herrick early in this century from his studies of vertebrate embryos. Caudal to the thalamus, in a transition zone between the diencephalon and the mesencephalon, are two additional regions: the **pretectum** dorsally and the **posterior tuberculum** ventrally.

Most structures within the telencephalon can be assigned to either its dorsal part, the **pallium,** or its ventral part, the **subpallium** (Fig. 19-1). In vertebrates with evaginated telencephalons, the pallium is divided into three major cortical regions: **medial** (limbic) **pallium, dorsal pallium,** and **lateral** (olfactory) **pallium.** In vertebrates with everted telencephalons, homologous divisions of the pallium are present but in a different topographic position due to the difference in development. The ventral part of the telencephalon contains two major areas, the **striatum** and the **septum,** that each contain a number of nuclei. An additional structure called the **claustrum** is present in the telencephalon of mammals. The claustrum consists of a strip of gray matter that lies in an intermediate position between the pallial cortex and the striatum. Whether its origin is pallial or subpallial has not yet been determined.

THE DIENCEPHALON

During development, the periventricular matrix gives rise to neurons that either remain *in situ* or migrate and form the nuclei of the various parts of the diencephalon. The degree of migration varies among Group I and Group II vertebrates. It also varies among the different parts of the diencephalon. Thus, different parts of the diencephalon are developed to a greater or lesser degree, depending on the particular group of vertebrates. In this section, we will survey the major sets of nuclei in each region of the diencephalon that are present when that region is relatively well developed.

Pretectum

The pretectum (Fig. 19-2) is a region that lies between the dorsal thalamus and the optic tectum. Its nuclei occupy **superficial, central,** and **periventricular** (or deep) **zones** that are in continuity with the superficial, central, and periventricular zones, respectively, of the optic tectum. A major commissural fiber tract, the **posterior commissure,** crosses the midline through the pretectal region, and a rostrocaudally running fiber tract, the fasciculus retroflexus, courses through the periventricular part of the pretectal region. The major inputs to the pretectum are from the retina and the optic tectum. The efferent pathways of the pretectum are involved in the regulation of eye and body movements in relation to prey and other features of the visual world. The pretectum acts in concert

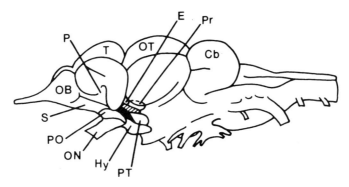

FIGURE 19-1. Lateral view of the brain of a longnose gar, showing the positions of the major divisions of the diencephalon. The extent of the dorsal thalamus is indicated with diagonal lines and that of the ventral thalamus with shading. Abbreviations: Cb, cerebellum; E, epithalamus; Hy, hypothalamus; OB, olfactory bulb; ON, optic nerve; OT, optic tectum; P, pallium of the telencephalon; PO, preoptic area; Pr, pretectum; PT, posterior tuberculum; S, striatum; T, telencephalon.

with another nucleus or pair of nuclei that lie in the tegmentum and constitute the **accessory optic system.** The relations between the pretectum and the accessory optic system are discussed in Chapter 20.

Posterior Tuberculum

Ventral to the area of the pretectum is another transitional zone between the diencephalon and mesencephalon called the posterior tuberculum (Figs. 19-2 and 19-3). Migrated nuclei of the posterior tuberculum are present only in fishes. The unimigrated part of the posterior tuberculum was discussed in Chapter 17, as its cells contain dopamine and may be homologous as a field to the substantia nigra and ventral tegmental area of cartilaginous fishes and/or amniote vertebrates. Migrated nuclei of the posterior tuberculum relay sensory information to the telencephalon from the lateral line and gustatory

systems. These pathways through the migrated nuclei of the posterior tuberculum are similar to sensory pathways through dorsal thalamic nuclei in their organization. The migrated tubercular nuclei are not homologous to any known nuclei in tetrapod vertebrates, however.

Epithalamus

The epithalamus (Fig. 19-1) is the most dorsal part of the diencephalon. In most vertebrates, it is composed of two major parts: the **pineal gland** and the **habenula** [Figs. 19-4(A) and 19-5]. In some vertebrates, the epithalamus also includes a **parietal eye.** The epithalamus is involved with the maintenance of circadian rhythms in response to the light cycle and is part of a major pathway for integration of limbic and striatal systems. The major afferent fiber tract to the habenula is the **stria medullaris** (Fig. 19-5), and its major efferent fiber tract is the **fasciculus retroflexus.**

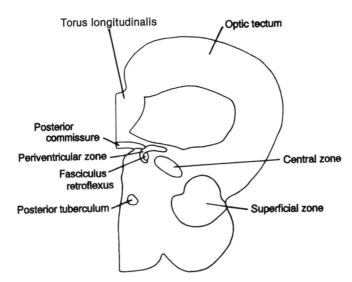

FIGURE 19-2. Drawing of a transverse hemisection through the right pretectum and posterior tuberculum of a herring (*Clupea harengus*). Data from Butler and Northcutt (1993). Note that in this figure and in figures in this and succeeding chapters on the forebrain, the ventricle is generally not labeled.

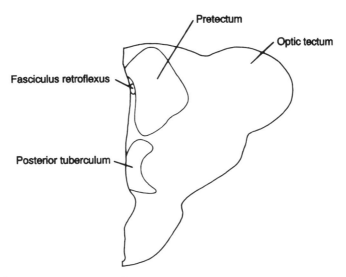

FIGURE 19-3. Drawing of a transverse hemisection through the right pretectum and posterior tuberculum of a bullfrog (*Rana catesbeiana*). Data from Neary and Northcutt (1983).

Dorsal Thalamus

The dorsal thalamus (Figs. 19-1, 19-4, and 19-5) is the gateway to the telencephalon. It includes nuclei that relay sensory information to the telencephalon and, in some vertebrates, nuclei that are primarily involved in circuits with pallial areas of the telencephalon. The greater part of the sensory information that reaches dorsal thalamic nuclei originates on the opposite, or contralateral, side of the body or of space. In anamniotes, a significant portion of the dorsal thalamic projections to the telencephalon are bilateral, while the rest are restricted to the telencephalon on the same, or ipsilateral, side. In amniotes, dorsal thalamic projections to the ipsilateral telencephalon predominate, but some bilateral pathways are present.

The **medial** and **lateral forebrain bundles** are a set of fiber tracts that course through the diencephalon, connecting telencephalic and diencephalic regions with each other and with more caudal parts of the brain. Another major fiber tract of the diencephalon is the **optic tract,** which is the continuation of the optic nerve past the **optic chiasm;** most or all of the

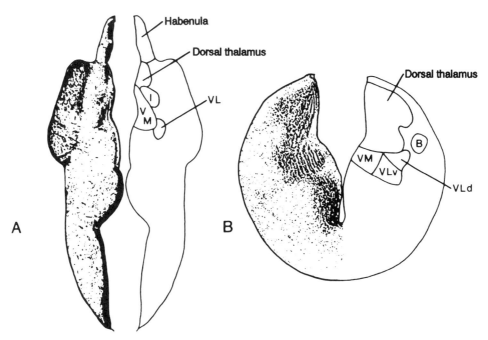

FIGURE 19-4. Transverse hemisections through the left diencephalons of (A) the bowfin (*Amia calva*) and (B) the bullfrog (*Rana catesbeiana*) with mirror-image drawings. Data from Butler and Northcutt (1992) and Neary and Northcutt (1983), respectively. Material from the former used with permission of S. Karger AG.

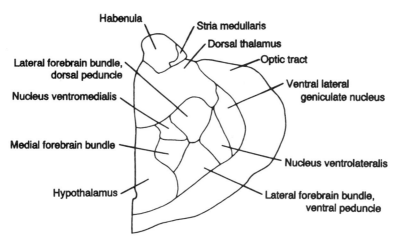

FIGURE 19-5. Drawing of a transverse hemisection through the right diencephalon of a lizard (*Iguana iguana*). Data from Butler and Northcutt (1973).

retinal ganglion cell axons in the optic nerve cross to the contralateral side of the brain via the optic chiasm. The axons in the optic tract project to sites in the diencephalon and to the optic tectum.

In the chapter below on the dorsal thalamus, we will introduce the various nuclei and nuclear groups of the dorsal thalamus. In most anamniote vertebrates, only three dorsal thalamic nuclei are present; in contrast, amniotes have a greatly expanded dorsal thalamus consisting of many nuclei and/or nuclear groups. A recent analysis (see Butler, 1994a) shows that the organizational plan of the dorsal thalamus in each of the major radiations of vertebrates, including mammals, is similar. The mammalian dorsal thalamus does have a number of derived (apomorphic) features, which, when identified and separated out, reveal the presence of the same basic features of the dorsal thalamus common to all vertebrates. Dorsal thalamic nuclei can be divided into two categories in all vertebrates:

- A **lemnothalamus** comprises nuclei predominantly in receipt of lemniscal retinal, somatosensory, and other afferents.

- A **collothalamus** comprises nuclei predominantly in receipt of sensory inputs relayed to them from the roof of the midbrain.

Ventral Thalamus

The ventral thalamus (Fig. 19-1) lies ventral to the dorsal thalamus and is involved with some sensory processing and with motor control systems. We will discuss some of these nuclei in detail in this chapter, since a separate chapter will not be devoted to the ventral thalamus. The other major nuclei within the ventral thalamus are related to motor control and will be discussed in Chapter 24 on the striatum.

In most anamniotes, at least three nuclei are present in the ventral thalamus. In cartilaginous and ray-finned fishes, these nuclei [Fig. 19-4(A)] have been termed **nucleus intermedius (I), nucleus ventromedialis (VM),** and **nucleus ventrolateralis (VL).** These nuclei all receive retinal projections, but most of their other connections remain to be investigated. In amphibians [Fig. 19-4(B)], a nucleus ventromedialis (VM) and dorsal **(VLd)**

and ventral **(VLv)** divisions of nucleus ventrolateralis have been recognized. Also present in a more lateral position is a nucleus that receives a heavy retinal projection and is called the **nucleus of Bellonci.** Its caudal continuation is a cell sparse area, the **neuropil of Bellonci** [B in Fig. 19-4(B)]. This nucleus was long thought to be a dorsal thalamic visual relay to the telencephalon, but it has no telencephalic projections and is now assigned to the ventral thalamus. Whether it might correspond to the nucleus intermedius of fishes is an open question.

Among amniotes, the ventral thalamus has been most extensively studied in mammals in which it is frequently referred to as the **subthalamus.** The ventral thalamus of mammals contains three nuclei that are interconnected primarily with the globus pallidus of the striatum and with the substantia nigra and play a role in motor control (see Chapter 24). In nonmammalian amniotes, the ventral thalamus also contains four or more nuclei, two of which are **nucleus ventromedialis** and **nucleus ventrolateralis** (Fig. 19-5). The ventrolateral nucleus receives a minor input from the retina, but the other connections of these two nuclei remain to be investigated.

Lateral Geniculate Nucleus Pars Ventralis. In all amniotes, the ventral thalamus contains a nucleus called the **ventral part** (or the **pars ventralis**) of the **lateral geniculate nucleus,** also referred to as the **ventral lateral geniculate nucleus.** This nucleus receives projections from the retina and optic tectum and is involved in visual functions. The ventral lateral geniculate nucleus does not project to the telencephalon as do the collothalamic and lemnothalamic visual nuclei of the dorsal thalamus. It participates instead in brainstem visual pathways. Among primates, this nucleus is known as the **pregeniculate nucleus** and is very variable in its size and location. In diapsid reptiles, birds, and turtles, the ventral lateral geniculate nucleus is of substantial size and forms an oval or lens-shaped nuclear mass that curves around the lateral edge of the diencephalon (Fig. 19-5).

In mammals, two divisions of the ventral lateral geniculate nucleus have been identified, each of which contains subdivisions with unique histochemical and connectional profiles. The subdivisions of the lateral division are involved in visuosensory processing; the lateral division receives projections from the retina, striate (visual) isocortex, pretectum, and superior colliculus. The lateral division, in turn, projects back to the superior colliculus and the pretectum and also projects to the pulvinar and dorsal lateral geniculate nucleus of the dorsal thalamus and to pontine nuclei and other brainstem visually related sites. In contrast, the subdivisions of the medial division of the ventral lateral geniculate nucleus are involved in visuomotor functions, with a role in the coordination of head and eye movements. This division projects to pontine nuclei, which in turn project to the cerebellum. The lateral cerebellar nucleus projects to the medial division of the ventral lateral geniculate nucleus, thus closing a ventral geniculate-ponto-cerebellar-ventral geniculate loop. The medial division also receives projections from pretectal nuclei and from visually related areas of parietal isocortex; in addition to pontine projections, it projects to the sensory-motor (deeper) part of the superior colliculus.

Thalamic Reticular Nucleus. A fifth ventral thalamic nucleus in mammals is the **thalamic reticular nucleus,** which is interconnected primarily with the nuclei of the dorsal thalamus and appears to be a rostral counterpart of the reticular formation of the hindbrain. The thalamic reticular nucleus is formed by a thin sheet of cells that enfolds the anterior, lateral, and ventral surfaces of the dorsal thalamus; the nucleus is traversed by axons that interconnect the telencephalon with the dorsal thalamus and the rest of the brainstem. The thalamic reticular nucleus receives projections from the isocortex and also a cholinergic input from nucleus cuneiformis in the midbrain (see Chapter 17), a part of the reticular formation. All of the neurons of the thalamic reticular nucleus are GABAergic and also positive for parvalbumin, a calcium-binding protein that frequently occurs in association with GABAergic neurotransmission. These neurons provide short latency, recurrent inhibitory input to the dorsal thalamic relay nuclei bilaterally.

A thalamic reticular nucleus has also been identified in crocodiles and may thus be a common feature of amniotes. This nucleus, however, appears to be more complexly organized than its mammalian counterpart. The neuron cell bodies of the nucleus lie in a location similar to that of the thalamic reticular nucleus of mammals, scattered among axons of the dorsal part, or peduncle, of a major fiber tract, the lateral forebrain bundle (Fig. 19-5), that interconnects the telencephalon with the rest of the brain. The neurons of the thalamic reticular nucleus project to the nuclei of the dorsal thalamus, as is the case in mammals, but they are not GABAergic. A small number of GABAergic neurons are present in the region but do not project to the dorsal thalamus. Some of the thalamic reticular neurons are positive for parvalbumin while others are not. Thus, several subpopulations of neurons are present in this nucleus in crocodiles, whereas only one population of neurons appears to constitute the homologous nucleus in mammals.

Hypothalamus and Preoptic Area

The hypothalamus (Fig. 19-1) lies in the most ventral part of the diencephalon. The region rostral to the hypothalamus is called the **preoptic area,** named for its position immediately dorsal to the optic chiasm. The preoptic area forms the rostral-most part of the diencephalon and is functionally related to the hypothalamus. The hypothalamus and preoptic area are primarily involved with the regulation of the internal organ systems of the body. The nuclei in these regions are connected with the hypophysis (pituitary) by fiber connections and hormonal systems. The preoptic area and hypothalamus are also widely connected with other parts of the brainstem and with the limbic system of the forebrain. A multitude of functions—including diurnal rhythms, reproductive behaviors, temperature regulation, feeding, drinking, sleeping, and emotional responses—are all under the control of or influenced by various parts of these regions.

THE TELENCEPHALON: PALLIUM

The telencephalic pallium in all vertebrates has three major divisions, although the question of whether these recognized divisions are homologous, each to each, as derivatives of embryonic fields or in other specified ways, across all groups of vertebrates has not been resolved. In Group I vertebrates with evaginated, laminar telencephalons (such as lampreys, squalomorph sharks, and amphibians) the three divisions are often referred to by geographical names: medial, dorsal, and lateral (Fig. 19-6). In vertebrates with simply everted telencephalons, such as some ray-finned fishes, the mediolateral position of these cortices is reversed (see Chapter 3), although whether all three of these regions are strict homologues of the medial, dorsal, and lateral pallia in vertebrates with evaginated telencephalons is in doubt. In the everted brain, the "medial" pallium is in a lateral position, and the "lateral" pallium is in a medial position. The pallial regions are designated as **P1, P2,** and **P3,** with P1 in the most medial position (Fig. 19-7). In teleosts, the migration of neurons away from the everted, periventricular surface during development results in a secondary alteration of the topographic relationships of both the pallial and the striatal areas.

In vertebrates with evaginated telencephalons, the medial pallium can also be referred to as the **limbic pallium,** a telencephalic part of the **limbic system** that is interconnected with the septum and the hypothalamus. The term limbic pallium can similarly be used to refer to the putatively homologous lateral pallium (P3) in vertebrates with everted telencephalons.

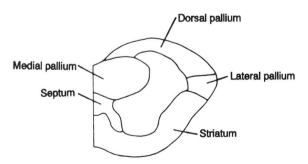

FIGURE 19-6. Drawing of a transverse hemisection through the right telencephalon of a squalomorph shark (*Notorynchus maculatus*). Data from Northcutt (1978).

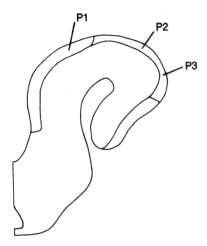

FIGURE 19-7. Drawing of a transverse hemisection through the right telencephalon of a non-teleost ray-finned fish, the bichir (*Polypterus palmas*). Data from Reiner and Northcutt (1992).

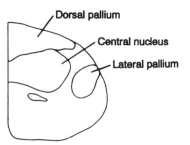

FIGURE 19-9. Drawing of a transverse hemisection through the right telencephalon of a shark (*Ginglymostoma cirratum*). Data from Luiten (1981).

includes a large region called the central nucleus, which at some levels extends all the way to the midline (Fig. 19-9). In amphibians (Fig. 19-10), medial, dorsal, and lateral pallial areas have been identified, but, as is the case with other anamniotes, questions persist regarding the homology of these areas with similarly designated pallial areas in amniotes.

Cerebral Cortex of Mammals

In mammals (Fig. 19-11), the limbic (medial) pallium, containing the **dentate gyrus** and a formation called **Ammon's horn**, as well as the olfactory (lateral) cortices, containing piriform cortex, are present in the telencephalic pallium, and each has three layers of cells and fibers. Such three-layered cortices are referred to as **allocortex.** The dorsal pallium is relatively greatly expanded. Most of the dorsal pallium forms a six-layered cortex called **isocortex** or **neocortex** (Figs. 19-11 and 19-12). Cortex that lies in the transition zone between allocortex and isocortex and has an intermediate number of layers is referred to as **transitional cortex.**

The outermost layer of isocortex, **layer I,** is the **molecular layer.** It is relatively cell sparse, consisting predominantly of afferent axons and the dendritic ramifications of neurons in deeper cortical layers. **Layer II** is called the **external granular layer,** in reference to its population of **granule cells,** relatively small neurons with short axons. **Layer III** is the **external pyramidal layer** and contains **pyramidal cells,** which have den-

Likewise, the adjectives **olfactory** or **piriform** can be used to refer to the lateral pallium in vertebrates with evaginated telencephalons and to the medially positioned pallium (P1) in those with everted telencephalons. The olfactory pallium, or primary olfactory cortex, is in receipt of projections relayed through the olfactory bulb. The olfactory bulb also projects directly to part of an associated group of nuclei that is most extensively developed in mammals and is collectively referred to as the **amygdala.**

In hagfishes (Fig. 19-8), the telencephalon is markedly expanded. The telencephalon undergoes evagination during development; the pallial areas are enlarged and laminated, and all receive direct olfactory projections. The pallial areas have been designated as **P1–P5,** with P1 being the most superficial, but no correspondence with the pallial areas P1–P3 in ray-finned fishes is meant to be implied by this terminology. The connections of the pallium in hagfishes need further study before comparisons can be made with other vertebrates.

Limbic (medial), dorsal, and olfactory (lateral) pallial divisions are present in sharks, skates and rays. The dorsal pallium

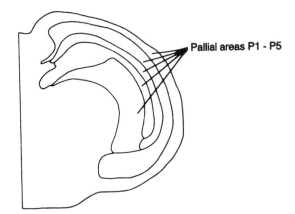

FIGURE 19-8. Drawing of a transverse hemisection through the right telencephalon of a hagfish. (*Eptatretus stouti*). Data from Wicht and Northcutt (1992). Used with permission of S. Karger AG.

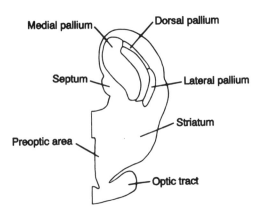

FIGURE 19-10. Drawing of a transverse hemisection through the right telencephalon of a bullfrog (*Rana catesbeiana*). Data from Northcutt and Kicliter (1980).

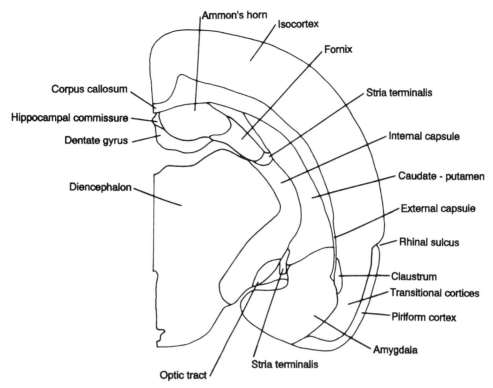

FIGURE 19-11. Drawing of a transverse hemisection through the right forebrain of a rat (*Rattus norvegicus*). Adapted from Pelligrino et al. (1979). Used with permission of Plenum Publishing Corp.

drites that extend from the apex of the cell soma towards the external surface of the cortex and are known as **apical dendrites.** Neurons in layers II and III give rise to projections to the deeper cortical layers, to other cortical areas, or to some subcortical structures.

Layer IV is the **internal granular layer** and contains a plethora of granule cells. Layers II and VI are usually well developed in cortical areas that receive ascending sensory projections from the dorsal thalamus, and this type of cortex is therefore called **granular cortex** or **koniocortex.** (The prefix "konio" is derived from the Greek word for dust, referring to the granulated appearance of the cortex.) Some mammals lack a well-developed layer IV, however, such as the hedgehog tenrec, an insectivore whose isocortex is illustrated in Figure 19-12. In contrast, a very prominent layer IV is shown in the visual cortex of a raccoon in Figure 26-6.

The **infragranular layers, layers V** and **VI,** give rise to the set of efferent cortical fibers that project to subcortical structures. Layer V is the **internal pyramidal layer,** and layer VI is called the **multiform layer.** Neurons in the latter are predominantly **fusiform cells,** which have elongated somata. Layer V neurons give rise to projections to the striatum, the superior colliculus, and the spinal cord, while layer VI neurons give rise to reciprocal projections to the dorsal thalamus. The pyramidal layers, layers III and V, are particularly well developed in cortical areas involved in motor control, and this type of cortex is therefore referred to as agranular cortex.

The isocortex has a columnar organization in that it is functionally organized as vertical **columns,** each about 300 μm in diameter, which extend perpendicularly through all six cortical layers. In sensory cortices, each column receives the input from a given peripheral receptive field, and the columns are arranged so that the topographic organization of the receptive fields is maintained.

In some mammals, the surface of the telencephalic hemispheres is relatively smooth (Fig. 19-13). This condition is referred to as **lissencephalic.** In other mammals (Fig. 19-14), particularly primates, the cerebral cortex is hypertrophied and is folded into **gyri** (singular = gyrus), or ridges, with **sulci** (singular = sulcus), or valleys, between them. This condition is referred to as **gyrencephalic.** During development in a gyrencephalic species, the crowns of the gyri grow outward, while the sulcal fundi remain relatively static (Fig. 19-15). The particular cortical areas defined by gyri in gyrencephalic species can be compared across species, whereas the sulci are more variable. Sulci are useful as guides to the location and identity of particular gyri.

Isocortex occupies the bulk of the cerebral hemispheres, which can be divided into **frontal, parietal, temporal,** and **occipital lobes** (Fig. 19-14). In gyrencephalic brains, two of the major sulci are the **central sulcus** (of Rolando), which divides the frontal from the parietal lobe, and the **lateral sulcus,** or Sylvian sulcus, which divides the temporal from the frontal and parietal lobes. An area of isocortex called **insular cortex** lies along the deep banks of the lateral sulcus. An additional sulcus to note here, the **rhinal sulcus,** separates isocortex from olfactory cortex (Fig. 19-11).

The motor cortex and the somatosensory cortex are located in adjacent or, in some mammals, overlapping parts of the frontal and parietal lobes, respectively. The frontal lobes also

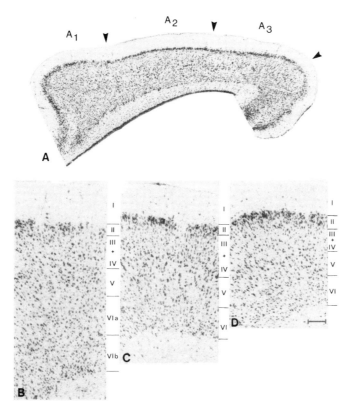

FIGURE 19-12. Medial part of isocortex in a hedgehog tenrec (*Echinops telfairi*). (A) Parasagittal section with rostral to the left showing areas A1, A2, and A3, with their boundaries indicated by arrowheads. Parts B, C, and D are transverse sections through areas A1, A2, and A3, respectively. The cortical lamination is less elaborate than in most other mammals, and a robust layer IV, the internal granular layer, is absent. From Künzle and Rehkämper (1992).

contain a related premotor cortical area and the frontal eye fields, which are involved in the control of eye movements. In humans, a part of the frontal lobe called Broca's area is involved in the motor control of speech. Auditory cortex lies in the temporal lobe, and visual cortex is located in the occipital lobe. The so-called primary, or striate, visual cortex occupies the banks of the **calcarine sulcus,** which is shown in Figure 19-16. Association cortices are present in addition to primary sensory cortices. A large part of the frontal lobe called the prefrontal

cortex is regarded as an association cortex, although it also receives a substantial ascending sensory input from the dorsal thalamus. Association cortical areas are also present in the temporal and parietal lobes, lying between the auditory, somatosensory, and visual sensory cortical areas.

The recognition of the various cortical areas and of their boundaries is an actively continuing topic of research. Boundaries are most often defined by changes in the cytoarchitecture (the pattern of cellular distribution) in the cortical layers and/or changes in maps of sensory representations as revealed by electrophysiology. A nineteenth century German neurologist, Korbinian Brodmann, was a pioneer of cytoarchitectonic studies. In 1908 Brodmann published a map of the human cortex, assigning a number to each of almost 50 identified regions. Many of these numbers are in the neuroscientific vernacular today, such as 3,1,2 for somatosensory cortex and 17 for the striate visual cortex.

A substantial range of variation exists between species in the pattern of gyri and sulci, where present, and in the extent and position of sensory and motor areas. These areas need to be identified in each species by studies tracing afferent connections or recording electrophysiological responses to stimulation. Sensory and motor areas identified in the brain of an echidna are shown in Figure 19-17. Variation in the pattern of gyri and sulci is shown among four mammals—a goat, a cat, an indri lemur, and a chimpanzee—in Figure 19-18.

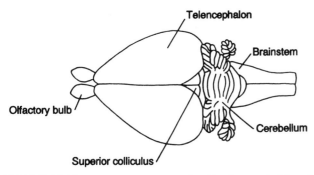

FIGURE 19-13. Dorsal view of a brain of a rabbit (*Lepus sp.*). Adapted from Northcutt (1979). Used with permission of The University of Chicago Press.

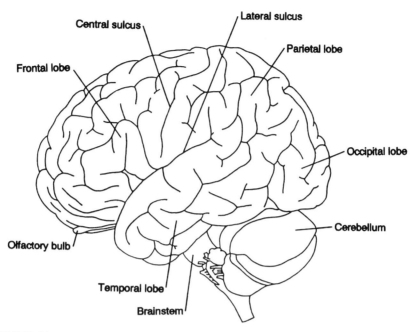

FIGURE 19-14. Lateral view of the brain of a primate (*Homo sapeins*). Data from Nieuwenhuys et al. (1978). Used with permission of Springer-Verlag.

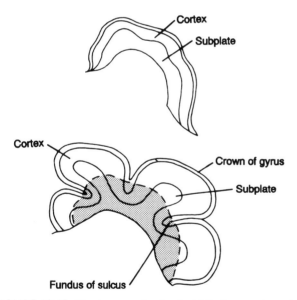

FIGURE 19-15. Development of gyri and sulci. Earlier stage in cortical development (top) showing the relative size and position of the developing subplate and cortex. Later stage (bottom) showing the subplate and cortex with gyral development and, indicated by shading, the earlier stage superimposed for comparison. Note that the sulcal fundi are near the surface of the earlier stage, while the gyral crowns have grown outward. Adapted from Welker (1990). Used with permission of Plenum Publishing Corp.

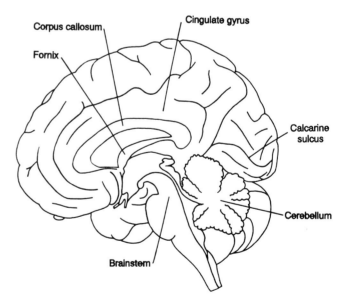

FIGURE 19-16. Medial mid-sagittal view of the brain of a primate (*Homo sapiens*), with rostral toward the left. Data from Nieuwenhuys et al. (1978). Used with permission of Springer-Verlag.

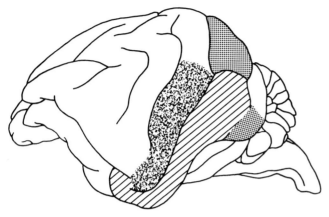

FIGURE 19-17. Lateral view of the brain of an echidna (*Tachyglossus aculeatus*), with rostral toward the left. The area of motor cortex is indicated by random stippling and that of somatosensory cortex by diagonal lines. Visual cortex is the shaded area dorsal to somatosensory cortex, and auditory cortex is the shaded area caudal to somatosensory cortex. Adapted from Rowe (1990). Used with permission of Plenum Publishing Corp.

The Olfactory Cortex of Mammals

The isocortex is expanded to such a degree in many mammals that parts of the limbic and olfactory cortices come into contact with each other ventrally in the temporal lobe. From rostral to caudal, the olfactory cortices, which receive direct input from the olfactory bulb, comprise the **anterior olfactory nucleus, piriform cortex,** the **olfactory tubercle** (to which

we will return below), a laminated part of a nuclear group called the **cortical,** or **corticomedial, amygdala,** and the **entorhinal cortex** (Figs. 19-11, 19-19, and 19-20). Additionally, the accessory olfactory bulb projects to a part of the cortical amygdala.

The Limbic Telencephalon of Mammals

The limbic cortex (Figs. 19-11 and 19-20) includes the **dentate gyrus,** the **hippocampus,** which is also referred to as **Ammon's horn,** and the **subicular cortices.** The hippocampal formation was divided into regions designated CA$_1$ to CA$_4$ (CA for *cornu Ammonis,* which is Ammon's horn in Latin) by Lorente de No in the early part of this century, and these designations are frequently used in the literature. The subicular cortices border entorhinal cortex, which in turn borders the piriform cortex. The **cingulate gyrus** (Figs. 19-16, 19-19, and 19-20), a part of the isocortex on the medial surface of the cerebral hemisphere, is connectionally and functionally included in the limbic system. The subicular cortices and entorhinal cortex are transitional cortices between the isocortex and the allocortical hippocampal and olfactory cortices.

Also a part of the limbic system is the collection of nuclei called the **amygdala** (Figs. 19-11 and 19-19). These nuclei are located in the rostral end of the temporal lobe. As referred to above, a laminated part of the amygdaloid nuclear complex—called the cortical or corticomedial amygdala—can be regarded as primary olfactory cortex since it receives direct projections from the olfactory bulb. The corticomedial amygdala is also

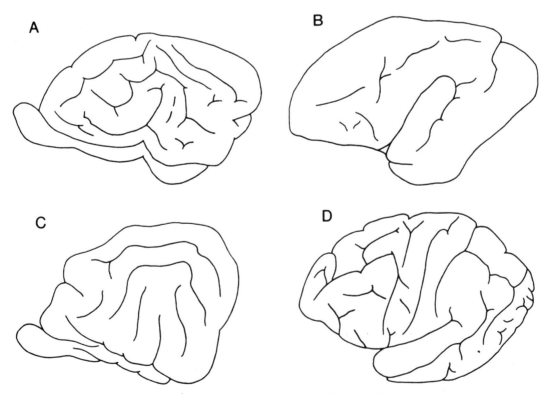

FIGURE 19-18. Lateral views of the cerebral hemispheres in (A) a goat, (B) an indri lemur, (C) a domestic cat, and (D) a chimpanzee. Data from Ariëns Kappers et al. (1967).

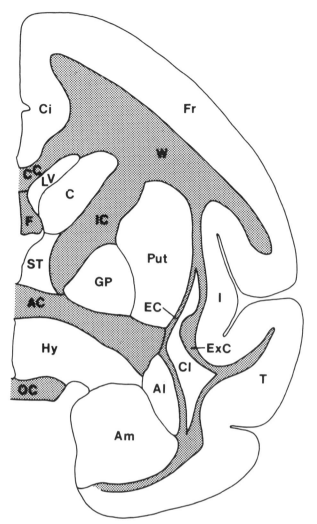

FIGURE 19-19. Drawing of a transverse hemisection through the right forebrain of a squirrel monkey (*Saimiri sciureus*). The major white matter fiber tracts are shaded. Abbreviations: AC, anterior commissure; Al, baso-lateral amygdala; Am, corticomedial amygdala; C, caudate nucleus; CC, corpus callosum; Ci, cingulate gyrus; Cl, claustrum; EC, external capsule; ExC, extreme capsule; F, fornix; Fr, frontal lobe isocortex; GP, globus pallidus; Hy, hypothalamus; I, insular isocortex; IC, internal capsule; LV, lateral ventricle; OC, optic chiasm; Put, putamen; ST, bed nucleus of the stria terminalis; T, temporal lobe isocortex; W, isocortical white matter. Data from Gergan and MacLean (1962).

connected with the hypothalamus, via a tract called the **stria terminalis,** and with the **bed nucleus of the stria terminalis** (Fig. 19-26). The **basal,** or **basolateral, amygdala** projects to the striatum and is widely interconnected with areas of associa-tion isocortex.

Fiber Tracts in the Telencephalon of Mammals

Many fiber tracts are present in the telencephalon. For orientation, several need to be introduced here. In mammals, the largest pathway connecting the telencephalon with the rest of the brain is a collection of afferent and efferent fibers that,

in the telencephalon, are called the **internal capsule** (Figs. 19-11 and 19-19). The internal capsule is a clearly demarcated fiber bundle as it passes through the region of the striatum. In fact, it is the fibers of the internal capsule that form the "stripes" of the striatum, giving it its name. The fibers in the internal capsule continue caudally through the brainstem, where they are referred to as the **cerebral peduncle** (Fig. 19-20). Superfi-cial to the internal capsule are two other, smaller fiber bundles: the **external capsule** and the **extreme capsule** (Fig. 19-19).

In all vertebrates, some fiber bundles are present in the telencephalon, as in the rest of the brain, that connect the two sides with each other. In mammals, the **hippocampal commissure** (Fig. 19-11) reciprocally interconnects the hippo-campal formations, and in primates, parts of isocortex. The **anterior commissure** (Fig. 19-19) reciprocally interconnects some of the more ventral and lateral parts of the telencephalon, including the olfactory cortices. In monotremes and marsupials, the anterior commissure also carries all of the commissural fibers that arise from isocortex, while in eutherian (placental) mammals, the anterior commissure carries only some of the isocortical commissural fibers, particularly those from the tem-poral isocortex.

Eutherian mammals have one additional commissural sys-tem that is unique among the vertebrate radiations. This unique commissural bundle is very large and is called the **corpus callosum** (Figs. 19-16, 19-19, and 19-20). Its fibers reciprocally connect wide areas of the isocortex with their respective, homo-topic (mirror-image) areas in the contralateral isocortex. The corpus callosum also carries commissural fibers from transi-tional cortex and the striatum. Several theories have been ad-vanced concerning the evolution of this commissure, including derivation from either the anterior commissure or the hippo-campal commissure, but a definitive consensus has not been achieved.

The Telencephalic Pallium of Nonmammalian Amniotes

As in mammals, the telencephalic pallium in diapsid rep-tiles, birds, and turtles is divisible into medial, dorsal, and lateral parts. In diapsid reptiles and turtles, neurons in the olfactory (lat-eral) pallium, part of the dorsal pallium, and the limbic (medial) parts of the pallium are migrated away from the periventricular surface so that a three-layered cortex—of two relatively cell-free zones with an intermediate cell-dense layer—is formed. The three-layered part of the dorsal pallium, the **dorsal cortex** (Fig. 19-21), receives ascending visual and somatosensory projections from nuclei in the dorsal thalamus that are in receipt of lemniscal pathways. As in most anamniote vertebrates, the surface of the cerebral hemisphere is relatively smooth (Fig. 19-22).

A second, substantial part of the dorsal pallium is formed by a large area, the **anterior dorsal ventricular ridge** (Fig. 19-21), which is a medially expanded bulge deep to the cortices. The anterior dorsal ventricular ridge (sometimes referred to simply as the dorsal ventricular ridge) receives ascending visual, auditory, and somatosensory projections from dorsal thalamic nuclei that receive their input from the midbrain tectum (the colliculi of mammals).

The medial pallium, which comprises the **medial** and **dor-somedial cortices** (Fig. 19-21), is thought to be homologous,

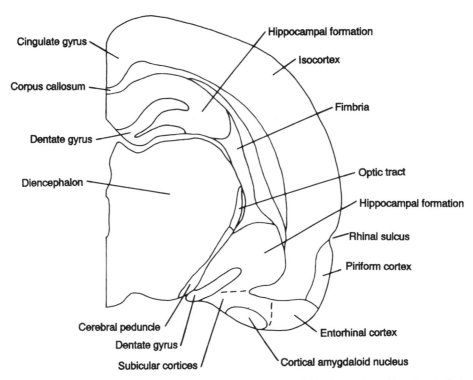

FIGURE 19-20. Drawing of a transverse hemisection through the right forebrain of a rat. Data from Pelligrino et al. (1979). Used with permission of Plenum Publishing Corp.

at least as a field, to the medial (limbic) pallium of mammals. The medial cortex has been homologized to the dentate gyrus as a specific set of neurons and the dorsomedial cortex likewise to Ammon's horn, although some data suggest a less discrete correspondence.

Some workers have proposed that the entire dorsal cortex of diapsid reptiles and turtles is homologous to transitional cortices, particularly the subiculum, rather than to any part of mammalian isocortex. Others have hypothesized that the dorsal cortex and the anterior dorsal ventricular ridge are homologous, as a field, to the isocortex (neocortex) of mammals. Still others have made a case for the hypothesis that the anterior dorsal ventricular ridge and those parts of the isocortex with similar ascending connections are independently evolved and therefore homoplaseous.

A recent analysis of the dorsal pallium in amniotes (see Butler, 1994b) somewhat bridges and reconciles these conflicting views. As in the case of the dorsal thalamus, which can be divided into a predominantly lemniscal-recipient lemnothalamus and a predominantly midbrain roof-recipient collothalamus, two divisions of the dorsal pallium appear to be present in all amniotes:

- A medial, **lemnothalamic pallium** receives its predominant dorsal thalamic input from the lemnothalamus.
- A lateral, **collothalamic pallium** receives its predominant dorsal thalamic input from the collothalamus.

This hypothesis holds that the dorsal cortex of nonmammalian amniotes is homologous as two or more fields to parts of isocortex (i.e., those parts that receive projections from the

lemnothalamus) and to the transitional cortices of mammals, while the anterior dorsal ventricular ridge is composed of several fields that are homologous to several respective fields in the rest of isocortex of mammals (i.e., those parts that receive projections from the collothalamus). Some features, such as the particular morphology of the dorsal ventricular ridge versus that of isocortex and particular cell phenotypes within the dorsal pallia of mammals and of some nonmammalian amniotes were evolved independently within the areas homologous as fields.

The **lateral cortex** (Fig. 19-21) in diapsid reptiles and turtles is thought to be homologous to at least part of the olfactory cortex of mammals. It is in receipt of direct olfactory projections. In the caudal part of the dorsal ventricular ridge is a large nucleus, called **nucleus sphericus** (Fig. 19-23), which consists of a laminar cell layer and a central neuropil and receives an input from the accessory olfactory bulb. Nucleus sphericus is homologous to part of the amygdalar nuclear group of mammals. Other amygdalar nuclei are also present, bordering the more rostral part of nucleus sphericus in the caudal part of the dorsal ventricular ridge.

In birds, as in diapsid reptiles, the pallium consists of medial (limbic), dorsal, and lateral (piriform) cortices and the dorsal ventricular ridge, which includes several different areas, including those called the **neostriatum** and **ectostriatum** (Fig. 19-24). The use of the term "striatum" in this nomenclature reflects the now discredited belief that the dorsal ventricular ridge in both reptiles and birds is part of the striatum. In the rostral part of the telencephalon in birds, an expanded part of the dorsal pallium lies in a dorsomedial position and corresponds to at least part of the dorsal cortex of diapsid reptiles. This area receives ascending visual and somatosensory projections and has been named the

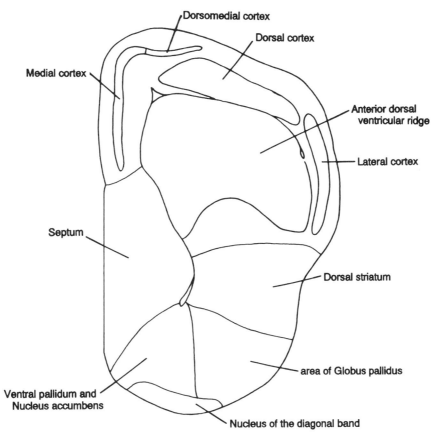

FIGURE 19-21. Drawing of a transverse hemisection through the right telencephalon of a lizard (*Gekko gecko*). Data from Ulinski (1983).

hyperstriatum accessorium, or **"Wulst"** (Fig. 19-13). Wulst is German for swelling, bulge, or hump, as is formed by this structure on the dorsal surface of the cerebral hemisphere (Fig. 19-24).

While the dorsal pallium of diapsid reptiles, birds, and turtles may contain regions that are homolgous to respective regions of mammalian isocortex as part or all of various fields, as discussed above, some specific features of given pallial areas have been independently derived in different lineages. Two specific phenotypes of dorsal pallial neurons, as defined by their neurotransmitter or neuroactive substance profile, appear to be present only in mammals, while a third appears to have been evolved independently in both mammals and birds.

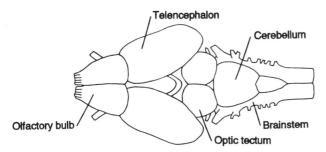

FIGURE 19-22. Dorsal view of the brain of a turtle (*Chrysemys picta*). Data from Northcutt (1979). Used with permission of The University of Chicago Press.

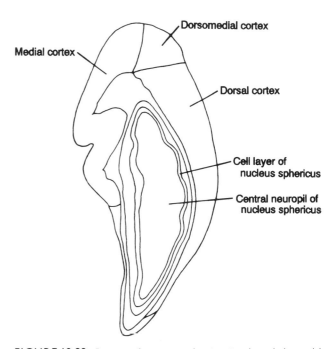

FIGURE 19-23. Drawing of a transverse hemisection through the caudal pole of the telencephalon of a lizard (*Tupinambis nigropunctatus*). Data from Ebbesson and Voneida (1969). Used with permission of S. Karger AG.

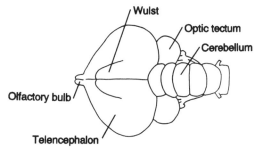

FIGURE 19-24. Dorsal view of the brain of a pigeon (*Columba livia*). Data from Northcutt (1979). Used with permission of The University of Chicago Press.

A variety of neuroactive substances characterize neurons in the infragranular layers (V and VI) of mammalian isocortex and are also present in the dorsal pallium of nonmammalian amniotes. In contrast, three substances—cholecystokinin-8 (CCK8), vasoactive intestinal polypeptide (VIP), and choline acetyltransferase (CHAT, an indicator of the presence of acetylcholine)—are specific for the neurons in layers II, III, and IV of mammalian isocortex. Neurons containing VIP or CHAT have not been found in the dorsal pallium of nonmammalian amniotes, while neurons containing CCK8 occur only in birds.

In mammals, neurons in layers II and III give rise to corticocortical connections, both ipsilaterally and interhemispherically, and granule cells in layer IV, which receive the major part of ascending dorsal thalamic sensory projections, are present in prodigious quantity. Some interhemispheric connections are present in diapsid reptiles, and some evidence of ipsilateral corticocortical connections exists, but such intercortical connections are sparse in nonmammalian amniotes as compared with mammals. Only in the dorsal pallium of birds do granule-like cells appear to exist in quantity.

Thus, some sets of neurons with specific neuroactive substances and connections, particularly some of those within layers II–IV of isocortex, may not have been present in the dorsal pallium of the common amniote ancestral stock. As these regions are considered to be homologous as fields, however, the presence of some specific apomorphic traits in given lineages is not an unlikely finding. The neuroactive substance or transmitter phenotype of a given neuron can be influenced during development by a variety of epigenetic factors. The presence of novel phenotypes of dorsal pallial neurons could thus occur as a result of minor shifts in the developmental process, and selective pressures favoring their maintenance would result in the establishment of such developmental shifts, themselves under genetic control, in the population.

As discussed above, only eutherian mammals have a massive interhemispheric commissure, the corpus callosum. An anterior commissure that carries olfactory and some isocortical interhemispheric fibers and a hippocampal commissure are also present in mammals. The best developed commissure in nonmammalian amniotes, the **anterior pallial commissure,** appears to correspond to the hippocampal commissure of mammals. It crosses the midline at the base of the medial cortex and primarily interconnects the medial pallial regions but has also been found to carry a minor component of dorsal pallial

interhemispheric fibers, at least in some diapsid reptiles. A more caudal and less well developed commissure, called the posterior pallial commissure or the anterior commissure, carries interhemispheric connections of the more caudal pallial regions. A large collection of ascending and descending fibers, consisting of the **medial and lateral forebrain bundles** (Fig. 19-5), is also present in the forebrain of nonmammalian amniotes, corresponding to the internal capsule/cerebral peduncle system and other projection systems of the forebrain in mammals. The lateral forebrain bundle is frequently divided into two parts called the **dorsal** and **ventral peduncles.**

THE TELENCEPHALON: SUBPALLIUM

The **striatum** and the **septum** are the two major divisions present in the ventral part of the telencephalon. The striatum lies ventrolaterally; it is concerned with motor functions and is also interconnected with the limbic system, while the septum lies ventromedially and is functionally and connectionally related to the limbic system.

The Striatum of Anamniotes

As we will discuss below in conjunction with the "striatum" in mammals, a variety of terms have been used for this region and its various structures, including corpus striatum, dorsal striatum, and so on. We continue to use the term striatum here for the entire region, based on general usage in studies of nonmammalian species. Striatum simply means "striped," and refers to the appearance of two of the nuclei in the ventrolateral telencephalon in mammals that are crossed by a fiber pathway and therefore have a striped appearance. The vagueness of this term from its general usage is an advantage that we wish to retain. Those nuclei in various radiations of vertebrates that are homologous to each other as discrete nuclei have other, well established, individual names that can be used in all such cases without causing confusion.

In anamniote vertebrates with evaginated telencephalons, the striatum occupies the ventrolateral part of the telencephalon, while the septal area lies in a more medial position, ventral to the medial pallium. Dorsal and ventral divisions of the striatum, a **nucleus accumbens,** and an **anterior** (or rostral) **entopeduncular nucleus** have been identified in amphibians. Similar regions are present in other anamniotes. One nucleus present in cartilaginous fishes—**area periventricularis ventrolateralis**—may correspond to the dorsal striatum of mammals (see below), and another nucleus—**nucleus superficialis basalis**—may correspond to the mammalian ventral striatum (see below). In ray-finned fishes, the eversion process during development results in at least part of the striatum occupying a dorsomedial position within the telencephalon.

The Ventrolateral Telencephalon of Mammals

In mammals, the terminology used for structures in the ventrolateral telencephalon needs to be reviewed before we discuss the structures themselves and their connections. Various nuclei have been referred to as the **basal ganglia,** in which three nuclei—the **caudate nucleus, putamen,** and **globus pallidus**—are most frequently included. The substantia nigra of the mesencephalon and other related structures are sometimes included in this term as well. The term basal ganglia is now waning in popularity in favor of newer terms that reflect a better understanding of the connections and neurotransmitters of the various cell groups. The ventrolateral telencephalon (Table 19-1) of mammals can be divided into three major parts:

- The "striatum", or **stiatopallidal complexes.**
- A ventral region referred to as the **extended amygdala.**
- A group of large, cholinergic neurons known as the basal nucleus of Meynert or the **magnocellular corticopetal cell complex.**

The Striatopallidal Complexes. The striatopallidal complexes have two divisions:

- A dorsal division, composed of the **dorsal striatum** and the **dorsal pallidum.**
- A ventral division, composed of the **ventral striatum** and the **ventral pallidum.**

The dorsal striatum comprises the caudate nucleus and the putamen (Fig. 19-19), which are in reality a single nucleus, artificially divided by fibers of the internal capsule at this level. The name caudate means "tail," in reference to the "C" shaped caudal extension of the caudate nucleus as seen in the sagittal plane (Fig. 19-25). The globus pallidus (Fig. 19-19) forms the dorsal pallidum and lies ventromedial to the caudate nucleus and putamen. Due to a lens-like shape in transverse section formed by the putamen and the globus pallidus, these two nuclei are sometimes referred to as the **lentiform nucleus,** although there is no connectional or functional basis for recognizing them as a single nucleus. The term lentiform nucleus is useful, however, as it is sometimes referred to as an anatomical landmark in relation to other structures. The caudate–putamen and the globus pallidus are sometimes referred to as the corpus striatum (striped body).

The globus pallidus has two segments. In primates, these segments are called **external** and **internal.** In nonprimate mammals, the external segment is simply called the globus pallidus, and the internal segment is called either the internal segment or the **entopeduncular nucleus.** This entopeduncular nucleus is the homologue of a cell group most often identified as the rostral entopeduncular nucleus in nonmammalian vertebrates. Some of the terms used for these entities in primates, nonprimate mammals, birds, and other vertebrates are listed in Table 19-2.

The term "extrapyramidal system" frequently has been used to inclusively describe descending pathways from the caudate nucleus, putamen, globus pallidus, substantia nigra, and other related nuclei and fiber tracts that ultimately terminate in the spinal cord. This term was derived as a contrast to the "pyramidal system," which is the efferent motor pathway from the motor cortex to the spinal cord. The pyramidal system is concerned with fine, voluntary motor movements, particularly of the distal parts of the extremities (such as in humans performing surgery, playing a piano, or writing). The extrapyramidal

TABLE 19-1. Ventrolateral Telencephalon of Mammals

			Caudate nucleus		
(1) Striatopallidal Complexes	Dorsal striatopallidal complex	Dorsal striatum	Putamen	Lentiform nucleus	Corpus striatum
		Dorsal pallidum	Globus pallidus		
	Ventral striatopallidal complex	Ventral striatum	Olfactory tubercle and nucleus accumbens		
		Ventral pallidum	Ventral pallidum (in rostral substantia innominata)		
(2) Extended Amygdala	Centromedial cortical part of amygdala				
	Sublenticular part of substantia innominata				
	Bed nucleus of stria terminalis				
	Caudomedial shell of nucleus accumbens				
(3) Nucleus Basalis					

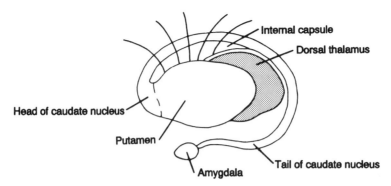

FIGURE 19-25. Reconstruction of a lateral view of the caudate nucleus and putamen, with part of the dorsal thalamus (shaded area) medial to the putamen. Rostral is toward the left. The fibers of the internal capsule are indicated by curved lines as they pass ventrocaudally between the caudate nucleus and the putamen.

system is concerned, in contrast, with the control of more gross motor movements and of stereotyped, species-typical display behaviors. The term extrapyramidal system has fallen out of favor in recent years, however.

Ventral to the globus pallidus is a region that was not named when it was first described and has hence become known as the **substantia innominata** (the "unnamed substance") (Fig. 19-26). Several different populations of neurons are present within the substantia innominata. More rostrally is a region that is anatomically within the olfactory cortex (as mentioned above) but is now recognized as part of the striatum: the **olfactory tubercle** (Fig. 19-26). In macrosmatic animals (those in which the sense of smell is well developed), cell bridges exist between the olfactory tubercle and the corpus striatum.

Deep to the olfactory tubercle and adjoining the ventromedial border of the caudate-putamen is a nucleus that can be divided into shell and core regions called **nucleus accumbens** (Fig. 19-26). This nucleus was originally named nucleus accumbens septi, but since this nucleus is now recognized as part of the striatopallidal complex rather than part of the septum, the "septi" is being dropped from its name in common usage. The core of nucleus accumbens surrounds the anterior commissure and is continuous cytoarchitectonically with the caudate-putamen. A distinct shell region encircles the core, except at the rostral pole where separate shell and core regions cannot be distinguished.

The ventral striatum has two main components: the olfactory tubercle and most of nucleus accumbens. The ventral pal-

lidum is composed of neurons that resemble those of the dorsal pallidum, that is, the globus pallidus, which are scattered within the substantia innominata, particularly in its more rostral part, extending rostroventrally between the globus pallidus and nucleus accumbens dorsally and the olfactory tubercle ventrally (Fig. 19-26).

The Extended Amygdala of Mammals. The second major part of the ventrolateral forebrain is the extended amygdala. The extended amygdala comprises four populations of neurons:

- The centromedial cortical part of the amygdalar nuclear complex.
- The more caudal, sublenticular part of the substantia innominata.
- The bed nucleus of the stria terminalis.
- The caudomedial part of the shell of nucleus accumbens.

The bed nucleus of the stria terminalis, a tract that connects the hypothalamus with the corticomedial amygdala as discussed above, lies between the septal region and nucleus accumbens (Fig. 19-26). The area of the extended amygdala, which extends caudally from the major part of nucleus accumbens, is shown in Figure 19-27, which shows structures that lie deep to the ventral surface of the brain.

The Magnocellular Corticopetal Cell Complex of Mammals. The third major part of the ventrolateral forebrain is a

TABLE 19-2. Globus Pallidus Nomenclature			
Primates	Nonprimate Mammals	Birds	A Number of Other Vertebrates
Globus pallidus, external segment	Globus pallidus or Globus pallidus proper or Globus pallidus, external segment	Paleostriatum primitivum	Other name, or included in the term striatum
Globus pallidus, internal segment	Entopeduncular nucleus or Globus pallidus, internal segment	(perhaps) Intra-peduncular nucleus	Rostral entopeduncular nucleus

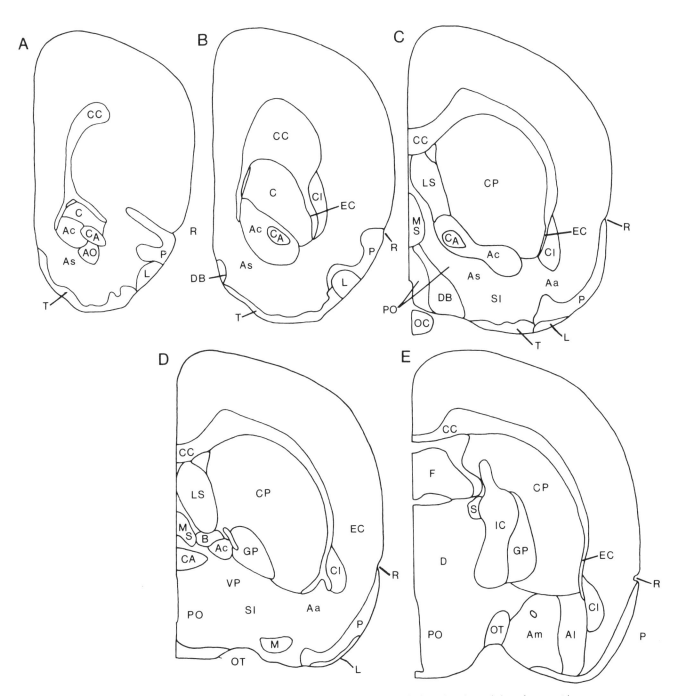

FIGURE 19-26. Drawings of a series of transverse sections through the right telencephalon of a rat, with A most rostral and E most caudal. Abbreviations: Aa, anterior amygdalar area; Ac, core of nucleus accumbens; Al, basolateral amygdala; Am, centromedial amygdala; Ao, anterior olfactory nucleus; As, shell of nucleus accumbens; B, bed nucleus of stria terminalis; C, caudate nucleus; CA, anterior commissure; CC, corpus callosum; Cl, claustrum; CP, caudate-putamen; D, diencephalon; DB, nucleus of the diagonal band; EC, external capsule; F, fornix; GP, globus pallidus; IC, internal capsule; L, lateral olfactory tract; LS, lateral septum; M, nucleus basalis (magnocellular corticopetal cell complex); MS, medial septum; OC, optic chiasm; OT, optic tract; P, piriform cortex; PO, preoptic area; R, rhinal sulcus; S, stria terminalis; SI, substantia innominata; T, olfactory tubercle; VP, ventral pallidum. Data from Pelligrino et al. (1979). Used with permission of Plenum Publishing Corp.

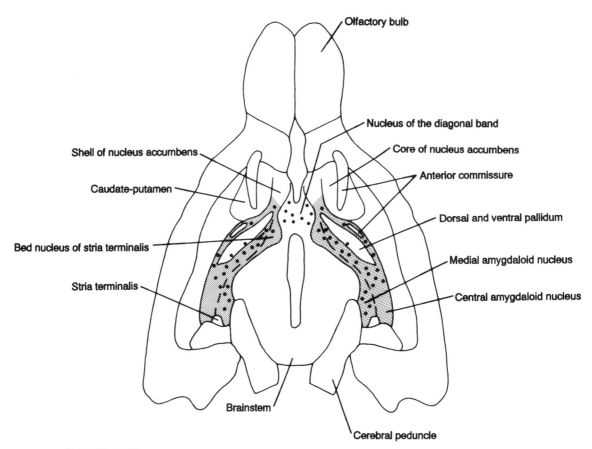

FIGURE 19-27. View of structures deep to the ventral surface of the brain in a rat. Rostral is toward the top. The area of the extended amygdala is indicated by shading, and the distribution of the large cholinergic neurons of nucleus basalis is indicated by the large dots. Data from Heimer and Alheid (1991). Used with permission of Plenum Publishing Corp.

single group of large neurons. These neurons lie within the substantia innominata, some scattered and some in aggregates, and constitute the **nucleus basalis** of Meynert, or the magnocellular corticopetal cell complex (Figs. 19-26 and 19-28). This cell group gives rise to a major cholinergic projection to all regions of the isocortex.

The Ventrolateral Telencephalon of Nonmammalian Vertebrates

In diapsid reptiles, birds, and turtles, the striatopallidal areas have been identified on the basis of their connections and histochemical profiles. Nomenclature varies, and a listing of some of the various terms used for the ventrolateral part of the telencephalon in nonmammalian amniotes is given in Table 19-3. The approximate boundaries of three of the major divisions of the striatopallidal complex—the striatum (dorsal striatum), globus pallidus (dorsal pallidum), and the nucleus accumbens part of the ventral striatum—are shown in a lizard in Figure 19-21. The ventral pallidum lies within the same general region of the basal forebrain as nucleus accumbens. A number of these nuclei are also shown in Figure 19-28 in the forebrain of a turtle. In turtles, the nucleus labeled paleostriatum augmentatum appears to correspond to the dorsal striatum and nucleus accumbens of mammals.

The connections and location of **area d** (and a medially related **area c** not shown in Fig. 19-28) are similar to those of the extended amygdala of mammals. Areas c and d are reciprocally connected with the caudal brainstem. They receive afferent inputs from diverse and numerous sites, including the dorsal column nuclei, raphe, locus coeruleus, substantia nigra, ventral tegmental area, hypothalamus, and, within the telencephalon, the olfactory cortex.

Large cholinergic neurons are present in the basal forebrain that are homologous to nucleus basalis of mammals. The distribution of these neurons is indicated by scattered large dots in Figure 19-28. It is possible that the unlabeled area in Figure 19-28(C) that is partly bordered by a dashed line medial to the more caudal parts of area d and the globus pallidus and that contains scattered large cholinergic neurons may correspond to the more caudal, sublenticular (i.e., ventral to the lentiform nucleus) part of the substantia innominata and thus to part of the extended amygdala of mammals. The main nuclei of the amygdala lie more caudally, within the caudal part of the dorsal ventricular ridge, as discussed above.

The Septum

The septum is the second major division of the ventral telencephalon. In anamniote vertebrates, it generally lies in the

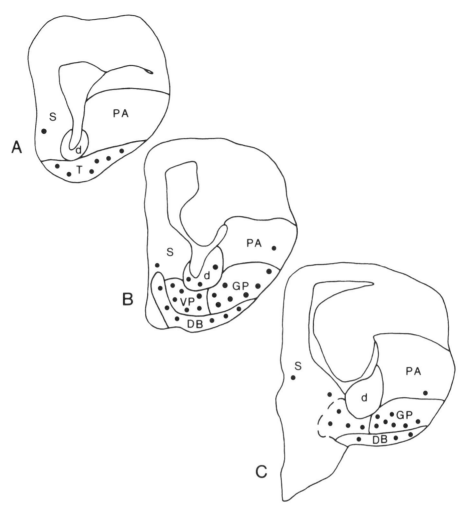

FIGURE 19-28. Drawings of a series of hemitransverse sections through the right telencephalon of a turtle, with A most rostral and C most caudal. The distribution of the large cholinergic neurons of nucleus basalis is indicated by the scattered large dots. Abbreviations: d, area d; DB, nucleus of the diagonal band; GP, globus pallidus; PA, paleostriatum augmentatum; S, septum; T, olfactory tubercle; VP, ventral pallidum. Data from Powers and Reiner, 1993. Used with permission of S. Karger AG.

TABLE 19-3. Comparative Striatal Nomenclature				
Mammals	Lizards	Crocodiles[a]	Birds	Turtles[a]
Caudate and putamen	Striatum	Dorsolateral area of ventrolateral area (VLA)	Paleostriatum augmentatum	Paleostriatum augmentatum and ?
Nucleus accumbens	Nucleus accumbens	Part of small celled field of VLA	Ventral part of lobus parolfactorius	
Globus pallidus	Globus pallidus	Ventrocaudal area of VLA	Paleostriatum primitivum	Globus pallidus
Ventral pallidum	Ventral pallidum	Ventral paleostriatum	Ventral pallidum	Ventral paleostriatum

[a] The terminology shown here for crocodiles and proposed homologies is based on Brauth (1988), which differs from that of Brauth and Kitt (1980). The terminology for turtles is that of Powers and Reiner (1993).

ventromedial wall of the telencephalon, ventral to the medial pallium, as shown in Figures 19-6 and 19-10. It is extensively interconnected with the hypothalamus and the medial pallium.

In mammals, the septal area receives the bulk of its afferent input from limbic cortex and the hypothalamus. The rostral part of the septum contains a lateral and a medial division. The lateral division contains the **lateral septal nuclei** (Fig. 19-26). The medial division contains **medial septal nuclei** and a nucleus called the **nucleus of the diagonal band of Broca** (Fig. 19-26). Two additional nuclei have been identified in the caudal part of the septum in mammals, called the **septofimbrial nucleus** and the **triangular nucleus.** In diapsid reptiles, birds, and turtles, the septum lies ventral to the medial pallium (Fig. 19-21), and is similarly interconnected with the hypothalamus and the medial pallium.

THE TELENCEPHALON: CLAUSTRUM

The **claustrum** (Figs. 19-11 and 19-19) is a thin layer of gray matter that lies between the pallial cortex of the temporal lobe and the putamen. It is present in mammals, but a homologue of the claustrum has yet to be identified in a nonmammalian vertebrate. The debate as to whether the claustrum is better categorized with the pallium or the striatum is unresolved.

In terms of its connections, the claustrum is primarily involved with the dorsal pallium. It has extensive reciprocal connections with all areas of the cerebral cortex, including the primary and association sensory cortices of the visual, auditory, and somatosensory systems and the motor and association cortices of the frontal lobe. The claustrum may thus play an integrative role in cortical functions. In addition, the claustrum receives afferent input from subcortical structures including the hypothalamus and a group of diffusely projecting nuclei in the dorsal thalamus called the intralaminar nuclei.

FOR FURTHER READING

Heimer, L. and Alheid, G. F. (1991) Piecing together the puzzle of basal forebrain anatomy. In T. C. Napier, P. W. Kalivas, and I. Hanin (eds.), *The Basal Forebrain: Anatomy to Function*. New York: Plenum, pp. 1–42.

Pritz, M. B. and Stritzel, M. E. 1990 A different type of vertebrate thalamic organization. *Brain Research*, 525, 330–334.

Welker, W. (1990) Why does cerebral cortex fissure and fold? A review of determinants of gyri and sulci. In E. G. Jones and A. Peters (eds.), *Cerebral Cortex, Vol. 8B: Comparative Structure and Evolution of Cerebral Cortex, Part II*. New York: Plenum, pp. 3–136.

Schwerdtfeger, W. K. and Germroth (eds.) (1990) *The Forebrain in Nonmammals: New Aspects of Structure and Development*. Berlin: Springer-Verlag.

Schwerdtfeger, W. K. and Smeets, W. J. A. J. (eds.) (1988) *The Forebrain of Reptiles: Current Concepts of Structure and Function*. Basel: Karger.

ADDITIONAL REFERENCES

Agarwala, S., Günlük, A. E., May, J. G., III, and Petry, H. M. (1992) Immunohistochemical organization of the ventral lateral genicu-late nucleus in the tree shrew. *Journal of Comparative Neurology*, 318, 267–276.

Agarwala, S., May, J. G., III, Moore, J. K., and Petry, H. M. (1992) Immunohistochemical organization of the ventral lateral geniculate nucleus in the ground squirrel. *Journal of Comparative Neurology*, 318, 255–266.

Amaral, D. G., Price, J. L., Pitkänen, A., and Carmichael, S. T. (1992) Anatomical organization of the primate amygdaloid complex. In J. P. Aggleton (ed.), *The Amygdala: Neurobiological Aspects of Emotion, Memory, and Mental Dysfunction*. New York: Wiley-Liss, Inc., pp. 1–66.

Andreu, M. J., Dávila, J. C., de la Calle, A., and Guirado, S. (1994) Monoaminergic innervation patterns in the anterior dorsal ventricular ridge of a lacertid lizard, *Psammodromus algirus*. *Brain, Behavior and Evolution*, 44, 175–186.

Ariëns Kappers, C. U., Huber, G. C. and Crosby, E. C. (1967) *The Comparative Anatomy of the Nervous System of Vertebrates, Including Man*. New York: Hafner Publishing Co.

Battaglia, G., Lizier, C., Colacitti, C., Princivalle, A. and Spreafico, R. (1994) A reticuloreticular commissural pathway in the rat thalamus. *Journal of Comparative Neurology*, 347, 127–138.

Braford, M. R., Jr. and Northcutt, R. G. (1983) Organization of the diencephalon and pretectum of the ray-finned fishes. In R. E. Davis and R. G. Northcutt (eds.), *Fish Neurobiology, Vol. 2: Higher Brain Areas and Functions*. Ann Arbor, MI: University of Michigan Press, pp. 117–163.

Brauth S. E. (1988) The organization and projections of the paleostriatal complex of *Caiman crocodilus*. In W. K. Schwerdtfeger and W. J. A. J. Smeets (eds.), *The Forebrain of Reptiles: Current Concepts of Structure and Function*. Basel: Karger, pp. 60–76.

Brauth, S. E. and Kitt, C. A. (1980) The paleostriatal system of *Caiman crocodilus*. *Journal of Comparative Neurology*, 189, 437–465.

Brog, J. S., Salyapongse, A., Deutch, A. Y. and Zahm, D. S. (1993) The patterns of afferent innervation of the core and shell in the "accumbens" part of the rat ventral striatum: immunohistochemical detection of retrogradely transported fluoro-gold. *Journal of Comparative Neurology*, 338, 255–278.

Bruce, L. L. and Butler, A. B. (1984) Telencephalic connections in lizards. I. Projections to cortex. *Journal of Comparative Neurology*, 229, 585–601.

Bruce, L. L. and Butler, A. B. (1984) Telencephalic connections in lizards. II. Projections to anterior dorsal ventricular ridge. *Journal of Comparative Neurology*, 229, 602–615.

Butler, A. B. (1994a) The evolution of the dorsal thalamus of jawed vertebrates, including mammals: cladistic analysis and a new hypothesis. *Brain Research Reviews*, 19, 29–65.

Butler, A. B. (1994b) The evolution of the dorsal pallium of amniotes: cladistic analysis and a new hypothesis. *Brain Research Reviews*, 19, 66–101.

Butler, A. B. and Northcutt, R. G. (1973) Architectonic studies of the diencephalon of *Iguana iguana* (Linnaeus). *Journal of Comparative Neurology*, 149, 439–462.

Butler, A. B. and Northcutt, R. G. (1992) Retinal projections in the bowfin, *Amia calva*: cytoarchitectonic and experimental analysis. *Brain, Behavior and Evolution*, 39, 169–194.

Butler, A. B. and Northcutt, R. G. (1993) The diencephalon of the Pacific herring, *Clupea harengus*: cytoarchitectonic analysis. *Journal of Comparative Neurology*, 328, 527–546.

Clascá, F., Avendaño, C., Román-Guindo, A., Llamas, A., and Reinoso-Suárez, F. (1992) Innervation from the claustrum of the frontal association and motor areas: axonal transport studies in the cat. *Journal of Comparative Neurology*, 326, 402–422.

Conley, M. and Friederich-Ecsy, B. (1993) Functional organization of the ventral lateral geniculate complex of the tree shrew (*Tupaia belangeri*): I. Nuclear subdivisions and retinal projections. *Journal of Comparative Neurology*, 328, 1–20.

Conley, M. and Friederich-Ecsy, B. (1993) Functional organization of the ventral lateral geniculate complex of the tree shrew (*Tupaia belangeri*): II. Connections with the cortex, thalamus, and brainstem. *Journal of Comparative Neurology*, 328, 21–42.

Crossland, W. J. and Uchwat, C. J. (1979) Topographic projections of the retina and optic tectum upon the ventral lateral geniculate nucleus in the chick. *Journal of Comparative Neurology*, 185, 87–106.

Díaz, C., Yanes, C., Trujillo, C. M., and Puelles, L. (1994) The lacertidian reticular thalamic nucleus projects topographically upon the dorsal thalamus: experimental study in *Gallotia galloti*. *Journal of Comparative Neurology*, 343, 193–208.

Ebbesson, S. O. E. and Voneida, T. J. (1969) The cytoarchitecture of the pallium in the tegu lizard (*Tupinambis nigropunctatus*). *Brain, Behavior and Evolution*, 2, 431–466.

Edwards, S. B., Rosenquist, A. C., and Palmer, L. A. (1974) An autoradiographic study of ventral lateral geniculate projections in the cat. *Brain Research*, 72, 282–287.

Finger, T. E. and Silver, W. L. (eds.), 1991, *Neurobiology of Taste and Smell*. Malabar, FL: Krieger Publishing Co.

Garrett, B., Osterballe, R., Slomianka, L., and Geneser, F. A. (1994) Cytoarchitecture and staining for acetylcholinesterase and zinc in the visual cortex of the Parma wallaby (*Macropus parma*). *Brain, Behavior and Evolution*, 43, 162–172.

Gergen, J. A. and MacLean, P. D. (1962) *A Stereotaxic Atlas of the Squirrel Monkey's Brain (Saimiri sciureus)*. Bethesda, MD: U.S. Dept of Health, Education, and Welfare.

Graybiel, A. M. (1974) Visuo-cerebellar and cerebello-visual connections involving the ventral lateral geniculate nucleus. *Experimental Brain Research*, 20, 303–306.

Haber, S. N., Lynd-Falta, E., and Mitchell, S. J. (1993) The organization of the descending ventral pallidal projections in the monkey. *Journal of Comparative Neurology*, 329, 111–128.

Harting, J. K., van Lieshout, D. P. and Feig, S. (1991) Connectional studies of the primate lateral geniculate nucleus: distribution of axons arising from the thalamic reticular nucleus of *Galago crassicaudatus*. *Journal of Comparative Neurology*, 310, 411–427.

Heimer, L., Switzer, R. D., and van Hoesen, G. W. (1982) Ventral striatum and ventral pallidum: components of the motor system? *Trends in Neurosciences*, 5, 83–87.

Henselmans, J. M. L. and Wouterlood, F. G. (1994) Light and electron microscopic characterization of cholinergic and dopaminergic structures in the striatal complex and the dorsal ventricular ridge of the lizard *Gekko gecko*. *Journal of Comparative Neurology*, 345, 69–83.

Isaacson, R. L. (1982) *The Limbic System*, 2nd ed. New York: Plenum.

Kicliter, E. and Bruce, L. L. (1983) Ground squirrel ventral lateral geniculate receives laminated retinal projections. *Brain Research*, 267, 340–344.

Künzle, H. and Rehkämper, G. (1992) Distribution of cortical neurons projecting to dorsal column nuclear complex and spinal cord in the hedgehog tenrec, *Echinops telfairi*. *Somatosensory and Motor Research*, 9, 185–197.

Lent, R. and Schmidt, S. L. (1993) The ontogenesis of the forebrain commissures and the determination of brain asymmetries. *Progress in Neurobiology*, 40, 249–276.

LeVay, S. and Sherk, H. (1981) The visual claustrum of the cat. I: Structure and connections. *Journal of Neuroscience*, 1, 956–980.

Luiten, P. G. M. (1981) Two visual pathways to the telencephalon in the nurse shark (*Ginglymostoma cirratum*). II. Ascending thalamo-telencephalic connections. *Journal of Comparative Neurology*, 196, 539–548.

Martinez-Guijarro, F. J., Soriano, E., Delrio, J. A., Blascoibanez, J. M., and Lopez-Garcia, C. (1993) Parvalbumin-containing neurons in the cerebral cortex of the lizard *Podarcis hispanica*: morphology, ultrastructure, and coexistence with GABA, somatostatin, and neuropeptide Y. *Journal of Comparative Neurology*, 336, 447–467.

Medina, L. and Reiner, A. (1994) Distribution of choline acetyltransferase immunoreactivity in the pigeon brain. *Journal of Comparative Neurology*, 342, 497–537.

Mitrofanis, J. and Guillery, R. W. (1993) New views of the thalamic reticular nucleus in the adult and the developing brain. *Trends in Neurosciences*, 16, 240–245.

Neary, T. J. and Northcutt, R. G. (1983) Nuclear organization of the bullfrog diencephalon. *Journal of Comparative Neurology*, 213: 262–278.

Nieuwenhuys, R., Voogd, J., and van Huijzen, C. (1978) *The Human Central Nervous System: A Synopsis and Atlas*. New York: Springer-Verlag.

Northcutt, R. G. (1978) Brain organization in the cartilaginous fishes. In E. S. Hodgson and R. F. Mathewson (eds.), *Sensory Biology of Sharks, Skates, and Rays*. Arlington, VA: Department of the Navy, pp. 117–193.

Northcutt, R. G. (1979) The comparative anatomy of the nervous system and the sense organs. In M. H. Wake (ed.), *Hyman's Comparative Vertebrate Anatomy*. Chicago: University of Chicago Press, pp. 615–769.

Northcutt, R. G. (1981) Evolution of the telencephalon in nonmammals. *Annual Review of Neuroscience*, 4, 301–350.

Northcutt, R. G. and Davis, R. E. (1983) Telencephalic organization in ray-finned fishes. In R. E. Davis and R. G. Northcutt (eds.), *Fish Neurobiology, Vol. 2: Higher Brain Areas and Functions*. Ann Arbor, MI: University of Michigan Press, pp. 203–236.

Northcutt, R. G. and Kicliter, E. (1980) Organization of the amphibian telencephalon. In S. O. E. Ebbesson (ed.), *Comparative Neurology of the Telencephalon*. New York: Plenum., pp. 203–255.

Paré, D. and Steriade, M. (1993) The reticular thalamic nucleus projects to the contralateral dorsal thalamus in macaque monkey. *Neuroscience Letters*, 154, 96–100.

Parent, A. (1986) *Comparative Neurology of the Basal Ganglia*. New York: Wiley.

Pellegrino, L. J., Pellegrino, A. S., and Cushman, A. J. (1979) *A Stereotaxic Atlas of the Rat Brain*. New York: Plenum.

Powers, A. S. and Reiner, A. (1993) The distribution of cholinergic neurons in the central nervous system of turtles. *Brain, Behavior and Evolution*, 41, 326–345.

Price, J. L. (1991) The central olfactory and accessory olfactory systems. In T. E. Finger and W. L. Silver (eds.), *Neurobiology of Taste and Smell*. Malabar, FL: Krieger Publishing Co., pp. 179–203.

Pritz, M. B. and Stritzel, M. E. (1993) Neuronal subpopulations in a reptilian thalamic reticular nucleus. *NeuroReport*, 4, 791–794.

Raos, V. and Bentivoglio, M. (1993) Crosstalk between the two sides of the thalamus through the reticular nucleus: a retrograde and anterograde tracing study in the rat. *Journal of Comparative Neurology*, 332, 145–154.

Reiner, A. (1991) A comparison of neurotransmitter-specific and neuropeptide-specific neuronal cell types present in the dorsal cortex in turtles with those present in the isocortex in mammals: implications for the evolution of isocortex. *Brain, Behavior and Evolution*, 38, 53–91.

Reiner, A. (1993) Neurotransmitter organization and connections of turtle cortex: implications for the evolution of mammalian isocortex. *Comparative Biochemistry and Physiology*, 104A, 735–748.

Reiner, A. and Northcutt, R. G. (1992) An immunohistochemical study of the telencephalon of the Senegal bichir (*Polypterus senegalus*). *Journal of Comparative Neurology*, 319, 359–386.

Rowe, M. (1990) Organization of the cerebral cortex in monotremes and marsupials. In E. G. Jones and A. Peters (eds.), *Cerebral Cortex, Vol. 8B: Comparative Structure and Evolution of Cerebral Cortex, Part II.* New York: Plenum, pp. 263–334.

Siemen, M. and Künzle, H. (1994) Connections of the basal telencephalic areas c and d in the turtle brain. *Anatomy and Embryology*, 189, 339–359.

Smeets, W. J. A. J., Nieuwenhuys, R., and Roberts, B. L. (1983) *The Central Nervous System of Cartilaginous Fishes: Structure and Functional Correlations.* Berlin: Springer-Verlag.

Ulinski, P. E. (1983) *Dorsal Ventricular Ridge: A Treatise on Forebrain Organization in Reptiles and Birds.* New York: Wiley.

Veenman, C. L., Albin, R. L., Richfield, E. K., and Reiner, A. (1994) Distributions of GABA$_A$, GABA$_B$, and benzodiazepine receptors in the forebrain and midbrain of pigeons. *Journal of Comparative Neurology*, 344, 161–189.

Voneida, T. J. and Ebbesson, S. O. E. (1969) On the origin and distribution of axons in the pallial commissures in the tegu lizard (*Tupinambis nigropunctatus*). *Brain, Behavior and Evolution*, 2, 467–481.

Wicht, H. and Northcutt, R. G. (1992) The forebrain of the Pacific hagfish: a cladistic reconstruction of the ancestral craniate forebrain. *Brain, Behavior and Evolution*, 40, 25–64.

Wilczynski, W. and Northcutt, R. G. (1983) Connections of the bullfrog striatum: afferent organization. *Journal of Comparative Neurology*, 214, 321–332.

20

Pretectum and Posterior Tuberculum

INTRODUCTION

The **pretectum** and the **posterior tuberculum** both lie in a transitional region between the diencephalon and the mesencephalon. The pretectum lies between the dorsal thalamus and the optic tectum. The posterior tuberculum is in the area that is ventral to the pretectum. The degree of development varies markedly in both the pretectum and the posterior tuberculum in different vertebrate radiations.

In this chapter, we will first consider the pretectum, a structure related primarily to the visual system and involved in visuomotor behaviors. We will then discuss the **accessory optic nuclei,** which are a set of nuclei that do not anatomically belong to the pretectum but are functionally interrelated with it. Although the accessory optic nuclei actually lie in the tegmentum, their connections and functions can best be understood in relation to the pretectal nuclei. In the final part of the chapter, we will discuss the migrated part of the posterior tuberculum, which is present in some vertebrates and is involved in relaying sensory information to the telencephalon.

PRETECTUM

The pretectal nuclei receive afferents primarily from the retina and the optic tectum and are involved in modulating motor behavior in response to visual input. In some Group I vertebrates as well as in Group II vertebrates, lateral migration of pretectal neurons occurs during development. In these animals, three pretectal zones, or sets of nuclei, can be identified: **superficial, central,** and **periventricular.** The periventricular zone curves around a caudally running fiber tract that originates in the habenula and is called the **fasciculus retroflexus.** This tract does not carry pretectal fibers but serves as a convenient landmark in the pretectal region.

Group I

In lampreys, the pretectum is an area of unmigrated neurons caudal to the dorsal thalamus (Fig. 20-1). It receives an input from the retina, but its other connections have not been studied. In squalomorph sharks, some migration of neurons occurs during the development of the pretectum. A **superficial pretectal nucleus** lies in a lateral position, and a **central pretectal nucleus** is present lateral to the posterior commissure. These two nuclei receive retinal projections (Fig. 20-2). A cell group medial to the central pretectal nucleus constitutes a **periventricular pretectal nucleus.** The optic tectum projects to all three pretectal nuclei.

In the pretectum of nonteleost ray-finned fishes, some neuronal migration also occurs during development. Pretectal nuclei are present in the superficial, central, and periventricular zones, that is, superficial to deep regions that are in continuity with similarly named zones of the optic tectum (see Chapter 18). In reedfishes and sturgeons, one pretectal nucleus is present in the superficial zone and is characterized by small cells: the **parvocellular superficial pretectal nucleus.** In gars and the bowfin *Amia,* and also in teleosts, as we will discuss below, several superficial nuclei are present (Fig. 20-3). It is possible that one or two of these nuclei (particularly nucleus corticalis) more properly belong in the central zone, but we will consider them all here due to their functional relationships and connections.

In gars and in *Amia,* the **parvocellular superficial pretectal nucleus,** a nucleus with large, widely scattered cells called the **magnocellular superficial pretectal nucleus,** and a more medially lying nucleus called the **posterior pretectal nucleus** are present in the superficial zone. In *Amia,* and possi-

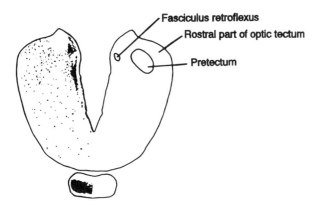

FIGURE 20-1. Transverse hemisection with mirror-image drawing through the pretectum of a lamprey (Petromyzon marinus). Adapted from Kennedy and Rubinson (1977).

bly in gars, an additional nucleus, called **nucleus corticalis,** is present in this region. The parvocellular superficial pretectal nucleus and nucleus corticalis receive retinal projections, and the magnocellular superficial pretectal nucleus receives tectal projections. Additional connections of these nuclei have been studied in teleosts, and it is known that they are involved in linking visual input to motor behaviors related to feeding.

A **central pretectal nucleus** lies in the central pretectal zone of nonteleost ray-finned fishes. **Dorsal** and **ventral periventricular pretectal nuclei** occupy the periventricular pretectal zone. The retina projects to the central pretectal nucleus and to the ventral and dorsal periventricular pretectal nuclei. The optic tectum projects to the central pretectal nucleus and the dorsal periventricular pretectal nucleus.

Among amphibians, the pretectum is somewhat more differentiated in anurans than in urodeles. In anurans (Fig. 20-4), at least five pretectal nuclei are present: the **posterior thalamic nucleus,** the **posterodorsal division of the lateral**

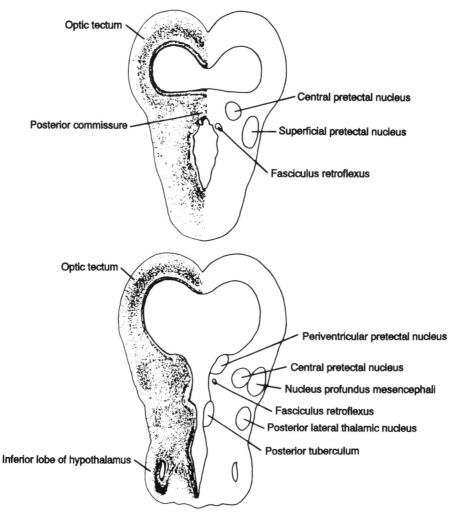

FIGURE 20-2. Transverse hemisections with mirror-image drawings through the pretectum of a shark (Squalus acanthias). The more rostral section is at the top. Adapted from Northcutt (1978) with additional data from Smeets and Northcutt (1987).

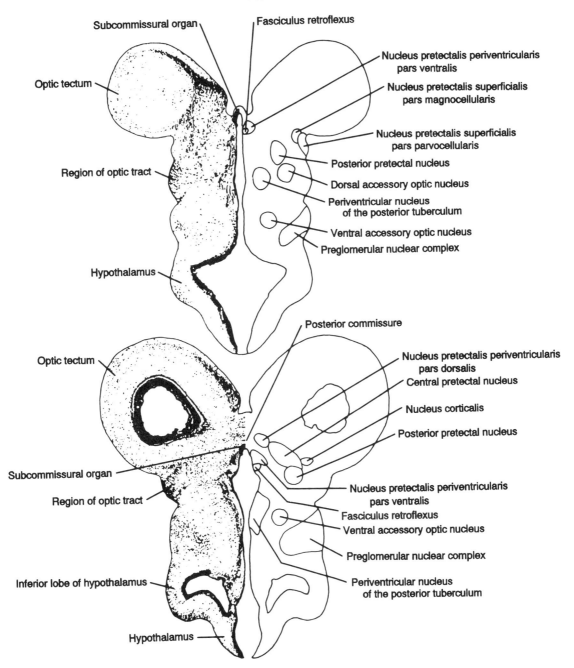

FIGURE 20-3. Transverse hemisections with mirror-image drawings through the pretectum of a bowfin *(Amia calva)*. The more rostral section is at the top. Adapted from Butler and Northcutt (1992). Used with permission of S. Karger AG.

nucleus, **nucleus lentiformis mesencephali,** the **nucleus of the posterior commissure,** and the **pretectal gray.** The first two of these nuclei have sometimes been grouped as a posterior part of the dorsal thalamus, but since they do not project to any part of the telencephalon, they are included in the pretectum here.

A number of the connections of the various pretectal nuclei have been described. Nucleus lentiformis mesencephali contains large cells and receives the heaviest retinal input. Additionally, it receives projections from the optic tectum, the dorsal

thalamus, and the accessory optic nucleus. Dendrites of neurons in the posterior thalamic nucleus extend into nucleus lentiformis mesencephali and may thus also receive these inputs. Nucleus lentiformis mesencephali projects reciprocally to the optic tectum and the accessory optic nucleus and projects to more caudal sites in the brainstem, including the ventral part of the hindbrain and the abducens (VI) oculomotor nucleus.

The posterior thalamic nucleus receives input from the retina, optic tectum, ventral hypothalamus, dorsal thalamus, striatum, and a variety of other sources. It projects to more

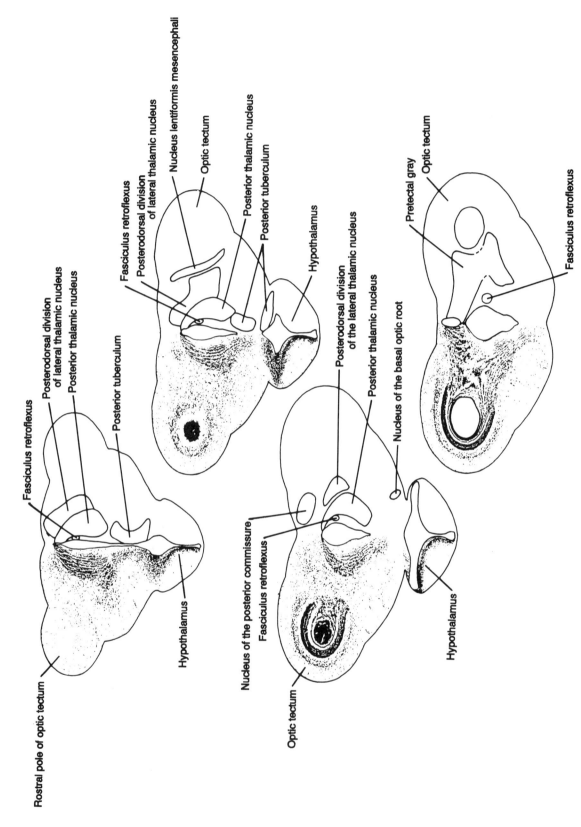

FIGURE 20-4. Transverse hemisections with mirror-image drawings through the pretectum of a frog (*Rana catesbeiana*). The most rostral section is at the top and the most caudal at the bottom. Adapted from Neary (1983).

Rostral pole of optic tectum

Fasciculus retroflexus

Posterodorsal division of lateral thalamic nucleus

Posterior thalamic nucleus

Posterior tuberculum

Hypothalamus

Nucleus lentiformis mesencephali

Optic tectum

Fasciculus retroflexus

Posterodorsal division of lateral thalamic nucleus

Posterior thalamic nucleus

Posterior tuberculum

Hypothalamus

Nucleus of the posterior commissure

Fasciculus retroflexus

Posterodorsal division of the lateral thalamic nucleus

Posterior thalamic nucleus

Nucleus of the basal optic root

Optic tectum

Hypothalamus

Pretectal gray

Optic tectum

Fasciculus retroflexus

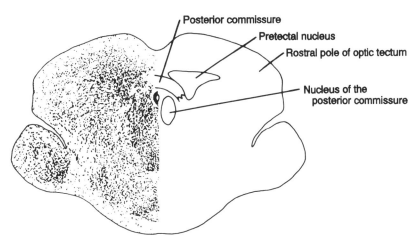

FIGURE 20-5. Transverse hemisection with mirror-image drawing through the pretectum of a hagfish *(Eptatretus stouti)*. Adapted from Wicht and Northcutt (1992). Used with permission of S. Karger AG.

rostral thalamic nuclei, including nucleus anterior, and to the optic tectum and torus semicircularis. It may project to the brainstem and spinal cord as well. The posterodorsal division of the lateral nucleus is known to project to the optic tectum and brainstem. The pretectal gray relays hypothalamic input to the brainstem.

Group II

In hagfishes, relatively limited migration of pretectal neurons occurs during development. Two nuclei (Fig. 20-5) have been identified in the pretectum and lie in the central to periventricular region: the **pretectal nucleus,** which lies lateral to the posterior commissure, and the **nucleus of the posterior commissure,** which lies ventral to the pretectal nucleus. The pretectal nucleus receives a substantial retinal projection.

In Group II cartilaginous fishes, the pretectum has superficial, central, and periventricular zones (Fig. 20-6). All three zones receive both retinal and tectal projections. The superficial pretectal zone (or nucleus) projects to the cerebellum. No part

of the pretectum gives rise to telencephalic projections, but most of the efferent projections of the pretectum to nuclei in the brainstem remain to be studied.

In teleosts (Fig. 20-7), the pretectal nuclei are more numerous and distinct than in other ray-finned fishes or in cartilaginous fishes. In fact, a marked hypertrophy of the pretectal and posterior tubercular regions is one of the hallmarks of teleosts, in direct contrast to the hypertrophy of the dorsal thalamus in amniotes that we will consider in Chapter 22. The periventricular pretectal zone contains two nuclei: a **ventral periventricular pretectal nucleus (PPv)** and a **dorsal periventricular pretectal nucleus (PPd).** (The dorsal periventricular pretectal nucleus is also frequently called the **optic nucleus of the posterior commissure.**) In some teleosts, including ostariophysans and percomorphs, a third periventricular pretectal nucleus is present: the **paracommissural nucleus.** When present, it lies dorsal to the dorsal periventricular pretectal nucleus.

The central pretectal zone is an area of scattered cells called the **central pretectal nucleus (PC).** The borders of this nucleus are often difficult to distinguish. Both the central and

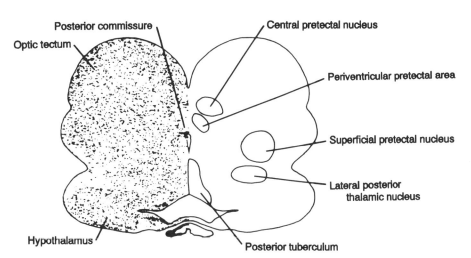

FIGURE 20-6. Transverse hemisection with mirror-image drawing through the pretectum of a skate *(Raja eglanteria)*. Adapted from Northcutt (1978) with additional data from Bodznick and Northcutt (1984).

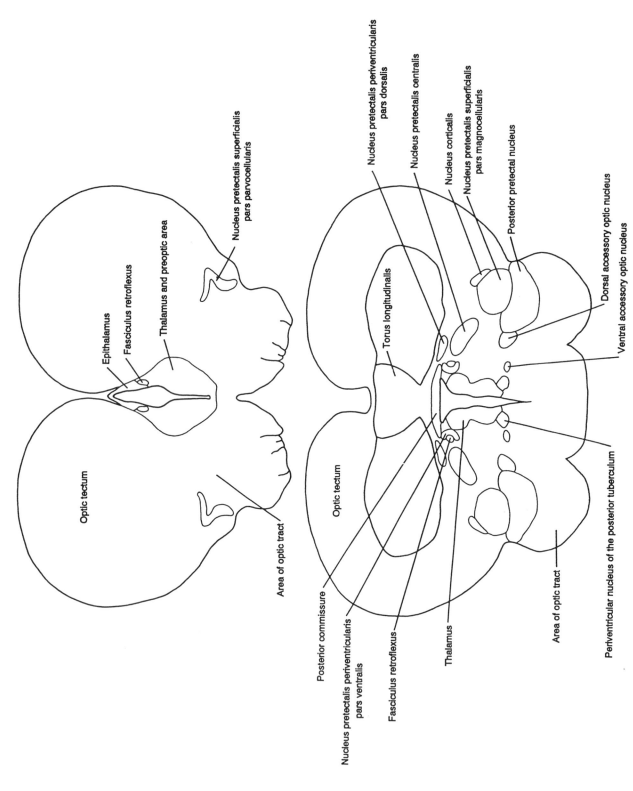

FIGURE 20-7. Drawings of transverse sections through the pretectum of a teleost (*Clupea harengus*). The more rostral section is at the top. Adapted from Butler and Northcutt (1993).

Nucleus pretectalis superficialis pars parvocellularis

Epithalamus

Fasciculus retroflexus

Thalamus and preoptic area

Optic tectum

Area of optic tract

Nucleus pretectalis periventricularis pars dorsalis

Nucleus pretectalis centralis

Nucleus corticalis

Nucleus pretectalis superficialis pars magnocellularis

Posterior pretectal nucleus

Dorsal accessory optic nucleus

Ventral accessory optic nucleus

Torus longitudinalis

Optic tectum

Posterior commissure

Nucleus pretectalis periventricularis pars ventralis

Fasciculus retroflexus

Thalamus

Area of optic tract

Periventricular nucleus of the posterior tuberculum

periventricular pretectal zones of teleosts are similar to those in other fishes.

The superficial pretectal zone is markedly expanded in teleosts and contains a number of nuclei. The most rostral of these nuclei is the **parvocellular superficial pretectal nucleus (PSp).** This nucleus lies at a point of bifurcation of the optic tract into dorsal and ventral optic tracts. It receives a substantial retinal projection and a tectal projection as well. The parvocellular superficial pretectal nucleus has frequently been identified incorrectly as the retino-recipient nucleus of the dorsal thalamus, and hence called the dorsal lateral geniculate nucleus.

The superficial zone of the pretectum in teleosts also contains a **magnocellular superficial pretectal nucleus (PSm),** which receives a tectal projection, and a **nucleus corticalis (NC),** which receives retinal and, probably, tectal projections. In three of the four groups of teleost fishes, one additional nucleus, the **posterior pretectal nucleus (PO),** is also present. In the fourth and largest group of teleosts, the euteleosts, two additional nuclei are present, which are homologous as a field to the posterior pretectal nucleus of the other ray-finned fishes. These nuclei are called the **intermediate superficial pretectal nucleus (PSi)** and **nucleus glomerulosus (G).**

Visual information is relayed through some of these superficial pretectal nuclei to part of the hypothalamus, and this pathway is thought to be involved in orienting the fish's body to the location of a food object. In most groups of ray-finned fishes, this pathway is from the parvocellular superficial pretectal nucleus and nucleus corticalis to the posterior pretectal nucleus, and hence to the **inferior lobe of the hypothalamus [LI,** Fig. 20-8(A)]. In euteleosts [Fig. 20-8(B)], the parvocellular superficial pretectal nucleus and nucleus corticalis project to the intermediate superficial pretectal nucleus (PSi), which, in

turn projects to nucleus glomerulosus. The latter projects to the inferior lobe of the hypothalamus. The parvocellular superficial pretectal nucleus also has more direct projections to motor system structures, including the trochlear nucleus **(IV)** and the inferior raphe **(IR),** the latter projecting, in turn, to the spinal cord.

The posterior pretectal nucleus, where present, varies greatly in its size among different species. In gars and in *Amia,* as discussed above, it is small. Among teleosts, the size of the nucleus varies from very small to extremely large. In euteleosts, two nuclei have differentiated in place of the single posterior pretectal nucleus, and the morphology of one of these nuclei, nucleus glomerulosus, is particularly complex, with concentric layers of cells and fibers. The particular demands of different feeding behaviors in various groups of ray-finned fishes have resulted in these marked differences in the nuclear structure of the pretectal region.

Another nucleus, the magnocellular superficial pretectal nucleus (PSm), may have different connections in two groups of euteleosts or, in one group or the other, may have been lost and replaced by a different cell population. This nucleus receives input from the optic tectum. In sunfishes *(Lepomis cyanellus)* and filefishes *(Navodon modestus),* which are percomorph euteleosts [Fig. 20-9(A)], it projects to two nuclei that, in turn, project back to the optic tectum. One of these nuclei is nucleus isthmi **(I;** see Chapter 16), which also projects to the corpus cerebelli **(Cb).** The second nucleus is called the lateral thalamic nucleus **(LT);** it is discussed below with the migrated nuclei of the posterior tuberculum. In goldfishes *(Carassius auratus),* which are ostariophysan euteleosts [Fig. 20-9(B)], PSm projects to two different nuclei—a cerebellar nucleus called nucleus lateralis valvulae **(NLV),** which projects to the cerebellum **(Cb)** and is also reciprocally connected with the inferior

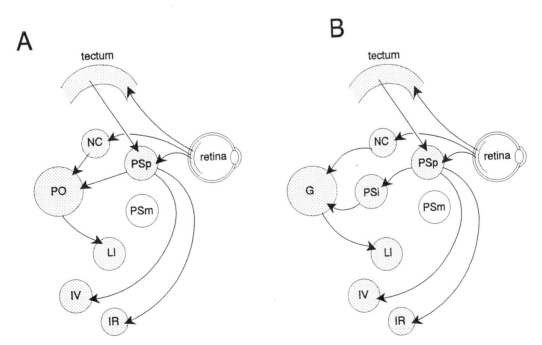

FIGURE 20-8. Diagram of pretectal pathways by which visual information reaches the inferior lobe of the hypothalamus in most groups of ray-finned fishes (A) and euteleosts (B). Abbreviations are given in the text.

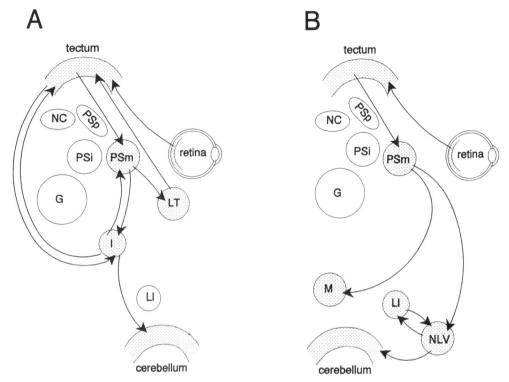

FIGURE 20-9. Diagram of the differing connections of nucleus pretectalis superficialis pars magnocellularis in percomorph euteleosts (A) and ostariophysan euteleosts (B). Abbreviations are given in the text.

lobe (LI) of the hypothalamus and another nucleus called the mammillary body (M), which is part of the posterior tuberculum. In both percomorphs and goldfishes, projections from PSm are relayed to the cerebellum, albeit by different routes. The pathways through PSm may modulate the visuomotor orienting behavior that is achieved with the pathways through PSp.

More studies are needed to determine which set of projections of PSm are present in most euteleosts and what the changes have been in one group or the other. Whatever the outcome, this part of the pretectum exhibits the potential for marked variation of central nervous system structure among species, whether in the cell populations present and/or in changes of their connections.

In monotremes, the pretectum is not a well differentiated area, in contrast to the pretectum in other mammals and in diapsid reptiles, birds, and turtles. Only a small pretectal nucleus is present in monotremes, called the **area pretectalis.**

In placental mammals, terminology for the pretectal nuclei varies, particularly between primate and nonprimate species. Six pretectal nuclei generally can be recognized: the **nucleus of the optic tract,** the **olivary pretectal nucleus,** the **nucleus of the posterior commissure,** and the **anterior, posterior,** and **medial pretectal nuclei.** Five of these nuclei are shown in a tree shrew in Figure 20-10 with other nuclei of the caudal part of the dorsal thalamus in the same region. The retina projects to the olivary pretectal nucleus, the nucleus of the optic tract, and the posterior pretectal nucleus in most placental mammals. It may also project to either the medial or the anterior pretectal nucleus, depending on the species. There is some indication of topographical order in the retinal projections to the pretectum.

In cats and monkeys, striate (visual) cortex projects to the nucleus of the optic tract, while in tree shrews, striate cortex projects to the posterior and anterior pretectal nuclei. Tectal projections also terminate in the pretectum; in cats, these reach the nucleus of the optic tract and olivary pretectal nucleus. Also in cats, somatosensory projections have been found to parts of the anterior and posterior pretectal nuclei that do not receive retinal input. These projections have two sources: the somatosensory cortex and the dorsal column nuclei. The anterior pretectal nucleus also receives a projection from the principal and spinal sensory trigeminal nuclei. Projections from the deep cerebellar nuclei and from the inferior colliculus terminate in the same areas of these nuclei as the somatosensory projections.

The pretectum in mammals is also involved in motor pathways. The nucleus of the posterior commissure receives a projection from the pallidum and projects to the superior colliculus, and some of the pretectal nuclei project to the oculomotor nuclear complex. Projections to the cerebellum, similar to those in diapsid reptiles, birds, and turtles as discussed below, have not been found, however. Among amniote vertebrates, the pretectum is relatively larger in nonmammalian amniotes than in mammals.

In most squamate reptiles and turtles, five to seven pretectal nuclei have been recognized, depending on the species (Fig. 20-11). The three pretectal nuclei that receive retinal input are **nucleus geniculatus pretectalis, nucleus lentiformis mesencephali,** and **nucleus posterodorsalis.** Nuclei lentiformis mesencephali and geniculatus pretectalis also receive tectal input, and telencephalic input has been found to nucleus pretectalis. Nucleus geniculatus pretectalis is known to project to the cerebellum.

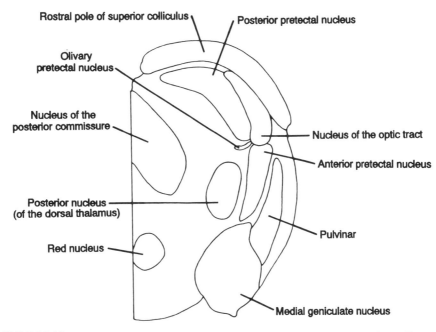

FIGURE 20-10. Drawing of a transverse hemisection through the pretectum of a tree shrew *(Tupaia glis)*. Lateral is toward the right, and the midline is on the left. Adapted from Weber (1985). Used with permission of S. Karger AG.

Additional pretectal nuclei lie more caudally and can be grouped as subdivisions of one larger area termed **nucleus lentiformis thalami.** The subdivisions include a **pars plicata,** a **pars extensa,** the **dorsal pretectal nucleus,** and the **medial pretectal nucleus.** The pars plicata and pars extensa of nucleus lentiformis thalami lie in the periventricular zone of the pretectum, while the dorsal and medial pretectal nuclei occupy the central zone. A **ventral pretectal nucleus** has also been recognized in some species. Where present, this nucleus lies in the superficial part of the pretectum. A nucleus called **nucleus pretectalis** in crocodiles, which may correspond to at least part of the nucleus lentiformis thalami of squamate reptiles, is known to project to a dorsal thalamic nucleus, nucleus rotundus, which receives a major ascending visual projection from the optic tectum.

In birds, pretectal nuclei similar to those in diapsid reptiles and turtles are present, although the nomenclature varies. Some of these nuclei are shown in Figure 20-12. Three nuclei receive a heavy retinal input: **nucleus lentiformis mesencephali,** the **tectal gray,** and the **area pretectalis.** Nucleus lentiformis mesencephali can be subdivided into a **pars medialis** and a **pars lateralis.** The tectal gray lies caudal to nucleus lentiformis mesencephali, at a level intermediate to the two levels shown in Figure 20-12. The retina has been found to project topographically to both parts of nucleus lentiformis mesencephali and to the tectal gray. A fourth nucleus, called **nucleus pretectalis diffusus** lies rostral to the area pretectalis at the level of the tectal gray and receives a sparser retinal input. Two more ventrally lying nuclei in the pretectal region (not included in Fig. 20-12), nucleus subpretectalis and nucleus interstitio-pretectalis-subpretectalis, which are collectively called the **subpretectal complex,** have been found to project to nucleus rotundus in the dorsal thalamus, as does the nucleus pretectalis in crocodiles.

ACCESSORY OPTIC SYSTEM

The movements of images across an animal's retina affect visuomotor behaviors, some of which are critical for survival. Such movements may be very similar whether they are produced (1) by the animal itself moving through a stationary space, (2) by movement of the animal's eyes while the animal is stationary, or (3) by movements of objects in the environment while both animal and eyes are stationary. Similar visual stimuli also result when the animal (stationary or moving) visually tracks an object, keeping the object's image in approximately the same location on the retina. Each of these similar retinal images may call for a different response; the visual system must therefore distinguish among them and send appropriate signals for action to the motor system in each case. Both pretectal and accessory optic structures make critical contributions to the solution of this problem.

The accessory optic system is a set of nuclei that is involved in regulating movements of the eyes in relation to the visual world. This system plays an important role in differentiating perceptual situations such as those listed above. It keeps an image stable on the retina as the eyes move, and it helps the animal to differentiate retinal image movement due to the animal moving from retinal image movement due to the object moving.

In teleosts, the functional partnership between the accessory optic nuclei and many of the pretectal nuclei is based on the common feature of a projection to the cerebellum, and thence to oculomotor nuclei and the vestibular system. The cerebellar projections of the pretectal nuclei in teleosts, which are present in addition to the projections discussed above, as well as the similar projections of the accessory optic nuclei,

A

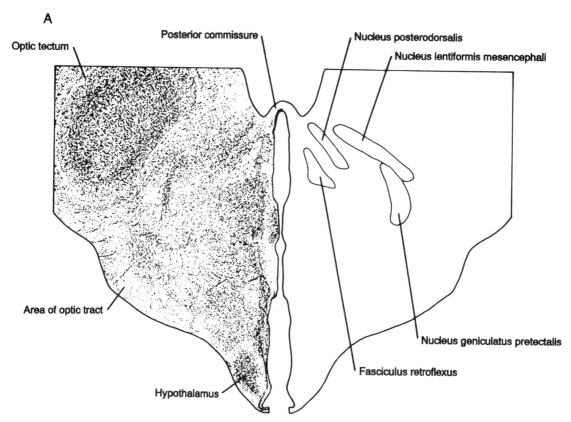

FIGURE 20-11. Transverse hemisections with mirror-image drawings through the pretectum of a lizard (*Tupinambis nigropunctatus*) in rostral-to-caudal order. Adapted from Cruce (1974).

will be discussed together here. Direct cerebellar projections from pretectal and/or accessory optic nuclei also are present in some but not all amniotes. Variation in the connections of these nuclei thus occurs to a marked degree in the different radiations of vertebrates.

The pretectal and accessory optic nuclei are involved in regulating movements of the eyes. A repeating pattern of eye movements, called **optokinetic nystagmus** (OKN), can be evoked by exposing an animal to patterns, such as black and white stripes, that are continuously moved across the visual field. The eyes first track the movement of the pattern in its direction and then return with a **saccade** (rapid movement) to fix on the next part of the pattern entering the visual field. The OKN is related to the processes of visual tracking and discriminating target movement from observer movement. Pretectal nuclei projecting to the cerebellum appear to be involved in OKN in the horizontal plane, while the accessory optic nuclei are involved in OKN in the vertical plane.

In most nontetrapod vertebrates, two accessory optic nuclei, dorsal and ventral, are present. These nuclei lie in the tegmentum and receive a retinal projection. Their functional interrelationship with the pretectum has been studied primarily in teleosts. In tetrapods, a single accessory, or basal, optic nucleus is present.

Group I

In lampreys, only a sparse retinal projection is present to the lateral part of the tegmentum. In squalomorph sharks, retinal projections terminate in two sites in the tegmentum (Fig. 20-13). The more dorsal site is an unnamed neuropil area ventral to the intercollicular nucleus. The more ventral site is called the **basal optic nucleus.** In three of the four radiations of nonteleost ray-finned fishes, only one tegmental retinal target is present, the **basal** (or ventral accessory) **optic nucleus.** In *Amia,* as in teleosts as we will discuss below, two accessory optic nuclei, dorsal and ventral, are present (Fig. 20-3).

In amphibians, a single basal optic nucleus is present that lies in the lateral part of the tegmentum (Fig. 20-4). This nucleus is usually referred to as the **nucleus of the basal optic root** (nBOR), in reference to the fascicle (or root) of optic fibers that innervates it. The nBOR receives a substantial retinal projection and has reciprocal connections with nucleus lentiformis mesencephali, as discussed above. It projects to the medial part of the tegmentum and to the oculomotor (III) and trochlear (IV) nuclei. Dendrites of the nucleus of the medial longitudinal fasciculus extend into nBOR and can thus relay information to the vestibular system and spinal cord. A projection from nBOR to the cerebellum is very sparse, if present at all. Similarly, pretectal nuclei do not project to the cerebellum in amphibians.

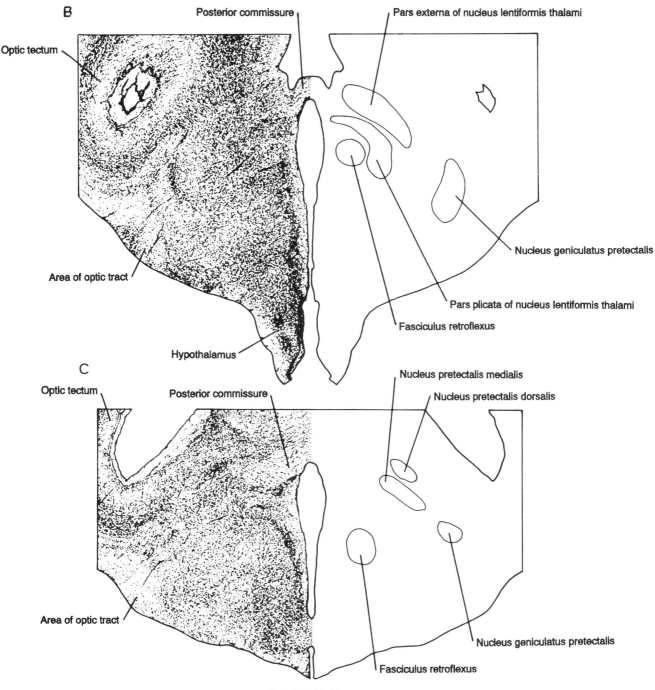

FIGURE 20-11. *(Continued)*

The roles of the pretectal and accessory optic nuclei in visuomotor behaviors have been studied with some thoroughness in amphibians. The regulation of eye movements in prey and predator detection, the detection of stationary objects such as barriers to escape, and the stabilization of gaze are among the functions of these nuclei.

Group II

In hagfishes, no retinal projections have been traced to the tegmentum. Retinal projections terminate in the pretectum as discussed above. In galeomorph sharks, two accessory optic nuclei have been found, one near the intercollicular nucleus and one more caudal and ventral at the level of the oculomotor nucleus. Only the former of these two nuclei has been found in a skate.

In teleosts, both **dorsal** and **ventral accessory optic nuclei** are present (Fig. 20-7) The connections of the accessory optic nuclei and the pretectum with the cerebellum and other brainstem sites are extensive in comparison with other anamniote vertebrates. These pathways are diagrammed in Figure 20-14. The nuclei that receive retinal and/or tectal inputs are indi-

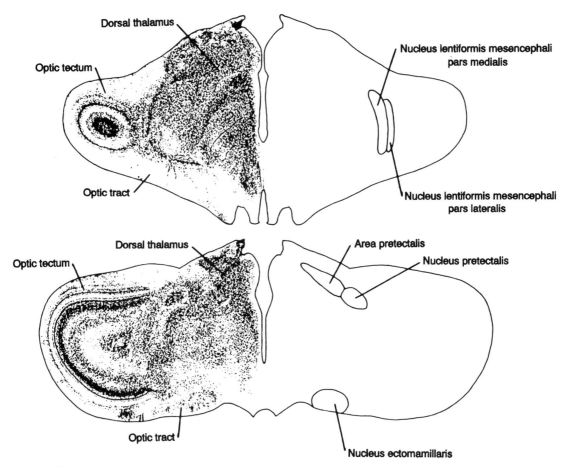

FIGURE 20-12. Transverse hemisections with mirror-image drawings through the pretectum of a bird (*Columba livia*). The more rostral section is at the top. The overlying telencephalon is not shown. Adapted from Karten and Hodos (1967). Used with permission of The Johns Hopkins University Press.

cated in the figure. Four of the nuclei project to the optic tectum—the central pretectal nucleus, the dorsal periventricular pretectal nucleus, the ventral periventricular pretectal nucleus, and the dorsal accessory optic nucleus—and other feedback circuits also exist.

As shown in the figure, the parvocellular superficial pretectal nucleus projects directly to the trochlear nucleus (IV) and is thus the only pretectal nucleus known to project to any of the oculomotor nuclei. The parvocellular superficial pretectal nucleus does not project to the cerebellum, however. Of all the pretectal and accessory optic nuclei, only the dorsal accessory optic nucleus projects to both an oculomotor nucleus (III in this case) and to the cerebellum. In the superficial pretectal zone, the magnocellular superficial pretectal nucleus projects either to nucleus isthmi (in percomorphs) or nucleus lateralis valvulae (in goldfishes), as discussed in detail above. Both the latter nuclei project, in turn, to the cerebellum. The central pretectal nucleus, the dorsal periventricular pretectal nucleus, and, where present, the paracommissural nucleus, project to the cerebellum, as does the ventral accessory optic nucleus.

The cerebellum (see Chapter 14) projects to oculomotor nuclei and to the nucleus of the medial longitudinal fasciculus, which is interconnected with the vestibular system. Hence, these multiple pathways allow visual input to be integrated with vestibular information. The motor output of the pretectal-accessory optic system is involved in smooth pursuit movements of the retina to track a moving image. Output to the motor systems of the body, such as that from the parvocellular superficial pretectal nucleus to the raphe and thence to the spinal cord, would allow for neck and body tracking of an image as well. Outputs to the inferior lobe of the hypothalamus from the posterior pretectal nucleus (or the intermediate superficial pretectal nucleus/nucleus glomerulosus in euteleosts) play a role in orientation and feeding responses, although further work on the efferent pathways is needed.

Mammals have three accessory optic nuclei, called the **dorsal, lateral,** and **medial terminal nuclei.** The dorsal terminal nucleus receives a projection from visual cortex, and the medial terminal nucleus is reciprocally connected with the nucleus of the optic tract in the pretectum. The mammalian accessory optic nuclei project to the oculomotor complex but not to the cerebellum.

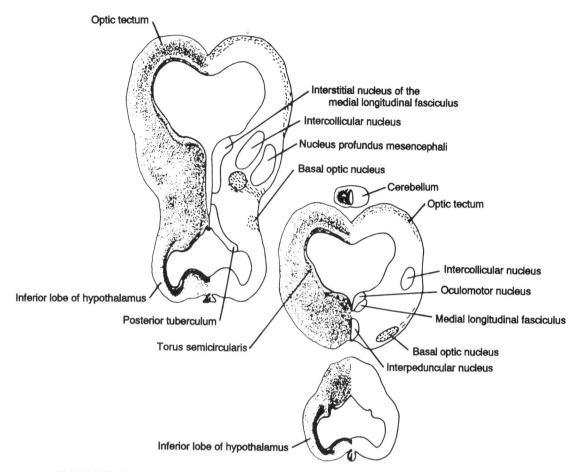

FIGURE 20-13. Transverse hemisections with mirror-image drawings through the midbrain of a shark (*Squalus acanthias*), with the retinal projections to tegmental nuclei and the optic tectum indicated by dashed lines and stippling. The more rostral section is at the upper left. Adapted from Northcutt (1979).

In diapsid reptiles and turtles a single **nucleus of the basal optic root** (nBOR) is present in the lateral part of the tegmentum. It receives retinal input and projects to the cerebellum and other sites in the caudal part of the brainstem. In birds (Fig. 20-12), this nucleus is also present and sometimes called **nucleus ectomamillaris.** In birds with frontally placed eyes, such as owls, a high proportion of nBOR neurons are binocularly responsive, as is the case in many mammals with frontally placed eyes. The high proportion of binocular neurons have evolved independently in these two groups in correlation with frontal vision and the utilization of depth cues in retinal image stabilization.

EVOLUTIONARY PERSPECTIVE

The pretectum is better differentiated in teleost fishes than in any other group of vertebrates. Among amniotes, it is better differentiated, that is, more elaborated, in diapsid reptiles, birds, and turtles than in placental mammals, and may have been secondarily reduced in monotremes. Teleosts have dorsal and ventral accessory optic nuclei, while nonmammalian amniotes have only the latter (nBOR).

Cerebellar projections from pretectal and accessory optic nuclei are present in two major groups of vertebrates, teleost fishes and nonmammalian amniotes. On the basis of current data, one must hypothesize that such cerebellar projections are apomorphic and were evolved independently in these two groups.

Pretectal and accessory optic pathways in teleosts and in nonmammalian amniotes are different in the degree to which they are elaborated. In nonmammalian amniotes, the pretectal and accessory optic nuclei project to the oculomotor complex and the cerebellum. In teleosts, the pretectal nuclei are characterized by reciprocal connections with a number of structures and by other multisynaptic, feedback pathways. These nuclei influence motor behavior via major efferent pathways to the hypothalamus, through the raphe to the spinal cord, to the oculomotor nuclei, and to the cerebellum.

As we will discuss in Chapter 22, placental mammals have the most highly differentiated and elaborate dorsal thalamus among the various radiations of vertebrates. In contrast, the pretectum and its associated accessory optic system is most highly differentiated and elaborated in teleosts. As we will now

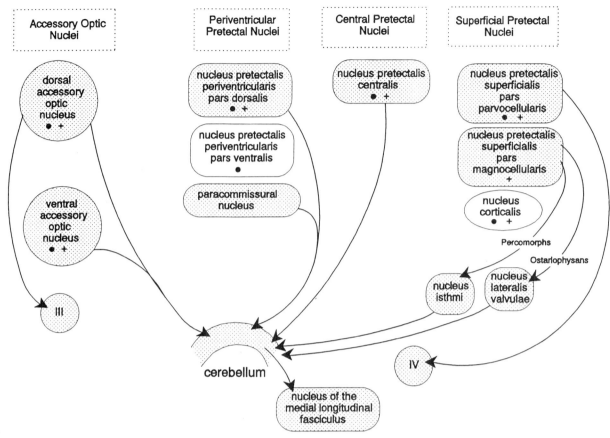

FIGURE 20-14. Diagram of pathways to the cerebellum and oculomotor nuclei in teleosts. Retinorecipient nuclei are indicated by a dot and tectorecipient nuclei by +.

discuss, teleosts also have migrated preglomerular nuclei that relay sensory information to the telencephalon and have no counterpart in mammals.

POSTERIOR TUBERCULUM

We have already encountered the unmigrated part of the posterior tuberculum in the tegmentum (Chapter 17). As discussed there, it lies in the medioventral part of the tegmentum, and some of its cells contain dopamine. The posterior tuberculum may thus be homologous as a field, at least in part, to the caudal diencephalic (A11) and/or rostral tegmental (rostral A9/A10) dopaminergic cell groups of mammals.

In most ray-finned fishes, migrated nuclei of the posterior tuberculum have been identified. The largest of these nuclei, the preglomerular nuclear complex, relays sensory information to the telencephalon. The pathways through the preglomerular nuclear complex mimic sensory relay pathways through the dorsal thalamus in their organization.

At least one migrated nucleus is also present in this caudal region of the diencephalon in cartilaginous fishes and will be discussed here. This nucleus has connections that are similar to some of the connections of the preglomerular nuclear complex of teleost fishes, and may be homologous as a field to it.

Group I

Migrated nuclei of the posterior tuberculum are not present in some of the Group I animals with laminar brains. In squalomorph sharks, a nucleus called the **posterior lateral thalamic nucleus** is present lateral to the periventricular part of the posterior tuberculum (Fig. 20-2). This nucleus is known to receive an electrosensory input in skates and rays. In a squalomorph shark, it has been found to project to the medial pallium in the telencephalon. In gars and *Amia*, a **preglomerular nuclear complex** is present lateral to the posterior tuberculum (Fig. 20-3).

Group II

In hagfishes, a relatively large nucleus called the **area lateralis of the posterior tuberculum** has been identified, but its connections remain to be studied. In Group II cartilaginous fishes, the **posterior lateral thalamic nucleus** (also called the **central thalamic nucleus**) lies lateral to the periventricular part of the posterior tuberculum (Fig. 20-6) and receives an electrosensory input. Both this nucleus and the posterior tuberculum project to the medial pallium in the telencephalon.

In teleosts, several migrated nuclei of the posterior tuberculum are present, the largest and most prominent of which is

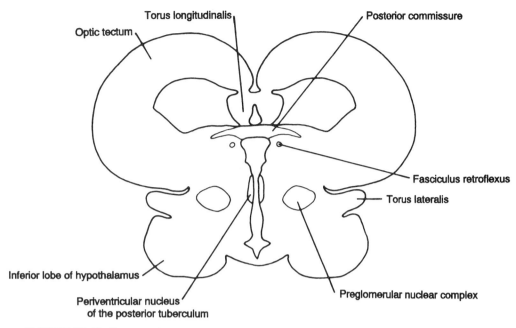

FIGURE 20-15. Drawing of a transverse section through the posterior tuberculum of a teleost (*Salmo gairdneri*). Data from Butler et al. (1991).

actually a group of nuclei that form the **preglomerular nuclear complex** (Fig. 20-15). *Caveat emptor:* unfortunately, the terminology in the literature on the preglomerular nuclear complex is particularly variable; many different terms, such as nucleus prethalamicus, nucleus rotundus, thalamus, and nucleus glomerulosus, have been used for all or part of the preglomerular nuclear complex. There is a similar array of terms for nucleus glomerulosus of the pretectum, which we discussed above. These terms have also been used for other nuclei in neighboring regions of the brain in other species. Until a more uniform terminology comes of age, the identity of any nucleus in this region must be determined by the content of the particular

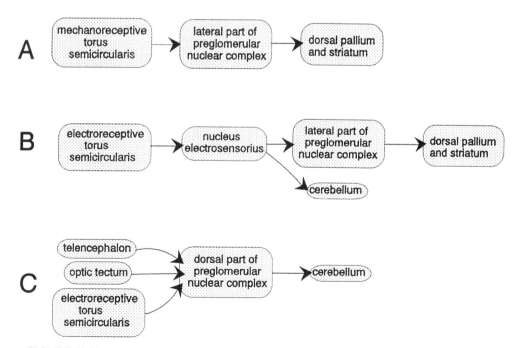

FIGURE 20-16. Diagram of some of the connections of the preglomerular nuclear complex. (A) Connections present in most teleosts, including electroreceptive teleosts. (B) Additional connections present in most electroreceptive teleosts. (C) Connections present in the electroreceptive mormyrid teleosts.

paper that one is reading, such as comments on the connections and location of the nucleus and on the terms others have used for it. With this *caveat* in mind, we will attempt here to offer a general survey of this area.

The cytoarchitecture and degree of development of the preglomerular nuclear complex varies among different species of teleosts. Boundaries between nuclei within the complex are not consistently easy to recognize, but lateral and medial divisions can be distinguished. Additional connectional studies are needed to define the number and boundaries of nuclei within the preglomerular complex in a variety of species.

Ascending sensory pathways that are relayed through parts of the preglomerular complex have been found in the lateral line and gustatory systems. While most teleosts have only a mechanosensory lateral line system, electroreception has been evolved independently in two groups of teleosts—some osteoglossomorphs and some ostariophysans. Differences in the development and connections of the preglomerular nuclear complex reflect this evolutionary history.

In most teleosts, including electroreceptive teleosts, the mechanosensory lateral line input to the torus semicircularis is relayed to the lateral part of the preglomerular nuclear complex (PGL). The PGL projects to the dorsal pallium and to the striatum in the telencephalon [Fig. 20-16(A)]. Auditory projections to PGL have also been found in some teleosts.

In electroreceptive teleosts, additional ascending pathways through PGL are present. The electrosensory part of the torus semicircularis in electroreceptive ostariophysans does not project directly to PGL but to a nucleus in the region of the pretectum called **nucleus electrosensorius** [Fig. 20-16(B)]. This nucleus projects to PGL, which, in turn, projects to the dorsal pallium and striatum in the telencephalon. Nucleus electrosensorius also projects to the cerebellum.

In mormyrids and notopterids, the electroreceptive osteoglossomorphs, the preglomerular nuclear complex is markedly enlarged and has more subdivisions than in other teleosts. In mormyrids [Fig. 20-16(C)], the electroreceptive part of the torus semicircularis projects to a dorsal part of the preglomerular complex (PGd), which also receives inputs from the telencephalon and the optic tectum. Instead of projecting to the telencephalon, however, PGd projects to the cerebellum. Within PGL, a ventral part (PGv) is the major source of ascending projections

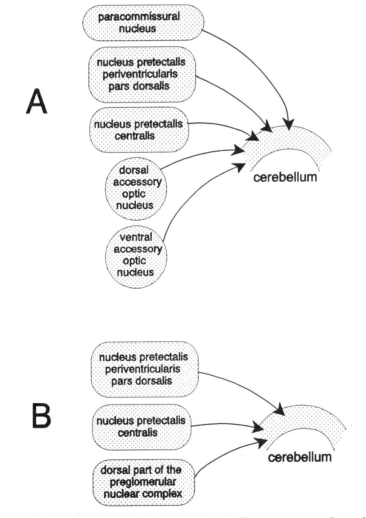

FIGURE 20-17. Diagram of the pretectal and accessory optic nuclear projections to the cerebellum in (A) most teleosts and (B) mormyrid teleosts.

to the telencephalon, but PGd does not project to PGv. Thus, electrosensory information does not appear to be relayed to the telencephalon in mormyrids.

In mormyrids and in some electroreceptive ostariophysans, the superficial pretectal and accessory optic nuclei are markedly reduced or absent. Electroreceptive inputs to the telencephalon and/or to the cerebellum have been selected for over the course of evolution in favor of visual pretectal/accessory optic pathways to the cerebellum. Five pretectal/accessory optic nuclei [Fig. 20-17(A)] project to the cerebellum in most teleosts, including the common goldfish (an ostariophysan)—in addition to cerebellar projections from nucleus electrosensorius where present. Only two of five cerebellar pathways of most teleosts are well developed in mormyrids, but a new pathway to the cerebellum from PGd is present [Fig. 20-17(B)]. This variation in electrosensory pathways is one of the more striking examples of the changes in nuclei and pathways that can occur in response to selective pressures.

In addition to relaying lateral line information to varying parts of the brain, the preglomerular nuclear complex relays gustatory information to the telencephalon. The secondary gustatory nucleus receives gustatory input from the facial lobes (see Chapter 11); in sunfishes, the secondary gustatory nucleus has been found to project to part of the preglomerular nuclear complex. In catfishes, the secondary gustatory nucleus projects to a nucleus called **nucleus lobobulbaris,** which lies just caudal to the major part of the preglomerular nuclear complex and is probably derived from it. Nucleus lobobulbaris projects to the striatum.

In most teleosts, the migrated nuclei of the posterior tuberculum include several additional, smaller nuclei in the vicinity of the preglomerular nuclear complex. These are **nucleus subglomerulosus,** the **lateral thalamic nucleus,** and the **posterior thalamic nucleus.** The lateral thalamic nucleus is known to receive afferent fibers from the magnocellular superficial pretectal nucleus and the secondary gustatory nucleus and to project to the optic tectum in percomorph euteleosts. The posterior thalamic nucleus projects to the telencephalon. More caudally and medially, a larger nucleus frequently called the mammillary body is present.

EVOLUTIONARY PERSPECTIVE

In amniote vertebrates, no migrated nuclei of the posterior tuberculum that relay ascending sensory information are present. Such migrated nuclei have been identified only in cartilaginous and ray-finned fishes. Whether the posterior lateral thalamic nucleus of cartilaginous fishes is homologous as a field to the preglomerular nuclear complex of teleosts or whether these nuclei have been evolved independently and are an example of parallelism or convergence remains to be determined.

The preglomerular nuclei in cartilaginous and ray-finned fishes appear to constitute an alternative to the dorsal thalamus as a relay of sensory information to the telencephalon. A marked degree of development of both the preglomerular nuclear complex and the pretectal nuclei combined with minimal development of the dorsal thalamus characterizes many fishes. This condition contrasts with the reverse situation in amniotes—the absence of migrated, sensory-relay nuclei derived from the posterior tuberculum, only a moderately developed pretectum, and an extensively developed dorsal thalamus.

FOR FURTHER READING

Braford, M. R., Jr. and Northcutt, R. G. (1983) Organization of the diencephalon and pretectum of the ray-finned fishes. In R. E. Davis and R. G. Northcutt (eds.), *Fish Neurobiology, Vol. 2: Higher Brain Areas and Functions.* Ann Arbor, MI: The University of Michigan Press, pp. 117–163.

Butler, A. B., Wullimann, M. F., and Northcutt, R. G. (1991) Comparative cytoarchitectonic analysis of some visual pretectal nuclei in teleosts. *Brain, Behavior and Evolution,* 38, 92–114.

Finger, T. E. and Karten, H. J. (1978) The accessory optic system in teleosts. *Brain Research,* 153, 144–149.

Fite, K. V. (1985) Pretectal and accessory-optic visual nuclei of fish, amphibia and reptiles: theme and variations. *Brain, Behavior and Evolution,* 26, 71–90.

Murakami, T., Fukuoka, T., and Ito, H. (1986) Telencephalic ascending acousticolateral system in a teleost *(Sebastiscus marmoratus),* with special reference to the fiber connections of the nucleus preglomerulosus. *Journal of Comparative Neurology,* 247, 383–397.

Striedter, G. F. and Northcutt, R. G. (1989) Two distinct visual pathways through the superficial pretectum in a percomorph teleost. *Journal of Comparative Neurology,* 283, 342–354.

Weber, J. T. (1985) Pretectal complex and accessory optic system of primates. *Brain, Behavior and Evolution,* 26, 117–140.

ADDITIONAL REFERENCES

Bangma, G. C. and ten Donkelaar, H. J. (1982) Afferent connections of the cerebellum in various types of reptiles. *Journal of Comparative Neurology,* 207, 255–273.

Bass, A. H. and Northcutt, R. G. (1981) Primary retinal targets in the Atlantic loggerhead sea turtle, *Caretta caretta. Cell and Tissue Research,* 218, 253–264.

Bass, A. H. and Northcutt, R. G. (1981) Retinal recipient nuclei in the painted turtle, *Chrysemys picta:* an autoradiographic and HRP study. *Journal of Comparative Neurology,* 199, 97–112.

Benevento, L. A. and Fallon, J. H. (1975) The ascending projections of the superior colliculus in the rhesus monkey *(Macaca mulatta). Journal of Comparative Neurology,* 160, 339–362.

Berkley, K. J. and Mash, D. C. (1978) Somatic sensory projections to the pretectum in the cat. *Brain Research,* 158, 445–449.

Bodnarenko, S. R., Rojas, X., and McKenna, O. C. (1988) Spatial organization of the retinal projection to the avian lentiform nucleus of the mesencephalon. *Journal of Comparative Neurology,* 269, 431–447.

Bodznick, D. and Northcutt, R. G. (1984) An electrosensory area in the telencephalon of the little skate, *Raja erinacea. Brain Research,* 298: 117–124.

Brecha, N. and Karten, H. J. (1979) Accessory optic projections upon oculomotor nuclei and vestibulocerebellum. *Science,* 203, 913–916.

Butler, A. B. and Northcutt, R. G. (1971) Retinal projections in *Iguana iguana* and *Anolis carolinensis. Brain Research,* 26, 1–13.

Butler, A. B. and Northcutt, R. G. (1971) Ascending tectal efferent projections in the lizard *Iguana iguana. Brain Research,* 35, 597–601.

Butler, A. B. and Northcutt, R. G. (1973) Architectonic studies of the diencephalon of *Iguana iguana* (Linnaeus). *Journal of Comparative Neurology*, 149, 439–462.

Butler, A. B. and Northcutt, R. G. (1978) New thalamic visual nuclei in lizards. *Brain Research*, 149, 469–476.

Butler, A. B. and Northcutt, R. G. (1992) Retinal projections in the bowfin, *Amia calva*: cytoarchitectonic and experimental analysis. *Brain, Behavior and Evolution*, 39, 169–194.

Butler, A. B. and Northcutt, R. G. (1993) The diencephalon of the Pacific herring, *Clupea harengus*: cytoarchitectonic analysis. *Journal of Comparative Neurology*, 328, 527–546.

Campbell, C. B. G. and Hayhow, W. R. (1971) Primary optic pathways in the echidna, *Tachyglossus aculeatus*: an experimental degeneration study. *Journal of Comparative Neurology*, 143, 119–136.

Campbell, C. B. G. and Hayhow, W. R. (1972) Primary optic pathways in the duckbill platypus *Ornithorynchus anatinus*: an experimental degeneration study. *Journal of Comparative Neurology*, 145, 195–208.

Cruce, J. A. F. (1974) A cytoarchitectonic study of the diencephalon of the tegu lizard, *Tupinambis nigropunctatus*. *Journal of Comparative Neurology*, 153, 215–238.

Fiebig, E. and Bleckmann, H. (1989) Cell groups afferent to the telencephalon in a cartilaginous fish *(Platyrhinoidis triseriata)*. *Neuroscience Letters*, 105, 57–62.

Finger, T. E. (1980) Nonolfactory sensory pathway to the telencephalon in a teleost fish. *Science*, 210, 671–673.

Fite, K. V. (1979) Optokinetic nystagmus and the pigeon visual system. In A. M. Granda and J. H. Maxwell (eds.), *Neural Mechanisms of Behavior In the Pigeon*. New York: Plenum, pp. 395–407.

Fite, K. V., Kwei-Levy, C., and Bengston, L. (1989) Neurophysiological investigation of the pretectal nucleus lentiformis mesencephali in *Rana pipiens*. *Brain, Behavior and Evolution*, 34, 164–170.

Fite, K. V., Reiner, A., and Hunt, S. P. (1979) Optokinetic nystagmus and the accessory optic system of pigeon and turtle. *Brain, Behavior and Evolution*, 16, 192–202.

Gamlin, P. D. R. and Cohen, D. H. (1988) Retinal projections to the pretectum in the pigeon *(Columba livia)*. *Journal of Comparative Neurology*, 269, 1–17.

Graham, J. (1977) An autoradiographic study of the efferent connections of the superior colliculus in the cat. *Journal of Comparative Neurology*, 173, 629–654.

Grover, B. G. and S. C. Sharma (1981) Organization of extrinsic tectal connections in goldfish *(Carassius auratus)*. *Journal of Comparative Neurology*, 196, 471–488.

Hutchins, B. and Weber, J. T. (1985) The pretectal complex of the monkey: a reinvestigation of the morphology and retinal terminations. *Journal of Comparative Neurology*, 232, 425–442.

Ito, H., Tanaka, H., Sakamoto, N., and Morita, Y. (1981) Isthmic afferent neurons identified by retrograde HRP method in a teleost, *Navodon modestus*. *Brain Research*, 207, 163–169.

Ito, H. and Vanegas, H. (1983) Cytoarchitecture and ultrastructure of nucleus prethalamicus, with special reference to degenerating afferents from optic tectum and telencephalon, in a teleost *(Holocentrus ascensionis)*. *Journal of Comparative Neurology*, 221, 401–415.

Ito, H. and Yoshimoto, M. (1990) Cytoarchitecture and fiber connections of the nucleus lateralis valvulae in the carp *(Cyprinus carpio)*. *Journal of Comparative Neurology*, 298, 385–399.

Kanwal, J. S., Finger, T. E., and Caprio, J. (1988) Forebrain connections of the gustatory system in ictalurid catfishes. *Journal of Comparative Neurology*, 278, 353–376.

Karten, H. J., Fite, K. V., and Brecha, N. (1977) Specific projection of displaced retinal ganglion cells upon the accessory optic system in the pigeon *(Columba livia)*. *Proceedings of the National Academy of Sciences USA*, 74, 1753–1756.

Karten, H. J. and Hodos, W. (1967) *A Stereotaxic Atlas of the Brain of the Pigeon (Columba livia)*. Baltimore, MD: The Johns Hopkins Press.

Kennedy, M. C. and Rubinson, K. (1977) Retinal projections in larval, transforming and adult sea lamprey, *Petromyzon marinus*. *Journal of Comparative Neurology*, 171, 465–479.

Lamb, C. F. and Caprio, J. (1993) Diencephalic gustatory connections in the channel catfish. *Journal of Comparative Neurology*, 337, 400–418.

Lamb, C. F. and Caprio, J. (1993) Taste and tactile responsiveness of neurons in the posterior diencephalon of the channel catfish. *Journal of Comparative Neurology*, 337, 419–430.

Luiten, P. G. M. (1981) Two visual pathways to the telencephalon in the nurse shark *(Ginglymostoma cirratum)*. II. Ascending thalamotelencephalic connections. *Journal of Comparative Neurology*, 196, 539–548.

Meek, J., Nieuwenhuys, R., and Elsevier, D. (1986) Afferent and efferent connections of cerebellar lobe C_1 of the mormyrid fish *Gnathonemus petersii*: an HRP study. *Journal of Comparative Neurology*, 245, 319–341.

Meek, J., Nieuwenhuys, R., and Elsevier, D. (1986) Afferent and efferent connections of cerebellar lobe C_3 of the mormyrid fish *Gnathonemus petersii*: an HRP study. *Journal of Comparative Neurology*, 245, 342–358.

Montgomery, N. M. and Fite, K. V. (1991) Organization of ascending projections from the optic tectum and mesencephalic pretectal gray in *Rana pipiens*. *Visual Neuroscience*, 7, 459–478.

Morita, Y., Ito, H., and Masai, H. (1980) Central gustatory paths in the crucian carp, *Carassius carassius*. *Journal of Comparative Neurology*, 191, 119–132.

Morita, Y., Murakami, T., and Ito, H. (1983) Cytoarchitecture and topographic projections of the gustatory centers in a teleost, *Carassius carassius*. *Journal of Comparative Neurology*, 218, 378–394.

Murakami, T., Morita, Y., and Ito, H. (1983) Extrinsic and intrinsic fiber connections of the telencephalon in a teleost, *Sebastiscus marmoratus*. *Journal of Comparative Neurology*, 216, 115–131.

Mustari, M. J., Fuchs, A. F., Kaneko, C. R. S., and Robinson, F. R. (1994) Anatomical connections of the primate pretectal nucleus of the optic tract. *Journal of Comparative Neurology*, 349, 111–128.

Neary, T. J. and Northcutt, R. G. (1983) Nuclear organization of the bullfrog diencephalon. *Journal of Comparative Neurology*, 213, 262–278.

Northcutt, R. G. (1978) Brain organization in the cartilaginous fishes. In E. S. Hodgson and R. F. Mathewson (eds.), *Sensory Biology Of Sharks, Skates, and Rays*. Arlington, VA: Office of Naval Research, pp. 117–193.

Northcutt, R. G. (1979) Retinofugal pathways in fetal and adult spiny dogfish, *Squalus acanthias*. *Brain Research*, 162, 219–230.

Northcutt, R. G. (1991) Visual pathways in elasmobranchs: organization and phylogenetic implications. *Journal of Experimental Zoology*, Suppl. 5, 97–107.

Northcutt, R. G. and Braford, M. R., Jr. (1984) Some efferent connections of the superficial pretectum in the goldfish. *Brain Research*, 296, 181–184.

Northcutt, R. G. and Butler, A. B. (1974) Evolution of reptilian visual systems: retinal projections in a nocturnal lizard, *Gekko gecko* (Linnaeus). *Journal of Comparative Neurology*, 157, 453–466.

Northcutt, R. G. and Butler, A. B. (1980) Projections of the optic tectum in the longnose gar, *Lepisosteus osseus*. *Brain Research*, 190, 333–346.

Northcutt, R. G. and Butler, A. B. (1993) The diencephalon and optic tectum of the longnose gar, *Lepisosteus osseus* (L.): cytoarchitectonics and distribution of acetylcholinesterase. *Brain, Behavior and Evolution,* 41, 57–81.

Northcutt, R. G. and Wullimann, M. F. (1988) The visual system in teleost fishes: morphological patterns and trends. In J. Atema, R. R. Fay, A. N. Popper, and W. N. Tavolga (eds.), *Sensory Biology of Aquatic Animals.* New York: Springer-Verlag, pp. 515–552.

Pritz, M. B. and Stritzel, M. E. (1992) Interconnections between the pretectum and visual thalamus in reptiles, *Caiman crocodilus. Neuroscience Letters,* 143, 205–209.

Reiner, A. and Karten, H. J. (1978) A bisynaptic retinocerebellar pathway in the turtle. *Brain Research,* 150, 163–169.

Repérant, J., Miceli, D., Rio, J.-P., Peyrichoux, J., Pierre, J., and Kirpitchnikova, E. (1986) The anatomical organization of retinal projections in the shark *Scyliorhinus canicula* with special reference to the evolution of the selachian primary visual system. *Brain Research Reviews,* 11, 227–248.

Rooney, D. J. and Szabo, T. (1991) Reciprocal connections between the "nucleus rotundus" and the dorsal lateral telencephalon in the weakly electric fish *Gnathonemus petersii. Brain Research,* 543, 153–156.

Scalia, F. (1972) The termination of retinal axons in the pretectal region of mammals. *Journal of Comparative Neurology,* 145, 223–258.

Scalia, F. (1972) Retinal projections to the olivary pretectal nucleus in the tree shrew and comparison with the rat. *Brain, Behavior and Evolution,* 6, 237–252.

Scalia, F. and Arango, V. (1979) Topographic organization of the projections of the retina to the pretectal region in the rat. *Journal of Comparative Neurology,* 186, 271–292.

Schmidt, M., Zhang, H.-Y., and Hoffmann, K.-P. (1993) OKN-related neurons in the rat nucleus of the optic tract and dorsal terminal nucleus of the accessory optic system receive a direct cortical input. *Journal of Comparative Neurology,* 330, 147–157.

Simpson, J. I. (1984) The accessory optic system. *Annual Review of Neuroscience,* 7, 13–41.

Smeets, W. J. A. J. (1981) Efferent tectal pathways in two chondrichthyans, the shark *Scyliorhinus canicula* and the ray *Raja clavata. Journal of Comparative Neurology,* 195, 13–23.

Smeets, W. J. A. J. (1981) Retinofugal pathways in two chondrichthyans, the shark *Scyliorhinus canicula* and the ray *Raja clavata. Journal of Comparative Neurology,* 195, 1–11.

Smeets, W. J. A. J. and Northcutt, R. G. (1987) At least one thalamotelencephalic pathway in cartilaginous fishes projects to the medial pallium. *Neuroscience Letters,* 78, 277–282.

Striedter, G. F. (1990) The diencephalon of the channel catfish, *Ictalurus punctatus.* I. Nuclear organization. *Brain, Behavior and Evolution,* 36, 329– 354.

Striedter, G. F. (1990) The diencephalon of the channel catfish, *Ictalurus punctatus.* II. Retinal, tectal, cerebellar and telencephalic connections. *Brain, Behavior and Evolution,* 36, 355–377.

Striedter, G. F. (1991) Auditory, electrosensory, and mechanosensory lateral line pathways through the forebrain in channel catfishes. *Journal of Comparative Neurology,* 312, 311–331.

Striedter, G. F. (1992) Phylogenetic changes in the connections of the lateral preglomerular nucleus in ostariophysan teleosts: a pluralistic view of brain evolution. *Brain, Behavior and Evolution,* 39, 329–357.

Voneida, T. J. and Sligar, C. M. (1979) Efferent projections of the dorsal ventricular ridge and the striatum in the tegu lizard, *Tupinambis nigropunctatus. Journal of Comparative Neurology,* 186, 43–64.

Wicht, H. and Northcutt, R. G. (1990) Retinofugal and retinopetal projections in the Pacific hagfish, *Eptatretus stouti* (Myxinoidea). *Brain, Behavior and Evolution,* 36, 315–328.

Wicht, H. and Northcutt, R. G. (1992) The forebrain of the Pacific hagfish: a cladistic reconstruction of the ancestral craniate forebrain. *Brain, Behavior and Evolution,* 40, 25–64.

Wullimann, M. F. (1988) The tertiary gustatory center in sunfishes is not nucleus glomerulosus. *Neuroscience Letters,* 86, 6–10.

Wullimann, M. F., Meyer, D. L., and Northcutt, R. G. (1991) The visually related posterior pretectal nucleus in the non-percomorph teleost *Osteoglossum bicirrhosum* projects to the hypothalamus: a DiI study. *Journal of Comparative Neurology,* 312, 415–435.

Wullimann, M. F. and Northcutt, R. G. (1988) Connections of the corpus cerebelli in the green sunfish and the common goldfish: a comparison of perciform and cypriniform teleosts. *Brain, Behavior and Evolution,* 32, 293–316.

Wullimann, M. F. and Northcutt, R. G. (1990) Visual and electrosensory circuits of the diencephalon in mormyrids: an evolutionary perspective. *Journal of Comparative Neurology,* 297, 537–552.

Wylie, D. R., Shaver, S. W., and Frost, B. J. (1994) The visual response properties of neurons in the nucleus of the basal optic root of the Northern saw-whet owl *(Aegolius acadicus). Brain, Behavior and Evolution,* 43, 15–25.

Yoshida, A., Sessle, B. J., Dostrovsky, J. O., and Chiang, C. Y. (1992) Trigeminal and dorsal column nuclei projections to the anterior pretectal nucleus in the rat. *Brain Research,* 590, 81–94.

Yoshimoto, M. and Ito, H. (1993) Cytoarchitecture, fiber connections, and ultrastructure of the nucleus pretectalis superficialis pars magnocellularis (PSm) in carp. *Journal of Comparative Neurology,* 336, 433–446.

21

Epithalamus

INTRODUCTION

The epithalamus is restricted to a medial position in all vertebrates and does not vary in the degree of its development as a function of Group I versus Group II vertebrates. We will therefore discuss the structures of the epithalamus in the various radiations of vertebrates without separately considering Groups I and II.

The epithalamus (Fig. 21-1) can be divided into two parts. The first part is immediately dorsal to the dorsal thalamus, composed of the **habenular nuclei** and any related nuclei. The second part is called the **epiphysis.** This part forms in the roof of the third ventricle between the habenula and the posterior commissure and is composed of the **pineal gland** and, in some cases, a **partietal eye** or a **frontal organ.** Fibers that originate from neurons within the epiphysis constitute the **epiphyseal nerve** (see Chapter 9), although they can be referred to as the **pineal tract** and/or the **parietal nerve.** The major functional roles of the epithalamus are the regulation of cyclic behavior, such as the circadian rhythm and reproductive cycles, in response to the day–night cycle and the modulation of other systems in the brain such as the limbic system, hypothalamus, raphe, and some motor-related pathways.

EPIPHYSIS

An epiphysis has not been identified in hagfishes. In lampreys, the epiphysis consists of two outgrowths from the roof of the third ventricle, called the pineal and parapineal organs, both of which have photoreceptive cells. In cartilaginous fishes, a single outgrowth, which has a glandular structure, is present in the epiphysis. Two outgrowths of a primarily glandular structure

are present in the epiphysis of some nonteleost ray-finned fishes, although only one has been identified in teleosts.

All amphibians have a pineal gland, and frogs additionally have a more rostral part of the epiphysis called the frontal organ, which is similar to the parietal eye of lizards. Most lizards and the tuatara *Sphenodon* have a parietal eye (see Chapter 2), with a pigmented retina, in addition to the pineal gland. Turtles and snakes have only a pineal gland, and adult crocodiles lack an epiphysis. Pineal and parapineal outgrowths are present in bird embryos, but only the glandular pineal persists in adults. A single epiphyseal outgrowth, the pineal gland, is present in most mammals, including monotremes.

Connections of the epiphyseal components have been studied to some extent in most vertebrate groups. In lampreys the epiphysis (pineal and parapineal organs together) contains photoreceptors and neurons, which have widespread projections to the diencephalon and midbrain. The efferent fibers course through the habenula and project to the dorsal thalamus, ventral thalamus, hypothalamus, pretectum, and posterior tuberculum in the diencephalon, as well as to a thickened part of the ependyma of the third ventricle ventral to the posterior commissure called the **subcommissural organ.** In the midbrain, epiphyseal projections reach the optic tectum, torus semicircularis, tegmentum, and oculomotor nucleus. A few neurons that lie in the dorsal tegmental area project epiphysopetally, that is, to the epiphysis.

Little is known about the projections of the pineal organ in cartilaginous fishes, but in ray-finned fishes, pineal projections are almost as widespread as they are in lampreys. As in lampreys, the pineal contains photoreceptor cells and neurons. Efferents from the pineal neurons project to the habenular nuclei, the subcommissural organ, and areas within the dorsal and ventral thalamus, the periventricular part of the hypothalamus, the pretectum, and the tegmentum. Neurons within the brain that project to the pineal have not yet been found in any jawed fishes.

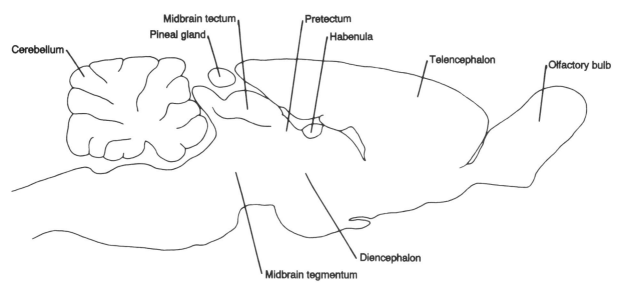

FIGURE 21-1. Drawing of a parasagittal section through the brain of a rat (*Rattus norvegicus*) showing the location of the pineal gland and the habenula. Adapted from Pellegrino et al. (1979). Used with permission of Plenum Publishing Corp.

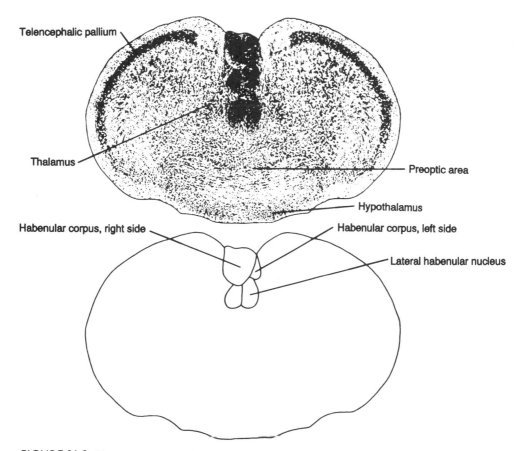

FIGURE 21-2. Transverse section with drawing through the habenula of a hagfish (*Eptatretus stouti*). Adapted from Wicht and Northcutt (1992). Used with permission of S. Karger AG.

In amphibians, projections of the epiphysis to sites in the diencephalon and midbrain are similar to those in agnathans and ray-finned fishes, but additional projections are present as well. In frogs, the pineal organ projects to the pretectum, the periventricular tegmental gray, and the region of the oculomotor nucleus. The frontal organ of the epiphysis in frogs projects to these areas and to the ventral thalamus. Additionally, the frontal organ projects to the medial part of the amygdala in the telencephalon. No epiphysopetal projections from the brain to the epiphysis are present, however.

Pineal projections, long thought to be absent in mammals, are present although relatively restricted. They terminate in the medial habenular nucleus and the pretectum in the region of the posterior commissure. Cells in the medial and lateral habenular nuclei project to the pineal organ, as do ependymal cells that line the pineal and parapineal recesses and are in contact with cerebrospinal fluid. A sympathetic input to the pineal is also present.

In diapsid reptiles and turtles, the pineal organ is known to project to parts of the diencephalon and mesencephalon, but the details of this projection remain to be worked out. The parietal eye gives rise to fibers that pass through and may terminate in the habenula. Parietal efferent fibers also project to the dorsolateral part of the dorsal thalamus, the pretectum, the periventricular part of the hypothalamus, the preoptic area, and the nucleus of the diagonal band in the septal area of the telencephalon. Additionally, a sparse epiphysopetal projection from scattered neurons in the periventricular part of the hypothalamus to the parietal eye is present.

Pineal projections are relatively restricted in birds. Avian pineal cells have irregular membrane segments that are similar to the outer segments of retinal photoreceptor cells and that contain opsin, the protein that combines with vitamin A to form the photosensitive pigment, rhodopsin (visual purple). Efferent fibers from the pineal project to the medial and lateral habenular nuclei and to the periventricular part of the hypothalamus. Epiphyseal projections do not reach parts of the dorsal or ventral thalamus, pretectum, or mesencephalon, however. As in diapsid reptiles and turtles, epiphysopetal projections to the pineal are present and originate from cells in the habenular nuclei and in the periventricular part of the hypothalamus. The pineal in birds is also known to receive a sympathetic nervous innervation from neurons in the superior cervical ganglion.

The role of the epiphysis in the regulation of circadian and other biological rhythms is well established, but the anatomical substrate for its activity has not been fully clarified. Photoreceptor cells are present within the epiphysis in most vertebrates. Cells within the pineal organ secrete melatonin and other regulatory hormones directly into the blood so that biological rhythms can be regulated by neuroendocrine control. The epiphysis also regulates biological rhythms via a number of connections of the epiphyseal cranial nerve.

The majority of projections from the epiphysis overlap regions to which the retina projects, indicating a complimentary role for some retinal and some epiphyseal inputs in regulating circadian rhythms. The epiphysis also may be involved in behavioral thermoregulation, such as some diapsid reptiles use to maintain body temperature during the daylight period. The epiphyseal projections to parts of the thalamus and hypothalamus, particularly the periventricular hypothalamus and preoptic

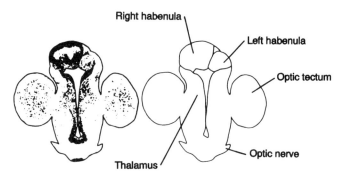

FIGURE 21-3. Transverse section with drawing through the habenula of a lamprey (*Petromyzon marinus*). Adapted from Kennedy and Rubinson (1977).

area, are to areas containing estradiol-concentrating neurons. These epiphyseal projections may affect gonadotropin secretion and mating behavior, which is seasonal and thus influenced by day–night periodicity. In birds, electrophysiological responses of pineal cells to artificially applied magnetic fields have been demonstrated, so a role for the pineal in the regulation and guidance of seasonal migrations is a possiblity.

Among jawed vertebrates, the presence of neurons in the brain that project epiphysopetally appears to be an apomorphic feature for amniotes. These neurons may be part of a multisynaptic pathway that relays information about the day–night cycle from the retina to the epiphysis. The retina projects to a nucleus in the preoptic area called the suprachiasmatic nucleus. Bilateral lesions of the suprachiasmatic nucleus result in disruptions of cyclic hormone secretions in the pineal organ, the circadian rhythms, and the estrous cycle. The possibility of a projection

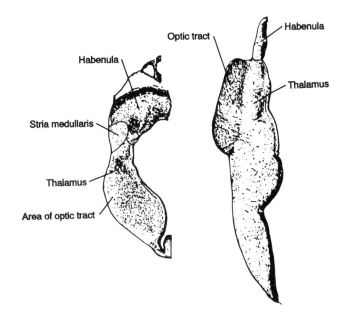

FIGURE 21-4. Transverse hemisections through the habenula of a squalomorph shark (*Squalus acanthias*), on the left, and the bowfin (*Amia calva*), on the right. Adapted from Northcutt (1979) and Butler and Northcutt (1992), respectively, and used with permission of Elsevier and S. Karger AG, respectively.

from the suprachiasmatic nucleus to the neurons that have been identified as projecting to the pineal, particularly those in the periventricular part of the hypothalamus, needs to be investigated.

HABENULA

One of the most striking morphological features of the habenula is that in many vertebrates it is markedly asymmetric in size on one side of the brain versus the other. In most, but not all, species with asymmetrical habenulae, the right habenula is larger than the left. The biological significance of this variation, if any, is unknown.

The habenular complex is well developed and prominent in hagfishes (Fig. 21-2), lampreys (Fig. 21-3), and cartilaginous and ray-finned fishes (Fig. 21-4), but only a limited amount of experimental work has been done on its connections in these vertebrates. The major efferent tract is the **fasciculus retroflexus** (or **habenulo-interpeduncular tract**) and is universally present in all fishes and in tetrapods. It is a discrete fiber bundle with a round shape in transverse section that courses through the dorsomedial parts of the dorsal thalamus and pretectum and thence to the **interpeduncular nucleus** in the brainstem. The position of the fasciculus retroflexus (FR) in

the pretectal region is shown in a number of the figures in Chapter 20.

The major afferent tract to the habenula from the telencephalon and other sources is the **stria medullaris** (**SM,** Fig. 21-4) and is recognizable in most fishes. An input to the habenula from a ventral area of the telencephalon, which is referred to as Vs (area ventralis telencephali pars supracommissuralis) has been found in teleosts. This ventral telencephalic region appears to be involved in the regulation of sexual behavior and may be the homologue of the basal amygdala of tetrapods.

In amphibians, the habenular complex is known to receive limbic afferents from the septal area and bed nucleus of the hippocampal commissure and striatal afferents from the entopeduncular nucleus (the homologue of the internal segment of the globus pallidus of primates). Input to the habenula also arises from part of the hypothalamus. The habenula projects to sites in the brainstem, including the interpeduncular nucleus.

The multiple connections of the epithalamus are most thoroughly worked out in mammals. In mammals, the epithalamus includes the habenular nuclei and two additional nuclei called the anterior and posterior paraventricular nuclei. The latter two nuclei are reciprocally connected with structures that are part of or related to the limbic system: the amygdala, the hippocampal formation, the lateral hypothalamus, and the preoptic area [Fig. 21-5(A)].

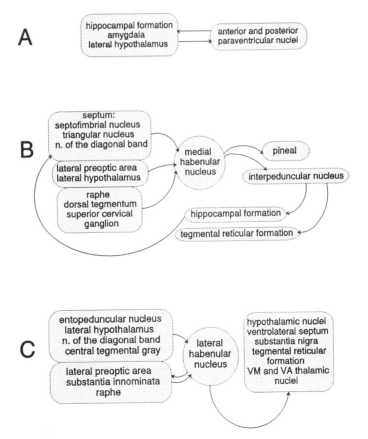

FIGURE 21-5. Diagram of connections of the habenular nuclei in mammals. (A) connections of the anterior and posterior periventricular nuclei; (B) connections of the medial habenular nucleus; and (C) connections of the lateral habenular nucelus.

The habenula in mammals contains medial and lateral habenular nuclei. The medial habenular nucleus is interconnected with the limbic system [Fig. 21-5(B)]. It receives inputs from the septofimbrial and triangular nuclei in the caudal part of the septum, the nucleus of the diagonal band in the medial part of the septum, the lateral preoptic and lateral hypothalamic areas, and the midbrain raphe. Like the pineal organ, the medial habenular nucleus receives a sympathetic input from the superior cervical ganglion. The medial habenular nucleus projects to the interpeduncular nucleus via the fasciculus retroflexus. The interpeduncular nucleus projects to the hippocampal formation, which in turn projects to the septum, thus forming a feedback loop for the integration of limbic-system information. In addition, the interpeduncular nucleus projects to the raphe and tegmental areas of the midbrain that receive input from the lateral habenular nucleus.

The medial habenular nucleus also contains neurons that project to the pineal organ, as discussed above. Due to the limbic afferents to the medial habenular nucleus, these epiphysopetal neurons would thus be subject to limbic system influence. Disruptions of the estrous cycle, sleeping disorders, and other cyclic behaviors altered as a result of stress may be influenced by this pathway.

The lateral habenular nucleus receives inputs from diverse sources related to both the striatum and the limbic system [Fig. 21-5(C)]: the entopeduncular nucleus (i.e., the internal segment of the globus pallidus), the lateral hypothalamus, the nucleus of the diagonal band, the substantia innominata, the lateral preoptic area, the midbrain raphe, and the central tegmental gray. The lateral habenular nucleus projects epiphysopetally to the pineal organ. The rest of the efferent projections of the lateral habenular nucleus are, like the afferent projections, related to both limbic and striatal structures. The lateral habenular nucleus projects to hypothalamic nuclei, the lateral preoptic area, substantia innominata, ventrolateral septum, the substantia nigra (pars compacta), the raphe and the adjoining tegmental reticular formation, and the ventral medial (VM) and ventral anterior (VA) thalamic nuclei (which are part of the ventral nuclear group and related to motor feedback pathways that will be discussed in Chapter 27).

The lateral habenular nucleus can be divided into three parts based on the predominance of particular connections. In rats, the input to the lateral part of the lateral habenular nucleus from the entopeduncular nucleus is substantial, while in cats and monkeys, the lateral part of the lateral habenular nucleus is dominated by input from the lateral hypothalamus [Fig. 21-6(A)]. The projections from this part of the lateral habenula are predominately to midbrain tegmental motor areas.

The inputs to the medial part of the lateral habenular nucleus in rats are predominantly from limbic sources, and efferent projections are predominantly to the raphe, which projects in turn to the hippocampal formation [Fig. 21-6(B)]. In cats, the two nonlateral parts of the lateral habenula are likewise dominated by limbic inputs—a central part dominated by inputs from the hypothalamus and raphe and a dorsomedial part dominated by inputs from the hypothalamus and the preoptic area.

Thus, the anterior and posterior paraventricular nuclei, the medial habenular nucleus, and the medial or central-dorsomedial part of the lateral habenular nucleus are all primarily interconnected with the limbic system. The lateral part of the lateral habenular nucleus is interconnected with both the striatal and limbic systems, although the striatal component is more prominent in rats than in cats or monkeys.

In diapsid reptiles (Fig. 21-7), birds, and turtles only limited information is available on habenular connections. Inputs to

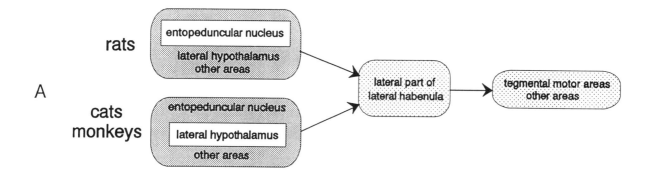

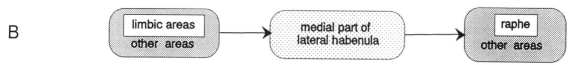

FIGURE 21-6. Diagram comparing the striatal and limbic connections of the lateral and medial parts of the lateral habenular nucleus. The highlighted names indicate major sources or major targets of the pathways.

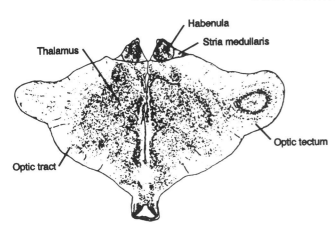

FIGURE 21-7. Transverse section through the habenula of a lizard (*Iguana iguana*). Adapted from Foster and Hall (1975).

the habenula from both striatal and limbic (septal) sources have been found, as well as from parts of the preoptic area and hypothalamus. The habenula projects via the fasciculus retroflexus to the interpeduncular nucleus and presumably to other sites in the brainstem as well.

EVOLUTIONARY PERSPECTIVE

The epithalamus itself and at least one of its components, the habenula, are present in all vertebrates and plesiomorphic for vertebrates. The habenula and its associated fiber tracts, particularly the fasciculus retroflexus, are very conservative features of vertebrate brains. Variation occurs in the presence or absence of the components of the epiphysis, however. Fossil evidence reveals that an epiphysis was present in most ostracoderm and placoderm fishes and is thus plesiomorphic for vertebrates. An epiphysis has not been identified in hagfishes and may have been secondarily lost in this group.

The pineal component of the epiphysis is present in lampreys and all groups of jawed vertebrates, except crocodiles. A second outgrowth of the epiphysis is present in lampreys, nonteleost ray-finned fishes, frogs, lizards, the tuatara, and bird embryos. The degree to which photoreceptor cells develop in one or both epiphyseal outgrowths is quite variable. In lampreys, both outgrowths contain photoreceptor cells, while in lizards and the tuatara, only the parietal eye is developed as a photoreceptive organ. This distribution suggests that the pineal component of the epiphysis is plesiomorphic for at least the common ancestral group of lampreys and jawed vertebrates and has been reduced and/or lost once in crocodiles. A second component of the epiphysis appears to have been gained multiple times and, at least in some cases, lost secondarily. Efferent projections to the pineal are present in lampreys and amniotes, but the variability in the location of the central neurons suggests that these projections were evolved independently.

FOR FURTHER READING

Díaz, C. and Puelles, L. (1992) Afferent connections of the habenular complex in the lizard *Gallotia galloti. Brain, Behavior and Evolution,* 39, 312–324.

Ekström, P. and van Veen, T. (1984) Pineal neural connections with the brain in two teleosts, the crucian carp and the European eel. *Journal of Pineal Research,* 1, 245–261.

Eldred, W. D., Finger, T. E., and Nolte, J. (1980) Central projections of the frontal organ of *Rana pipiens,* as demonstrated by the anterograde transport of horseradish peroxidase. *Cell and Tissue Research,* 211, 215–222.

Kemali, M., Guglielmotti, V., and Gioffré, D. (1980) Neuroanatomical indentification of the frog habenular connections using peroxidase (HRP). *Experimental Brain Research,* 38, 341–347.

Korf, H.-W., Oksche, A., Ekström, P., Gery, I., Zigler, J. S., Jr., and Klein, D. C. (1986) Pinealocyte projections into the mammalian brain revealed with S-antigen antiserum. *Science,* 231,735–737.

Puzdrowski, R. L. and Northcutt, R. G. (1989) Central projections of the pineal complex in the silver lamprey *Ichthyomyzon unicuspis. Cell and Tissue Research,* 225, 269–274.

ADDITIONAL REFERENCES

Berk, M. L. and Butler, A. B. (1981) Efferent projections of the medial preoptic nucleus and medial hypothalamus in the pigeon. *Journal of Comparative Neurology,* 203, 379–399.

Butler, A. B. and Northcutt, R. G. (1992) Retinal projections in the bowfin, *Amia calva:* cytoarchitectonic and experimental analysis. *Brain, Behavior and Evolution,* 39, 169–194.

Butler, A. B. and Northcutt, R. G. (1993) The diencephalon of the Pacific herring, *Clupea harengus:* cytoarchitectonic analysis. *Journal of Comparative Neurology,* 328, 527–546.

Cragg, B. G. (1961) The connections of the habenula in the rabbit. *Experimental Neurology,* 3, 388–409.

Ekström, P. (1984) Central neural connections of the pineal organ and retina in the teleost *Gasterosteus aculeatus* L. *Journal of Comparative Neurology,* 226, 321–335.

Ekström, P. and van Veen, T. (1983) Central connections of the pineal organ in the three-spined stickleback, *Gasterosteus aculeatus* L. (Teleostei). *Cell and Tissue Research,* 232, 141–155.

Foster, R. E. and Hall, W. C. (1975) The connections and laminar organization of the optic tectum in a reptile (*Iguana iguana*). *Journal of Comparative Neurology,* 163, 397–426.

Hafeez, M. A. and Zerihun, L. (1974) Studies on central projections of the pineal nerve tract in rainbow trout, *Salmo gairdneri* Richardson, using cobalt chloride iontophoresis. *Cell and Tissue Research,* 154, 485–510.

Herkenham, M. and Nauta, W. J. H. (1977) Afferent connections of the habenular nuclei in the rat. A horseradish peroxidase study, with a note on the fiber-of-passage problem. *Journal of Comparative Neurology,* 173, 123–146.

Herkenham, M. and Nauta, W. J. H. (1979) Efferent connections of the habenular nuclei in the rat. *Journal of Comparative Neurology,* 187, 19–48.

Jones, E. G. (1985) *The Thalamus.* New York: Plenum.

Kennedy, M. C. and Rubinson, K. (1977) Retinal projections in larval, transforming and adult sea lamprey, *Petromyzon marinus. Journal of Comparative Neurology,* 171, 465–479.

Klein, D. C. and Moore, R. Y. (1979) Pineal *N*-acetyltransferase and hydroxyindole-*O*-methyltransferase: control by the retinohypothalamic tract and the suprachiasmatic nucleus. *Brain Research,* 174, 245–262.

Korf, H.-W. and Wagner, U. (1981) Nervous connections of the parietal eye in adult *Lacerta* s. *sicula* Rafinesque as demonstrated by anterograde and retrograde transport of horseradish peroxidase. *Cell and Tissue Research,* 219, 567–583.

Korf, H.-W., Zimmerman, N. H., and Oksche, A. (1982) Intrinsic neurons and neural connections of the pineal organ of the house sparrow, *Passer domesticus,* as revealed by anterograde and retrograde transport of horseradish peroxidase. *Cell and Tissue Research,* 222, 243–260.

Krayniak, P. F. and Siegel, A. (1978) Efferent connections of the septal area in the pigeon. *Brain, Behavior and Evolution,* 15, 389–404.

Kudo, M., Yamamoto, M. and Nakamura, Y. (1991) Suprachiasmatic nucleus and retinohypothalamic projections in moles. *Brain, Behavior and Evolution,* 38, 332–338.

McBride, R. L. (1981) Organization of afferent connections of the feline lateral habenular connections. *Journal of Comparative Neurology,* 198, 89–99.

Møller, M. and Korf, H.-W. (1987) Neural connections between the brain and pineal gland of the golden hamster (*Mesocricetus auratus.*) *Cell and Tissue Research,* 247, 145–153.

Nauta, H. J. W. (1979) Projections of the pallidal complex: an autoradiographic study in the cat. *Neuroscience,* 4, 1853–1874.

Northcutt, R. G. (1979) Retinofugal pathways in fetal and adult spiny dogfish, *Squalus acanthias. Brain Research,* 162, 219–230.

Parent, A. (1986) *Comparative Neurobiology Of the Basal Ganglia.* New York: Wiley.

Pellegrino, L. J., Pellegrino, A. S., and Cushman, A. J. (1979) *A Stereotaxic Atlas of the Rat Brain.* New York: Plenum.

Russchen, F. T. and Jonker, A. J. (1988) Efferent connections of the striatum and the nucleus accumbens in the lizard *Gekko gecko. Journal of Comparative Neurology,* 276, 61–80.

Saper, C. B., Swanson, L. W., and Cowan, W. M. (1979) An autoradiographic study of the efferent connections of the lateral hypothalamic area in the rat. *Journal of Comparative Neurology,* 183, 689–706.

Shiga, T., Oka, Y., Satou, M., Okumoto, N., and Ueda, K. (1985) Efferents from the supracommissural ventral telencephalon in the hime salmon (landlocked red salmon, *Oncorhynchus nerka*): an anterograde degeneration study. *Brain Research Bulletin,* 14, 55–61.

van der Kooy, D. and Carter, D. A. (1981) The organization of the efferent projections and striatal afferents of the entopeduncular nucleus and adjacent areas in the rat. *Brain Research,* 211, 15–36.

Villani, L., Dipietrangelo, L., Pallotti, C., Pettazzoni, P., Zironi, I., and Guarnieri, T. (1994) Ultrastructural and immunohistochemical study of the telencephalo-habenulo-interpeduncular connections of the goldfish. *Brain Research Bulletin,* 34, 1–5.

Wicht, H. and Northcutt, R. G. (1992) The forebrain of the Pacific hagfish: a cladistic reconstruction of the ancestral craniate forebrain. *Brain, Behavior and Evolution,* 40, 25–64.

Wyss, J. M., Swanson, L. W. and Cowan, W. M., (1979) A study of subcortical afferents to the hippocampal formation in the rat. *Neuroscience,* 4, 463–476.

Yañez, J. and Anadón, R. (1994) Afferent and efferent connections of the habenula in the larval sea lamprey (*Petromyzon marinus* L.): an experimental study. *Journal of Comparative Neurology,* 345, 148–160.

Yañez, J., Anadón, R., Holmqvist, B. I., and Ekström, P. (1993) Neural projections of the pineal organ in the larval sea lamprey (*Petromyzon marinus* L.) revealed by indocarbocyanine dye tracing. *Neuroscience Letters,* 164, 213–216.

22

Dorsal Thalamus

INTRODUCTION

The dorsal thalamus is a collection of nuclei within the diencephalon that relays ascending sensory and related information to the telencephalon and is thus a crucial link between the external world and an animal's sensory experience of it. The dorsal thalamus is the key to the telencephalon in more ways than one; these two regions have extensive neuroanatomical connections and corresponding functional interactions, and the evolution of these two regions is similarly intertwined. In this sense, the dorsal thalamus is a Rosetta stone for deciphering the history of forebrain evolution.

The way in which the organization of the dorsal thalamus of vertebrates is presented in this chapter represents a significant departure from previous views. The present weight of evidence indicates that many previously held ideas about the evolution of the dorsal thalamus are no longer plausible. The approach to dorsal thalamic organization taken in this chapter is based on a new interpretation of forebrain evolution (see Butler, 1994a,b). This new interpretation encompasses the major body of literature now available on the organization and connections of the dorsal thalamus of vertebrates. It represents the most parsimonious reconstruction of dorsal thalamic evolutionary history that the data allow. We recognize, however, that not everyone will accept this interpretation of dorsal thalamic evolution. Some may feel that such a relatively unestablished interpretation does not belong in a textbook. We have therefore endeavored to be as objective as possible in writing this chapter and to keep separate the data from our interpretation.

The first part of this chapter is a brief introduction to dorsal thalamic nuclei followed by a synopsis of their organization and derivation within the conceptual framework of the new interpretation. In succeeding parts, data on the nuclei are dis-

cussed in more depth, and only the grouping of the nuclei reflects the bias of the new interpretation: those nuclei that receive their predominant input from the roof of the midbrain are discussed before the nuclei that have other inputs predominating. In the final two parts of this chapter, the phylogenetic implications of the data and the defining characteristics of the dorsal thalamus are discussed.

OVERVIEW OF DORSAL THALAMIC NUCLEI

The nuclei of the dorsal thalamus lie in a periventricular position or are only slightly migrated away from the periventricular matrix in most anamniote vertebrates. Three nuclei can be distinguished in the dorsal thalamus of anamniotes. In contrast, the dorsal thalamus in amniotes is characterized by lateral migration of the neurons and by a multiplicity of distinct nuclei. This elaboration of the dorsal thalamus in amniotes contrasts with the elaboration of the pretectal and posterior tubercular regions in teleost fishes, as pointed out in Chapter 20.

Group I

In lampreys (Fig. 22-1), a dorsal thalamic region is present, but individual nuclei remain to be identified. In squalomorph sharks (Fig. 22-2) and nonteleost ray-finned fishes (Fig. 22-3), the dorsal thalamus contains three nuclei: the **central posterior nucleus**, the **dorsal posterior nucleus**, and **nucleus anterior.** Nucleus anterior lies rostral to the other two nuclei and predominantly receives retinal projections. The dorsal posterior nucleus receives the bulk of the optic tectal projections,

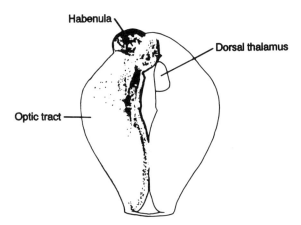

FIGURE 22-1. Transverse hemisection with mirror-image drawing through the diencephalon of a lamprey (*Petromyzon marinus*). Adapted from Kennedy and Rubinson (1977).

and the central posterior nucleus receives auditory projections from the torus semicircularis. In the dorsal thalamus of amphibians, similar nuclei receiving predominantly retinal, optic tectal, and toral inputs are present. Slight migration, enlargement, and subdividing of the nuclei occurs in frogs, however, and the nomenclature is different in some respects.

In frogs, the dorsal thalamus (Fig. 22-4) has been divided into three rostral-to-caudal zones. The most rostral zone contains **nucleus anterior,** which receives retinal and a number of other afferent projections. The middle zone contains the toral-recipient **central posterior nucleus** (also called the **central nucleus**) and two nuclei—the **anterior lateral nucleus** and the **posterior lateral nucleus**—that receive most of the optic tectal projections. Whether only the anterior lateral nucleus or both of these nuclei are homologous to the dorsal posterior nucleus of other Group I anamniotes remains to be resolved. Somatosensory projections relayed through the midbrain roof also project to the middle thalamic zone, but their exact locus of termination is not yet known. The posterior zone may be a

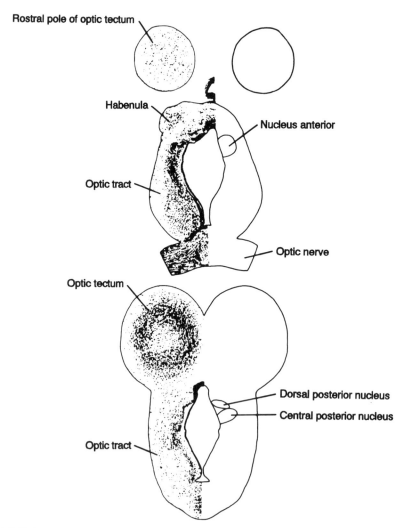

FIGURE 22-2. Transverse hemisections with mirror-image drawings through the diencephalon of a squalomorph shark (*Squalus acanthias*). The upper section is more rostral. Adapted from Northcutt (1979). Used with permission of Elsevier.

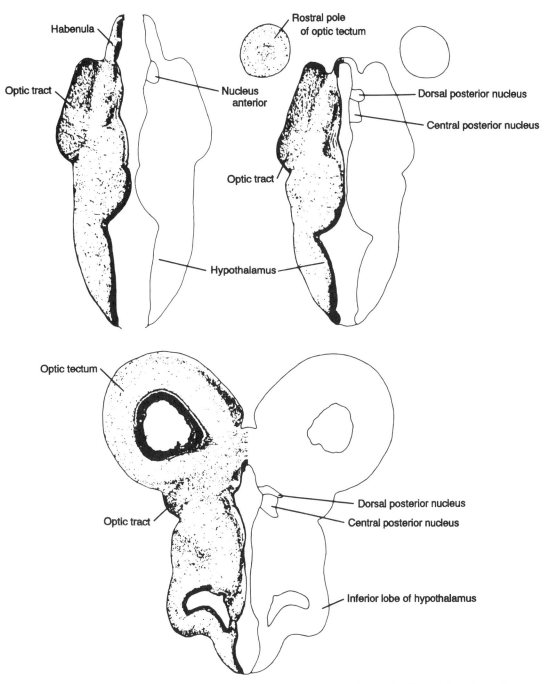

FIGURE 22-3. Transverse hemisections with mirror-image drawings through the diencephalon of a bowfin (*Amia calva*). The upper left section is most rostral. Adapted from Butler and Northcutt (1992). Used with permission of S. Karger AG.

part of the pretectum rather than the dorsal thalamus and is treated as such here (see Chapter 20).

Group IIA

Some migration of neurons in the dorsal thalamus occurs in hagfishes. Four nuclei have been recognized (Fig. 22-5), one of which—**nucleus anterior**—appears to be homologous to the same-named nucleus in other anamniotes and receives reti-

nal projections. The homologies of the three remaining nuclei—the **internal nucleus,** the **external nucleus,** and the **subhabenular nucleus**—are uncertain.

In galeomorph sharks, skates, and rays, some migration of neurons also occurs in the dorsal thalamus (DTh, Fig. 22-6). A **nucleus anterior** and a **dorsal posterior nucleus** have been recognized in some species on the basis of their major inputs from the retina and the optic tectum, respectively, as homologues of the like-named nuclei in other anamniotes, but

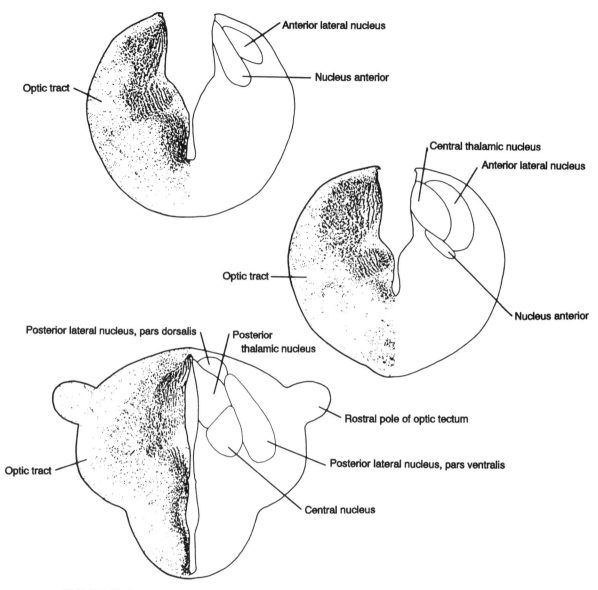

FIGURE 22-4. Transverse hemisections with mirror-image drawings through the diencephalon of a bullfrog (*Rana catesbeiana*). The upper left section is most rostral. Adapted from Neary and Northcutt (1983).

additional work needs to be done to determine the homologies of other nuclei present.

In teleosts, three nuclei—**nucleus anterior,** which lies most rostrally and receives predominantly retinal input, and the **central posterior** (toral-recipient) and **dorsal posterior** (optic tectal-recipient) **nuclei**—are present in the dorsal thalamus (Fig. 22-7). Migration of neurons does not occur to any marked degree, although the two more caudal nuclei have laterally extending parts.

Group IIB

In most mammals, a number of groups of nuclei (Fig. 22-8) are present. The major superior colliculus (optic tectum)-recipient nucleus is called either the **lateral posterior nucleus** (LP) or the **pulvinar,** depending upon the species, and is usually referred to as the **LP/pulvinar complex (LP/pul)**

for simplicity. The inferior colliculus (torus semicircularis)-recipient nucleus is the **medial geniculate body.** Both the LP/pulvinar complex and the medial geniculate body have subdivisions. Somatosensory and multisensory projections from the midbrain roof terminate in various parts of a group of nuclei called the **posterior nuclear group.** The **dorsal lateral geniculate nucleus** predominantly receives a substantial retinal input, and a set of nuclei called the **ventral nuclear group** receives lemniscal somatosensory and related motor feedback inputs. The most rostral group in the dorsal thalamus, the **anterior nuclear group,** receives input from part of the hypothalamus and several other sources. The **intralaminar** and **medial nuclear groups** lie near the anterior nuclear group. The intralaminar nuclei receive diverse inputs. The most prominent component of the medial nuclear group, the **mediodorsal nucleus,** receives a substantial projection from the olfactory cortex.

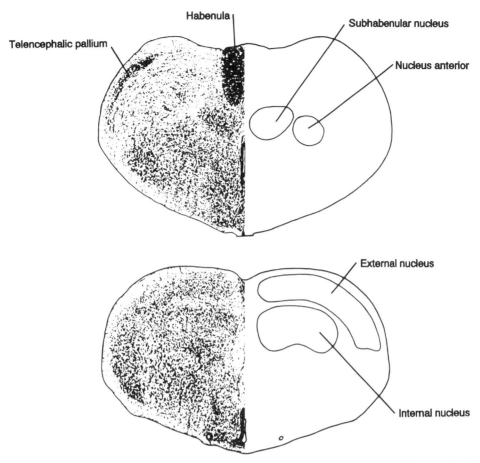

FIGURE 22-5. Transverse hemisections with mirror-image drawings through the diencephalon of a hagfish (*Eptatretus stouti*). The upper section is more rostral. Adapted from Wicht and Northcutt (1992). Used with permission of S. Karger AG.

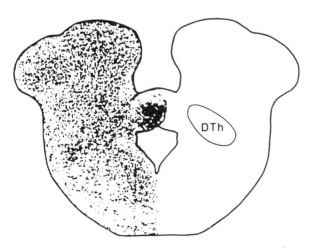

FIGURE 22-6. Transverse hemisection with mirror-image drawing through the diencephalon of a thornback skate (*Platyrhinoidis triseriata*). DTh = dorsal thalamus. Adapted from Northcutt (1978).

In diapsid reptiles, birds, and turtles, as in mammals, a greater number of distinct nuclei are present in the dorsal thalamus than in anamniotes. A number of these nuclei are shown in Figures 22-9 and 22-10. Various names have been used for these nuclei, resulting in terminology that is cumbersome at best and that can be confusing and intimidating. Only selected terms are used here, and they are listed in Table 22-1. Two midbrain roof-recipient nuclei are **nucleus rotundus (R),** which receives visual projections from the optic tectum, and **nucleus reuniens pars compacta [RE,** also called **nucleus medialis** of lizards and **nucleus ovoidalis (O)** of birds], which receives auditory projections from the torus semicircularis. Two nuclei also receive somatosensory and/or multisensory inputs from the midbrain roof: (1) the **medialis complex (M,** also called **nucleus medialis posterior** of lizards, **nucleus caudalis** of turtles and the **caudal part of nucleus dorsolateralis posterior,** or **cDLP,** of birds) and (2) the **pars diffusa of nucleus reuniens** of diapsid reptiles and turtles (**nucleus semilunaris parovoidalis** of birds).

In the more rostral part of the thalamus, a nucleus that we will call the **dorsal lateral optic nucleus (DLON),** which has been variably named in the literature, receives predominantly retinal projections. In birds, a neighboring nucleus, **nucleus**

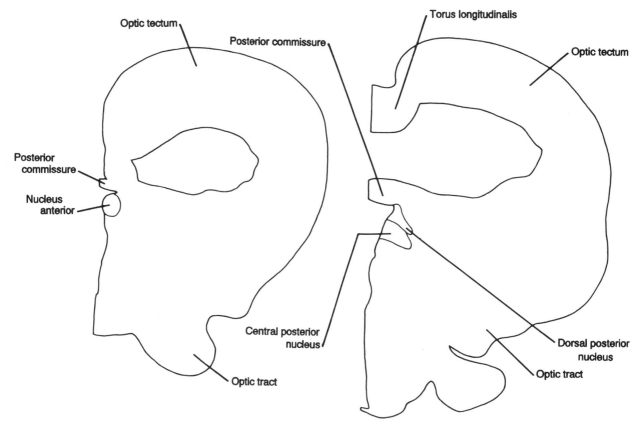

FIGURE 22-7. Drawings of transverse hemisections through the right diencephalon of a teleost (*Clupea harengus*). The section on the left is more rostral. Adapted from Butler and Northcutt (1993).

dorsalis intermedius ventralis anterior (DIVA), receives a lemniscal somatosensory input. A discrete nucleus with corresponding afferent somatosensory connections has not been found in diapsid reptiles or turtles. Two other nuclei—**nucleus dorsolateralis anterior** (DLA) and **nucleus dorsomedialis** (DM)—receive a more diverse set of connections.

SYNOPSIS OF THE DUAL ELABORATION HYPOTHESIS OF DORSAL THALAMIC EVOLUTION

In anamniotes, two basic divisions of the dorsal thalamus can be recognized: (1) a rostral part predominantly in receipt of direct retinal projections and, in some cases, other projections including somatosensory lemniscal projections; (2) a more caudal part predominantly in receipt of inputs relayed to it from the midbrain roof. The rostral part is called the **lemnothalamus** in reference to the lemniscal nature of its inputs, that is, the sensory input pathways to it are direct, as opposed to being routed through an "extra" synapse in the midbrain roof. The caudal part is called the **collothalamus** in reference to its predominant inputs being relayed through the midbrain roof— the colliculi, as its components are called in mammals. The lemnothalamus comprises a single nucleus in anamniotes, nucleus anterior, and projects bilaterally to the medial and dorsal

pallia in the telencephalon. The collothalamus comprises at least two nuclei, which are most frequently called the dorsal posterior and central posterior nuclei, and probably a third nuclear area related to the somatosensory system. The collothalamic nuclei project ipsilaterally to the striatum.

In amniotes, a number of distinct nuclei are present in the dorsal thalamus. Several nuclei (or nuclear groups) receive their predominant input from the midbrain roof. A number of additional nuclei (or nuclear groups) are present, some of which predominantly receive lemniscal sensory inputs. All of these latter additional nuclei are referred to here collectively as the **rostral dorsal thalamic nuclei.**

During development in some vertebrates, neurons migrate away from the region of the periventricular matrix where they are generated and form multiple, migrated nuclei in various parts of the brain. In the dorsal thalamus, this process occurs to a marked degree only in amniotes and is the single most important difference between the dorsal thalamus of amniotes and that of anamniotes. The central problem has been to identify where all of the nuclei that are present in the dorsal thalamus of adult amniotes come from—both in embryological and evolutionary terms.

The new interpretation of the evolution of the dorsal thalamus (Butler, 1994a), on which this chapter is based, encompasses the idea that the set of nuclei in each radiation of amniote vertebrates that do *not* receive their predominant input from the midbrain roof are all homologous as a field to the nucleus anterior of anamniotes; that is, these nuclei constitute the lemnothalamus. In other words, the embryonic anlage that gives rise to nucleus anterior in anamniotes is hypothesized to be the same as that which gives rise to the rostral dorsal thalamic nuclei in amniotes. The set of nuclei in the dorsal thalamus of amniotes that receive visual, auditory, somatosensory, and multisensory inputs from the midbrain roof likewise constitute the collothalamus, as do the nuclei with similar inputs in the dorsal thalamus of anamniotes. While the possibility of parallelism or convergence cannot be dismissed in the elaboration of the dorsal thalamus of mammals and the elaboration of the dorsal thalamus in nonmammalian amniotes, the specified homologies as hypothesized here represent the most parsimonious interpretation of the known developmental and morphological similarities.

The key event that is hypothesized to have occurred in the development of the dorsal thalamus in the ancestral stock of extant amniote vertebrates, that is, in the captorhinomorphs, was an increase in the number of neurons produced by the dorsal thalamic germinal matrix in conjunction with migration of the earlier-produced neurons away from the area of the matrix. This developmental change occurred for both the lemnothalamic and collothalamic divisions, resulting in the formation of multiple, discrete, migrated nuclei within both divisions: a dual elaboration of the dorsal thalamus.

With the more lateral position of a number of nuclei in captorhinomorphs, additional changes occurred as well. The collothalamic nuclei gained projections to the pallium in addition to those to the striatum, some of the lemnothalamic projections to the contralateral pallium were lost, and diffuse projections to widespread areas of both the pallium and striatum from some of the lemnothalamic nuclei were gained. More changes then occurred within the separate radiations of mammals (synapsids) and of the nonsynapsid amniotes, including a shift to a relatively caudal position of the visual and somatosensory lemnothalamic nuclei in mammals due to extensive secondary expansion of the rest of the lemnothalamus. In the following sections, the visual, auditory, and somatosensory–multisensory collothalamic nuclei are discussed first, and the remaining nuclei assigned to the lemnothalamus—the rostral dorsal thalamic nuclei—are then considered as a group.

MIDBRAIN-VISUAL DORSAL THALAMIC RELAY NUCLEI

As we discussed in Chapter 18, the optic tectum projects to multiple nuclei in the diencephalon. Within the dorsal thalamus, two visual tecto-recipient nuclei project to the telencephalon. The tectal input to one of these nuclei (nucleus anterior in anamniotes) is minor, the major input to nucleus anterior being from the retina. The tectal input to the other nucleus is substantial, and this nucleus is discussed in this section.

Group I

Studies have not yet been done in lampreys or squalomorph sharks to identify a particular set of neurons in receipt of tectal projections. In nonteleost ray-finned fishes, the optic

tectum projects to a nucleus in the dorsal thalamus, the **dorsal posterior nucleus** (Fig. 22-3), which projects to the striatum (called area ventralis in these fishes) in the telencephalon. Whether this pathway is topographically organized has not been determined.

In amphibians, the optic tectum projects to two neighboring nuclei, called the **anterior lateral nucleus** and the **posterior lateral nucleus** (Fig. 22-4). These projections are topographic only to the dorsal part of the posterior lateral nucleus, which does not project to the telencephalon. The anterior lateral nucleus receives nontopographic tectal input and projects to the striatum.

Group II

Tectal efferent projections have not yet been studied in hagfishes. In Group II cartilaginous fishes, the optic tectum projects to the **dorsal posterior nucleus.** The dorsal posterior nucleus projects to the striatum and/or to the dorsal pallium. More studies are needed to clarify the details of this pathway. The dorsal posterior nucleus in teleost fishes (Fig. 22-7) receives tectal projections and projects to the putative striatum (area ventralis).

In placental mammals, a nuclear group with a number of subdivisions, referred to as the **LP/pulvinar complex** [the lateral posterior nucleus and the pulvinar, Fig. 22-8(C–F)], receives topographic projections from the superior colliculus and projects primarily to extrastriate visual cortex in the telencephalon. It also projects to the striatum. A homologue of the LP/pulvinar of placental mammals has not yet been identified in monotremes.

In diapsid reptiles, birds, and turtles, the thalamic nucleus receiving substantial tectal visual projections is called **nucleus rotundus** (Figs. 22-9 and 22-10). It is a large nucleus and has a round shape in transverse section. Nucleus rotundus projects to a circumscribed area within the dorsal ventricular ridge (called the ectostriatum in birds) and to the striatum. This pathway is topographically organized in birds but not in diapsid reptiles or turtles.

MIDBRAIN-AUDITORY DORSAL THALAMIC RELAY NUCLEI

Like the ascending visual pathway from the roof of the midbrain, an ascending auditory pathway from the torus semicircularis (inferior colliculus) to a dorsal thalamic nucleus and thence to the telencephalon is a widespread feature of vertebrates.

Group I

Ascending projections of the torus semicircularis, which receives ascending auditory projections, have yet to be investigated in lampreys and in squalomorph sharks. In nonteleost ray-finned fishes, the torus semicircularis projects to the **central posterior nucleus** (Fig. 22-3), and this nucleus projects to the striatum. A similarly located nucleus in the dorsal thalamus in

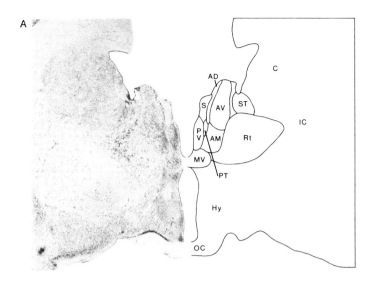

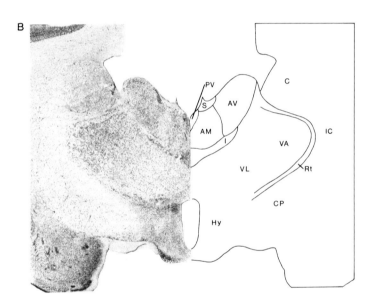

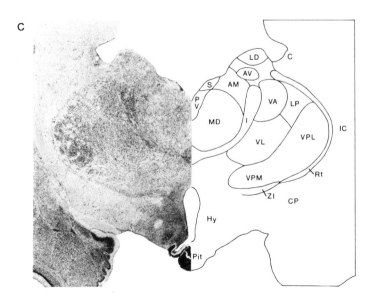

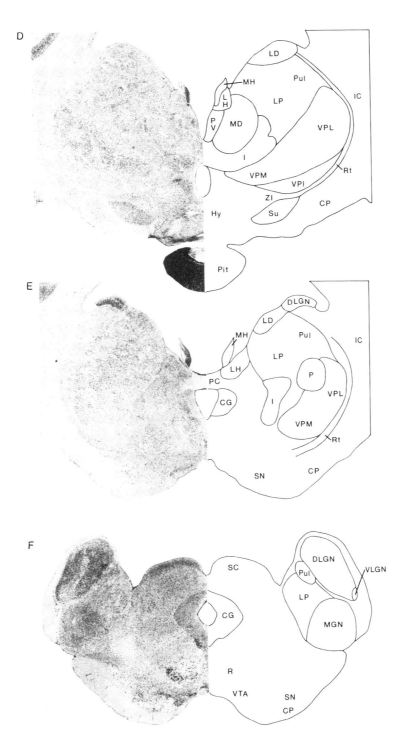

FIGURE 22-8. Transverse hemisections with mirror-image drawings through the diencephalon of a mammal (*Procyon lotor*) from rostral (A) to caudal (F). Abbreviations: AD, anterodorsal nucleus of the anterior nuclear group; AM, anteromedial nucleus of the anterior nuclear group; AV, anteroventral nucleus of the anterior nuclear group; C, caudate nucleus; CG, central gray; CP, cerebral peduncle; DLGN, dorsal lateral geniculate nucleus; Hy, hypothalamus; I, intralaminar nuclear group; IC, internal capsule; LD, lateral dorsal nucleus, which is associated with the anterior nuclear group; LH, lateral habenula; LP, lateral posterior nucleus; MD, mediodorsal nucleus; MGN, medial geniculate nucleus; MH, medial habenula; MV, medioventral (reuniens) nucleus of the medial nuclear group; OC, optic chiasm; P, posterior nuclear group; PC, posterior commissure; Pit, pituitary; PT, parataenial nucleus of the medial nuclear group; Pul, pulvinar; PV, paraventricular nuclear group of epithalamus; R, red nucleus; Rt, thalamic reticular nucleus (of the ventral thalamus); S, stria medullaris; SC, superior colliculus; SN, substantia nigra; ST, stria terminalis; Su, subthalamic nucleus; VA, ventral anterior nucleus of the ventral nuclear group; VL, ventral lateral complex of the ventral nuclear group; VLGN, ventral lateral geniculate nucleus; VPI, ventral posterior inferior nucleus of the ventral nuclear group; VPL, ventral posterolateral nucleus of the ventral nuclear group; VPM, ventral posteromedial nucleus of the ventral nuclear group; VTA, ventral tegmental area; ZI, zona incerta. Photomicrographs and data on nuclear boundaries courtesy of Wally Welker.

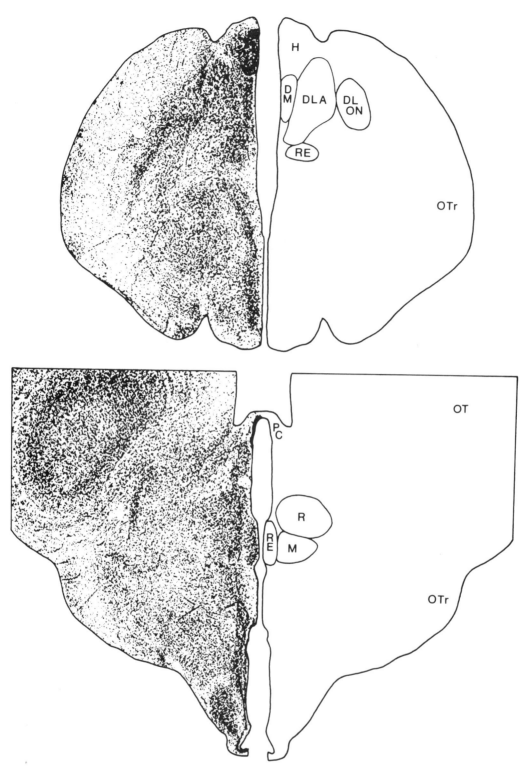

FIGURE 22-9. Transverse hemisections with mirror-image drawings through the diencephalon of a lizard (*Tupinambis nigropunctatus*). The upper section is the more rostral. Abbreviations: DLA, nucleus dorsolateralis anterior; DLON, dorsal lateral optic nucleus; DM, nucleus dorsomedialis; H, habenula; M, medialis complex; OT, optic tectum; OTr, optic tract; PC, posterior commissure; R, nucleus rotundus; RE, nucleus reuniens pars compacta. Adapted from Cruce (1974) with additional data from Pritz (1974) and Pritz and Stritzel (1990, 1994).

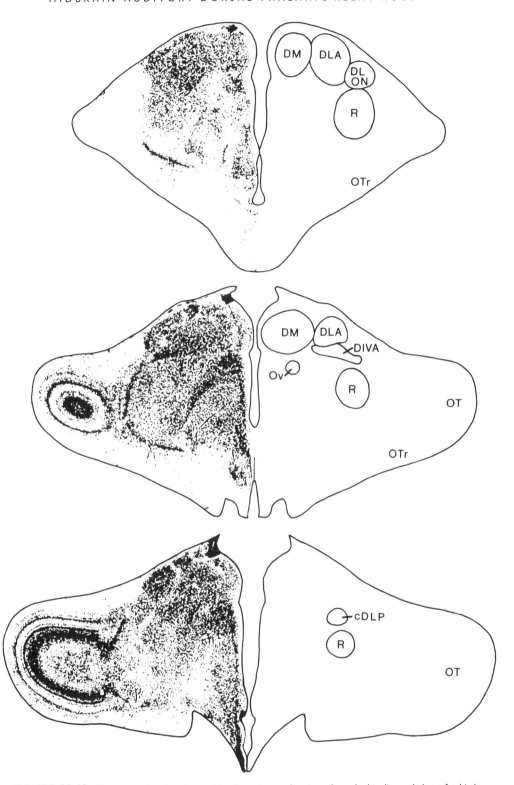

FIGURE 22-10. Transverse hemisections with mirror-image drawings through the diencephalon of a bird (*Columba livia*). The uppermost section is the most rostral. Abbreviations: cDLP, caudal part of nucleus dorsolateralis posterior; DIVA, nucleus dorsalis intermedius ventralis anterior; DLA, nucleus dorsolateralis anterior; DLON, dorsal lateral optic nucleus; DM, nucleus dorsomedialis; OT, optic tectum; OTr, optic tract; Ov, nucleus ovoidalis; R, nucleus rotundus. Adapted from Karten and Hodos (1967) with additional data from Arends and Zeigler (1991) and Wild (1987b). Used with permission of The Johns Hopkins University Press.

TABLE 22-1. Terminology Used for Dorsal Thalamic Nuclei in Amniotes

	Mammals	Diapsid Reptiles and Turtles	Birds
Lemniscal Visual Nuclei	Dorsal lateral geniculate nucleus (DLGN)	Dorsal lateral optic nucleus (DLON)	Dorsal lateral optic nucleus (DLON)
Lemniscal Somatosensory and Related Nuclei	Ventral nuclear group (and medial part of posterior nuclear group: Pom)	?	Nucleus dorsalis intermedius ventralis anterior (DIVA)
Rostral Nuclei with Diverse Inputs	Anterior, medial, and intralaminar nuclear groups	Nucleus dorsolateralis anterior (DLA), nucleus dorsomedialis (DM), and perirotundal nuclei	Nucleus dorsolateralis anterior (DLA), nucleus dorsomedialis (DM), and perirotundal nuclei
Mibrain Visual Relay Nuclei	Lateral posterior–pulvinar complex (LP/pulvinar)	Nucleus rotundus (R)	Nucleus rotundus (R)
Midbrain Auditory Relay Nuclei	Medial geniculate body	Nucleus reuniens pars compacta (RE)	Nucleus ovoidalis (O)
Midbrain Somatosensory-Multisensory Relay Nuclei	Limitans/suprageniculate complex and lateral part of posterior nuclear group (Pol)	Medialis (M) complex and nucleus reuniens pars diffusa	Caudal part of nucleus dorsolateralis posterior (cDLP) and nucleus semilunaris parovoidalis

amphibians, called the **central nucleus** (Fig. 22-4), likewise receives toral input and projects to the striatum.

Group II

Information on toral projections and the related ascending pathways is lacking for hagfishes. In galeomorph sharks, skates, and rays, forebrain auditory pathways also remain to be studied, but auditory evoked potentials have been recorded from a part of the dorsal pallium, called the central nucleus, in the telencephalon. Teleost fishes, like other ray-finned fishes, have ascending toral projections to the **central posterior nucleus** (Fig. 22-7) in the dorsal thalamus, and the central posterior nucleus projects to the putative striatum.

In mammals, the inferior colliculus projects to the **medial geniculate body** [Fig. 22-8(F)] in the dorsal thalamus, and the medial geniculate body projects to auditory cortex and to the striatum. In diapsid reptiles (Fig. 22-9) and turtles, ascending auditory projections from the torus semicircularis terminate in **nucleus reuniens pars compacta** (Fig. 22-9) in the dorsal thalamus, which, in turn, projects to a circumscribed part of the dorsal ventricular ridge (separate from the visual projection area) and to the striatum. In birds, the dorsal thalamic, auditory relay nucleus, **nucleus ovoidalis** (Fig. 22-10), projects to the auditory part of the dorsal ventricular ridge, called Field L, and to the striatum.

MIDBRAIN-SOMATOSENSORY AND MIDBRAIN-MULTISENSORY DORSAL THALAMIC RELAY NUCLEI

Part of the midbrain roof receives somatosensory inputs. Within the midbrain roof, some neurons also integrate multimodal information: somatosensory and auditory or somatosensory, auditory, and visual. Ascending pathways carry this information to the telencephalon via the dorsal thalamus.

Group I

Somatosensory afferents to the midbrain roof, including somesthetic projections, have not been studied in most anamniotes. In frogs, ascending projections from the spinal cord and the region of the obex, where the dorsal column nuclei are located, terminate in the midbrain roof. It has been postulated that somatosensory (and possibly multisensory) projections are present from the midbrain roof to a region within the area of the central and lateral nuclei (Fig. 22-4) in the dorsal thalamus.

Group II

In placental mammals, the dorsal thalamus contains a posterior group of nuclei [Fig. 22-8(E)]. (The medial part of the posterior group, or Pom, receives lemniscal trigeminal input; it therefore appears to belong instead to the ventral nuclear group.) The predominant input to two areas of the posterior nuclear group that are referred to collectively as the **limitans/suprageniculate complex** and to the **lateral part of the posterior nuclear group,** or Pol, is from the roof of the midbrain and includes somatosensory projections and multisensory (i.e., particularly auditory–somatosensory) projections. These parts of the posterior group project to areas of the isocortex associated with these sensory modalities. Among mammals, the posterior nuclear group is best developed in primates.

The somatosensory-recipient part of the midbrain roof in diapsid reptiles and turtles projects to an area of the dorsal thalamus that lies between the midbrain–visual and midbrain–auditory recipient nuclei and has been given a variety of names. It is referred to here as the **medialis complex** (Fig. 22-9), as it has been recently designated in crocodiles. The medialis complex projects both to the striatum and to a discrete region within the dorsal ventricular ridge, a region that lies between the visual- and auditory-recipient areas of the ridge. A neighboring nucleus, **nucleus reuniens pars diffusa** (not shown in Fig. 22-9) also receives midbrain roof input and projects to an area of the dorsal ventricular ridge neighboring that of the medialis complex projection and also interposed between the visual and auditory recipient areas.

In birds, similarly located and connected nuclei are present in the dorsal thalamus. One of these nuclei is referred to as the **caudal part of nucleus dorsolateralis posterior,** or cDLP (Fig. 22-10). The second nucleus (not shown in Fig. 22-10) borders nucleus ovoidalis. It is called either **nucleus semilunaris parovoidalis** or the **ventromedial part of nucleus ovoidalis.** These nuclei predominantly receive somatosensory and multisensory inputs from the midbrain roof and project to the dorsal ventricular ridge.

ROSTRAL DORSAL THALAMIC NUCLEI

One nucleus in anamniotes, **nucleus anterior,** and a diverse array of nuclei in amniotes, particularly mammals, receive their predominant input from structures other than the midbrain roof. Multiple inputs have been found for these nuclei, including a prominent retinal projection and lemniscal somatosensory projections. Other inputs include minor ones from the midbrain roof and those related to the hypothalamus and limbic system. These nuclei also appear to have a significant serotoninergic input from the raphe as a common character.

Group I

In lampreys, the optic tract lies on the lateral (superficial) edge of the diencephalon and terminates on the dendrites of dorsal thalamic neurons in the rostral diencephalon (Fig. 22-1), but thalamotelencephalic projections remain to be studied. In squalomorph sharks, retinal axons terminate on the dendrites of a nucleus in the rostral part of the dorsal thalamus, nucleus anterior (Fig. 22-2), which also receives a minor tectal projection. As we will discuss below, nucleus anterior projects to the dorsal and/or medial pallium in some Group II sharks and in skates, but it has only a sparse projection to the medial pallium in squalomorph sharks. In nonteleost ray-finned fishes, the retina projects to a dorsal thalamic nucleus, nucleus anterior (Fig. 22-3). This nucleus also receives minor tectal projections, and it projects bilaterally to the medial pallium (and perhaps part of the dorsal pallium) in the telencephalon. Nucleus anterior does not project to the striatum.

In salamaders and lungfishes, in which almost no migration of neurons occurs, retinal axons terminate on the laterally extended dendrites of dorsal thalamic neurons. Most connections of the dorsal thalamus with the telencephalon have not yet been studied, however. In frogs, nucleus anterior (Fig. 22-4) receives multiple sensory inputs from a large variety of sources, including visual input from the retina, tectal input relayed via a nucleus in the posterior thalamus, ascending somatosensory input from the region of the obex, ascending auditory input, and projections from the ventral part of the hypothalamus and

the medial pallium. Nucleus anterior projects bilaterally to the medial pallium and the medial part of the dorsal pallium. It also projects to the optic tectum, the posterior dorsal thalamus, and the hypothalamus. It does not project to the striatum, however.

Group IIA

In hagfishes (Fig. 22-5), group II cartilaginous fishes, and most teleosts (Fig. 22-7), nucleus anterior receives a major projection directly from the retina. Minor tectal projections have also been found to nucleus anterior in sharks and teleosts (Fig. 22-7). Nucleus anterior is known to project to the central nucleus of the dorsal pallium in sharks (primarily contralaterally) and bilaterally to the medial pallium in skates, although its total pattern of projections has yet to be studied in any cartilaginous fish. In teleosts, nucleus anterior projects bilaterally to the medial pallium, but it does not project to the striatum.

Group IIB

Mammals. In placental mammals, not only multiple nuclei but multiple groups of nuclei characterize the dorsal thalamus. The major groups in the rostral dorsal thalamus (Fig. 22-8) are the

- Anterior nuclear group.
- Medial nuclear group.
- Intralaminar nuclear group.
- Ventral nuclear group.
- Dorsal lateral geniculate nucleus.

The anterior nuclear group—the **anterodorsal, anteroventral,** and **anteromedial nuclei** [Fig. 22-8(A–C)]—and part of a closely related nucleus, the **lateral dorsal nucleus** [Fig. 22-8(C)], projects heavily and topographically to parts of the ipsilateral medial (limbic) cortex, particularly the cingulate cortex. In at least some mammals, the anteromedial nucleus additionally projects to the contralateral cortex. Afferent input to the anterior nuclear group arises primarily in the medial pallium and in a part of the ventral hypothalamus called the mammillary bodies. This anterior nuclear group has also been found to receive a sparse input from the retina in a variety of species.

One nucleus traditionally assigned to the medial nuclear group, the **medioventral nucleus** [Fig. 22-8(A), sometimes called the **reuniens nucleus**], has reciprocal connections with the medial pallium in addition to an input from the lateral pallium (olfactory cortex). It can be regarded as related more to the anterior group than the medial group because of the similarity of its connections with the medial pallium.

The most prominent nucleus within the medial nuclear group is the **mediodorsal nucleus** [Fig. 22-8(C and D)]. This nucleus receives a major afferent input from the olfactory cortex. It also receives projections from the mesencephalic reticular formation and the amygdala. The mediodorsal nucleus projects to the ipsilateral prefrontal and prelimbic cortices. Recently, it also has been found to project to transitional orbital and medial cortices in the contralateral telencephalon. Thus, the mediodorsal nucleus provides a bilateral relay of olfactory information to nonolfactory areas of the dorsal pallium.

The intralaminar nuclear group [Fig. 22-8(B–E], which comprises six nuclei in eutherian mammals, is characterized by giving rise to diffuse projections that terminate in widespread areas of the telencephalon, including bilateral projections to transitional cortices. The third nucleus traditionally assigned to the medial nuclear group, the **parataenial nucleus,** can be grouped with the intralaminar nuclei as it also gives rise to diffuse telencephalic projections. The intralaminar nuclei project ipsilaterally to isocortex and limbic cortex and to the striatum. The afferent input to the intralaminar nuclei is from diverse sources, including the cortex, the superior colliculus (optic tectum), the pretectum, the spinal cord, and a number of sites in the brainstem. The latter are primarily related to the motor system and include nucleus cuneiformis, the reticular formation, and the substantia nigra.

Low-frequency stimulation of the intralaminar nuclei produces a set of repetitive, negative electrical waves in the cortex, as monitored by electroencephalographic (EEG) recordings, called the **cortical recruiting response.** Spontaneous cortical rhythms are suppressed during the response, and the EEG is like that taken during sleep. Conversely, higher frequency stimulation of the intralaminar nuclei produces an EEG pattern similar to a state of wakefulness. Thus, the intralaminar nuclei modulate and regulate the state of arousal.

While placental mammals have anterior, medial, and intralaminar nuclear groups that are each composed of multiple nuclei, and hence referred to as groups, monotremes do not. In monotremes, an **anterior nucleus** is present as is a **medioventral (reuniens) nucleus. A mediodorsal nucleus** that projects to the prefrontal cortex is also present. Monotremes lack a parataenial nucleus and intralaminar nuclei, however; no nucleus has been found that gives rise to diffuse telencephalic projections.

Somatosensory information from the spinal cord, the principal nucleus of V, and the dorsal column nuclei in the region of the obex ascends to the ventral nuclear group [Fig. 22-8(B–E)] in the dorsal thalamus of placental mammals. This group also receives, in its different nuclei, inputs that convey feedback information on motor activities from the cerebellar and vestibular nuclei, part of the corpus striatum, and the substantia nigra. An ascending gustatory projection also terminates in part of the ventral nuclear group. The somatosensory portion of the ventral nuclear group projects topographically to the ipsilateral somatosensory cortex. Part of the nuclear area that receives input from the corpus striatum, substantia nigra, and gustatory system projects bilaterally to frontal and medial cortices, while the nucleus receiving cerebellar and striatal inputs projects to ipsilateral motor cortex. A large area of the dorsolateral thalamus has been identified as the somatosensory relay area in monotremes.

In monotremes and placental mammals, a substantial retinal projection terminates in the **dorsal lateral geniculate nucleus** [Fig. 22-8(F)]. This nucleus also receives a minor input from the optic tectum. The dorsal lateral geniculate nucleus, along with the lemniscal somatosensory nuclei, lies in a more caudal position in the dorsal thalamus of placental mammals than it does in monotremes or does the comparable dorsal lateral optic nucleus in diapsid reptiles, birds, and turtles; we

will consider a possible reason for this difference in position below. The dorsal lateral geniculate nucleus projects topographically to the ipsilateral striate (visual) cortex.

Nonmammalian Amniotes. In diapsid reptiles (Fig. 22-9), birds (Fig. 22-10), and turtles, a number of nuclei are present in the dorsal thalamus in addition to the midbrain–visual, –auditory, and –somatosensory–multisensory relay nuclei discussed above. Two of the nuclei in the rostral part of the dorsal thalamus are **nucleus dorsolateralis anterior (DLA)** of diapsid reptiles and turtles—and a pair of corresponding nuclei in birds called **nucleus dorsolateralis anterior thalami** and **nucleus dorsolateralis anterior thalami pars medialis** (shown as one nuclear area, DLA, in Fig. 22-10)—and **nucleus dorsomedialis (DM).** These nuclei partially encircle nucleus rotundus and thus form part of a "shell" region around it. Other nuclei complete the shell around the more ventral aspects of nucleus rotundus, and these nuclei along with DLA and DM are referred to collectively as the **perirotundal nuclei.**

A variety of structures project to DLA, including the spinal cord, torus semicircularis, hypothalamus, other thalamic nuclei, and the septum. It also receives a serotoninergic input from the raphe and a sparse retinal projection. Widespread areas of the telencephalon receive projections from DLA, including the medial cortices, the dorsal cortex including its lateral part, the pallium thickening, the lateral cortex, and, in turtles, the striatum. The projection of DLA to the pallial cortices is bilateral. Nucleus dorsomedialis also receives a variety of inputs, including afferents from the spinal cord and the raphe. A direct projection to it from the olfactory cortex recently has been found in birds. Like DLA, DM has widespread telencephalic projections to cortical areas, the dorsal ventricular ridge, and the striatum.

In birds, a lemniscal, somatosensory relay nucleus, **nucleus dorsalis intermedius ventralis anterior (DIVA),** is present in the same region of the dorsal thalamus. DIVA (Fig. 22-10) receives ascending somatosensory projections and projects bilaterally to part of the Wulst in the dorsal pallium. A separate nucleus with similar connections has not yet been found in diapsid reptiles or turtles; a ventromedial part of the perirotundal nuclei (DLA/DM) may be homologous to DIVA as part of a field, however, since ascending somatosensory projections have been found to this area.

A more laterally lying nucleus that lies in the rostral part of the dorsal thalamus receives the bulk of the retinal projection. This nucleus projects to the dorsal pallium in the telencephalon. It has been referred to by many different terms in the literature, due to the fact that multiple retinal targets are present in this region of the diencephalon, most of which are part of the ventral thalamus. For example, the retino-recipient dorsal thalamic nucleus has been called nucleus intercalatus in lizards and the dorsal optic nucleus in turtles. In birds, a group of five nuclei in this region receive retinal projections and project to the telencephalic pallium. For purposes of clarity and objectivity, we will call the retinal target (of one or more nuclei) in the dorsal thalamus that in turn projects to the dorsal pallium the **dorsal lateral optic nucleus** in diapsid reptiles (Fig. 22-9), birds (Fig. 22-10), and turtles.

The dorsal lateral optic nucleus receives a substantial retinal input and other lesser inputs, including a minor input from the optic tectum. In diapsid reptiles and turtles, this nucleus projects ipsilaterally to an area in the dorsolateral part of the dorsal ventricular ridge and/or pallial thickening (in lizards, snakes, and crocodiles) or to the lateral part of the dorsal cortex (in turtles). In birds, the dorsal lateral optic nucleus projects bilaterally to the visual Wulst, with the projection overlapping to some degree the somatosensory input discussed above. In birds, this thalamotelencephalic pathway is known to be topographically organized.

EVOLUTIONARY PERSPECTIVE

A sensory relay nucleus of the dorsal thalamus that receives input from the optic tectum (superior colliculus) and projects to the telencephalon has been found in all vertebrates studied except monotremes. Variously called the dorsal posterior nucleus, the anterior lateral nucleus, nucleus rotundus, the LP/pulvinar complex, this nucleus projects ipsilaterally to the striatum in all vertebrates studied. In some cartilaginous fishes and in amniotes, this nucleus also projects to the dorsal pallium. A dorsal thalamic nucleus receiving tectal projections and projecting to the ipsilateral striatum is thus plesiomorphic for at least jawed vertebrates. The addition of projections of the nucleus to the dorsal pallium has occurred twice independently, in sharks and in amniotes.

A sensory relay nucleus of the dorsal thalamus that receives input from the torus semicircularis (inferior colliculus) and projects to the telencephalon is present in all vertebrates studied. This nucleus is called the central posterior nucleus, nucleus reuniens pars compacta, nucleus ovoidalis, or the medial geniculate body. Like the optic tectal relay nucleus, this nucleus projects to the ipsilateral striatum. Additionally, it projects to the dorsal pallium in amniotes. The presence of this nucleus and its projection to the ipsilateral striatum is thus plesiomorphic for at least jawed vertebrates, and its projection to the dorsal pallium in amniotes is apomorphic.

Relay nuclei of the dorsal thalamus that receive somatosensory and multisensory input from the optic tectum and project to the telencephalon have been found in all amniotes. An area of the dorsal thalamus with similar connections is postulated to be present in frogs. Among amniotes, these nuclei are called the limitans/suprageniculate nucleus and the lateral part of the posterior nuclear group, the medialis complex and the pars diffusa of nucleus reuniens, and nucleus dorsolateralis posterior and nucleus semilunaris parovoidalis or the ventromedial part of nucleus ovoidalis. Such nuclei appear to be plesiomorphic for at least amniotes and also appear to be homologous, at least as a field, to a part of the dorsal thalamus of amphibians in the region of the central and lateral nuclei.

A single nucleus in the rostral thalamus, nucleus anterior, is present in anamniote vertebrates. This nucleus is known to receive a substantial retinal input and a lesser tectal input in fishes. In amphibians, it receives multiple inputs, including projections from the retina, the posterior thalamus relaying tectal input, the ventral hypothalamus, the medial pallium, and somatosensory and auditory structures. Nucleus anterior projects bilaterally to the medial pallium. This nucleus and its major retinal and minor tectal afferent projections are plesiomorphic features of at least jawed vertebrates and probably all vertebrates. The bilateral projections of the nucleus to the medial pallium may also be plesiomorphic for vertebrates. The presence of multiple inputs from diverse, nonvisual sources, including a somatosensory input, to nucleus anterior is apomorphic for tetrapods.

The nucleus dorsolateralis anterior of diapsid reptiles, birds, and turtles projects bilaterally to the medial pallium and to other cortical areas as well. It receives input from diverse sources, including a sparse retinal input. Its bilateral projection to the medial pallium and its diverse inputs are similar to the connections of the nucleus anterior of anamniotes. Nucleus dorsomedialis projects diffusely to the ipsilateral pallium and striatum, and it receives a spinal input. The spinal input to nucleus dorsomedialis is similar to that of nucleus anterior of anamniotes, and its diffuse projections include the medial pallium to which nucleus anterior likewise projects. Both nucleus dorsolateralis anterior and nucleus dorsomedialis lie in the rostral thalamus. These two nuclei may therefore be homologous as part of a field to the nucleus anterior of anamniotes.

Nucleus dorsalis intermedius ventralis anterior (DIVA) of birds receives ascending, lemniscal somatosensory input. As in the case of nucleus dorsomedialis, this input is similar to that of the nucleus anterior of amphibians. The projections of DIVA to the dorsal pallium are bilateral in birds, similar to the bilaterality of the projections of the nucleus anterior of anamniotes. Likewise, the dorsal lateral optic nucleus of diapsid reptiles, birds, and turtles receives a substantial lemniscal retinal input, as does the nucleus anterior of anamniotes. Its projections to the pallium are bilateral in birds, although only ipsilateral in diapsid reptiles and turtles.

On the basis of these data, it would appear that the dorsal lateral optic nucleus is also homologous as part of a field to the nucleus anterior of anamniotes. If a lemniscal somatosensory nucleus is found to be present in diapsid reptiles and turtles, as in birds, this nucleus would likewise be part of the field that is homologous to the nucleus anterior of anamniotes. In other words, the embryonic anlage that gives rise to nucleus anterior in anamniotes gives rise to nucleus dorsolateralis anterior, nucleus dorsomedialis, the dorsal lateral optic nucleus, and any lemniscal somatosensory nucleus in nonsynapsid amniotes. The lateral migration of neurons during the development of the dorsal thalamus and the presence of multiple nuclei as opposed to a single nucleus anterior are apomorphic for amniotes.

The anterior nuclear group of placental mammals [and the anterior and medioventral (reuniens) nucleus of monotremes] receives inputs from the medial cortex and the hypothalamus, as well as a sparse retinal input. This group projects to the ipsilateral medial cortex, and part of it additionally projects bilaterally to the cortex. The intralaminar nuclear group has diverse inputs and projects diffusely to the pallium and striatum; this diffuse projection is predominantly ipsilateral, but some projections to the contralateral telencephalic cortex are present as well.

Connections similar to the olfactory afferent connections of the mammalian mediodorsal nucleus, which projects to prefrontal isocortex, resemble the projection to nucleus dorsomedialis from olfactory cortex in birds. The target of the mediodorsal nucleus in mammalian prefrontal cortex is hallmarked by a very dense innervation of ascending dopaminergic fibers, and regions that are similarly densely innervated by dopaminergic fibers have been identified in the dorsal pallium of both lizards and birds. Thus, an olfacto-thalamo-dorsal pallial pathway may be present in most or all amniotes. Taken as a whole, the connections of the anterior, medial, and intralaminar nuclei of mammals resemble those of the dorsolateral and dorsomedial nuclei of nonsynapsid amniotes and those of nucleus anterior in amphibians.

Based on these data and on the location of nuclei in the rostral part of the dorsal thalamus, we can thus hypothesize that the nuclei of the rostral dorsal thalamus in mammals and in nonsynapsid amniotes are homologous as a field to the nucleus anterior of anamniotes. Projections to the telencephalon from these nuclei that are only ipsilateral, instead of bilateral, are apomorphic where present, although the possibilty exists that some of the bilateral projections present in amniotes may have been first lost and then reevolved. Projections from these nuclei to the striatum are apomorphic for amniotes.

The ventral nuclear group and the dorsal lateral geniculate nucleus lie in the caudal part of the dorsal thalamus in placental mammals. As we will discuss below, this caudal position may be secondarily due to the relative expansion of the anterior, medial, and intralaminar nuclear groups. The connections of a large part of the ventral nuclear group and the dorsal lateral geniculate nucleus are similar to those of the somatosensory relay nuclei (where present) and the dorsal lateral optic nucleus of nonsynapsid amniotes, respectively. The afferent connections of these nuclei are also similar to some of the afferent connections of the nucleus anterior of anamniotes. These data suggest that the dorsal lateral geniculate nucleus of mammals is homologous as a discrete nucleus to the dorsal lateral optic nucleus of nonsynapsid amniotes. The mammalian ventral nuclear group and the dorsal lateral geniculate nucleus are also homologous as a field, along with the anterior, medial, and intralaminar nuclear groups, to the nucleus anterior of anamniotes.

In the captorhinomorph amniotes that gave rise to all extant amniote vertebrates, increased cell proliferation during development resulted in an abundance of migrated nuclei in the dorsal thalamus. In the ancestral line of the placental mammals, continued expansion of the lemnothalamus occurred, and a number of nuclear groups, rather than single nuclei, were formed. The relative expansion of the anterior, medial, and intralaminar groups of nuclei resulted in a relative, caudal displacement of the ventral nuclear group and the dorsal lateral geniculate nucleus within the dorsal thalamus. The evolution of the dorsal thalamus in tetrapods is summarized in Figure 22-11.

TOWARDS A NEW DEFINITION OF THE DORSAL THALAMUS IN VERTEBRATES

Some previous definitions of the dorsal thalamus of vertebrates have been based primarily on features of the mammalian dorsal thalamus. One of the frequently used criteria for a dorsal thalamic nucleus has been that it must project to a specified cytoarchitectonic area within the isocortex or limbic cortex or

to their homologues. As we discussed above, collothalamic nuclei project only to the striatum in anamniotes. The projections of lemnothalamic nuclei to the pallium are frequently bilateral, rather than just ipsilateral, in anamniotes and amniotes alike. Projections to the pallium, in addition to projections to the striatum, are an apomorphy of collothalamic nuclei in amniotes, as are projections to the striatum from some of the lemnothalamic nuclei in amniotes. The distributions of topographic organization of various pathways through the dorsal thalamus and of reciprocal connections from the pallium to the dorsal thalamus need further investigation in a variety of species, but neither can be said to be a defining feature of a dorsal thalamic nucleus.

Difficulties in recognizing the cytoarchitectonic areas of the dorsal thalamus must also be considered. The neurons of nuclei derived from one embryonic anlage may migrate or be displaced to a different region within the adult brain. For example, the preglomerular nuclear complex lies near the area of the dorsal thalamus in fishes, as discussed in Chapter 20. Thus, to distinguish nuclei in the region of the dorsal thalamus from other diencephalic nuclei, such as ventral thalamic, pretectal, or preglomerular nuclei, on the basis of the presence or absence of a projection to the telencephalon is hazardous at best. Nuclei derived from other sources, such as the hypothalamus in some fishes, also serve as relay nuclei to the telencephalon (as will be discussed in Chapter 23). The pulvinar in humans, but not in other primates or any other vertebrate, contains a population of neurons that have migrated from the telencephalon.

The definition of the dorsal thalamus that is proposed here is as follows: *the dorsal thalamus of jawed vertebrates comprises those nuclei that are derived from the embryonic dorsal thalamic anlage, belong to one of two fundamental divisions that are based on the source of the predominant afferent input, receive input predominantly from the contralateral sensory world, and project to the striatum or to the pallium or to both.* Dorsal thalamic nuclei may have either specifically or nonspecifically (diffusely) organized projections, may or may not have projections that are topographically organized, may or may not receive reciprocal projections from the telencephalon, and may project to the telencephalon bilaterally or only ipsilaterally. Diencephalic nuclei other that those of the dorsal thalamus may also project to the telencephalon.

A number of apomorphic features of dorsal thalamic nuclei were acquired in ancestral amniotes, as were additional ones in marsupial and placental mammals. The selective pressures encountered in a terrestrial habitat, as opposed to an aquatic one, favored the establishment of these features. Lateral migration of the neurons of the sensory relay nuclei is correlated with a marked increase in the number of subdivisions of these nuclei and the importance of visual, auditory, and somatosensory cues in the terrestrial world.

The elaboration of the anterior nuclear group of mammals into multiple nuclei and the shift of this system from a multisensory relay to, in effect, a cortical-thalamo-cortical circuit (with the additional relay through the mammillary bodies) is correlated with the increased use of the medial (limbic) cortex in learning and memory. All vertebrates have ascending sensory system relays, but in mammals in particular, the analysis and

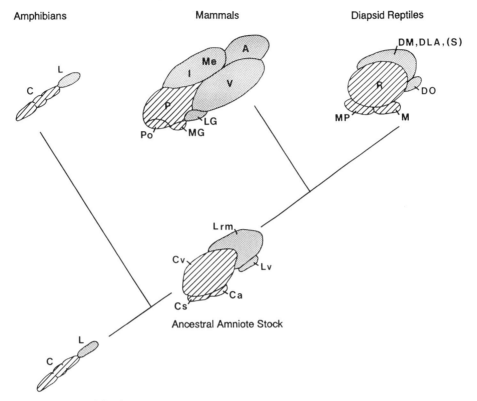

FIGURE 22-11. Summary of hypothesis (Butler, 1994a) of dorsal thalamic evolution in tetrapods in dendrogram form; oblique dorsolateral views of the whole dorsal thalamus, with rostral toward the upper right and medial toward the upper left. Lemnothalamic nuclei are indicated by shading and collothalamic nuclei by diagonal lines. A single lemnothalamic nucleus (nucleus anterior) and two to three caudal collothalamic nuclei (visual, auditory, and possibly somatosensory–multisensory) were present in ancestral tetrapods and are present in extant amphibians. In the ancestral amniote stock, both collothalamic and lemnothalamic divisions of the dorsal thalamus were elaborated; three collothalamic nuclei and at least two lemnothalamic nuclear areas were present. In the synapsid line that gave rise to extant mammals, a marked additional elaboration of the lemnothalamus occurred, resulting in a relatively larger expansion of this division of the dorsal thalamus than occurred in most nonmammalian amniotes. Among nonmammalian amniotes, birds (not represented in this figure) appear to have independently further elaborated their lemnothalamus. Abbreviations: A, anterior nuclear group; C, collothalamus; Ca, auditory collothalamic nucleus; Cs, somatosensory–multisensory collothalamic nucleus; Cv, visual collothalamic nucleus; DLA, nucleus dorsolateralis anterior; DM, nucleus dorsomedialis; DO, dorsal lateral optic nucleus; I, intralaminar nuclear group; L, lemnothalamus; LG, dorsal lateral geniculate nucleus; Lrm, rostromedial part of the lemnothalamus; Lv, visual (dorsal lateral) lemnothalamic nucleus; M, nucleus medialis (also called nucleus reuniens pars compacta); Me, medial nuclear group; MG, medial geniculate body; MP, nucleus medialis posterior (also called the medialis complex); P, lateral posterior/pulvinar complex; Po, posterior nuclear group; R, nucleus rotundus; (S), postulated lemniscal somatosensory nuclear area; V, ventral nuclear group.

use of this sensory information by the limbic system is vastly increased. The expansion and elaboration of medially lying nuclei in amniotes, such as the medial nuclear group and the intralaminar nuclei, is correlated with the increased complexity of the terrestrial environment.

As is the case in many regions of the brain (such as the pretectum, the posterior tuberculum, the hypothalamus, and various parts of the telencephalon) the dorsal thalamus has evolved differently in different radiations of vertebrates. Within the diencephalon of nontetrapods, the pretectum and posterior tuberculum are among those regions that are most elaborated and expanded. In contrast, the dorsal thalamus is most elabo-

rated and expanded in amniotes and, within amniotes, in placental mammals. The differential elaboration of different parts of the central nervous system reflects the variety of adaptations in the various radiations of vertebrates that have evolved in response to the demands of new and changing habitats.

FOR FURTHER READING

Braford, M. R., Jr. and Northcutt, R. G. (1983) Organization of the diencephalon and pretectum of the ray-finned fishes. In R. E.

Davis and R. G. Northcutt (eds.), *Fish Neurobiology, Vol. 2*. Ann Arbor, MI: University of Michigan Press, pp. 117–164.

Butler, A. B. (1994a) The evolution of the dorsal thalamus of jawed vertebrates, including mammals: cladistic analysis and a new hypothesis. *Brain Research Reviews*, 19, 29–65.

Harting, J. K., Glendenning, K. K., Diamond, I. T., and Hall, W. C. (1973) Evolution of the primate visual system: anterograde degeneration studies of the tecto-pulvinar system. *American Journal of Physical Anthropology*, 38, 383–392.

Jones, E. G. (1985) *The Thalamus*. New York: Plenum.

Neary, T. J. and Northcutt, R. G. (1983) Nuclear organization of the bullfrog diencephalon. *Journal of Comparative Neurology*, 213, 262–278.

Northcutt, R. G. (1991) Visual pathways in elasmobranchs: organization and phylogenetic implications. *Journal of Experimental Zoology*, Suppl. 5, 97–107.

ADDITIONAL REFERENCES

Abramson, B. P. and Chalupa, L. M. (1988) Multiple pathways from the superior colliculus to the extrageniculate visual thalamus of the cat. *Journal of Comparative Neurology*, 271, 397–418.

Arends, J. J. A. and Zeigler, H. P. (1991) Organization of the cerebellum in the pigeon (*Columba livia*): II. Projections of the cerebellar nuclei. *Journal of Comparative Neurology*, 306, 245–272.

Baisden, R. H. and Hoover, D. B. (1979) Cells of origin of the hippocampal afferent projection from the nucleus reuniens thalami: a combined Golgi-HRP study in the rat. *Cell and Tissue Research*, 203, 387–391.

Balaban, C. D. and Ulinski, P. S. (1981) Organization of thalamic afferents to anterior dorsal ventricular ridge in turtles. I. Projections of thalamic nuclei. *Journal of Comparative Neurology*, 200, 95–129.

Bass, A. H. and Northcutt, R. G. (1981) Retinal recipient nuclei in the painted turtle, *Chrysemys picta*: an autoradiographic and HRP study. *Journal of Comparative Neurology*, 199, 97–112.

Beckstead, R. M. (1984) The thalamostriatal projection in the cat. *Journal of Comparative Neurology*, 223, 313–346.

Benevento, L. A. and Fallon, J. A. (1975) The ascending projections of the superior colliculus in the rhesus monkey (*Macaca mulatta*). *Journal of Comparative Neurology*, 160, 339–362.

Bennis, M., Repérant, J., Rio, J.-P., and Ward, R. (1994) An experimental re-evaluation of the primary visual system of the European chameleon, *Chamaeleo chameleon*. *Brain, Behavior and Evolution*, 43, 173–188.

Berson, D. M. and Graybiel, A. M. (1978) Parallel thalamic zones in the LP/pulvinar complex of the cat identified by their afferent and efferent connections. *Brain Research*, 147, 139–148.

Bingman, V. P., Casini, G., Nocjar, C., and Jones, T.-J. (1994) Connections of the piriform cortex in homing pigeons (*Columba livia*) studied with fast blue and WGA–HRP. *Brain, Behavior and Evolution*, 43, 206–218.

Bonke, B. A., Bonke, D., and Scheich, H. (1979) Connectivity of the auditory forebrain nuclei in the guinea fowl (*Numida meleagris*). *Cell and Tissue Research*, 200, 101–121.

Brauth, S. E., McHale, C. M., Brasher, C. A. and Dooling, R. J. (1987) Auditory pathways in the budgerigar. I. Thalamo-telencephalic projections. *Brain, Behavior and Evolution*, 30, 174–199.

Bruce, L. L. and Butler, A. B. (1984a) Telencephalic connections in lizards. I. Projections to cortex. *Journal of Comparative Neurology*, 229, 585–601.

Bruce, L. L. and Butler, A. B. (1984b) Telencephalic connections in lizards. II. Projections to anterior dorsal ventricular ridge. *Journal of Comparative Neurology*, 229, 602–615.

Bullock, T. H. and Corwin, J. T. (1979) Acoustic evoked activity in the brain in sharks. *Journal of Comparative Physiology*, 129, 223–234.

Butler, A. B. (1994b) The evolution of the dorsal pallium in the telencephalon of amniotes: cladistic analysis and a new hypothesis. *Brain Research Reviews*, 19, 66–101.

Butler, A. B. and Northcutt, R. G. (1973) Architectonic studies of the diencephalon of *Iguana iguana* (Linnaeus). *Journal of Comparative Neurology*, 149, 439–462.

Butler, A. B. and Northcutt, R. G. (1978) New thalamic visual nuclei in lizards. *Brain Research*, 149, 469–476.

Butler, A. B. and Northcutt, R. G. (1992) Retinal projections in the bowfin, *Amia calva*: cytoarchitectonic and experimental analysis. *Brain, Behavior and Evolution*, 39, 169–194.

Butler, A. B. and Northcutt, R. G. (1993) The diencephalon of the Pacific herring, *Clupea harengus*: cytoarchitectonic analysis. *Journal of Comparative Neurology*, 328, 527–546.

Butler, A. B. and Saidel, W. M. (1991) Retinal projections in the freshwater butterfly fish, *Pantodon buchholzi* (Osteoglossoidei). I. Cytoarchitectonic analysis and primary visual pathways. *Brain, Behavior and Evolution*, 38, 127–153.

Cropper, E. C., Eisenman, J. S., and Asmitia, E. C. (1984) An immunocytochemical study of the serotoninergic innervation of the thalamus of the rat. *Journal of Comparative Neurology*, 224, 38–50.

Cruce, J. A. F. (1974) A cytoarchitectonic study of the diencephalon of the tegu lizard, *Tupinambis nigropunctatus*. *Journal of Comparative Neurology*, 153, 215–238.

Dermon, C. R. and Barbas, H. (1994) Contralateral thalamic projections predominantly reach transitional cortices in the rhesus monkey. *Journal of Comparative Neurology*, 344, 508–531.

Diamond, M. E., Armstrong-Jones, M., and Ebner, F. F. (1992) Somatic sensory responses in the rostral sector of the posterior group (POm) and in the ventral posterior medial nucleus (VPM) of the rat thalamus. *Journal of Comparative Neurology*, 318, 462–476.

Dinopoulos, A. (1994) Reciprocal connections of the motor neocortical area with the contralateral thalamus in the hedgehog (*Erinaceus europaeus*) brain. *European Journal of Neuroscience*, 6, 374–380.

Divac, I. and Mogensen, J. (1985) The prefrontal "cortex" in the pigeon: catecholamine histofluorescence. *Neuroscience*, 15, 677–682.

Divac, I. and Öberg, R. G. E. (1990) Prefrontal cortex: the name and the thing. In W. K. Schwerdtfeger and P. Germroth (eds.), *The Forebrain in Nonmammals: New Aspects of Structure and Development*. Berlin: Springer-Verlag, pp. 213–223.

Druga, R., Rokyta, R., and Benes, V., Jr. (1991) Thalamocaudate projections in the macaque monkey (a horseradish peroxidase study). *Journal für Hirnforschung*, 32, 765–774.

Ebbesson, S. O. E. and Schroeder, D. M. (1971) Connections of the nurse shark's telencephalon. *Science*, 173, 254–256.

Echteler, S. E. (1985) Organization of the central auditory pathways in a teleost fish, *Cyprinus carpio*. *Journal of Comparative Physiology A*, 156, 267–280.

Echteler, S. E. and Saidel, W. M. (1981) Forebrain connections in the goldfish support telencephalic homologies with land vertebrates. *Science*, 212, 683–685.

Fernald, R. D. and Shelton, L. C. (1985) The organization of the diencephalon and the pretectum in the cichlid fish *Haplochromis burtoni*. *Journal of Comparative Neurology*, 238, 202–217.

Foster, R. E. and Hall, W. C. (1978) The organization of central auditory pathways in a reptile, *Iguana iguana*. *Journal of Comparative Neurology*, 178, 783–832.

Gamlin, P. D. R. and Cohen, D. H. (1986) A second ascending visual pathway from the optic tectum to the telencephalon in the pigeon (*Columba livia*). *Journal of Comparative Neurology*, 250, 296–310.

Graybiel, A. M. (1972) Some fiber pathways related to the posterior thalamic region in the cat. *Brain, Behavior and Evolution*, 6, 363–393.

Goffinet, A. M. (1990) Cortical architectonic development: a comparative study in reptiles. In W. K. Schwerdtfeger and P. Germroth (eds.), *The Forebrain in Nonmammals: New Aspects of Structure and Development*. Berlin: Springer-Verlag, pp. 135–144.

Goffinet, A. M., Daumerie, Ch., Langewerf, B., and Pieau, C. (1986) Neurogenesis in reptilian cortical structures: ³H-thymidine autoradiographic analysis. *Journal of Comparative Neurology*, 243, 106–116.

Hall, W. C. and Ebner, F. F. (1970) Parallels in the visual afferent projections of the thalamus in the hedgehog (*Paraechinus hypomelas*) and the turtle (*Pseudemys scripta*). *Brain, Behavior and Evolution*, 3, 135–154.

Hall, W. C. and Ebner, F. F. (1970) Thalamotelencephalic projections in the turtle (*Pseudemys scripta*). *Journal of Comparative Neurology*, 140, 101–122.

Herkenham, M. (1979) The afferent and efferent connections of the ventromedial thalamic nucleus in the rat. *Journal of Comparative Neurology*, 183, 487–518.

Höhl-Abrahão, J. C. and Creutzfeldt, O. D. (1991) Topographical mapping of the thalamocortical projections in rodents and comparison with that in primates. *Experimental Brain Research*, 87, 283–294.

Hoogland, P. V. (1982) Brainstem afferents to the thalamus in a lizard, *Varanus exanthematicus*. *Journal of Comparative Neurology*, 210, 152–162.

Itaya, S. K., Van Hoesen, G. W., and Benevento, L. A. (1986) Direct retinal pathways to the limbic thalamus of the monkey. *Experimental Brain Research*, 61, 607–613.

Itoh, K., Kaneko, T., Kudo, M., and Mizuno, N. (1984) The intercollicular region in the cat: a possible relay in the parallel somatosensory pathways from the dorsal column nuclei to the posterior complex of the thalamus. *Brain Research*, 308, 166–171.

Kaas, J. H., Huerta, M. F., Weber, J. T., and Harting, J. K. (1978) Patterns of retinal terminations and laminar organization of the lateral geniculate nucleus of primates. *Journal of Comparative Neurology*, 182, 517–554.

Karten, H. J. (1967) The ascending auditory pathway in the pigeon (*Columba livia*). I. Diencephalic projections of the inferior colliculus (nucleus mesencephali lateralis, pars dorsalis). *Brain Research*, 6, 409–427.

Karten, H. J. (1968) The ascending auditory pathway in the pigeon (*Columba livia*). II. Telencephalic projections of the nucleus ovoidalis thalami. *Brain Research*, 11, 134–153.

Karten, H. J. and Hodos, W. (1967) *A Stereotaxic Atlas of the Brain of the Pigeon (Columba livia)*. Baltimore, MD: The Johns Hopkins Press.

Karten, H. J. and Hodos, W. (1970) Telencephalic projections of the nucleus rotundus in the pigeon (*Columba livia*). *Journal of Comparative Neurology*, 140, 35–52.

Karten, H. J., Hodos, W., Nauta, W. J. H., and Revzin, A. M. (1973) Neural connections of the "visual Wulst" of the avian telencepahlon. Experimental studies in the pigeon (*Columba livia*) and owl (*Speo-*

tyto cunicularia). *Journal of Comparative Neurology*, 150, 253–278.

Karten, H. J., Konishi, M., and Pettigrew, J. (1978) Somatosensory representation in the anterior Wulst of the owl (*Speotyto cunicularia*). *Society for Neuroscience Abstracts*, 4, 554.

Karten, H. J. and Revzin, A. M. (1966) The afferent connections of the nucleus rotundus in the pigeon. *Brain Research*, 2, 368–377.

Kennedy, M. C. and Rubinson, K. (1977) Retinal projections in larval, transforming and adult sea lamprey, *Petromyzon marinus*. *Journal of Comparative Neurology*, 171, 465–479.

Kicliter, E. (1979) Some telencephalic connections in the frog, *Rana pipiens*. *Journal of Comparative Neurology*, 185, 75–86.

Kitt, C. A. and Brauth, S. E. (1982) A paleostriatal-thalamic-telencephalic path in pigeons. *Neuroscience*, 7, 2735–2751.

Korzeniewska, E. and Güntürkün, O. (1990) Sensory properties and afferents of the n. dorsolateralis posterior thalami of the pigeon. *Journal of Comparative Neurology*, 292, 457–479.

Kosareva, A. A. (1980) Retinal projections in lamprey (*Lampetra fluviatilis*). *Journal für Hirnforschung*, 21, 243–256.

Krettek, J. E. and Price, J. L. (1977) The cortical projections of the mediodorsal nucleus and adjacent thalamic nuclei in the rat. *Journal of Comparative Neurology*, 171, 157–191.

Krubitzer, L. A. and Kaas, J. H. (1992) The somatosensory thalamus of monkeys: cortical connections and a redefinition of nuclei in marmosets. *Journal of Comparative Neurology*, 319, 123–140.

Kudo, M. and Niimi, K. (1980) Ascending projection of the inferior colliculus in the cat: an autoradiographic study. *Journal of Comparative Neurology*, 191, 545–556.

Lin, C.-S., May, P. J., and Hall, W. C. (1984) Nonintralaminar thalamostriatal projections in the gray squirrel (*Sciurus carolinensis*) and tree shrew (*Tupaia glis*). *Journal of Comparative Neurology*, 230, 33–46.

Lohman, A. H. M. and van Woerden-Verkley, I. (1978) Ascending connections to the forebrain in the tegu lizard. *Journal of Comparative Neurology*, 182, 555–594.

Luiten, P. G. M. (1981) Two visual pathways to the telencephalon in the nurse shark (*Ginglymostoma cirratum*). II. Ascending thalamotelencephalic connections. *Journal of Comparative Neurology*, 196, 539–548.

Martinet, L., Servière, J., and Peytevin, J. (1992) Direct retinal projections of the "non-image forming" system to the hypothalamus, anterodorsal thalamus and basal telencephalon of mink (*Mustela vison*) brain. *Experimental Brain Research*, 89, 373–382.

Martinez-Garcia, F. and Lorente, M.-J. (1990) Thalamo-cortical projections in the lizard *Podarcis hispanica*. In W. K. Schwerdtfeger and P. Germroth (eds.), *The Forebrain in Nonmammals: New Aspects of Structure and Function*. Berlin: Springer-Verlag, pp. 93–102.

McCormick, C. A. (1992) Evolution of the central auditory pathways in anamniotes. In D. B. Webster, R. R. Fay, and A. N. Popper (eds.), *The Evolutionary Biology of Hearing*. New York: Springer-Verlag, pp. 323–350.

Miceli, D., Marchand, L., Repérant, J., and Rio, J.-P. (1990) Projections of the dorsolateral anterior complex and adjacent thalamic nuclei upon the visual Wulst in the pigeon. *Brain Research*, 518, 317–323.

Miceli, D., Peyrichoux, J., and Repérant, J. (1975) The retino-thalamo-hyperstriatal pathway in the pigeon (*Columba livia*). *Brain Research*, 100, 125–131.

Mogensen, J. and Divac, I. (1982) The prefrontal 'cortex' in the pigeon: behavioral evidence. *Brain, Behavior and Evolution*, 21, 60–66.

Montgomery, N. M. and Fite, K. V. (1989) Retinotopic organization of central optic projections in *Rana pipiens*. *Journal of Comparative Neurology,* 283, 526–540.

Northcutt, R. G. (1978) Brain organization in the cartilaginous fishes. In E. S. Hodgson and R. F. Mathewson (eds.), *Sensory Biology of Sharks, Skates, and Rays.* Arlington, VA: Office of Naval Research, pp. 117–193.

Northcutt, R. G. (1979) Retinofugal pathways in fetal and adult spiny dogfish, *Squalus acanthias. Brain Research,* 162, 219–230.

Northcutt, R. G. (1981) Localization of neurons afferent to the telencephalon in a primitive bony fish, *Polypterus palmas. Neuroscience Letters,* 22, 219–222.

Northcutt, R. G. and Butler, A. B. (1980) Projections of the optic tectum in the longnose gar, *Lepisosteus osseus. Brain Research,* 190, 333–346.

Northcutt, R. G. and Butler, A. B. (1991) Retinofugal and retinopetal projections in the green sunfish, *Lepomis cyanellus. Brain, Behavior and Evolution,* 37, 333–354.

Oliver, D. L. and Hall, W. C. (1978) The medial geniculate body of the tree shrew *Tupaia glis.* II. Connections with the neocortex. *Journal of Comparative Neurology,* 182, 459–494.

Parent, A. (1976) Striatal afferent connections in the turtle (*Chrysemys picta*) as revealed by retrograde axonal transport of horseradish peroxidase. *Brain Research,* 108, 25–36.

Pritz, M. B. (1974) Ascending connections of a thalamic auditory area in a crocodile, *Caiman crocodilus. Journal of Comparative Neurology,* 153, 199–214.

Pritz, M. B. (1975) Anatomical identification of a telencephalic visual area in crocodiles: ascending connections of nucleus rotundus in *Caiman crocodilus. Journal of Comparative Neurology,* 164, 323–338.

Pritz, M. B. and Stritzel, M. E. (1990) Thalamic projections from a midbrain somatosensory area in a reptile, *Caiman crocodilus. Brain, Behavior and Evolution,* 36, 1–13.

Pritz, M. B. and Stritzel, M. E. (1992) A second auditory area in the non-cortical telencephalon of a reptile. *Brain Research,* 569, 146–151.

Pritz, M. B. and Stritzel, M. E. (1994) Anatomical identification of a telencephalic somatosensory area in a reptile, *Caiman crocodilus. Brain, Behavior and Evolution,* 43, 107–127.

Raczkowski, D. and Rosenquist, A. C. (1981) Retinotopic organization in the cat lateral posterior-pulvinar complex. *Brain Research,* 221, 185–191.

Rakic, P. (1991) Critical cellular events during cortical evolution: radial unit hypothesis. In B. L. Finlay, G. Innocenti and H. Scheich (eds.), *The Neocortex: Ontogeny and Phylogeny.* New York: Plenum, pp. 21–32.

Regidor, J. and Divac, I. (1987) Architectonics of the thalamus in the echidna (*Tachyglossus aculeatus*): search for the mediodorsal nucleus. *Brain, Behavior and Evolution,* 30, 328–341.

Regidor, J. and Divac, I. (1992) Bilateral thalamocortical projection in hedgehogs: evolutionary implications. *Brain, Behavior and Evolution,* 39, 265–269.

Reiner, A. (1986) Is prefrontal cortex found only in mammals? *Trends in Neurosciences,* 9, 298–300.

Repérant, J., Miceli, D., Rio, J.-P., Peyrichoux, J., Pierre, J. and Kirpitchnikova, E. (1986) The anatomical organization of retinal projections in the shark *Scyliorhinus canicula* with special reference to the evolution of the selachian primary visual system. *Brain Research Reviews,* 11, 227–248.

Saunders, P. T. and Ho, M.-W. (1984) The complexity of organisms. In J. W. Pollard (ed.), *Evolutionary Theory: Paths Into the Future.* Chichester: John Wiley pp. 121–139.

Scalia, F., Knapp, H., Halpern, M., and Riss, W. (1968) New observations on the retinal projections in the frog. *Brain, Behavior and Evolution,* 1, 324–353.

Schneider, A. and Necker, R. (1989) Spinothalamic projections in the pigeon. *Brain Research,* 484, 139–149.

Smeets, W. J. A. J. (1981) Efferent tectal projections in two chondrichthyans, the shark *Scyliorhinus canicula* and the ray *Raja clavata. Journal of Comparative Neurology,* 195, 13–23.

Smeets, W. J. A. J., Hoogland, P. V. and Lohman, A. H. M. (1986) A forebrain atlas of the lizard *Gekko gecko. Journal of Comparative Neurology,* 254, 1–19.

Smeets, W. J. A. J., Hoogland, P. V., and Voorn, P. (1986) The distribution of dopamine immunoreactivity in the forebrain and midbrain of the lizard *Gekko gecko:* an immunohistochemical study with antibodies against dopamine. *Journal of Comparative Neurology,* 253, 46–60.

Springer, A. D. and Mednick, A. S. (1985) Retinofugal and retinopetal projections in the cichlid fish *Astronotus ocellatus. Journal of Comparative Neurology,* 236, 179–196.

Striedter, G. F. The diencephalon of the channel catfish, *Ictalurus punctatus:* II. Retinal, tectal, cerebellar, and telencephalic connections. *Brain, Behavior and Evolution,* 36, 355–377.

Symonds, L. L., Rosenquist, A. C., Edwards, S. B., and Palmer, L. A. (1981) Projections of the pulvinar-lateral posterior complex to visual cortical areas in the cat. *Neuroscience,* 6, 1995–2020.

Takada, M. (1992) The lateroposterior thalamic nucleus and substantia nigra pars lateralis: origin of dual innervation over visual system and basal ganglia. *Neuroscience Letters,* 139, 153–156.

Thompson, S. M. and Robertson, T. R. (1987) Organization of subcortical pathways for sensory projections to the limbic cortex. II. Afferent projections to the thalamic lateral dorsal nucleus in the rat. *Journal of Comparative Neurology,* 265, 189–202.

Ulinski, P. S. (1984) Thalamic projections to the somatosensory cortex of the echidna, *Tachyglossus aculeatus. Journal of Comparative Neurology,* 229, 153–170.

Ulinski, P. S. (1986) Organization of corticogeniculate projections in the turtle, *Pseudemys scripta. Journal of Comparative Neurology,* 254, 529–542.

Ulinski, P. S. and Nautiyal, J. (1988) Organization of retinogeniculate projections in turtles of the genera *Pseudemys* and *Chrysemys. Journal of Comparative Neurology,* 276, 92–112.

Van Groen, T. and Wyss, J. M. (1992) Projections from the laterodorsal nucleus of the thalamus to the limbic and visual cortices in the rat. *Journal of Comparative Neurology,* 324, 427–448.

Waldmann, C. and Güntürkün, O. (1993) The dopaminergic innervation of the pigeon caudolateral forebrain: immunocytochemical evidence for a 'prefrontal cortex' in birds? *Brain Research,* 600, 225–234.

Welker, W. I. (1973) Principles of organization of the ventrobasal complex in mammals. *Brain, Behavior and Evolution,* 7, 253–336.

Wicht, N. and Himstedt, W. (1988) Topologic and connectional analysis of the dorsal thalamus of *Triturus alpestris* (Amphibia, Urodela, Salamandridae). *Journal of Comparative Neurology,* 267, 545–561.

Wicht, N. and Northcutt, R. G. (1992) The forebrain of the Pacific hagfish: a cladistic reconstruction of the ancestral craniate forebrain. *Brain, Behavior and Evolution,* 40, 25–64.

Wilczynski, W. and Northcutt, R. G. (1983) Connections of the bullfrog striatum: afferent organization. *Journal of Comparative Neurology,* 214, 321–332.

Wild, J. M. (1987a) Thalamic projections to the paleostriatum and neostriatum in the pigeon (*Columba livia*). *Neuroscience*, 20, 305–327.

Wild, J. M. (1987b) The avian somatosensory system: connections of regions of body representation in the forebrain of the pigeon. *Brain Research*, 412 205–223.

Winer, J. A., Wenstrup, J. J., and Larue, D. T. (1992) Patterns of GABAergic immunoreactivity define subdivisions of the mustached bat's medial geniculate body. *Journal of Comparative Neurology*, 319, 172–190.

Winer, J. A. and Wenstrup, J. J. (1994) Cytoarchitecture of the medial geniculate body in the mustached bat (*Pteronotus parnellii*). *Journal of Comparative Neurology*, 346, 161–182.

Winer, J. A. and Wenstrup, J. J. (1994) The neurons of the medial geniculate body in the mustached bat (*Pteronotus parnellii*). *Journal of Comparative Neurology*, 346, 183–206.

23

The Visceral Brain: The Hypothalamus and the Autonomic Nervous System

INTRODUCTION

The viscera are the contents of the four main body cavities: the cranium, the thorax, the abdomen, and the pelvis. Used in a strict sense, the term "visceral" thus refers to the internal body organs. Indeed, elsewhere in this book, we have used the term visceral in its strict sense and will do so again later in this chapter when we discuss the autonomic nervous system. We now, however, must introduce the use of "visceral" in a more general sense; that is, referring to a wide variety of biological processes that psychologists would classify as "emotion" and "motivation" and that zoologists would classify as survival behavior for the individual and the species. Individual survival behaviors include finding and obtaining the basic necessities of life: food, water, maintenance of body temperature, and shelter from unfavorable environmental situations. Species-survival behaviors include courtship, defense of territory, mate selection, and parental care of the young. The visceral brain, both in the narrow sense and in the wider sense of the term, includes the **hypothalamus.** In addition to interacting with the autonomic nervous system, the hypothalamus has important connections with sensory systems such as the olfactory system and the visual system. In addition to the hypothalamus and the autonomic nervous system, we also will discuss the **area preoptica** (sometimes spelled **praeoptica**) or **preoptic area,** which is a transitional area between the diencephalon and the telencephalon.

In this chapter, we will first discuss the hypothalamus in anamniotes. We have not divided anamniotes into Group I and Group IIA for two reasons. First, the cytoarchitecture of the hypothalamus does not vary between these groups as much as do a number of other parts of the central nervous system.

Second, less information about the connections and functions of the hypothalamus is available than is the case for many other brain regions in anamniotes. We then will discuss the hypothalamus of amniotes, and will end with a description of the autonomic nervous system.

The Hypothalamus

In the majority of the brain regions we have encountered thus far, we have found nuclear groups that have more or less discrete boundaries. Some of these boundaries are formed by a sheath of myelinated or sometimes unmyelinated axons that sharply delimits a neuronal population from the surrounding tissue. In other cases, the neurons' sizes, shapes, or other characteristics distinguish a cell population from its neighbors. In the hypothalamus, however, some of the neuronal populations do not have a discrete appearance, and therefore merge with or overlap their neighboring neuronal groups; hence they are known as "areas" rather than "nuclei." Two such examples are the lateral hypothalamic area and the preoptic area. In addition, like the thalamus, the hypothalamus consists of a large number of nuclei with similar sounding names. Some are variations on the directional terms, "medial," "lateral," "dorsal," and "ventral." Other names are descriptive of location or appearance, such as "supraoptic" (meaning "above the optic tract"), or "arcuate" (meaning "having an arc-like shape"). A further complication, as in the thalamus and elsewhere in the central nervous system, is that various scientists have tended to use somewhat different nomenclature to identify these cell groups. Where differing nomenclatures have been used, we have tried to use the most common name and to mention alternative names that refer to the same group of neurons.

The Hypothalamus and the Endocrine System

The hypothalamus is closely associated with the **endocrine system** because of its intimate relationship with the **pituitary** or **hypophysis,** which is located just below the floor of the hypothalamus. The zone of attachment between the hypothalamic floor and the hypophysis is the **eminentia medialis** or **median eminence.** The hypophysis itself is divisible into two regions: the more posterior, which is called the **neurohypophysis** or **pars nervosa,** and the more anterior, which is called the **adenohypophysis.** The neurohypophysis is in continuity with the hypothalamus proper, and, as its name implies, it consists of neural tissue. The neurohypophysis receives the terminals of peptide-secreting neurosecretory cells that have their somata in the hypothalamus. The secretions of these neurosecretory cells are picked up by the circulatory system and transported to target organs such as the kidney and the reproductive system.

The adenohypophysis is composed of three subdivisions: the most posterior region, known as the **pars intermedia** or intermediate part, which is immediately adjacent to the neurohypophysis; a middle region called the **pars tuberalis** or **tuberal region;** and the most anterior division, the **pars distalis** or **anterior lobe.** The pars tuberalis is sometimes known as the **pars infundibularis** because of its association with the **infundibulum,** which is Latin for "little funnel" and is the stalk by which the pituitary is attached to the hypothalamus. The adenohypophysis, unlike the neurohypophysis, is glandular tissue. The andenohypophyseal or anterior pituitary hormones travel through the general circulation to specific target organs, some of which are endocrine glands, such as the thyroid or the gonads, others of which are nonendocrine glands, such as the mammary glands of mammals, and still others that are general organ systems such as the skeleton or the pigment cells of the skin.

The pituitary serves as the link between the hypothalamus and the general circulatory system so that neuropeptides and monoamines produced in the hypothalamus and median eminence can affect the various peripheral target organs. Two mechanisms of linkage exist: a direct linkage and an indirect linkage. As shown in Figure 23-1, the linkage to the neurohypophysis or posterior pituitary is direct; neurosecretory neurons in the hypothalamus send their axons through the median eminence and infundibulum to terminate in the neurohypophysis where their secretions of **oxytocin** and **vasopressin** are introduced into the general circulation via a network of capillaries. These hormones have affects on the kidney in all vertebrates, on the transport of ova and reproductive behavior in amniotes in general, and on contractions of the uterus and the flow of milk in mammals. These hormones also are important in certain aspects of social behavior in mammals and perhaps other vertebrate classes as well.

Also shown in Figure 23-1 is the indirect linkage in which the neurosecretory neurons terminate in the median eminence. The secretions of these neurons are known as **releasing hormones** or **releasing factors.** These secretions are picked up by capillaries and transported via an intrinsic hypophyseal circulatory network known as the **hypophyseal portal system.**

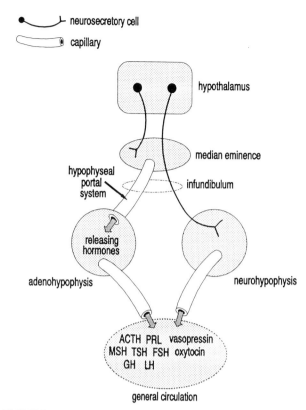

FIGURE 23-1. A schematic diagram of the hypothalamic hypophyseal portal system.

These releasing hormones enter the adenohypophysis where they affect the synthesis and release of a number of anterior pituitary hormones. Some releasing hormones stimulate the release of anterior pituitary hormones, and others inhibit their release. The hormones of the anterior pituitary, their releasing hormones, and their target tissues are listed in Tables 23-1 and 23-2. The pituitary hormones, in addition to their well-known affects on peripheral organs of the body, also are neuroactive substances at a number of locations in the brain. The reader might find a comparison of Tables 23-1 and 23-2 with Table 2-1, which lists some of the many neuroactive substances, to be useful at this point. In addition to serving as neuroactive substances in the brain, many of these hypothalamic and pituitary hormones serve more than one function. For example, **prolactin** is involved in electrolyte balance in ray-finned fishes, metamorphosis in amphibians, and milk production in mammals, and it also plays a role in parental care in many vertebrates.

Circumventricular Organs

The third ventricle, like the ventricular system in general, is lined with a thin layer of ependymal cells. In certain regions, however, this layer is thicker and forms specialized organs. These organs, called **circumventricular organs,** contact the cerebrospinal fluid that fills the ventricle and are also innervated by axons, often from hypothalamic and limbic structures. These

TABLE 23-1. Summary of Major Hypothalamic and Anterior Pituitary Hormones that Regulate Endocrine Glands

Hypothalamic Releasing Hormone	Pituitary Hormone	Target Endocrine Gland
Gonadotropic releasing hormone (GnRH)	Follicle stimulating hormone (FSH), luteinizing hormone (LH)	Testis, ovary
Corticotropin releasing hormone (CRH)	Adrenocorticotropic hormone (ACTH)	Adrenal cortex
Thyrotropin releasing hormone (TRH)	Thyroid stimulating hormone (TSH)	Thyroid, oviduct (reptiles, birds)

organs secrete substances into the cerebrospinal fluid, and they can also detect the presence of substances in this fluid. Circumventricular organs are present in all vertebrates, although their number can vary considerably. Anamniotes have four or five; diapsid reptiles, mammals, and turtles typically have six; and birds have nine.

The functions of these organs are not well understood. They do, however, seem to play a role in a variety of activities of the visceral brain, such as the regulation of water balance and blood pressure, the monitoring of levels of various tropic hormones and their hypothalamic releasing factors in the cerebrospinal fluid, ingestive behavior, nausea and vomiting, and biological rhythms. Table 23-3 lists the most frequently described organs and gives some indication of their functions as they are presently understood.

The Hypothalamus and the Limbic System

The hypothalamus functions in close association with the limbic system, which will be discussed in detail in Chapter 30. In brief, the limbic system is involved in emotional and motivational behavior, as well as certain aspects of memory. It is mainly a telencephalic system, but has strong connections with the thalamus, hypothalamus, and midbrain tegmentum. The main telencephalic limbic structures that have major connections with the hypothalamus are the **septum,** the **hippocampal formation,** and the **amygdala.** Many structures in the limbic system have receptors for gonadal hormones and are involved in social behaviors that depend on such hormone levels for their timing and expression, such as courtship, territorial defense, aggression, and parental behavior. The limbic sys-

TABLE 23-2. Summary of Major Hypothalamic and Pituitary Hormones that Regulate Nonendocrine Glands and Organs

Hypothalamic Releasing Hormone	Pituitary Hormone	Target Gland or Organ
Prolactin releasing factor (PRF—also known as vasoactive intestinal polypeptide or VIP), prolactin release-inhibiting hormone (PIH)	Prolactin (PRL)	Ovary (mammals) Mammary gland (mammals) Skin (reptiles) Crop (pigeons and doves) Kidney (ray-finned fishes)
Melanocyte stimulating hormone releasing factor (MRF), melanocyte stimulating hormone release-inhibiting factor (MIF)	Melanocyte stimulating hormone (MSH)	Pigment cells
Growth hormone releasing hormone (GHRH), somatostatin (SOM)	Growth hormone (GH)	Skeleton
These hormones are produced in the hypothalamus and are transported to the posterior pituitary where they are secreted. No hypothalamic releasing hormones are involved.	Arginine vasotocin (AVT—called vasopressin in reptiles and birds)	Kidney and aorta (ray-finned fishes, reptiles, and birds) Oviduct (ray-finned fishes, reptiles, birds)
	Isotocin (called mesotocin in reptiles and birds)	Blood vessels and oviduct (ray-finned fishes)
	Oxytocin	Uterus (mammals) Mammary gland (mammals), oviduct (reptiles, birds) Kidney (ray-finned fishes) Social behavior (mammals and possibly birds)
	Vasopressin	Kidney Social behavior (mammals)

TABLE 23-3. The Major Circumventricular Organs

Organ	Function	Reported in
Vascular organ of the lamina terminalis	Water balance Blood pressure Reproduction	Birds and mammals
Lateral septal organ	Secretes vasoactive intestinal polypeptide (VIP)	Birds
Subfornical organ (called the subseptal organ in non-mammals)	Water balance Blood pressure	Birds, turtles, and mammals
Paraventricular organ	Secretes tropic hormones Blood pressure Water balance	Ray-finned fishes, diapsid reptiles, birds, and turtles
Median eminence and neurohypophysis	See Tables 23-1 and 23-2	Amphibians, diapsid reptiles, birds, mammals, and turtles
Pineal	Secretes serotonin and melatonin Biological rhythms	All vertebrates
Subcommissural organ	Secretes glycoproteins Biological rhythms	All vertebrates
Subtrochlear organ	?	Birds
Area postrema	Food aversions Nausea and vomiting Blood pressure	Ray-finned fishes, diapsid reptiles, mammals, birds, and turtles

tem also affects the endocrine system by means of its connections with the hypothalamus.

The Preoptic Area

The preoptic area is a region at the junction of the telencephalon and diencephalon. In keeping with its transition-zone location between two major brain subdivisions, this area plays an important role in the activities of both subdivisions. It is closely related to both the hypothalamus and the limbic system. In particular, the preoptic area is associated with reproductive behavior. It contains many receptor sites for the estrogenic and androgenic sex hormones produced by the testes and ovaries. The preoptic area is one of the regions of the brain that shows **sexual dimorphism,** which means a structural difference between males and females. Other brain areas that also have receptor sites for androgens and estrogens are the hypothalamus and the amygdala.

Maintenance of body temperature within a specific range is important for vertebrates because too low a temperature may slow down the rate at which vital biochemical reactions occur, and too warm a temperature can make these reactions occur at too high a rate so that the body's reserves of nutrients and energy are too quickly depleted. Further extremes of heat or cold can result in death. Endothermic animals (mammals and birds) can generate heat internally and regulate temperature by panting, by adjusting the amount of insulating air trapped under feathers or hair, sweating, and other physiological mechanisms. Ectothermic animals, on the other hand, have few or no physiological mechanisms for regulation of temperature, and therefore they regulate their temperature behaviorally by changing their location when the ambient temperature becomes unsatisfactory for their needs. Reptiles posses some physiological mechanisms, such as panting and modification of blood flow at the skin, but

must still supplement these with behavioral methods to maintain their body temperatures in the optimal range. Whether the thermoregulation is physiological or behavioral, the preoptic area, in conjunction with the posterior hypothalamus, is concerned with regulation of body temperature in amniotes and ray-finned fishes. Insufficient data exist to know whether this area plays an important role in this vital function in amphibians.

THE HYPOTHALAMUS IN ANAMNIOTES

Two major parts of the hypothalamic region are present in all vertebrates: a preoptic area and the hypothalamus proper. The preoptic area is the rostral most part of the hypothalamic region and is named for the fact that it lies rostral to, or in front of, the optic chiasm. It receives a modest input from the retina in most vertebrates. In anamniotes, the hypothalamus proper can usually be divided into dorsal and ventral parts and may also have several other distinct nuclei.

Jawless and Cartilaginous Fishes

As discussed in Chapter 21, projections to the preoptic area and the hypothalamus from the epiphysis have been identified in lampreys, ray-finned fishes, and amphibians. These projections along with those from the retina are involved in the control of biological rhythms. Interconnections of the hypothalamus and preoptic area with the habenula are probably also present in anamniotes, although these pathways have been studied in detail only in amniotes.

The preoptic area and hypothalamus in agnathans are poorly understood regions. In lampreys, a **preoptic area** and

dorsal and ventral divisions of the hypothalamus have been recognized, but few data are available on cytoarchitecture or connections. In hagfishes (Fig. 23-2), the preoptic area contains four nuclei: the **periventricular, dorsal, external,** and **intermediate preoptic nuclei.** The hypothalamus is poorly differentiated, consisting of densely packed cells medially and a more diffuse field of cells laterally that is called the **infundibular nucleus.** In cartilaginous fishes (Fig. 23-3), a preoptic area is present, and the hypothalamus has a **medial part** surrounding the third ventricle and **inferior lobes.** Telencephalic pro-

jections are present to the preoptic area and the inferior lobes. The retina is known to project to the preoptic area in both hagfishes and cartilaginous fishes, as it probably also does in lampreys.

Actinopterygians

In ray-finned fishes, the cytoarchitecture, connections, and functions of the preoptic area and hypothalamus have been studied in better detail than in other aquatic anamniotes. Both of

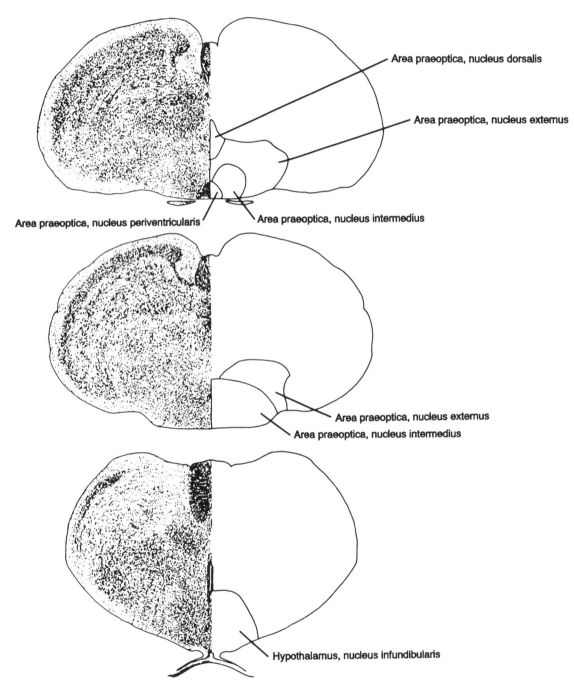

FIGURE 23-2. Transverse hemisections with mirror-image drawings through the diencephalon of a hagfish (*Eptatretus stouti*). The upper hemisection is the most rostral. Adapted from Wicht and Northcutt (1992). Used with permission of S. Karger AG.

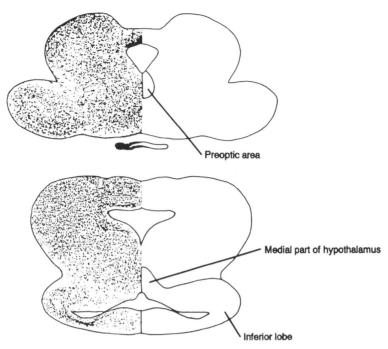

FIGURE 23-3. Transverse hemisections with mirror-image drawings through the diencephalon of a skate (*Raja eglanteria*) and a guitarfish (*Platyrhinoidis triseriata*). Adapted from Northcutt (1978).

these regions contain a number of distinct nuclei. The preoptic region contains an anterior nucleus, the **nucleus preopticus parvocellularis anterioris,** formed by small cells and, as are other nuclei in this region, named for its position and cell size. As shown in Figure 23-4, three nuclei lie caudal to the latter nucleus. **Nucleus preopticus magnocellularis,** named for its large cells, actually contains both large and small celled areas referred to, respectively, as the **pars magnocellularis** and **pars parvocellularis. A suprachiasmatic nucleus** is present in the preoptic region and is named for its position dorsal to the optic chiasm. A **nucleus preopticus parvocellularis posterioris** lies ventral to the thalamus caudally.

The part of the hypothalamus that borders the third ventricle is generally divided into dorsal and ventral parts, which are referred to as the **dorsal hypothalamus** and the **ventral hypothalamus.** The hypothalamus also contains other nuclei, some of which are laterally migrated. Three nuclei—the **nuclei ventralis tuberis, lateralis tuberis,** and **anterior tuberis**—consist of small aggregations of very large neurons. In most groups of ray-finned fishes, a laterally lying lobe is present in the hypothalamus, called the **inferior lobe,** into which a hypothalamic ventricle extends. This lobe contains neurons diffusely scattered through it, called **nucleus diffusus of the inferior lobe,** and a smaller, more compact nucleus called **nucleus centralis of the inferior lobe** [Figure 23-5(A)].

Endocrine Functions. A number of the endocrine functions of the preoptic area and hypothalamus have been studied in ray-finned fishes. Neurosecretory neurons in the preoptic area project to the posterior part of the pituitary gland, the pars nervosa. The pars nervosa interdigitates with a part of the adenohypophysis called the pars intermedia, which releases **melanocyte stimulating hormone,** a hormone that causes disper-

sion of melanin in melanocytes and consequent darkening of the skin. Neurons in both the preoptic area and hypothalamus are believed to regulate the release of melanocyte stimulating hormone within the pars intermedia. Other preoptic axons that terminate in the pars nervosa release **arginine vasotocin,** a diuretic hormone that increases glomerular filtration in the kidneys and increases blood pressure by causing vasoconstriction in the ventral aorta, or **isotocin,** which acts with arginine vasotocin to increase blood pressure and also causes contraction of the smooth-muscle fibers in the ovary and oviduct during parturition or oviposition.

In teleosts, the part of the adenohypophysis that is known as the pars distalis is directly innervated by axons from the hypothalamus. This situation is different from that in other vertebrates in which a vascular portal system transports hormones from the hypothalamus to the pars distalis. The teleostean pars distalis can be divided into a **rostral zone** that has **prolactin** and **corticotropin** containing terminals and a more **proximal zone** with **gonadotropin** and **growth hormone** containing terminals. **Thyrotropin** containing terminals occur throughout the pars distalis. Prolactin is important for survival of teleosts in fresh water since it enhances ion concentration and maintains plasma sodium levels. Corticotropin controls blood corticosteroid levels in their normal daily fluctuations and acutely in response to stress. Gonadotropin controls ovulation and ovarian recrudescence, which is a regrowth of the ovaries that occurs following the release of all mature oocytes (eggs) during spawning, and it also influences sperm production in males. Growth hormone functions as its name implies, and thyrotropin regulates the function of the thyroid gland in controlling metabolic rate.

The release of thyrotropin is regulated by a release-inhibitory hormone secreted by neurons in the anteromedial

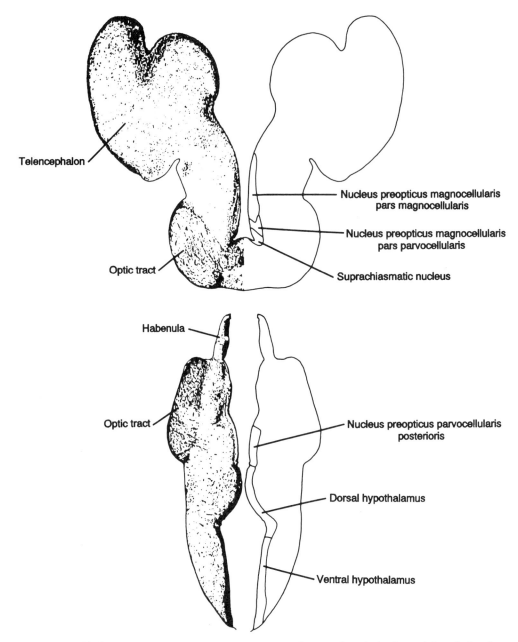

Telencephalon

Nucleus preopticus magnocellularis
pars magnocellularis

Nucleus preopticus magnocellularis
pars parvocellularis

Optic tract

Suprachiasmatic nucleus

Habenula

Optic tract

Nucleus preopticus parvocellularis
posterioris

Dorsal hypothalamus

Ventral hypothalamus

FIGURE 23-4. Transverse hemisections with mirror-image drawings through the diencephalon of a bowfin (*Amia calva*). The upper section is more rostral. Adapted from Butler and Northcutt (1992). Used with permission of S. Karger AG.

part of the ventral hypothalamus, while the release of the other hormones of the pars distalis is regulated primarily by the tuberal nuclei. Neurons in the nucleus lateralis tuberis secrete hormones that regulate the release of gonadotropin, prolactin, and corticotropin, the latter also being affected by neurons in the preoptic area. Neurons in the nucleus anterior tuberis may secrete a hormone that affects the release of growth hormone.

Connections of the Preoptic Area and Hypothalamus. In addition to playing a significant role in the regulation of hormone release, the preoptic area and hypothalamus have widespread interconnections within the central nervous system. Most of the

major projection systems of these regions will be discussed in amphibians, where a more complete picture is available, but one set of hypothalamic connections that is involved in feeding behavior may be unique to ray-finned fishes. This pathway involves a visual input relay through the pretectum, as discussed in Chapter 20. The retina projects to the parvocellular superficial pretectal nucleus and to nucleus corticalis. In most groups of ray-finned fishes, these two nuclei project to the posterior pretectal nucleus, which in turn projects to the inferior lobe of the hypothalamus. In euteleost fishes, the relay from these nuclei to the inferior lobe is through two nuclei—the intermediate superficial pretectal nucleus and nucleus glomerulosus—rather

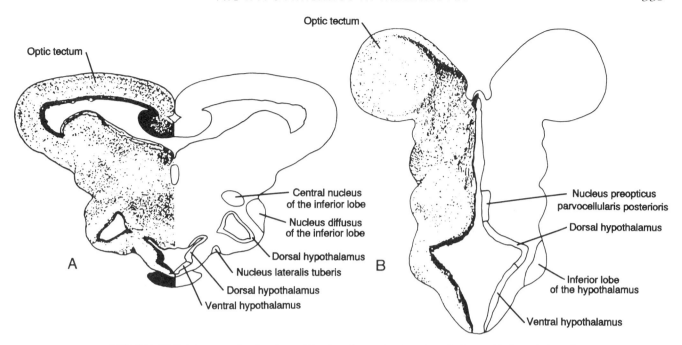

FIGURE 23-5. Transverse hemisections with mirror-image drawings through the caudal diencephalon of (A) a gar (*Lepisosteus osseus*) and (B) a bowfin (*Amia calva*). Adapted from Northcutt and Butler (1980) and Butler and Northcutt (1992), respectively. Material used with permission of Elsevier and S. Karger AG, respectively.

than through the posterior pretectal nucleus. The inferior lobe receives a descending input from the telencephalon, and in channel catfish, which have an exceptionally well-developed gustatory system, the inferior lobe has been found to project directly to the telencephalon, although it does not appear to do so in other fishes.

Functions of the Preoptic Area and Hypothalamus. Studies involving electrical stimulation of the preoptic area and the hypothalamus shed some light on a number of the behavioral repertoires in which these areas participate. Preoptic area stimulation can result in courtship behaviors and nest building, sperm release, egg release and spawning movements, aggressive behavior, or the display behavior of vertical banding, the latter apparently involving the release of melanocyte stimulating hormone. Stimulating the dorsomedial part of the hypothalamus results in feeding and agonistic behaviors, vertical banding, sound production, or sperm release. Stimulation of the inferior lobe results in feeding behavior, for which the pretectal afferent pathways were discussed above. Biting attacks, chasing, vertical movements, and circling are also behaviors that have been elicited by inferior lobe stimulation.

Descending projections from the preoptic area to the spinal cord and from the hypothalamus to the medulla may have a role in controlling heart rate in teleosts. Different parts of the hypothalamus appear to have antagonistic roles in increasing or decreasing heart rate in a wide range of vertebrates. In teleosts, stimulation of the preoptic area and the dorsal part of the dorsal hypothalamus results in heart rate decrease, while stimulation of some sites in the inferior lobe results in an increase in the heart rate. Stimulation of other sites in the inferior lobe results in an initial decrease of the heart rate followed by

an increase in rate when the stimulation is terminated. Both cardiac regulatory functions and hormonal control systems are correlated with hypothalamic involvement in behavioral displays and reactions as discussed above.

Sarcopterygians

In lungfishes, preoptic and hypothalamic regions are present (Fig. 23-6), but most neurons in these and other regions of the brain are in a periventricular position. The same is true

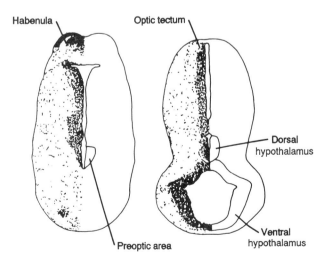

FIGURE 23-6. Transverse hemisections with mirror-image drawings through the diencephalon of a lungfish (*Protopterus annectens*). The hemisection on the left is more rostral. Adapted from Northcutt (1977).

of these regions in salamanders. In frogs, the preoptic region contains a rostral nucleus, called the **anterior preoptic area,** and two nuclei caudally, a **magnocellular nucleus** and a **suprachiasmatic nucleus.** Within the hypothalamus, two periventricular nuclei are present that comprise the **dorsal hypothalamus** and the **ventral hypothalamus.** A **lateral hypothalamic nucleus** is also present. It is composed of scattered cells and on topographic grounds may be homologous to part or all of the cells in the inferior lobe of the hypothalamus in fishes.

Various nuclei within the preoptic area in amphibians receive multiple inputs, including a major input from the medial pallium. The septum, lateral amygdala, nucleus anterior of the dorsal thalamus, an auditory relay nucleus called the secondary isthmal nucleus, hypothalamus, retina, and epiphysis also project to the preoptic area. The preoptic area projects to the hypothalamus as well as to the striatum and the spinal cord.

The hypothalamus receives a set of afferents similar to those to the preoptic area—from the medial pallium, lateral amygdala, and preoptic area—and additional afferents from the striatum, nucleus anterior and the central nucleus of the dorsal thalamus, the secondary isthmal nucleus, the tegmentum, the secondary visceral nucleus, and the raphe. Efferent targets of hypothalamic projections include the lateral amygdala, part of the subpallium, the preoptic area, nucleus anterior and the posterior nucleus of the dorsal thalamus, the periventricular parts of the pretectum and optic tectum, the torus semicircularis, and the tegmentum. Thus, the preoptic area and hypothalamus comprise a multifaceted interface between the limbic system, sensory systems, and neuroendocrine systems.

THE HYPOTHALAMUS IN AMNIOTES

In this introductory survey, we will not attempt a complete description of the neuronal groups and their various subdivisions in the hypothalamus of amniotes. Instead, we will concentrate on the main nuclei that are common to most mammals, diapsid reptiles, birds, and turtles. In general, the hypothalamus can be divided roughly into medial and lateral regions, an anterior region, and a posterior region. Many of the nuclei in these regions are shown schematically in Figure 23-7.

The anterior region consists of a cluster of nuclei, some of which are quite small. These include the **anterior nucleus,** the **preoptic area,** which has been subdivided into medial and lateral zones, the **supraoptic nucleus,** the **suprachiasmatic nucleus,** and the **paraventricular nucleus.** The posterior hypothalamus comprises the **mammillary bodies,** which are twin, hemispherical bulges in the floor of the mammalian hypothalamus that bear resemblance (at least in some minds) to human, female breasts. Also in the posterior zone is the **posterior nucleus,** which is located dorsal to the mammillary bodies. The mammillary bodies are subdivided into **medial** and **lateral mammillary nuclei.** In some mammals, an **intermediate mammillary nucleus** is present as well. The main constituents of the lateral region are the **lateral hypothalamic area** (LHA), which is a diffuse region without sharp boundaries. The medial region is sometimes known as the **tuberal region** for the **tuber cinereum,** which is a swelling of the floor of the diencephalon between the mammillary bodies. The middle or tuberal zone

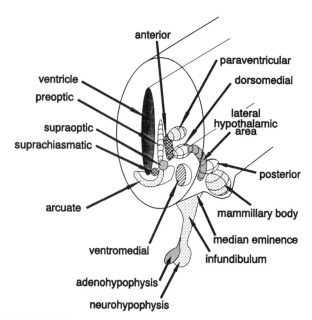

FIGURE 23-7. A schematic representation of major nuclei of the hypothalamus of a mammal. Rostral is to the left.

consists of a **dorsomedial nucleus** and a **ventromedial nucleus.** Figures 23-8 and 23-9 show some of the preoptic and hypothalamic nuclei in a diapsid reptile (a lizard), and a bird (a pigeon).

Connections of the Hypothalamus

The hypothalamus has extensive connections with the diencephalic and mesencephalic components of the limbic system and with the autonomic neuronal groups of the brainstem and spinal cord. Although we discuss them here separately as afferent and efferent, the reader should be aware that many of the connections of the hypothalamus are reciprocal; that is, major pathways of the hypothalamus, such as the medial forebrain bundle or mammillothalamic tract, have axons that travel in both directions.

Retinal Connections and the Epiphysis. Many of the functions of the hypothalamus depend on the daily cycle of light that begins at dawn and ends at dusk. Others depend on the seasonal increases (winter and spring) and decreases (summer and autumn) in the length of the day. The dawn–dusk light cycles are important for some types of **circadian biological rhythms,** which are daily rhythms with a duration of 24 h, ± 3 h. The longer cycles are important for seasonal rhythms, such as the annual onset of the mating season or annual migrations. Two groups of photoreceptors that have direct connections with the hypothalamus are the retina and the epiphysis or pineal (see Chapters 2 and 21). The axons of the retinal photoreceptors leave the optic tract (see Chapter 26) and enter the hypothalamus where they terminate in the nucleus suprachiasmaticus in the anterior hypothalamus. Electrophysiological and other studies of the suprachiasmatic nucleus suggest that it may serve

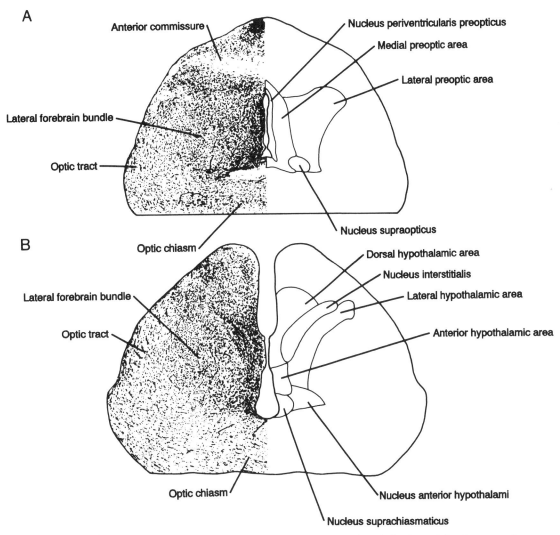

FIGURE 23-8. Transverse hemisections with mirror-image drawings through the diencephalon, from rostral to caudal, of a lizard (*Tupinambis nigropunctatus*). Adapted from Cruce (1974).

(continued)

as the pacemaker for circadian rhythms in conjunction with the pineal.

In diapsid reptiles (as in anamniotes), the pineal functions as a light gathering organ. Its photoreceptors have axons that terminate in the pretectum and tegmentum of the midbrain. In mammals and birds, the pineal's influence on biological rhythms is via the hormone **melatonin,** which is produced in the pineal during the dark phase of the daily light cycle. Concentrations of melatonin also have been reported in the suprachiasmatic nucleus. No evidence of pineal photoreceptors has yet been found in birds or mammals.

Other Afferent Connections. The main afferent connections of the hypothalamus are from the limbic system of the forebrain; that is, the septum, the amygdala, and the hippocampus of the telencephalon and the limbic thalamus, which is the anterior nuclear group in mammals. Because feeding, which depends so much on taste and smell, is central to the functions

of the hypothalamus, the olfactory areas of the telencephalon and the gustatory region of the medulla provide inputs to the hypothalamus as well. Olfaction also is important to social and reproductive behaviors, which likewise are important hypothalamic functions. Finally, inputs arrive at the hypothalamus from the midbrain tegmentum, which serves as the contact point between the limbic system and the reticular formation. The following are the main "highways" by which inputs from these regions arrive at the hypothalamus:

- The **medial forebrain bundle,** which contains axons that arise in the olfactory areas of the telencephalon, the septum, the hippocampal formation, and parts of the amygdala and terminate in the lateral hypothalamic area and the lateral preoptic area.

- The **fornix,** which arises in the hippocampal formation and terminates in the medial and lateral mammillary nuclei and the lateral preoptic area.

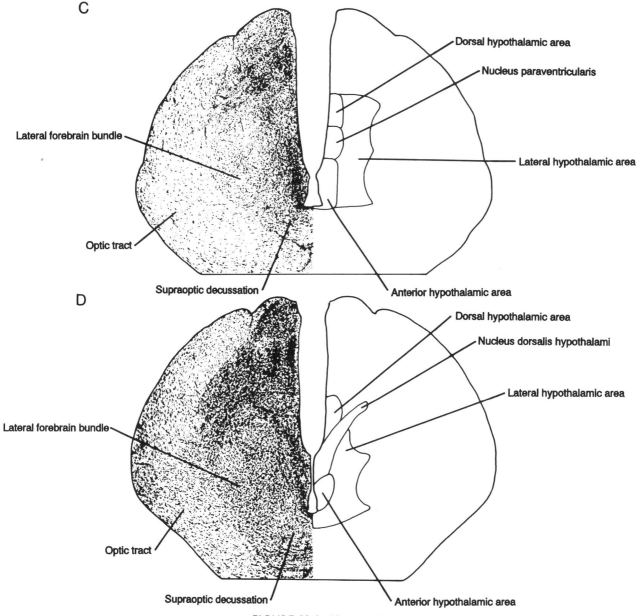

FIGURE 23-8. (Continued)

- The **mammillary peduncle,** which originates in the dorsal and ventral tegmental nuclei of the mesencephalon and terminates in the mammillary body.
- The **dorsal longitudinal fasciculus,** which arises in the periaqueductal gray region of the midbrain and terminates in the periventricular region of the hypothalamus.
- The **ascending serotoninergic axons** from the raphe group of nuclei (see Chapter 13) also travel in the medial forebrain bundle and terminate in the suprachiasmatic and mammillary nuclei as well as in the preoptic area.

Efferent Connections. An important efferent system of the hypothalamus is formed by the pathways that control the sympathetic autonomic neurons of the thoracic and lumbar regions

of the spinal cord and the mostly parasympathetic autonomic cranial nerve nuclei of the caudal hindbrain and sacral spinal cord. The autonomic system will be discussed in greater detail later in this chapter. In addition, as we mentioned above, many of the fiber connections of the hypothalamus are reciprocal. Thus a number of the efferent pathways are the same as those of the afferent pathways. These are:

- The **medial forebrain bundle,** which sends axons to the septum, the hippocampal formation, parts of the amygdala, and the raphe group of the reticular formation.
- The **mammillary efferent pathways,** which include the **mammillotegmental tract,** which is the reciprocal pathway from the mammillary body to the dorsal and ventral

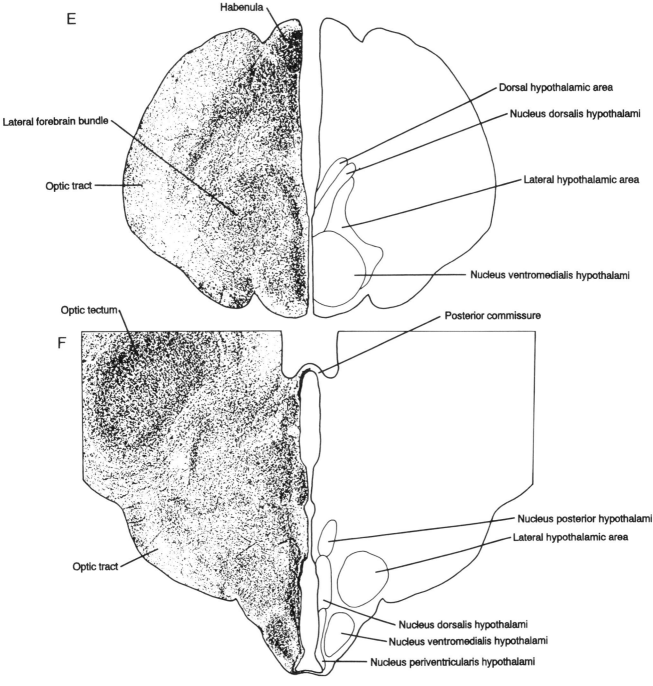

FIGURE 23-8. (*Continued*)

tegmental nuclei, and the **mammillothalamic tract,** from the mammillary body to the anterior or limbic nucleus of the thalamus.

- The **dorsal longitudinal fasciculus,** which arises in the periventricular region of the hypothalamus and terminates in the periaqueductal gray region of the midbrain.
- The **descending autonomic pathways** from the paraventricular nucleus and the lateral and posterior hypothalamic areas to the visceromotor cranial nerve nuclei of the medulla, such as the dorsal motor nucleus of the vagus

and the nucleus ambiguus, as well as traveling as far as the intermediolateral column of the spinal cord.

- The **posterior hypophyseal pathways,** which carry the axons of the neurosecretory cells of the supraoptic nucleus and the paraventricular nucleus into the posterior lobe of the pituitary (neurohypophysis) and which are the source of vasopressin and oxytocin.
- The **anterior hypophyseal pathways,** which originate in the medial or tuberal hypothalamus, especially from the arcuate nucleus and terminate in the median eminence

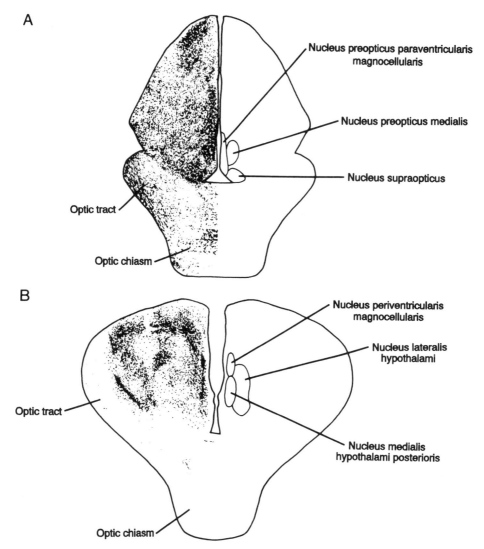

FIGURE 23-9. Transverse hemisections with mirror-image drawings through the diencephalon, from rostral to caudal, of a bird (*Columba livia*). Adapted from Karten and Hodos (1967). Used with permission of the Johns Hopkins University Press.

where they stimulate the production of the hypothalamic releasing hormones or factors that travel via the hypophyseal portal system to the adenohypophysis (anterior pituitary) to regulate the synthesis and production of a variety of anterior pituitary hormones.

Functions of the Hypothalamus

The anterior-posterior axis and the medial-lateral axis of the hypothalamus correspond roughly to a functional subdivision of this very complex structure. In a general sense, the anterior and posterior regions of the hypothalamus regulate water balance through their effects on the kidney via the neurohypophysis or posterior pituitary. In addition, these regions are involved in virtually all aspects of reproduction and parental care. These include control of the production and synthesis of the adenohypophyseal (anterior pituitary) hormones, the movement of ova in the oviduct, contractions of the muscles

of the organs of reproduction, the ejection of fertilized ova, the contractions of the uterus in mammals and viviparous (live bearing) reptiles, and the production of milk in mammals and "crop milk" in pigeons and doves. Through the neurohypophyseal hormone oxytocin, the flow of milk is controlled. In addition, the hormones of both the anterior and posterior pituitary are involved in the regulation of the social aspects of reproduction such as courtship, territorial defense, competition for mates, care of the young, etc. In addition, these regions contain a major pacemaker for biological rhythms, an area that is involved in temperature regulation, and an area that controls the autonomic nervous system. The lateral and medial hypothalamic zones are involved in ingestive behavior (feeding and drinking) as well as aggressive behavior and control of the autonomic nervous system. A number of the functions of the major hypothalamic nuclei and areas are presented in Table 23-4.

The hypothalamus is an important integrating region of the brain; it is a crossroads for influences from the pallial regions

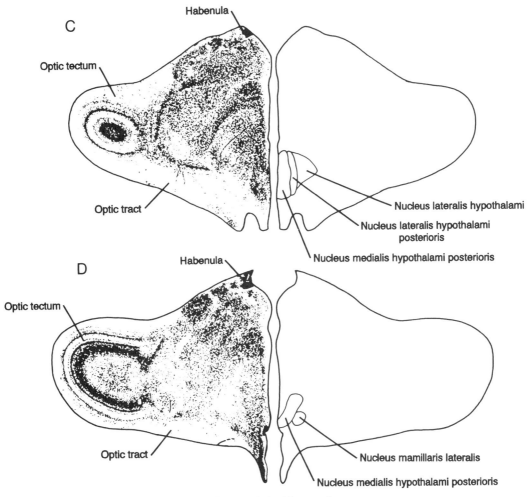

FIGURE 23-9. (Continued)

of the telencephalon, the limbic system, the reticular formation, sensory systems (especially the gustatory, olfactory, visual, and viscerosensory systems), and control pathways to the endocrine and limbic systems. Some of its nuclear groups, such as the paraventricular and supraoptic nuclei that are in direct control of the contractions of the uterus and the flow of milk in mammals, are essentially motor in the sense that the motor nuclei of the cranial nerves or of the spinal cord are motor; that is, they directly control effector organs. Somewhat less direct control of effectors is exerted by the rest of the neuroendocrine hypothalamus. The majority of hypothalamic neurons, however, are involved in the integration of the many afferent influences and the initiation and maintenance of complex motivations and behaviors that cannot be considered "motor" in any direct sense of that term.

THE AUTONOMIC NERVOUS SYSTEM

At many points in this book we have focused on the somatic motor system, which is the system that moves the skeleton and thus the body. These somatic movements can be reflexive, such as those produced by the cerebellum and vestibular system to maintain posture and muscle tone, or they can be voluntary.

We now turn to a motor system that is visceral rather than somatic and that is entirely involuntary; that is, the autonomic nervous system. We have already encountered the autonomic nervous system elsewhere, such as in Chapters 7 and 8.

The autonomic nervous system is composed of two divisions: the **sympathetic division** and the **parasympathetic division.** These two divisions work together to provide a smooth and efficient neural control system for the viscera. In general, the parasympathetic system tends to "down-regulate" or reduce the functions of an organ and the sympathetic tends to "up-regulate" or increase these functions. Thus, in the case of the heart, the sympathetic division increases heart rate and the parasympathetic division decreases heart rate. Likewise, the parasympathetic division relaxes certain smooth muscles, such as those of the anal and bladder sphincters, while the sympathetic division contracts them. This relationship, however, is not entirely consistent; for example, the sympathetic division relaxes the ciliary muscle inside the eye whereas the parasympathetic division contracts it. Likewise, the sympathetic division relaxes the muscles of the intestinal tract whereas the parasympathetic division constricts them.

The sympathetic system becomes powerfully activated during times of strenuous exertion, such as sudden escape from a predator or leaping out of the way of an automobile. During

TABLE 23-4. Functions of Some of the Major Hypothalamic Nuclei and Areas

Hypothalamic Nucleus or Area	Functions
Preoptic area Anterior nucleus	Temperature regulation—cooling of the body Reproduction and parental behavior Control of the autonomic nervous system
Supraoptic nucleus Paraventricular nucleus	Control of water balance via kidney Control of reproductive organs during ovulation and copulation Control of milk flow in mammals Contraction of uterus in mammals and viviparous reptiles Social and parental behavior Control of the autonomic nervous system
Posterior nucleus Mammillary body	Temperature regulation—warming of the body Control of the autonomic nervous system
Suprachiasmatic nucleus	Pacemaker for biological rhythms
Arcuate nucleus	Control of synthesis and release of anterior pituitary hormones
Lateral hypothalamic area	Feeding behavior Drinking behavior Aggression Control of the autonomic nervous system
Ventromedial nucleus Dorsomedial nucleus	Feeding behavior Aggression Control of the autonomic nervous system

Autonomic Neurochemistry

Neither the sympathetic nor the parasympathetic division terminates directly on the target organs. Both systems send their efferent axons to ganglia located outside of the central nervous system. The axons from the central nervous system to the ganglia are known as the **preganglionic axons,** and those that leave the ganglion are known as **postganglionic axons.** The neurotransmitter found in the synaptic vesicles of the preganglionic axons of both divisions is **acetylcholine.** In addition, acetylcholine is the neurotransmitter for the postganglionic axons of the parasympathetic division. The postganglionic neurotransmitter of the sympathetic division is **norepinephrine,** except in the adrenal medulla, where it is **epinephrine.** In addition to acetylcholine and norepinephrine, a number of neuropeptides and other neuroactive substances have been found in the autonomic nervous system. These include **vasoactive intestinal peptide, enkephalin, somatostatin, substance P, cholecystokinin,** and **nitric oxide.** Some of these substances have sympathetic-like affects and others have parasympathetic-like affects.

Amniotes

The sympathetic division is located in the thoracic and lumbar levels of the spinal cord and therefore is sometimes referred to as the **thoracolumbar division.** In-coming visceral afferent axons terminate in the **intermediomedial column** of the spinal cord. The outgoing visceral efferent neurons have their cell bodies in the **intermediolateral column.** These sympathetic efferent neurons have relatively short axons and soon after they leave the spinal cord, they terminate in the sympathetic ganglion (see Chapters 7 and 8). Thus they are preganglionic axons. The postganglionic axons originate in the sympathetic ganglion and terminate in the peripheral organ systems such as the liver, the heart, gut, and glands. Because the sympathetic ganglia are located close to the spinal cord, the postganglionic axons are rather long in order to reach their target organs. The neurotransmitter of the postganglionic sympathetic axon, norepinephrine, is sometimes referred to as noradrenalin; thus the sympathetic postganglionic axons also are known as **noradrenergic axons.**

The visercal motor neurons of the intermediolateral column terminate not only on sympathetic ganglia at their own level of the spinal cord, they also terminate on ganglia that are above or below that level. These ascending and descending connections between adjacent sympathetic ganglia form a longitudinal "chain," much like beads on a string. This longitudinal chain is known as the **sympathetic chain of ganglia,** or simply as the **sympathetic chain.**

The parasympathetic division differs in many respects from the sympathetic division. First, its neurons are located in the hindbrain and in the sacral levels of the spinal cord. Thus it is sometimes referred to as the **craniosacral division.** Second, its incoming visceral afferent axons terminate in the sensory nuclei of the cranial nerves in the visceral afferent column (see Chapter 10) as well as in the sacral levels of the spinal cord. The visceral efferent neurons of the parasympathetic division are located in the visceral motor nuclei of the hindbrain, especially that of the **vagus nerve.** Third, the parasympathetic gan-

such exertions, the sympathetic division greatly increases heart rate, blood pressure, and respiration as well as stimulates the liver to release extra amounts of glucose into the circulatory system. At the same time, certain other visceral activities are inhibited such as contraction of gut musculature and secretion of digestive enzymes. Because of these dramatic effects, the sympathetic division has the reputation of being an "emergency" system, while the parasympathetic system has the reputation of being the system that operates during nonemergency times. Indeed the word "sympathetic" comes from the Greek word *pathos,* which means "feeling" or "emotion." This summary, however, is not a complete representation of the role of the sympathetic division, which works constantly, even under nonemergency situations, in partnership with the parasympathetic division to maintain a smoothly functioning internal environment.

TABLE 23-5. Summary of the Autonomic Nervous System

	Sympathetic	Parasympathetic
Location in CNS	Thoracic and lumbar spinal cord	Hindbrain and sacral spinal cord
Location of ganglia	Close to spinal cord	Close to target organ
Length of preganglionic axons	Short	Long
Length of postganglionic axons	Long	Short
Preganglionic neurotransmitter	Acetylcholine	Acetylcholine
Postganglionic neurotransmitter	Norepinephrine	Acetylcholine
Effect on heart rate, blood pressure, respiration	Increases	Decreases
Effects on visceral blood vessels, cutaneous blood vessels	Constriction	
Effects on anal and bladder sphincters	Constriction	Relaxation
Effects on ciliary muscle and intestinal muscles	Relaxation	Constriction
Effect on dilator pupillae muscle	Contraction	
Effect on constrictor pupillae muscle		Contraction
Effect on muscles that elevate hairs and feathers	Contraction	
Effect on liver	Release glucose	Secretion of bile
Effect on pancreas, stomach, tear glands, and salivary glands		Secretion
Effect on sweat glands and adrenal medulla	Secretion	

glia are located near the target organs rather than close to the central nervous system, which means that the preganglionic axons are rather long and the postganglionic axons are relatively short. Finally, since the neurotransmitter of the postganglionic axon terminals, like that of the preganglionic terminals of both divisions, is acetylcholine, these axons often are referred to as **cholinergic.** Table 23-5 summarizes the main differences between the sympathetic and parasympathetic divisions. Table 23-6 summarizes the target organs of the various components of the autonomic nervous system.

Anamniotes

The autonomic nervous system has not received the same attention by scientists in anamniotes as it has in amniotes, especially mammals and some birds. In general, the autonomic nervous system of anamniotes is not so neatly divided into sympathetic and parasympathetic divisions as it is in amniotes. Moreover, the pattern of dual innervation by both divisions of the autonomic system is not a typical feature of the organs of anamniotes. In general anamniotes, especially nontetrapod anamniotes, have fewer glands in the head than do terrestrial animals, because lubrication of the oral cavity and eyes usually is not a problem in the aquatic environment. Jawless vertebrates, for example, have no sympathetic ganglia; axons that are probably sympathetic pass from the spinal cord directly to target organs, such as blood vessels. Jawless vertebrates, do have a parasympathetic system that originates in the motor nucleus of the vagus. Cartilaginous fishes, like jawless vertebrates, also lack a well-developed sympathetic system. They do possess sympathetic ganglia, but these are not located in a chain along the side of the spinal cord. Parasympathetic innervation is supplied mainly by the vagus nerve. The stomachs of these animals receive dual sympathetic and parasympathetic innervation.

In ray-finned teleost fishes, a sympathetic chain is present, and dual innervation of additional organs can be observed. A similar pattern can be found in amphibians except that in these tetrapod anamniotes, the heart is dually innervated by sympathetic and parasympathetic axons. In addition, because amphibians have developed both tear glands and salivary glands, these organs have parasympathetic innervation that originates in the motor nuclei of cranial nerves VII and IX. Some frogs have the sacral component of the parasympathetic division, but this appears lacking in salamanders.

TABLE 23-6. Summary of the Major Target Organs of the Autonomic Nervous System

Target Organ	Sympathetic	Parasympathetic
Ciliary muscle, dilator pupillae, constrictor pupillae		Nucleus III complex
Salivary glands, tear glands	Thoracic spinal cord	Nucleus VII and nucleus IX
Adrenal medulla		
Heart, lungs, stomach, liver, pancreas, kidney		Nucleus X
Small intestine	Thoracic and lumbar spinal cord	Nucleus X
Large intestine		Nucleus X and sacral cord
Bladder and rectal sphincters, gonads, genitals	Lumbar spinal cord	Sacral cord

EVOLUTIONARY PERSPECTIVE

The preoptic area and hypothalamus are quite large and well developed in teleosts and cartilaginous fishes, but are relatively small and unelaborated in amphibians. The hypothalamus appears to have undergone an independent expansion in mammals, diapsid reptiles, birds, and turtles. Although some variation exists in the structure and organization of these areas, in general the hypothalamus and preoptic area have been fairly conservative in their evolution. The relationship between the hypothalamus and the endocrine system has undergone some interesting changes, however, in the functions of some of the hormones in different classes of vertebrates. Even though the hormones themselves have remained largely unmodified chemically, their functions in particular classes have changed. The evolution of the autonomic nervous system has been quite conservative, especially in the tetrapod lineage.

FOR FURTHER READING

Crews, D. and Gans, C. (1992) The interaction of hormones, brain, and behavior: an emerging discipline in herpetology. In C. Gans and D. Crews (eds.), *Hormones, Brain, and Behavior. Biology of the Reptilia, Volume 18E*. Chicago: University of Chicago Press, pp. 1–23.

Demski, L. S. (1983) Behavioral effects of electrical stimulation of the brain. In R. E. Davis and R. G. Northcutt (eds.), *Fish Neurobiology, Vol. 2: Higher Brain and Functions*. Ann Arbor, MI: University of Michigan Press, pp. 317–359.

Gans, C. and Crews, D. (eds.), *Hormones, Brain, and Behavior. Biology of the Reptilia, Volume 18E*. Chicago: University of Chicago Press.

Gorbman, A. and Davey, K. (1991) Endocrines. In C. L. Prosser (ed.) *Neural and Integrative Animal Physiology*. New York: Wiley, pp. 693–754.

Hastings, J. W., Rusak, B., and Boulos, Z. (1991) Circadian rhythms: the physiology of biological timing. In C. L. Prosser (ed.), *Neural and Integrative Animal Physiology*. New York: Wiley, pp. 435–536.

Kuenzel, W. J. and van Tienhoven, A. (1982) Nomenclature and location of avian hypothalamic nuclei and associated circumventricular organs. *Journal of Comparative Neurology*, 206, 293–313.

Neary, T. J. and Northcutt, R. G. (1983) Nuclear organization of the bullfrog diencephalon. *Journal of Comparative Neurology*, 213, 262–278.

Nelson, D. O., Heath, J. C., and Prosser, C. L. (1984) Evolution of temperature regulatory mechanisms. *American Zoologist*, 24, 798–807.

Northcutt, R. G. and Kicliter, E. (1980) Organization of the amphibian telencephalon. In S. O. E. Ebbesson (ed.), *Comparative Neurology of the Telencephalon*. New York: Plenum, pp. 203–255.

Ottinger, M. A. (1989) The brain-pituitary-gonad axis in homeotherms. In C. G. Scanes and M. Schreibman (eds.), *Development, Maturation, and Senescence of Neuroendocrine Systems: A Comparative Approach*. New York: Academic, pp. 135–153.

Peter, R. E. and Fryer, J. N. (1983) Endocrine functions of the hypothalamus of actinopterygians. In R. E. Davis and R. G. Northcutt (eds.), *Fish Neurobiology, Vol. 2: Higher Brain Areas and Functions*. Ann Arbor, MI: University of Michigan Press, pp. 165–201.

Shepherd, G. M. (1988) *Neurobiology*. New York: Oxford University Press.

ADDITIONAL REFERENCES

Allison, J. D. and Wilczynski, W. (1991) Thalamic and midbrain auditory projections to the preoptic area and ventral hypothalamus in the green tree frog (*Hyla cinerea*). *Brain, Behavior and Evolution*, 38, 322–331.

Allison, J. D. and Wilczynski, W. (1994) Efferents from the suprachiasmatic nucleus to basal forebrain nuclei in the green tree frog (*Hyla cinerea*). *Brain, Behavior and Evolution*, 43, 129–139.

Bauer, D. H. and Demski, L. S. (1980) Vertical banding evoked by electrical stimulation of the brain in anesthetized green sunfish, *Lepomis cyanellus*, and bluegill, *L. macrochirus*. *Journal of Experimental Biology*, 84, 149–160.

Berk, M. L. and Butler, A. B. (1981) Efferent projections of the medial preoptic nucleus and medial hypothalamus in the pigeon. *Journal of Comparative Neurology*, 203, 379–399.

Bradford, M. R., Jr. and Northcutt, R. G. (1983) Organization of the diencephalon and pretectum of the ray-finned fishes. In R. E. Davis and R. G. Northcutt (eds.), *Fish Neurobiology, Vol. 2: Higher Brain Areas and Functions*. Ann Arbor, MI: University of Michigan Press, pp. 117–163.

Butler, A. B. and Northcutt, R. G. (1973) Architectonic studies of the diencephalon of *Iguana iguana* (Linneaus). *Journal of Comparative Neurology*, 149, 439–462.

Butler, A. B. and Northcutt, R. G. (1992) Retinal projections in the bowfin, *Amia calva*: cytoarchitectonic and experimental analysis. *Brain, Behavior and Evolution*, 39, 169–194.

Cooper, W. E., Jr. and Greenberg, N. (1992) Reptilian coloration and behavior. In C. Gans and D. Crews (eds) *Hormones, Brain, and Behavior. Biology of the Reptilia, Volume 18E*. Chicago: University of Chicago Press, pp. 298–422.

Cruce, J. A. F. (1974) A cytoarchitectural study of the diencephalon of the tegu lizard, *Tupinabis nigropunctatus*. *Journal of Comparative Neurology*, 153, 215–238.

Demski, L. S. (1973) Feeding and aggressive behavior evoked by hypothalamic stimulation in a cichlid fish. *Comparative Biochemistry and Physiology*, 44A, 685–692.

Demski, L. S., Bauer, D. H., and Gerald, J. W. (1975) Sperm release evoked by electrical stimulation of the fish brain: a functional-anatomical study. *Journal of Experimental Zoology*, 191, 215–232.

Demski, L. S. and Gerald, J. W. (1972) Sound production evoked by electrical stimulation of the brain in toadfish, *Opsanus tau*. *Animal Behavior*, 20, 507–513.

Demski, L. S. and Knigge, K. M. (1971) The telencephalon and hypothalamus of the bluegill (*Lepomis macrochirus*): evoked feeding, aggressive and reproductive behavior with representative frontal sections. *Journal of Comparative Neurology*, 143, 1–16.

Demski, L. S. and Sloan, H. E. (1985) A direct magnocellular-preoptico-spinal pathway in goldfish: implications for control of sex behavior. *Neuroscience Letters*, 55, 383–388.

Demski, L. S. (1984) The evolution of neuroanatomical substrates of reproductive behavior: sex steroid and LHRH-specific pathways including the terminal nerve. *American Zoologist*, 24, 809–830.

Hoogland, P. V. and Vermeulen-Van der Zee, E. (1989) Efferent connections of the dorsal cortex of the lizard *Gecko gecko* studied with *Phaseolus vulgaris*-leucoagglutinin. *Journal of Comparative Neurology*, 285, 298–303.

Hornby, P. J. and Demski, L. S. (1988) Functional-anatomical studies of neural control of heart rate in goldfish. *Brain, Behavior and Evolution*, 31, 181–192.

Johnston, S. A. and Maler, L. (1992) Anatomical organization of the hypophysiotropic systems in the electric fish, *Apteronotus leptorhynchus*. *Journal of Comparative Neurology*, 317, 421–437.

Karten, H. J. and Hodos, W. (1967) *A Stereotaxic Atlas of the Brain of the Pigeon (Columba livia)*. Baltimore, MD: Johns Hopkins Press.

Larsen, P. J., Hay-Schmidt, A., and Mikkelsen, J. D. (1994) Efferent connections from the lateral hypothalamic region and lateral preoptic area to the hypothalamic paraventricular nucleus of the rat. *Journal of Comparative Neurology*, 342, 299–319.

Molist, P., Rodriguez-Moldes, I., and Anadón, R. (1993) Organization of catecholaminergic systems in the hypothalamus of two elasmobranch species, *Raja undulata* and *Scyliorhinus canicula*. A histofluorescence study. *Brain, Behavior and Evolution*, 41, 290–302.

Murakami, T., Morita, Y., and Ito, H. (1983) Extrinsic and intrinsic fiber connections of the telencephalon in a teleost, *Sebastiscus marmoratus*. *Journal of Comparative Neurology*, 216, 115–131.

Northcutt, R. G. (1977) Retinofugal projections in the lepidosirenid lungfishes. *Journal of Comparative Neurology*, 174, 553–574.

Northcutt, R. G. (1978) Brain organization in the cartilaginous fishes. In E. S. Hodgson and R. F. Mathewson (eds.), *Sensory Biology of Sharks, Skates, and Rays*. Arlington, VA: Office of Naval Research, pp. 117–193.

Northcutt, R. G. (1979) Retinofugal pathways in fetal and adult spiny dogfish, *Squalus acanthias*. *Brain Research*, 162, 219–230.

Northcutt, R. G. and Butler, A. B. (1980) Projections of the optic tectum in the longnose gar, *Lepisosteus osseus*. *Brain Research*, 190, 333–346.

Northcutt, R. G. and Ronan, M. (1992) Afferent and efferent connections of the bullfrog medial pallium. *Brain, Behavior and Evolution*, 40, 1–16.

Kuenzel, W. J. and Blähser, S. (1993/4) The visceral forebrain system in birds: its proposed anatomical components and functions. *Poultry Science Review*, 5, 29–36.

Prasada Rao, P. D., Job, T. C., and Schreibman, M. P. (1993) Hypophysiotropic neurons in the hypothalamus of the catfish *Clarias batrachus*: a cobaltous lysine and HRP study. *Brain, Behavior and Evolution*, 42, 24–38.

Puzdrowski, R. L. and Northcutt, R. G. (1989) Central projections of the pineal complex in the silver lamprey *Ichthyomyzon unicuspis*. *Cell and Tissue Research*, 225, 269–274

Sakamoto, N. and Ito, H. (1982) Fiber connections of the corpus glomerulosus in a teleost, *Navodon modestus*. *Journal of Comparative Neurology*, 205, 291–298.

Shiga, T., Oka, Y., Satou, M., Okumoto, N., and Ueda, K. (1985) An HRP study of afferent connections of the supracommissural ventral telencephalon and the medial preoptic area in hime salmon (landlocked red salmon, *Oncorhynchus nerka*). *Brain Research*, 361, 162–177.

Striedter, G. F. (1990) The diencephalon of the channel catfish, *Ictalurus punctatus*. II. Retinal, tectal, cerebellar and telencephalic connections. *Brain, Behavior and Evolution*, 36, 355–377.

Streider, G. F. and Northcutt, R. G. (1989) Two distinct visual pathways through the superficial pretectum in a percomorph teleost. *Journal of Comparative Neurology*, 283, 342–354.

Walker, W. F. (1987) *Functional Anatomy of the Vertebrates*. Philadelphia: Saunders.

Wicht, H. and Northcutt, R. G. (1990) Retinofugal and retinopetal projections in the Pacific hagfish, *Eptatretus stouti* (Myxinoidea). *Brain, Behavior and Evolution*, 36, 315–328.

Wicht, H. and Northcutt, R. G. (1992) The forebrain of the Pacific hagfish: a cladistic reconstruction of the ancestral craniate forebrain. *Brain, Behavior and Evolution*, 40, 25–64.

Whittier, J. M. and Tokarz, R. R. (1992) Physiological regulation of sexual behavior in female reptiles. In C. Gans and D. Crews (eds.), *Hormones, Brain, and Behavior. Biology of the Reptilia, Volume 18E*. Chicago: University of Chicago Press, pp. 24–69.

Wilczynski, W., Allison, J. D., and Marler, C. A. (1993) Sensory pathways linking social and environmental cues to endocrine control regions of amphibian forebrains. *Brain, Behavior and Evolution*, 42, 252–264.

Wullimann, M. F., Meyer, D. L., and Northcutt, R. G. (1991) The visually related posterior pretectal nucleus in the non-percomorph teleost *Osteoglossum bicirrhosum* projects to the hypothalamus: a Di-I study. *Journal of Comparative Neurology*, 312, 415–435.

Part Five

THE FOREBRAIN: TELENCEPHALON

24

Striatum

INTRODUCTION

The striatopallidal complexes, extended amygdala, and magnocellular corticopetal cell complex (nucleus basalis) are located in the ventral part of the telencephalon and are involved with the control of movements. In this chapter, we will primarily focus on the striatopallidal complexes. In mammals (Fig. 24-1), the **dorsal striatopallidal complex** is composed of the **caudate nucleus, putamen,** and **globus pallidus,** while the **ventral striatopallidal complex** is predominantly composed of **nucleus accumbens,** the **olfactory tubercle,** and the **ventral pallidum** within the rostral part of the substantia innominata, as was discussed in Chapter 19.

A number of neurotransmitters and neuromodulators have been identified in these complexes in mammals, and similar substances have been identified in some nonmammalian vertebrates in the same ventral region of the telencephalon. These substances are important because of their roles in integrating the various inputs and in transmitting the various outputs of the striatopallidal complexes. They also fortuitously provide markers by which homologues of the mammalian striatopallidal complexes can be identified in nonmammals. Some of the striatopallidal neurochemicals of mammals are illustrated in Figure 24-2, in relation to the neurons in which they have been found (see Table 2-1). Finding a particular neuroactive substance in a particular nucleus or fiber bundle in a nonmammalian vertebrate brain not only helps to identify homologous structures but, as information accumulates, will help unravel the functional evolution of this system, particularly the substantial increase in the complexity of motor pathways that occurred within the radiation of land vertebrates.

In this chapter, we will include a minor digression into the nuclei of the ventral thalamus (subthalamus). These nuclei are discussed here, rather than in a separate chapter, since their connections are so intimately related to the striatum. Nuclei believed to be part of the ventral thalamus, due to their position, have been identified in most groups of vertebrates, but very little information is yet available on their connections in anamniotes. Thus, the ventral thalamic nuclei will be introduced in conjunction with the amniote striatopallidal complexes. Ascending sensory projections from collothalamic nuclei in the dorsal thalamus to the striatum in anamniotes were discussed in Chapter 22 and will not be repeated in any detail here.

The magnocellular corticopetal cell complex, which comprises the cholinergic cells distributed through various structures in the ventral part of the telencephalon, will also be discussed in this chapter. This population of cells interconnects the limbic-related part of the striatum with the isocortex and appears to be involved in cortical arousal and attentional functions.

THE STRIATAL SYSTEM

Group I

In lampreys, the striatum is present in the ventral part of the telencephalon and has densely packed cells in its caudal part (Fig. 24-3). In squalomorph sharks, several striatal components have been tentatively identified (Fig. 24-4). A densely packed lamina of cells in the ventral telencephalon is called the **area superficialis basalis.** This area is densely innervated by the dopaminergic neurons of the midbrain and may thus be the homologue of the nucleus accumbens of mammals and possibly, as a field, also of the olfactory tubercle. A group of cells dorsal to this lamina is called the **area periventricularis ventrolateralis** and appears to be the homologue of the caudate

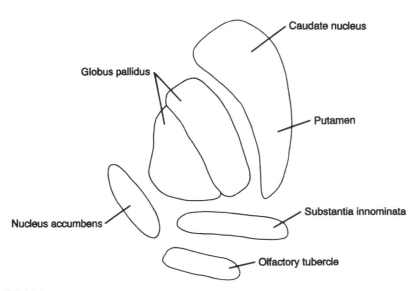

FIGURE 24-1. Rostrocaudally foreshortened diagram of major components of the striatopallidal complexes in mammals, schematically shown on the right side of the brain with lateral toward the right.

and putamen of mammals. A smaller medial group of cells, the **interstitial nucleus of the basal forebrain bundle,** may correspond to the ventral pallidum.

The area periventricularis ventrolateralis projects to the posterior tuberculum (Fig. 24-5). The dopaminergic neurons of the posterior tuberculum in anamniotes may be homologous to some of the caudal diencephalic (A11) and/or rostral tegmental (A9/A10) dopaminergic neurons in mammals (see Chapter 17). Some of the fibers in this projection contain the neuromodu-

lator substance P (SP), and others contain the neuromodulator, enkephalin (ENK). The posterior tuberculum dopaminergic (DA) cells project back to area periventricularis ventrolateralis, and this area also receives a projection of serotonin-containing (5HT) fibers from the superior raphe (SR).

In nonteleost ray-finned fishes, a number of different groups have been identified in the ventral part of the telencephalon that may be homologous to various striatal and septal forebrain components of amniotes. In the bichir, *Polypterus,* histochemical and connectional evidence suggests that a large, periventricular cell group, called the **dorsal nucleus of the ventral telencephalic area (Vd),** may be the homologue of the caudate-putamen (Fig. 24-6). It contains fibers rich in substance P, enkephalin, serotonin, and tyrosine hydroxylase (an indicator of catecholamines) and is high in cholinesterase. This nucleus may also be homologous as a field to the globus pallidus.

A nucleus dorsal to Vd, whimsically called **'nother nucleus of the ventral telencephalic area (Vn),** is similar to the caudate-putamen in histochemistry but of uncertain homology due to its location dorsal to Vd. The nucleus ventral to Vd, the **ventral nucleus of the ventral telencephalic area (Vv),** has reciprocal connections with the limbic part of the pallium.

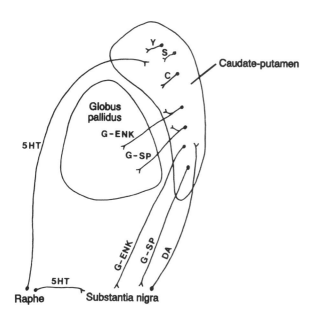

FIGURE 24-2. Some of neurotransmitters and neuromodulators of the dorsal striatopallidal complex and related brainstem nuclei in mammals. Abbreviations: C, acetylcholine; DA, dopamine; G-ENK, GABA (gamma-aminobutyric acid) co-localized with enkephalin; G-SP GABA co-localized with substance P; S, somatostatin; Y, neuropeptide Y; 5HT, serotonin (5-hydroxytryptamine).

FIGURE 24-3. Transverse hemisection with mirror-image drawing through the telencephalon of a lamprey. Adapted from Northcutt (1981). Redrawn, with permission, from the *Annual Review of Neuroscience,* Volume 4, © 1981, by Annual Reviews Inc.

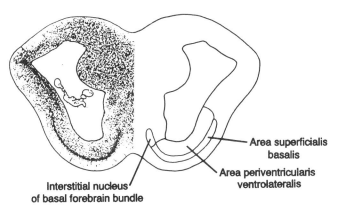

FIGURE 24-4. Transverse hemisection with mirror-image drawing through the telencephalon of a squalomorph shark (*Squalus acanthias*). Adapted from Northcutt et al. (1988).

The ventral nucleus maybe homologous as a field to septal nuclei, nucleus accumbens, and possibly part of the substantia innominata. A laterally lying nucleus, known as the **lateral nucleus of the ventral telencephalic area (Vl),** has connections and histochemistry similar to the medial septal nucleus and the olfactory tubercle. Additional studies of connections are needed to clarify the relationships of these nuclei.

In lungfishes, the ventral part of the telencephalon can be divided into a medial and a lateral region [Fig. 24-7(A)]. Histochemical findings indicate that the lateral region, called the **lateral subpallium (LS),** may be homologous as a field to all or most of the striatum of mammals, that is, the caudate-putamen, globus pallidus, and, in its caudal part, nucleus accumbens. Substance P-containing fibers have been traced from the lateral subpallium to a ventrolateral part of the tegmentum. The cells of this tegmental nucleus contain catecholamine (presumably dopamine) and project reciprocally to the lateral subpallium. The lateral subpallium is also innervated by serotonin-containing fibers from somewhere in the brainstem—presumably the raphe complex of the reticular formation.

In newts, serotonin- and dopamine-containing fibers are present in the striatum, and afferent projections arise from the putative homologue of the amygdala, dorsal thalamus, posterior tuberculum (dopaminergic), locus coeruleus (dopaminergic),

and raphe (serotoninergic). Projections to the striatum from a structure in the hypothalamus, called the periventricular organ, have also been found. Lesions of the periventricular organ disrupt stereotyped courtship behavior.

In frogs [Fig. 24-7(B and C)], dorsal and ventral parts of the striatum have been identified, as has a nucleus accumbens and an olfactory tubercle. A more caudal nucleus, the **anterior entopeduncular nucleus,** may also be part of the striatal region. The connections and histochemistry of the dorsal and ventral striatum and the anterior entopeduncular nucleus do not allow for strict correspondence with particular, respective parts of the striatum in mammals, however.

Unlike the situation in amniotes, the heaviest input to the striatum in frogs arises from the dorsal thalamic nuclei that are in receipt of sensory information from the midbrain roof. The anterior entopeduncular nucleus also projects to the striatum, particularly to its ventral part. Other inputs are much sparser and arise from the amygdala, the preoptic area, the posterior tuberculum, and the superficial isthmal reticular formation.

The dorsal and ventral parts of the striatum project to the anterior entopeduncular nucleus and, with nucleus accumbens and the anterior entopeduncular nucleus, to the tegmentum. The ventral striatum and anterior entopeduncular nucleus also project to the pretectum. The anterior entopeduncular nucleus projects to the optic tectum and the torus semicircularis in the midbrain roof. Areas in receipt of striatal projections in the tegmentum include the nucleus profundus mesencephali, the superficial isthmal reticular formation, and several nuclei within the tegmentum.

These efferent projections are the basis for feedback pathways to the major afferent source of the striatum, the midbrain roof-recipient dorsal thalamic nuclei (Fig. 24-8). The anterior entopeduncular nucleus, the pretectum, and areas in the tegmentum project to both the optic tectum and the torus semicircularis. Additionally, the isthmal reticular nucleus projects to the torus semicircularis, and nucleus profundus mesencephali projects to the optic tectum. The torus semicircularis and optic tectum project to the dorsal thalamic nuclei that, in turn, project to the striatum.

In frogs, the catecholaminergic (presumably dopaminergic) cells in the posterior tuberculum project to the striatum (Fig. 24-9). Efferent targets of the striatum, in turn, include two nuclei—nucleus profundus mesencephali and the superficial isthmal reticular formation—that may be homologous to the noncatecholaminergic pars reticulata of the substantia nigra. The efferent projections of these two nuclei to the optic tectum support this idea. The striatum in frogs does not receive any projection from the pallium, so no corticostriatal system like that in mammals is present. The striatum in frogs, and presumably in other amphibians, is thus somewhat the opposite of the striatum in amniotes, in which sensory projections to the striatum are minor and motor circuits predominate.

Group IIA

In the enlarged and elaborate telencephalon of hagfishes, the striatum is believed to be a cell group that lies deep to the pallial areas and rostrolateral (rather than ventrolateral as in other vertebrates) to the lateral ventricle (Fig. 24-10). This cell group receives projections from the posterior tuberculum, and

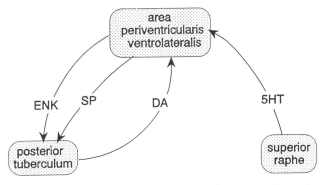

FIGURE 24-5. Diagram of some of the striatal connections in squalomorph sharks. Abbreviations: DA, dopamine; ENK, enkephalin; SP, substance P; 5HT, serotonin.

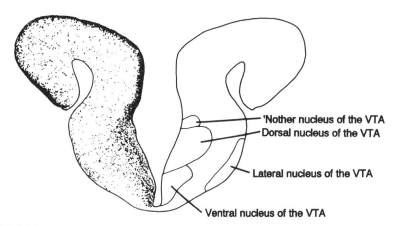

FIGURE 24-6. Transverse hemisection with mirror-image drawing through the telencephalon of a bichir (*Polypterus palmas*). Adapted from Reiner and Northcutt (1992). VTA = ventral telencephalic area.

fibers of a basal forebrain bundle arise and/or terminate within it.

In galeomorph sharks, skates, and rays, the distinctive lamina of cells that forms nucleus superficialis basalis in squalomorph sharks is present in the ventral part of the telencephalon and may be the homologue of nucleus accumbens, and possibly also the olfactory tubercle, of mammals. A more diffuse mass of cells dorsal to this lamina has been identified as the striatum and corresponds to the cell group named area periventricularis ventrolateralis in squalomorph sharks, the probable homologue

of the caudate-putamen of mammals. In skates (Fig. 24-11), but not in sharks, the striatum (area periventricularis ventrolateralis) has been found to receive ascending visual projections relayed from the midbrain roof through the dorsal thalamus and may receive other sensory input as well. Dopamine-containing fibers have been found in the striatum and nucleus superficialis basalis in rays. Somatostatin and neuropeptide Y have also been found in the basal telencephalon in sharks.

In teleost fishes (Fig. 24-12), hypertrophy of the telencephalon combined with the developmental process of eversion

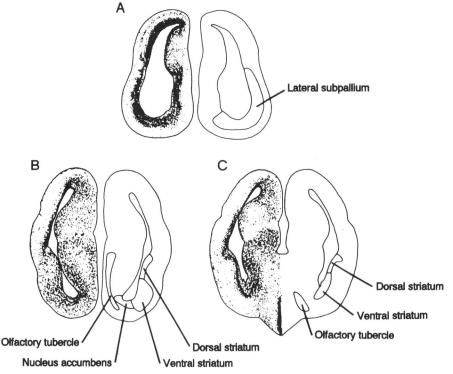

FIGURE 24-7. Transverse hemisections with mirror-image drawings through the telencephalon in (A) a lungfish (*Protopterus annectens*) and (B and C) a bullfrog (*Rana catesbeiana*). Adapted from Reiner and Northcutt (1987) and Northcutt and Kicliter (1980), respectively. Material from the latter used with permission of Plenum Publishing Corp.

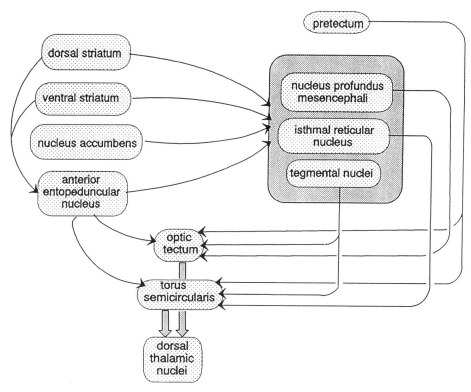

FIGURE 24-8. Some of the feedback pathways from the striatum to dorsal thalamic nuclei in frogs.

(see Chapter 3) makes topological comparisons difficult. Histochemical and connectional data are not yet adequate to completely clarify this situation. Three cell groups in the ventral telencephalon are present that correspond to cell groups in the bichir *Polypterus*: the dorsal (Vd), ventral (Vv), and lateral (Vl) nuclei of the ventral telencephalic area.

The dorsal nucleus of the ventral telencephalic area (Vd) may be homologous to the caudate-putamen (and possibly the globus pallidus) of land vertebrates in all ray-finned fishes.

In teleosts, it contains dopaminergic- and serotoninergic-containing fibers, substance P- and enkephalin-containing fibers and neurons, and GABAergic neurons. It is also rich in cholinesterase.The ventral nucleus (Vv) may be similarly homologous as a field to the septal nuclei, nucleus accumbens, and perhaps part of the substantia innominata. It has catecholamine-, serotonin-, substance P-, and enkephalin-containing fibers. The lateral nucleus (Vl) resembles the olfactory tubercle of mammals in having an olfactory input. Cholinergic neurons,

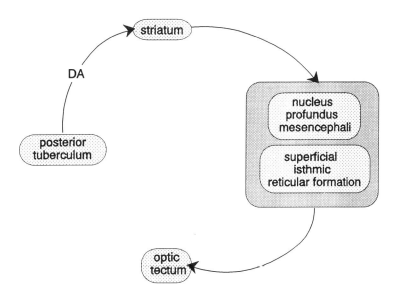

FIGURE 24-9. A motor pathway to the optic tectum in frogs.

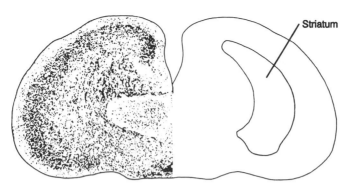

FIGURE 24-10. Transverse hemisection with mirror-image drawing through the telencephalon in a hagfish (*Eptatretus stouti*). Adapted from Wicht and Northcutt (1992). Used with permission of S. Karger AG.

somatostatin-containing neurons, and neuropeptide Y- containing neurons are present within Vl.

Two other parts of the telencephalon in teleosts have also been tentatively identified as part of the striatum on the basis of histochemistry: the **central zone (Dc)** and **medial zone (Dm) of the dorsal telencephalic area** (Fig. 24-12). Evidence is, however, ambiguous. These parts of the telencephalon contain high levels of cholinesterase in some teleosts but not in others. The medial zone has only low levels of substance P- and enkephalin-containing fibers. The central zone is known to project to the optic tectum. Both zones receive ascending sensory projections from the dorsal thalamus. These two telencephalic zones possibly are unique to teleosts, having been evolved solely within this radiation. They appear to be some type of elaboration of the dorsolateral pallium and/or striatal region. Whether they are homologues of part of the striatum or part of the dorsal ventricular ridge/lateral isocortex of amniotes or parts of both remains an unsolved problem for the present.

Group IIB

With the invasion of the land, changes occurred in the motor system to accommodate a new, more expanded set of movements. The striatal motor system governs the movements of the trunk and limbs. Much of its output is inhibitory in order to prevent all but the desired movements. As we discussed in Chapter 19, the striatopallidal complex has four major divisions in all amniote vertebrates. The dorsal part of the complex contains the dorsal striatum (caudate nucleus and putamen of mam-

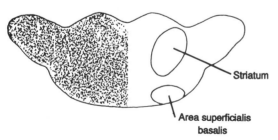

FIGURE 24-11. Transverse hemisection with mirror-image drawing through the telencephalon of a skate (*Raja eglanteria*). Adapted from Northcutt (1978).

mals) and the dorsal pallidum (globus pallidus of mammals). The ventral part of the complex contains the ventral striatum (olfactory tubercle and nucleus accumbens of mammals) and the ventral pallidum (a rostral part of the substantia innominata of mammals). Table 19-3 shows some of the terms that have been used in various amniotes.

The striatopallidal complexes are involved with circuits that interconnect pallial motor and premotor areas with the dorsal thalamus and with the subthalamus (i.e., the ventral thalamus), mesencephalic dopaminergic nuclei (substantia nigra and ventral tegmental area), and other related nuclei. The connections of the striatum with ventral thalamic nuclei are best understood in mammals, but some information on this circuit is also available for nonmammalian amniotes. We will briefly digress to cover the ventral thalamus in amniotes here and then return to a comparative view of the striatal systems.

The Ventral Thalamus. In mammals, three ventral thalamic nuclei are closely related to the striatum. Two of these nuclei, the **subthalamic nucleus** (of Luys) and the **zona incerta**, are shown in Figure 24-13. The third nucleus, the **nucleus of field H of Forel** (also known as the **subthalamic reticular nucleus**), lies at the level of the mesodiencephalic junctional region, immediately rostral to the red nucleus and ventrolateral to the ventral tegmental area. Forel's field H is composed of cells scattered among fiber tracts.

Specific portions of fiber tracts, most of which arise in the globus pallidus and loop through the subthalamic region, are also named. The **subthalamic fasciculus** consists of fibers that reciprocally connect the globus pallidus with the subthalamic nucleus. Fibers of the **lenticular fasciculus** (Fig. 24-13), which is also called **field H₂ of Forel,** arise from the most medial part of the globus pallidus and pass between the zona incerta and the subthalamic nucleus. Fibers of the **ansa lenticularis** arise from the inner segment of the globus pallidus and pass medial to the subthalamic nucleus. The latter two fascicles join and enter the **thalamic fasciculus** (Fig. 24-13), which is also called **field H₁ of Forel,** to project to the ventral nuclear group of the dorsal thalamus.

The subthalamic nucleus primarily serves as a relay from the globus pallidus to the substantia nigra and vice versa. It also projects back to the caudate-putamen. The zona incerta

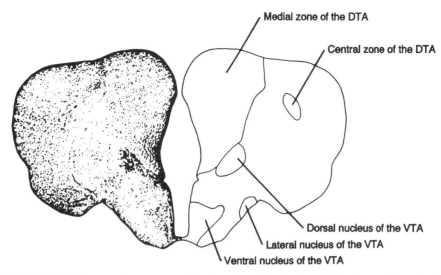

FIGURE 24-12. Transverse hemisection with mirror-image drawing through the telencephalon of a teleost (*Lepomis cyanellus*). Adapted from Northcutt and Davis (1983). DTA = dorsal telencephalic area; VTA = ventral telencephalic area.

receives projections from sensorimotor cortex, the ventral lateral geniculate nucleus, the trigeminal nuclear complex, cerebellar nuclei, and the spinal cord. Its projections are not well understood but include efferents to the pretectum, the posterior pituitary lobe, and the spinal cord. Forel's field H (the subthalamic reticular nucleus) receives input from the globus pallidus, spinal cord, and reticular formation. It projects to reticular nuclei in the brainstem and to neck motoneurons in the cerivcal spinal cord. As a whole, the ventral thalamic nuclei serve to link the globus pallidus and cortex with nuclei related to motor output or to motor feedback circuits.

The Mammalian Striatopallidal Complex. A transverse hemisection through part of the striatum in a primate (*Homo sapiens*) is shown in Figure 24-14. The caudate nucleus and putamen are pierced by fascicles of the internal capsule (I). The nuclei that comprise the ventral striatum, nucleus accumbens and the olfactory tubercle, lie in a medioventral position. At a more

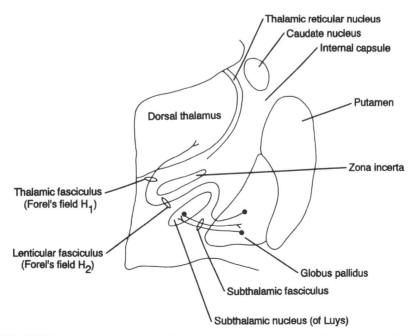

FIGURE 24-13. Semischematic drawing of a transverse hemisection through the right ventral thalamus of a primate (*Homo sapiens*). Dorsal is toward the top and lateral toward the right.

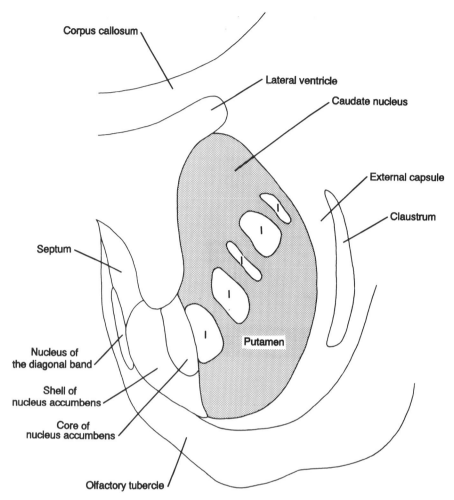

FIGURE 24-14. Semischematic drawing of a right transverse hemisection through the striatal area of a primate (*Homo sapeins*). Lateral is toward the right, and the midline is at the left. I = internal capsule. In this and Figure 25–15, the caudate nucleus and putamen are indicated by shading.

caudal level, as shown in Figure 24-15, the two segments of the globus pallidus are present, as is the ansa lenticularis, which carries efferent fibers from the internal segment of the globus pallidus to the diencephalon.

The dorsal striatum (also sometimes referred to as the **neo-striatum**) has a bicompartmental organization in that it contains a mosaic of **patches** within a **matrix,** as first defined in rats. The patches are regions of high μ opiate receptor binding within the matrix of lower receptor binding. Neurons within the patches receive input from prelimbic cortical areas and have reciprocal connections with the ventral part of the A9 (substantia nigra) dopaminergic cells in the tegmentum. Neurons within the matrix receive input from isocortical areas and from dopaminergic cells in the intermediate part of A9. They project to GABAergic neurons in the pars reticulata of the substantia nigra. A similar bicompartmental organization identified in cats and monkeys has been referred to as a **striosome/matrix** organization. Discrete zones within the matrix that receive somatotopically organized projections from sensory and motor cortices have been identified as **matrisomes.** Some of the neurons within the face and arm region of the putamen respond to visual as well as tactile stimuli and thus provide a bimodal map of space within the dorsal striatum.

Nucleus accumbens has two major parts: a **core** that is continuous cytoarchitectonically with the caudate-putamen and a **shell** region around the core. A rostral pole region is also recognized as a separate entity. While a bicompartmental organization (patch/matrix) has been identified in the dorsal striatum, the connectivity and neurochemical patterns within different areas of nucleus accumbens suggest a much more complex, multicompartmental organization.

The dorsal and ventral striatopallidal complexes of mammals both receive input from the diffusely projecting intralaminar nuclei of the dorsal thalamus. The dorsal striatum also receives projections from dorsal thalamic sensory relay nuclei. In addition to these inputs, both the dorsal and ventral complexes are characterized by involvement in three major sets of pathways that serve to regulate motor output.

The dorsal striatopallidal complex has one set of pathways that interconnect the dorsal striatum and pallidum with the substantia nigra, ventral tegmental area, and subthalamic nuclei (Fig. 24-16). Both the substantia nigra and the ventral tegmental area project to the caudate-putamen, and the caudate-putamen and globus pallidus project to the substantia nigra. The globus pallidus is reciprocally connected with the subthalamic nucleus.

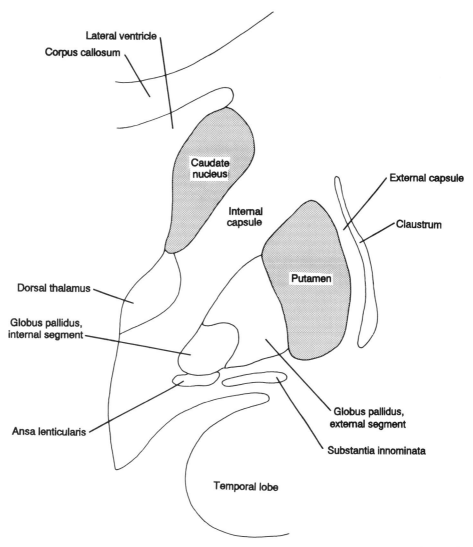

FIGURE 24-15. Semischematic drawing of a right transverse hemisection through the striatal area of a primate (*Homo sapeins*), similar to that shown in Figure 24–14 but at a more caudal level.

The amygdala is reciprocally connected with the substantia nigra and the ventral tegmental area and projects to both the globus pallidus and the caudate-putamen.

The second set of pathways of the dorsal striatopallidal complex is from isocortex through the dorsal striatum and dorsal pallidum to dorsal thalamic nuclei and then back to motor isocortex (Fig. 24-17), which gives rise to the long descending corticospinal tract. This feedback circuit of the dorsal striatopallidal complex is not only in one direction; the globus pallidus projects back to the caudate-putamen.

The third set of pathways involves descending projections via the tectum, substantia nigra, and nucleus tegmenti pedunculopontinus pars compacta (Fig. 24-16). The latter nucleus has widespread projections to the brainstem reticular formation, as well as to the thalamus and other parts of the forebrain. Some of the connections of the dorsal striatopallidal complex are shown in more detail in Figure 24-18, where the connections of the two parts of the substantia nigra are differentiated.

Of the three major circuits that involve the ventral striatopallidal complex, one is involved with interconnections with dopamine-containing, tegmental cell groups and the subthalamus (Fig. 24-19). The ventral tegmental area and substantia nigra both have reciprocal connections with the ventral striatum, and the ventral pallidum is reciprocally connected with the subthalamic nucleus. The ventral pallidum projects to the substantia nigra, the habenula, and the amygdala, which has reciprocal connections with the subthalamic nucleus as well as with both the ventral tegmental area and substantia nigra.

The second set of pathways of the ventral striatopallidal complex is from prefrontal, olfactory, and limbic cortices through the ventral striatum and ventral pallidum to a dorsal thalamic nucleus (the mediodorsal nucleus), and then back to limbic-related isocortex (Fig. 24-20). Like the cortical circuit involving the dorsal striatopallidal complex, this circuit is not only one way; the ventral pallidum projects reciprocally back to the ventral striatum.

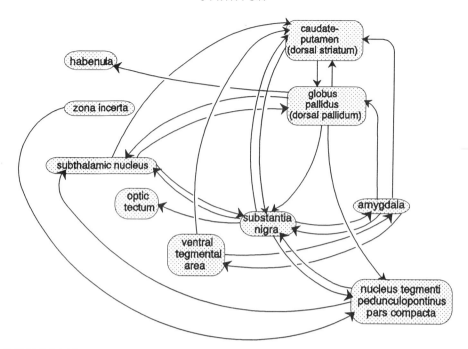

FIGURE 24-16. Some connections of the dorsal striatum and dorsal pallidum with brainstem structures in mammals.

The third set of pathways are descending projections from the ventral pallidum through nucleus tegmenti pedunculopontinus pars compacta (Fig. 24-19). The latter nucleus projects to the tectum via a relay in the substantia nigra (Fig. 24-16) and to the reticular formation.

The dorsal striatum and pallidum are thus more involved with isocortical pathways, while the ventral striatum and pallidum are more involved with limbic pathways. Nonetheless, inputs from the two systems interface in a number of different nuclei, including the habenula, amygdala, subthalamic nucleus, and tegmental nuclei, so that integration of information can occur. One of the most important of such integrative pathways is a projection from nucleus accumbens (primarily its core) to the dopaminergic neurons of the substantia nigra pars compacta, which in turn project directly to the caudate-putamen. Via this route, the ventral, more limbic parts of the striatum, which receive limbic cortical inputs, can affect the dorsal, more motor parts of the striatum.

The habenula receives projections from the ventral tegmental area (with its input from the ventral striatum), and directly from the dorsal pallidum. The amygdala receives projections from the ventral pallidum, has reciprocal connections with the ventral tegmental area and substantia nigra, and projects to both the dorsal striatum and the ventral striatum. The subthalamic nucleus has reciprocal connections with the dorsal pallidum, substantia nigra, and amygdala, and it projects to the dorsal striatum. In the tegmentum, the substantia nigra, ventral tegmental area, and nucleus tegmenti pedunculopontinus pars compacta are all involved in multiple interconnections of the dorsal and ventral striatopallidal complexes.

Note that the projections shown here for both striatopallidal complexes are only some of those involved in this complex system. For example, isocortical projections to brainstem nuclei such as the ventral tegmental area have not been touched on, cortical projections of the dorsal striatum have been omitted, and a number of other connections and interconnections are also not included. We have chosen only some of the major pathways to illustrate the basic organization of the system here and have taken the risk of giving only a partial picture with the hope of not becoming lost in too much detail.

A multitude of various excitatory and inhibitory neurotransmitters and neuromodulators is present in the striatopallidal complexes, allowing for complex interactions in processing and integrating afferent information. Some of the various neurons and connections for which particular neurochemicals have been identified in the dorsal striatopallidal complex and related

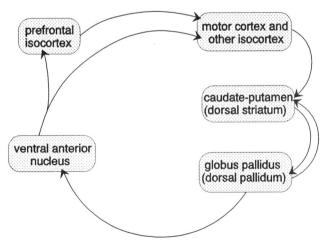

FIGURE 24-17. Some connections of the dorsal striatum and dorsal pallidum with isocortex in mammals.

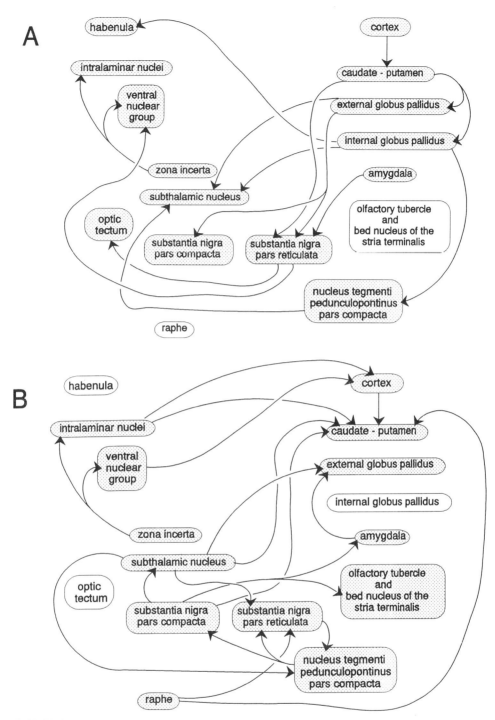

FIGURE 24-18. Some connections of the dorsal striatum, dorsal pallidum, substantia nigra, and related structures in mammals. (A) emphasizes output pathways from the striatopallidal structures and (B) input pathways to them.

nuclei are shown in Figure 24-21. Isocortical afferent fibers contain the excitatory transmitter glutamate (Gl), and afferents from the raphe contain serotonin (5HT). Various striatal neurons contain GABA, enkephalin, substance P, neuropeptide Y, somatostatin, and acetylcholine. More than one of these transmitters may be present (co-localized) within a single neuron. As shown in Figure 24-2, the ascending fibers projecting from

the substantia nigra and the ventral tegmental area contain dopamine, while the reciprocal, striatonigral fibers contain GABA co-localized with either enkephalin or substance P.

The Striatopallidal Complex in Nonmammalian Amniotes. In diapsid reptiles, birds, and turtles, studies of connections allow for tentative identification of homologues of mammalian sub-

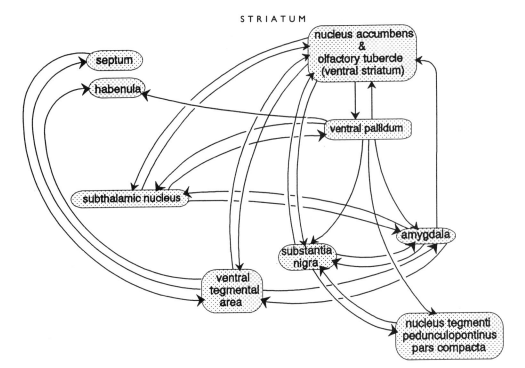

FIGURE 24-19. Some connections of the ventral striatum and ventral pallidum with brainstem structures in mammals.

thalamic nuclei in some cases. In lizards and crocodiles, a nucleus called the **anterior entopeduncular nucleus** is probably the homologue of the subthalamic nucleus of mammals. In birds, this nucleus is called the **anterior nucleus of the ansa lenticularis.** In birds, a possible homologue of the nuclei of the fields of Forel has also been identified and is called the **area ventromedialis thalami.**

The striatopallidal complex of diapsid reptiles and turtles occupies the ventrolateral part of the telencephalon, ventral to the dorsal ventricular ridge (Figs. 24-22 and 24-23). The dorsal striatum is often referred to simply as the striatum, as shown in the figures, and is homologous to the caudate nucleus and putamen of mammals. Structures that have been identified as the globus pallidus (dorsal pallidum) and the ventral pallidum

lie ventral to part of the dorsal striatum, while nucleus accumbens (part of the ventral striatum) lies in a more rostromedial position.

The striatopallidal complexes both receive input from a nucleus in the dorsal thalamus, called nucleus dorsomedialis, that has diffuse projections to many parts of the telencephalon. The dorsal striatum receives input from dorsal thalamic sensory relay nuclei. In addition to these two sets of inputs, each of the striatopallidal complexes is characterized by involvement in three major sets of pathways that all serve to regulate motor output.

The dorsal striatopallidal complex has one set of pathways that interconnect the complex with tegmental dopamine-containing nuclei and with the subthalamus [Fig. 24-24(A)]. Both the ventral tegmental area and the substantia nigra, which

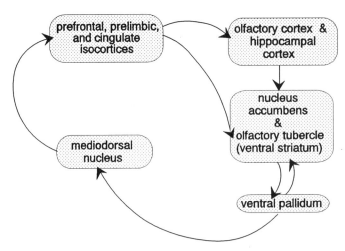

FIGURE 24-20. Some connections of the ventral striatum and ventral pallidum with cortical areas in mammals.

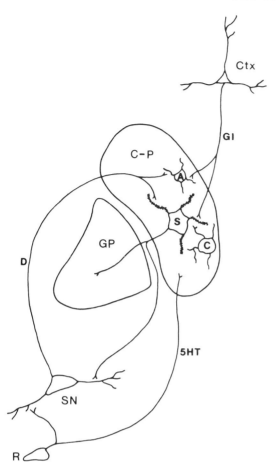

FIGURE 24-21. Cerebral cortical neurons (Ctx) with the excitatory transmitter glutamate (Gl) synapse on two populations of caudate-putamen (C-P) neurons—those with spiny dendrites (S) and those with smooth dendrites. The smooth-dendrite (aspiny) neurons (A) contain the inhibitory transmitter GABA or the modulators neuropeptide Y or somatostatin. Dopamine-containing axons (D) from the substantia nigra (SN) synapse on both spiny- and aspiny-dendrite neurons. Large neurons (C) contain the transmitter acetylcholine and regulate the balance of GABA and dopamine. The neurons of the raphe (R) contain serotonin (5HT) and project to the substantia nigra as well as to the caudate-putamen. Cells with spiny dendrites in the caudate-putamen project to the globus pallidus (GP) and substantia nigra. These cells contain GABA and/or the modulators enkephalin or substance P.

have dopamine-containing neurons, project to the dorsal striatum. The dorsal striatum and the globus pallidus project to the substantia nigra, which in turn projects to the optic tectum. Descending projections arise from the latter to motor centers in the brainstem and to the upper part of the spinal cord. The dorsal striatopallidal complex is also interconnected with the anterior entopeduncular nucleus, the homologue of the subthalamic nucleus of mammals.

Nonmammalian amniotes do not have cortical neurons that project directly to the spinal cord as found in mammals. In these animals, telencephalic control of voluntary movements is via descending pathways through the reticular formation and other brainstem sites. Nonetheless, a second set of pathways from the dorsal striatum through the dorsal thalamus to part of the dorsal cortex and back to the dorsal striatum may be present

in diapsid reptiles, although further work is needed to verify it. The globus pallidus projects to a nucleus called area ventromedialis thalami [Fig. 24-24(A)], which also receives cerebellar input and lies immediately next to the nucleus that receives dorsal column somatosensory input. The area ventromedialis thalami projects to part of the telencephalon near or within the somatosensory region of the dorsal ventricular ridge or to the dorsal cortex or to both. Cells in the same part of dorsal cortex then project to the dorsal striatum.

The third set of pathways from the dorsal striatopallidal complex is through a nucleus in the pretectum, called the dorsal nucleus of the posterior commissure, and thence to the optic tectum [Fig. 24-24(A)] and through other brainstem nuclei to the optic tectum and the reticular formation. The pathway through the pretectum is present in turtles, crocodiles, and some—but not all—lizards.

One of the three major sets of pathways that characterize the ventral striatopallidal complex in diapsid reptiles and turtles is the interconnections of the complex with tegmental dopamine-containing nuclei and with the subthalamus. Nucleus accumbens receives projections from the ventral tegmental area, and both nucleus accumbens and the ventral pallidum project back to the ventral tegmental area [Fig. 24-24(B)]. The substantia nigra is not interconnected with the ventral striatopallidal complex, however. The ventral pallidum projects to the ventromedial nucleus, which may be a homologue of field H of Forel, and to the habenula in the epithalamus.

The ventral striatopallidal complex in diapsid reptiles and turtles is not known to be part of a pathway from the ventral pallidum through the dorsal thalamus to the dorsal pallium, as is present in mammals and may be present in birds. Instead, nucleus accumbens receives projections from the same part of the dorsal cortex as does the dorsal striatum. Additional studies are needed on possible dorsal pallial and descending projections of this system.

The dorsal striatopallidal complex and ventral striatopallidal complex interface in a number of nuclei that receive projections from both complexes. The ventral tegmental area is the best analyzed of these points of interface in diapsid reptiles and turtles. It receives projections from the ventral striatopallidal complex and projects to both the dorsal and ventral striatum (Fig. 24-24). The fibers that arise in the substantia nigra and ventral tegmental area and project to the paleostriatal complexes in diapsid reptiles and turtles contain dopamine, and ascending projections of the raphe are serotoninergic. Enkephalin, GABA, dopamine, acetylcholinesterase, neuropeptide Y, and other neurochemicals similar to those in mammals are present within the paleostriatal complexes. The distribution of these neurochemicals correlates with the homologies of the various parts of these complexes indicated by connectional studies. Similar histochemical results have been found for this system in birds and will not be discussed separately below.

In birds, the four components of the striatopallidal complexes are the **paleostriatum augmentatum** (dorsal striatum), the **paleostriatum primitivum** (dorsal pallidum), the **ventral part of lobus parolfactorius** (ventral striatum), and the ventral pallidum (Figs. 24-25 and 24-26). A nucleus called **nucleus intrapeduncularis** lies in the central part of the paleostriatum primitivum. Whether this nucleus is a rostral thalamic cell group

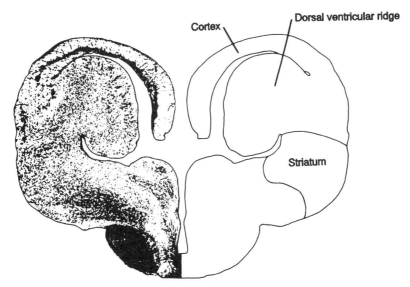

FIGURE 24-22. Transverse hemisection with mirror-image drawing through the telencephalon of a turtle. Adapted from Balaban and Ulinski (1981).

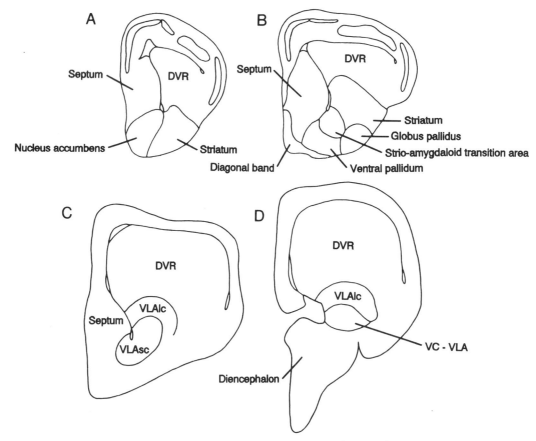

FIGURE 24-23. Semischematic drawings of transverse hemisections through the right telencephalon in (A and B) a lizard and (C and D) a crocodile. VC-VLA = ventrocaudal portion of VLAlc; VLAlc = large-celled part of ventrolateral area; VLAsc = small-celled part of ventrolateral area. Adapted from Russchen and Jonker (1988) and Brauth and Kitt (1980), respectively.

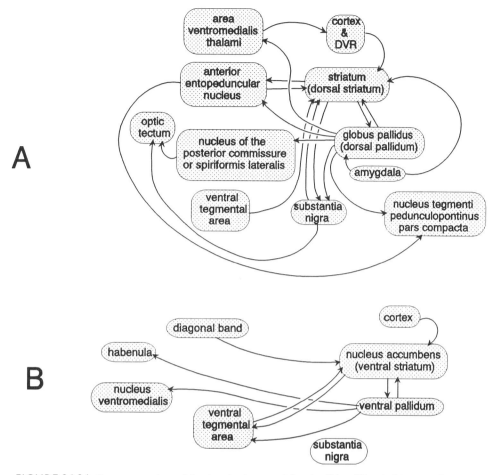

A

B

FIGURE 24-24. Some connections of the dorsal striatum and dorsal pallidum (A) and of the ventral striatum and ventral pallidum (B) in diapsid reptiles and turtles.

or part of the dorsal pallidum, as its projection to the habenula (Fig. 24-25) suggests, has not been resolved.

The dorsal and ventral striatopallidal complexes in birds both receive input from a medially lying group of cells in the dorsal thalamus that gives rise to diffuse projections to wide areas of the telencephalon. This group of cells appears to be homologous to the nucleus dorsomedialis of diapsid reptiles and turtles and the intralaminar group of nuclei in mammals. The dorsal striatum also receives projections from dorsal thalamic sensory relay nuclei. As in diapsid reptiles and turtles, the striatopallidal complexes of birds are each characterized by involvement in three major sets of pathways.

The dorsal striatopallidal complex is involved in one set of pathways that interconnect the complex with tegmental dopaminergic cell groups and with the subthalamus (Fig. 24-25). The substantia nigra (confusingly sometimes called nucleus tegmenti pedunculopontinus in birds) receives projections from both the dorsal striatum (paleostriatum augmentatum) and the dorsal pallidum (paleostriatum primitivum) and projects back to the dorsal striatum. The dorsal pallidum has reciprocal connections with the anterior nucleus of the ansa lenticularis (homologous to the subthalamic nucleus of mammals).

The second major set of pathways of the dorsal striatopallidal complex of birds is a circuit through the dorsal thalamus

and dorsal pallium. This circuit is from the dorsal pallidum to a nucleus in the dorsal thalamus, nucleus dorsalis intermedius posterior. The latter nucleus projects to part of the dorsal pallium called the **neostriatum intermediale,** which relays through two more dorsal pallial cell groups: the **neostriatum intermedium laterale** and **archistriatum intermedium.** The archistriatum intermedium projects back to the dorsal striatopallidal complex.

The third set of pathways of the dorsal striatopallidal complex is a projection system to the optic tectum (Fig. 24-25) and brainstem via the substantia nigra and other brainstem nuclei. A nucleus in the dorsal thalamus, called **nucleus spiriformis lateralis,** receives projections from the paleostriatum primitivum (dorsal pallidum), the anterior nucleus of the ansa lenticularis, and the substantia nigra and projects to the optic tectum. Nucleus spiriformis lateralis appears to be homologous to the dorsal nucleus of the posterior commissure of diapsid reptiles.

One of the three sets of pathways of the ventral striatopallidal complex involves interconnections with dopamine-containing cell groups in the tegmentum and with the subthalamus. Subthalamic connections of the complex remain to be studied in birds, but reciprocal connections of nucleus accumbens with both the ventral tegmental area and the substantia nigra are known (Fig. 24-26).

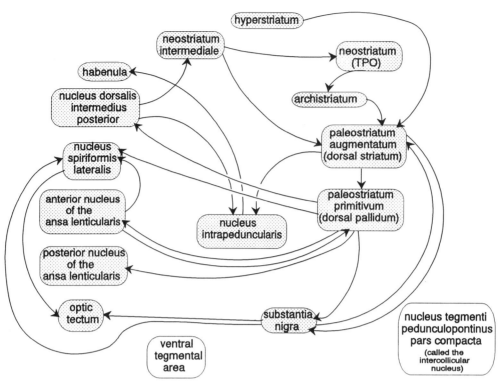

FIGURE 24-25. Some connections of the dorsal striatum and dorsal pallidum in birds.

The second major set of pathways of the ventral striatopallidal complex—a pathway through the dorsal thalamus and dorsal pallium—has not yet been found in birds, but some findings indicate that it may be present. As we discussed above, a nucleus in the dorsal thalamus in mammals receives projections from the ventral pallidum and projects to prefrontal cortex, which also receives a dopaminergic input from the tegmentum. A dorsal pallial area, called the **posterodorsolateral telencephalic field** (Fig. 24-26), that is similar to the mammalian prefrontal cortex in its dopaminergic innervation has been found in pigeons, but a

possible pathway to this area from the ventral pallidum through the dorsal thalamus has not yet been investigated.

The third set of pathways of the ventral striatopallidal complex would be expected to consist of descending pathways from the complex to brainstem nuclei other than the homologues of the ventral tegmental area and substantia nigra. These projections have not yet been investigated in birds.

The dorsal and ventral striatopallidal complexes of birds interface in a number of nuclei. One of these interfaces is the habenula, which receives projections from the paleostriatum

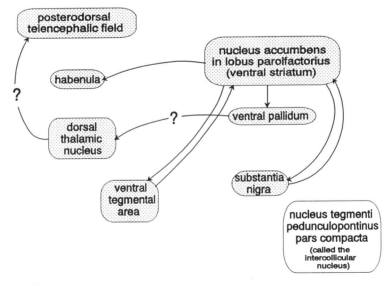

FIGURE 24-26. Some connections of the ventral striatum and ventral pallidum in birds.

augmentatum (the dorsal striatum), relayed through nucleus intrapeduncularis (Fig. 24-25), and from nucleus accumbens (the ventral striatum) (Fig. 24-26). A second interface is the homologue of the substantia nigra, which has reciprocal connections with both nucleus accumbens (Fig. 24-26) and the paleostriatum augmentatum (Fig. 24-25), in addition to an input from the paleostriatum primitivum (the dorsal pallidum, Fig. 24-25).

MAGNOCELLULAR CORTICOPETAL CELL COMPLEX

The magnocellular corticopetal cell complex (or nucleus basalis) is one of three major groupings of cholinergic neurons in the basal forebrain of amniotes. In addition to nucleus basalis, cholinergic neurons are distributed within the medial septum and the diagonal band. These three groups of cholinergic neurons project widely to cortical areas, with the septal and diagonal band neurons projecting to limbic and olfactory cortices and the basalis neurons projecting to isocortex and its dorsal pallial homologues. Those in the medial septum project to the hippocampus in mammals and to the medial cortex in diapsid reptiles and turtles. In mammals, the diagonal band has a vertical limb, and the cholinergic cells located there project to the hippocampus. The diagonal band also has a horizontal limb, and the cholinergic cells located there project to olfactory cortex. In lizards, cholinergic cells in the diagonal band have been found to project to the lateral (olfactory) cortex, but a diagonal band projection to the medial cortex appears to be absent.

Cholinergic neurons within nucleus basalis project to all areas of isocortex in mammals. In turtles, similarly located cholinergic neurons that lie within the ventral paleostriatum and globus pallidus have been found to project to the dorsal cortex. This group of cholinergic neurons receives inputs from nucleus accumbens in mammals and the corresponding area d in turtles (see Fig. 19-29).Cholinergic cell groups that correspond to all three basal forebrain areas in mammals have been identified in lizards as well.

While less data are available on cholinergic neurons in the basal telencephalon of anamniotes, such neurons may be widely distributed among vertebrates. Whether they tend to occur in three major regions, as in amniotes, has not yet been determined. Cholinergic neurons have been identified in the ventrolateral zone of the area ventralis (Vv) in the telencephalon of a cyprinid teleost. The ventrolateral zone may be homologous to the olfactory tubercle of land vertebrates. Cholinergic neurons are also present in the striatal and septal regions in frogs.

EVOLUTIONARY PERSPECTIVE

A striatopallidal area, with dorsal and ventral divisions, in the ventral part of the telencephalon has been identified in most major radiations of vertebrates. Some neuroscientists refer to part or all of the dorsal striatopallidal complex of mammals as the "neostriatum" and part or all of the ventral striatopallidal complex as the "paleostriatum," in the belief that the former was evolved more recently than the latter. On the contrary, the wide distribution of both dorsal and ventral divisions of the

complex indicate that both are plesiomorphic for at least jawed vertebrates and equally old in phylogenetic terms.

Even though only fragmentary information is available on the presence of various neurochemicals in the striatal area of some groups of vertebrates, the data currently available suggest that the transmitters and modulators identified in mammals may all be present in most or all vertebrate groups. Thus, the evolution of the complex roles in motor control that the striatal system plays in land vertebrates may not be explicable as simply the addition of new pathways with new transmitters for new interactions within the system. The pathways and nuclei associated with various parts of the striatal area in anamniotes need much additional research for us to be able to identify the specific changes that have occurred in the striatal system within and between the major radiations of anamniotes and to compare that variation with striatal system organization in various land vertebrates.

The set of dopaminergic projections from either the posterior diencephalon and/or the ventral part of the midbrain to the striatum is widespread among vertebrates, as discussed above in Chapter 17. In Group I vertebrates, dopamine-containing neurons are present in the posterior part of the diencephalon within the posterior tuberculum but not farther caudally in any part of the mesencephalon. In frogs, the dopamine-containing tubercular neurons remain in a relatively medial position but are known to give rise to ascending projections to the telencephalon. In amniotes, populations of mesencephalic dopamine-containing cells migrate laterally to form the pars compacta of the substantia nigra, while others remain more medially in the ventral tegmental area. A similar lateral migration of dopamine-containing cells occurs in the mesencephalon of Group II cartilaginous fishes. The question of the evolutionary origin of the midbrain dopaminergic cell groups in both Group II cartilaginous fishes and amniotes awaits further embryological and connectional data.

Cholinergic neurons are present in three regions of the basal telencephalon in amniotes—the medial septum, diagonal band, and nucleus basalis—and these cell groups are thus plesiomorphic for at least amniotes. Nucleus basalis, also known as the magnocellular corticopetal cell complex, has wide projections to isocortex in mammals and projects to dorsal cortex in the nonmammalian amniotes that have been studied. Cholinergic cells that may correspond to one or more of the basal telencephalic cholinergic groups of amniotes have also been found in some anamniotes, but more data is needed on their distribution and projection patterns before their evolutionary history among all of the extant vertebrate groups can be clarified.

FOR FURTHER READING

Baskerville, K. A., Chang, H. T., and Herron, P. (1993) Topography of cholinergic afferents from the nucleus basalis of Meynert to representational areas of sensorimotor cortices in the rat. *Journal of Comparative Neurology,* 335, 552–562.

Gerfen, C. R. (1992) The neostriatal mosaic: multiple levels of compartmental organization. *Trends in Neurosciences,* 15, 133–138.

Heimer, L., Switzer, R. D., and Van Hoesen, G. W. (1982) Ventral striatum and ventral pallidum: components of the motor system? *Trends in Neurosciences,* 5, 83–87.

Parent, A. (1986) *Comparative Neurology of the Basal Ganglia*. New York: Wiley.

Powers, A. S. and Reiner, A. (1993) The distribution of cholinergic neurons in the central nervous system of turtles. *Brain, Behavior and Evolution, 41,* 326–345.

Smeets, W. J. A. J. and Reiner, A. (eds.) (1994) *Phylogeny and Development of Catecholamine Systems in the CNS of Vertebrates*. Cambridge, England: Cambridge University Press.

ADDITIONAL REFERENCES

Anderson, K. D. and Reiner, A. (1991) Striatonigral projection neurons: a retrograde labeling study of the percentages that contain substance P or enkephalin in pigeons. *Journal of Comparative Neurology, 303,* 658–673.

Balaban, C. D. and Ulinski, P. S. (1981) Organization of thalamic afferents to anterior dorsal ventricular ridge in turtles. I. Projections of thalamic nuclei. *Journal of Comparative Neurology, 200,* 95–129.

Beckstead, R. M. (1984) The thalamostriatal projection in the cat. *Journal of Comparative Neurology, 223,* 313–346.

Beckstead, R. M., Domesick, V. B., and Nauta, W. J. H. (1979) Efferent connections of the substantia nigra and ventral tegmental area in the rat. *Brain Research, 175,* 191–217.

Berendse, H. W., Groenewegen, H. J., and Lohman, A. H. M. (1992) Compartmental distribution of ventral striatal neurons projecting to the mesencephalon in the rat. *Journal of Neuroscience, 12,* 2079–2103.

Bigl, V., Woolf, N. J., and Butcher, L. L. (1982) Cholinergic projections from the basal forebrain to frontal, parietal, temporal, occipital, and cingulate cortices: a combined fluorescent tracer and acetylcholinesterase analysis. *Brain Research Bulletin, 8,* 727–749.

Bissoli, R., Contestabile, A., Niso, R., and Szabo, T. (1989) Regional levels of cholinergic, GABAergic and excitatory amino acidic transmitters in fish telencephalon. *Comparative Biochemistry and Physiology, 93C,* 317–320.

Brauth, S. E. (1988) The organization and projections of the paleostriatal complex of *Caiman crocodilus*. In W. K. Schwerdtfeger and W. J. A. J. Smeets (eds.), *The Forebrain of Reptiles: Current Concepts of Structure and Function*. Basel: Karger, pp. 60–76.

Brauth, S. E. (1990) Histochemical strategies in the study of neural evolution. *Brain, Behavior and Evolution, 36,* 100–115.

Brauth, S. E., Ferguson, J. L., and Kitt, C. A. (1978) Prosencephalic pathways related to the paleostriatum of the pigeon (*Columba livia*). *Brain Research, 147,* 205–221.

Brauth, S. E. and Kitt, C. A. (1980) The paleostriatal system of *Caiman crocodilus*. *Journal of Comparative Neurology, 189,* 437–465.

Brog, J. S., Salyapongse, A., Deutch, A. Y., and Zahm, D. S. (1993) The pattern of afferent innervation of the core and shell in the "accumbens" part of the rat ventral striatum: immunohistochemical detection of retrogradely transported fluoro-gold. *Journal of Comparative Neurology, 338,* 255–278.

Chiba, A. and Honma, Y. (1992) Distribution of neuropeptide Y-like immunoreactivity in the brain and hypophysis of the cloudy dogfish, *Scyliorhinus torazame*. *Cell and Tissue Research, 268,* 453–461.

Chiba, A., Honma, Y., Ito, S., and Homma, S. (1989) Somatostatin-immunoreactivity in the brain of the gummy shark, *Mustelus manazo* Bleeker, with special regard to the hypothalamo-hypophyseal system. *Biomedical Research, 10,* Suppl. 3, 1–12.

Ciani, F., Franceschini, V., and Del Grande, P. (1988) Histochemical and biochemical study on the acetylcholinesterase and choline acetyltransferase in the brain and spinal cord of frog, *Rana esculenta*. *Journal für Hirnforschung, 29,* 157–163.

Corio, M., Thibault, J., and Peute, J. (1992) Distribution of catecholaminergic and serotoninergic systems in forebrain and midbrain of the newt, *Triturus alpestris* (Urodela). *Cell and Tissue Research, 268,* 377–387.

Domesick, V. B. (1988) Neuroanatomical organization of dopamine neurons in the ventral tegmental area. *Annals of the New York Academy of Sciences, 537,* 10–26.

Druga, R., Rokyta, R., and Benes, V., Jr. (1991) Thalamocaudate projections in the macaque monkey (a horseradish peroxidase study). *Journal für Hirnforschung, 32,* 765–774.

Dube, L., Clairambault, P. and Malacarne, G. (1990) Striatal afferents in the newt *Triturus cristatus*. *Brain, Behavior and Evolution, 35,* 212–226.

Dunnett, S. B., Everitt, B. J., and Robbins, T. W. (1991) The basal forebrain—cortical cholinergic system: interpreting the functional consequences of excitotoxic lesions. *Trends in Neurosciences, 14,* 494–501.

Echteler, S. M. and Saidel, W. M. (1981) Forebrain connections in the goldfish support telencephalic homologies with land vertebrates. *Science, 212,* 683–685.

Ekström, P. (1987) Distribution of choline acetyltransferase immunoreactive neurons in the brain of a cyprinid teleost (*Phoxinus phoxinus* L.). *Journal of Comparative Neurology, 256,* 494–515.

Flaherty, A. W. and Graybiel, A. M. (1993) Two input systems for body representations in the primate striatal matrix: experimental evidence in the squirrel monkey. *Journal of Neuroscience, 13,* 1120–1137.

Goldman, P. S. and Nauta, W. J. H. (1977) An intricately patterned prefrontal-caudate projection in the rhesus monkey. *Journal of Comparative Neurology, 171,* 369–386.

Goldman-Rakic, P. S. and Selemon, L. D. (1990) New frontiers in basal ganglia research. *Trends in Neurosciences, 13,* 241–244.

Gonzalez, A. and Russchen, F. T. (1988) Connections of the basal ganglia in the lizard *Gekko gecko*. In W. K. Schwerdtfeger and W. J. A. J. Smeets (eds.), *The Forebrain of Reptiles: Current Concepts of Structure and Function*. Basel: Karger, pp. 50–59.

Gonzalez, A., Russchen, F. T., and Lohman, A. H. M. (1990) Afferent connections of the striatum and the nucleus accumbens in the lizard *Gekko gecko*. *Brain, Behavior and Evolution, 36,* 39–58.

Graybiel, A. M. (1978) Organization of the nigrotectal connection: an experimental study in the cat. *Brain Research, 143,* 339–348.

Graybiel, A. M. (1990) Neurotransmitters and neuromodulators in the basal ganglia. *Trends in Neurosciences, 13,* 243–254.

Graziano, M. S. A. and Gross, C. G. (1993) A bimodal map of space: somatosensory receptive fields in the macaque putamen with corresponding visual receptive fields. *Experimental Brain Research, 97,* 96–109.

Greenberg, N., MacLean, P. D., and Ferguson, J. L. (1979) Role of the paleostriatum in species-typical display behavior of the lizard (*Anolis carolinensis*). *Brain Research, 172,* 229–241.

Groenewegen, H. J. and Berendse, H. W. (1990) Connections of the subthalamic nucleus with limbic-innervated parts of the basal ganglia. A neuroanatomical tracing study in the rat. *Journal of Comparative Neurology, 294,* 607–622.

Groenewegen, H. J., Meredeth, G. E., Berendse, H. W., Voorn, P., and Wolters, J. G. (1989) The compartmental organization of the ventral striatum in the rat. In A. R. Crossman and M. A. Sambrook (eds.), *Neural Mechanisms in Disorders of Movement*. London: John Libbey, pp. 45–54.

Haber, S. N., Lynd-Balta, E., and Mitchell, S. J. (1993) The organization of the descending ventral pallidal projections in the monkey. *Journal of Comparative Neurology*, 329, 111–128.

Hallanger, A. E. and Wainer, B. H. (1988) Ascending projections from the pedunculopontine tegmental nucleus and the adjacent mesopontine tegmentum in the rat. *Journal of Comparative Neurology*, 274, 483–515.

Hallanger, A. E., Levey, A. I., Lee, H. J., Rye, D. B., and Wainer, B. H. (1987) The origins of cholinergic and other subcortical afferents to the thalamus in the rat. *Journal of Comparative Neurology*, 262, 105–124.

Hay-Schmidt, A. and Mikkelsen, J. D. (1992) Demonstration of a neuronal projection from the entopeduncular nucleus to the substantia nigra of the rat. *Brain Research*, 576, 343–347.

Hedreen, J. C. (1977) Corticostriatal cells identified by the peroxidase method. *Neuroscience Letters*, 4, 1–7.

Herkenham, M. and Nauta, W. J. H. (1977) Afferent connections of the habenular nuclei in the rat. A horseradish peroxidase study with a note on the fiber-of-passage problem. *Journal of Comparative Neurology*, 173, 123–146.

Herkenham, M. and Pert, C. B. (1981) Mosaic distribution of opiate receptors, parafascicular projections and acetylcholinesterase in rat striatum. *Nature (London)*, 291, 415–418.

Hoogland, P. V. (1977) Efferent connections of the striatum in *Tupinambis nigropunctatus. Journal of Morphology*, 152, 229–246.

Hoogland, P. V. and Vermeulen-Van der Zee, E. (1990) Distribution of choline acetyltransferase immunoreactivity in the telencephalon of the lizard *Gekko gecko. Brain, Behavior and Evolution*, 36, 378–390.

Hoover, J. E. and Strick, P. L. (1993) Multiple output channels in the basal ganglia. *Science*, 259, 819–821.

Hornby, P. J. and Piekut, D. T. (1988) Immunoreactive dopamine β-hydroxylase in neuronal groups in the goldfish brain. *Brain, Behavior and Evolution*, 32, 252–256.

Isa, T. and Sasaki, S. (1992) Descending projections of Forel's field H neurones to the brain stem and the upper cervical spinal cord in the cat. *Experimental Brain Research*, 88, 563–579.

Isa, T. and Sasaki, S. (1992) Mono- and disynaptic pathways from Forel's field H to dorsal neck motoneurones in the cat. *Experimental Brain Research*, 88, 580–593.

Jones, E. G. (1985) *The Thalamus*. New York: Plenum.

Jongen-Rêlo, A. L., Groenewegen, H. J., and Voorn, P. (1993) Evidence for a multi-compartmental histochemical organization of the nucleus accumbens in the rat. *Journal of Comparative Neurology*, 337, 267–276.

Karten, H. J. and Dubbeldam, J. L. (1973) The organization and projections of the paleostriatal complex in the pigeon (*Columba livia*). *Journal of Comparative Neurology*, 148, 61–90.

Kita, H. and Kitai, S. T. (1986) Two distinct enkephalin afferents to the substantia nigra in the rat: light and electron microscopic studies of immunohistochemically labeled terminals. *Society for Neuroscience Abstracts*, 12, 653.

Kitt, C. A. and Brauth, S. E. (1981) Projections of the paleostriatum upon the midbrain tegmentum in the pigeon. *Neuroscience*, 6, 1551–1566.

Kitt, C. A. and Brauth, S. E. (1982) A paleostriatal-thalamic-telencephalic path in pigeons. *Neuroscience*, 7, 2735–2751.

Kitt, C. A. and Brauth, S. E. (1986) Telencephalic projections from midbrain and isthmal cell groups in the pigeon. II. The nigral complex. *Journal of Comparative Neurology*, 247, 92–110.

Kubota, Y. and Kawaguchi, Y. (1993) Spatial distributions of chemically identified intrinsic neurons in relation to patch and matrix compartments of rat neostriatum. *Journal of Comparative Neurology*, 332, 499–513.

Kuo, H. and Chang, H. T. (1992) Ventral pallido-striatal pathway in the rat brain: a light and electron microscopic study. *Journal of Comparative Neurology*, 321, 626–636.

Lapper, S. R., Smith, Y., Sadikot, A. F., Parent, A., and Bolam, J. P. (1992) Cortical input to parvalbumin-immunoreactive neurones in the putamen of the squirrel monkey. *Brain Research*, 580, 215–224.

Lin, C.-S., May, P. J., and Hall, W. C. (1984) Nonintralaminar thalamostriate projections in the gray squirrel (*Sciurus carolinensis*) and tree shrew (*Tupaia glis*). *Journal of Comparative Neurology*, 230, 33–46.

Martinoli, M.-G., Dubourg, P., Geffard, M., Calas, A., and Kah, O. (1990) Distribution of GABA-immunoreactive neurons in the forebrain of the goldfish, *Carassius auratus. Cell and Tissue Research*, 260, 77–84.

Medina, L., Martí, E., Artero, C., Fasolo, A., and Puelles, L. (1992) Distribution of neuropeptide Y-like immunoreactivity in the brain of the lizard *Gallotia galloti. Journal of Comparative Neurology*, 319, 387–405.

Medina, L. and Smeets, W. J. A. J. (1991) Comparative aspects of the basal ganglia-tectal pathways in reptiles. *Journal of Comparative Neurology*, 308, 614–629.

Meredith, G. E., Blank, B., and Groenewegen, H. J. (1989) The distribution and compartmental organization of the cholinergic neurons in nucleus accumbens of the rat. *Neuroscience*, 31, 327–345.

Meredith, G. E. and Smeets, W. J. A. J. (1987) Immunocytochemical analysis of the dopamine system in the forebrain and midbrain of *Raja radiata*: evidence for a substantia nigra and ventral tegmental area in cartilaginous fish. *Journal of Comparative Neurology*, 265, 530–548.

Mesulam, M.-M. and Van Hoesen, G. W. (1976) Acetylcholinesterase-rich projections from the basal forebrain in the rhesus monkey to neocortex. *Brain Research*, 109, 152–157.

Nauta, H. J. W. (1974) Evidence of a pallidohabenular pathway in the cat. *Journal of Comparative Neurology*, 156, 19–28.

Nauta, H. J. W. (1979) Projections of the pallidal complex: an autoradiographic study in the cat. *Neuroscience*, 4, 1853–1874.

Nauta, H. J. W. and Cole, M. (1978) Efferent projections of the subthalamic nucleus: an autoradiographic study in the monkey and cat. *Journal of Comparative Neurology*, 180, 1–16.

Nauta, W. J. H., Smith, G. P., Faull, R. M., and Domesick, V. B. (1978) Efferent connections and nigral afferents of the nucleus accumbens septi in the rat. *Neuroscience*, 3, 385–401.

Nieuwenhuys, R. (1963) The comparative anatomy of the actinopterygian forebrain. *Journal für Hirnforschung*, 6, 171–192.

Nieuwenhuys, R. (1985) *Chemoarchitecture of the Brain*. Berlin: Springer-Verlag.

Northcutt, R. G. (1978) Brain organization in the cartilaginous fishes. In E. S. Hodgson and R. F. Mathewson (eds.), *Sensory Biology of Sharks, Skates, and Rays*. Arlington, VA: Office of Naval Research, pp. 117–193.

Northcutt, R. G. (1981) Evolution of the telencephalon in nonmammals. *Annual Review of Neuroscience*, 4, 301–350.

Northcutt, R. G. (1991) Visual pathways in elasmobranchs: organization and phylogenetic implications. *Journal of Experimental Zoology*, Suppl. 5, 97–107.

Northcutt, R. G. and Butler, A. B. (1988) Projections of the olfactory bulb and nervus terminalis in the silver lamprey. *Brain, Behavior and Evolution*, 32, 96–107.

Northcutt, R. G. and Davis, R. E. (1983) Telencephalic organization in ray-finned fishes. In R. E. Davis and R. G. Northcutt (eds.), *Fish Neurobiology, Vol 2: Higher Brain Areas and Functions*. Ann Arbor, MI: The University of Michigan Press, pp. 203–236.

Northcutt, R. G. and Kicliter, E. (1980) Organization of the amphibian telencephalon. In S. O. E. Ebbesson (ed.), *Comparative Neurology of the Telencephalon*. New York: Plenum pp. 203–255.

Northcutt, R. G., Reiner, A. and Karten, H. J. (1988) Immunohistochemical study of the telencephalon of the spiny dogfish, *Squalus acanthias*. *Journal of Comparative Neurology*, 277, 250–267.

Parent, A. (1973) Demonstration of a catecholaminergic pathway from the midbrain to the strio-amygdaloid complex in the turtle (*Chrysemys picta*). *Journal of Anatomy* (*London*), 114, 379–387.

Parent, A. (1975) The monoaminergic innervation of the telencephalon of the frog, *Rana pipiens*. *Brain Research*, 99, 35–47.

Parent, A. (1976) Striatal afferent connections in the turtle (*Chrysemys picta*) as revealed by retrograde axonal transport of horseradish peroxidase. *Brain Research*, 108, 25–36.

Parent, A. (1979) Identification of the pallidal and peripallidal cells projecting to the habenula in monkey. *Neuroscience Letters*, 15, 159–164.

Parent, A. (1990) Extrinsic connections of the basal ganglia. *Trends in Neurosciences*, 13, 254–258.

Parent, A. and Hazrati, L.-N. (1993) Anatomical aspects of information processing in primate basal ganglia. *Trends in Neurosciences*, 16, 111–116.

Parent, A., Mackey, A. and De Bellefeuille, L. (1983) The subcortical afferents to caudate nucleus and putamen in primate: a fluorescence retrograde double labeling study. *Neuroscience, 10*, 1137–1150.

Pickavance, L. C., Staines, W. A., and Fryer, J. N. (1992) Distributions and colocalization of neuropeptide Y and somatostatin in the goldfish brain. *Journal of Chemical Neuroanatomy*, 5, 221–233.

Pontet, A., Danger, J. M., Dubourg, P., Pelletier, G., Vaudry, H., Calas, A., and Kah, O. (1989) Distribution and characterization of neuropeptide Y-like immunoreactivity in the brain and pituitary of the goldfish. *Cell and Tissue Research*, 255, 529–538.

Reiner, A. (1986) Is prefrontal cortex found only in mammals? *Trends in Neurosciences*, 9, 298–300.

Reiner, A. (1986) The co-occurrence of substance P-like immunoreactivity in striatopallidal and striatonigral projection neurons in birds and reptiles. *Brain Research,* 371, 155–161.

Reiner, A. (1986) Transmitter-specific projections from the basal ganglia to the tegmentum in pigeons. *Society for Neuroscience Abstracts,* 12, 873.

Reiner, A. (1991) A comparison of neurotransmitter-specific and neuropeptide-specific neuronal cell types present in the dorsal cortex in turtles with those present in the isocortex in mammals: implications for the evolution of isocortex. *Brain, Behavior and Evolution,* 38, 53–91.

Reiner, A. (1993) Neurotransmitter organization and connections of turtle cortex: implications for the evolution of mammalian isocortex. *Comparative Biochemistry and Physiology,* 104A, 735–748.

Reiner, A., Brauth, S. E., Kitt, C. A., and Karten, H. J. (1980) Basal ganglionic pathways to the tectum: studies in reptiles. *Journal of Comparative Neurology,* 193, 565–589.

Reiner, A., Brecha, N. C., and Karten, H. J. (1982) Basal ganglia pathways to the tectum: the afferent and efferent connections of the lateral spiriform nucleus of pigeon. *Journal of Comparative Neurology,* 208, 16–36.

Reiner, A. and Northcutt, R. G. (1987) An immunohistochemical study of the telencephalon of the African lungfish, *Protopterus annectens*. *Journal of Comparative Neurology,* 256, 463–481.

Reiner, A. and Northcutt, R. G. (1992) An immunohistochemical study of the telencephalon of the Senegal bichir (*Polypterus senegalus*). *Journal of Comparative Neurology,* 319, 359–386.

Royce, G. J. (1978) Cells of origin of subcortical afferents to the caudate nucleus: a horseradish peroxidase study in the cat. *Brain Research,* 153, 465–475.

Royce, G. J. and Laine, E. (1984) Efferent connections of the caudate nucleus, including cortical projections of the striatum and other basal ganglia: an autoradiographic and horseradish peroxidase investigation in the cat. *Journal of Comparative Neurology,* 226, 28–49.

Russchen, F. T. and Jonker, A. J. (1988) Efferent connections of the striatum and the nucleus accumbens in the lizard *Gekko gecko*. *Journal of Comparative Neurology,* 276, 61–80.

Selemon, L. D., Gottlieb, J. P., and Goldman-Rakic, P. S. (1994) Islands and striosomes in the neostriatum of the rhesus monkey: non-equivalent compartments. *Neuroscience,* 58, 183–192.

Sharma, S. C., Berthoud, V. M., and Breckwoldt, R. (1989) Distribution of substance P-like immunoreactivity in the goldfish brain. *Journal of Comparative Neurology,* 279, 104–116.

Sesack, S. R. and Pickel, V. M. (1992) Prefrontal cortical efferents in the rat synapse on unlabeled neuronal targets of catecholamine terminals in the nucleus accumbens septi and on dopamine neurons in the ventral tegmental area. *Journal of Comparative Neurology,* 320, 145–160.

Shinonaga, Y., Takada, M., and Mizuno, N. (1992) Direct projections from the central amygdaloid nucleus to the globus pallidus and substantia nigra in the cat. *Neuroscience,* 51, 691–703.

Smeets, W. J. A. J. (1988) The monoaminergic systems of reptiles investigated with specific antibodies against serotonin, dopamine, and noradrenaline. In W. K. Schwerdtfeger and W. J. A. J. Smeets (eds.), *The Forebrain of Reptiles: Current Concepts of Structure and Function*. Basel: Karger, pp. 97–109.

Smeets, W. J. A. J., Nieuwenhuys, R., and Roberts, B. L. (1983) *The Central Nervous System of Cartilaginous Fishes: Structure and Functional Correlations*. Berlin: Springer-Verlag.

Smeets, W. J. A. J. and Reiner, A., Catecholamines in the CNS of vertebrates: current concepts of evolution and functional significance. In W. J. A. J. Smeets and A. Reiner (eds.), *Phylogeny and Development of Catecholamine Systems in the CNS of Vertebrates*. Cambridge, England: Cambridge University Press, pp. 463–481.

Steininger, T. L., Rye, D. B., and Wainer, B. H. (1992) Afferent projections to the cholinergic pedunculopontine tegmental nucleus and adjacent midbrain extrapyramidal area in the albino rat. I. Retrograde tracing studies. *Journal of Comparative Neurology,* 321, 515–543.

Takada, M. (1992) The lateroposterior thalamic nucleus and substantia nigra pars lateralis: origin of dual innervation over the visual system and basal ganglia. *Neuroscience Letters,* 139, 153–156.

ten Donkelaar, H. J. and de Boer-van Huizen, R. (1981) Basal ganglia projections to the brain stem in the lizard *Varanus exanthematicus* as demonstrated by retrograde transport of horseradish peroxidase. *Neuroscience,* 6, 1567–1590.

Ulinski, P. S. (1986) Neurobiology of the therapsid-mammal transition. In N. Hotton, III., P. D. MacLean, J. J. Roth and E. C. Roth (eds.), *The Ecology and Biology of Mammal-like Reptiles*. Washington, DC: Smithsonian Institution Press, pp. 149–171.

van der Kooy, D. and Carter, D. A. (1981) The organization of the efferent projections and striatal afferents of the entopeduncular nucleus and adjacent areas in the rat. *Brain Research,* 211, 15–36.

Van Hoesen, G. W., Yeterian, E. H., and Lavizzo-Mourey, R. (1981) Widespread corticostriate projections from temporal cortex of rhesus monkey. *Journal of Comparative Neurology,* 199, 205–219.

Wicht, H. and Himstedt, W. (1990) Brain stem projections to the telencephalon in two species of amphibians, *Triturus alpestris* (Urodela) and *Ichthyophis kohtaoensis* (Gymnophiona). In W. K. Schwerdtfeger and P. Germroth (eds.), *The Forebrain in Nonmammals: New Aspects of Structure and Development.* Berlin: Springer-Verlag, pp.43–55.

Wicht, H. and Northcutt, R. G. (1992) The forebrain of the Pacific hagfish: a cladistic reconstruction of the ancestral craniate forebrain. *Brain, Behavior and Evolution,* 40, 25–64.

Wilczynski, W. and Northcutt, R. G. (1983) Connections of the bullfrog striatum: afferent organization. *Journal of Comparative Neurology,* 214, 321–332.

Wilczynski, W. and Northcutt, R. G. (1983) Connections of the bullfrog striatum: efferent organization. *Journal of Comparative Neurology,* 214, 333–343.

Wild, J. M. (1987) Thalamic projections to the paleostriatum and neostriatum in the pigeon (*Columba livia*). *Neuroscience,* 20, 305–327.

Záborszky, L. and Cullinan, W. E. (1992) Projections from the nucleus accumbens to cholinergic neurons of the ventral pallidum: a correlated light and electron microscopic double-immunolabeling study in rat. *Brain Research,* 570, 92–101.

25

Dorsal Pallium

INTRODUCTION

In all groups of vertebrates, the telencephalic pallium can be divided into at least three areas. In vertebrates with evaginated telencephalons, these areas are most commonly called the medial, dorsal, and lateral pallia. Comparable pallial areas are present in transposed order in vertebrates with everted telencephalons. In amniotes, the Group IIB vertebrates, the dorsal pallium forms a large part of the telencephalon and receives ascending sensory projections from the dorsal thalamus.

While a dorsal part of the pallium lying between the medial and lateral pallia has been identified in all vertebrate groups, the question of whether any or all of the dorsal pallia in various anamniotes are homologous to the dorsal pallial areas in amniotes is one of the most important unresolved questions in comparative neurobiology. Put simply, did a dorsal pallium homologous to that in extant amniotes exist in the earliest vertebrates or has a dorsal pallial, ascending-sensory receptive part of the telencephalon been evolved separately in several vertebrate lineages?

In attempting to resolve this question, comparative neuroanatomists have focused on several defining features of the dorsal pallium in amniotes, including the receipt of ascending sensory projections from the dorsal thalamus and the absence of afferent projections from the olfactory bulb. This chapter begins with a review of areas identified as dorsal pallium in anamniotes and the somewhat fragmentary data available, continues with an overview of afferent sensory and related pathways to the dorsal pallium in amniotes, and concludes with a new interpretation of dorsal pallial evolution.

THE DORSAL PALLIUM IN GROUP I VERTEBRATES

In lampreys, a small region has been tentatively identified as the dorsal pallium [Fig. 25-1(A)], based solely on its position in an unevaginated part of the telencephalon. This region receives olfactory input, but ascending projections from the diencephalon have not been investigated. In squalomorph sharks, electrophysiological evidence indicates that both the medial and dorsal pallia [Fig. 25-1(B)] receive visual input, but anatomical tracing studies of thalamic projections remain to be done.

In nonteleost ray-finned fishes, such as the bichir *Polypterus*, the pallial areas designated **P2** and **P3** [Fig. 25-1(C)] topologically correspond to the dorsal and medial pallia of vertebrates with evaginated telencephalons, respectively. Both of these pallial areas may receive dorsal thalamic and posterior tubercular input, the medial part of P2 receiving more of this ascending input than the rest of P2 and P3. In lungfishes, the dorsal pallium [Fig. 25-2(A)] receives olfactory bulb input, but possible ascending thalamic projections to the medial and/or dorsal pallia have not yet been studied.

In frogs, the rostral part of the dorsal thalamus (nucleus anterior) relays visual input to both the medial and dorsal pallia [Fig. 25-2(B)]. Olfactory bulb projections are not restricted to the lateral pallium but also terminate in the ventral part of the dorsal pallium. The possibility that the dorsal pallium and part of the lateral pallium in frogs are homologous to the dorsal pallium of amniotes has been raised on the basis of similarities in histochemistry, cellular continuity, and embryological data. We will return to this question in the last section of this chapter.

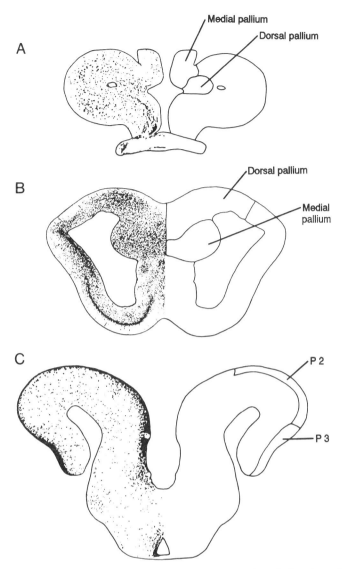

A

Medial pallium
Dorsal pallium

B

Dorsal pallium
Medial pallium

C

P 2
P 3

FIGURE 25-1. Transverse hemisections with mirror-image drawings through the telencephalons of (A) a lamprey, (B) a squalomorph shark (*Squalus acanthias*), and (C) a bichir, (*Polypterus palmas*). Part A is adapted from Northcutt (1981). Redrawn, with permission, from the *Annual Review of Neuroscience*, Volume 4 © 1981, by Annual Reviews Inc. Part B is adapted from Northcutt et al. (1988), and Part C is from Northcutt and Davis (1983).

THE DORSAL PALLIUM IN GROUP IIA VERTEBRATES

Neuroanatomical Organization

In hagfishes, the entire pallium (Fig. 25-3) is in receipt of olfactory projections, and ascending projections from the diencephalon have not been studied. In Group II cartilaginous fishes, some data are available on visual, auditory, and lateral line projections to the telencephalon. Auditory responses have been recorded in the telencephalon of sharks, the best responses being in a medial area within the middle part of the hemisphere. Within the telencephalon of Group II (galeo-

morph) sharks [Fig. 25-4(A)], a large nucleus that is called the **central nucleus** is present in a central location. An area designated as dorsal pallium lies dorsal to the central nucleus, but the central nucleus is believed also to be a part of the dorsal pallium. The dorsal pallium is richly populated with a substantial variety of different neuronal cell types.

In galeomorph sharks, ascending visual projections from the rostral part of the dorsal thalamus (nucleus anterior) terminate within the central nucleus. Visual projections from the tecto-recipient dorsal posterior nucleus in the dorsal thalamus may likewise terminate in the central nucleus, as well as in the striatum. Visual evoked potentials have also been recorded in both the dorsal and medial pallia. Visual potentials have simi-

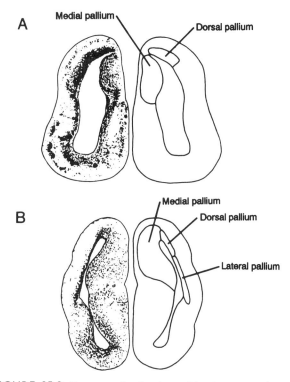

FIGURE 25-2. Transverse hemisections with mirror-image drawings through the telencephalons of (A) a lungfish (*Protopterus annectens*) and (B) a bullfrog (*Rana catesbeiana*). Part A is adapted from Reiner and Northcutt (1987), and part B is from Northcutt and Kicliter (1980). The latter used with permission of Plenum Publishing Corp.

larly been recorded in the medial pallium of skates, to which nucleus anterior projects. Behavioral studies have shown that sharks can discriminate patterns following surgical removal of their optic tecta, indicating that their telencephalic visual areas mediate this function.

In skates, a telencephalic area has been identified electrophysiologically in which both electrosensory and visual responses can be recorded. This area is within the medial pallium, and nucleus anterior of the dorsal thalamus, as well as the posterior tuberculum and the electrosensory lateral posterior nucleus are known to project to it. The evolutionary relationship of the dorsal and medial pallia to each other in cartilaginous fishes and of both to the dorsal pallium of amniotes remains unclear.

The telencephalon in teleosts is a complex and poorly understood part of the brain. An area in the telencephalon of the common goldfish that may be the homologue of the dorsal pallium of other vertebrates is known to receive projections from the retino-recipient nucleus anterior in the dorsal thalamus. This region in the brain, the **dorsal zone of area dorsalis telencephali (Dd),** is shown in the brain of the green sunfish in Figure 25-4(B). A more lateral region, the **lateral zone of area dorsalis telencephali (Dl),** thought to be the homologue of the medial pallium of other vertebrates (see Chapter 30), may also receive projections from nucleus anterior.

Other ascending sensory pathways are present to the telencephalon in teleosts that are relayed through the preglomerular nuclear complex, a migrated group of the posterior tuberculum (see Chapter 20), rather than through the dorsal thalamus. For example, gustatory information is relayed through the preglomerular complex to a centrally lying part of the telencephalon, the **central zone of area dorsalis telencephali (Dc).** Likewise, acousticolateral information is relayed from the torus semicircularis through the preglomerular complex to Dd and a more medial area, the **medial zone of area dorsalis telencephali (Dm).** The more ventral part of Dm may be a homologue of the medial pallium, while the more dorsal part may be an independently evolved portion of the telencephalon that resembles the caudate and putamen of amniotes.

Behavioral Issues

The possible roles of an elaborated telencephalon in Group IIA vertebrates or of the pallial areas in Group I vertebrates for a number of complex behaviors, such as intraspecific communication, reproduction, feeding, and learning, have yet to be explored. For example, no part of the telencephalon of cartilaginous fishes, ray-finned fishes, or amphibians has yet been identified as having a role in acoustic or electric social communication. Such communication behaviors have evolved independently multiple times within various lineages of vertebrates and involve differing peripheral organs in the production of communicative signals.

As will be discussed in Chapter 28, the motor nuclei in the brainstem that are involved in communication have direct inputs from the telencephalon and the midbrain in birds and some mammals. Studies are needed in fishes and amphibians on possible descending pathways, whether direct or multisynaptic, that could affect the output of the motor nuclei involved in producing communication signals. For example, electrical stimulation of the preoptic area and of the supracommissural nucleus of the ventral telencephalon (Vs, Fig. 30-9) recently has been found to evoke acoustic communication signals in toadfishes. Toadfishes produce such sounds by controlled contraction of sonic muscles in the lateral walls of the swim bladder. Other sound producing mechanisms used by various fishes include grinding of pharyngeal teeth and electric organ discharges. Amphibians use the larynx for vocalization. The motor nuclei controlling the production of communication signals variously receive input from the reticular formation and/or pacemaker nuclei in the medulla.

In a gymnotiform fish, some of the supramedullary pathways involved in the control of electric discharges have been elucidated. A diencephalic nucleus, nucleus electrosensorius, receives ascending electrosensory and octavolateral input and projects to multiple targets that include a nearby prepacemaker nucleus. The latter nucleus projects to the pacemaker nucleus in the medulla and is involved in the control of the frequency of electric organ discharge. Investigating possible descending telencephalic inputs to electrosensory-related nuclei in the diencephalon of gymnotids and to other nuclei in other anamniotes that are involved in the production of communication signals will add to our understanding of motor system and telencephalic evolution.

Some other motor behavior patterns in anamniotes are related to mate selection or territorial defense, but forebrain projections that might be involved in the control of these behaviors, which are based on visual recognition and the

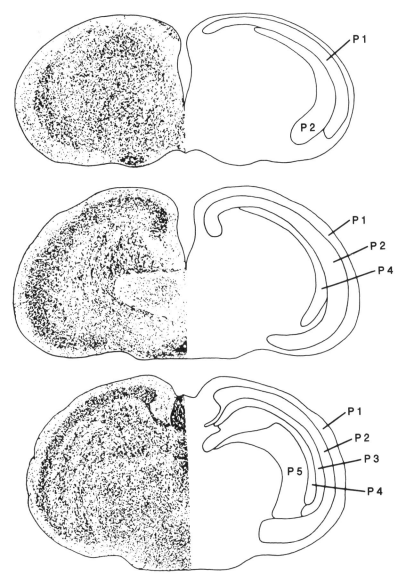

FIGURE 25-3. Transverse hemisections with mirror-image drawings through the telencephalon of a hagfish (*Eptatretus stouti*). The uppermost section is most rostral. P1–P5 are pallial areas. Adapted from Wicht and Northcutt (1992). Used with permission of S. Karger AG.

response behaviors of conspecifics, remain to be explored. For example, some species of perciform teleost fishes—Siamese fighting fishes, paradise fishes, sunfishes, and various cichlids—erect their gill covers as part of a frontal display behavior during aggressive encounters with male conspecifics. The trigeminal motor nucleus, which controls the movement of the gill covers, is known to receive inputs from the reticular formation and other hindbrain sites. In addition, some other brainstem inputs to the reticular formation are known, such as those from octavolateralis nuclei. Possible descending telencephalic projections to these brainstem nuclei remain to be investigated.

Teleosts have elaborate telencephalons with multiple migrated nuclear groups, as shown in the green sunfish in Figure 25-4(B). Involvement of some part of the telencephalic

pallium in complex behavioral patterns has been demonstrated by using electrical stimulation. Stimulation of the telencephalic hemisphere evokes arousal and swimming behavior. Stimulation restricted to the medioventral part of the telencephalon produces feeding behavior, while stimulation of the dorsal central part (Dc) produces sweeping or nest building behaviors.

Some studies of learning have been done on various teleosts following surgical ablation of the telencephalon. In such cases, instrumental conditioning is impaired when the reinforcement is remote from the correct response. Avoidance learning tasks and successive reversal learning tasks are also severely impaired. Thus, the telencephalon in teleosts appears to play a significant role in some of the integrative functions necessary for learning.

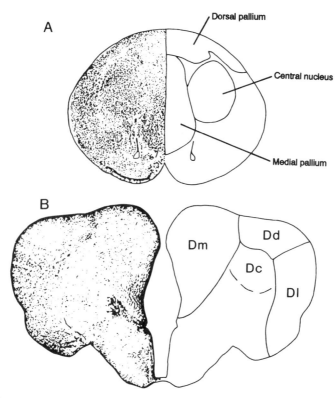

FIGURE 25-4. Transverse hemisections with mirror-image drawings through the telencephalons of (A) a galeomorph shark (*Ginglymostoma cirratum*) and (B) a teleost fish (*Lepomis cyanellus*). Dc, Dd, Dl, and Dm = central, dorsal, lateral, and medial zones of the area dorsalis telencephali, respectively. Part A is adapted from Ebbesson (1980) and part B is from Northcutt and Davis (1983). Material adapted from Ebbesson (1980) used with permission of Plenum Publishing Corp.

AFFERENT SYSTEMS TO THE DORSAL PALLIUM IN AMNIOTES

As introduced in Chapter 19, the dorsal pallium in diapsid reptiles, birds, and turtles consists of the dorsal cortex and the dorsal ventricular ridge. In mammals, the dorsal pallium consists of the isocortex, which forms the bulk of the cerebral hemispheres, and transitional cortices. In contrast, the dorsal ventricular ridge of nonsynapsid amniotes consists of either a laminar cell plate with some centrally scattered cells or an enlarged area of nuclear groups. The dorsal pallia of amniotes share a number of common organizational features, including the organization of afferent sensory pathways. These afferent pathways, which were introduced in Chapter 22, will be reviewed and summarized here. In the following three chapters (Chapters 26–28), the major visual, auditory, and somatosensory-motor regions of the telencephalon in amniotes will be discussed in detail.

It should be noted that Chapters 26–28 address the dorsal pallium in Group IIB vertebrates and, with this chapter as part of the set, maintain the same order of presentation as in many other midbrain and forebrain chapters, that is, Group I, Group IIA, and, within Group IIB, (1) mammals, (2) diapsid reptiles, and (3) birds. This sequence represents the order in which these groups diverged from the common ancestral amniote stock (see Chapter 4). Material on turtles, which appear in the fossil record after diapsids, is included with that on diapsid reptiles in these chapters only as a matter of convenience due to the similarities.

Visual Pathways

Two major visual pathways to the telencephalon are present in all amniotes and are diagramatically shown for mammals and diapsid reptiles in Figure 25-5. The retina projects to the dorsal lateral geniculate nucleus in mammals and the dorsal lateral optic nucleus in diapsids. This dorsal thalamic (lemnothalamic) nucleus projects to striate cortex in mammals and to a lateral part of dorsal cortex in diapsids. The retina also projects to the superior colliculus in mammals and to the optic tectum in diapsids. This midbrain visual structure projects to the lateral posterior/pulvinar complex (LP/pulvinar) in mammals and to nucleus rotundus in diapsids. This dorsal thalamic (collothalamic) nucleus projects to extrastriate cortex in mammals and to a part of the dorsal ventricular ridge (DVR) in diapsids. A number of association visual cortical areas are also known to be present in mammals that receive ascending visual input from one or both of these thalamic visual nuclei.

Somatosensory Pathways

Two major somatosensory pathways to the telencephalon are known in mammals, but among nonsynapsid amniotes, these pathways have only been fully identified in birds. These pathways are diagramatically shown in Figure 25-6. The dorsal column nuclei in the region of the obex project to a dorsal thalamic nucleus, the ventral posterolateral nucleus within the ventral nuclear group in mammals and nucleus dorsalis inter-

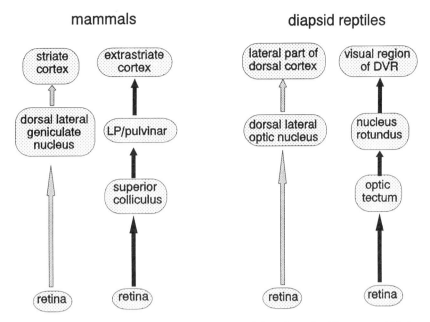

FIGURE 25-5. Diagram of visual pathways in mammals and diapsid reptiles. The gray and black arrows indicate lemnothalamic and collothalamic pathways, respectively.

medius ventralis anterior (DIVA) in birds. From the dorsal thalamus (lemnothalamic somatosensory nucleus), this information is relayed to the primary and related somatosensory cortices in mammals and to a part of the Wulst in birds. The dorsal column nuclei also project to the somesthetic part of the tectum, which, along with auditory and other sensory midbrain areas, projects to the limitans–suprageniculate complex and the lateral part of the posterior nuclear group (Pol) in mammals and to the caudal part of nucleus dorsolateralis posterior (cDLP) and nucleus semilunaris parovoidalis in birds. These dorsal thalamic (collo-

thalamic) nuclei project to an area of somatosensory–multisensory cortex in mammals and to parts of the dorsal ventricular ridge in birds.

Auditory Pathway

An ascending auditory pathway has been found in all amniotes and is diagramatically shown in mammals and diapsids in Figure 25-7. The auditory part of the midbrain, the inferior colliculus in mammals and the torus semicircularis

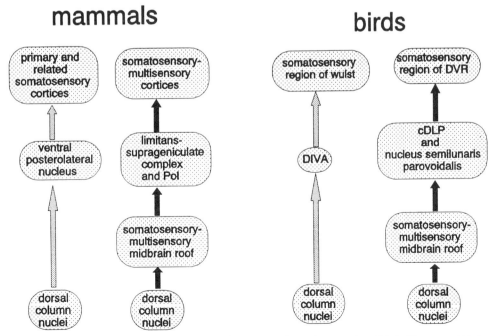

FIGURE 25-6. Diagram of somatosensory–multisensory pathways in mammals and birds. The gray and black arrows indicate lemnothalamic and collothalamic pathways, respectively.

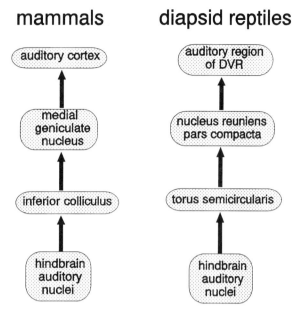

FIGURE 25-7. Diagram of auditory pathway in mammals and diapsid reptiles.

in diapsids, projects to the medial geniculate nucleus in mammals and to nucleus reuniens pars compacta in diapsids. This dorsal thalamic (collothalamic) nucleus projects to auditory cortex in mammals and to a part of the dorsal ventricular ridge in diapsids.

Olfactory Pathway

In mammals, an olfactory pathway (Fig. 25-8) is present from the olfactory cortex to the mediodorsal nucleus in the dorsal thalamus. The mediodorsal nucleus, a part of the lemnothalamus, projects to prefrontal and prelimbic cortical areas.

The cortical terminal area of mediodorsal afferent fibers is characterized by a large quantity of dopaminergic fibers. In birds, the olfactory cortex similarly projects to part of nucleus dorsomedialis (Fig. 25-8). An area of the telencephalon has been found in birds, in a lateral part of the dorsal ventricular ridge, that is also in continuity with the dorsal cortical area, which is similarly rich in dopaminergic fibers. A possible projection from the olfactory-recipient part of nucleus dorsomedialis to this dopaminergic-rich part of the telencephalon remains to be investigated, however, as do possible similar pathways in diapsid reptiles or turtles. Behavioral studies have revealed that this dopaminergic-rich telencephalic region in birds has functions similar to that of mammalian prefrontal cortex.

Other Afferent Pathways

The final major thalamopallial pathways that we need to outline here arise from lemnothalamic nuclei and consist of the projections of the anterior, medial, and intralaminar nuclear groups in mammals and of the dorsomedial and anterior dorsolateral nuclei in nonsynapsid amniotes. In mammals, some of these nuclei project specifically to limbic and prelimbic cortices, including the subicular cortices, and some project diffusely to wide areas of the cortex. In nonsynapsid amniotes, the dorsomedial and anterior dorsolateral nuclei give rise to specific projections to the dorsal cortex as well as diffuse projections to wide areas of the dorsal cortex and dorsal ventricular ridge.

SYNOPSIS OF THE DUAL EXPANSION HYPOTHESIS OF DORSAL PALLIAL EVOLUTION

The extensiveness of connections between the dorsal thalamus and the dorsal pallium does not allow for either structure to have undergone significant evolutionary change alone. For

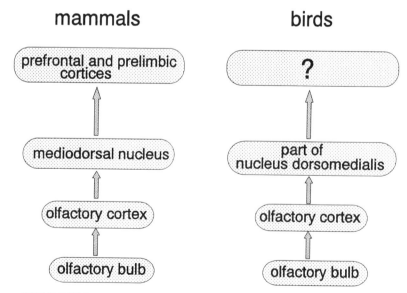

FIGURE 25-8. Diagram of olfactory dorsal thalamic pathway in mammals and birds.

this reason, the dorsal thalamus serves as a Rosetta stone for deciphering the evolutionary history of the dorsal pallium. The relationships of the lemnothalamus and collothalamus (see Chapter 22) with the dorsal pallium provide a key to telencephalic evolution, particularly within amniotes. Elucidation of evolutionary history of the telencephalon in the Group II members of various radiations of anamniotes must await additional data.

As discussed above, the question of what part or parts of the pallium in various anamniotes is or are homologous to the dorsal pallium in amniotes remains open. The presence of a dorsal pallial region, not all of which is in receipt of olfactory bulb projections, in most if not all jawed vertebrates implies that at least some part of the amniote dorsal pallium may have a homologue in jawed anamniotes. The anamniote dorsal pallial area might be homologous to all parts of the amniote dorsal pallium. Alternatively, both the dorsal pallium and part or all of the lateral pallium in anamniotes might be so. A further possibility is that the dorsal pallium in anamniotes may be the homologue of only part or all of the more medial portion of the amniote dorsal pallium, and the more lateral portion of the amniote dorsal pallium (which receives ascending projections from the collothalamus) may be a *de novo* structure in amniotes.

For our present purposes, the term "dorsal pallium" is used for amniotes simply to refer to the pallial region between the medial pallium and the lateral, piriform, olfactory-recipient pallium. The presence of such a dorsal pallium comprising two divisions as discussed below, is hypothesized to have been a feature of the common ancestral stock of amniotes. The possibility exists that similarities in dorsal pallial development and morphology could have arisen independently in mammals and in nonmammalian amniotes, that is, as a result of parallelism or convergence. However, an hypothesis of a specified homology, as accepted here, represents the most parsimonious interpretation of the data.

In amniotes, two basic divisions of the dorsal pallium can be recognized (Butler, 1994b): (1) a **medial lemnopallial division** predominantly (or plesiomorphically) in receipt of projections from lemnothalamic nuclei, and (2) a **lateral collopallial division** predominantly (or plesiomorphically) in receipt of projections from collothalamic nuclei. In the ancestral captorhinomorphs (Fig. 25-9), a *dual expansion* of both of these divisions is hypothesized to have occurred as a result of an increase in the duration and/or rate of cell proliferation in the germinal matrix, particularly during the symmetrical division stage and thus resulting in an exponential increase in the number of progenitor cells. Projections from the elaborated lemnothalamus to the expanded medial division of the dorsal pallium were increased, and projections from the elaborated collothalamus

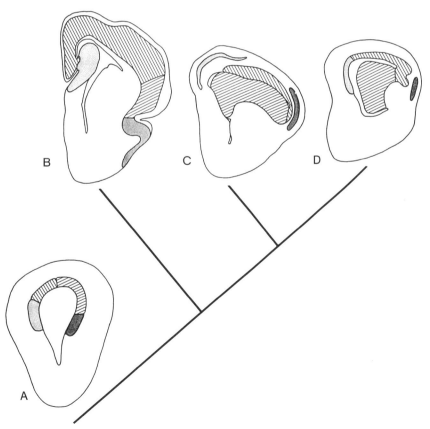

FIGURE 25-9. Evolution of the dorsal pallium in amniotes. The medial lemnopallial division of the dorsal pallium is indicated by diagonal lines slanting down to the right, and the lateral collopallial division of the dorsal pallium is indicated by diagonal lines slanting up to the right. Medial pallium is indicated by light shading and lateral pallium by dark shading. (A) captorhinomorph ancestral amniote; (B) extant mammal; (C) extant diapsid reptile; (D) extant turtle. From Butler (1994b). Used with permission of Elsevier.

to the expanded lateral division of the dorsal pallium were gained at this time.

In the synapsid line that led to mammals, additional expansions of both pallial divisions occurred, while in the line leading to both diapsid amniotes (diapsid reptiles and birds) and to extant anapsid amniotes (turtles), further dorsal pallial expansions were primarily confined to the lateral division. In the synapsid line, a second change involving an increase in the degree of radial organization resulted in an "outside-in" migration pattern during development (see Chapter 3) in most of the dorsal pallium, as opposed to the "inside-out" pattern retained by nonsynapsid amniotes.

New neuronal phenotypes, particularly the granule cells of the main thalamic-recipient layers of sensory cortices, layers II and IV, also appeared in the synapsid line in conjunction with the outside-in migration pattern. Several neuronal phenotypes have been identified in mammals by their neurotransmitter histochemistry and their pattern of intracortical and interhemispheric projections that are absent in diapsid reptiles and turtles (see Chapter 19). These neuronal phenotypes thus may be apomorphic for mammals. In considering the evolution of various areas within the dorsal pallium, respective areas in mammals and nonmammalian amniotes therefore appear to be homologous as fields in receipt of specified dorsal thalamic projections rather than as discrete, neurotransmitter-specific, laminar cell populations. Some dorsal pallial neuronal populations with neurotransmitter phenotypes similar to those in layers II–IV of mammalian isocortex are also present in birds and thus appear to be the result of homoplasy.

The medial lemnopallial division of the dorsal pallium comprises medial and lateral parts. In mammals, the medial part forms the subicular, cingulate, prefrontal, primary sensorimotor, and related cortices, and in nonsynapsid amniotes, it forms the medial part of the dorsal cortex. The lateral part of the medial lemnopallial division forms the striate cortex in mammals and the lateral part of the dorsal cortex, or pallial thickening, of diapsid reptiles and turtles and the visual Wulst in birds. The lateral collopallial division of the dorsal pallium forms the extrastriate, auditory, somatosensory–multisensory, and related cortices in mammals and the areas of the dorsal ventricular ridge with similar collothalamic afferents in nonsynapsid amniotes.

These two new hypotheses of dorsal thalamic and dorsal pallial evolution are consistent with a principle called **minimum increase in complexity** (see Saunders and Ho, 1981 and 1984). This concept notes that while selection pressures may act on any given change, the change itself must first occur randomly within a population. The phenotype is of course determined and produced by both the genotype and the complex, interactive processes of embryological development, which are themselves under genotypic control. Most changes increase complexity, and, thus, the simpler the change, the greater is its probability of occurring. The relative fitness of the change is then determined by the degree to which it is adaptive. The changes in the development of the dorsal thalamus and dorsal pallium hypothesized here would involve relatively few alleles and thus constitute a relatively minimal increase in complexity. The myriad selection pressures involved in the conquest of the land then favored the resultant elaborations of the dorsal thalamus and expansions of the dorsal pallium.

FOR FURTHER READING

Butler, A. B. (1994b) The evolution of the dorsal pallium in the telencephalon of amniotes: cladistic analysis and a new hypothesis. *Brain Research Reviews*, 19, 65–101.

Ebbesson, S. O. E. (1980) *Comparative Neurology of the Telencephalon*. New York: Plenum.

Jones, E. G. and Peters, A. (eds.) (1990) *Cerebral Cortex, Vol. 8A: Comparative Structure and Evolution of Cerebral Cortex, Part I*. New York: Plenum.

Karten, H. J. and Shimizu, T. (1989) The origins of neocortex: connections and lamination as distinct events in evolution. *Journal of Congitive Neuroscience*, 1, 291–301.

Northcutt, R. G. and Davis, R. E. (1983) Telencephalic organization in ray-finned fishes. In R. E. Davis and R. G. Northcutt (eds.), *Fish Neurobiology, Vol. 2: Higher Brain Areas and Functions*. Ann Arbor, MI: The University of Michigan Press, pp. 203–236.

Reiner, A. (1991) A comparison of neurotransmitter-specific and neuropeptide-specific neuronal cell types present in the dorsal cortex in turtles with those present in the isocortex in mammals: implications for the evolution of isocortex. *Brain, Behavior and Evolution*, 38, 53–91.

Ulinski, P. S. (1983) *Dorsal Ventricular Ridge: A Treatise on Forebrain Organization in Reptiles and Birds*. New York: Wiley.

ADDITIONAL REFERENCES

Bass, A. H. (1989) Evolution of vertebrate motor systems for acoustic and electric communication: peripheral and central elements. *Brain, Behavior and Evolution*, 33, 237–247.

Bingman, V. P., Casini, G., Nocjar, C., and Jones, T.-J. (1994) Connections of the piriform cortex in homing pigeons (*Columba livia*) studied with fast blue and WGA–HRP. *Brain, Behavior and Evolution*, 43, 206–218.

Bodznick, D. and Northcutt, R. G. (1984) An electrosensory area in the telencephalon of the little skate, *Raja erinacea. Brain Research*, 298, 117–124.

Bullock, T. H. and Corwin, J. T. (1979) Acoustic evoked activity in the brain in sharks. *Journal of Comparative Physiology*, 129, 223–234.

Butler, A. B. (1994a) The evolution of the dorsal thalamus of jawed vertebrates, including mammals: cladistic analysis and a new hypothesis. *Brain Research Reviews*, 19, 29–65.

Clairambault, P. and Timmel, J. F. (1990) Developmental organization of the amphibian pallium. In W. K. Schwerdtfeger and P. Germroth (eds.), *The Forebrain in Nonmammals: New Aspects of Structure and Development*. Berlin: Springer-Verlag, pp. 29–41.

Demski, L. S. (1983) Behavioral effects of electrical stimulation of the brain. In R. E. Davis and R. G. Northcutt (eds.), *Fish Neurobiology, Vol. 2: Higher Brain Areas and Functions*. Ann Arbor, MI: University of Michigan Press, pp. 317–359.

Ebbesson, S. O. E. (1980) On the organization of the telencephalon in elasmobranchs. In S. O. E. Ebbesson (ed.), *Comparative Neurology of the Telencephalon*. New York: Plenum, pp. 1–16.

Echteler, S. M. and Saidel, W. M. (1981) Forebrain connections in the goldfish support telencephalic homologies with land vertebrates. *Science*, 212, 683–685.

Fine, M. L. and Perini, M. A. (1994) Sound production evoked by electrical stimulation of the forebrain in the oyster toadfish. *Journal of Comparative Physiology A*, 174, 173–185.

Finger, T. E. (1980) Nonolfactory sensory pathway to the telencephalon in a teleost fish. *Science*, 210, 671–673.

Gorlick, D. L. (1990) Neural pathway for aggressive display in *Betta splendens*: midbrain and hindbrain control of gill-cover erection behavior. *Brain, Behavior and Evolution*, 36, 227–236.

Graeber, R. C., Ebbesson, S. O. E., and Jane, J. A. (1973) Visual discrimination in sharks without optic tectum. *Science*, 180, 413–415.

Ito, H., Morita, Y., Sakamoto, N., and Ueda, S. (1980) Possibility of telencephalic visual projection in teleosts, Holocentridae. *Brain Research*, 197, 219–222.

Ito, H., Murakami, T., Fukuoka, T., and Kishida, R. (1986) Thalamic fiber connections in a teleost (*Sebastiscus marmoratus*): visual, somatosensory, octavolateral, and cerebellar relay region to the telencephalon. *Journal of Comparative Neurology*, 250, 215–227.

Karten, H. J. (1969) The organization of the avian telencephalon and some speculations on the phylogeny of the amniote telencephalon. In C. R. Noback and J. M. Petras (eds.), *Comparative and Evolutionary Aspects of the Vertebrate Central Nervous System, Annals of the New York Academy of Sciences*, 167, 146–179.

Karten, H. J. (1991) Homology and evolutionary origins of the 'neocortex.' *Brain, Behavior and Evolution*, 38, 264–272.

Keller, C. H., Maler, L., and Heiligenberg, W. (1990) Structural and functional organization of a diencephalic sensory-motor interface in the gymnotiform fish, *Eigenmannia*. *Journal of Comparative Neurology*, 293, 347–376.

Kawasaki, M., Maler, L., Rose, G. J., and Heiligenberg, W. (1988) Anatomical and functional organization of the prepacemaker nucleus in gymnotiform electric fish: the accommodation of two behaviors in one nucleus. *Journal of Comparative Neurology*, 276, 113–131.

Manso, M. J. and Anadón, R. (1993) Golgi study of the telencephalon of the small-spotted dogfish *Scyliorhinus canicula* L., *Journal of Comparative Neurology*, 333, 485–502.

Mogensen, J. and Divac, I. (1993) Behavioral effects of ablation of the pigeon-equivalent of the mammalian prefrontal cortex. *Behavioural Brain Research*, 55, 101–107.

Morita, Y., Murakami, T., and Ito, H. (1983) Cytoarchitecture and topographic projections of the gustatory centers in a teleost, *Carassius carassius*. *Journal of Comparative Neurology*, 218, 378–394.

Murakami, T., Fukuoka, T. and Ito, H. (1986) Telencephalic ascending acousticolateral system in a teleost (*Sebastiscus marmoratus*), with special reference to the fiber connections of the nucleus preglomerulosus. *Journal of Comparative Neurology*, 247, 383–397.

Murakami, T., Morita, Y., and Ito, H. (1983) Extrinsic and intrinsic fiber connections of the telencephalon in a teleost, *Sebastiscus marmoratus*. *Journal of Comparative Neurology*, 216, 115–131.

Northcutt, R. G. (1974) Some histochemical observations on the telencephalon of the bullfrog, *Rana catesbeiana* Shaw. *Journal of Comparative Neurology*, 157, 379–390.

Northcutt, R. G. (1981) Evolution of the telencephalon in nonmammals. *Annual Review of Neuroscience*, 4, 301–350.

Northcutt, R. G. (1991) Visual pathways in elasmobranchs: organization and phylogenetic implications. *Journal of Experimental Zoology*, Suppl. 5, 97–107.

Northcutt, R. G. and Kicliter, E. (1980) Organization of the amphibian telencephalon. In S. O. E. Ebbesson (ed.), *Comparative Neurology of the Telencephalon*. New York: Plenum, pp. 203–255.

Northcutt, R. G. and Puzdrowski, R. L. (1988) Projections of the olfactory bulb and nervus terminalis in the silver lamprey. *Brain, Behavior and Evolution*, 32, 96–107.

Northcutt, R. G., Reiner, A., and Karten, H. J. (1988) Immunohistochemical study of the telencephalon of the spiny dogfish, *Squalus acanthias*. *Journal of Comparative Neurology*, 277, 250–267.

Overmier, J. B. and Hollis, K. L. (1983) The teleostean telencephalon in learning. In R. E. Davis and R. G. Northcutt (eds.), *Fish Neurobiology, Vol. 2: Higher Brain Areas and Functions*. Ann Arbor, MI: University of Michigan Press, pp. 265–284.

Reiner, A. (1993) Neurotransmitter organization and connections of turtle cortex: implications for the evolution of mammalian isocortex. *Comparative Biochemistry and Physiology*, 104A, 735–748.

Reiner, A. and Northcutt, R. G. (1987) An immunohistochemical study of the telencephalon of the African lungfish, *Protopterus annectens*. *Journal of Comparative Neurology*, 256, 463–481.

Reiner, A. and Northcutt, R. G. (1992) An immunohistochemical study of the telencephalon of the Senegal bichir (*Polypterus senegalus*). *Journal of Comparative Neurology*, 319, 359–386.

Saunders, P. T. and Ho, M.-W. (1981) On the increase in complexity in evolution. II. The relativity of complexity and the principle of minimum increase. *Journal of Theoretical Biology*, 90, 515–530.

Saunders, P. T. and Ho, M.-W. (1984) The complexity of organisms. In J. W. Pollard (ed.), *Evolutionary Theory: Paths Into the Future*. New York: Wiley, pp. 121–139.

Smeets, W. J. A. J. and Northcutt, R. G. (1987) At least one thalamotelencephalic pathway in cartilaginous fishes projects to the medial pallium. *Neuroscience Letters*, 78, 277–282.

Striedter, G. F. (1990) The diencephalon of the channel catfish, *Ictalurus punctatus*. II. Retinal, tectal, cerebellar and telencephalic connections. *Brain, Behavior and Evolution*, 36, 355–377.

Wicht, H. and Northcutt, R. G. (1992) The forebrain of the Pacific hagfish: a cladistic reconstruction of the ancestral craniate forebrain. *Brain, Behavior and Evolution*, 40, 25–64.

Wullimann, M. F. and Northcutt, R. G. (1990) Visual and electrosensory circuits of the diencephalon in mormyrids: an evolutionary perspective. *Journal of Comparative Neurology*, 297, 537–552.

26

Visual Forebrain in Amniotes

INTRODUCTION

A fox stealthily moves through the tall grass of a meadow in search of a meal. A noise somewhat to the animal's left attracts its attention. The large, sound-gathering pinnae of its ears localize the source; the eyes and head reflexly turn to this point in space. Something is moving there. The fox fixates both eyes in the direction of the sound source. The image of a ground squirrel comes into focus on the fox's retina. So too does the image of the tall grass in front of the squirrel as well as images of the earth, twigs, and leaves behind the squirrel. The fox's visual system must sort out this jumble of lines, textures, colors, and shapes to present the fox with a representation of the critical features of its visual environment. The target must be sharply differentiated from the foreground and the background. Other factors need to be evaluated: How far away is it? It this something to pounce on for an easy meal? Or is it something that is too big to handle? Is it something with formidable defenses like a porcupine or snake? Or is it something that might make a meal of the fox? Finally, the fox must consider what the consequences were of its last encounter with something that looked like that. The neuron populations in the fox's visual system perform these and related analyses continuously as the fox moves through the visual world.

In previous chapters, we discussed the visual system at the receptor level and at the levels of the midbrain and thalamus. We pointed out that the tectum of the midbrain was a region that permitted the central nervous system to coordinate "maps" of the body with "maps" of the surrounding sensory worlds of vision, hearing, and some other distance senses. These midbrain maps allowed for our fox's eyes and head to orient accurately towards the source of the sound. Information about other properties of the external visual world are sent to the dorsal thalamus directly from the retina via the optic nerve and tract. From the dorsal thalamus, visual pathways ascend to the telencephalon, which processes the visual information that will be used to guide the fox's decisions.

Much of our understanding of the visual forebrain comes from the study of mammals. Until the last several decades, the mammalian pattern of thalamic nuclei and telencephalic areas was regarded as a unique achievement of mammals. Comparative anatomists and physiologists now recognize that the mammalian pattern also exists in diapsid reptiles, birds, and turtles and indeed has certain features in common with amphibians, cartilaginous fishes, and ray-finned fishes as well. A unifying framework for discussion of these features is provided by the concept of dual ascending sensory pathways to the dorsal thalamus, the **lemnothalamic** and **collothalamic** pathways, which we introduced in Chapters 22 and 25. The lemnothalamic pathways bring sensory information to the dorsal thalamus directly (i.e., without a synapse in the midbrain roof) from the spinal cord, brainstem, and retina. The collothalamic pathways also carry sensory information to the dorsal thalamus from the spinal cord, brainstem, and retina, but they do so via the midbrain roof, which is known as the colliculi in mammals and the tectum in nonmammals.

Both the collothalamic and lemnothalamic nuclei send their efferents to separate areas of the **dorsal pallium** in the telencephalon. In mammals, these areas are known as **sensory isocortex.** The presence of thick granule cell layers (II and IV) is one of the hallmarks of sensory cortex. The axons of the dorsal thalamic neurons terminate in the granule cell layers of the sensory cortical areas (see Chapter 22). In this chapter, we will explore the visual pathways to the telencephalon and their regions of termination.

IPSILATERAL RETINAL PATHWAYS AND STEREOSCOPIC VISION

The typical relationship between the retina and the brain in vertebrates is for each retina to send most of its axons to the contralateral half of the brain. Thus, the right eye sends its optic nerve across the midline to the left side of the brain. The place where the two optic nerves cross is known as the **optic chiasm.** The term "chiasm" is derived from the Greek word for the crossing of two lines, as in the Greek letter *chi,* which has the same shape as our letter X. For most vertebrates, the majority of optic axons project to the contralateral side of the brain, although some ipsilateral axons have been reported in many vertebrates. In mammals, however, the situation is somewhat different; considerable numbers of retinal axons do not cross but enter the brain on the same side. One theory of the heavy contralateral termination of optic axons holds that this crossing is related to the right–left reversal (as well as inversion) of the images of objects when they pass through the optics of the eye and form on the retina.

The marked increase in the percentage of ipsilateral retinal axons may be a consequence of a shift in the location of the eyes from a more lateral position in the skull to a more frontal position. In general, the more frontal the location of the eyes, the greater is the proportion of ipsilateral retinal axons. In rodents, which have relatively laterally located eyes, the ipsilateral retinal axons are approximately 10% of the total number. In carnivores and primates, which have frontal eyes, the ipsilateral retinal axons are in the range of 40–50% of the total retinal projection to the brain.

Binocularity and Stereoscopic Vision. Frontally located eyes have a considerable degree of overlap in the regions of visual space that each eye can survey (binocular vision). If the extent of overlap is sufficient, it permits the development of stereoscopic (three-dimensional) vision, which offers the animal a highly precise estimate of the distances of objects. Such an ability is of enormous advantage to predators, which must accurately assess the distance of prey for the final assault. Stereoscopic vision is also of value to animals that are not typically predators; animals that have become adapted to life in the trees, such as squirrels, tree shrews, and primates. Accurate determination of distance is vital for life in the arboreal environment as the animal moves from branch to branch, sometimes for executing feats of acrobatics, as well as for gathering food. Similar considerations apply to the diurnal bats that make use of vision rather than echolocation when foraging for food.

Frontal Vision and Visual System Evolution. The shift of the location of the eyes in the skull appears to have begun in the therapsid ancestors of mammals. The shift may have occurred as a consequence of the development of the **zygomatic arch** of the skull. The zygomatic arch is a bony arc formed by three bones of the skull of mammals. Under this arc passes the **temporalis muscle,** which is one of the major muscles that close the jaw. The space under the zygomatic arch through which this masticatory muscle passes is known as the **temporal opening.** The chewing of food (rather than just swallowing it whole or

in large chunks) and the resultant increased efficiency of food-resource utilization are characteristics that have only evolved fully in mammals, allowed for by the increased development of the zygomatic arch and temporalis muscle. In addition, the structure of the skull changed, permitting greater mechanical strength to withstand the enormously increased stresses that were placed on the skull due to the greater pull of the increasingly more powerful muscles. With the enlargement of the zygomatic arch and strengthening of the skull, the **orbits** (eye sockets) were pushed forward to the front of the skull and closer to the midline.

Figure 26-1 shows skulls that represent four stages in the evolution of the therapsid skull. These animals vary in their use of jaws for chewing and also in the location of the orbits in the skull. In this figure, the early therapsid from the late Permian era shows the typical early amniote pattern of a small zygomatic arch and small temporal openings through which the temporalis muscle passed en route to the lower jaw. This skull also has orbits that are located in a lateral position. In the two Lower Triassic animals, one herbivorous and one carnivorous, enlarged temporal openings are present, and the orbits are located more frontally. The Lower and Middle Triassic carnivores show the pattern that is typical of modern carnivores, primates, and other frontal-eyed mammals, that is, a large zygomatic arch, large temporal openings, and frontally located orbits.

As may be seen in Figure 26-2, the shift to frontal eyes resulted in a greater overlap of the visual fields of each eye, producing a large binocular field, which in turn set the stage for the development of effective stereoscopic vision. In general, the greater the degree of overlap, the greater the distance over which stereoscopic vision is functional. One should note, however, that the shift of the orbits to the front and towards the midline may not have been driven by selective pressures related to vision, but rather by selective pressures related to feeding. Once in their new location, however, the relatively few ipsilateral retinal axons now took on a more important role, and a new set of selective pressures on the visual system came into operation, effecting a new evolutionary trend towards greater ipsilateral representation.

Figure 26-2 also illustrates the point that only predators, or those with a low risk of predation because they are adapted for life in the trees, can afford the luxury of stereoscopic vision; the price paid for frontal eyes is the loss of panoramic (360°) vision, which is an effective early warning system for the approach of predators from the rear. Some therian mammals, such as rabbits, have lateral eyes, located high on the skull, which afford panoramic vision, but this may have evolved secondarily, that is, after the development of a substantial ipsilateral projection of the retina on the central visual system.

The shift to more frontal eyes and the consequent increase in ipsilateral retinal input to the brain had a profound impact on the evolution of the mammalian central visual system. At the dorsal thalamic level, it is correlated with an elaborate organization of the visual lemnothalamus into **laminae** or plates that are organized to keep ipsilateral and contralateral inputs separate. Indeed, the visual information coming from each eye is kept separate until it reaches the level of the isocortex where interaction then permits the fusion of the two retinal

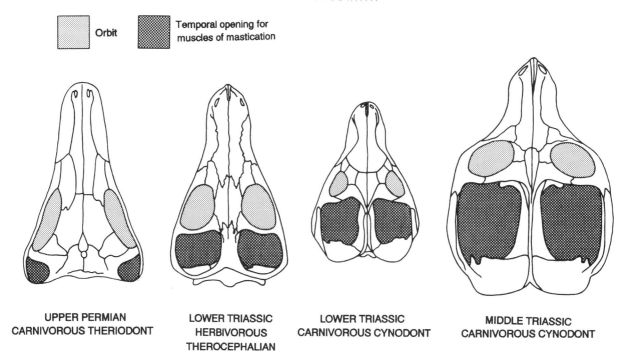

▦ Orbit	▨ Temporal opening for muscles of mastication

UPPER PERMIAN CARNIVOROUS THERIODONT　　**LOWER TRIASSIC HERBIVOROUS THEROCEPHALIAN**　　**LOWER TRIASSIC CARNIVOROUS CYNODONT**　　**MIDDLE TRIASSIC CARNIVOROUS CYNODONT**

FIGURE 26-1. Drawings of skulls of four synapsids to show the relationship between the development of the muscles used for chewing, the zygomatic arch (the arc of bone that surrounds the temporal opening), and the shift from lateral to frontal orbits. The Upper Permian carnivorous theriodont has distinctly lateral orbits, small muscles of mastication, and a small temporal opening. The Lower Triassic herbivorous therocephalian and Lower Triassic carnivorous cynodont, later forms than A, have more frontal orbits, larger muscles of mastication, larger temporal openings, and larger zygomatic arches. The Middle Triassic carnivorous cynodont, similar to the lower Triassic carnivore, has distinctly frontal eyes, massive muscles of mastication, large temporal openings, and a large zygomatic arch. From *Vertebrate Paleontology and Evolution* by R. L. Carroll. Copyright © 1988 by W. H. Freeman and Company. Used with permission.

images, resulting in our perception of the three dimensionality of objects and visual space.

VISUAL PATHWAYS TO THE TELENCEPHALON IN MAMMALS

Lemnothalamic Visual Forebrain

The visual lemnothalamic route is from the ganglion cell layer of the retina (see Chapter 2) to its target cell population in the dorsal thalamus. The mammalian term for this structure the **nucleus geniculatus lateralis, pars dorsalis,** which translates to the dorsal part (or division) of the lateral geniculate nucleus, or, more simply, the dorsal lateral geniculate nucleus. This structure is often abbreviated as GLd or LGd. The word "geniculate" comes from the Latin word *genu*, which means a "knee" and refers to the knee-like appearance of this structure in humans and other primates. The modifier "lateral" serves to distinguish it from the medial geniculate nucleus, which is an auditory dorsal thalamic relay nucleus. The similarity of names and the close proximity of the two nuclei is unfortunate, however, because it suggests to beginners that the medial geniculate nucleus is the auditory equivalent of the dorsal lateral geniculate nucleus. As discussed in greater detail in Chapter 28, the two are quite different; the dorsal lateral geniculate nucleus is a

lemnothalamic nucleus, whereas the medial geniculate nucleus is a collothalamic nucleus.

In mammals, the pars dorsalis of the lateral geniculate nucleus is composed of two or more subdivisions. Of these subdivisions, some receive the ipsilateral retinal axons, and the others receive the contralateral retinal axons. In mammals with laterally located eyes, such as many ground-dwelling rodents, rabbits, and insectivores, the dorsal lateral geniculate nucleus consists of a single column of cells in the caudal dorsal thalamus. The axons of the ipsilateral and contralateral components of the **optic tract** (the name for the bundle of retinal axons after they pass the optic chiasm) terminate in separate segments within this cell column, with the contralateral segment being the larger. The locations of the ipsilateral and contralateral termination fields can only be determined experimentally by chemically or otherwise labeling the axons that follow each route. The distributions of labeled and unlabeled axons thus reveal the "covert" or hidden segregation pattern. In some mammals, such as rats and opossums, segregation of the ipsilateral and contralateral retinal axons is incomplete, that is, the ipsilateral and contralateral axons terminate in overlapping fields. Figure 26-3 shows examples of two types of retinal termination patterns in the dorsal lateral geniculate nucleus in two lateral-eyed mammals, a rodent (rat) and an insectivore (hedgehog). At the bottom, the figure shows the right and left retinas, and the right and left optic nerves emerging from their respective retinas. At the optic chiasm, the nerves partially decussate

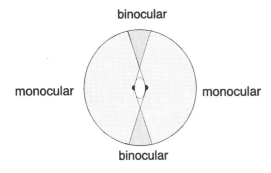

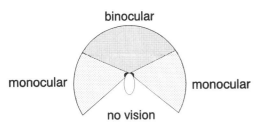

FIGURE 26-2. Diagram of the effects of a shift from lateral orbits to frontal orbits on the size and location of the monocular and binocular visual fields. Top: A mammal with lateral eyes has panoramic vision, a small field of binocular overlap in the front, another small binocular field in the back, and large monocular fields at the sides. Bottom: A mammal with frontal orbits has a large binocular field in the front, moderate monocular fields at the sides, and no vision in a substantial sector of the visual field at the rear.

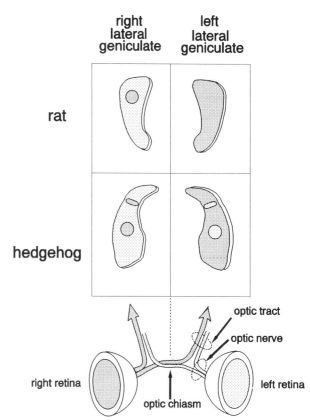

FIGURE 26-3. Drawing of ipsilateral and contralateral terminations of the optic tract in the pars dorsalis of the lateral geniculate nucleus in a rat and a hedgehog. The darkly shaded areas indicate the fields of termination from the right retina. In both mammals, the ipsilateral projections in the right dorsal lateral geniculate nucleus are much smaller than are the contralateral projections to the left dorsal lateral geniculate nucleus. Note that in the dorsal lateral geniculate nucleus of the hedgehog, the zones where the ipsilateral projections from the left retina would fall are free of contralateral axons from the right retina, that is, a complete segregation of ipsilateral and contralateral terminations exists. In contrast, in the rat dorsal lateral geniculate nucleus, no zone is present that is solely dedicated to ipsilateral fibers; the ipsilateral zone overlaps the contralateral zone.

(cross the midline) and form the right and left optic tracts. The figure points out an important distinction between the optic nerve and the optic tract, that is, the retinal axons contained in the left optic nerve all originate in the left retina, whereas the left optic tract contains axons from both retinas.

The panels above the optic tracts in Figure 26-3 contain representations of transverse sections through the right and left dorsal lateral geniculate nuclei of a rat and a hedgehog. The darkly shaded areas indicate the regions where axons of the right optic tract terminate. In the rat, the left dorsal lateral geniculate nucleus is completely filled with terminals of axons originating in the right retina. In the right dorsal lateral geniculate nucleus, the axon terminations from the right retina are confined to a small region. Note that the left dorsal lateral geniculate nucleus gives no clue as to where the retinal axons from its ipsilateral (left eye) would terminate. Contrast the rat dorsal lateral geniculate nucleus with that of the hedgehog. First, the hedgehog has two zones of ipsilateral termination in the right dorsal lateral geniculate nucleus. Second, the region of ipsilateral termination in the left dorsal lateral geniculate nucleus has no contralateral terminals in it, that is, the hedgehog dorsal lateral geniculate nucleus exhibits complete segregation of ipsilateral and contralateral regions by means of specialized regions that are devoted solely to ipsilateral axons.

In mammals with frontal eyes, such as tree shrews, cats, and monkeys, the dorsal lateral geniculate nucleus consists of a number of plate-like cell columns or laminae. Figure 26-4 shows the dorsal lateral geniculate nucleus of three frontal-eyed mammals: an insectivore that makes extensive use of vision (tree shrew), a carnivore (cat), and a primate (rhesus monkey). In such animals, the individual laminae receive exclusively ipsilateral or contralateral fibers, that is, an "overt" segregation. Within an overtly segregated dorsal lateral geniculate nucleus, however, some laminae may have a covert segregation as well. For example, in the tree shrew shown in Figure 26-4, the ipsilateral terminations are in layers 1 and 5, and the contralateral terminations are in layers 2, 3, 4, and 6, which is a complete overt segregation. In the rhesus monkey, the ipsilateral terminations are in layers 2, 3, and 5, and the contralateral terminations are in layers 1, 4, and 6. The S layer, however, shows covert segregation with a small region of it receiving ipsilateral axons and the larger region receiving contralateral axons. A similar pattern is seen in cats, in which the ipsilateral terminal field is in laminae A1, C1, and the rostral portion of M and the contralateral projection is to laminae A, C, C2, and the caudal portion of M.

Cell Sizes in the Dorsal Lateral Geniculate Nucleus. In primates, the cell columns or laminae consist of three types: **mag-**

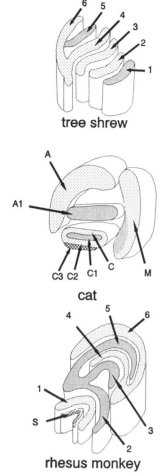

tree shrew

cat

rhesus monkey

FIGURE 26-4. Drawings of various forms of the laminated pars dorsalis of the lateral geniculate nucleus in three species of mammals: a tree shrew, a cat, and a rhesus monkey. The contralateral projections are shown in light stipple and the ipsilateral projections in dark stipple. Rostral is towards the left and caudal towards the right. Adapted from Casagrande and Norton (1991).

As summarized in Table 26-1, the X pathway or channel leaves the retina via medium sized ganglion cells and terminates in the parvocellular laminae of the dorsal lateral geniculate nucleus. These ganglion cells are mainly found in the foveal region of the retina, which is the region of highest visual acuity (sharpest vision). The Y pathway is associated with large retinal ganglion cells and terminates on the magnocellular laminae of the dorsal lateral geniculate nucleus. The Y ganglion cells are typically found in the periphery and are quite responsive to stimuli that move. The W cells have been the least studied. They terminate on the koniocellular components of the dorsal lateral geniculate nucleus. The W channel cells are the most variable in their properties and indeed may actually consist of several different cell types. These three channels differ in the sizes of their receptive fields, their ability to resolve small differences between stimuli, their responsiveness to temporal changes in stimuli and contrast changes, as well as other properties. The X and Y cells of the dorsal lateral geniculate nucleus (parvocellular and magnocellular) terminate in different levels of layer IV of the visual region of the isocortex: the so-called "striate" cortex. The W cells terminate in layers I and II. Figure 26-5 is a schematic representation of the X, Y, and W channels in the lemnothalamic visual pathway of a primate. Cells with X-, Y-, and W-like properties have also been described in non-primate mammals, such as cats, ferrets, rats, tree shrews, opossums, and rabbits. Although widely distributed throughout mammals, the X, Y, and W channels have not been widely reported in nonmammals. Whether this lack is due to the uniqueness of these neuronal systems in mammals or to the paucity of electrophysiological studies in nonmammalian preparations is not clear at present. Certain characteristics of X and Y cells have been reported in nonmammals, especially in birds, which have independently evolved visual systems that are the equal of (and in some instances superior to) the best visual systems of mammals, including the primates. For the present, however, the X, Y, and W channels that have been fully characterized in mammals by morphological, chemical, and electrophysiological criteria are not known in nonmammalian vertebrates.

Species Differences in the Dorsal Lateral Geniculate. What is the phyletic relationship among the individual laminae of the dorsal lateral geniculate nucleus in different species of mammals? Is lamina 6 of a tree shrew homologous with lamina 6 of a rhesus monkey? Which lamina of a carnivore dorsal lateral geniculate nucleus is homologous with layer 6 of a primate dorsal lateral geniculate nucleus? The answer to this question is "all and none." The reason for this seemingly paradoxical answer is that we are dealing here with a **field homology** (see Chapter 1). Each component of the dorsal lateral geniculate nucleus of mammals is derived from the single dorsal lateral optic nucleus (as will be discussed below) of their early amniote ancestors, as is the dorsal lateral optic nucleus of all other descendants of the early amniote stock (diapsid reptiles, birds, and turtles). When this unitary population of neurons became segregated into specialized ipsilateral and contralateral divisions, as in different species of mammals, or into separate subdivisions as we will see below in birds, these subdivisions retained their homologous relationship as a field to the ancestral condition. Thus, we cannot say that any single component of the

nocellular (large celled), **parvocellular** (small celled), and **koniocellular** (very small celled). The magnocellular component comprises two layers (layers 1 and 2 in the monkey in Fig. 26-4), in which one layer receives axons from the ipsilateral eye and the other from the contralateral eye. Laminae 3–6 are parvocellular. Diurnal primates have a greater proportion of the parvocellular part of the dorsal lateral geniculate nucleus than do nocturnal primates, especially in the areas of this nucleus that are devoted to the central part of the visual field, where acuity is highest. The koniocellular components of the monkey dorsal lateral geniculate nucleus are found in the S layer and in the spaces between the laminae (the **intralaminar zone**). Each of these cell types is associated with a different class of physiological characteristics known as *X, Y,* and *W*. These classes of characteristics are considered to be "channels" that are presumed to bring different aspects of the visual world to the visual parts of the telencephalon. Since these channels operate concurrently, they are sometimes referred to as "parallel pathways" or are said to perform "parallel processing."

TABLE 26-1. Classes of Cells in the Visual Lemnothalamus of Primates

Cell Class	Retinal Ganglion Cell Type	Lateral Geniculate Cell Type	Cortical Layer of Termination	Physiological Properties
X	Medium (mostly in fovea)	Parvocellular	IV (also III and VI)	Small receptive field High spatial resolution Low temporal resolution Low sensitivity to contrast and motion
Y	Large (mostly in periphery)	Magnocellular	IV	Medium receptive field Moderate spatial resolution High temporal resolution High sensitivity to contrast and motion
W	Small (variable location)	Koniocellular	I and II	Medium receptive field Moderate spatial resolution Variable sensitivity to movement

dorsal lateral geniculate nucleus of one mammal is uniquely homologous to any other component in any other mammal; rather, all components of one species are homologous to all components of another as a field.

Pars Ventralis of the Lateral Geniculate. A **pars ventralis (ventral division)** of the lateral geniculate nucleus is also present in the thalamus of mammals, but it is a part of the ventral thalamus (see Chapter 19). This structure, which is a typical feature of vertebrate visual systems, receives axons from the retina but does not send its efferents to the telencephalon as do the lemnothalamic and collothalamic visual nuclei of the dorsal thalamus. In primates, the ventral part of the lateral

geniculate nucleus is known as the **pregeniculate nucleus** and it is very variable in its size and location. For example, in prosimians it is located on the lateral surface of the pars dorsalis, whereas in anthropoids, it has shifted so far dorsally that one author has described it as an "eyebrow" over the dorsal division.

Striate Visual Cortex. The lemnothalamic visual efferents from the dorsal lateral geniculate nucleus terminate in the most posterior region of the cerebral hemisphere, in part of the **occipital lobe.** This area of isocortex is known as the **striate cortex** because of the prominent stripe or band located within layer IV. This stripe is called various names after the scientists who first described it, such as the **stripe** (or **line**) **of Gennari,**

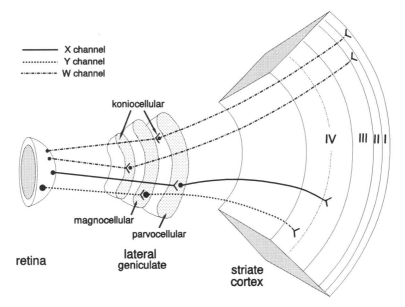

FIGURE 26-5. The X, Y, and W channels in the retinogeniculate-striate visual pathway of a mammal. The X channel begins with small ganglion cells in the central retina and terminates in the parvocellular portion of the dorsal lateral geniculate nucleus. The Y channel originates in large ganglion cells located in the peripheral regions of the retina and terminates in the magnocellular region of the dorsal lateral geniculate nucleus. The W channel begins in the smallest ganglion cells at various retinal locations and terminates in the koniocellular regions of the dorsal lateral geniculate nucleus, which are in a specialized lamina and also in the space between the magnocellular and parvocellular laminae.

after the eighteenth-century Italian physician, Francesco Gennari. It also is known as the **outer stripe of Baillargier,** after the nineteenth-century French physician, Jules Baillargier, who observed this band and one other while he was describing the lamination of the cerebral cortex. This stripe is formed by the heavily myelinated axons of dorsal lateral geniculate neurons as they pass through layer IV en route to their destinations within this region of cortex. Figure 26-6 shows a low-power photomicrograph of striate cortex with a prominent stripe in a raccoon. Figure 26-7 shows the location of the striate cortex in six mammals: a marsupial (opossum), a rodent (rat), a lagomorph (rabbit), an arboreal insectivore (tree shrew), and a prosimian primate (a bushbaby).

Most maps of striate cortex show its location on the lateral surface of the brain, which in many instances gives a misleading underestimate of the total area, because this cortical region

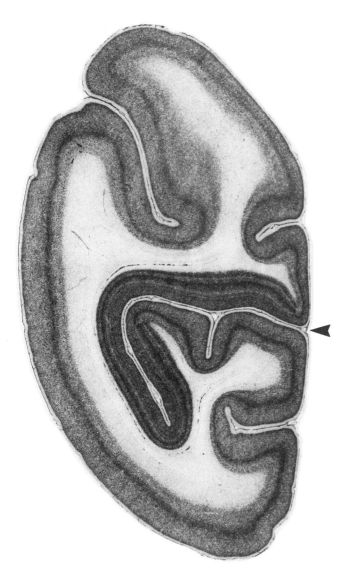

FIGURE 26-6. Photomicrograph of a section through the striate cortex—in the involuted area internal to the arrowhead—in a mammal (*Procyon lotor*). Note the particularly dark stripe in this region that marks layer IV. Photomicrograph courtesy of Wally Welker.

continues onto the medial surface. Figure 26-8 shows the striate cortex of an owl monkey on both the lateral and medial surfaces. Note on the medial surface the presence of a deep sulcus, the **calcarine** (claw-like) **sulcus**, which is a prominent feature of the telencephalon in primates. A considerable amount of striate cortex is "buried" within the depths of this sulcus. Buried striate cortex (as well as buried cortex with other specializations) occurs commonly in gyrencephalic mammals. The efferents of the striate cortex are predominantly to extrastriate cortex, but this cortical area also projects to sites including the dorsal lateral geniculate nucleus, superior colliculus, and other targets of the optic tract in the diencephalon and mesencephalon.

Retinotopic Organization. An important characteristic of the central visual system is its retinotopic organization, which is a form of topographic organization found in sensory and motor systems (see Chapter 2). In a retinotopic organization, the relationship between each point on the retinal surface, which corresponds to a point in visual space is preserved in the subsequent cell populations of the visual pathway, for example, the dorsal lateral geniculate nucleus and striate cortex. Although the topology and topography are preserved, the proportional representations of each area are not generally maintained. Thus, the central retina, which in mammals is the area of highest visual acuity, is represented by a greater proportion of neural tissue in the striate cortex than it is on the retina. This increase is known as the **magnification factor.**

Figure 26-9 shows an example of retinotopic organization in the **middle temporal area** or **area MT** of the **extrastriate cortex** of an owl monkey. The extrastriate cortex is the collothalamic cortex, which we will discuss below; we turn to it here simply for illustrating retinotopic organization. The upper diagram in Figure 26-9 represents location in visual space. The plus numbers represent the degrees of elevation of the target in the upper visual field; the minus numbers represent the degrees of elevation of the target in the lower visual field. The lower diagram indicates the positions of these visual field locations within area MT. As we will see below, the extrastriate visual cortical areas contain a large number of such retinotopically organized areas.

Columnar Organization. In the 1960s, David Hubel and Torsten Wiesel documented the existence of a vertical (i.e., columnar) organization of cat visual cortex; that is, that the visual cortex is organized in a series of vertical columns, with each column conveying information about a variety of stimulus properties such as direction, orientation, and color. Moreover, each eye has its own set of columns in the cortex, that is, the segregation into ipsilateral and contralateral subdivisions persists at the cortical level in a covert form. Thus, if an eye is removed surgically, the physiologist finds that certain columns are active while neighboring columns are "silent." Since a given eye is said to dominate a particular column, this distribution of right and left eye inputs to the right or left visual cortex is known as **ocular dominance columns.** These covert columns can be visualized by a variety of chemical techniques. One method involves the mitochondrial enzyme, cytochrome oxidase, which increases in concentration when cells are metabolically active, such as when they are being stimulated. Staining for cytochrome oxidase in the cortex of an animal that has had

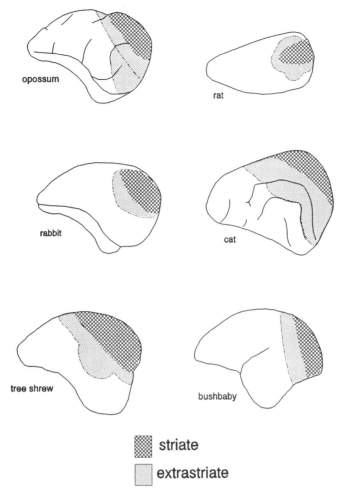

FIGURE 26-7. Variations in the extent of striate and extrastriate visual cortex in six mammals. In each drawing, rostral is towards the left.

one eye open and one eye closed reveals a series of perpendicular bands of staining within striate cortex that are particularly dark in layer IV, the granule cell layer that receives its predominant input from the dorsal lateral geniculate nucleus. Other techniques reveal the existence of these bands in other layers. Indeed, many anatomists and physiologists now believe that the columns are continuous throughout the full depth of the cortex.

Figure 26-10 illustrates ocular dominance columns in a small region of striate cortex on the right side. The darker shaded regions receive input from the contralateral (left) eye, and the lighter shaded regions receive input from the ipsilateral (right) eye. Notice that the ocular dominance columns extend throughout the entire thickness of the cortex (layers I–VI). The ocular dominance columns themselves are further subdivided into functionally different, smaller columns, such as **orientation columns** in which the cells in one such column may prefer vertical lines, while some of its neighboring columns may prefer diagonal lines, and still other columns would selectively respond to horizontal lines.

Within the ocular dominance columns revealed by cytochrome oxidase staining, the staining is not uniform; instead; it is patchy with lighter and darker regions, especially within layers I–III. The darker regions (with more intense activity) have been described as **blobs, puffs,** or **patches** by various workers. These differences in staining density represent differences in functional organization. The cells found within the darker regions, called **blob cells,** differ from the cells found in the lighter regions, **interblob cells,** in several characteristics. A greater proportion of the blob cells respond differently to colors but do not respond very well to differences in the orientation of the stimulus (horizontal, vertical, or diagonal). The interblob cells tend to respond to stimulation of either eye, whereas blob cells are more monocular, and the blob cells are more responsive to the overall shape of an object rather than to its fine details.

Figure 26-11 shows a pair of adjacent ocular dominance columns in the left striate cortex. One column is dominated by the right eye and one by the left. The figure shows that the left-eye dominated column receives its stimulation from the left half of the retina of the left eye, while the right-eye dominated column receives its input from the left half of the right retina. Note the laminar segregation of ipsilateral and contralateral axons in the dorsal lateral geniculate nucleus. Within each ocular dominance column is a series of orientation columns, each

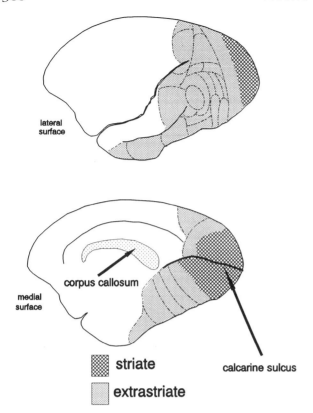

FIGURE 26-8. Drawing of the striate and extrastriate visual cortices of an owl monkey. The subdivisions within the extrastriate visual cortex indicate multiple representations of the visual world. Based on data from Allman and McGuinness (1988) and Sereno and Allman (1991).

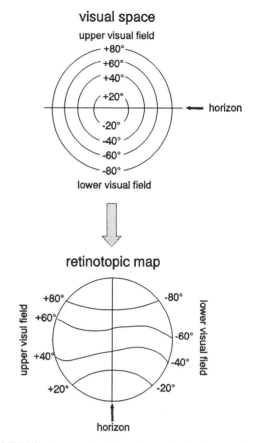

FIGURE 26-9. Diagram of retinotopic organization in extrastriate area MT (middle temporal area) in an owl monkey. The positive numbers in the upper figure indicate degrees of elevation into the upper visual field. The negative numbers indicate degrees of elevation into the lower visual field. The lower figure shows how points in space are represented in area MT.

of which prefers a different stimulus orientation. Within the orientation columns in layers I–III are the blobs. Pairs of adjacent ocular dominance columns—with their integral orientation columns, blobs, and interblobs—are known as **hypercolumns** and appear to be a kind of basic organizational unit for analyzing the composition of a particular region of visual space.

Clear instances of ocular dominance columns have been reported in a variety of Old World monkeys and apes and in carnivores. The columns are less clear in prosimians and New World monkeys and appear to be absent in rodents and lagomorphs. This distribution across species suggests that the presence of ocular dominance columns may be related to the process of stereoscopic vision.

Collothalamic Visual Forebrain

The collothalamic visual nucleus in mammals is called **nucleus lateralis posterior** or, in primates, the **pulvinar,** which derives from the Latin word for a pillow or cushion. Primates also have a nucleus called lateralis posterior, but it is not a collothalamic visual nucleus. In cats, on the other hand, both nucleus lateralis posterior and the pulvinar are collothalamic visual nuclei. The pulvinar in both cats and primates consists of a number of subdivisions of which most, but not all, are collothalamic visual nuclei.

In the nomenclature of the German neurologist Korbinian Brodmann (1868–1918), the striate cortex is **area 17,** and the visual collothalamic termination area in the extrastriate cortex

is **area 18** and **area 19.** The extrastriate visual cortex begins immediately adjacent to the rostral boundary of the striate cortex (Fig. 26-7). Each of the subdivisions within the extrastriate visual cortex is a representation of the visual field. The principal source of afferents to the extrastriate visual cortex is via the collothalamic visual pathway from the retina to the superior colliculus to the lateralis posterior/pulvinar complex and hence to the cortex. The termination field of neurons from the visual collothalamus is the region of cortex adjoining striate cortex. A secondary source of afferents to extrastriate cortex is the dorsal lateral geniculate nucleus, as shown in Figure 26-12, although this projection is considerably more modest than the projection of the dorsal lateral geniculate nucleus to striate cortex. Likewise, the lateralis posterior/pulvinar complex sends some axons to the striate cortex.

The relationship between the lateralis posterior/pulvinar complex and the extrastriate cortex is intricate and involves reciprocal connections between this collothalamic cell group and both extrastriate and striate cortices. Moreover, the nucleus appears to be divided into striate–recipient and extrastriate–recipient zones in addition to the zone that receives axons from the superior colliculus. Like the extrastriate cortex that it innervates, the lateralis posterior/pulvinar complex contains multiple representations of the visual field with

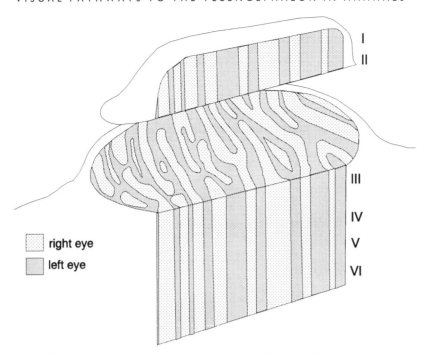

FIGURE 26-10. Diagram of ocular dominance columns in a small region of the striate cortex of a mammal. A slab of tissue has been removed at the top to reveal the three-dimensional structure of the columns. The Roman numerals indicate the layers of the cortex.

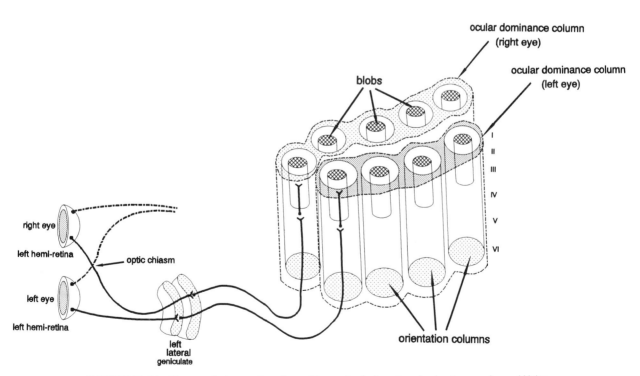

FIGURE 26-11. Diagram of a hypercolumn formed by a pair of adjacent ocular dominance columns. Within each ocular dominance column can be seen other forms of columnar organization such as orientation columns and "blob" columns.

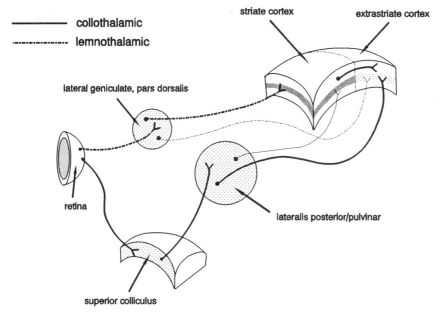

FIGURE 26-12. Summary of collothalamic and lemnothalamic visual pathways in mammals. Heavy lines indicate major projections. Thin lines indicate minor projections.

retinotopic organization. Again, like the extrastriate visual cortex, the retinotopic organization is not as precise as is found in the dorsal lateral geniculate nucleus and striate cortex. The number of representations of the visual field in the lateralis posterior/pulvinar complex is, however, far fewer than are found in the extrastriate visual cortex, which suggests that the lemnothalamic input to the extrastriate cortex may be more of an organizing force for the multiple visual areas than is the collothalamic input.

The high degree of dependence of many (and perhaps most) species of mammals on vision may be seen in the large proportion of cortex devoted to vision. Some of the cortical areas that respond to visual stimulation overlap areas traditionally regarded as somatosensory or motor. Typically, the total area of dorsal pallium devoted to both striate and extrastriate cortices is 45–50%, but when diurnal mammals are compared to nocturnal mammals, the percentages are higher for the diurnal by 10–15%. In diurnal Old World monkeys, the total area of cortex responsive to visual stimuli to at least some degree is now estimated to be approximately 75% and may be even higher in great apes and humans.

The striate visual cortex has a highly retinotopic organization. In the extrastriate cortex, the situation is considerably more complicated because not one, but a multiplicity of retinotopic organizations is present. Figure 26-8 shows the subdivisions of the extrastriate visual cortex in an owl monkey, which is a New World monkey. The figure depicts 24 separate regions in which retinotopic maps can be found. For the sake of simplicity, we have not provided the names of the individual regions. Moreover, by the time you read this chapter, half a dozen additional regions may have been found. The middle temporal area, referred to above, is one of these regions, located in the middle part of the temporal region as its name states. In Old World monkeys, 36 regions have been found. Prosimian pri-

mates, as well as carnivores, rodents, and lagomorphs have at least 13 or 14 extrastriate visual areas. The interconnections of these multiple extrastriate areas are very complex, but in general, those closest to the striate cortex send their efferents to those extrastriate areas that are more remote in a kind of cascade manner.

The maps within various parts of extrastriate cortex differ from one another in a number of respects; for example, adjacent maps may be mirror images of one another or they may only map part of the visual field rather than the whole field. Indeed, a number of these newly discovered visual areas overlap areas of cortex traditionally considered to be motor or somatosensory, as mentioned above. Some of these areas seem to be more involved with color, others more with the shape of objects, and still others more with stimuli that move.

To return to our fox in the meadow who was processing the retinal image of the ground squirrel, the fox located the squirrel with its earlier stages of processing, that is, its superior colliculus, which has a precise retinotopic organization. It interpreted the retinal images with the lemnothalamic, collothalamic, intercortical, and intracortical pathways of a number of its multiple visual cortical areas. Other areas of extrastriate cortex, along with multimodal association cortices and limbic cortices involved in memory functions, as will be discussed in Chapter 30, allowed the fox to process whether encounters with similar stimuli in the past were satisfactory or not so that it could come to a decision as to whether or not to pounce. Exactly how and where the information that is processed in these multiple cortical regions comes together to form perceptions of whole objects, with the qualities of form, color, position, and motion all integrated, and how the cortical regions organize unfamiliar patterns of stimuli into meaningful visual images are the subjects of intensive investigation by vision scientists.

TABLE 26-2. Thalamic Sources of Afferents to the Visual Telencephalon of Amniotes

Amniotes	Lemnothalamic	Collothalamic
Mammals	Lateral geniculate nucleus, pars dorsalis (various numbers of laminae)	N. lateralis posterior and/or pulvinar
Diapsid Reptiles	Dorsal lateral optic nucleus (nucleus intercalatus thalami)	Nucleus rotundus
Birds	Dorsal lateral optic nucleus (nucleus opticus principalis thalami or OPT complex), which includes: Nucleus dorsolateralis, pars lateralis Nucleus dorsolateralis anterior, pars magnocellularis Nucleus suprarotundus Dosolateral nucleus of OPT Nucleus superficialis parvocellularis	Nucleus rotundus
Turtles	Dorsal lateral optic nucleus (dorsal optic nucleus)	Nucleus rotundus

PATHWAYS TO THE VISUAL TELENCEPHALON IN DIAPSID REPTILES, BIRDS, AND TURTLES

The collothalamic and lemnothalamic visual pathways that were described for mammals above and in Chapters 22 and 25 also occur in diapsid reptiles, birds, and turtles. The recent uncovering of information about the visual system of nonmammals, however, poses a number of nomenclatural problems for students of visual system comparative anatomy and evolution. These difficulties stem from the fact that many of the neuronal groups in the dorsal thalamus already were named before their affiliation with the visual system was understood. This has led to a variety of different names for structures that have the same relationship to the components of the visual pathway. The task

of the student would be simplified if the neuronal population in the dorsal thalamus that receives axons directly from the retina and in turn sends its axons directly to the visual pallium had the same name in all vertebrates; regrettably this is not the case. In an attempt to ease the pain, Table 26-2 lists the comparable collothalamic and lemnothalamic visual nuclei in these four groups of amniotes. Figure 26-13 schematically summarizes the main visual pathways.

Lemnothalamic Visual Pathways

In diapsid reptiles, birds, and turtles, the optic tract terminates in a cluster of nuclei, some of which appear to be within the dorsal thalamus. One or more of these nuclei were initially thought to correspond in some way to components of the dorsal lateral geniculate nucleus of mammals

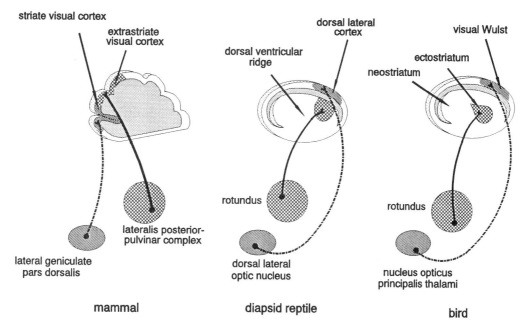

FIGURE 26-13. Summary of the collothalamic and lemnothalamic visual pathways to the telecephalon in three amniotes; a mammal, a diapsid reptile, and a bird. Rostral is toward the right and caudal toward the left.

and were so named. To complete the picture, in each case, a prominent, plate-like pars ventralis of the lateral geniculate nucleus was identified as well. More recent studies, however, have revealed that not all of the retinal target nuclei besides the pars ventralis are in the dorsal thalamus or send their efferents to the telencephalon; they therefore cannot all be considered as comparable to the pars dorsalis of the lateral geniculate nucleus of mammals. Some of these other nuclei, for example, send their efferents to the optic tectum and pretectum. Indeed, in diapsid reptiles and turtles, only one dorsal thalamic, retinorecipient nucleus sends its efferents to the telencephalon. This projection is ipsilateral.

Because of the very different names assigned to the various dorsal thalamic visual nuclei in different animals, we recommend the use of a more general name, the **dorsal lateral optic nucleus** for the lemnothalamic visual nucleus in nonmammalian amniotes. In some diapsid reptiles, the dorsal lateral optic nucleus has been called **nucleus intercalatus thalami.** In some turtles, it is known as the **dorsal optic nucleus.** In birds, the retina also projects to a number of discrete thalamic nuclei. Unlike the case in diapsid reptiles and turtles, however, five of the avian visual nuclei are in the dorsal thalamus and project to the telencephalon. Unlike their nonavian amniote counterparts, these telencephalic projections are mostly bilateral. This group of lemnothalamic visual nuclei in birds has been called the **nucleus opticus principalis thalami,** or **OPT complex.**

As in mammals, in diapsid reptiles, birds, and turtles, ascending axons from the lemnothalamus project to part of the dorsal pallium in the telencephalon, that is, in the lateral part of dorsal cortex in turtles [Fig. 26-14(A)] and in a corresponding cortical area called the **pallial thickening**

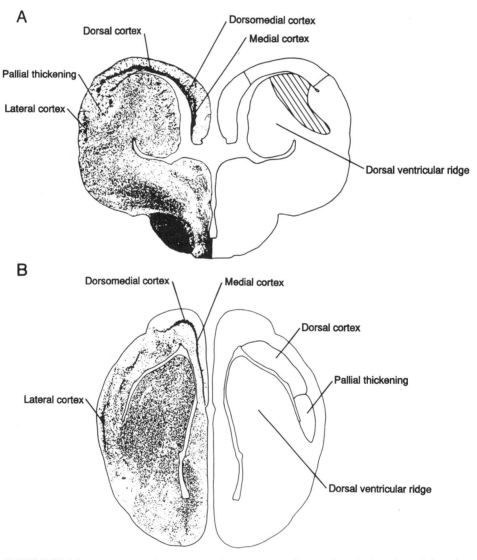

FIGURE 26-14. (A) Transverse hemisection with mirror-image drawing through the telencephalon of a turtle, showing the dorsal cortex and the visual part (diagonal lines) of the anterior dorsal ventricular ridge. Adapted from Balaban and Ulinski (1981). (B) Transverse hemisection with mirror-image drawing through the telencephalon of a diapsid reptile (*Tupinambis nigropunctatus*), showing the dorsal cortex. Adapted from Ebbesson and Voneida (1969). Used with permission of S. Karger AG.

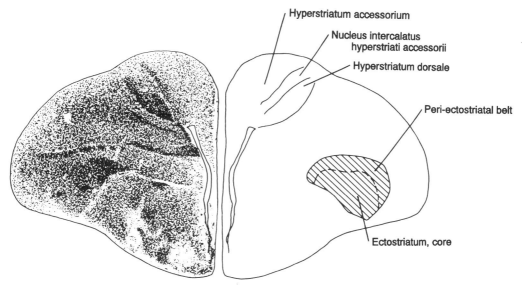

FIGURE 26-15. Transverse hemisection with mirror-image drawing through the telencephalon of a bird (*Columba livia*), showing the visual area of the anterior dorsal ventricular ridge, the ectostriatum (diagonal lines), and the parts of the visual Wulst. Adapted from Karten and Hodos (1967) with additional data from Karten and Hodos (1970). Used with permission of the Johns Hopkins University Press.

in lizards [Fig. 26-14(B)]. In birds, the termination of the lemnothalamic visual projections is in a dorsal region of the telencephalon known as the **visual Wulst.** The German word *Wulst* means a bulge or bump. The visual Wulst contains several layers of granule cells, which are similar in morphology and in some electrophysiological characteristics, including retinotopic organization and columnar organization, to the granule cells in the striate cortex of mammals. These avian granule cells are thought to have evolved independently in birds, convergent with the similar granule cells in mammals, since such cell populations appear to be absent in diapsid reptiles and turtles.

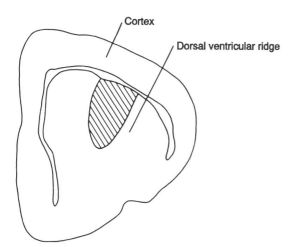

FIGURE 26-16. Drawing of a transverse hemisection through the right telencephalon of a diapsid reptile (*Iguana iguana*), showing the visual area (diagonal lines) of the anterior dorsal ventricular ridge. Adapted from Bruce and Butler (1984b).

The layers within the visual Wulst that receive the lemnothalamic visual projections are the **hyperstriatum dorsale,** the **hyperstriatum intercalatus superior,** and the **nucleus intercalatus hyperstriati accessorii** (Fig. 26-15). The densest terminations are to the hyperstriatum dorsale and nucleus intercalatus hyperstriati accessorii. Like the striate cortex of mammals, the telencephalic components of the lemnothalamic visual pathway send their efferent projections to the optic tectum (superior colliculus), the lemnothalamic visual nucleus, and other targets of the optic tract.

Collothalamic Visual Pathways

Table 26-2 summarizes the collothalamic visual nuclei in mammals, diapsid reptiles, birds, and turtles. Here the nomenclature is relatively simple; in all three nonmammalian groups, the collothalamic visual nucleus is called **nucleus rotundus.** This nucleus, like its mammalian counterpart, projects to a region of the visual telencephalon separate from that of the lemnothalamic visual projection. In diapsid reptiles [Fig. 26-14(B)] and turtles [Fig.26-14(A)], this region is known as the **visual area of the anterior dorsal ventricular ridge.** In birds (Fig. 26-15), the telencephalic target of nucleus rotundus is called the **ectostriatum,** which is a well-demarcated zone within the neostriatum (anterior dorsal ventricular ridge) region. The ectostriatum is composed of two divisions: the **ectostriatal core** and the **peri-ectostriatal belt.** The collothalamic fibers terminate in the core region, which in turn sends its axons to the belt region. A further similarity between the avian nucleus rotundus and the lateralis posterior/pulvinar complex of mammals is that nucleus rotundus itself consists of several subdivisions, each projecting to a different region of the ectostriatum. In addition, another collothalamic nucleus, **nucleus triangularis,** which sits like a pointed hat atop nucleus rotundus, also receives tectal projections and projects to the entirety of the ecto-striatum. Efferents from the ectostriatum are to other cell groups within the hyperstriatum and neostriatum regions of the telencephalon.

EVOLUTIONARY TRENDS IN THE VISUAL SYSTEM OF AMNIOTES

An examination of the ascending visual pathways to the telencephalon in mammals, diapsid reptiles, birds, and turtles reveals a striking pattern of similarities, which almost certainly represents the heritage of these groups from their common ancestral, amniote stock. In addition, striking differences can be seen as well. Figure 26-13 shows a number of the common features of the pattern among mammals and diapsids, which also apply to turtles, in the collothalamic and lemnothalamic visual projections to the telencephalon. The collothalamic visual route is from nucleus rotundus, or nucleus lateralis posterior/pulvinar, to the visual area of the anterior dorsal ventricular ridge, which in birds is known as the ectostriatum and which in mammals is the extrastriate cortex. The lemnothalamic visual route is from the dorsal lateral optic nucleus, or dorsal lateral geniculate nucleus of mammals, to the lemnothalamic visual part of the dorsal pallium, that is, the pallial thickening in diapsid reptiles, visual Wulst in birds, lateral part of the dorsal cortex in turtles, and striate cortex in mammals.

Overlying this pattern of similarities are certain dramatic differences that have emerged in mammals and birds. The most dramatic differences may be seen in mammals, probably as a consequence of their development of substantial ipsilateral projections from the retina to the visual lemnothalamus. These differences include: a dorsal lateral geniculate nucleus that is segregated into ipsilateral and contralateral subdivisions; magnocellular, koniocellular, and parvocellular subdivisions; X, Y, and W channels; and a high degree of retinotopic organization. In mammals, the large number of representations of the visual field is unique among amniotes. The variation among the multiple representations suggests that multiple areas have been evolved independently in the various groups of mammals.

Birds, by a process of convergent evolution, independently have evolved certain similarities to mammals: multiple components of the visual lemnothalamus (OPT complex) as well as granule cells and a columnar organization in their dorsal pallial equivalent to striate cortex, the visual Wulst. Some birds with an extreme frontal position to their eyes, such as owls, have a highly developed binocular organization to the visual Wulst. A major difference, however, between the avian and mammalian visual systems is that the mammalian system has developed ipsilateral retinal projections to the dorsal thalamus to an extent unparalleled among other vertebrates, whereas the avian visual system is contralateral from the retina to the lemnothalamic diencephalon but bilateral from the diencephalon to the visual Wulst. Among all vertebrate groups, birds have the best color vision as well as outstanding visual acuity that would be the envy of any mammalian predator.

FOR FURTHER READING

Albright, T. D., Desimone, R. and Gross, C. G. (1984) Columnar organization of directionally selective cells in visual area MT of the macaque. *Journal of Neurophysiology,* 51, 16–31.

Allman, J. (1990) Evolution of neocortex. In A. Peters and E. G. Jones (eds.), *Cerebral Cortex, Vol. 8A: Comparative Structure and Evolution of Cerebral Cortex.* New York: Plenum, pp. 269–283.

Allman, J. and McGuinness, E. (1988) Visual cortex in primates. In H. D. Steklis and J. Erwin (eds.), *Comparative Primate Biology, Vol. 4: The Neurosciences.* New York: Liss, pp. 279–326.

Bagnoli, P., Francesconi, W., and Magni, F. (1982) Visual Wulst-optic tectum relationships in birds: a comparison with the mammalian corticotectal system. *Archives of Italian Biology,* 120, 212–235.

Balaban, C. D. and Ulinski, P. S. (1981) Organization of thalamic afferents to anterior dorsal ventricular ridge in turtles. I. Projections of thalamic nuclei. *Journal of Comparative Neurology,* 200, 95–129.

Bass, A. and Northcutt, R. G. (1981) Retinal recipient nuclei in the painted turtle, *Chrysemys picta:* an autoradiographic and HRP study. *Journal of Comparative Neurology,* 199, 97–112.

Benowitz, L. J. and Karten, H. J. (1976) Organization of the tectofugal visual pathway in the pigeon: a retrograde transport study. *Journal of Comparative Neurology,* 167, 503–520.

Blake, R. and Wilson, H. R. (1991) Neural models of stereoscopic vision. *Trends in Neurosciences,* 14, 445–452.

Bruce, L. L. and Butler, A. B. (1984a) Telencephalic connections in lizards. I. Projections to cortex. *Journal of Comparative Neurology,* 229, 585–601.

Bruce, L. L. and Butler, A. B. (1984b) Telencephalic connections in lizards. II. Projections to anterior dorsal ventricular ridge. *Journal of Comparative Neurology,* 229, 602–619.

Butler, A. B. (1994a) The evolution of the dorsal thalamus of jawed vertebrates, including mammals: cladistic analysis and a new hypothesis. *Brain Research Reviews,* 19: 29–65.

Butler, A. B. (1994b) The evolution of the dorsal pallium in the telencephalon of amniotes: cladistic analysis and a new hypothesis. *Brain Research Reviews,* 19: 66–101.

Butler, A. B. and Northcutt, R. G. (1978) New thalamic visual nuclei in lizards. *Brain Research,* 149, 469–476.

Campbell, C. B. G. (1972) Evolutionary patterns in mammalian diencephalic visual nuclei and their fiber connections. *Brain, Behavior and Evolution,* 6, 216–236.

Carroll, R. L. (1988) *Vertebrate Paleontology and Evolution.* San Francisco: Freeman.

Casagrande, V. A. and Norton, T. T. (1991) Lateral geniculate nucleus: a review of its physiology and function. In A. G. Leventhal (ed.), *Vision and Visual Dysfunction, Vol. 4: The Neural Basis of Visual Function.* London: Macmillan, pp. 41–84.

Chalupa, L. M. (1991) Visual function of the pulvinar. In A. G. Leventhal (ed.), *Vision and Visual Dysfunction, Vol. 4: The Neural Basis of Visual Function.* London: Macmillan, pp. 140–159.

Engelage, J. and Bischof, H.-J. (1993) The organization of the tectofugal pathway in birds: a comparative review. In H. P. Zeigler and H.-J. Bischof (eds.), *Vision, Brain and Behavior in Birds.* Cambridge, MA: MIT Press, pp. 137–158.

Güntürkün, O., Miceli, D. and Watanabe, M. (1993) Anatomy of the avian thalamofugal pathway. In H. P. Zeigler and H.-J. Bischof (eds.), *Vision, Brain and Behavior in Birds.* Cambridge, MA: MIT Press, pp. 115–135.

Hodos, W. (1993) The visual capabilities of birds. In H. P. Zeigler and H.-J. Bischof (eds.), *Vision, Brain and Behavior in Birds.* Cambridge, MA: MIT Press, pp. 63–76.

Kaas, J. H., Guillery, R. W., and Allman, J. M. (1972) Some principles of organization in the dorsal lateral geniculate nucleus. *Brain, Behavior and Evolution,* 6, 253–299.

Kaas, J. H. and Huerta, M. F. (1988) The subcortical visual system of primates. In H. D. Steklis and J. Erwin (eds.), *Comparative Primate Biology, Vol. 4: The Neurosciences.* New York: Liss, pp. 327–391.

Karten, H. J. and Hodos, W. (1970) Telencephalic projections of the nucleus rotundus in pigeons (*Columba livia*). *Journal of Comparative Neurology,* 140, 33–52.

Karten, H. J., Hodos, W., Nauta, W. J. H., and Revzin, A. M. (1973) Neuronal connections of the 'visual Wulst' of the avian telencephalon. Experimental studies in the pigeon (*Columba livia*) and the owl (*Speotyto cunicularia*). *Journal of Comparative Neurology,* 150, 253–278.

LeVay, S. and Nelson, S. B. (1991) Columnar organization of the visual cortex. In A. G. Leventhal (ed.), *Vision and Visual Dysfunction, Vol. 4: The Neural Basis of Visual Function.* London: Macmillan, pp. 266–315.

Lohman, A. H. M. and Smeets, W. J. A. J. (1990) The dorsal ventricular ridge and cortex of reptiles in historical and phylogenetic perspective. In B. L. Finlay, G. Innocenti, and H. Scheich (eds.), *The Neocortex.* New York: Plenum, pp. 59–74.

Reiner, A. (1991) A comparison of neurotransmitter-specific and neuropeptide-specific neuronal cell types present in dorsal cortex in turtles with those present in isocortex in mammals: implications for the evolution of isocortex. *Brain, Behavior and Evolution,* 38, 53–91.

Repérant, J., Rio, J.-P., Ward, R., Hergueta, S., Miceli, D., and Lemire, M. (1992) Comparative analysis of the primary visual system of reptiles. In C. Gans and P. S. Ulinski (eds.), *Biology of the Reptilia, Vol. 17, Neurology C: Sensorimotor Integration.* Chicago: The University of Chicago Press, pp. 175–240.

Robinson, D. L. and Petersen, S. E. (1992) The pulvinar and visual salience. *Trends in Neurosciences,* 15, 127–132.

Sereno, M. I. and Allman, J. M. (1991) Cortical visual areas in mammals. In A. G. Leventhal (ed.), *Vision and Visual Dysfunction, Vol. 4: The Neural Basis of Visual Function.* London: Macmillan, pp. 160–172.

Shimizu, T. and Karten, H. J. (1993) The avian visual system and the evolution of neocortex. In H. P. Zeigler and H.-J. Bischof (eds.), *Vision, Brain and Behavior in Birds.* Cambridge, MA: MIT Press, pp. 103–135.

Spear, P. D. (1991) Functions of extrastriate cortex in non-primate species. In A. G. Leventhal (ed.), *Vision and Visual Dysfunction, Vol. 4: The Neural Basis of Visual Function.* London: Macmillan, pp. 339–370.

Ulinski, P. S. (1990) The cerebral cortex of reptiles. In A. Peters and E. G. Jones (eds.), *Cerebral Cortex, Vol. 8A: Comparative Structure and Evolution of Cerebral Cortex.* New York: Plenum, pp. 139–215.

Ulinski, P. S. and Margoliash, D. (1990) Neurobiology of the reptile-bird transition. In A. Peters and E. G. Jones (eds.), *Cerebral Cortex, Vol. 8A: Comparative Structure and Evolution of Cerebral Cortex.* New York: Plenum, 217–265.

Van Essen, D. C. (1985) Functional organization of primate visual cortex. In A. Peters and E.G. Jones (eds.), *Cerebral Cortex, Vol. 3: Visual Cortex.* New York: Plenum, pp. 259–329.

Zeigler, H. P. and Bischof, H.-J. (eds.) (1993) *Vision, Brain and Behavior in Birds.* Cambridge, MA: MIT Press.

ADDITIONAL REFERENCES

Bonhoeffer, T. and Grinvald, A. (1993) The layout of iso-orientation domains in area 18 of cat visual cortex: optical imaging reveals a pinwheel-like organization. *The Journal of Neuroscience,* 13, 4157–4180.

Ebbesson, S. O. E. and Voneida, T. J. (1969) The cytoarchitecture of the pallium in the tegu lizard (*Tupinambis nigropunctatus*). *Brain, Behavior and Evolution,* 2, 431–466.

Kaas, J. H. and Krubitzer, L. A. (1992) Area 17 lesions deactivate area MT in owl monkeys. *Visual Neuroscience,* 9, 399–407.

Kaas, J. H. and More, A. (1993) Connections of visual areas of the upper temporal lobe of owl monkeys: the MT crescent and dorsal and ventral subdivisions of FST. *Journal of Neuroscience,* 13, 534–546.

Karten, H. J. and Hodos, W. (1967) *A Stereotaxic Atlas of the Brain of the Pigeon (Columba livia).* Baltimore, MD: The Johns Hopkins University Press.

Mark, R. F. and Marotte, L. R. (1992) Australian marsupials as models for the developing mammalian visual system. *Trends in Neurosciences,* 15, 51–57

Obermayer, K. and Blasdel, G. G. (1993) Geometry and ocular dominance columns in monkey striate cortex. *Journal of Neuroscience,* 13, 4114–4129.

Payne, B. R. (1993) Evidence for visual cortical area homologs in cat and macaque monkey. *Cerebral Cortex,* 3, 1–25.

27

Somatosensory and Motor Forebrain in Amniotes

INTRODUCTION

In this chapter, we will discuss the somatosensory (or somesthetic) forebrain and the motor forebrain of mammals, diapsid reptiles, birds, and turtles. The somatosensory worlds of these animals consist of the consequences of mechanical and thermal stimulation of the body's external and internal surfaces. These include the heating and cooling, and indentation and stretching of surfaces, as well as the bending of joints, the stretching of tendons and muscles, and other consequences of the activity of the musculoskeletal system. In addition, the bending of hairs, feathers, and scales serve as important indicators of events in the external environment. Chapter 2 summarized many of these stimuli and their somatosensory receptors.

As in Chapter 26, most of our discussion will center on mammals because they have the most elaborate telencephalic representations of these systems and because they have been more intensively studied than nonmammalian amniotes. Many of the principles of organization that we saw in the visual isocortex of mammals will apply as well to their somatosensory and motor isocortices. These include topographic organization, multiple representations, and collothalamic and lemnothalamic input. In the somatosensory and motor systems, however, the topography that is represented is that of the body of the animals, that is, a **somatotopic organization,** rather than the topography of the visual world as in the visual cortex. Because of the similarity of their somatotopic organizations, the somatosensory and motor systems are convenient to discuss in the same chapter.

Both the extent and the details of the somatotopic organization of the somatosensory regions and the motor regions of the telencephalon can be measured physiologically, but by different methods. Somatosensory mapping is performed by placing a recording electrode on (or in) the telencephalon and then stimulating various body parts by touch, temperature, pressure, and so on, and noting which body regions affect electrical activity at the telencephalic electrode location. Motor mapping also uses a telencephalic electrode, but in this instance it is a stimulating electrode. The experimenters observe which regions of the body move when a particular telencephalic site is stimulated electrically. These physiological maps are then correlated with morphological maps that are in turn correlated with cytoarchitectural criteria, areas of termination of thalamic input, and areas of origin of descending pathways to the brainstem and spinal cord. An important consideration in evaluating these maps is that the size and shape of the map can change according to the methods used to elicit it; for example, the size and shape of the maps can be affected by the type and depth of anesthesia used or by the type of recording or stimulating electrode. In general, a reasonably good congruence exists between maps obtained by different methods. Whatever the method used, however, the resulting maps are of the contralateral body half because of the crossing, or decussation, of the ascending secondary sensory axons on route to the dorsal thalamus and of the descending motor pathways on route to the spinal cord, whether directly or via relay through parts of the brainstem.

THE SOMATOSENSORY AND MOTOR FOREBRAIN OF MAMMALS

One of the defining characteristics of mammals is the presence of hair. When we think about hair, we tend to think of temperature regulation and protection. Hairs, however, also have an important sensory function (see Chapter 2) in that they can be deflected by objects near to the body surface. The hair follicle receptors at the bases of the hairs then report the fact of this close encounter with something to the somatosensory system. Hairy skin, like a gloved hand, cannot be used for determining the fine surface details of an object. For this function, direct contact between the skin itself and the objects being inspected is required. The hairless skin on the palms of hands and forepaws and on the soles of feet and hindpaws is known as **glabrous** skin and has a high degree of sensitivity for touch and fine detail. Mammals with a highly developed sense of touch typically have separate representations of the hairy and glabrous regions of the skin in their somatosensory cortices.

The Ventral Tier Nuclei of the Dorsal Thalamus

Anatomists sometimes find it convenient to divide the dorsal thalamus into two layers or tiers (as in the tiers of a stadium). A **dorsal tier** and a **ventral tier** of the dorsal thalamus have thus been recognized. The nuclei of the ventral tier have names that begin with "ventralis," such as ventralis anterior, ventralis lateralis, ventralis posteromedialis, or ventralis posterolateralis. The ventral tier nuclei have in common that they are the somatosensory and motor nuclei of the dorsal thalamus. Please do not confuse the ventral tier of the dorsal thalamus with the ventral thalamus, which is an entirely different structure, discussed in Chapters 19 and 24.

Just as the somatosensory cortex and the motor cortex are located more or less in close proximity to each other, so are their associated dorsal thalamic nuclei. The more anterior of the ventral tier nuclei have their principal connections with the cortical components of the motor system as well as with noncortical components, such as the cerebellum, corpus striatum (i.e., the dorsal striatum and dorsal pallidum), and substantia nigra. The more caudal group of ventral tier nuclei receive the ascending somatosensory axons from the spinal cord, the dorsal column nuclei, and the trigeminal nuclei of the brainstem.

Somatosensory Lemnothalamus

The somatosensory nuclei of the dorsal thalamus are known as the **ventrolateral** (or **ventrobasal**) **complex.** This group of nuclei is located ventral to the lateralis posterior/pulvinar complex that we encountered in the previous chapter and consists of two principal nuclei: **nucleus ventralis posteromedialis (VPM)** and **nucleus ventralis posterolateralis (VPL).** Nucleus ventralis posterolateralis receives its input from the somatosensory cell groups in the dorsal horn of the spinal cord and also from the dorsal column nuclei (nuclei gracilis and cuneatus) via the **medial lemniscus,** which thus carries information about somatosensory stimulation of the body. Nucleus ventralis posteromedialis is the comparable thalamic nucleus for somatosensory projections from the head and face regions. It receives ascending axons from the sensory nuclei of the trigeminal nerve in the pons. Together, nuclei ventralis posterolateralis and ventralis posteromedialis contain a complete, somatotopically organized representation of the body, including the head and face. In primates, several additional components of the ventrolateral complex exist as well. Within the ventrolateral complex are a superior region that receives input from muscle and joint receptors, a central region that contains a somatotopic representation of both slowly adapting

and rapidly adapting skin receptors, and an inferior region that receives input from the spinothalamic tract with pain and temperature information. The ventrolateral complex sends its efferents to the major area of somatosensory isocortex, often referred to as **S1** and **S2,** or **SI** and **SII.**

Related to the ventrolateral complex in terms of the organization of its afferent and efferent projections is a more caudal nucleus, the **medial part of the posterior nuclear group (Pom).** The Pom receives somatosensory lemniscal projections from the dorsal column nuclei in the caudal medulla. In turn, Pom projects to somatosensory cortex (SII). We thus include Pom with the ventrolateral complex in the somatosensory lemnothalamus.

Somatosensory Collothalamus

Caudal to the ventrolateral complex is another group of nuclei—the **posterior nuclear group,** which includes the **nucleus limitans** and the **suprageniculate nucleus.** The suprageniculate/limitans nuclear complex receives input from the somatosensory-multisensory part of the tectum, that is, the superior colliculus, of the midbrain, which receives ascending axons from the dorsal column nuclei. The suprageniculate/limitans complex thus constitutes the collothalamic somatosensory thalamus. The efferents of the suprageniculate/nucleus limitans complex are to isocortex located in the posterior region of the insula. This **posterior insular cortex** is comparable to extrastriate visual cortex of mammals in that it is the target of a collothalamic rather than a lemnothalamic pathway.

Motor Lemnothalamus

The rostral (or oral) end of the ventral tier comprises the motor division of the dorsal thalamus. Unfortunately, this is another of those regions with confusing nomenclature due to subdivisions into separate nuclei of nuclei that originally were thought to be a single entity and due also to the differences in thalamic organization between species that are correlated with differences in cortical organization. The motor nuclei of the ventral tier are considered to be part of the lemnothalamus due to their lemniscal-like inputs from motor related structures, that is, direct afferent projections without relay through the roof of the midbrain, and their projections to motor cortices (M1 and M2). We will briefly digress to note that if motor collothalamic pathways exist, one might consider the suprageniculate/limitans complex in this context, in that the somatosensory part of the superior colliculus receives motor-related input from the substantia nigra and via relay through the pretectum (see Chapters 17 and 24). This pathway thus could be considered motor as well as sensory.

Two points should be kept in mind about the motor thalamus. The first is that unlike motor cortex, the motor thalamus has no direct connections with motor nuclei of the brainstem or spinal cord; it is "motor" only in the sense that its major connections (both afferent and efferent) are with motor cortex, the cerebellum, and the corpus striatum. It thus serves as a major input pathway to motor cortex as well as being an important feedback pathway from cortex to thalamus and back to cortex. The second point is that the description of motor thalamus presented here is based largely on studies of cats and primates, which are the subjects of the overwhelming majority of research studies in this area, and which seem to have some of the most elaborately developed forebrain motor systems.

The most rostral component of the motor thalamus is **nucleus ventralis anterior (VA),** which is involved in pathways associated with the noncortical components of the motor system: the corpus striatum (especially the internal segment of the globus pallidus) and the pars reticulata of the substantia nigra of the midbrain. Moving caudally, we next encounter **nucleus ventralis lateralis (VL),** which receives axons from the cerebellum by way of relay through the deep cerebellar nuclei. Still farther caudally, we come to **nucleus ventralis posterolateralis (VPL),** which also receives cerebellar input via the deep cerebellar nuclei.

The alert reader may now be justifiably confused, for the moment, since we just said above that VPL is part of the somatosensory system. The problem lies in the fact that VPL is one of those instances of a nucleus that was originally thought to be a single structure and was later deemed by some workers to be two separate entities that were difficult to separate on cytoarchitectonic criteria alone. Thus, we now refer to a **nucleus ventralis posterolateralis, pars oralis (VPLo),** which is the motor component of VPL, and a **nucleus ventralis posterolateralis, pars caudalis (VPLc),** which is the somatosensory component. To make matters worse, scientists who study the somatosensory system tend to refer to VPLc simply as VPL, whereas those who study the motor system usually distinguish the motor component, VPLo, from the somatosensory component, VPLc. Furthermore, some workers consider VPLo to be a subdivision of nucleus ventralis lateralis, rather than of nucleus ventralis posterolateralis.

A small consolation for this nomenclatural confusion may be taken from the fact that, like the motor/somatosensory arrangement in isocortex, the motor components of the ventral tier are rostral (oral) to all of the somatosensory components, which facilitates remembering that VPLo is motor while VPLc is somatosensory. One additional nucleus remains to be added to the cast of characters in the motor thalamus; this is a collection of neurons situated medial to VPLo and known by the slightly mysterious name of **nucleus X.** Nucleus X, along with VPLo, may also be a subdivision of nucleus ventralis lateralis. The nuclei of the motor thalamus and their relationships are diagrammed in Figure 27–8(A).

Somatotopic Organization

A principle of central nervous system organization that has emerged in many of the preceding chapters is topographic organization. This principle can be observed in its most dramatic form in somatotopic maps. Moreover, the greater use an animal makes of a sensory or motor system, the greater the amount of neural tissue that is dedicated to processing the functions of that system. Thus, those body parts that have the greatest sensitivity have the largest representation in the somatosensory cortex, and those body parts that have the greatest dexterity of movement have the biggest representation in the somatotopic, cortical motor map. The structure with the largest representation on the somatosensory map may not necessarily have the largest representation on the motor map, however. For example, in humans, the hand and especially the fingers are very well repre-

sented in both the somatosensory and motor maps. In the somatosensory map, the index finger has the largest amount of cortex dedicated to it, whereas in the motor map, the greatest representation is that of the thumb. In general, animals that use their face and snout region to explore the environment have large representations of these areas. The duckbilled platypus, for example, which uses its bill to feel about in muddy waters for food, has an enormous area of somatosensory cortex dedicated to the sensations from the bill. Other examples are insectivores, which have a large representation of the snout, and bats, which have a large cortical area dedicated to sensations from the wings.

The somatotopic maps of both the somatosensory cortex and motor cortex are laid out in the same manner with the most caudal structures being represented most medially, sometimes spreading over onto the medial surface of the hemisphere. The mid-body structures are located on the lateral surface, and the most rostral structures have their representation in the ventrolateral part of the somatosensory and motor cortical areas. Taste and vestibular sensitivity are represented here as well. The most ventrolateral representations in the somatosensory map are those of the viscera of the abdomen and thorax.

Figure 27-1 shows the cerebral hemispheres of a primate that has well-developed forepaws and hindpaws, including good manipulative ability of the digits. Note that the sequence of the cortical areas devoted to each body part corresponds to the sequence in which those parts are arranged in the animal's body. Note also the correspondence of the somatosensory and motor areas. In general, the somatosensory and motor maps are located in adjacent areas of cortex, and the maps often are in register with one another so that corresponding somatosensory and motor areas are across from one another. Because of the differences between the amounts of cortex dedicated to a particular body part, the correspondence between the somatosensory and motor maps cannot be perfect; nevertheless, the alignment of the two maps is a striking feature of the organiza-

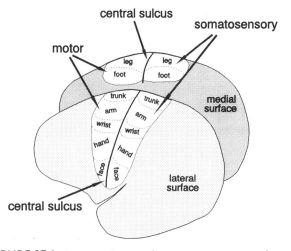

FIGURE 27-1. Somatotopic maps of somatosensory cortex and motor cortex represented on opposite sides of the central sulcus in a primate brain. The maps begin on the medial surface and continue onto the lateral surface. The maps are arranged with the caudal body parts represented most medially and the rostral parts represented ventrally and laterally.

tion of the somatosensory-motor cortices, especially in primates.

The precise alignment of the somatosensory map and the motor map might lead one to conclude that the two maps have a high degree of interconnectedness. In fact, relatively few direct connections between somatosensory cortex and motor cortex have been reported. Instead, the principal efferent pathways from somatosensory cortex are back to the regions of the ventrolateral complex from which they originated, as well as some projections to the corpus striatum, the dorsal column nuclei, and the spinal cord.

Overlap and Separation of Somatosensory and Motor Maps

A frequently observed mammalian pattern is for the somatosensory map to abut on the motor map or to have a strip of isocortex separating the two maps. This intermediate strip is sometimes known as "silent" cortex because it does not respond to mechanical or thermal stimulation of the body, and electrical stimulation of this strip does not result in movement of a body part. A number of exceptions to this rule have been described, however. In the platypus, for example, somatosensory and motor cortices have been reported to partially overlap each other, with about one third of motor cortex being in the somatosensory area and about one quarter of somatosensory cortex being in the motor area. Like their fellow monotreme the platypus, echidnas have a combined somatosensory-motor cortical area; in addition, however, they also have a separate motor area, which is totally without a corresponding somatosensory map. Among the North American marsupials, such as the American opossum, complete overlap is the normal condition, whereas in Australian marsupials, such as the brush-tailed possum, wombats, and wallabies, the situation is similar to that found in echidnas, that is, a combined somatosensory-motor area plus an additional purely motor area. Other examples of this pattern may be seen in hedgehogs. In contrast to these, other mammals such as tree shrews, some rodents, some primates, and carnivores have a strip of cortex that separates the somatosensory map from the motor map.

The point of this survey is to illustrate that considerable variability has been reported within and between various groups of mammals in the degree of separation of the somatosensory and motor cortices. Some examples of these relationships can be seen in Figure 27-2. This body of data must be regarded with some caution, however, as the distinction between sensory and motor areas can be a difficult one to make. Depending on the stimulus parameters, movements can be evoked from somatosensory cortical areas, and motor cortical areas can respond to somatosensory stimulation. The degree to which various groups of mammals actually vary in the organization of their somatosensory-motor cortices and the amount of overlap of the various areas needs to be reevaluated in further studies.

Somatosensory Barrels

Figure 2-22(B) shows an example of the somatotopic representation of the forepaw area in a mammal with well-developed digits. Such an animal would make extensive use

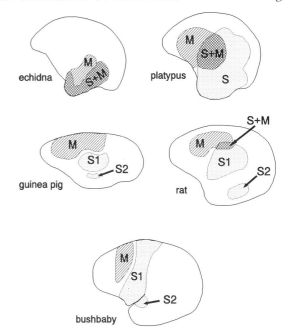

FIGURE 27-2. Examples of the variety of locations of somatosensory and motor cortices in five different mammals. M = motor cortex; S = somatosensory cortex; S + M indicates areas of overlap of somatosensory and motor cortices. Many mammals have more than one somatosensory map. SI = primary somatosensory map; S2 = secondary somatosensory map.

of the forepaws in manipulation of objects. Examples of animals of this sort are primates, raccoons, and many rodents. In this map, each individual digit and each individual palm pad has its own dedicated region of cortex. Even more dramatic than this punctate representation are the **barrel fields** of the face areas of the somatosensory cortex of mammals that make extensive use of their vibrissae (whiskers) for exploration of the external world. Because the vibrissae extend beyond the width of the face, they serve as a tactile early warning system, that is, they indicate that the face is about to touch something just before the contact is made. This information is extremely useful to animals that move about in tunnels or narrow burrows or that squeeze their way through narrow apertures. Nocturnal animals also find the extensions of the face's tactile surface to be helpful in feeling their way through dark or dimly lit environments. Carnivores and rodents usually have well-developed vibrissae. What makes somatosensory representation of the vibrissae so extraordinary is that each whisker has its own dedicated cortical area. Because of the cylindrical shape of these cortical areas through the depth of cortical layer IV, they are known as **barrels.** Thus, each whisker has its own barrel. Moreover, the layout of the barrels in the cortical map is virtually a picture of the arrangement of the whiskers on the face. Figure 27-3 shows the vibrissae on the snout of a small rodent and its corresponding cortical barrel field.

Perhaps even more dramatic than the organization of the barrel field is the somatotopic organization of the snout of the star-nosed mole, which has a star-shaped arrangement of 22 fleshy, finger-like projections on its rostral end. Each nostril is thus surrounded by a hemistar of 11 of these fleshy "fingers."

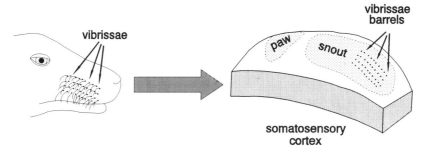

FIGURE 27-3. The somatotopic representation of the snout region of a rodent. Each vibrissa on the snout has its own representation as a "barrel" in the snout region of the map. The topographic distribution of vibrissae on the snout is the same as in the barrel field of the somatosensory cortex.

These fingers are used by the mole to manipulate and probe the soil surrounding the snout for food objects. When the somatosensory cortex of this creature was mapped, the experimenters found that not only does each finger have its own representation in the face area, but that the array of finger representations in the map exactly mirrors the relative size and shape of the fingers of the hemistar surrounding the contralateral nostril. In addition to an acute sense of touch, a recent report suggests that these fingers may also possess the capacity for electroreception that could detect the muscle action potentials of the mole's earthworm prey.

Multiple Representations of the Sensory Somatotopic Map

The somatosensory cortex has multiple representations in a number of mammalian species. In the visual system, the lemnothalamic visual nucleus predominantly projects to one visual representation, striate cortex, that is thus considered to be the visual lemnocortex, while the collothalamic visual nucleus predominantly projects to multiple extrastriate visual representations that thus constitute the visual collocortex. In the somatosensory system, at least two representations, S1 and S2 (or SI and SII), receive projections from the somatosensory lemnothalamic nuclei and are thus considered to be somatosensory lemnocortex. The posterior insular cortex, which receives projections from the somatosensory collothalamic nuclei, is considered to constitute the somatosensory collocortex.

In primates, S1 occupies Brodmann's areas 3b, 1, and 2. S2 usually is located ventral to S1. A full map of the contralateral body half exists in each of these areas. Usually the maps are mirror images of one another, that is, nose to nose. Typically, the S2 map is smaller and less detailed that the S1 map. Moreover, the S2 maps tend to have larger receptive fields than the S1 maps. This is the arrangement that has been described for rats and mice, insectivores, and bats. In addition to the S1 and S2 maps, multiple representations of certain body parts that the animal uses extensively, such as vibrissae or hands, may have a second representation within S1.

Three complete somatotopic representations have been found in the cortex of monotremes, and among marsupials, opossums also have three complete representations. Other marsupials, such as wallabies and wombats, lack the dual represen-

tation of the vibrissae but do have a dual representation of the hairy skin of the upper lip. Some prosimians have a dual representation of the glabrous skin of the hand within S1. A similar dual representation of the forepaw has been reported in squirrels. Raccoons, which also make extensive use of the forepaw, have only a single representation of the glabrous forepaw, but it is enormous in size and contains many highly specialized subregions as well as very small receptive fields. In addition, representations of the hairy part of the forepaw are present as well. Examples of the relative size and location of S1 and S2 in various mammals are shown in Figure 27-2.

A number of species of mammals have more than two somatosensory cortical areas. An additional complete representation of the body half has been reported in squirrels, which gives them a total of three maps. Cats not only have S1 and S2, but an S3, S4, and S5 as well. Old World and New World monkeys, in addition to an S2 map, have two complete maps of the contralateral body half within the central strip of S1. In addition, on either side of the dual contralateral maps are a pair of matching maps that represent deeper tissues (such as muscles and joints) of the same body parts as in the central maps. While these four maps are not completely identical to each other, the degree of correspondence between them is striking. Additional regions of somatotopic representation are the **retroinsular** (in back of the insula) cortex in primates, which may correspond to S4 in cats, and certain areas of the insula, the **granular** and **dysgranular insular areas,** which appear to be similar to S2, but rather more specialized for the face and hand. Finally, we come to the **posterior parietal cortex,** which is located caudal to the somatosensory cortex and comprises Brodmann's areas 5 and 7. This area receives its inputs from S1 (especially Brodmann's area 2) and sends its efferents to S2.

Afferents to Somatosensory Cortex

Figure 27-4 summarizes the lemnothalamic and collothalamic afferents to the somatosensory cortex. As shown in this figure, somatosensory axons from receptors for pain, touch, temperature, and other somesthetic sensations from the body enter the dorsal horn of the spinal cord. Some lemnothalamic axons of dorsal horn cells pass directly to the ventrolateral complex of the dorsal thalamus and terminate in VPL (i.e.,

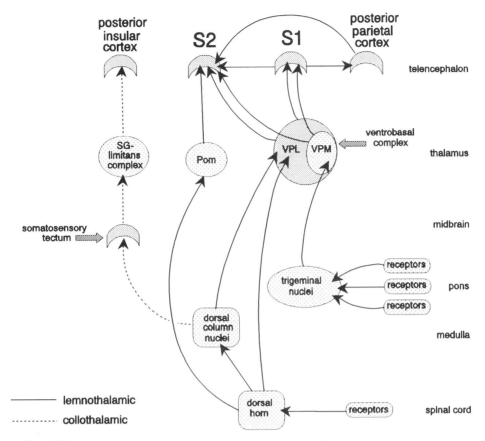

FIGURE 27-4. A schematic representation of the lemnothalamic and collothalamic somatosensory pathways in mammals. SI = primary somatosensory cortex; and S2 = secondary somatosensory cortex; SG = suprageniculate nucleus; Pom = medial division of the posterior nuclear group; VPL = nucleus ventralis posterolateralis; VPM = nucleus ventralis posteromedialis. The receptors shown at the level of the pons represent the three branches of the trigeminal nerve. The receptor shown at the spinal cord level represents somatosensory input from the body apart from the head.

VPLc), which contains the representation of the body. Other somatosensory axons from the dorsal horn project to the dorsal column nuclei from which lemnothalamic efferents also pass to VPL. The corresponding somatosensory axons from receptors in the face and head regions of the body enter by way of the three branches of the trigeminal nerve in the pons and terminate in the principal and descending nuclei of the trigeminal nerve. Efferent axons of these trigeminal nuclei join those from the spinal cord and medulla to terminate in VPM, which is specialized for the somatotopic representation of the head. The efferents of the ventrolateral complex are to S1 and S2. A second lemnothalamic pathway originates in the dorsal horn. These axons terminate in the medial part of the posterior nuclear group (Pom). The efferent axons of Pom project to S2.

Also shown in Figure 27-4 is the collothalamic somatosensory pathway. This pathway arises in the dorsal column nuclei with its first termination in the somatosensory–multisensory region of the midbrain tectum, that is, the superior colliculus. From there, tectothalamic axons project to the suprageniculate (SG)/nucleus limitans group in the posterior thalamus, the efferents of which terminate in the posterior insular cortex.

Efferents of Somatosensory Cortex

Both S1 and S2 are the major sources of efferents from the somatosensory cortex, with S4 (or its equivalent) also serving as an additional output route in those mammals that possess a fourth representation of the body. Figure 27-5 summarizes the efferents of S1, which may be divided into two categories: **cortico–cortical** and **descending.** The cortical targets of S1 are motor cortex, S2, and the posterior parietal cortex. The descending efferents of the somatosensory cortex are to the corpus striatum and to the brainstem and spinal cord components of the somatosensory system: the dorsal horn of the spinal cord, the dorsal column nuclei, and the ventrolateral complex.

The efferents of S2 and S4 are summarized in Figure 27-6. The efferents of S2 are to S4 (when present), as well as to motor cortex, posterior parietal cortex, frontal cortex—including the **frontal eye fields** (which are a region of the frontal lobe that are specialized for movements of the eyes)—**perirhinal cortex** (the area around the rhinal fissure in the ventrolateral frontal lobe), and the pontine nuclei. The projections of S4 are more limited and are to motor cortex, posterior parietal cortex, and the frontal eye fields, which will be discussed below. If

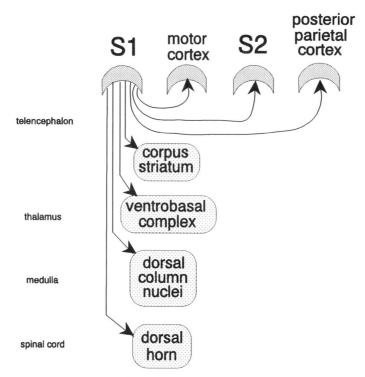

FIGURE 27-5. Schematic representation of the efferents of somatosensory cortical area S1.

we compare the afferent and efferent connections of S2 in Figures 27-4 and 27-6, we see that S2 is a major player in the somatosensory story, with inputs from S1, posterior parietal cortex, the thalamic ventrolateral complex and Pom, and with efferents to a variety of cortical and subcortical areas.

Motor Cortex

The motor cortex has played a key role in experimental support for the concept that different biological and behavioral functions are localized in specific regions of the brain and, specifically, in the idea of topographic organization within the brain. This support initially came from studies of damage to the motor cortex (usually war injuries and animal experiments)

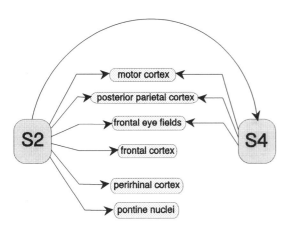

FIGURE 27-6. Schematic representation of the efferents of somatosensory cortical areas S2 and S4.

and electrical stimulation studies of epileptic patients during surgery. Studies that related small loci of damage in the motor cortex to localized regions of paralysis of body parts revealed the somatotopic motor map. These studies, in addition to the stimulation experiments, led to the view that voluntary movement is controlled by the motor cortex.

Multiple Motor Representations of the Body

The somatosensory system is not the only region of the central nervous system in which multiple representations of the body exist. The cortical component of the motor system also has multiple body representations. For example, in some primates, up to six motor areas are present, each of which has a representation of the body. The **primary motor area,** which has been designated as **M1,** can be divided into rostral and caudal subdivisions; it is located in Brodmann's area 4, as shown in Figure 27-7. Other motor areas include a **supplementary motor area, M2,** and **dorsal and ventral premotor areas.** An additional premotor area has been reported on the medial surface of the cerebral hemisphere in the cingulate gyrus. This region corresponds to Brodmann's area 24. Because of its connections to other cortical motor areas, spinal cord, and motor thalamus, as well as electrophysiological criteria, the dorsal portion of area 24, known as area 24c, which is located within the depths of the cingulate sulcus, has been proposed as **M3.** Its connections with limbic cortex suggest that it may serve as one of the routes by which the limbic system can influence voluntary motor behavior.

If each of these areas contains a more or less complete representation of the body, how can we tell them apart? The answer is that the supplementary motor, premotor cortical areas

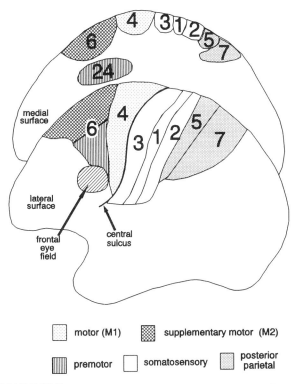

motor (M1)   supplementary motor (M2)

premotor   somatosensory   posterior parietal

FIGURE 27-7. The location of some of the somatosensory and motor cortical areas of a monkey. The numbers refer to Brodmann's cortical areas. The primary motor cortex (M1) corresponds to area 4. The supplementary motor area (M2) is located in the dorsal and medial regions of area 6. The ventral region of area 6 and area 24 on the medial surface are known as premotor areas. A portion of area 24 contains an M3 region. Areas 5 and 7 constitute the posterior parietal motor areas. Also shown is the S1 somatosensory area, which comprises areas 3, 1, and 2.

are distinguished from primary motor cortex, and from each other, by the types of movements elicited by stimulation within each map, the complexity of the movements, and the amount of stimulation required to produce a movement. Figure 27-7 shows the locations of some of these areas in a monkey brain.

The Cortical Eye Fields

Ventral to areas 4 and 6 is area 8. Parts of area 8 and the surrounding tissue constitute an area known as the **frontal eye field.** Stimulation of this cortical field results in strong movements of the eyes to the opposite side of the body. This area is regarded as containing command neurons for voluntary movements of the eyes and probably plays a role in the coordination of voluntary eye movements as well. The efferents of this area are to the oculomotor complex of the midbrain and the pontine reticular formation. The location of the frontal eye field may be seen in Figure 27-7.

An additional eye field is located in the occipital lobe, within the visual cortical areas, especially in V1 (striate cortex). This field is called the **occipital eye field.** In contrast to the frontal eye field, however, this region subserves involuntary eye movements of the sort that would result from the eyes pursuing some moving object in the visual field. The efferents of the occipital eye field are to the midbrain tectum.

Efferents of Motor Cortex

In addition to its cortico-cortical connections with M2 and the premotor cortical areas, the principal efferent pathway from M1 is the **corticospinal tract** (or **pyramidal tract**) from motor cortex to the contralateral spinal cord. This pathway is involved in voluntary movements of the body. Also included with the corticospinal tract is the **corticobulbar tract,** which controls voluntary movement of the face, head, and neck. The reader might find that a review of the material in Chapter 8 on the spinal cord of tetrapods would be helpful at this point. In many mammals, including many primates, rodents, and some carnivores (cats and raccoons), the corticospinal tract extends the entire length of the spinal cord, or nearly so. In other mammals, such as monotremes, marsupials, tree shrews, some insectivores (moles), anteaters, and elephants, the corticospinal tract only penetrates the spinal cord as far as the thoracic region. Finally, in many hoofed mammals (goats, cows, deer, and sheep), some insectivores (hedgehogs), and aardvarks, the corticospinal tract extends only to the cervical level of the spinal cord.

Area M1 is not the only contributor to the corticospinal tract, although it may contribute as much as 50% of its axons. The balance of the axons forming the tract are from M2, the premotor area, S1, and area 5 of the posterior parietal cortex. In addition to cortico-cortical, corticospinal, and corticobulbar efferents, M1 also sends axons to the **pontine nuclei,** which send their efferents to the cerebellum, and to the VPLo and ventralis lateralis nuclei of the motor thalamus. Area M2 sends its efferents to the other cortical motor areas, the corpus striatum, and nucleus ventralis lateralis of the motor thalamus, as well as making a contribution to the corticospinal tract. Finally, axons originating in the premotor areas terminate in M1, M2, the spinal cord, and nucleus X of the motor thalamus. The afferents and efferents of motor cortex are diagrammed in Figure 27-8, which is based largely on studies of monkeys.

THE SOMATOSENSORY AND MOTOR FOREBRAIN OF NONMAMMALIAN AMNIOTES

The somatosensory and motor representations in the forebrains of diapsid reptiles, birds, and turtles are much more limited than in mammals. In particular, the elaborate and highly detailed somatic and motor topographic maps are not a feature of the dorsal pallium in the telencephalons of these brains. Likewise, multiple representations of the body map, so typical of mammals, have not been reported in these animals. In addition, pathways for direct dorsal pallial control of brainstem and spinal motor neurons have not been observed. Instead, control of motor functions by the telencephalon is largely limited to the striatum (see Chapters 19 and 24) by way of brainstem coordinating systems such as the reticular formation and optic tectum (see Chapters 13 and 18).

Somatosensory System

The organization of ascending pathways via the dorsal thalamus does not follow quite the same pattern in the somato-

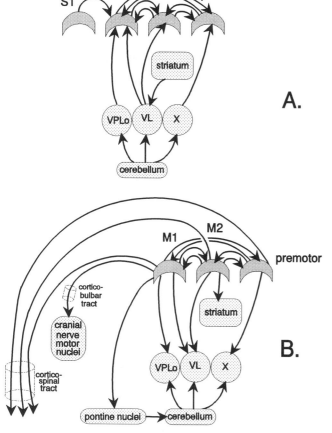

FIGURE 27-8. (A) schematic representation of the afferent pathways to cortical motor areas M1, M2, and the premotor areas. The inputs to these three cortical areas are mainly from VPLo, VL, and nucleus X of the ventral tier of the dorsal thalamus, from S1, and from one another. (B) schematic representation of the efferents of M1, M2, and the premotor cortical areas. The main efferents of the three motor cortical areas are to brainstem and spinal cord motor areas via the corticospinal and corticobulbar tracts, to pontine nuclei, which have connections with the cerebellum, and to the striatum. VPLo = nucleus ventralis posterolateralis, pars oralis; VL = nucleus ventralis lateralis.

sensory system of diapsids and turtles as it does in mammals. In mammals, as we have discussed, both collothalamic and lemnothalamic somatosensory pathways are present, as they are in the visual system. In addition, a lemnothalamic pathway from the trigeminal nuclei of the pons, which innervate the face and head, is juxtaposed to the lemnothalamic pathway from the body at both the dorsal thalamic and telencephalic levels. In diapsids and turtles, some evidence indicates that both lemnothalamic and collothalamic somatosensory pathways via the dorsal thalamus are present; trigeminal lemnothalamic input bypasses the dorsal thalamus, however.

Lemnothalamic Somatosensory System

Among nonmammalian amniotes, the lemnothalamic somatosensory system has been most clearly elucidated in birds. The avian lemnothalamic pathway is very much like that in

mammals. The ascending axons from the dorsal column nuclei in the caudal medulla terminate in a nucleus in the dorsal thalamus called **nucleus dorsalis intermedius ventralis anterior (DIVA).** The DIVA projects to a region of the rostral pallium, within the hyperstriatum accessorium, which is located anterior to the visual Wulst (the termination area of the visual lemnothalamic pathway) and could be considered therefore as a "somatosensory Wulst" [Fig. 27-9(A)]. In lizards, a lemniscal somatosensory projection to part of nucleus dorsolateralis anterior in the dorsal thalamus and then to the dorsal cortex [Fig. 27-9(B)] has been found. Whether or not this pathway corresponds to that in birds via DIVA to the somatosensory Wulst requires further study to be resolved.

Trigeminal Lemnothalamic Somatosensory System

The avian trigeminal lemnothalamic pathway is not "thalamic" in the usual sense of the term because it has no thalamic component. The pathway, known as the **quintofrontal tract,** originates in the principal nucleus of the trigeminal in the pons and terminates in the basal, frontal telencephalon in a region that is known as **nucleus basalis,** but should more properly be called **nucleus trigeminalis telencephali,** or the **telencephalic trigeminal nucleus.** Why is this nuclear group in the telencephalon and not in the dorsal thalamus, like its mammalian counterpart, nucleus ventralis posteromedialis (VPM)? Some embryological evidence suggests that part of the ventrolateral complex of mammals originates in the telencephalon and eventually migrates to the diencephalon. If this is the case, then the telencephalic trigeminal nucleus of birds may be the unmigrated homolog of VPM. Whatever the reason for its location, the pattern of efferents from this nucleus is what one would expect for telencephalic projections from a dorsal thalamic nucleus. The efferents are to two zones in the neostriatum of the anterior dorsal ventricular ridge. The first is directly dorsal to the telencephalic trigeminal nucleus in the frontal (rostral) part of the neostriatum and has been designated **neostriatum frontale, pars trigeminale;** the second projection field is in the caudal neostriatum and is known as **neostriatum caudale, pars trigeminale.** These two areas are reminiscent of the face and snout areas of S1 and S2 of mammals. A detailed topographic mapping study of the somatosensory representation of the body is yet to be done. There is, however, within this region a topographical representation of the head and bill that resembles the bill area of the platypus described above. This ascending trigeminal system has been implicated in the sensory control of pecking and grasping of food objects and the manipulation of them in the mouth to a point where they can be swallowed.

Although the avian trigeminal somatosensory pathway has been well documented, we are unable to find any studies in the literature that unequivocally report the presence of this pathway in diapsid reptiles and turtles. A report of research on ranid frogs, however, indicates the presence of a similar pathway from the trigeminal region of the caudal brainstem directly to an area in the ventrolateral part of the telencephalon. In addition, several studies of diapsid reptiles have found suggestions of a quintofrontal pathway. The presence of this arrangement in amphibians, diapsid reptiles, and birds would strongly

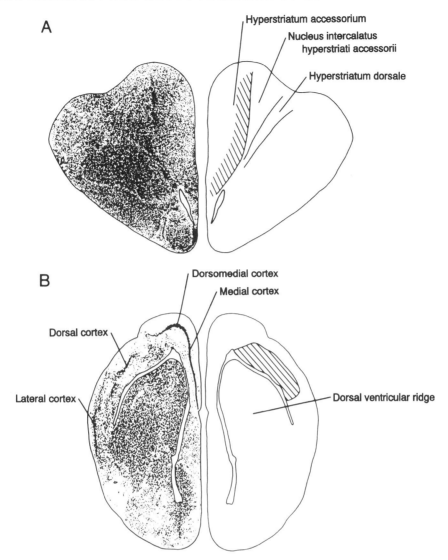

FIGURE 27-9. (A) Transverse hemisection with mirror-image drawing through the somatosensory Wulst in the telencephalon of a bird (*Columba livia*). The area of termination of the somatosensory lemnothalamic projection is indicated by diagonal lines. Adapted from Karten and Hodos (1967) with additional data from Wild (1987b). Used with permission of The Johns Hopkins University Press. (B) Transverse hemisection with mirror-image drawing through the telencephalon of a lizard (*Tupinambis nigropunctatus*). The location of the putative lemnothalamic telencephalic somatosensory area in the dorsal cortex is indicated by diagonal lines. Adapted from Ebbesson and Voneida (1969) with the somatosensory area inferred from Bruce and Butler (1984a). Used with permission of S. Karger AG.

suggest that a telencephalic target for the ascending trigeminal pathway is a common feature of at least tetrapods and that a dorsal thalamic location of part of this system is unique to mammals.

Collothalamic Somatosensory System

In diapsid reptiles and turtles, ascending axons from the spinal cord and dorsal column nuclei ascend to the midbrain somatosensory area in the tectum. From the tectum, somatosensory projections ascend to terminate in the **medialis complex** in the dorsal thalamus. The medialis complex sends its efferents to a region of the anterior dorsal ventricular ridge that lies

between the target region of projections from nucleus rotundus (the collothalamic visual counterpart to the medialis complex) and the target region of axons from nucleus reuniens, pars centralis, which is the auditory collothalamic nucleus that we will encounter in Chapter 28. Examples of the telencephalic termination field of the medialis complex are shown in a turtle in Figure 27-10 and in a diapsid reptile (an iguana) in Figure 27-11.

The thalamic counterpart of the collothalamic somatosensory pathway in birds comprises two nuclear areas: **nucleus semilunaris parovoidalis** and the caudal part of **nucleus dorsolateralis posterior (cDLP).** These nuclei send their efferent axons to the neostriatum intermedium, which is a region

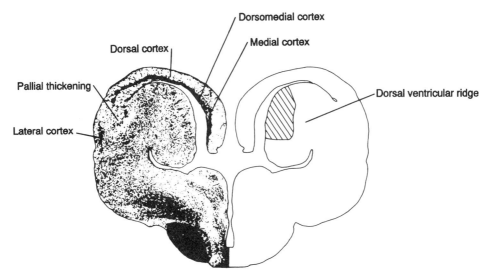

FIGURE 27-10. Transverse hemisection with mirror-image drawing through the telencephalon of a turtle. The location of the collothalamic telencephalic somatosensory area in the dorsal ventricular ridge is indicated by diagonal lines. Adapted from Balaban and Ulinski (1981).

within the avian anterior dorsal ventricular ridge (Fig. 27-12). The rostral region of nucleus dorsolateralis posterior appears more closely related to the collothalamic visual system. The collothalamic and nontrigeminal lemnothalamic pathways to the telencephalon in diapsid reptiles and birds are diagrammed schematically in Figure 27-13.

Motor System

The relatively limited somatosensory representation in the telencephalon in diapsid reptiles, birds, and turtles may be

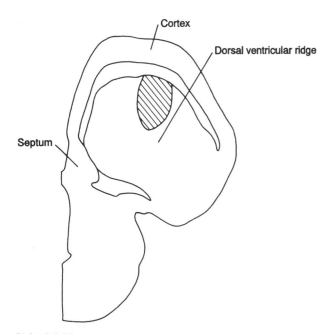

FIGURE 27-11. Drawing of a transverse hemisection through the telencephalon of a lizard (*Iguana iguana*). The location of the collothalamic telencephalic somatosensory area in the dorsal ventricular ridge is indicated by diagonal lines. Adapted from Bruce and Butler (1984b).

related to the relative dearth of descending motor pathways from the telencephalon to central pattern generators associated with motor nuclei of the brainstem and the spinal cord. Neurons in a lateral part of the telencephalon of birds, called the **tempero-parieto-occipital area** (Fig. 27-12), project to the paleostriatum and may constitute a cortical-like motor area. Similarly, neurons in the medial-to-central part of the dorsal cortex in gecko lizards have been found to project to the striatum. While some descending pathways from the telencephalon to the brainstem reticular formation have been noted—particularly a pathway from the intermediate archistriatum to the parvocellular reticular formation in the medulla of birds—we can only speculate that these regions are motor in the absence of detailed stimulation studies to determine that some form of motor somatotopy exists. Other descending pathways arise in regions that are sensory or are closely related to sensory structures. Because a pathway is descending does not guarantee that it is motor. Sensory systems may require descending pathways to the reticular formation for such behavioral functions as focusing attention on a particular sensory target, which may have little or nothing to do with motor function in the sense that motor cortex is related to movement in mammals.

Another difficulty is the apparent absence of an obvious motor thalamus that has reciprocal connections with pallial or dorsal ventricular ridge motor cell groups as exist in mammals. The few studies that have been done report little or no projections of cerebellar nuclei or other brainstem groups to the dorsal thalamus in diapsid reptiles or turtles. Birds have a dorsal thalamic nucleus, **nucleus dorsointermedius posterior (DIP)** that receives input from the globus pallidus equivalent of the corpus striatum and sends its efferent axons to the neostriatum of the dorsal ventricular ridge. This region, in turn, sends efferents to the **archistriatum intermedium,** which is the source of a descending pathway to the reticular formation. Another dorsal thalamic nucleus, **nucleus dorsolateralis intermedius (DLI)** receives projections from a variety of sources, including vestibular and cerebellar nuclei, and projects to part of the neostriatum and to the dorsal and ventral striatum. Given

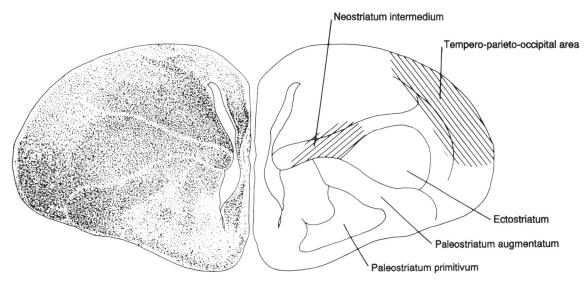

FIGURE 27-12. Transverse hemisection with mirror-image drawing through the telencephalon of a bird (*Columba livia*). The location of the collothalamic telencephalic somatosensory area in the neostriatum intermedium, within the dorsal ventricular ridge, is indicated by the more medial area of diagonal lines. The location of putative motor neurons that project to the paleostriatum is indicated by the area of diagonal lines in the lateral part of the telencephalon. Adapted from Karten and Hodos (1967) with additional data from Wild (1987b) and Brauth et al. (1978). Used with permission of the Johns Hopkins University Press.

the current absence of evidence for a definitive avian motor cortex, however, one can only speculate that DIP and DLI might be the avian equivalents of nuclei ventralis anterior and ventralis lateralis of mammals.

The main routes to the control of central pattern generators and motor neurons in nonmammalian amniotes appear to be via the striatum and its connections to a variety of descending spinal pathways, and via the tectum with its tectospinal route. A substantial amount of additional research is needed in diapsid

reptiles, birds, and turtles before we can develop a full understanding of the evolution of motor systems in amniotes.

EVOLUTIONARY PERSPECTIVE

In contrast to the elaborate telencephalic somatosensory and motor systems of mammals, those of diapsid reptiles, birds, and turtles seem modest indeed. What factors might have led

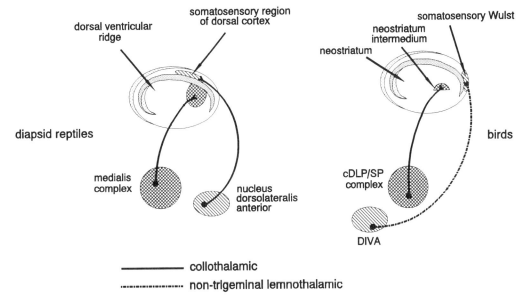

FIGURE 27-13. A schematic diagram of the nontrigeminal lemnothalamic somatosensory pathway to the telencephalon possibly in diapsid reptiles via part of nucleus dorsolateralis anterior and in birds, and the collothalamic somatosensory pathway to the telencephalon in diapsid reptiles and birds.

to the elaboration of these systems in mammals while leaving them relatively undeveloped in the other amniote groups? One possible factor might have been the evolution of hair as the mammalian skin covering. Because of the vastly greater numbers of hairs on mammalian skin than there are scales on diapsid reptiles or feathers on birds, the skin has the capacity to be a much more sensitive receptor sheet in mammals than in the other amniote taxa. The increased numbers of hairs also allows for a heightened sensitivity to precisely where on the body surface some stimulation occurs and to how large is the area of contact.

A second consideration is the development in mammals of direct telencephalic control over motor nuclei of the spinal cord, both in lamina VIII as well as directly onto the motor neurons of lamina IX. Increased precision in the use of digits, especially in great apes, simian monkeys, raccoons, and rodents such as rats, squirrels, chipmunks, and hamsters, and to a lesser extent in prosimians, dogs, tree shrews, insectivores, monotremes, and marsupials, has led to the development of the elaborate corticospinal motor system that is characteristic of mammals. Many diapsid reptiles and birds have the capacity to manipulate their digits and can perform prehensile motor acts in grasping objects such as tree limbs, seizing prey as do predatory birds, or manipulating their eggs or food objects; the digital abilities of these animals, however, are very limited in comparison with those of a squirrel, for example, deftly manipulating an acorn as it chews first on one side and then on the other, a monkey delicately lifting a tasty insect held between thumb and index finger to its mouth, or a human typing a manuscript. The latter precision motor behaviors require the collaboration of a sophisticated somatosensory system to provide the feedback necessary to permit the motor system to calibrate its commands precisely in accordance with the immediate consequences of those commands.

Finally, while a lemnothalamic somatosensory pathway may be a common feature of amniotes, discrete lemnothalamic relay nuclei for this system may be absent in diapsid reptiles and turtles, even though they are present in birds and mammals. The possibility exists that birds and mammals independently evolved the discrete relay nuclei for this system by a process of parcellation of the lemnothalamus, rather than possessing them as a heritage from their common amniote ancestors. With regard to the trigeminal component of the lemnothalamic somatosensory system, mammals may be unique among vertebrates in having the target of the ascending trigeminal lemniscal pathway located in the dorsal thalamus, in close juxtaposition to the somatic lemniscal pathway.

FOR FURTHER READING

Balaban, C. D. and Ulinski, P. S. (1981) Organization of thalamic afferents to anterior dorsal ventricular ridge in turtles. *Journal of Comparative Neurology*, 200, 95–129.

Bates, J. F. and Goldman-Rakic, P. (1993) Prefrontal connections of medial motor areas in the rhesus monkey. *Journal of Comparative Neurology*, 336, 211–228.

Bruce, L. L. and Butler, A. B. (1984a) Telencephalic connections in lizards. I. Projections to cortex. *Journal of Comparative Neurology*, 229, 585–601.

Bruce, L. L. and Butler, A. B. (1984b) Telencephalic connections in lizards. II. Projections to the anterior dorsal ventricular ridge. *Journal of Comparative Neurology*, 229, 602–615.

Butler, A. B. (1994a) The evolution of the dorsal thalamus of jawed vertebrates, including mammals: cladistic analysis and a new hypothesis. *Brain Research Reviews*, 19, 29–65.

Butler, A. B. (1994b) The evolution of the dorsal pallium in the telencephalon of amniotes: cladistic analysis and a new hypothesis. *Brain Research Reviews*, 19, 66–101.

Ebbesson, S. O. E. and Voneida, T. J. (1969) The cytoarchitecture of the pallium in the tegu lizard (*Tupinambis nigropunctatus*). *Brain, Behavior and Evolution*, 2, 431–466.

Ghez, C. (1991) Voluntary movement. In E. R. Kandel, J. H. Schwartz, and T. J. Jessell (eds.), *Principles of Neural Science*. Norwalk, CT: Appleton, pp. 609–625.

Heffner, R. and Masterton, R. B. (1975) Variation in the form of the pyramidal tract and its relationship to digital dexterity. *Brain, Behavior and Evolution*, 12, 161–200.

Hepp-Reymond, M.-C. (1988) Functional organization of motor cortex and its participation in voluntary movements. In H. D. Steklis and J. Erwin (eds.), *Comparative Primate Biology, Vol. 4: The Neurosciences*. New York: Liss, pp. 501–624.

Johnson, J. I. (1990) Comparative development of somatic sensory cortex. In E. G. Jones and A. Peters (eds.), *Cerebral Cortex, Vol. 8B: Comparative Structure and Evolution of the Cerebral Cortex, Part II*. New York: Plenum, pp. 335–449.

Jones, E. G. (1986) Connectivity of primate sensory–motor cortex. In E. G. Jones and A. Peters (eds.), *Cerebral Cortex, Vol. 5: Sensory-Motor Areas and Aspects of Connectivity*. New York: Plenum, pp. 115–183.

Kaas, J. H. (1982) The segregation of function in the nervous system: why do sensory systems have so many subdivisions? *Contributions to Sensory Physiology*, 7, 201–240.

Kaas, J. H. and Pons, T. P. (1988) The somatosensory system of primates. In H. D. Steklis and J. Erwin (eds.), *Comparative Primate Biology, Vol. 4: The Neurosciences*. New York: Liss, pp. 421–468.

Northcutt, R. G. and Kicliter, E. (1980) Organization of the amphibian telencephalon. In S. O. E. Ebbesson (ed.), *Comparative Neurology of the Telencephalon*. New York: Plenum, pp. 203–255.

Olson, C. R., Musil, S. Y., and Goldberg, M. E. (1993) Posterior cingulate cortex and visuospatial cognition: properties of single neurons in the behaving monkey. In B. A. Vogt and M. Gabriel (eds.), *Neurobiology of Cingulate Cortex and Limbic Thalamus*. Boston: Birkhauser, pp. 366–380.

Pritz, M. B. and Northcutt, R. G. (1980) Anatomical evidence for an ascending somatosensory pathway to the telencephalon in crocodiles (*Caiman crocodilus*). *Experimental Brain Research*, 40, 324–345.

Pritz, M. B. and Stritzel, M. E. (1990) Thalamic projections from a midbrain somatosensory area in a reptile, *Caiman crocodilus*. *Brain, Behavior and Evolution*, 36, 1–13.

Pritz, M. B. and Stritzel, M. E. (1994) Anatomical identification of a telencephalic somatosensory area in a reptile, *Caiman crocodilus*. *Brain, Behavior and Evolution*, 43, 107–127.

Reiner, A., Brauth, S. E., and Karten, H. J. (1984) Evolution of the amniote basal ganglia. *Trends in Neurosciences*, 7, 320–325.

Rowe, M. (1990) Organization of the cerebral cortex in monotremes and marsupials. In E. G. Jones and A. Peters (eds.), *Cerebral Cortex, Vol. 8B: Comparative Structure and Evolution of Cerebral Cortex, Part II*. New York: Plenum, pp. 263–334.

Ulinski, P. S. (1990) The cerebral cortex of reptiles. In E. G. Jones and A. Peters (eds.), *Cerebral Cortex, Vol. 8A: Comparative Structure* and Evolution of the Cerebral Cortex, Part I. New York: Plenum, pp. 139–215.

Ulinski, P. S. and Margoliash, D. (1990) Neurobiology of the reptile-bird transition. In E. G. Jones and A. Peters (eds.), *Cerebral Cortex, Vol. 8A: Comparative Structure and Evolution of the Cerebral Cortex, Part I*. New York: Plenum, pp. 217–265.

Wild, J. M. (1987a) Thalamic projections to the paleostriatum and neostriatum in the pigeon (*Columba livia*). *Neuroscience*, 20, 305–327.

Wild, J. M. (1987b) The avian somatosensory system: connections of regions of body representation in the forebrain of the pigeon. *Brain Research*, 412, 205–223.

Wild, J. M., Arends, J. J. A., and Zeigler, H. P. (1985) Telencephalic connections of the trigeminal system in the pigeon (*Columba livia*): a trigeminal sensorimotor circuit. *Journal of Comparative Neurology*, 234, 441–464.

Wise, S. P. and Donoghue, J. P. (1986) Motor cortex of rodents. In E. G. Jones and A. Peters (eds.), *Cerebral Cortex, Vol. 5: Sensory-Motor Areas and Aspects of Connectivity*. New York: Plenum, pp. 243–270.

Zeigler, H. P. and Karten, H. J. (1973) Brain mechanisms and feeding behavior in the pigeon (*Columba livia*). I. Quintofrontal structures. *Journal of Comparative Neurology*, 152, 59–82.

ADDITIONAL REFERENCES

Arends, J. J. A. and Zeigler, H. P. (1991) Organization of the cerebellum in the pigeon (*Columba livia*). II. Projections of the cerebellar nuclei. *Journal of Comparative Neurology*, 306, 245–272.

Brauth, S. E., Ferguson, J. L., and Kitt, C. A. (1978) Prosencephalic pathways related to the paleostriatum of the pigeon (*Columba livia*). *Brain Research*, 147, 205–221.

Catania, K. C., Northcutt, R. G., Kaas, J. H. and Beck, P. D. (1993) Nose stars and brain stripes. *Nature (London)*, 364, 493.

Favorov, O. V. and Kelly, D. G. (1994) Mimicolumn organization within somatosensory cortical segregates: I. Development of afferent connections. *Cerebral Cortex*, 4, 408–427.

Gonzalez, A., Russchen, F. T., and Lohman, A. H. M. (1990) Afferent connections of the striatum and the nucleus accumbens in the lizard *Gekko gecko*. *Brain, Behavior and Evolution*, 36, 39–58.

Karten, H. J. and Hodos, W. (1967) *A Stereotaxic Atlas of the Brain of the Pigeon (Columba livia)*. Baltimore: The Johns Hopkins University Press.

Morecraft, R. J. and Van Hoesen, G. W. (1992) Cingulate input to the primary and supplementary motor cortices of the rhesus monkey: evidence for somatotopy in areas 24c and 23c. *Journal of Comparative Neurology*, 322, 471–489.

Lu, S.-M. and Lin, R. C.-S. (1993) Thalamic afferents of the rat barrel cortex: a light- and electron-microscopic study using *Phaseolus vulgaris* leucoagglutinin as an anterograde tracer. *Somatosensory and Motor Research*, 10, 1–16.

Musil, S. Y. and Olson, C. R. (1993) The role of cat cingulate cortex in sensorimotor integration. In B. A. Vogt and M. Gabriel (eds.), *Neurobiology of Cingulate Cortex and Limbic Thalamus*. Boston: Birkhauser, pp. 345–365.

Stepniewska, I., Preuss, T. M., and Kaas, J. H. (1993) Architectonics, somatotopic organization, and ipsilateral cortical connections of the primary motor area (M1) of owl monkeys. *Journal of Comparative Neurology*, 330, 238–271.

Weller, W. L. (1993) SmI cortical barrels in an Australian marsupial, *Trichosurus vulpecula* (brush-tailed possum): structural organization, patterned distribution, and somatotopic relationships. *Journal of Comparative Neurology*, 337, 471–492.

28

Auditory and Vocal Forebrain in Amniotes

INTRODUCTION

In this chapter, we will continue our discussion of sensory and motor aspects of the forebrain. This time, however, we will be discussing the world of sound: how its detection and production are processed in the dorsal thalamus and telencephalon. Sound waves are a form of rhythmic disturbance of the surrounding medium. For animals that live in air, such as humans, this means that some rhythmic event produces a rhythmic pattern of movement of air molecules sufficient to cause our tympanic membrane (eardrum) to vibrate. This pattern of vibration is detected by the hair cells in the cochlea (see Chapter 2) and is transmitted to the brain. For animals that live in water, the stimulus to be detected is a disturbance of the water molecules surrounding it. Some animals, like moles, however, move through tunnels of earth; the earth conducts vibrations to the air in the tunnel surrounding the animal where these vibrations are detected by its auditory system. In this chapter we also will describe the telencephalic pathways for the control of vocal production.

Vocal productions (and their detection) are important in the lives of animals as social signals to identify individuals, to keep animals in contact with each other when they are out of sight, to establish and maintain territories, to attract mates, to warn of predators, and to threaten intruders. In some special cases, such as owls, bats, and dolphins, hearing is used to detect the location of prey by accurately localizing the source of sounds in space. In the case of owls, the salient sounds are those made by the prey as it moves through the surrounding environment of leaves, grass, and twigs. Bats, however, use the entire vocal-auditory system to produce sounds that reflect off the prey (or other objects); these reflected signals are detected by the bat. Bats can then "compute" the location in space from which they were reflected. Indeed no sound signal has any value unless it can be detected and interpreted. Neuronal

populations and pathways in the telencephalon play a role in the production, detection, and interpretation of vocal signals and other sounds.

In this chapter, we will make frequent reference to auditory structures and axonal pathways of the hindbrain and midbrain that were described in Chapters 11, 12, 15, and 17. The reader might want to briefly review those before going further in this chapter.

Location of Sound Sources

The location of a sound source is one of the most important aspects of the acoustic world that an animal must process. In order to make an accurate determination of source location, an animal must make use of the differences in the signals that arrive at its two ears. If we had only one ear, we would have to turn 360° in order to determine the direction from which the sound seemed to be the loudest, which would be the direction of the source. Two ears, however, permit source direction to be determined without moving the head at all, although head movements can improve directional accuracy. The acoustical stimuli that enter the two ears are known as **binaural** signals (bi = two, and aural = ear). The differences between the acoustical signals that reach each ear, because of their different locations in space, are called **binaural cues** to direction. These binaural cues are largest when the sound source is directly in line with the entrance to one of the two ears. Localization is generally most accurate when the target is in front of the listener.

Three types of binaural differences can occur when the source is lined up directly with the entrance to the ear and are illustrated in Figure 28-1.

- **Binaural time difference** is the difference in time required for the sound wave to reach the second ear after arriving at the first ear. This difference is very short and is determined partly by the speed of sound (which can vary slightly in different atmospheric conditions) and partly by the size of the head (which determines the distance between the two ears). This time difference is less than 1 ms for humans and shorter for animals with smaller heads. Owls make exceptional use of this cue in their superb ability to detect the source of a sound made by a prey animal, such as a small rodent. Figure 28-1(A) shows the sound arriving at the near ear at time 1 (t1) and later at the far ear at time 2 (t2).

- **Binaural intensity difference** results from the second ear being in the sound "shadow" of the head. Just as the head can cast a shadow when a beam of light is shined on it, so it can cast a sound shadow to an acoustical wave because some of the sound is reflected off the first side of the head, which reduces intensity at the second ear, and some of the sound gets absorbed by the head, which further reduces the intensity at the second ear. Figure 28-1(B) shows some of the sound being reflected from the near side of the head, which makes the intensity at the near ear (i1) greater than the intensity at the far ear (i2).

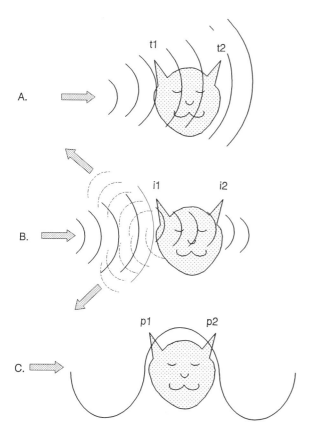

FIGURE 28-1. Binaural cues to the localization of sounds in space. The arrows indicate the direction of the sound wave. (A) Binaural time difference. The sound arrives at the near ear at time 1 (t1), which is sooner than when it arrives at the far ear at time 2 (t2). (B) Binaural intensity difference. Sound is reflected from the head on the near side so that the far ear is in the sound "shadow" of the head. Thus the intensity at the near ear (i1) is greater than the intensity at the far ear (i2). (C) Binaural phase difference. The near ear receives the rising phase of the sound wave (p1) at the same instant that the far ear receives the falling phase (p2). Binaural phase differences depend on the frequency of the sound wave and the spacing between the ears.

- **Binaural phase difference** occurs when the acoustical wave that arrives at the first ear is out of phase with the wave arriving at the second ear; that is, the first ear may be receiving a rising phase of the wave at the same instant that the second ear is receiving a falling phase of the wave. The availability of this cue depends on the relationship between the distance between adjacent peaks of the acoustical wave, which is inversely proportional to their frequency, and the distance between the two ears. This binaural cue is of particular importance to small animals that have ears so close together that the time difference is too short for the nervous system to process. Likewise, their small heads do not produce enough of a shadow for a detectable binaural intensity difference. Since others of their own species (and occasionally potential predators) typically are the most frequent target of vocalizations, small animals have become adapted to vocalize in frequency ranges that maximize the effectiveness of the binaural phase-difference cues of their intended listeners. A limitation, however, is that the phase-difference cue is ineffective at high frequencies when the diameter of the head is greater than the peak-to-peak distance of the acoustical wave; this would permit the first ear to be receiving the rising phase of one wave while the second ear was receiving the falling phase of a different wave. Fortunately for the animals, these high frequencies are just the ones that tend to be most reflected and absorbed, resulting in a sound shadow for the second ear, and hence produce an intensity difference that assists in the localization process. In Figure 28-1(C), the near ear is receiving the rising phase (p1) of the sound wave at the same instant that the far ear is receiving the falling phase (p2).

Although we have discussed three types of binaural cues, the auditory system actually computes the binaural time difference from the binaural phase difference rather than detecting time differences directly.

Echolocation

Many prey objects make virtually no sound as they move through their environment, such as a moth flying through the night sky. Others have the sounds that they make masked by other sounds from the environment, such as fishes swimming in the sea. Certain predators have overcome these disadvantages by producing vocalizations and listening for the echoes of these vocalizations as they are reflected off objects in the environment. The capabilities of these animals are so extraordinary that they can detect the size, shape, speed, direction, and even surface features of the objects that reflect back the acoustic waves of these vocal emissions. The two groups of animals that have become highly adapted for the skillful use of these mechanisms are the microchiropteran bats and dolphins.

Auditory Channels for Time (Phase) and Intensity

To the extent that an animal uses the binaural time difference to localize the source of a sound, whether the prey is the primary source as in the case of owls, or whether prey is reflecting an acoustical wave generated by the listener, as for bats and dolphins, the auditory system must be capable of detecting and processing minute time differences. Likewise, the phase and intensity differences must be processed as well. In some diapsid reptiles, birds, and turtles, two nuclear groups in the hindbrain form a timing channel. These are the **nucleus magnocellularis,** which is one of the cochlear nuclei that receive the axons of the auditory nerve, and the **nucleus laminaris,** which is a target of the magnocellular nucleus. Nucleus laminaris receives axons from both the ipsilateral and contralateral nucleus magnocellularis and computes the time difference between the signals at the two ears. It is able to do so because the action potentials (or spike potentials) of the auditory nerve are **phase locked;** that is, each action potential or spike is generated at the same phase of the acoustical wave. When the spikes tend to occur at approximately the same location (also known as the "phase angle") on each successive wave, they are said to be "phase locked." Nucleus laminaris thus forms a map of auditory space based on arrival-time differences of phase-locked auditory action potentials. In owls, the nucleus laminaris is very well developed. The phase locking also allows this channel to be responsive to phase. Mammals lack a nucleus laminaris. On the other hand, their **medial superior olivary nucleus** of the hindbrain performs the phase-locked encoding of binaural time differences and may be homologous to the nucleus laminaris of nonmammalian amniotes. In addition to time and phase, the auditory nerve also provides information about intensity. This latter information is processed in the other cochlear nucleus, **nucleus angularis,** which is the origin of the intensity channel. This brief discussion of the coding of binaural differences and the central representation of auditory space is intended only to give the reader a flavor of the structure-function relationships that exist in the auditory system. A fuller treatment of this rich topic may be found in the references to review articles and research reports at the end of this chapter.

DESIGN FEATURES OF THE AUDITORY SYSTEM

Topographic Organization

Topographic organization of sensory systems has been a major theme of Chapters 26 and 27 on vision and somesthesis and the motor system. In this chapter, we will encounter this mapping principle again, but in a quite different form than before. In the topographic maps that we have already discussed, the topography that was mapped was of space, either the visual space around the animal or the spatial map of the body. In the case of the auditory system, the topography is one of the frequency, time delay, and other parameters of the sound waves that make up the acoustical signals to be detected. Thus, instead of finding maps of visual space or maps of the body, here we will find maps that represent the physical properties of sound waves that are laid out in an orderly, systematic fashion, such as high tones at one end of the map, low tones at the other end, and intermediate tones located in the middle. A topographic organization based on the frequencies of tones is known as a **tonotopic organization.** Similar orderly maps can be found for other acoustic parameters depending on the particular prop-

erty of the sound that the animal must process, such as the location in space of the sound source, or its speed and direction, if it is a moving source. Tonotopic organization can be found at each level of the auditory pathway from the peripheral auditory organs to the telencephalon.

Bilateral Interaction in the Auditory Pathway

Another important design feature of the auditory system that makes it different from the visual or somatosensory systems is the rich network of interconnections between the right and left auditory pathways at each level. In general, the visual and somesthetic pathways of the right and left sides are kept separated until the telencephalon, where the right and left topographic representations can interact via commissural axons. In contrast, commissural axons are present at every level of the auditory system from the pons and medulla, where the auditory axons of the cochlear branch of cranial nerve VIII enter and go through their initial stages of processing, through to the telencephalon. The only exception is at the level of the auditory thalamus, which appears to lack any commissural connections. The high degree of bilateral interaction probably has been selected for due to the importance of making constant comparisons of the signals at the two ears for the crucial information about directionality. This constant comparison of stimuli on the right and left is not a feature of somesthesis, and the closest counterpart in vision would be the comparison of stimulation of corresponding points on the two retinas that is the basis of stereoscopic vision.

NOMENCLATURE OF THE AUDITORY SYSTEM

A great deal of research has been done on the auditory systems of amniotes, but the majority of studies have concentrated on a remarkably small number of species. Maps of the auditory telencephalon have been made in a number of species of mammals, some songbirds and nonsongbirds, and a few species of diapsid reptiles. Detailed studies, however, of the auditory cortex or auditory dorsal ventricular ridge and their relationships with the auditory dorsal thalamus have mainly been carried out in cats, several species of monkeys, and several species of birds. A further complicating factor has been the variation in the nomenclature of the telencephalic and diencephalic components of this system in different species and by different investigators studying the same species. In this chapter we will use as the mammalian nomenclature the names that seem to be the most common, which are the ones used most often (but not always) for cats and a number of primate species. The avian nomenclature is relatively consistent among species and researchers. Diapsid reptiles and turtles seem to be the most neglected among the amniotes. This may be due to their relatively unsophisticated auditory systems in contrast to those of birds and mammals, which in turn may be related to the relative paucity of sounds that they are capable of producing in comparison to birds and mammals.

AUDITORY THALAMUS

Collothalamic Auditory Pathway

In the two preceding chapters, we saw that the somatosensory system and the visual system each have both collothalamic and lemnothalamic pathways to the telencephalon. The auditory system, however, only has collothalamic pathways to the telencephalon; that is, all ascending routes to the dorsal thalamus are via the auditory midbrain. To date, no lemnothalamic auditory pathways have been reported in any vertebrate.

In lizards and snakes the thalamic auditory nucleus is **nucleus medialis;** in crocodiles and turtles this medial nucleus is known as **nucleus reuniens,** which means the "reuniting" nucleus, because in crocodiles the two nuclei on either side are fused across the midline. In crocodiles, the nucleus reuniens has two divisions: **nucleus reuniens, pars centralis** and nucleus reuniens, pars diffusa. Both divisions of nucleus reuniens receive ascending axons from the auditory midbrain, but the more substantial termination field is within the central division. The central division projects auditory input to a discrete part of the dorsal ventricular ridge.

The equivalent auditory nucleus in birds is called **nucleus ovoidalis,** because of its ovoid shape. Like the nucleus reuniens, pars centralis of crocodiles, the nucleus ovoidalis receives the main ascending auditory pathway from the auditory midbrain. Nucleus ovoidalis projects to the auditory Field L region of the telencephalon in the dorsal ventricular ridge. Field L will be discussed in greater detail later in this chapter.

The mammalian counterpart of these nuclei is called **nucleus geniculatus medialis,** or **medial geniculate nucleus,** or often **medial geniculate body.** The medial geniculate nucleus has several subdivisions that vary both with species and the names assigned to them. In general, however, three main subdivisions, sometimes with additional subunits, seem to be the most common description. These three subdivisions are **medial, ventral,** and **magnocellular.** The magnocellular division of the medial geniculate nucleus should not be confused with the magnocellular cochlear nucleus, which is located in the hindbrain and receives axons of the cochlear branch of nerve VIII.

We will take this opportunity to remind you of a point made in Chapter 26; namely, that the medial geniculate nucleus (or body) is not the auditory equivalent of the dorsal lateral geniculate nucleus (or body) of the visual system because the medial geniculate nucleus is collothalamic and the lateral geniculate nucleus is lemnothalamic. This caveat having been noted, we must now mention that the ventral division of the medial geniculate nucleus nevertheless has certain features in common with the pars dorsalis of the lateral geniculate nucleus; that is, it has a laminar organization with a tonotopic organization within each lamina. The other divisions of the medial geniculate have tonotopic organization as well.

AUDITORY TELENCEPHALON

Columnar Organization

Like the visual and somatosensory cortices that we discussed in the two preceding chapters, the auditory cortex also has a columnar organization. The avian auditory telencephalon, which has been the most intensively studied among the nonmammalian amniotes, also has a columnar organization with different acoustical frequency bands of the tonotopic organization being represented as vertical columns through the individual laminae that are the structural organization of this area. In mammals, an additional type of organization, similar to the hypercolumns of visual cortex, has been described, which consists of alternating pairs of horizontal bands through the cortex. The neurons in one band of the pair are excited when both ears are stimulated; the neurons in the adjacent band are excited when one ear is stimulated but are inhibited when the other ear is also stimulated. These auditory "hypercolumns" (reminiscent of the visual hypercolumns described in Chapter 26) provide the telencephalic representation of auditory space. Figure 28-2 shows a schematic representation of tonotopic and columnar organization in a mammal. Figure 28-7 shows tonotopic and columnar organization in the avian Field L.

Mammals

A feature of the visual, somatosensory, and motor cortices is multiple representation of either the spatial body map or the map of visual space. Mammals also have multiple representations of the auditory map, but it is a tonotopic map, not a representation of auditory space. The main tonotopic map is referred to as **A1.** In the Brodmann nomenclature of human cortical areas, the primary auditory cortex is found in areas 41 and 42, which are located on the medial surface of the temporal lobe and into the depths of the lateral sulcus. The region containing the secondary map is designated as **A2.** A1 and A2 typically exist as a pair, with the sequence of tones in A1 and A2 reversed so that the high-frequency end of A1 is adjacent to the high-frequency end of A2. In addition, a number of other tonotopic maps have been found in different species of mammals. Differences both in organization and nomenclature make comparisons difficult. In general, however, these addi-

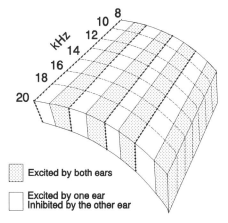

FIGURE 28-2. Columnar and tonotopic organization in the auditory cortex of a cat. The tonotopic organization is the orderly representation of acoustical frequencies. Adjacent pairs of horizontal columns are excited by acoustical stimulation of both ears or are inhibited by stimulation of one ear and are excited by stimulation of the other. These pairs of columns are similar in some respects to the hypercolumns of the visual cortex.

tional areas form a semicircular or circular **auditory belt** around A1 and A2. Marsupials appear to have only a primary or A1 area. Most eutherian mammals examined thus far have at least A1 and A2. In addition, other cortical areas in which the neurons respond to acoustical stimuli have been reported, but these are not tonotopically organized. In Figure 28-3, we present examples of the auditory cortex in several mammalian species with their individual tonotopically organized regions indicated. For simplicity of discussion, however, we will refer to three regions: A1, A2, and the auditory belt.

Bats. The auditory cortex of the microchiropteran bats, which are nocturnal and make use of echolocation to identify prey and other objects, including their distance, speed, and direction of movement, is organized in a highly complex manner. Because bats are moving through space as they emit their high-frequency vocalizations, the speed of the movement of the bat is added to the speed of the acoustical wave as it leaves the bat and is subtracted from the speed of the returning echo signal as it bounces off a stationary object. This is the **Doppler shift** phenomenon that affects the pitch of sounds. A common example of the Doppler shift can be observed in the change of pitch in the siren of an emergency vehicle as it approaches you and then passes you. As the siren passes you, the pitch of its sound drops as the speed of the vehicle is first added to the speed of the sound wave and then subtracted from it. Bats must compensate for the Doppler shift in order to make accurate distance, speed, and direction estimates. In addition, the reflections from a moving target, such as a fluttering moth, are much more complex than those from a stationary tree and produce complex Doppler shifts in the frequencies of the returning echoes. Not only can bats make these compensations, but they can use the Doppler-shift information to identify different types of insects. These abilities are represented in the organization of their auditory cortex.

Figure 28-4 shows a schematic representation of some of the many specialized areas of the auditory cortex of a mustache bat that have been described by Nabuo Suga and co-workers. The diagram shows a large tonotopic map that extends well into the ultrasonic range with a very large area of cortex dedicated to the 61-kHz frequency, which is the frequency of the animal's echolocation vocalization. This animal also has a number of additional areas that respond either to constant-frequency stimuli or to stimuli in which the frequency is rapidly changed (frequency-modulated stimuli). These correspond to the types of acoustical signals that are reflected from various surfaces in its environment including its prey. By taking advantage of differences in the amplitudes, delays, Doppler shifts, and other characteristics of these reflected signals, the bat can make extremely accurate estimates of the size, distance, velocity, azimuth, elevation, and even fine details of the surfaces that reflect its vocalizations.

The auditory systems of bats have been the subject of intense investigation, and this brief description does not do justice to the rich detail and the amazingly sophisticated organization that has been elucidated by these studies. Readers with an interest in the bat auditory system will find references to detailed review articles in the reading list at the end of the chapter.

Blind Mole Rats. The mole rat is a rodent that spends its entire life in totally dark, underground tunnels. Such an animal has no use for a light detection system and its eyes are covered with furry skin. In spite of its covered eyes, it does possess a dorsal division of the lateral geniculate nucleus and striate cortex. At birth, the retina projects to the lateral geniculate nucleus, but this pathway soon degenerates and stimulation of the eyes with light does not result in the generation of electrical activity in its striate cortex. Blind mole rats, however, depend heavily on the auditory system for information about the external environment. The remarkable characteristic of blind mole rats is that neurons from the auditory system invade the idle visual system and convert its functions to auditory processing. Thus, the pars dorsalis of the lateral geniculate as well as the striate cortex of mole rats are responsive to auditory stimulation. This

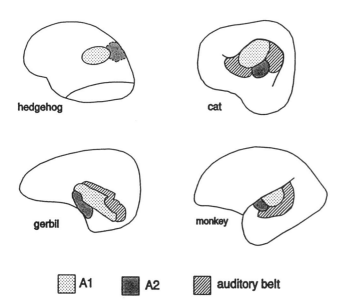

A1   A2   auditory belt

FIGURE 28-3. The major subdivisions of the auditory cortex in four mammals: a hedgehog, a cat, a gerbil, and monkey.

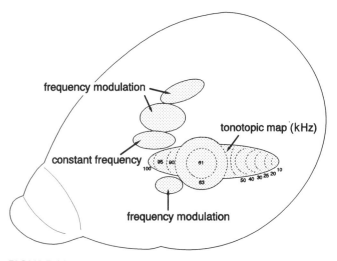

FIGURE 28-4. Some of the major areas of the auditory cortex of the mustache bat including a large tonotopic map and specialized areas for the detection of constant-frequency and frequency-modulated tones. Adapted from Suga (1984).

"invasion" of a neighboring sensory system may offer an important clue to one of the mechanisms of sensory system evolution.

Connections of the Auditory Cortex. The connections between the auditory cortex and the auditory dorsal thalamus of a cat are shown schematically in Figure 28-5. The figure shows three regions of auditory cortex: A1, A2, and a region described as the auditory belt, which varies considerably between species and contains a number of subdivisions that are not always contiguous. In general, however, the principal source of afferents to A1 is the ventral division of the medial geniculate. The magnocellular division has a very widespread projection and sends its efferents to virtually all auditory cortical areas. Axons originating in the dorsal division of the medial geniculate mainly terminate in the auditory belt region that surrounds A1 and A2. The auditory cortical areas, like those of the somatosensory and visual cortices are interconnected with each as shown in the figure. In addition, A1 sends efferent axons to the ventral division of the medial geniculate as well as to the inferior colliculus, which constitutes the auditory midbrain of mammals.

Diapsid Reptiles, Birds, and Turtles

As with somatosensory and visual collothalamic pathways from the dorsal thalamus to the telencephalon in diapsid reptiles, birds, and turtles, the target of the auditory collothalamic pathway is the dorsal ventricular ridge. In the case of the auditory pathway, the termination field is in (or very near to) the medial wall of the dorsal ventricular ridge, as illustrated in an iguana (Fig. 28-6), a turtle (Fig. 28-7), and a pigeon (Fig.28-8).

The telencephalic auditory area in the dorsal ventricular ridge in birds is more complex than in diapsid reptiles or turtles. This area is known as **Field L,** based on the telencephalic nomenclature of the avian telencephalon proposed by Maximillian Rose, in which he designated the various regions of the telencephalon with letters of the alphabet rather than names such as ectostriatum or neostriatum. The merit of Rose's system is that it is completely neutral and implies nothing about homologies with structures in the brains of mammals or other vertebrates. Its disadvantage is that names (even if they later may turn out to be inappropriate or misleading) are easier to associate with a region of the brain than are letters of the alphabet. This is particularly true when some notion of homology or equivalence is inherent in the name. As a result, Rose's nomenclature never achieved popularity among researchers except for Field L, for which there was no specific preexisting name.

Field L consists of three laminae, **L1, L2,** and **L3** (Fig. 28-8). The thalamorecipient layer of Field L is L2; different regions within Field L2 receive the ascending axons from nucleus ovoidalis and from nucleus semilunaris parovoidalis. Fields L1 and L3 are interconnected with L2. Figure 28-9 shows a schematic representation of the columnar and tonotopic organization in Field L. The figure indicates that each lamina of Field L contains its own tonotopic organization.

Figure 28-10 summarizes the pathways of the auditory forebrain in amniotes. The figure shows the pathways in a mammal, a diapsid reptile, and a bird. In each, the axonal pathway from the thalamic auditory nucleus (the medial geniculate, the reuniens–medialis complex, or the nucleus ovoidalis) is shown to its respective telencephalic target (auditory cortex, the auditory area of the dorsal ventricular ridge, or Field L). In each class, only a collothalamic pathway has been reported.

VOCAL TELENCEPHALON

The emission of sounds from the oral cavity is a characteristic of many vertebrate species, especially those animals that live

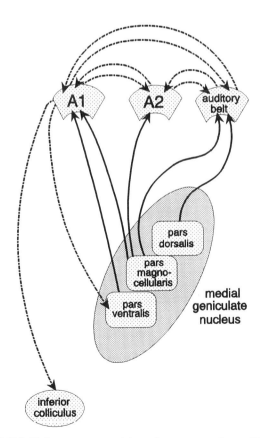

FIGURE 28-5. Connections of the auditory cortex of a cat. The solid lines indicate afferents from the subdivisions of the thalamic medial geniculate nucleus. Broken lines indicate efferents from auditory cortical areas.

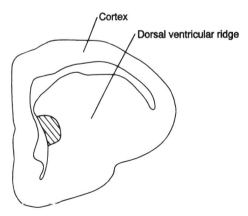

FIGURE 28-6. Drawing of a transverse hemisection through the telencephalon of a lizard. The auditory projection area within the dorsal ventricular ridge is indicated by diagonal lines. Adapted from Bruce and Butler (1984).

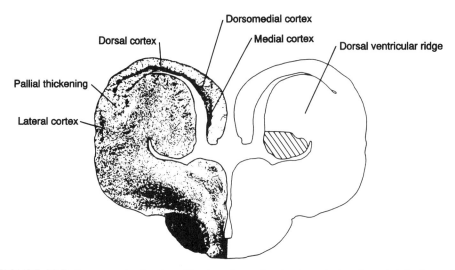

FIGURE 28-7. Transverse hemisection with mirror-image drawing through the telencephalon of a turtle. The auditory projection area within the dorsal ventricular ridge is indicated by diagonal lines. Adapted from Balaban and Ulinski (1981).

in air. Air-breathing animals are well adapted for this behavior because of the relative ease of manipulation of the air column that is moving in and out of the body for respiration. This is much the way the performer on a musical wind instrument manipulates the air column within the instrument by blowing and operating various valves or tube lengths to produce different musical notes. The sounds made by air breathing animals, known as **vocalizations,** serve a variety of purposes such as establishment and defense of territories, attraction of mates, intimidation of rivals, contact with mates or young when not in view, and warning others of the species of the presence of predators and other dangers. Although these forms of communication carry considerable information for the listener, they do not constitute language because they lack grammatical rules of sentence structure such as subject and an object. Only humans

are capable of *vocal* language. Until recently, humans were thought to be the only animals that are capable of language of any sort, vocal or otherwise. Recent studies of chimpanzees have shown that these animals, which are generally believed to be the closest primate relatives to humans, have the capacity to learn and use nonvocal language in the form of manipulation of symbols or through the use of American sign language, such as is used by hearing or speech impaired humans. These animals also demonstrate an ability to understand the grammatical content of human speech, rather than merely responding to spoken commands such as we might use with a pet dog or some other domestic animal. Although these observations are impressive, some controversy surrounds the interpretation of this research, and some experts in human language remain unconvinced.

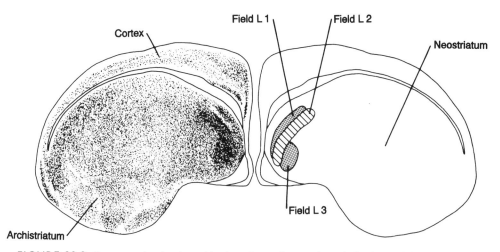

FIGURE 28-8. Transverse hemisection with mirror-image drawing through the diencephalon of a pigeon. The auditory area (L2) and its related regions (L1 and L3) in the dorsal ventricular ridge are indicated by diagonal lines and shading respectively. Adapted from Karten and Hodos (1967) with additional data from Wild et al. (1993). Used with permission of The Johns Hopkins University Press.

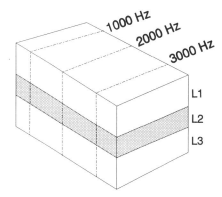

FIGURE 28-9. Tonotopic and columnar organization in Field L of a bird.

Diapsid Reptiles and Turtles

Relatively little research has been done on the vocal control mechanisms of diapsid reptiles and turtles. This is perhaps due to reptilian dependence on visual and olfactory signals for communication. Some diapsid reptiles and turtles do emit vocalizations, but they are rather limited in their variety in contrast to the rich variety of vocal signals produced by mammals and the extraordinary variety and complexity of vocalizations of which humans and birds are capable.

Vocalization and Hearing

Hearing is closely related to vocalization for several reasons. First, affecting the behavior of another animal is almost invariably the purpose of vocalization; if the vocalization is not heard, it cannot have an effect. Second, hearing is important for the development and maintenance of vocalization in many animals. The vocalizer hears its own vocal productions and therefore can modulate or modify them in accordance with the situation. In addition, many species of birds learn their species

typical calls during their early posthatch development by listening to their parents vocalize. Likewise, humans learn articulate speech not only by listening to other humans speaking, but also by listening to their own vocal productions.

Birds

The vocal-control system of birds, especially in songbirds, such as canaries and finches, has been the subject of considerable attention by neuroscientists during the past two decades. Substantial progress has been made in mapping the axonal pathways from the telencephalon to the nucleus of the hypoglossal nerve (cranial nerve XII) in the caudal hindbrain, a portion of which controls the organs of vocalization. These studies also have illuminated the relationship between the auditory system and the vocal control system. Figure 28-11 summarizes the major pathways of the avian auditory and vocal-control systems. In Fig. 28-11(A) the avian auditory system is represented in a sagittal perspective. In the inner ear, near the caudal end of the brainstem, the basilar papilla (see Chapter 2) is shown as the peripheral auditory organ that contains the auditory hair-cell receptors. Axons from these hair cells form the acoustical branch of cranial nerve VIII and terminate in the cochlear nuclei. Ascending axons from the cochlear nuclei terminate in the auditory midbrain in the **nucleus mesencephalicus lateralis, pars dorsalis (MLd).** Further links in the pathway are from MLd to nucleus ovoidalis in the dorsal thalamus and then to field L in the telencephalon.

The organ of vocalization in birds is called the **syrinx,** which is the Greek word for "whistle," and which is located at the bifurcation of the trachea. The nucleus in the hindbrain that controls the syrinx is the **tracheosyringeal division of the hypoglossal nucleus** of cranial nerve XII, otherwise known as **nucleus XIIts.** This division of nucleus XII contains the motor neurons that control the muscles of the trachea and syrinx.

Figure 28-11 (B and C) show the forebrain pathways that link the auditory system with the vocal-control pathway. Two

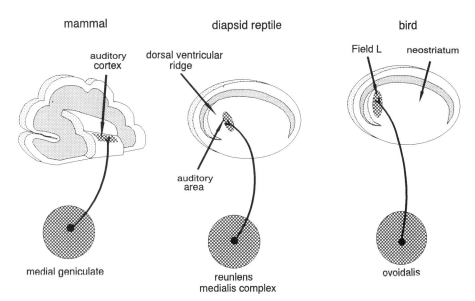

FIGURE 28-10. Summary of the auditory forebrain in mammals, diapsid reptiles, and birds. Rostral is to the left.

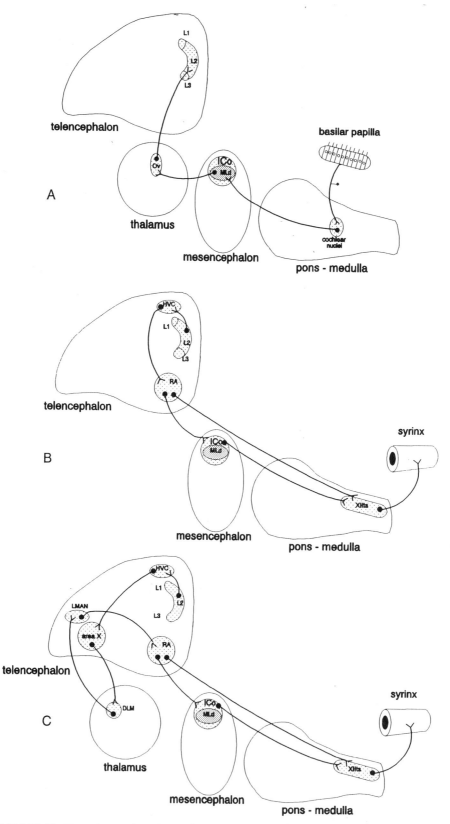

FIGURE 28-11. The auditory and vocal-control pathways in a songbird. (A) The auditory pathway. (B) The direct vocal-control pathway. (C) The indirect vocal-control pathway. See text for abbreviations.

main forebrain pathways have been described. The first, which is illustrated in Figure 28-11(B), is the direct pathway from field L to a region of the caudal hyperstriatum ventrale known as the **higher vocal center** or **HVC.** The importance of HVC for vocal behavior may be seen in the strong relationship between the volume of HVC and the number of song types that a songbird species typically has in its repertoire; the more song types, the greater the volume of HVC. Axons from HVC pass to the **robust nucleus of the archistriatum (RA),** which is located in the region of the telencephalon that gives rise to other descending pathways that are presumed to be motor. Most of the axons of the RA pass directly to nucleus XIIts. Some, however, terminate in the midbrain in the **nucleus intracollicularis (ICo),** which surrounds the MLd. Axons from nucleus intracollicularis also descend to nucleus XIIts in the hindbrain. This pathway from HVC to RA is referred to by some investigators as the **direct pathway** for control of nucleus XIIts. This pathway appears to control the initiation and production of vocalization.

The **indirect pathway,** which is shown in Figure 28-11(C), is considerably more complex and passes from field L to HVC and then to a nucleus in the rostral telencephalon known as the vocal-control area or **telencephalic area X,** which is not to be confused with nucleus X of the motor thalamus of some mammals. Area X is unusual since it demonstrates a **sexual dimorphism** (a morphological difference between males and females), such that it is well developed in male song birds, but does not appear as a distinct morphological entity in females. In males, the projection from HVC to area X is substantial; in females a modest projection from HVC terminates on neurons in the region in which area X is found in males, even though area X does not stand out from the surrounding neurons in females as it does in males. This sexual dimorphism is consistent with the observation that male songbirds have a greater number and a richer variety of songs in their repertoires than do female songbirds. Axons from area X descend to the dorsal thalamus to the **pars medialis of nucleus dorsolateralis (DLM).** Efferent axons from DLM return to the telencephalon to the **pars lateralis of the nucleus magnocellularis of the anterior neostriatum (LMAN).** The efferents of LMAN are to RA, which is the source of descending telencephalic control over the nucleus XIIts. The indirect pathway appears to be related to the learning and memory of song.

Mammals

Although the great majority of mammals vocalize to a greater or lesser extent, the mammals most studied in this regard are humans because of the importance of speech and language disturbances after various types of head injuries, strokes, and other types of neurological disorders. An area at the caudal end of the left inferior frontal gyrus was identified by the nineteenth-century French surgeon Pierre Paul Broca as an area of cortex, now known as **Broca's area,** which is responsible for motor control of speech. This area corresponds to Brodmann's area 44 and the adjacent part of area 45. Broca's area is adjacent to the region of motor area M1 that controls movements of the mouth, face, lips, and tongue.

Broca referred to patients with damage to this region as having **aphasia,** or the inability to express themselves by speech. Broca's aphasia, which is a motor aphasia, should be distinguished from Wernicke's aphasia (named for Karl Wernicke, a nineteenth-century German psychiatrist), which is a sensory aphasia that is manifested by an inability to understand speech. The cortical area in which damage results in Wernicke's aphasia, known as **Wernicke's area,** is located in the left temporal lobe, immediately caudal to the A1 auditory cortex. This area is equivalent to the posterior third of Brodmann's area 22. Damage to these areas also can affect both the comprehension and production of written language. Wernicke's area and Broca's area are interconnected by a major intratelencephalic axon bundle known as the **arcuate fasciculus,** which is so named because of the arc that it forms as it first runs caudally to leave the temporal lobe and then swings rostrally to enter the ventral frontal lobe. Other areas that have been reported to be involved in the production of vocalization are the supplementary motor area, the anterior cingulate cortex, and the periaquaductal gray. That language functions and certain other psychological functions are located predominantly in the right or left hemisphere rather than being distributed with bilateral symmetry, as are the majority of neural functions, has led to the concept of **hemispheric dominance (specialization)** or **cerebral dominance.** A discussion of this interesting concept would take us too far afield from the purposes of this book. Thus we can do no more here than to call it to your attention and refer you to the list of further readings at the end of this chapter.

In mammals the production of vocal sounds is carried out by the **larynx,** which is a chamber composed of cartilage and several sets of muscles. The movement of the respiratory air column is modulated by various muscles and also resonates in the mouth cavity to produce vocal sounds. Vocalizations can also be varied by changing the size and shape of the mouth cavity by adjustment of the jaws, lips, and tongue. The muscles of the larynx are innervated by motor neurons of the **nucleus ambiguus,** which is located in the caudal hindbrain. The jaw, lip, and tongue muscles receive their innervation from the motor nuclei of the trigeminal, facial, and hypoglossal nerves. The actions of these four sets of hindbrain motor nuclei are coordinated by the reticular formation of the pons and medulla. The vocal control pathway of humans is thus not nearly as direct as in birds; it originates in Broca's area and other areas of motor cortex that include the supplementary motor area (M2) and the medial premotor area in the cingulate cortex (Brodmann's area 24). Neurons in these areas send their axons via the corticobulbar tract to widespread regions of the reticular formation of the pons and medulla. These reticular formation areas in turn send efferents to motor V, motor VII, nucleus XII, and the nucleus ambiguus. The sensory and motor pathways for speech and vocalization in humans are shown in Figure 28-12.

An important distinction must be made between speech and vocalization. Virtually all mammals are capable of vocalization, but only humans posses that special type of vocalization that we refer to as speech. Only humans use Broca's area and Wernicke's area for producing and understanding speech. In nonhuman mammals, telencephalic control of vocalization is carried out by the supplementary motor area (M2) and the anterior cingulate cortex (area 24.) Electrical stimulation of the region of M1 in nonhuman primates that corresponds to Broca's area in humans results in movements of the larynx muscles, but rarely are such movements accompanied by actual vocalization. In contrast, in many mammals and other tetrapods as

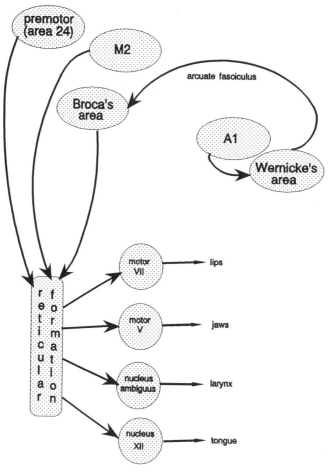

FIGURE 28-12. The vocal-control pathway in humans.

EVOLUTIONARY PERSPECTIVE

Auditory pathways in diapsid reptiles, birds, and turtles follow a consistent pattern of a collothalamic pathway from the auditory midbrain to a dorsal thalamic auditory nucleus to the medial dorsal ventricular ridge. The mammalian pattern is consistent with this overall plan; the thalamic auditory nucleus has several divisions that have different terminal fields in auditory isocortex. A lemnothalamic auditory pathway that passes from the auditory receptive nuclei of the hindbrain directly to the thalamus, thus bypassing the auditory midbrain, has not been described in mammals, diapsid reptiles, birds, or turtles. The absence of such a pathway may be related to a major difference between the topographic organization of the auditory pathway and that of the somatosensory and visual pathways; namely, that the auditory topography is tonotopic, that is, each region of the topographic map represents a certain number of cycles of the acoustic wave per second. In contrast, the visual and somatosensory topographies are spatial topographies based on visual space and space on the body surface. A speculative possibility is that spatial topographies require a lemnothalamic organization plus a collothalamic organization, whereas a tonal topography can make do with only a collothalamic organization. Perhaps future research will shed light on this possibility.

A consistent pattern also can be seen in the evolution of the vocal-control pathways. The portion of the pathway that seems common to mammals, diapsid reptiles, birds, and turtles is control of vocalization by the periaquaductal gray region of the dorsal tegmentum in the region of the junction between the midbrain and pons. This area probably acts via the reticular formation of the pons and medulla to influence the motor nuclei of the hindbrain that control vocalization. Two groups of amniotes have evolved additional neuronal networks that control the vocalization-control system. In mammals, because the size and shape of the oral cavity affects vocalization, those motor nuclei that affect oral-cavity size and shape (trigeminal, facial, and hypoglossal) are participants in this system. Mammals have evolved specialized regions of isocortex that each have a detailed topography of the body including the face and oral region. These motor areas exert control over the motor nuclei of the brainstem via the reticular formation. Humans, which are the only animals that are capable of molding their vocalizations into speech, have an additional specialized motor-speech area near the face region of M1 that is important for the production of speech sounds.

well, vocalization results from stimulation of regions in the **periaquaductal gray** area of the dorsal midbrain tegmentum.

Birds differ somewhat from mammals in that they lack lips and are relatively incapable of affecting oral cavity size and shape, except by opening and closing the beak. Birds have a specialized vocal organ, the syrinx (in addition to a larynx), but only the hypoglossal nucleus, a portion of which controls the muscles of the syrinx, is involved. Although avian vocalizations do not have the cognitive and linguistic content of human speech, they nevertheless have an extremely rich complexity and variability, especially among songbirds. The avian vocal-control pathways have evolved independently of those in mammals, with some similarities and some differences.

One might be tempted to compare the higher vocal center (HVC), for example, with Wernicke's area since they both receive their inputs from the auditory system and send their efferents to structures that are involved in the initiation and production of vocalizations (Broca's area in the case of humans and the robust nucleus, or RA, of the archistriatum in the case of birds). Such speculations are interesting, but should be undertaken cautiously and considered together with the differences. For example, a dorsal thalamic relay is important in the indirect or long pathway from HVC to RA, whereas the dorsal thalamus is not regarded as a major participant in mammalian vocalization.

FOR FURTHER READING

Balaban, C. D. and Ulinski, P. S. (1981) Organization of thalamic afferents to anterior dorsal ventricular ridge in turtles. I. Projections of thalamic nuclei. *Journal of Comparative Neurology,* 200, 95–129.

Brugge, J. F. and Reale, R. A. (1985) Auditory cortex. In A. Peters and E. G. Jones (eds.), *Cerebral Cortex. Volume 4. Association and Auditory Cortices.* New York: Plenum, pp. 229–271.

Carr, C. E. (1992) Evolution of the central auditory system in reptiles and birds. In D. B. Webster, R. R. Fay, and A. N. Popper (eds.), *The Evolutionary Biology of Hearing.* New York: Springer, pp. 511–543.

Carr, C. E. (1993) Processing of temporal information in the brain. *Annual Review of Neuroscience,* 16, 223–243.

Foster, R. E. and Hall, W. C. (1978) The organization of the central auditory pathways in a reptile, *Iguana iguana. Journal of Comparative Neurology,* 178, 738–832.

Heffner, R. S. and Heffner, H. E. (1992) Evolution of sound localization in mammals. In D. B. Webster, R. R. Fay, and A. N. Popper (eds) *The Evolutionary Biology of Hearing.* New York: Springer, pp. 691–715.

Hall, W. S. and Brauth, S. E. (eds.) (1994) Avian auditory-vocal interfaces. Fifth Annual Karger Workshop. *Brain, Behavior and Evolution,* 44, 187–286.

Karten, H. J. (1967) Organization of the ascending auditory pathway in the pigeon (*Columba livia*): I. Diencephalic projections of the inferior colliculus (nucleus mesencephali lateralis pars dorsalis). *Brain Research,* 6, 409–427.

Karten, H. J. (1968) Organization of the ascending auditory pathway in the pigeon (*Columba livia*): II. Telencephalic projections of the nucleus ovoidalis thalami. *Brain Research,* 7, 134–153.

Merzenich, M. M. and Schreiner, C. E. (1992) Mammalian auditory cortex–some comparative observations. In D. B. Webster, R. R. Fay, and A. N. Popper (eds.), *The Evolutionary Biology of Hearing.* New York: Springer, pp. 673–689.

Newman, J. D. (1988) Primate hearing mechanisms. In H. D. Steklis and J. Erwin (eds.), *Comparative Primate Biology. Volume 4. Neurosciences.* New York: Liss, pp. 469–499.

Nottebohm, F., Kelley, D. B., and Paton, J. A. (1982) Connections of vocal control nuclei in the canary telencephalon. *Journal of Comparative Neurology,* 207, 344–357.

Pollak, G. (1992) Adaptations of basic structures and mechanisms in the cochlea and central auditory pathway of the mustache bat. In D. B. Webster, R. R. Fay, and A. N. Popper (eds.), *The Evolutionary Biology of Hearing.* New York: Springer, pp. 751–778.

Pritz, M. B. (1974) Ascending connections of a thalamic auditory area in a crocodile, *Caiman crocodilus. Journal of Comparative Neurology,* 153, 179–198.

Pritz, M. B. (1980) Parallels in the organization of auditory and visual systems in crocodiles. In S. O. E. Ebbesson (ed.), *Comparative Neurology of the Telencephalon.* New York: Plenum, pp. 331–342.

Scheich, H. (1990) Representational geometries of telencephalic auditory maps in birds and mammals. In Finlay, B., Innocenti, G., and Scheich, H. (eds.), *The Neocortex.* New York: Plenum, pp. 119–138.

Shepherd, G. M. (1988) *Neurobiology.* Chapter 23. Communication and speech. New York: Oxford University Press, pp. 468–483.

Suga, N. (1984) The extent to which biosonar information is represented in the bat auditory cortex. In G. M. Edelman, W. E. Gall, and W. M. Cowan (eds.), *Dynamic Aspects of Neocortex.* New York: Wiley, pp. 315–373.

Suga, N. (1988) Auditory neuroethology and speech processing: complex-sound processing by combination-sensitive neurons. In G. M. Edelman, W. E. Gall, and W. M. Cowan (eds.), *Auditory Function: Neurobiological Bases of Hearing.* New York: Wiley, pp. 679–720.

Sutton, D. and Jürgens, U. (1988) Neural control of vocalization. In H. D. Steklis and J. Erwin (eds.), *Comparative Primate Biology. Volume 4. Neurosciences.* New York: Liss, pp. 625–647.

Webster, D. B., Fay, R. R., and Popper, A. N. (eds.) (1992) *The Evolutionary Biology of Hearing.* New York: Springer.

ADDITIONAL REFERENCES

Batzri-Israeli, R., Kelly, J. B., Glendenning, K. K., Masterton, R. B., and Wollberg, Z. (1990) Auditory cortex of the long-eared hedgehog (*Hemiechinus auritus*). *Brain, Behavior and Evolution,* 36, 237–248.

Bottjer, S. W., Halsema, K. A., Brown, S. A., and Meisner, E. A. (1989) Axonal connections of a forebrain nucleus concerned with vocal learning. *Journal of Comparative Neurology,* 379, 312–326.

Brauth, S. E. and Reiner, A. (1991) Calcitonin-gene related peptide is an evolutionarily conserved marker within the amniote thalamo-telencephalic auditory pathway. *Journal of Comparative Neurology,* 313, 227–239.

Bruce, L. L. and Butler, A. B. (1984) Telencephalic connections of lizards. II. Projections to anterior dorsal ventricular ridge. *Journal of Comparative Neurology,* 229, 602–619.

Devoogd, T. J., Krebs, J. R., Healy, S. D., and Purvis, A. (1993) Relations between song repertoire size and the volume of brain nuclei related to song: comparative evolutionary analyses amongst oscine birds. *Proceedings of the Royal Society* (*London*) B, 254, 75–82.

Durand, S. E., Tepper, J. M., and Cheng, M.-F. (1992) The shell region of the nucleus ovoidalis: a subdivision of the avian auditory thalamus. *Journal of Comparative Neurology,* 323, 495–511.

Hall, W. S., Brauth, S. E., and Heaton, J. T. (1994) Comparison of the effects of lesions in nucleus basalis and Field 'L' on vocal learning and performance in the budgerigar (*Melopsittachus undulatus*). *Brain, Behavior and Evolution,* 44, 133–148.

Heil, P., Bronchti, G., Wollberg, Z., and Scheich, H. (1991) Invasion of visual cortex by the auditory system in the naturally blind mole rat. *NeuroReport,* 2, 735–738.

Karten, H. J. and Hodos, W. (1967) *A Stereotaxic Atlas of the Pigeon Brain* (*Columba livia*). Baltimore, MD: The Johns Hopkins University Press.

Kupferman, I. (1991) Localization of higher cognitive and affective functions: the association cortices. In E. R. Kandel, J. H. Schwartz, and T. M. Jessell (eds.), *Principles of Neural Science.* Norwalk, CT: Appleton and Lange, pp. 823–838.

Nottebohm, F., Stokes, T. M., and Leonard, C. M. (1976) Central control of song in the canary. *Journal of Comparative Neurology,* 165, 457–468.

Simpson, H. B. and Vicario, D. S. (1991) Early estrogen treatment of female zebra finches masculinizes the brain pathway for learned vocalizations. *Journal of Neurobiology,* 22, 777–793.

Striedter, G. F. (1994) The vocal control pathways in budgerigars differ from those in songbirds. *Journal of Comparative Neurology,* 343, 35–56.

Wild, J. M. (1994) The auditory-vocal-respiratory axis in birds. *Brain, Behavior and Evolution,* 44, 192–209.

Wild, M. J., Karten, H. J., and Frost, B. J. (1993) Connections of the auditory forebrain in the pigeon (*Columba livia*). *Journal of Comparative Neurology,* 337, 32–62.

29

Terminal Nerve and Olfactory Forebrain

INTRODUCTION

Most vertebrates have two rostral nerves: the **olfactory nerve** and the **terminal nerve.** Some tetrapods have a third rostral nerve, the **vomeronasal nerve**, in addition to the olfactory and terminal nerves. The olfactory and vomeronasal nerves, along with the sense of taste, constitute the chemosensory system. The modality of the terminal nerve has not yet been established.

The rostral chemosensory nerves are composed of the axons of bipolar receptor cells that, along with secretory supporting cells, form the olfactory epithelium of the nasal cavities. Their axons terminate within the most rostral parts of the brain. The olfactory nerve terminates within the **olfactory bulb** in fishes. In tetrapods, the site of termination of the olfactory nerve is called the **main olfactory bulb** to distinguish it from the site of termination of the vomeronasal nerve, the **accessory olfactory bulb.** The collections of axons that originate in the olfactory bulb(s) and project to areas within the rest of the brain are referred to as tracts. Thus, in diapsid reptiles, for example, the olfactory nerve terminates in the main olfactory bulb, and the main olfactory bulb gives rise to the **olfactory tract**, which terminates in the **olfactory cortex** in the telencephalon.

In this chapter, we will first discuss the organization of the olfactory nerve system in Group I and Group II vertebrates. Then we will discuss the vomeronasal nerve system in tetrapods. The system that involves the terminal nerve differs somewhat in its organization and functional role from the other two rostral nerve systems and will be discussed last. The terminal nerve system will first be discussed in general terms for vertebrates, and then its special features in teleosts will be considered.

OLFACTORY SYSTEM

The olfactory nerve arises from neurons in the olfactory epithelium that have dendritic endings specialized for the detection of chemical stimuli and that project into the olfactory bulb. The paired olfactory bulbs are formed by evagination during development, whether the telencephalic hemispheres are formed by evagination, as in most vertebrate groups, or by eversion, as in ray-finned fishes.

The olfactory input through the olfactory bulbs is a two-neuron relay system in which the **receptor cells** project to cells called **mitral cells** (for their resemblance to a bishop's mitre) (Fig. 29-1). The multiple ramifications of both the axon of the receptor cell (R) and the dendrite of the mitral cell (M) form a complex sphere, called a **glomerulus** (G), where they meet, like the joining of two hands with the fingers partially flexed and intermingled. **Granule cells** (Gr) are also present in the olfactory bulb, and along with centrifugal inputs from the forebrain, influence the processing of olfactory information in the bulb.

The olfactory bulbs relay olfactory information to the olfactory pallium via the olfactory tracts. The relative lengths of both the olfactory nerves and the olfactory tracts are variable. In many vertebrates, including some cartilaginous and bony fishes, the olfactory bulbs are sessile in position, that is, they are juxtaposed to the telencephalon. In these cases, the olfactory nerves may be of shorter or longer length, depending on the distance of the olfactory epithelium from the olfactory bulbs. In other cartilaginous and bony fishes, the olfactory bulb lies close to the olfactory epithelium, and the olfactory tract is an elongated structure running between the olfactory bulb and the telencephalon.

The olfactory pallium is usually thought of as a distinct cytoarchitectonic area in the telencephalon that has afferent

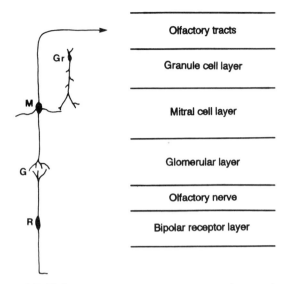

FIGURE 29-1. Olfactory bulb organization in vertebrates: schematic drawing of an olfactory bipolar receptor cell (R) with its axon terminating in a glomerulus (G) formed by the receptor axon terminals and the distal branches of the dendrite of a mitral cell (M), which in turn projects to the olfactory cortex (arrow). A granule cell (Gr) is also shown terminating on a dendrite of the mitral cell.

olfactory innervation, but such a working definition is acceptable only as a first approximation. Other criteria, such as topology, cytoarchitecture, efferent projections, and histochemical characteristics, are also important in defining and distinguishing pallial areas, but data on these aspects of the olfactory system in anamniotes are limited. Olfactory fibers terminate in more than one distinct pallial area in a number of anamniotes, and some of these areas appear to be part or all of the dorsal pallium on topological, cytoarchitectonic, and/or histochemical grounds. However, due to the relative dearth of information on the latter features and on efferent connections of these pallial areas, identifying them as homologues of specific pallial areas in amniote vertebrates is somewhat hazardous.

Among the telencephalic regions in amniotes that receive olfactory inputs relayed from the olfactory pallium is a subset of a group of nuclei, collectively called the **amygdala.** In amphibians and many amniotes, another part of the amygdala receives direct vomeronasal input from the accessory olfactory bulb. The latter system of projections will be discussed below in this chapter. The amygdala's other connections and its relation to the limbic system, of which it is considered a part, will be discussed in Chapter 30.

Group I

In lampreys, the olfactory bulbs contain a **glomerular layer** that lies deep to the olfactory nerve fibers, a **layer of mitral cells,** and a **granular cell layer** that contains cells of varying morphology. Axons from cells in the granular cell layer contribute to the olfactory tracts along with the mitral cell axons. Olfactory bulb projections are dense within a large area in the telencephalon, which has been identified as the **lateral pallium** (Fig. 29-2) partly because of this heavy olfactory input. A sparser olfactory projection is also present to the area identified as the dorsal pallium and possibly to the medial pallium as well.

In squalomorph sharks, the olfactory bulbs [Fig. 29-3] are covered by a layer of olfactory nerve fibers, deep to which are a **glomerular layer** and an **internal cellular layer.** The latter layer contains both mitral and granular cells. The layers of the olfactory peduncle can be traced into a **medial** as well as a **lateral pallial area,** which are thus interpreted as subdivisions of the lateral pallium in receipt of olfactory projections [Fig. 29-3(B)].

In nonteleost ray-finned fishes, the olfactory bulbs [Fig. 29-4] are covered by an outer layer of olfactory nerve fibers over the **glomerular layer.** Deep to the glomerular layer is an **external cellular layer** that contains both mitral cells and smaller, **periglomerular cells.** The axons of the periglomerular cells project peripherally to the receptor cells in the olfactory epithelium. A layer primarily of **secondary olfactory fibers** lies deep to the external cellular layer. This layer is primarily composed of the axons of mitral cells and divides into the medial and lateral olfactory tracts. The deepest layer of neurons is the **internal cellular layer** and contains granule cells.

Due to the development of the telencephalon by eversion instead of evagination, the most medial pallial field, sometimes designated **P1** [Fig. 29-4], is the part of the pallium in receipt of olfactory bulb projections via the lateral olfactory tract. The olfactory fibers terminate on the more distal parts of the dendrites of the pallial cells, which lie in a periventricular position. This pallial area is thought to be homologous to the lateral, olfactory pallium of vertebrates with evaginated telencephalic hemispheres.

In the coelacanth *Latimeria*, the olfactory bulb is covered by olfactory nerve fibers and contains a glomerular layer, a cellular zone that is presumed to contain mitral cells, and a more internal zone of granular cells. During the development of the telencephalon in *Latimeria*, the ventral, subpallial part evaginates. The dorsal, pallial part does not undergo an evagination, however, but is thickened and similar to the everted, pallial part of the telencephalon in ray-finned fishes. The olfactory tract lies in a central position within the pallial part of the telencephalon (Fig. 29-5).

In lungfishes, the olfactory bulbs contain a superficial olfactory nerve layer, a **glomerular layer,** an **external plexiform layer** of fibers, a **mitral cell layer,** an **internal plexiform layer,** and an **internal granular cell layer.** The transition zone between the olfactory bulb and the pallium is marked by a loss of the glomerular layer. Olfactory fibers project to areas identified, on topological, cytoarchitectonic, and histochemical grounds, as both the lateral and dorsal pallia.

In some tetrapods, both a main olfactory bulb (in receipt of olfactory nerve fibers) and an accessory olfactory bulb (in receipt of vomeronasal nerve fibers) are present. In amphibians the main olfactory bulb [Figs. 29-6 and 29-7] comprises a superficial layer of olfactory afferent fibers (rostral to the level shown in the figure), a **glomerular layer,** an **external plexiform layer,** a **mitral cell layer,** and an **internal granular cell layer.** As is also the case in lungfishes, the granular cell layer is relatively thick and densely packed with cells. In amphibians, mitral cell axons enter the lateral olfactory tract [Fig. 29-7] and

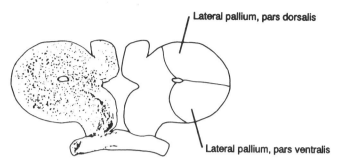

Lateral pallium, pars dorsalis

Lateral pallium, pars ventralis

FIGURE 29-2. Transverse hemisection with mirror-image drawing through the telencephalon of a lamprey. Adapted from Northcutt (1981) with additional data from Northcutt and Puzdrowski (1988). Redrawn, with permission, from the *Annual Review of Neuroscience,* Volume 4, © 1981, by Annual Reviews Inc.

terminate in the **lateral pallium** and the ventral half of the dorsal pallium. In frogs, some fibers from the medial olfactory tract reach part of the rostral medial pallium, and in salamanders, some medial olfactory tract fibers terminate in rostral parts of both the medial and dorsal pallia.

A number of connections of the lateral pallium, in addition to its input from the olfactory bulb, have been studied in amphibians. The lateral pallium also receives afferent projections from the medial pallium and from the septum, amygdala, entopeduncular nucleus, preoptic area, anterior and central parts of the dorsal thalamus, and infundibular hypothalamus. Thus, the olfactory processing done by the lateral pallium is modulated by a number of forebrain inputs. The output of the lateral pallium in amphibians is predominantly within the telencephalon—to the olfactory bulb, medial pallium, dorsal pallium, diagonal band, septum, and striatum. While the lateral pallium also projects directly to the infundibular hypothalamus, the main route by which the lateral pallium influences the hypothalamus is through the medial pallium.

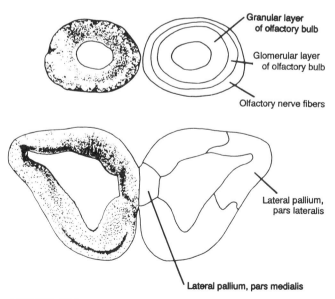

Granular layer of olfactory bulb

Glomerular layer of olfactory bulb

Olfactory nerve fibers

Lateral pallium, pars lateralis

Lateral pallium, pars medialis

FIGURE 29-3. Transverse hemisections with mirror-image drawings through the olfactory bulb (top) and telencephalon of a squalomorph shark *(Squalus acanthias).* After Northcutt et al. (1988).

Group II

In hagfishes, the olfactory nerves enter the olfactory bulb [Fig. 29-8] and terminate on the dendrites of mitral cells within the **glomerular layer.** A **periglomerular layer** and the **mitral cell layer** lie deep to the glomerular layer rostrally. In a more caudal part of the olfactory bulb, an **internal granular cell layer** is present. Both mitral and granule cells contribute axons to the olfactory tract.

The pallium of the telencephalon in hagfishes can be divided into five major divisions, three of which are shown in Figure 29-8, and the lateral olfactory tract projects massively to all five divisions. The lateral olfactory tract forms deep and superficial divisions that course within the deepest (**P5**) and the most superficial (**P1**) pallial divisions, respectively. The deep division of the lateral olfactory tract is unique to hagfishes. Whether the entire pallium in hagfishes is homologous to the lateral, piriform pallium of other vertebrates or is a field homologue of two or more of the pallial areas present in other vertebrates is an unresolved question. Studies of the efferent projections of the various divisions of the pallium and of other afferent projections may help to clarify telencephalic organization in these animals.

In Group II cartilaginous fishes, the olfactory bulb is superficially covered by a layer of olfactory receptor cell axons and contains **glomerular, mitral,** and **granular cell layers.** Fibers from the olfactory bulb divide into lateral and medial olfactory tracts in Group II sharks and terminate in the dorsal and lateral pallia. In skates (Fig. 29-9), only a single olfactory tract can be distinguished. This tract terminates in the lateral pallium and may also innervate a region on the medial border of the lateral pallium called **nucleus a.**

In teleosts, the olfactory nerve fibers form a superficial layer over the olfactory bulb. The olfactory bulb [Fig. 29-10] contains **glomerular, mitral, secondary olfactory fiber,** and **internal granular cell layers.** An example of a mitral cell in a carp is shown in Figure 29-11, although the morphology varies to some extent. A second type of neuron is present in the mitral cell layer in at least some teleosts and is called a **ruffed cell** in reference to the presence of many pedunculated protrusions on the initial portion of the axon. The axons of ruffed cells are believed to project into the telencephalon along with the mitral cell axons. The ruffed cells are unusual neurons, however, in that their dendrites surround the dendrites of the mitral cells

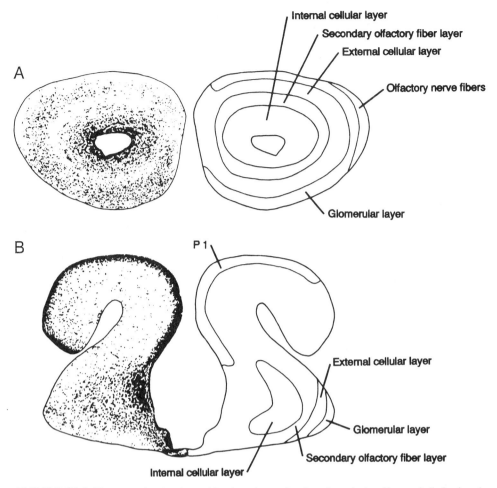

FIGURE 29-4. Transverse hemisections with mirror-image drawings through the olfactory bulb (top) and rostral telencephalon of a bichir *(Polyterus palmas).* After Northcutt and Davis (1983).

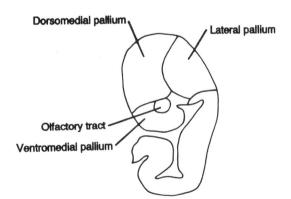

FIGURE 29-5. Drawing of a transverse hemisection through the right telencephalon of the coelacanth *(Latimeria chalumnae).* Adapted from Nieuwenhuys and Meek (1990b). Used with permission of Plenum Publishing Corp.

like glial processes, rather than receiving an array of synaptic inputs, while the ruffed portion of their axons have reciprocal synapses with granule cells. Their function in olfactory processing and their evolutionary origin remain to be determined. The granular cells of teleosts have smooth dendrites; they lack axons and appear to play a major role in the integration of olfactory information within the olfactory bulb.

The olfactory tract in teleosts comprises medial and lateral divisions. The olfactory fibers terminate predominantly within the **posterior zone of area dorsalis telencephali** [Fig. 29-10]. Recently, evidence of direct projections from the olfactory receptor cells to the posterior zone of area dorsalis and to the ventral and commissural nuclei of area ventralis (see Fig. 30-9) has also been found. Receptor cell axons that bypass the glomeruli of the mitral cells and continue to the telencephalon along with the axons of the mitral cells account for these projections.

The olfactory system has been studied in most detail in mammals. The neuronal organization of the mammalian olfactory bulb is shown semischematically in Figure 29-12. The chemosensory receptor cells terminate on mitral cell dendrites within the glomeruli in the **glomerular layer.** Centrifugal fibers, which arise in sites in the basal forebrain and project back to the olfactory bulb within the olfactory tract, synapse on

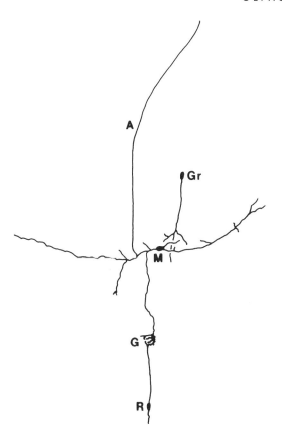

FIGURE 29-6. Drawing of a mitral cell (M) in the olfactory bulb of a frog (*Rana pipiens*) as seen in a Golgi preparation. An olfactory receptor cell (R) synapses on a glomerulus (G) on the distal end of a mitral cell dendrite. A granule cell (Gr) also synapses on the mitral cell. The axon (A) of the mitral cell projects to olfactory cortex. Adapted from Scalia et al. (1991).

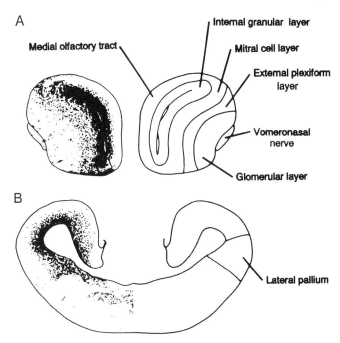

FIGURE 29-7. Transverse hemisections with mirror-image drawings through the olfactory bulb (top) and telencephalon of a tiger salamander (*Ambystoma tigrinum*). Adapted from Northcutt and Kicliter (1980). Used with permission of Plenum Publishing Corp.

the dendrites of granule cells, which, in turn, have reciprocal synapses with the mitral cell dendrites.

As in a variety of other vertebrates, the olfactory bulb in mammals gives rise to medial and lateral olfactory tracts. These tracts project to the **olfactory cortex,** which lies on the ventral surface of the telencephalon, and to neighboring areas including an **anterior olfactory nucleus** and the **olfactory tubercle** (see Fig. 29-19). These latter structures are sometimes considered to be parts of the olfactory cortex itself. The olfactory cortex comprises a **frontal olfactory cortex** lying next to the lateral olfactory tract and a more caudal, **temporal olfactory cortex.** The latter encompasses the **cortical part of the amygdala,** which receives vomeronasal projections.

The olfactory cortex in mammals projects to a variety of sites within the forebrain. It projects to some neighboring isocortical areas, to nucleus accumbens (ventral striatum), to limbic areas including the amygdaloid complex and the hippocampal formation, and to the lateral part of the hypothalamus. The olfactory cortex also has a substantial efferent projection to a nucleus in the dorsal thalamus called the **mediodorsal nucleus.**

The mediodorsal nucleus relays olfactory information to **prefrontal cortex,** and prefrontal cortex in humans (Fig. 29-13) has been implicated as the site of many higher, conscious cortical functions and as being dysfunctional in disorders such as schizophrenia. The cortical area of the projection of the mediodorsal nucleus is also richly innervated by dopaminergic fibers that arise in the brainstem. Among mammals, primates have an exceptionally large prefrontal cortical area. (Echidnas also have a very extensive prefrontal cortex but may have acquired this expansion independently.)

In diapsid reptiles (Fig. 29-14), the main olfactory bulb is covered by olfactory nerve fibers and contains a **glomerular layer** of glomeruli and periglomerular cells, an **external granular layer** of small cells, an **external plexiform layer** with a few displaced mitral cells, a **mitral cell layer,** an **internal plexiform layer,** and a densely packed **internal granular layer.** The olfactory tract projects to the **lateral,** or **piriform, cortex** in the lateral part of the telencephalic pallium, shown in a lizard in Figure 29-15. The olfactory bulbs of birds and turtles are organized in a like manner, as are their projections to lateral cortex. Efferent projections of the lateral cortex have been studied to some extent in nonsynapsid amniotes; lateral cortical efferents to nucleus dorsomedialis in the dorsal thalamus of birds recently have been described.

Studies of the distribution of dopaminergic fibers in the forebrain of birds have shown that a lateral and caudal part of the dorsal ventricular ridge, called the **posterodorsolateral neostriatum** (Fig. 29-16), is markedly rich in dopaminergic fiber innervation. While this site has been assigned to the dorsal ventricular ridge, its continuity with the dorsal cortical part of the telencephalon is important to note. A similarly located part of the dorsal ventricular ridge that is rich in dopamine has been found in lizards, although not in snakes or turtles. Behavioral studies in birds have shown that ablation of this part of the

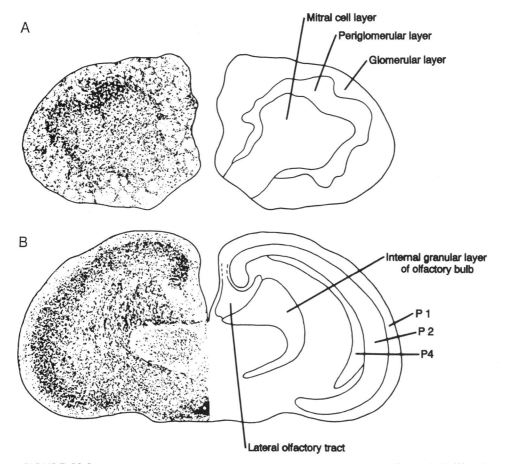

FIGURE 29-8. Transverse hemisections with mirror-image drawings through the olfactory bulb (A) and telencephalon of a hagfish *(Eptatretus stouti)*. Adapted from Wicht and Northcutt (1992). Used with permission of S. Karger AG.

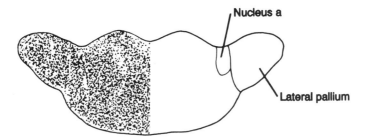

FIGURE 29-9. Transverse hemisection with mirror-image drawing through the telencephalon of a skate *(Raja eglanteria)*. Adapted from Northcutt (1978).

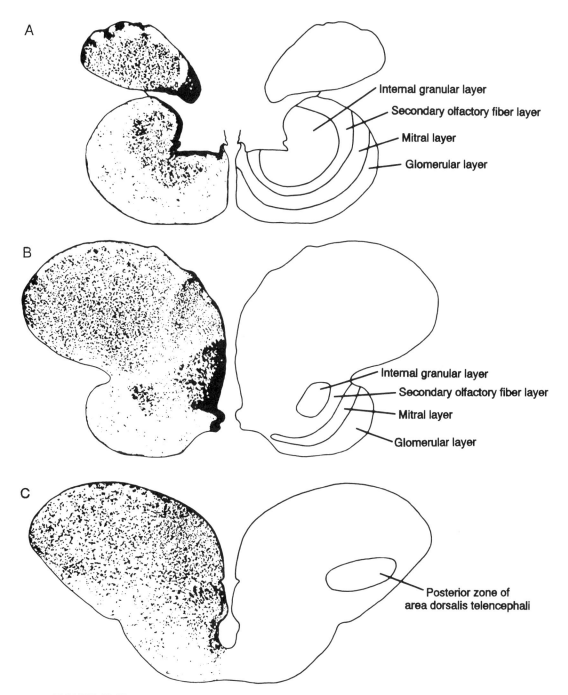

FIGURE 29-10. Transverse hemisections with mirror-image drawings through the olfactory bulb (A) and telencephalon in a teleost *(Salmo gairdneri)*. Adapted from Northcutt and Davis (1983).

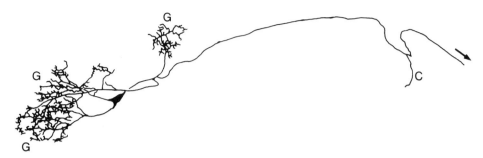

FIGURE 29-11. Drawing of a mitral cell as reconstructed from a Golgi preparation in the olfactory bulb of a carp. The distal part of the mitral cell dendrite ramifies to form several glomerular tufts (G). The axon gives off a collateral branch (C) before continuing into the telencephalon (arrow). Data from Satou (1990).

brain results in deficits in tasks requiring delayed alternation, which are similarly impaired by prefrontal cortical ablations in mammals.

The possible homology of the posterodorsolateral neostriatum in birds and of the corresponding part of the dorsal ventricular ridge in other nonsynapsid amniotes with the prefrontal cortex in mammals would be strongly supported if an olfactory input projection system through the dorsal thalamus could be identified. Nucleus dorsomedialis receives an afferent projection from piriform cortex, but a possible pathway from this

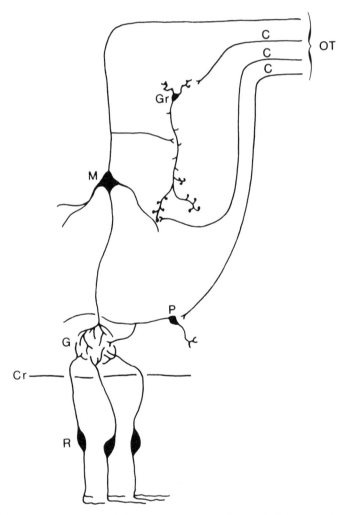

FIGURE 29-12. Schematic drawing of the organization of the olfactory bulb in mammals. Abbreviations: C: centrifugal fibers; Cr, cribiform plate; G, glomerulus; Gr, granule cell; M, mitral cell; OT, olfactory tract; P, periglomerular cell; R, receptor cell. Adapted from Heimer (1995, p. 271). Used with permission of Springer-Verlag.

nucleus to the dopamine-rich area of the neostriatum has not yet been described. Alternatively, the dopamine-rich part of the dorsal pallium may be plesiomorphic for amniotes, but the dorsal thalamic olfactory relay to this part of the pallium may have evolved only in mammals.

The role of a strong olfactory input to the prefrontal cortex in mammals is still somewhat enigmatic, as some mammals, such as primates, are generally considered to have reduced their use of the olfactory system in favor of other senses, particularly vision. Olfactory input may be more important than previously realized in both limbic and neocortical functions—including emotion, memory, learning, awareness of self, social interactiveness, and the set of distinctive behavioral characteristics unique to the individual.

VOMERONASAL SYSTEM

A vomeronasal nerve is present in amphibians and, among amniotes, in mammals and most squamate reptiles (lizards and snakes). It is absent in the thecodonts (crocodiles and birds) and in turtles. Whether it is present in the rhynchocephalian *Sphenodon* is unclear. Until recently, this nerve was believed to be absent in some mammals, including some of the primates, but recent evidence suggests that it is probably present in all

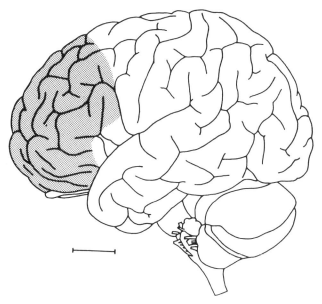

FIGURE 29-13. Drawing of the brain of a primate (*Homo sapiens*). Prefrontal cortex is indicated by shading. Adapted from Nieuwenhuys et al. (1978). Used with permission of Springer-Verlag.

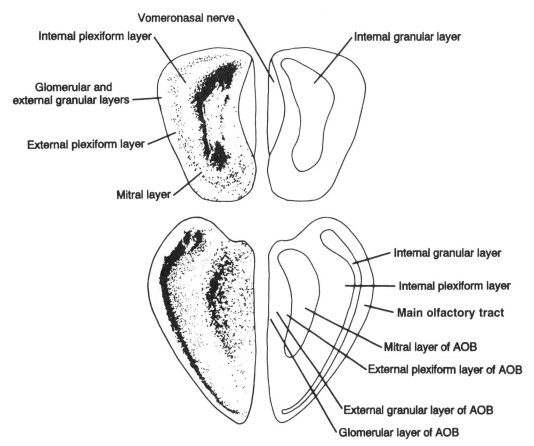

FIGURE 29-14. Transverse sections with mirror-image drawings through the main olfactory bulb (top) and the main and accessory olfactory bulbs in a snake (*Elaphe obsoleta rosalleni*). Adapted from Halpern (1980). AOB = accessory olfactory bulb. Used with permission of Plenum Publishing Corp.

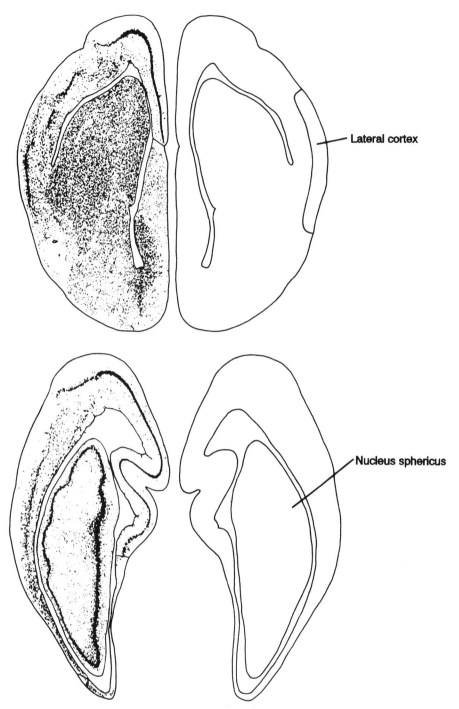

FIGURE 29-15. Transverse sections with mirror-image drawings through the telencephalon of a lizard (*Tupinambis nigropunctatus*). The upper section is more rostral. Adapted from Ebbesson and Voneida (1969). Used with permission of S. Karger AG.

mammals. The nerve has bipolar cells with dendrites ending in the olfactory epithelium, either in a restricted part of the olfactory epithelial area, as in urodele amphibians, or in a blind pouch next to the epithelial surface and called the **vomeronasal** (or **Jacobson's**) **organ,** as in other amphibians and amni-

otes. The vomeronasal nerve projects to the accessory olfactory bulb.

In amphibians, the accessory olfactory bulb is present as a thickened part of the caudal wall of the main olfactory bulb. Neurons in the accessory olfactory bulb project to the ventrolat-

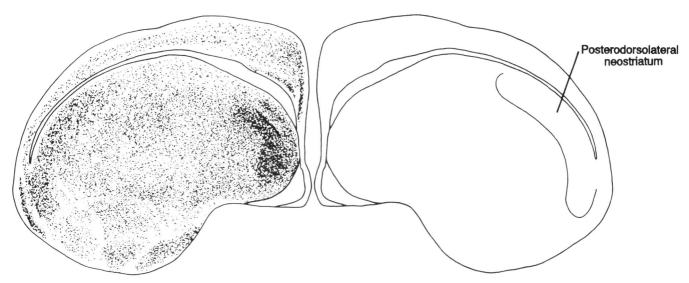

FIGURE 29-16. Transverse hemisection with mirror-image drawing through the telencephalon of a bird (*Columba livia*). Adapted from Karten and Hodos (1967) with additional data from Waldmann and Güntürkün (1993) and Divac and Mogensen (1985). Used with permission of The Johns Hopkins University Press.

eral tip of the lateral pallium, an area that is called the lateral prominence and may in fact be a part of the amygdala rather than of the lateral pallium. The accessory olfactory bulb projects most heavily to a more ventral part of the forebrain, called the **pars lateralis of the amygdala** (Fig. 29-17).

In amniotes, a number of subdivisions of the amygdala are present. In mammals, the amygdala (see Chapter 30) is frequently divided into **corticomedial** and **basolateral** divisions. Within the corticomedial division, the anterior cortical nucleus receives projections from the main olfactory bulb, while the posterior cortical and medial nuclei receive projections from the accessory olfactory bulb. Nuclei in the basolateral amygdala receive indirect olfactory input relayed through part of the olfactory cortex.

In lizards and snakes, in which both olfactory and vomeronasal systems are present, the amygdala has three divisions, two of which receive direct olfactory input. The **olfactory amygdala** receives input from the main olfactory bulb, and the **vomeronasal amygdala,** which is most often called **nucleus sphericus** (Fig. 29-15), receives input from the accessory olfac-

tory bulb (Fig. 29-14). In all amniotes and amphibians studied to date, the areas of termination of fibers from the main and accessory olfactory bulbs are separate. The pars lateralis of the amygdala in frogs, the posterior cortical and medial amygdaloid nuclei of mammals, and the nucleus sphericus in lizards and snakes appear to be homologous as pallial, vomeronasal-receptive cell groups.

The vomeronasal system plays an important role in mediating unconditioned and reinforcing properties of natural chemicals, such as those used for prey trailing by snakes. In mammals, the pheromonal cues detected by the vomeronasal system are essential for some components of normal copulatory behavior and affect the regulation of the onset of puberty, female cyclicity, the release of some hormones, the maintenance of pregnancy, maternal behavior, and other related functions.

TERMINAL NERVE

The function of the terminal nerve is uncertain, although it is currently suspected of having a neuromodulatory role, regulating neuronal excitability in the various sites to which it projects, including areas within the more ventral and medial parts of the forebrain. The nerve has been found to project to the retina and to the optic tectum in teleost fishes. It was described in cartilaginous fishes by Fritsch in 1878 and by Locy in 1905 and is now known to be widely distributed among vertebrates. The dendrites of terminal nerve neurons are distributed on the nasal septum and are in close proximity to the blood vessels in that area. The cell bodies of these neurons are most frequently located within a nearby ganglion, but their position is variable, particularly within bony fishes.

The cell bodies of the terminal nerve contain **luteinizing hormone-releasing hormone** (LHRH), which is sometimes alternatively called **gonadotropin-releasing hormone** (GnRH). In mammals, most terminal nerve cell bodies lie in

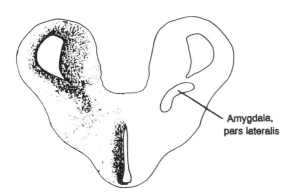

FIGURE 29-17. Transverse hemisection with mirror-image drawing through the telencephalon of a frog (*Rana catesbeiana*). Adapted from Wilczynski and Northcutt (1983).

the ganglion of the nerve, but some additional LHRH-containing cells lie along the course of the nerve as it enters the brain. The LHRH-containing fibers of the terminal nerve project to the preoptic area, the medial part of the hypothalamus, the main and accessory olfactory bulbs, olfactory cortex, parts of isocortex, and the midbrain. A possible role for the terminal nerve in pheromonal responses for reproductive behavior, including the promotion of ejaculation by copulating males, has been suggested; however, the contribution of the terminal nerve in this and other behaviors, versus the contribution of the medial part of the olfactory system, remains to be clarified.

In lampreys, neurons that resemble the terminal nerve cell bodies of other vertebrates lie along the olfactory nerve as it enters the olfactory bulb, and the axons have been traced to the ventral thalamus, the ventral part of the hypothalamus, and the mesencephalic tegmentum. Unlike other vertebrates, however, terminal nerve fibers in lampreys do not appear to contain LHRH. In cartilaginous fishes, terminal nerve cell bodies are positive for LHRH and lie within as many as three ganglia as well as along the distal part of the nerve itself. The nerve appears to project mainly to the preoptic area and the septum; fibers also terminate directly on cerebral blood vessels. In sturgeons, a number of GnRH positive cell bodies, presumably at least some of which are terminal nerve cell bodies, are distributed within the ventral part of the forebrain.

In lungfishes, the terminal nerve has two components: an anterior root (or **terminal nerve proper**) and a posterior root (also called the **nervus praeopticus**). Terminal/preoptic nerve fibers have been traced to the septum and to the preoptic nucleus, but the terminal sites of more caudally passing fibers that enter the lateral forebrain bundle have yet to be identified. Terminal nerve projections to septal, preoptic, and hypothalamic sites have been found in amphibians, where, as in other vertebrates, LHRH appears to play a role in reproductive behaviors via the hypothalamic-pituitary system. The central projections of the terminal nerve need further study in nonsynapsid amniotes, but are probably to septal and preoptic-hypothalamic sites. The involvement of LHRH in various aspects of reproductive behavior has been established.

In teleost fishes, the terminal nerve cell bodies contain LHRH and project to medial, ventral sites in the forebrain. In teleosts in which the olfactory bulb is sessile, the majority of cell bodies of the terminal nerve do not lie in a ganglion but migrate into the medial and ventral parts of the forebrain. That intracerebral, LHRH-positive cells can originate from the medial olfactory placode and migrate into the forebrain has recently been confirmed with experimental embryological findings in mice. In teleosts, the migrated terminal nerve neurons project to visual structures, particularly the retina. Diencephalic neurons that project to the retina also have recently been found in sturgeons, but whether these neurons are LHRH (GnRH) positive and thus terminal nerve neurons remains to be confirmed.

In teleosts, the migrated neurons of the terminal nerve form a broken rostrocaudal column through the ventral parts of the telencephalon and diencephalon. They have been referred to collectively as the **nucleus olfactoretinalis** and have been found to receive an ascending projection from a set of neurons in the area of the isthmus located rostral to the locus coeruleus. Axons from the cell bodies of the terminal nerve project to the retina via the optic nerve. In the retina, many of the fibers synapse on dopamine-containing **interplexiform cells.** The effect of the terminal nerve innervation of the retina on visual functions remains to be elucidated.

The terminal nerve in fishes is believed to play a role in reproductive behaviors due to the LHRH content of its neurons. The olfactory system may, however, play a more prominent role in reproduction. Exposure to sex pheromones has been found to increase the activity of olfactory nerve neurons in the medial part of the olfactory bulb but not of terminal nerve neurons. Tactile stimulation, on the other hand, was found to result in a decrease of terminal nerve neuron activity, and the contribution of the terminal nerve to reproductive behavior may thus be related to the physical contact that occurs during spawning behavior.

EVOLUTIONARY PERSPECTIVE

The olfactory nerve component of the rostral chemosensory system is present in all groups of vertebrates and is plesiomorphic for vertebrates. Olfactory projections to the pallium are restricted, mostly to the lateral pallium, in jawed vertebrates but are more extensive in lampreys and are to the entire pallium in hagfishes. From this distribution, extensive olfactory innervation of the pallium appears to be the plesiomorphic condition for vertebrates. Likewise, restricted pallial olfactory projections are the plesiomorphic condition for jawed vertebrates.

The presence of a terminal nerve in jawless vertebrates is questionable, but a terminal nerve is at least plesiomorphic for jawed vertebrates. In ray-finned fishes with sessile olfactory bulbs, the majority of cell bodies of the terminal nerve are in a migrated position within the brain. The projection of terminal nerve neurons to visual structures such as the retina is apomorphic in some or all ray-finned fishes. While both olfactory and terminal nerve systems are present in most vertebrates, variation does occur. In toothed whales, for example, both the olfactory and vomeronasal systems have been lost, but the terminal nerve system has been markedly expanded.

The vomeronasal system is present in some tetrapods but not in nontetrapod vertebrates and is apomorphic for tetrapods. It has been secondarily lost in thecodonts (crocodiles and birds) and turtles and may also have been reduced or lost in *Sphenodon*. Although previously believed to be absent in some mammals, particularly primates, recent evidence suggests that a vomeronasal system may have been retained in all mammals.

FOR FURTHER READING

Braford, M. R., Jr. and Northcutt, R. G. (1974) Olfactory bulb projections in the bichir, *Polypterus. Journal of Comparative Neurology*, 156, 165–178.

Demski, L. S. and Schwanzel-Fukuda, M. (eds.) (1987) *The Terminal Nerve (Nervus Terminalis): Structure, Function, and Evolution. Annals of the New York Academy of Sciences*, 159.

Lohman, A. H. M. and Smeets, W. J. A. J. (1993) Overview of the main and accessory olfactory bulb projections in reptiles. *Brain, Behavior and Evolution*, 41, 147–155.

Münz, H., Claas, B., Stumpf, W. E., and Jennes, L. (1982) Centrifugal innervation of the retina by luteinizing hormone releasing hormone (LHRH)-immunoreactive telencephalic neurons in teleostean fishes. *Cell and Tissue Research*, 222, 313–323.

Northcutt, R. G. and Puzdrowski, R. L. (1988) Projections of the olfactory bulb and nervus terminalis in the silver lamprey. *Brain, Behavior and Evolution*, 32, 96–107.

Reiner, A. (1986) Is prefrontal cortex found only in mammals? *Trends in Neurosciences*, 9, 298–300.

ADDITIONAL REFERENCES

Bass, A. H. (1981) Olfactory bulb efferents in the channel catfish, *Ictalurus punctatus. Journal of Morphology*, 169, 91–111.

Bazer, G. T., Ebbesson, S. O. E., Reynolds, J. B., and Bailey, R. P. (1987) A cobalt-lysine study of primary olfactory projections in king salmon fry (*Oncorhynchus tshawytscha* Walbaum). *Cell and Tissue Research*, 248, 499–503.

Becerra, M., Manso, M. J., Rodriguez-Moldes, I., and Anadón, R. (1994) Primary olfactory fibers project to the ventral telencephalon and preoptic region in trout (*Salmo trutta*): a developmental immunocytochemical study. *Journal of Comparative Neurology*, 342, 131–143.

Bingman, V. P., Casini, G., Nocjar, C., and Jones, T.-J. (1994) Connections of the piriform cortex in homing pigeons (*Columba livia*) studied with fast blue and WGA–HRP. *Brain, Behavior and Evolution*, 43, 206–218.

Brown, J. W. (1987) The nervus terminalis in insectivorous bat embryos and notes on its presence during human ontogeny. *Annals of the New York Academy of Sciences*, 519, 184–200.

Burghardt, G. M. (1993) The comparative imperative: genetics and ontogeny of chemoreceptive prey responses in natricine snakes. *Brain, Behavior and Evolution*, 41, 138–146.

Carmichael, S. T., Clugnet, M. C., and Price, J. L. (1994) Central olfactory connections in the macaque monkey. *Journal of Comparative Neurology*, 346, 403–434.

Demski, L. S., Fields, R. D., Bullock, T. H., Schreibman, M. P., and Margolis-Nunno, H. (1987) The terminal nerve of sharks and rays. *Annals of the New York Academy of Sciences*, 519, 15–32.

Demski, L. S. and Northcutt, R. G. (1983) The terminal nerve: a new chemosensory system in vertebrates? *Science*, 220, 435–437.

Demski, L. S., Ridgway, S. H., and Schwanzel-Fukuda, M. (1990) The terminal nerve of dolphins: gross structure, histology and lunteinizing-hormone-releasing hormone immunohistochemistry. *Brain, Behavior and Evolution*, 36, 249–261.

Divac, I., Holst, M.-C., Nelson, J., and McKenzie, J. S. (1987) Afferents of the frontal cortex in the echidna (*Tachyglossus aculeatus*). Indication of an outstandingly large prefrontal area. *Brain, Behavior and Evolution*, 30, 303–320.

Divac, I. and Mogensen, J. (1985) The prefrontal "cortex" in the pigeon: catecholamine histofluorescence. *Neuroscience*, 15, 677–682.

Divac, I. and Öberg, R. G. E. (1990) Prefrontal cortex: the name and the thing. In Schwerdtfeger, W. K. and Germroth, P. (eds.), *The Forebrain in Nonmammals: New Aspects of Structure and Development*. Berlin: Springer-Verlag, pp. 213–220.

Dulka, J. G. (1993) Sex pheromone systems in goldfish: comparisons to vomeronasal system in tetrapods. *Brain, Behavior and Evolution*, 42, 265–280.

Ebbesson, S. O. E. and Voneida, T. J. (1969) The cytoarchitecture of the pallium in the tegu lizard (*Tupinambis nigropunctatus*). *Brain, Behavior and Evolution*, 2, 431–466.

Eisthen, H. L., Sencglaub, D. R., Schroeder, D. M., and Alberts, J. R. (1994) Anatomy and forebrain projections of the olfactory and vomeronasal organ in axolotls (*Ambystoma mexicanum*). *Brain, Behavior and Evolution*, 44, 108–124.

Fujita, I., Sorensen, P. W., Stacey, N. E., and Hara, T. J. (1991) The olfactory system, not the terminal nerve, functions as the primary chemosensory pathway mediating responses to sex pheromones in male goldfish. *Brain, Behavior and Evolution*, 38, 313–321.

Fuster, J. M. (1989) *The Prefrontal Cortex: Anatomy, Physiology, and Neuropsychology of the Frontal Lobe*. New York: Raven.

Graves, B. M. (1993) Chemical delivery to the vomeronasal organs and functional domain of squamate chemoreception. *Brain, Behavior and Evolution*, 41, 198–202.

Greenberg, N. (1993) Central and endocrine aspects of tongue-flicking and exploratory behavior in *Anolis carolinensis. Brain, Behavior and Evolution*, 41, 210–218.

Grober, M. S. and Bass, A. H. (1991) Neuronal correlates of sex/role change in labrid fishes: LHRH-like immunoreactivity. *Brain, Behavior and Evolution*, 38, 302–312.

Halpern, M. (1980) The telencephalon of snakes. In S. O. E. Ebbesson (ed.), *Comparative Neurology of the Telencephalon*. New York: Plenum, pp. 257–295.

Halpern, M. (1988) Vomeronasal system functions: role in mediating the reinforcing properties of chemical stimuli. In W. K. Schwerdtfeger and W. J. A. J. Smeets (eds.), *The Forebrain of Reptiles: Current Concepts of Structure and Function*. Basel: Karger, pp. 142–150.

Hatanaka, T. and Matsuzaki, O. (1993) Odor responses of the vomeronasal system in Reeve's turtle, *Geoclemys reevesii. Brain, Behavior and Evolution*, 41, 183–186.

Heimer, L. (1995) *The Human Brain and Spinal Cord: Functional Neuroanatomy and Dissection Guide*, Second Edition. New York: Springer-Verlag.

Hofmann, M. H. and Meyer, D. L. (1989a) Central projections of the nervus terminalis in four species of amphibians. *Brain, Behavior and Evolution*, 34, 301–307.

Hofmann, M. H. and Meyer, D. L. (1989b) The nervus terminalis in larval and adult *Xenopus laevis. Brain Research*, 498, 167–169.

Hofmann, M. H., Piñuela, C., and Meyer, D. L. (1993) Retinopetal projections from diencephalic neurons in a primitive actinopterygian fish, the sterlet *Acipenser ruthenus. Neuroscience Letters*, 161, 30–32.

Holtzman, D. A. (1993) The ontogeny of nasal chemical senses in garter snakes. *Brain, Behavior and Evolution*, 41, 163–170.

Honkanen, T. and Ekström, P. (1990) An immunocytochemical study of the olfactory projections in the three-spined stickleback, *Gasterosteus aculeatus*, L. *Journal of Comparative Neurology*, 292, 65–72.

Inouchi, J., Wang, D., Jiang, X. C., Kubie, J., and Halpern, M. (1993) Electrophysiological analysis of the nasal chemical sense in garter snakes. *Brain, Behavior and Evolution*, 41, 171–182.

Karten, H. J. and Hodos, W. (1967) *A Stereotaxic Atlas of the Brain of the Pigeon (Columba livia)*. Baltimore: The Johns Hopkins Press.

Kosaka, T. and Hama, K. (1979) Ruffed cell: a new type of neuron with a distinctive initial unmyelinated portion of the axons in the olfactory bulb of the goldfish (*Carassius auratus*). I. Golgi impregnation and serial thin sectioning studies. *Journal of Comparative Neurology*, 186, 301–319.

Leprêtre, E., Anglade, I., Williot, P., Vandesande, F., Tramu, G., and Kah, O. (1993) Comparative distribution of mammalian GnRH (gonadotrophin-releasing hormone) and chicken GnRH-II in the brain of the immature Siberian sturgeon (*Acipenser baeri*). *Journal of Comparative Neurology*, 337, 568–583.

Levine, R. L. and Dethier, S. (1985) The connections between the olfactory bulb and the brain in the goldfish. *Journal of Comparative Neurology*, 237, 427–444.

Martínez-Garcia, F., Olucha, F. E., Teruel, V., and Lorente, M. J. (1993) Fiber connections of the amygdaloid formation of the lizard *Podarcis hispanica*. *Brain, Behavior and Evolution*, 41, 156–162.

Mendoza, A. S., Küderling, I., Kuhn, H. J., and Kühnel. W. (1994) The vomeronasal organ of the New World monkey *Saguinus fuscicollis* (Callitrichidae). A light and transmission electron microscopic study. *Annals of Anatomy*, 176, 217–222.

Meyer, D. L., von Bartheld, C. S., and Lindörfer, H. W. (1987) Evidence for the existence of a terminal nerve in lampreys and in birds. *Annals of the New York Academy of Sciences*, 519, 385–391.

Mogensen, J. and Divac, I. (1982) The prefrontal 'cortex' in the pigeon: behavioral evidence. *Brain, Behavior and Evolution*, 21, 60–66.

Mogensen, J. and Divac, I. (1993) Behavioural effects of ablation of the pigeon-equivalent of the mammalian prefrontal cortex. *Behavioural Brain Research*, 55, 101–107.

Moore, F. L., Muske, L., and Propper, C. R. (1987) Regulation of reproductive behaviors in amphibians by LHRH. *Annals of the New York Academy of Sciences*, 519, 108–116.

Muske, L. E. (1993) Evolution of gonadotropin-releasing hormone (GnRH) neuronal systems. *Brain, Behavior and Evolution*, 42, 215–230.

Muske, L. E. and Moore, F. L. (1988) The nervus terminalis in amphibians: anatomy, chemistry and relationship with the hypothalamic gonadotropin-releasing hormone system. *Brain, Behavior and Evolution*, 32, 141–150.

Neary, T. J. (1990) The pallium of anuran amphibians. In E. G. Jones and A. Peters (eds.), *Cerebral Cortex, Vol. 8A: Comparative Structure and Evolution of Cerebral Cortex, Part I*. New York: Plenum, pp. 107–138.

Nieuwenhuys, R. (1965) The forebrain of the crossopterygian *Latimeria chalumnae* Smith. *Journal of Morphology*, 117, 1–24.

Nieuwenhuys, R. (1967) Comparative anatomy of olfactory centers and tracts. *Progress in Brain Research, 23, 1–64*.

Nieuwenhuys, R. and Meek, J. (1990a) The telencephalon of actinopterygian fishes. In E. G. Jones and A. Peters (eds.), *Cerebral Cortex, Vol. 8A: Comparative Structure and Evolution of Cerebral Cortex, Part I*. New York: Plenum, pp. 31–73.

Nieuwenhuys, R. and Meek, J. (1990b) The telencephalon of sarcopterygian fishes. In E. G. Jones and A. Peters (eds.), *Cerebral Cortex, Vol. 8A: Comparative Structure and Evolution of Cerebral Cortex, Part I*. New York: Plenum, pp. 75–106.

Nieuwenhuys, R., Voogd, J., and van Huijzen, C. (1978) *The Human Nervous System: A Synopsis and Atlas*. Berlin: Springer-Verlag.

Northcutt, R. G. (1978) Brain organization in the cartilaginous fishes. In E. S. Hodgson and R. F. Mathewson (eds.), *Sensory Biology of Sharks, Skates, and Rays*. Arlington, VA: Office of Naval Research, pp. 117–193.

Northcutt, R. G. (1981) Evolution of the telencephalon in nonmammals. *Annual Review of Neuroscience*, 4, 301–350.

Northcutt, R. G. and Butler, A. B. (1991) Retinofugal and retinopetal projections in the green sunfish, *Lepomis cyanellus*. *Brain, Behavior and Evolution*, 37, 333–354.

Northcutt, R. G. and Davis, R. E. (1983) Telencephalic organization in ray-finned fishes. In R. G. Northcutt and R. E. Davis (eds.), *Fish Neurobiology, Vol. 2: Higher Brain Areas and Functions*. Ann Arbor, MI: The University of Michigan Press, pp. 203–236.

Northcutt, R. G. and Kicliter, E. (1980) Organization of the amphibian telencephalon. In S. O. E. Ebbesson (ed.), *Comparative Neurology of the Telencephalon*. New York: Plenum, pp. 203–255.

Northcutt, R. G., Reiner, A., and Karten, H. J. (1988) Immunohistochemical study of the telencephalon of the spiny dogfish, *Squalus acanthias*. *Journal of Comparative Neurology*, 277, 250–267.

Northcutt, R. G. and Royce, G. J. (1975) Olfactory bulb projections in the bullfrog *Rana catesbeiana*. *Journal of Morphology*, 145, 251–268.

Nunez Rodriguez, J., Kah, O., Breton, B., and Le Menn, F. (1985) Immunocytochemical localization of GnRH (gonadotropin releasing hormone) systems in the brain of a marine teleost fish, the sole. *Experientia*, 41, 1574–1576.

Oelschläger, H. A. (1988) Persistence of the nervus terminalis in adult bats: a morphological and phylogenetic approach. *Brain, Behavior and Evolution*, 32, 330–339.

Oeth, K. M. and Lewis, D. A. (1992) Cholecystokinin- and dopamine-containing mesencephalic neurons provide distinct projections to monkey prefrontal cortex. *Neuroscience Letters*, 145, 87–92.

Price, J. L. (1991) The central olfactory and accessory olfactory systems. In T. E. Finger and W. L. Silver (eds.), *Neurobiology of Taste and Smell*. Malabar, Florida: Krieger Publishing Co., pp. 179–203.

Ray, J. P. and Price, J. L. (1993) The organization of projections from the mediodorsal nucleus of the thalamus to orbital and medial prefrontal cortex in macaque monkeys. *Journal of Comparative Neurology*, 337, 1–31.

Reiner, A. and Northcutt, R. G. (1987) An immunohistochemical study of the telencephalon of the African lungfish, *Protopterus annectens*. *Journal of Comparative Neurology*, 256, 463–481.

Reiner, A. and Northcutt, R. G. (1992) An immunohistochemical study of the telencephalon of the senegal bichir (*Polypterus senegalus*). *Journal of Comparative Neurology*, 319, 359–386.

Rooney, D., Døving, K. B., Ravaille-Veron, M., and Szabo, T. (1992) The central connections of the olfactory bulbs in cod, *Gadus morhua* L. *Journal für Hirnforschung*, 33, 63–75.

Rooney, D. J., New, J. G., Szabo, T., and Ravaille-Veron, M. (1989) Central connections of the olfactory bulb in the weakly electric fish, *Gnathonemus petersii*. *Cell and Tissue Research*, 257, 423–436.

Rooney, D. J. and Szabo, T. (1990) Secondary olfactory projections in *Gnathonemus petersii* (Mormyridae): a comparative perspective. In W. K. Schwerdtfeger and P. Germroth (eds.), *The Forebrain in Nonmammals: New Aspects of Structure and Development*. Berlin: Springer-Verlag, pp. 1–15.

Satou, M. (1990) Synaptic organization, local neuronal circuitry, and functional segregation of the teleost olfactory bulb. *Progress in Neurobiology*, 34, 115–142.

Scalia, F. (1972) The projection of the accessory olfactory bulb in the frog. *Brain Research*, 36, 409–411.

Scalia, F. (1976) Structure of the olfactory and accessory olfactory systems. In R. Llinás and W. Precht (eds.), *Frog Neurobiology*. Berlin: Springer-Verlag, pp. 213–233.

Scalia, F. and Ebbesson, S. O. E. (1971) The central projections of the olfactory bulb in a teleost (*Gymnothorax funebris*). *Brain, Behavior and Evolution*, 4, 376–399.

Scalia, F., Gallousis, G., and Roca, S. (1991) A note on the organization of the amphibian olfactory bulb. *Journal of Comparative Neurology*, 305, 435–442.

Scalia, F., Halpern, M., Knapp, H., and Riss, W. (1968) The efferent connexions of the olfactory bulb in the frog: a study of degenerating unmyelinated fibers. *Journal of Anatomy*, 103, 245–262.

Scalia, F., Halpern, M., and Riss, W. (1969) Olfactory bulb projections in the South American caiman. *Brain, Behavior and Evolution*, 2, 238–262.

Schober, A., Meyer, D. L., and von Bartheld, C. S. (1994) Central projections of the nervus terminalis and the nervus praeopticus in the lungfish brain revealed by nitric oxide synthase. *Journal of Comparative Neurology*, 349, 1–19.

Schwanzel-Fukuda, M. and Pfaff, D. W. (1989) Origin of luteinizing hormone-releasing hormone neurons. *Nature (London)*, 338, 161–164.

Schwenk, K. (1993) The evolution of chemoreception in squamate reptiles: a phylogenetic approach. *Brain, Behavior and Evolution*, 41, 124–137.

Scott, J. W. and Harrison, T. A. (1991) The olfactory bulb: anatomy and physiology. In T. E. Finger and W. L. Silver (eds.), *Neurobiology of Taste and Smell*. Malabar, FL: Krieger, pp. 151–178.

Smeets, W. J. A. J. (1983) The secondary olfactory connections in two chondrichthians, the shark *Scyliorhinus canicula* and the ray *Raja clavata*. *Journal of Comparative Neurology*, 218, 334–344.

Smeets, W. J. A. J. (1990) The telencephalon of cartilaginous fishes. In E. G. Jones and A. Peters (eds.), *Cerebral Cortex, Vol 8A: Comparative Structure and Embryology of Cerebral Cortex, Part I*. New York: Plenum, pp. 3–30.

Smeets, W. J. A. J., Hoogland, P. V., and Voorn, P. (1986) The distribution of dopamine immunoreactivity in the forebrain and midbrain of the lizard *Gekko gecko*: an immunohistochemical study with antibodies against dopamine. *Journal of Comparative Neurology*, 253, 46–60.

Springer, A. D. (1983) Centrifugal innervation of goldfish retina from ganglion cells of the nervus terminalis. *Journal of Comparative Neurology*, 214, 404–415.

Stell, W. K., Walker, S. E., and Ball, A. K. (1987) Functional-anatomical studies on the terminal nerve projection to the retina of bony fishes. *Annals of the New York Academy of Sciences*, 519, 80–96.

Szabo, T., Blähser, S., Denizot, J.-P. and Ravaille-Véron, M. (1991) Projection olfactif primaire extrabulbaire chez certains poissons téléostéens. *Comptes Rendus d'Académie des Sciences*, Paris, Série III, 312, 555–560.

Uchiyama, H. (1990) Immunohistochemical subpopulations of retinopetal neurons in the nucleus olfactoretinalis in a teleost, the whitespotted greenling (*Hexagrammos stelleri*). *Journal of Comparative Neurology*, 293, 54–62.

Ulinski, P. S. (1990) The cerebral cortex of reptiles. In E. G. Jones and A. Peters (eds.), *Cerebral Cortex, Vol. 8A: Comparative Structure and Evolution of Cerebral Cortex, Part I*. New York: Plenum, pp. 139–215.

von Bartheld, C. S. and Meyer, D. L. (1986) Central connections of the olfactory bulb in the bichir, *Polypterus palmas*, reexamined. *Cell and Tissue Research*, 244, 527–535.

von Bartheld, C. S., Rickmann, M. J., and Meyer, D. L. (1986) A light- and electron-microscopic study of mesencephalic neurons projecting to the ganglion of the nervus terminalis in the goldfish. *Cell and Tissue Research*, 246, 63–70.

Waldmann, C. and Güntürkün, O. (1993) The dopaminergic innervation of the pigeon caudolateral forebrain: immunocytochemical evidence for a 'prefrontal cortex' in birds? *Brain Research*, 600, 225–234.

Waters, R. M. (1993) Odorized air current trailing by garter snakes, *Thamnophis sirtalis*. *Brain, Behavior and Evolution*, 41, 219–223.

Weiss, O. and Meyer, D. L. (1988) Odor stimuli modulate retinal excitability in fish. *Neuroscience Letters*, 93, 209–213.

Weldon, P. J. and Ferguson, M. W. J. (1993) Chemoreception in crocodilians: anatomy, natural history, and empirical results. *Brain, Behavior and Evolution*, 41, 239–245.

Wenzel, B. M. (1987) The olfactory and related systems in birds. *Annals of the New York Academy of Sciences*, 519, 137–149.

Wicht, H. and Northcutt, R. G. (1992) The forebrain of the Pacific hagfish: a cladistic reconstruction of the ancestral craniate forebrain. *Brain, Behavior and Evolution*, 40, 25–64.

Wicht, H. and Northcutt, R. G. (1993) Secondary olfactory projections and pallial topography in the Pacific hagfish, *Eptatretus stouti*. *Journal of Comparative Neurology*, 337, 529–542.

Wilczynski, W. and Northcutt, R. G. (1983) Connections of the bullfrog striatum: afferent organization. *Journal of Comparative Neurology*, 214, 321–332.

Wirsig, C. R. (1987) Effects of lesions of the terminal nerve on mating behavior in the male hamster. *Annals of the New York Academy of Sciences*, 519, 241–251.

Wysocki, C. J. and Meredith, M. (1991) The vomeronasal system. In T. E. Finger and W. L. Silver (eds.), *Neurobiology of Taste and Smell*. Malabar, FL: Krieger, pp. 125–150.

Zucker, C. L. and Dowling, J. E. (1987) Centrifugal fibers synapse on dopaminergic interplexiform cells in the teleost retina. *Nature (London)*, 330, 166–168.

30

Limbic Telencephalon

INTRODUCTION

In most vertebrates, the medial part of the telencephalon comprises the telencephalic component of one of the brain's major coordinating systems, the limbic system. The main component of the telencephalic limbic system is the **medial pallium,** which contains the **hippocampal formation** and related pallial areas. The **septal nuclei** and the **amygdala,** which have a more ventral position, are additional major telencephalic components of the limbic system. The amygdala has two divisions in most vertebrates: a so-called **cortical division,** which is a ventral continuation of the lateral pallium, and a **basal division,** which is subpallial in origin. Both divisions contain nuclear groups but no acutal cortical (i.e., layered) areas. A region called the **extended amygdala** includes the centromedial amygdala (central and medial nuclei of the cortical amygdala) and several other structures of the basal forebrain that are unified by similar connections and histochemical characteristics. Other structures outside the telencephalon are also considered to be part of the limbic system and include parts of the epithalamus, dorsal thalamus, and hypothalamus.

Generally speaking, in cartilaginous and ray-finned fishes and in amphibians, both the dorsal and medial pallia receive sensory projections relayed from the rostral part of the dorsal thalamus, and the medial pallium receives the majority of these projections. In some cases, the medial pallium also receives sensory information relayed through nuclei outside of the dorsal thalamus as well. In amniotes, in contrast, most of the retinal and somatosensory input to the rostral thalamus is relayed to the expanded dorsal pallium, with only a minor amount being relayed directly to the medial pallium. Pathways from the dorsal pallium, as well as from the lateral pallium, secondarily relay sensory information to the medial pallium. The rostral thalamic nuclei that relay a minor sensory input to the medial pallium are primarily involved in cortical–subcortical–cortical circuits with the medial pallium.

In mammals, behavioral studies indicate that the limbic system is involved in motivation, memory, learning, emotion, sexual behavior, and many integrative functions necessary for appropriate responses to stimuli. The hippocampal formation is essential for short-term memory and is thus the gateway for the storage in long-term memory of new events. The limbic system also interacts with other major systems throughout the brain, particularly the olfactory system, dorsal pallial areas, and the striatum. Both the septal nuclei and amygdala receive direct inputs from the olfactory bulb, while the hippocampal formation does not. The amygdala is also interconnected with the striatum, the hypothalamus, and widespread areas of isocortex, including the prefrontal cortex. Relatively few behavioral studies of these structures have been done in nonmammalian vertebrates. Those that have been done suggest that the telencephalic limbic structures play a similar role in these animals.

THE LIMBIC STRUCTURES OF THE TELENCEPHALON IN ANAMNIOTES

Group I

In lampreys, a relatively small part of the telencephalon forms the medial pallium (Fig.30-1). A septal nucleus lies farther rostrally between the rest of the telencephalon and the rostral part of the diencephalon and receives an input from the olfactory bulb. A nuclear area homologous to the amygdala of other vertebrates has not yet been identified. A cell-poor area bordering the striatum may be a candidate for an amygdalar homologue, however, as olfactory fibers terminate there.

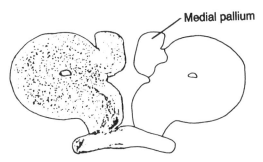

FIGURE 30-1. Transverse hemisection with mirror-image drawing through the telencephalon of a lamprey. Adapted from Northcutt (1981). Redrawn, with permission, from the *Annual Review of Neuroscience,* Volume 4, © 1981, by Annual Reviews Inc.

In squalomorph sharks (Fig. 30-2), the medial pallium is also relatively small and is fused across the midline with the contralateral medial pallium. A septal nucleus lies in the ventromedial part of the telencephalon. A nucleus (called **nucleus A** in some cases and **nucleus X** or **nucleus N** in others) in the lateroventral part of the telencephalon may be homologous to the cortical amygdala of other vertebrates. This nucleus receives projections from the olfactory bulb. The medial pallium in squalomorph sharks is known to receive afferent projections from the dorsal pallium and the septal nucleus in the telencephalon and the posterior tubercle and a nucleus called the posterior lateral thalamic nucleus in the diencephalon. The latter nucleus relays input from the lateral line electrosensory system to the medial pallium.

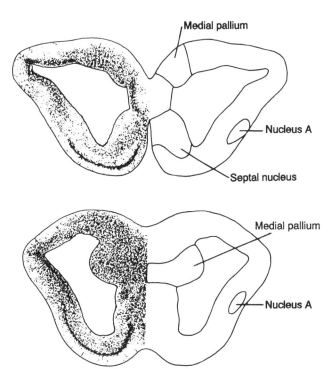

FIGURE 30-2. Transverse hemisections with mirror-image drawings through the telencephalon of a squalomorph shark (*Squalus acanthias*). The upper section is more rostral. Adapted from Northcutt et al. (1988).

In nonteleost ray-finned fishes (Fig. 30-3), the "medial" pallium lies in the lateral part of the telencephalon, due to the development of the telencephalon by eversion (see Chapter 3). This pallial area is referred to as the **P3** part of the pallium. Septal nuclei, called the **dorsal (Vd)** and **ventral (Vv) nuclei of the ventral telencephalic area,** are relatively large. A more caudally lying nucleus, called the **supracommissural nucleus of the ventral telencephalic area (Vs),** receives olfactory projections via the medial olfactory tract and has been identified as the amygdala.

In lungfishes (Fig. 30-4), **dorsal, intermediate,** and **ventral divisions** of the medial pallium can be identified on cytoarchitectural and histochemical grounds, but connections remain to be studied. A septal area, called the **medial subpallium,** lies ventral to the medial pallium. Caudally, an area that lies ventral to the septal area, called the **central subpallium,** is present and may be homologous to part or all of the amygdala of other vertebrates.

In frogs (Fig. 30-5), the medial pallium is enlarged relative to its condition in lungfishes and to the dorsal and lateral pallia. It is possible to divide the medial pallium into three parts on cytoarchitectonic criteria, but these parts cannot be assumed to correspond 1:1 to the three parts of the medial pallium in lungfishes or to specific parts of the medial pallium in other vertebrates. **Medial** and **lateral septal nuclei** lie ventral to the medial pallium. A nucleus called the **medial amygdala** lies ventral to the septal nuclei, and, farther caudally, a **lateral amygdala** has been identified. The medial amygdala corresponds topographically to the nucleus identified as the amygdala in lungfishes. The lateral amygdala receives projections from the olfactory bulb and thus may be homologous to the cortical amygdala of other vertebrates.

The medial pallium in frogs receives projections [Fig. 30-6(A)] from the dorsal and lateral pallia, the nucleus of the diagonal band, the septal nuclei, the medial and lateral amygdalar nuclei, nucleus accumbens, and a bed nucleus of the pallial commissure. The latter nucleus is a small cell group in the medial part of the caudal telencephalon. The rostral part of the medial pallium also receives an input from the olfactory bulb. Afferent projections to the medial pallium from outside the telencephalon arise from the anterior nucleus of the dorsal thalamus, which relays visual and somatosensory information, and from the serotonin-containing cells within the raphe division of the brainstem reticular formation.

The medial pallium in frogs has reciprocal connections with the contralateral medial pallium. Within the telencephalon it projects to a number of areas, including the septal nuclei, the bed nucleus of the pallial commissure, the diagonal band nucleus, the olfactory bulb, both parts of the amygdala, and the dorsal and lateral pallia. Outside the telencephalon it projects to structures including the habenula, nucleus anterior of the dorsal thalamus, ventral thalamus, preoptic area, and hypothalamus [Fig. 30–6(B)].

Group IIA

The telencephalon in hagfishes is large and complex (Fig. 30-7). The olfactory bulb is known to project to all parts of what appear to be pallial areas, whereas most portions of both the dorsal and medial pallia do not receive olfactory input in

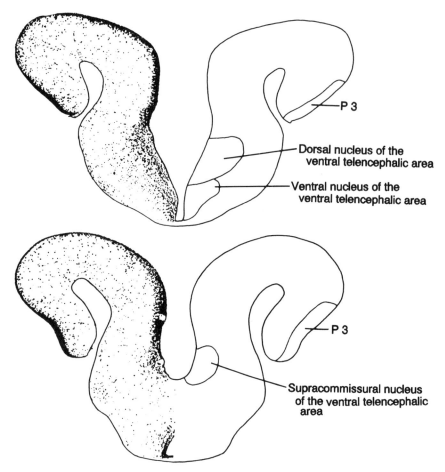

FIGURE 30-3. Transverse hemisections with mirror-image drawings through the telencephalon of a bichir (*Polypterus palmas*). The upper section is more rostral. Adapted from Reiner and Northcutt (1992).

most vertebrates. Thalamic projections to the telencephalon have not yet been studied in hagfishes, so their distribution cannot yet be compared with the olfactory projections. A medioventrally lying **septal area** has been tentatively identified, but a possible amygdalar area remains unknown.

Among galeomorph sharks, skates, and rays, the degree of elaboration of cell masses within the telencephalon varies from moderate to high. In skates (Fig. 30-8), the migration of cell masses in the telencephalon obscures relationships. A medial pallium has been identified in the dorsomedial telencephalon, and **medial** and **lateral septal nuclei** lie ventral to it. A possible amygdalar homologue has not been found, however, as olfactory projections do not extend beyond the lateral pallium. Both electrosensory and visual inputs are known to reach the medial pallium in skates. Lateral line electrosensory input is relayed through a caudal diencephalic nucleus, the lateral posterior thalamic nucleus and visual input through nucleus anterior of the dorsal thalamus. The posterior tuberculum also projects to the medial pallium. Lateral line mechanosensory information also reaches the telencephalon, but the anatomy of this pathway is not yet known.

In both teleost and nonteleost ray-finned fishes, a pallial part of the telencephalon, called the area dorsalis, and a subpallial part, called the area ventralis, are present in the telencephalon. A part of the pallium lying in a caudal and lateral position

(called the **posterior zone of area dorsalis** or **Dp**) has been identified as the homologue of the piriform, (or lateral,) cortex of other vertebrates since it receives projections from the olfactory bulb. The remaining areas of the pallium are thus thought to be homologous to the medial and dorsal pallia of other vertebrates and/or, in part, to be independently evolved pallial areas unique to teleosts. Two candidates for a homologue of the medial cortex of other vertebrates have been proposed, neither of which is particularly satisfactory.

A part of the pallium, referred to as the **lateral zone of area dorsalis (Dl,** Fig. 30-9), has been proposed as a homologue of the medial pallium of other vertebrates since it lies in a lateral position in the everted telencephalon and does not receive direct projections from the olfactory bulb. The relationships among telencephalic cell groups in teleosts are secondarily complicated, however, by the enlargement and migration that occurs during development following eversion.

A medially lying part of the pallium, the **medial zone of area dorsalis (Dm),** has been compared previously to part of the striatum of other vertebrates, as discussed above (Chapter 24), but current evidence for this comparison is not convincing. The ventral part of Dm, **Dm-v,** does not receive direct olfactory bulb input, gives rise to projections to the hypothalamus that might be similar to limbic-hypothalamic pathways in other vertebrates, and, like limbic pallium in other vertebrates, has com-

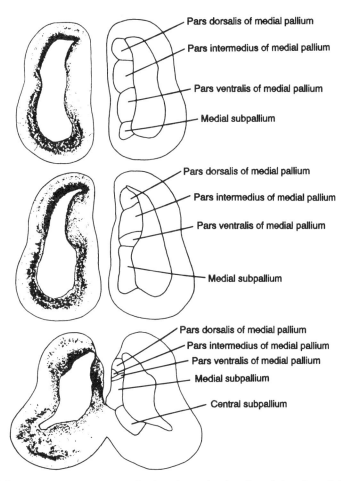

Pars dorsalis of medial pallium
Pars intermedius of medial pallium
Pars ventralis of medial pallium
Medial subpallium

Pars dorsalis of medial pallium
Pars intermedius of medial pallium
Pars ventralis of medial pallium
Medial subpallium

Pars dorsalis of medial pallium
Pars intermedius of medial pallium
Pars ventralis of medial pallium
Medial subpallium
Central subpallium

FIGURE 30-4. Transverse hemisections with mirror-image drawings through the telencephalon of a lungfish (*Protopterus annectens*). The uppermost section is most rostral. Adapted from Reiner and Northcutt (1987).

missural connections with the contralateral Dm-v. Thus, Dm-v has also been proposed as a homologue of the medial pallium of other vertebrates.

The pallial target of the olfactory bulb, Dp, projects to the medial cortex in most vertebrates, but it does not project to either Dl or Dm-v in teleosts. Both Dl and the more dorsal part of Dm, Dm-d, receive sensory inputs relayed through dorsal thalamic nuclei as well as through diencephalic nuclei outside the dorsal thalamus, that is, various parts of the preglomerular nuclear complex. The latter pathways appear similar to sensory relay pathways through the caudal diencephalon of sharks that project to the medial pallium, as discussed above. Thus, whether Dl, part of Dm, both, or neither are homologous to the medial pallium of other vertebrates remains an unsolved puzzle.

Fortunately, we can be somewhat more certain in identifying ventral parts of the telencephalon in teleosts (30–9), due to topological, connectional, and histochemical data that are less in conflict with the corresponding data for this part of the telencephalon in other vertebrates. Two medially lying areas, the **dorsal** (Vd) and **ventral** (Vv) **nuclei of area ventralis** are considered to be homologous, respectively, to the lateral and medial septal nuclei of other vertebrates. As discussed in Chapter 24, Vv may be homologous as a field to part of the septum as well as to nucleus accumbens and the substantia innominata. Five more caudal areas—**NT** (**nucleus taenia**) and

Vs, Vp, Vc, and **Vi** (the **supracommissural, postcommissural, commissural,** and **intermediate nuclei of area ventralis,** respectively)—receive projections from the olfactory bulb and are considered to be homologous to all or part of the amygdala of other vertebrates.

LIMBIC STRUCTURES OF THE TELENCEPHALON IN AMNIOTES (GROUP IIB)

While nuclei form the pallial parts of the telencephalon in Group IIA vertebrates, both nuclei and cortices are present in the pallium in Group IIB vertebrates. In this section, we will first compare the pallial parts of the limbic system among mammals, diapsid reptiles, birds, and turtles. We will then consider the subpallial limbic structures in these three groups of amniotes.

Nomenclature

Before discussing the hippocampal formation and limbic system of mammals, we will pause briefly to consider the bases of their rather creative nomenclature. The term hippocampal formation is used here to designate the **hippocampus proper**

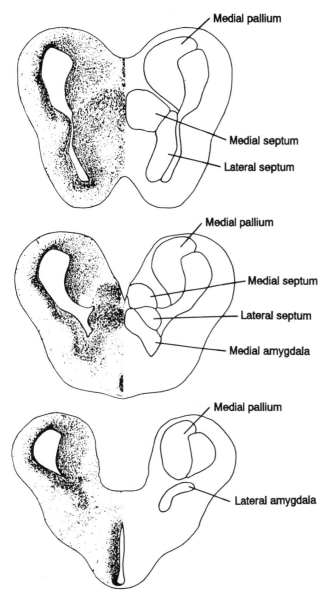

FIGURE 30-5. Transverse hemisections with mirror-image drawings through the telencephalon of a bullfrog (*Rana catesbeiana*). The uppermost section is most rostral. Adapted from Wilczynski and Northcutt (1983).

(**Ammon's horn**), the **dentate gyrus**, and their related transitional cortices, that is, the **subicular** and **entorhinal cortices**. The term **limbic lobe** was introduced by the French neuroscientist Pierre Paul Broca in 1878 to refer to the hippocampal formation and the cingulate gyrus, parahippocampal gyrus, and related cortices, which medially border the surrounding isocortical cortices. The French noun *limbe* means border or halo, and the adjective *limbique* thus means bordering. The term **limbic system** was introduced by the American neuroscientist Paul MacLean in 1952 to encompass the limbic lobe and related subcortical structures, including the amygdala (meaning "almond"), septal nuclei (septum meaning "wall"), and related parts of the striatum and diencephalon.

The term hippocampus is of much earlier origin. It was introduced by a student of Andreas Versalius, named Julius

Caesar Arantius, in 1587. The same name is used for the teleost fish with the common name of seahorse. Both the part of the brain and the fish were named hippocampus, meaning "horse-caterpillar," for their supposed resemblance to such a chimerid creature. Unfortunately, the likeness has proved to be obvious only in the case of the fish. For the hippocampus in the brain, the debate over which end should be the head and which the tail is as unresolvable as it is esoteric.

The hippocampus is also called Ammon's horn, for its equally poorly perceived likeness to a ram's horn. The term ram's horn was in fact introduced for the structure by Winslow in 1732 before being superseded in 1742 by a more pretentious term introduced by Garengeot, Ammon's horn, which was derived from the name of an Egyptian god, Amun Kneph, a god the Egyptians depicted either as a ram or as a human with ram's horns. Entering into an argument that creativity is better left to artists than scientists would be well beyond the scope of our subject, so we will now return to the *terra firma* of simple neuroanatomy.

Limbic Pallium of Mammals

The hippocampal formation of monotremes and marsupial mammals is similar topologically to that in nonmammalian amniotes in that it lies along the medial wall of the hemisphere (Fig. 30-10). In adult placental mammals, the hippocampal formation also occupies the medial wall of the hemisphere, but its relationships with other structures are distorted by the greatly expanded isocortical areas. It also folds inwardly during development so that the dentate gyrus, a prominent layer of closely packed neuron cell bodies, is innermost. Although not obvious in transverse sections, in sagittal sections the cellular layer of the dentate gyrus appears as a series of waves or notches, like a row of teeth, and the gyrus is named for this attribute.

A horizontal section through the hippocampal formation in a rodent is shown in Figure 30-11. The dentate gyrus caps the end of the curved hippocampus, or Ammon's horn. The hippocampus is divided into four "fields" on cytoarchitectural grounds, and the fields are referred to as **CA1–CA4**, "CA" being the abbreviation for the Latin for Ammon's horn: *Cornu Ammonis*. The cortical region adjoining the CA1 field of the hippocampus is called the subicular region and has three subdivisions that are, in medial-to-lateral order, the **subiculum**, the **presubiculum**, and the **parasubiculum**. Entorhinal cortex adjoins the parasubiculum. The two major pathways to and from the hippocampal formation are also shown in Figure 30-11. The superficial pathway is called the **perforant path**, while the pathway bordering the lateral ventricle is the **alveus**.

Figure 30-12(A) shows the approximate level of a transverse section through the forebrain of a primate for the drawings of the hippocampal formation in the rest of the figure. In Figure 30-12(B) a diagram at this level of a transverse section of the developing forebrain is shown to illustrate how the hippocampal formation (HF) rotates inwardly upon itself (arrows). An enlargement of the area that includes the hippocampal formation on the left side is shown for the adult brain in Figure 30-12(C), and enlargements of the hippocampal formation itself, as shown in Figure 30-12(C), are shown in Figures 30-12(D and E). In an unfolded form, the dentate gyrus would be the most medial part of the hippocampal formation.

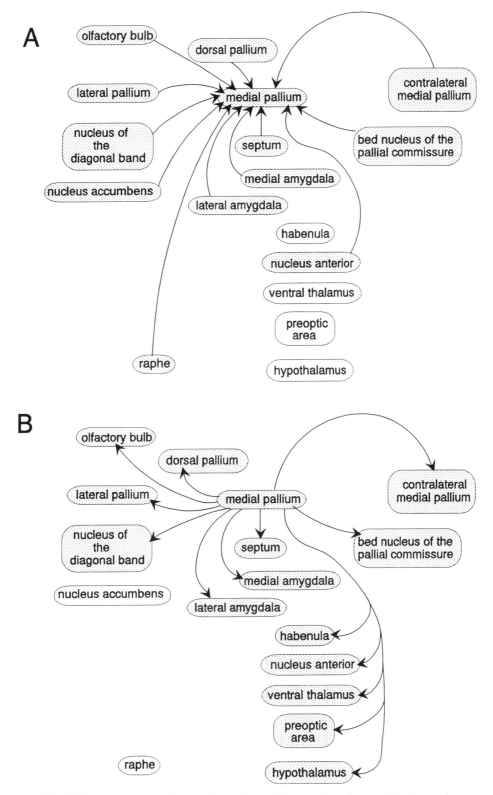

FIGURE 30-6. Diagrams of the afferent (A) and efferent (B) connections of the medial pallium in frogs.

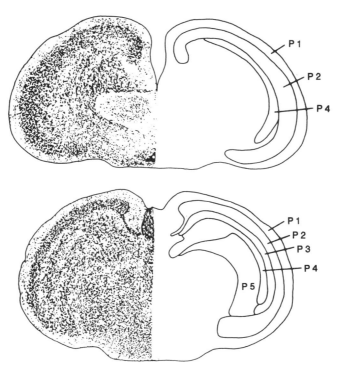

FIGURE 30-7. Transverse hemisections with mirror-image drawings through the telencephalon of a hagfish (*Eptatretus stouti*). The upper section is more rostral. PI–P5 are pallial areas. Adapted from Wicht and Northcutt (1992). Used with permission of S. Karger AG.

In the dentate gyrus, the majority of neurons have spherical to oval shaped cell bodies and are called **granule cells.** These cells are cytologically similar to the **candelabra cells** of the medial cortex of squamate reptiles and turtles, which we will discuss below. **Pyramidal cells** are also present, as are other neurons called **basket cells.** Ammon's horn is composed primarily of large and small pyramidal cells, some of which are **double pyramidal cells.** Smaller interneurons are also present. The large pyramidal cells give rise to the efferent pathway of the hippocampus. The axons that form this efferent pathway travel within the alveus and then coalesce to form a tract. The area of their coalescence is called the **fimbria** (from the Latin for "fringe"), and the tract from that point on is called the **fornix** (Figs. 30-11 and 30-12). The word "fornix" is Latin for "arch," which is the shape of this tract in the sagittal plane over the region of the diencephalon.

The hippocampus receives inputs from two major sources, one of which is relayed through the entorhinal cortex (Fig. 30-13). Entorhinal cortex receives an array of inputs from parts of the ventral portion of the temporal lobe, part of the frontal lobe called the orbitofrontal cortex, cortices neighboring the entorhinal cortex, primary sensory cortices, the septal nuclei, and numerous subcortical structures. Entorhinal cortex projects, via the perforant path, directly to some of the CA fields of the hippocampus and to the dentate gyrus, which also projects to the hippocampus. Some of the details of the connections involved in this pathway are included in Figure 30-16. The CA fields then project heavily to the subicular cortices, which also receive direct projections from entorhinal cortex and give rise to the majority of efferent projections from the hippocampal formation.

The major efferent projections of the subicular cortices are to the cingulate cortex, to the amygdala, and to widespread areas of association cortices in the frontal, parietal, temporal, and occipital lobes. The subicular cortices also project to nucleus accumbens and the olfactory tubercle, that is, the ventral

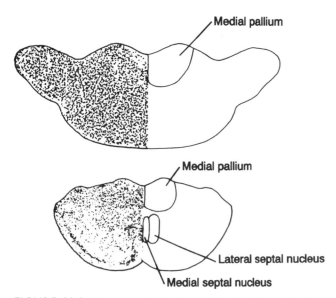

FIGURE 30-8. Transverse hemisections with mirror-image drawings through the telencephalon of a skate (*Raja eglanteria*). The upper section is more rostral. Adapted from Northcutt (1978).

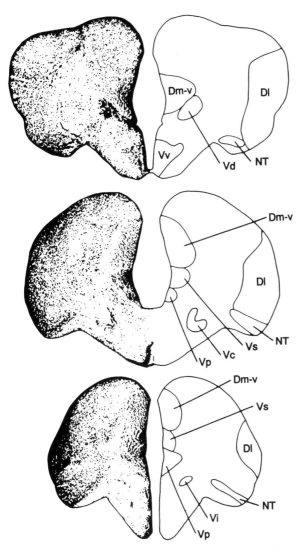

FIGURE 30-9. Transverse hemisections with mirror-image drawings through the telencephalon of a teleost (*Lepomis cyanellus*). The uppermost section is most rostral. Adapted from Northcutt and Davis (1983). Abbreviations: Dl, lateral zone of area dorsalis; Dm-v, ventral part of medial zone of area dorsalis; NT, nucleus taenia; Vc, commissural nucleus of area ventralis (V); Vd, dorsal nucleus of V; Vi, intermediate nucleus of V; Vp, postcommissural nucleus of V; Vs, supracommissural nucleus of V; Vv, ventral nucleus of V.

striatum, the site of an important interface between the limbic and striatal systems (see Chapter 24). Entorhinal cortex thus serves as a gateway for input to the hippocampus, while the subicular cortices serve as a gateway for its output. The entorhinal cortex also plays a role as a gateway for output, since it has reciprocal projections with the subicular cortices and projects back to isocortex.

The second major set of afferent connections to the hippocampal formation involves the fornix system, which also carries efferent fibers. The fornix interconnects the hippocampal formation with nuclei in the septal area, diencephalon, and more caudal brainstem. A circuit within this system (Fig. 30-14), recognized by and named for the American neuroscientist James

Papez, involves projections from the mammillary body in the hypothalamus to the anterior nuclear group of the dorsal thalamus, and thence to the cingulate gyrus. The cingulate gyrus then projects to entorhinal cortex, which in turn projects to the subicular cortices and Ammon's horn, as well as to the dentate gyrus. Ammon's horn and the subicular cortices project back, via the fornix, to the mammillary body. Papez proposed that this circuit might be involved in functions related to both the experience and expression of emotions. Numerous additional interconnections have subsequently been identified within this circuit, including direct projections from the mammillary body and supramammillary nucleus to the hippocampal formation (not shown in Fig. 30-14), from the anterior nuclear group of the dorsal thalamus to the subiculum, and from the subiculum and Ammon's horn to the anterior nuclear group.

The hippocampal formation receives additional afferent connections, primarily via the fornix. Dorsal thalamic nuclei of the medial nuclear group, particularly nucleus reuniens, and the paraventricular nuclei of the epithalamus also project to the hippocampal formation. As discussed in Chapter 21, the paraventricular nuclei have reciprocal connections with the amygdala and with lateral hypothalamic and preoptic areas as well as with the hippocampal formation. Similarly, the medial habenula is involved in a limbic circuit, discussed below in conjunction with the septal nuclei. Monoaminergic projections from the locus coeruleus and serotoninergic inputs from the midbrain raphe also reach the hippocampal formation.

An important input to the limbic telencephalon arises from the dopamine-containing cells of the ventral tegmental area (VTA) in the midbrain. This system has been referred to as the mesolimbic dopamine system, in contrast to the nigrostriatal dopamine system of projections from the substantia nigra (SN) to the striatum. Recent findings show that significant overlap exists in the projections of both the substantia nigra and the ventral tegmental area to both the limbic system and the striatum, and thus the two sets of projections should be considered as a unitary, **mesotelencephalic dopamine system.** Various combinations of neurons in the VTA or the SN or both project to limbic cortices including the cingulate cortex, entorhinal cortex, and the hippocampal formation. They also project to related parts of isocortex, including the prefrontal cortex, as well as to subpallial limbic structures as discussed below. This ascending dopamine system has been implicated in a wide range of functions, including reward aspects of learning and various cognitive functions. In humans, abnormally high levels of dopamine are correlated with symptoms of schizophrenia.

Damage to the hippocampal formation in mammals results in a number of changes in behavior, including increased locomotor activity, decreased distractibility, and severe impairments of learning and memory. Hippocampal memory deficits prevent the accumulation of new knowledge, while memory of events prior to the damage is preserved.

Limbic Pallium in Nonmammalian Amniotes

In diapsid reptiles (Figs. 30-15 and 30-16) and turtles (Fig. 30-17), the medial pallium is a cortical lamina of cells that is divided into two parts: **medial cortex (MC)** and **dorsomedial cortex (DMC).** These two parts of the medial pallium are also referred to as the **small-celled** and **large-celled parts,**

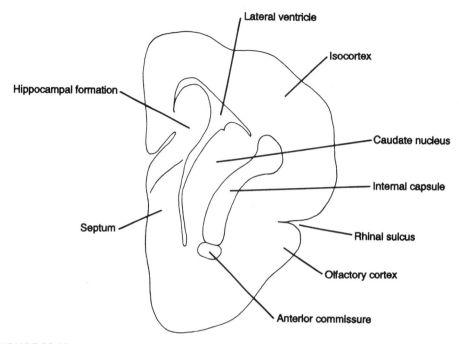

FIGURE 30-10. Drawing of a transverse hemisection through the right telencephalon of a marsupial (*Hypsiprymnus rufescens*). Data from Ariëns Kappers et al. (1967).

respectively. A **dorsal cortex (DC)** is also present and divisible into two parts. The dorsal cortex has been traditionally recognized as the homologue of at least part of the nonlimbic isocortex of mammals, but some of its features need to be considered here along with those of the medial pallium in regard to the limbic system of amniotes. The lateral part of the dorsal cortex receives visual input relayed by the dorsal lateral optic nucleus from the retina, and has been discussed in Chapter 26. Some of

the connections of the medial part of the dorsal cortex resemble limbic system connections of mammals and will be included here.

In turtles and in squamate reptiles (i.e., the tuatara, lizards, and snakes) the medial pallium is a three-layered cortex. The majority of neuron cell bodies lie in the middle layer, **layer 2** (Fig. 30-15), and most of their dendrites branch in the most superficial layer, **layer 1.** A few neuron cell bodies are present in both layer 1 and the deepest layer, **layer 3.** In some snakes

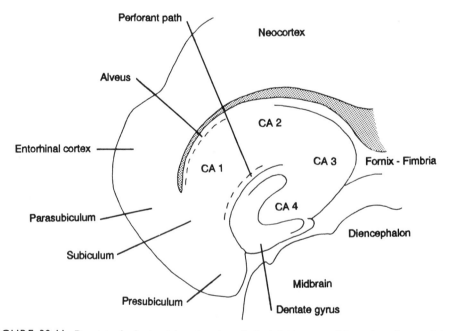

FIGURE 30-11. Drawing of a horizontal section through the left hippocampal formation of a rat. Fields CA1–CA4 constitute the hippocampus proper. The part of the lateral ventricle present in this region is indicated by shading. Adapted from Isaacson (1982). Used with permission of Plenum Publishing Corp.

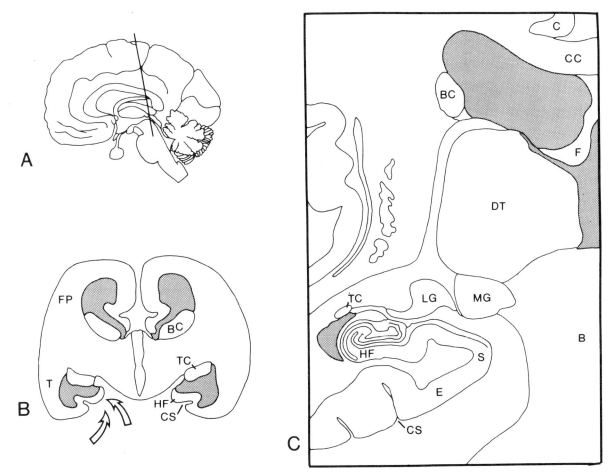

FIGURE 30-12. The hippocampal formation in a primate brain (*Homo sapiens*). (A) Sagittal section with line indicating the level and angle of the transverse hemisection through the left hippocampal formation shown in (C). The internal rotation (arrows) of the hippocampal formation within the temporal lobe during development is shown in transverse section in (B). (D) Enlarged drawing of a transverse section of the hippocampal formation on the left. (E) Drawing of left hippocampal formation similar to that shown in (D), to show fiber pathways. In (B, C, and D) ventricular areas are indicated by shading. Abbreviations: B, brainstem; BC, body of caudate nucleus; C, cingulate gyrus; CC, corpus callosum; CS, collateral sulcus; DT, dorsal thalamus; E, entorhinal cortex; F, fornix; FP, frontal-parietal isocortex; HF, hippocampal formation; LG, dorsal lateral geniculate nucleus; MG, medial geniculate body; S, subicular cortices; T, temporal isocortex; TC, tail of caudate nucleus. Adapted from DeArmond et al. 1989.

(continued)

(Fig. 30-16), an additional layer of densely packed cell bodies is present ventral to the dorsomedial cortex and is called the **ventral cell plate.**

The majority of cells in layer 2 of the medial cortex are densely packed and characterized as small **double pyramidal cells** (Fig. 30-18), similar to the double pyramidal cells of mammals. The word "double" refers to the fact that dendrites of these cells extend in both superficial and deep (diametrically opposite) directions. Their morphology varies somewhat, indicating that several types of double pyramidal cells are present. One type has been identified as **candelabra cells.** Most dendrites of the candelabra cells ramify extensively in layer 1, the major afferent fiber layer of the cortex, while the basal dendrites ramify in the superficial part of layer 3. Axons of candelabra and other double pyramidal cells enter a major limbic fiber tract that runs through the deep part of layer 3 called the alveus. The axons bifurcate, and one branch runs to the septal nuclei while the other runs dorsally toward the dorsomedial cortex.

Layer 1 of medial cortex contains few cells, most of which are either stellate or horizontal. A moderate number of cells are present in layer 3 and are of variable morphology, including spiny multipolar, pyramidal, and horizontal types.

Neurons in the dorsomedial cortex are primarily concentrated in layer 2 and consist of double pyramidal cells that are larger than those in the medial cortex. Their apical dendrites also ramify extensively in layer 1, and their basal dendrites ramify in the superficial part of layer 3. The axons of these cells enter the alveus in the deep part of layer 3. Bipolar cells are present in layer 1 of the dorsomedial cortex, and several cell types, including double pyramidal, multipolar, and horizontal, are present in layer 3.

The medial cortex in squamate reptiles and turtles receives afferent projections from several other cortices in the telencephalon, while the dorsomedial cortex is primarily interconnected with just the medial cortex. The lateral (olfactory) cortex, the dorsal cortex, and the dorsomedial cortex project to the medial

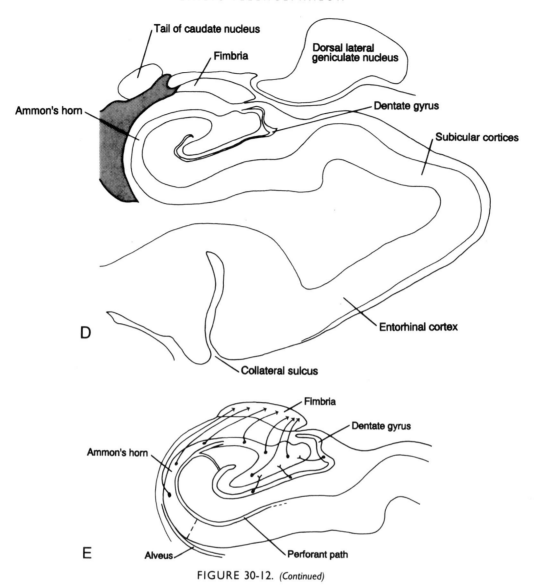

FIGURE 30-12. *(Continued)*

cortex. The medial cortex projects back reciprocally to the dorsal and dorsomedial cortices and is also reciprocally connected with the septum. The dorsomedial cortex has a substantial commissural connection with the contralateral dorsomedial cortex, in addition to its reciprocal connection with the medial cortex, and receives an input from the septum. Ascending inputs to the medial and dorsomedial cortices are similar and include projections from the anterior dorsolateral nucleus of the dorsal thalamus, the mammillary and periventricular parts of the hypothalamus, the raphe nuclei, the locus coeruleus, and the reticular formation.

Connections of the medial part of the dorsal cortex that resemble those of the limbic system of other vertebrates include efferent projections to the septal nuclei, amygdalar nuclei, nucleus accumbens, striatum, and hypothalamus. Like the medial and dorsomedial cortices, the dorsal cortex also receives afferent projections from the anterior dorsolateral nucleus of the dorsal thalamus, the mammillary nuclei of the hypothalamus, the raphe, and the locus coeruleus, and, as discussed above, is reciprocally connected with the medial cortex.

Based on similarities in cytological characteristics of the neurons, some of the afferent connections, and the laminar arrangement of the sets of terminal fields in layer 1 from various afferent sources, the medial cortex has been compared with the dentate gyrus in the hippocampal formation of mammals. Likewise, the dorsomedial cortex has been compared with Ammon's horn. These comparisons are supported by recent immunohistochemical findings on the presence of similar subpopulations of GABA neurons in the dorsomedial cortex of lizards and in Ammon's horn in mammals. The medial part of the dorsal cortex has been compared with parts of the mammalian limbic system that border Ammon's horn, that is, the subiculum, the entorhinal cortex, and/or adjacent cortices that lie in a transitional position between the hippocampal formation and isocortex. The topological order of these cortices in squamate reptiles and turtles on one hand and in mammals on the other may thus be the same: the sequence of medial cortex to dorsomedial cortex to the medial part of dorsal cortex corresponding to the sequence of dentate gyrus to Ammon's horn to subicular and entorhinal cortices.

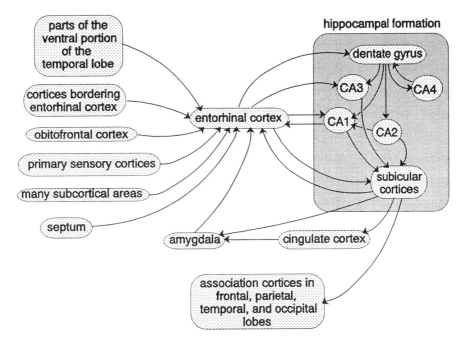

FIGURE 30-13. Diagram of some of the connections of the hippocampal formation in mammals.

The telencephalon of crocodiles appears to be cytoarchitectonically similar to that in squamate reptiles and turtles, but studies of connections of many of the cortical areas and subpallial structures have not yet been done. Likewise, little is known about the histochemical characteristics or development of the crocodilian telencephalon. These studies are needed to resolve an apparent riddle concerning the evolution of the cortical limbic structures encountered in birds and calls into question the existence of a 1:1 correspondence in parts of the medial pallium in mammals versus other amniote vertebrates referred to above.

In the telencephalon of birds, the medial pallium contains two laminae of densely packed neurons, which form a "V" shape and are collectively referred to as the **hippocampus,** and a region of more diffusely scattered neurons called the **parahippocampal area.** The degree of development of the avian hippocampal region has been found to be related to spatial memory requirements engendered by food-storing behaviors. Birds that store and then retrieve food have significantly larger hippocampal formations than related species that do not store food. This hippocampal enlargement has occurred at least three times independently in different groups of birds: titmice, corvids, and nuthatches.

The hippocampus and parahippocampal area of birds are homologous, at least as a field, to the medial pallium of squamate reptiles and turtles. While little information is available on the intracortical connections in birds, detailed analysis of multiple immunohistochemical markers—including those for acetylcholine, catecholamine, GABA, serotonin, and a variety of neuropeptides—indicates that the hippocampus of birds is homologous to Ammon's horn of mammals and that the parahippocampal area of birds is homologous to the dentate gyrus of mammals. Presumably the adjoining parts of the dorsal cortex

are homologous to the subicular and entorhinal cortices of mammals.

Unfortunately, this interpretation has the "wrong" topological order (Fig. 30-19). Instead of a topology of dentate gyrus to Ammon's horn to subicular and entorhinal cortices as is present in mammals, the topological order in birds would be Ammon's horn to dentate gyrus to subicular and entorhinal cortices. As discussed above the latter topology for birds is also in contrast to that present in squamate reptiles and turtles.

Hence, the riddle: if a 1:1 correspondence exists between individual parts of the medial pallium in mammals and in other amniotes, has this order been reversed in birds? What topological order will be found in the medial pallium of crocodiles (Fig. 30-19), which appears cytoarchitectonically similar to that in squamate reptiles? What are the connections of the various parts of the medial pallium in birds? How is the function of the medial pallium of birds affected by the apparently different arrangement of its component parts?

Limbic Subpallium: Septum

The major subpallial parts of the limbic telencephalon in amniotes are the septal area and the amygdalar complex, which we will now consider. In mammals, the septal area lies in the medial wall of the telencephalon. The lateral division of the septum contains a pair of **lateral septal nuclei.** The medial division of the septum contains the **medial septal nucleus.** The nucleus of the **diagonal band of Broca** is sometimes included within the medial division of the septum, as is done here, and sometimes is referred to as a separate entity. An additional posterior division of the septum is also present and contains the **septofimbrial** and **triangular nuclei.**

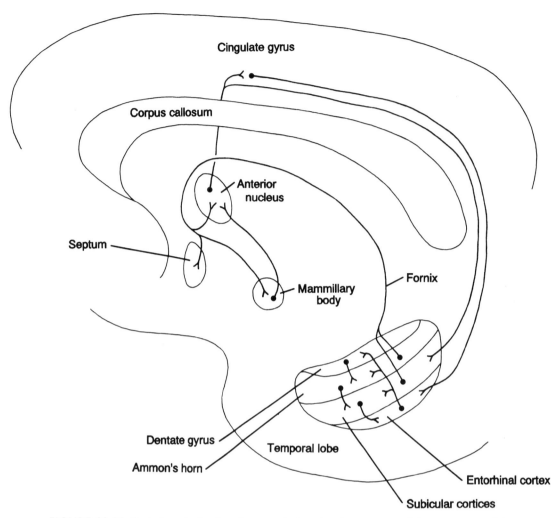

FIGURE 30-14. Semischematic drawing of a parasagittal section showing some of the connections of the hippocampal formation. Rostral is toward the left.

The major connections of the septum are reciprocal connections with telencephalic and diencephalic limbic structures and a pathway that forms part of a limbic circuit through the habenula. The septal nuclei (Fig. 30-20) are reciprocally connected with the hippocampus and the subicular cortices, with the preoptic area and a variety of nuclei in the hypothalamus, including the mammillary complex, and with the dopamine-containing, mesotelencephalic SN–VTA. The septal nuclei also receive inputs from the amygdala, particularly its basolateral part, and from the locus coeruleus. Additional efferents are to the dentate gyrus and entorhinal cortex, the habenula, and the dorsal raphe nucleus.

The limbic circuit that encompasses the septum and the habenula (see Chapter 21) involves projections from the hippocampus and subiculum to septal nuclei. The medial division of the septal area, particularly the nucleus of the diagonal band, projects to the medial habenular nucleus, which in turn projects to the interpeduncular nucleus. The interpeduncular nucleus projects back to the hippocampal formation.

Damage to the septal area results in a number of behavioral changes, although, since the results of septal lesions vary

significantly among different species, generalizations are somewhat hazardous. In some mammals, including humans, increased rage reactions and hyperemotionality occur immediately following septal lesions, although these behaviors return to normal over time. Animals with septal lesions show hyper-responsiveness to stimuli and also increased aggressive responses, which may in fact be a manifestation of an increase in defensiveness. Such animals also increase water intake and decrease open-field locomotion. In normal animals, the septum is believed to assist in attending to the less salient stimuli in the environment, rather than only to the more salient stimuli.

In the subpallial part of the telencephalon (Figs. 30-15 and 30-16) in diapsid reptiles, birds, and turtles, septal nuclei are present and lie ventral to the medial cortex. Only some of the connections of the septal area are known in nonmammalian amniotes. In diapsid reptiles, the septal area (including the nucleus of the diagonal band) has been found to project to the medial and lateral habenular nuclei. In birds, the medial septal nucleus is known to receive hypothalamic projections. The septal area projects to the hippocampal and parahippocampal

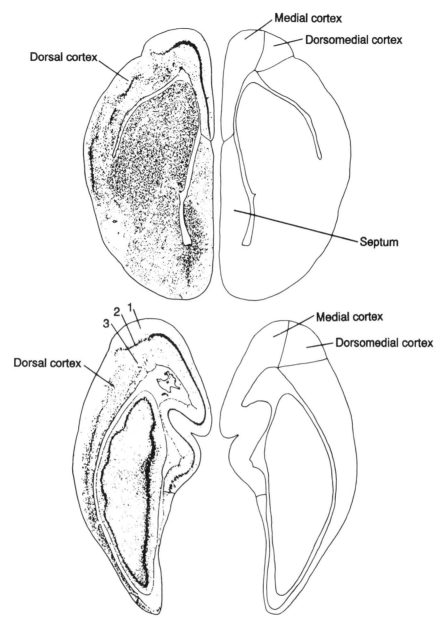

FIGURE 30-15. Transverse hemisections with mirror-image drawings through the telencephalon of a lizard (*Tupinambis nigropunctatus*). The upper section is more rostral. The numbers 1, 2, and 3 indicate layers of cortex. Adapted from Ebbesson and Voneida (1969). Used with permission of S. Karger AG.

cortices, portions of the hypothalamus, the dorsomedial thalamus, the lateral habenular nucleus, and the midbrain tegmentum and reticular formation.

Limbic Subpallium: Amygdala

The amygdaloid complex lies in the ventrolateral part of the caudal telencephalon in most amniotes. In adult mammals, the amygdala lies in the tip of the temporal lobe (Fig. 30-14). The amygdala has striatal, olfactory, limbic, and isocortical connections and is thus an important point of interface for multiple systems. As discussed in Chapter 24, the amygdalostriatal system pathways include reciprocal connections with the substantia nigra, ventral tegmental area, and subthalamic

nucleus. A circuit of amygdalar projections to the ventral striatum, ventral striatal projections to the ventral pallidum, and ventral pallidal projections back to the amygdala is also present, as are amygdalar efferents to the dorsal striatum, i.e., the caudate nucleus and putamen.

The amygdala in mammals has two major divisions, referred to as cortical (or **corticomedial**) and basal (or **basolateral**), each of which contains multiple nuclei. The corticomedial amygdala contains a nucleus, the **anterior cortical nucleus,** that receives direct olfactory projections from the main olfactory bulb and two other nuclei (the **posterior cortical nucleus** and **medial nucleus**) that receive vomeronasal projections from the accessory olfactory bulb. A **central nucleus** is also present in the corticomedial amygdala. The baso-

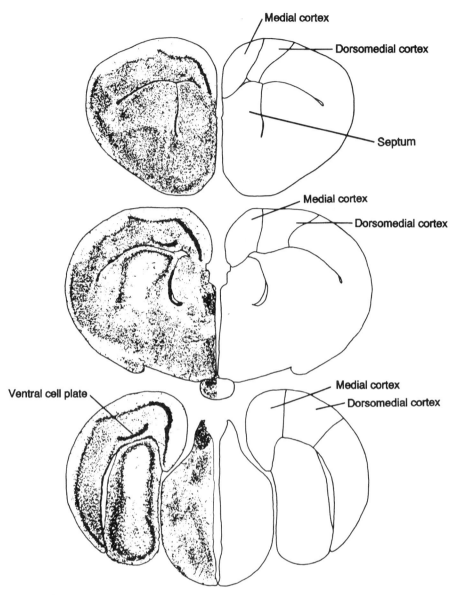

FIGURE 30-16. Transverse hemisections with mirror-image drawings through the telencephalon of a snake (*Eryx johni*). Adapted from Ulinski (1974).

lateral amygdala includes **basal, lateral, accessory basal,** and **paralaminar nuclei.** Olfactory input from the main olfactory bulb to the basolateral amygdala is indirect, being relayed to it through part of olfactory cortex.

The **extended amygdala** is a group of nuclei that include the central and medial cortical amygdalar nuclei, the bed nucleus of the stria terminalis, the sublenticular part of the substantia innominata, and the caudomedial shell of nucleus accumbens. The bed nucleus of the stria terminalis is viewed by many as an extended part of the central and medial amygdalar nuclei. As a group, the nuclei of the extended amygdala have similar afferent connections from cortical areas including the hippocampus, entorhinal cortex, frontal isocortex, and olfactory cortex. They also have similar efferent projections to the magnocellular basal forebrain, hypothalamus, and other brainstem areas.

The corticomedial amygdala primarily serves as a relay of olfactory information to hypothalamic and brainstem structures [Fig. 30-21(A)]. A major efferent tract from the amygdala is the stria terminalis, and the corticomedial amygdala has reciprocal connections with the bed nucleus of the stria terminalis. Both the corticomedial amygdala and the bed nucleus of the stria terminalis are connected with the medial habenula, various thalamic nuclei, the hypothalamus, the ventral tegmental area, central gray, and locus coeruleus. The corticomedial amygdala also has reciprocal connections with nucleus basalis (the magnocellular basal forebrain), projects to the globus pallidus and the substantia nigra, and has relatively sparse connections with the entorhinal cortex and parts of isocortex.

In studies of hamsters, the vomeronasal-recipient medial nucleus of the corticomedial amygdala has been found to play a crucial role in male copulatory behavior. Lesions restricted

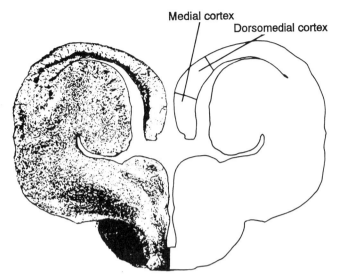

FIGURE 30-17. Transverse hemisection with mirror-image drawing through the telencephalon of a turtle. Adapted from Balaban and Ulinski (1981).

to this nucleus markedly diminish the male's licking and sniffing of the female's anogenital region and eliminate the male's mating behavior. The medial nucleus is thus believed to be crucial for the relay of chemosensory information to preoptic and hypothalamic regions involved in the mediation of sexual behaviors.

While the corticomedial amygdala is primarily related to the hypothalamus and the rest of the brainstem, the basolateral amygdala is primarily related to the telencephalic cortices. The basolateral amygdala [Fig. 30-21(B)] receives indirect olfactory input relayed through olfactory cortex, has reciprocal connections with nucleus basalis, and projects to the dorsal striatum and other parts of the striopallidal complexes, the preoptic area,

and the hypothalamus, but it is also heavily interconnected with widespread isocortical areas. It has reciprocal connections with some unimodal sensory cortices, polysensory association cortices, and prefrontal, insular, and cingulate cortices. It also has reciprocal projections with the entorhinal cortex, hippocampus, and the part of the subiculum that immediately adjoins the hippocampus (sometimes called the **prosubiculum**), although the amygdalofugal projections are much more pronounced than the reciprocal projections from the hippocampal formation.

Additional influence of the basolateral amygdala on the prefrontal cortex is mediated by a pathway through the medio-

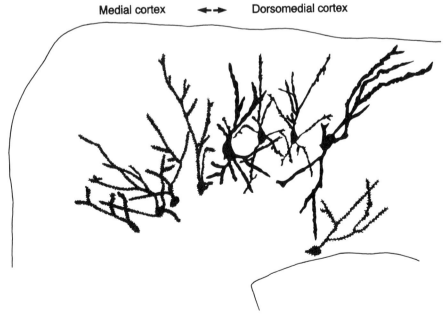

FIGURE 30-18. Neurons of the medial pallium in a snake (*Natrix sipedon*). Adapted from Ulinski (1974).

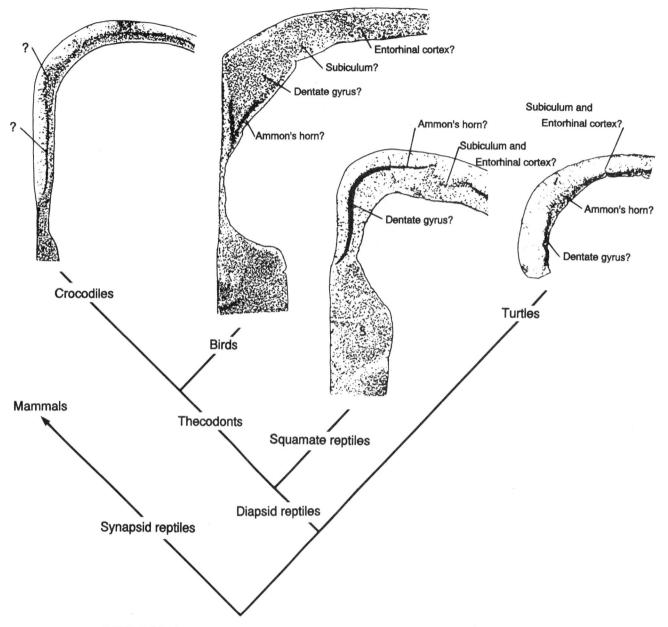

FIGURE 30-19. Comparison of medial pallial organization in extant nonmammalian amniote groups. S = septum.

dorsal nucleus of the dorsal thalamus, and inputs from the basolateral nucleus to the mediodorsal nucleus are carried by several routes (Fig. 30-22). The basolateral amygdala projects to nucleus accumbens. The ventral pallidum receives a projection from nucleus accumbens in addition to a sparse direct projection from the basolateral amygdala and, in turn, projects to the mediodorsal nucleus. The mediodorsal nucleus also receives a direct projection from the basolateral amygdala and a sparser direct projection from the corticomedial amygdala (not shown in the figure). The mediodorsal nucleus projects robustly to prefrontal cortex, as does the basolateral amygdala itself.

Lesions of the amygdala produce a variety of deficits and alterations in behavior. Orienting towards unfamiliar stimuli and responding to novel events are decreased, although visual discrimination learning is not impaired. Stimulation of the rostral part of the amygdala evokes flight and fear responses, while stimulation of the caudal part evokes defensive and/or aggressive reactions. The amygdala and areas of the prefrontal and temporal cortices all appear to be essential for the maintenance of normal social behavior. Animals with amygdalar lesions show a marked loss of aggressive behaviors. Subtle behaviors used in social interactions are also impaired by amygdalar lesions; monkeys high in social rank drop to the bottom of the social hierarchy following lesions of the amygdala.

Amygdalar lesions disrupt some learning tasks and the acquisition of conditioned emotional responses. Its widespread

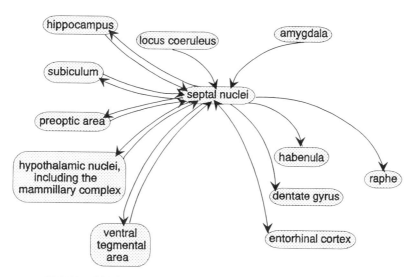

FIGURE 30-20. Diagram of some of the connections of the septal nuclei.

connections with striatal, hypothalamic, limbic, and isocortical structures allow for involvement in somatic and visceral responses to novel, emotional, and social stimuli. In humans, an increase in local synaptic activity in the amygdala, as measured by cerebral blood flow, is correlated with the alterations of mood that occur with affective disorders. Likewise, the anxiety that accompanies a state of panic or the expectation of pain involves the amygdala.

Multiple subdivisions of the amygdala have been recognized in diapsid reptiles, birds, and turtles. In squamate reptiles, the amygdalar nuclear group has three main divisions. The division called the **olfactory amygdala** receives direct projections from the main olfactory bulb, and the **vomeronasal amygdala** (or **nucleus sphericus**) receives projections from the accessory olfactory bulb. The third division, called the **dorsal amygdalar group,** receives projections from the other two divisions of the amygdala and from multiple other sites in the telencephalon. As discussed in Chapter 29, the main olfactory bulb is innervated by the olfactory nerve, the input for which is derived from a large part of the olfactory epithelium, while the accessory olfactory bulb is innervated by the vomeronasal nerve, the input for which is derived from the olfactory epithelium in Jacobson's organ. Nucleus sphericus is a distinctive nucleus in the caudal part of the amygdalar region that is formed by an oval lamina of cells (as seen in transverse section). The neurons of nucleus sphericus concentrate sex hormones, as do some of the other amygdaloid nuclei. In snakes, nucleus sphericus is known to project robustly to the hypothalamus.

In birds, amygdalar nuclei lie within a nuclear group in the ventrolateral part of the telencephalon referred to as the **archistriatum.** Some of the nuclei of the archistriatum are connectionally part of the somatomotor system (see Chapters 27 and 28). Three nuclei, **archistriatum mediale, archistriatum posterior,** and **nucleus taeniae,** appear to be homologous to the mammalian amygdala. Afferents include those from the parahippocampal area, the locus coeruleus, and a nucleus in the ventral tegmentum. The amygdalar nuclei of the archistriatum project to the hypothalamus via the **tractus occipitomesencephalicus pars hypothalami** (HOM). The hypothalamus,

via multisynaptic descending pathways, influences heart rate and related functions of the sympathetic nervous system. In accord with results of amygdalar lesions in mammals, as discussed above, lesions of the medial, amygdalar part of the archistriatum in wild birds, which are normally fearful in captivity, result in tameness. Stimulation of the medial archistriatum, in contrast, produces escape behavior.

EVOLUTIONARY PERSPECTIVE

Medial pallial, septal, and amygdalar areas are present in all jawed vertebrates. Their possible presence in agnathans needs to be addressed in further studies. The medial pallium that has been tentatively identified in lampreys and that is present in cartilaginous fishes is relatively small. Problems have arisen in identifying the part(s) of the telencephalon in teleosts that is (are) homologous to the medial pallium of other vertebrates, although a medial pallial, nonolfactory recipient homologue has been identified in nonteleost ray-finned fishes in a lateral position. A medial pallium and septal and amygdalar areas are clearly present in amphibians and in all amniote vertebrates. The presence of a medial pallium and associated subpallial structures is thus a feature that was present in at least the ancestral stock of jawed vertebrates and may have been present in the earliest vertebrates.

MacLean, who introduced the term limbic system, hypothesized that the brain of mammals can be understood in terms of a tripartite organization that he has called the "triune brain." As discussed in Chapter 6, he proposed that the limbic system of most extant mammals constitutes the middle of three components that were acquired progressively over evolution. According to the hypothesis, the limbic system was gained in early mammals and layered over a striatal "reptile brain." At a later stage of mammalian evolution, the third component, the isocortical (neocortical) "new mammal brain," was added. MacLean calls the limbic system the "old mammal brain." The findings of comparative neuroanatomical investigations, as we have reviewed here, do not support the evolutionary history of the

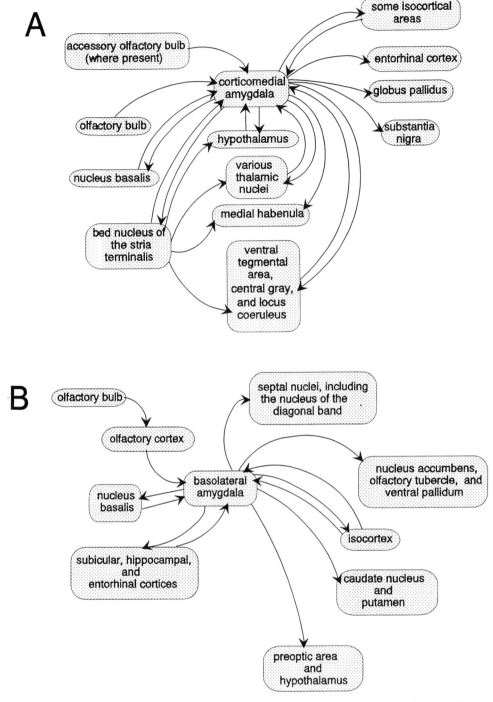

FIGURE 30-21. Diagrams of some of the connections of (A) the corticomedial amygdala and (B) the basolateral amygdala.

limbic system as described by the triune brain hypothesis. The present evidence indicates that the limbic system evolved long before the advent of any amniote vertebrates, let alone early mammals.

The limbic system is, in fact, significantly enlarged in all amniote vertebrates as compared with its condition in anamniotes. In the common ancestral amniote stock, the limbic system underwent considerable enlargement and differentiation. Dis-

tinctly different parts of the medial pallium can be identified in mammals, diapsid reptiles, birds, and turtles, even though their specific homologous relationships are not yet entirely understood. Likewise, expansion and subdivision has occurred within the septal and amygdalar areas.

A marked increase in the interconnections of the limbic system and the striatal motor system may also have occurred in amniotes. The currently known interconnections of the sys-

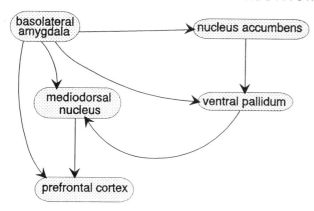

FIGURE 30-22. Diagram of connections involving the ventral striatum and ventral pallidum and the mediodorsal nucleus of the dorsal thalamus.

tems are most numerous in mammals, lesser in nonmammalian amniotes, and least in anamniotes. Many more studies of connections are needed in anamniotes, however, to test this possibility, as the apparent trend may simply be a reflection of the relative amount of available data.

An hypothesis more in keeping with currently available data is that a major change in the organization of afferent connections to the pallium occurred in ancestral amniotes. In anamniotes, in which dorsal thalamic nuclei are in a relatively periventricular position, sensory input from the rostral part of the thalamus (i.e., the lemnothalamus) is relayed directly to the medial pallium. In amniotes, the medial pallium receives little direct sensory input. Dorsal thalamic sensory relay nuclei project to expanded dorsal pallial areas instead, and the latter areas in turn project to the medial pallium.

Thus, the medial pallium in amniotes is relatively expanded while at the same time being freed of the task of analyzing primary sensory input. Instead the medial pallium receives so-called "higher order inputs of an associational nature" and thus functions at a more complex level of information analysis. In mammals, the medial pallium is the largest and most highly differentiated relative to other amniotes, and it plays a major role in learning, memory, and the associated functions of motivation and emotion.

FOR FURTHER READING

Erichsen, J. T., Bingman, V. P., and Krebs, J. R. (1991) The distribution of neuropeptides in the dorsomedial telencephalon of the pigeon (*Columba livia*): a basis for regional subdivisions. *Journal of Comparative Neurology*, 314, 478–492.

Hoogland, P. V. and Vermeulen-Van der Zee, E. (1993) Medial cortex of the lizard *Gekko gecko*: a hodological study with emphasis on regional specialization. *Journal of Comparative Neurology*, 331, 326–338.

Isaacson, R. L. (1982) *The Limbic System*. New York: Plenum.

Lehman, M. N. and Winans, S. S. (1982) Vomeronasal and olfactory pathways to the amygdala controlling male hamster sexual behavior: autoradiographic and behavioral analyses. *Brain Research*, 240, 27–41.

Schwerdtfeger, W. K. and Germroth, P. (1990) Archicortical and periarchicortical areas in the vertebrate forebrain. In W. K. Schwerdtfeger and P. Germroth (eds.), *The Forebrain in Nonmammals: New Aspects of Structure and Development*. Berlin: Springer-Verlag, pp. 197–212.

Thompson, S. M. and Robertson, R. T. (1987) Organization of subcortical pathways for sensory projections to the limbic cortex. I. Subcortical projections to the medial limbic cortex in the rat. *Journal of Comparative Neurology*, 265, 175–188.

ADDITIONAL REFERENCES

Amaral, D. G., Price, J. L., Pitkänen, A., and Carmichael, S. T. (1992) Anatomical organization of the primate amygdaloid complex. In Aggleton, J. P. (ed.), *The Amygdala: Neurobiological Aspects of Emotion, Memory, and Mental Dysfunction*. New York: Wiley-Liss, pp. 1–66.

Ariëns Kappers, C. U., Huber, G. C. and Crosby, E. C., (1967) *The Comparative Anatomy of the Nervous System of Vertebrates, Including Man*. New York: Hafner Publishing Co.

Balaban, C. D. and Ulinski, P. S. (1981) Organization of thalamic afferents to anterior dorsal ventricular ridge in turtles. I. Projections of thalamic nuclei. *Journal of Comparative Neurology*, 200, 95–129.

Benowitz, L. (1980) Functional organization of the avian telencephalon. In S. O. E. Ebbesson (ed.), *Comparative Neurology of the Telencephalon*. New York: Plenum, pp. 389–421.

Berbel, P. J. (1988) Cytology of medial and dorso-medial cerebral cortices in lizards: a Golgi study. In W. K. Schwerdtfeger and W. J. A. J. Smeets (eds.), *The Forebrain of Reptiles: Current Concepts of Structure and Function*. Basel: Karger, pp. 12–19.

Berk, M. L. and Butler, A. B. (1981) Efferent projections of the medial preoptic nucleus and medial hypothalamus in the pigeon. *Journal of Comparative Neurology*, 203, 379–399.

Bingman, V. P. (1993) Vision, cognition, and the avian hippocampus. In H. P. Zeigler and H.-J. Bischof (eds.), *Vision, Brain, and Behavior in Birds*. Cambridge, MA: The MIT Press, pp. 391–408.

Bruce, L. L. and Butler, A. B. (1984) Telencephalic connections in lizards. I. Projections to cortex. *Journal of Comparative Neurology*, 229, 585–601.

Bodznick, D. and Northcutt, R. G. (1984) An electrosensory area in the telencephalon of the little skate, *Raja erinacea*. *Brain Research*, 298, 117–124.

Cohen, D. H. (1975) Involvement of the avian amygdalar homologue (archistriatum posterior and mediale) in defensively conditioned heart rate change. *Journal of Comparative Neurology*, 160, 13–36.

Cruce, J. A. F. (1975) An autoradiographic study of the projections of the mammillothalamic tract in the rat. *Brain Research*, 85, 211–219.

Dávila, J. C., Megas, M., de la Calle, A., and Guirado, S. (1993) Subpopulations of GABA neurons containing somatostatin, neuropeptide Y, and parvalbumin in the dorsomedial cortex of the lizard *Psammodromus algirus*. *Journal of Comparative Neurology*, 336: 161–173.

DeArmond, S. J., Fusco, M. M., and Dewey, M. M. (1989) *Structure of the Human Brain: A Photographic Atlas, Third Edition*. New York: Oxford University Press.

Desan, P. H. (1988) Organization of the cerebral cortex in turtle. In W. K. Schwerdtfeger and W. J. A. J. Smeets (eds.), *The Forebrain of Reptiles: Current Concepts of Structure and Function*. Basel: Karger, pp. 1–11.

Díaz, C. and Puelles, L. (1992) Afferent connections of the habenular complex in the lizard *Gallotia galloti*. *Brain, Behavior and Evolution*, 39, 312–324.

Domesick, V. B. (1972) Thalamic relationships of the medial cortex in the rat. *Brain, Behavior and Evolution*, 6, 457–483.

Ebbesson, S. O. E. and Voneida, T. J. (1969) The cytoarchitecture of the pallium in the tegu lizard (*Tupinambis nigropunctatus*). *Brain, Behavior and Evolution*, 2, 431–466.

Echteler, S. M. and Saidel, W. M. (1981) Forebrain connections in the goldfish support telencephalic homologies with land vertebrates. *Science*, 212, 683–685.

Fallon, J. H. (1988) Topographic organization of ascending dopaminergic projections. *Annals of the New York Academy of Sciences*, 537, 1–9.

Fibiger, H. C. and Phillips, A. G. (1988) Mesocorticolimbic dopamine systems and reward. *Annals of the New York Academy of Sciences*, 537, 206–215.

Gaffan, D., Murray, E. A., and Fabre-Thorpe, M. (1993) Interaction of the amygdala with the frontal lobe in reward memory. *European Journal of Neuroscience*, 5, 968–975.

Gasbarri, A., Packard, M. G., Campana, E., and Pacitti, C. (1994) Anterograde and retrograde tracing of projections from the ventral tegmental area to the hippocampal formation in the rat. *Brain Research Bulletin*, 33, 445–452.

Halpern, M. (1980) The telencephalon of snakes. In S. O. E. Ebbesson (ed.), *Comparative Neurology of the Telencephalon*. New York: Plenum, pp. 257–293.

Halpern, M. (1988) Vomeronasal system functions: role in mediating the reinforcing properties of chemical stimuli. In W. K. Schwerdtfeger and W. J. A. J. Smeets (eds.), *The Forebrain of Reptiles: Current Concepts of Structure and Function*. Basel: Karger, pp. 142–150.

Halpern, M., Morrell, J. I., and Pfaff, D. W. (1982) Cellular [³H]estradiol localization in the brains of garter snakes: an autoradiographic study. *General and Comparative Endocrinology*, 46, 211–224.

Healy, S. D. and Krebs, J. R. (1993) Development of hippocampal specialization in a food-storing bird. *Behavioural Brain Research*, 53, 127–131.

Heimer, L. and Alheid, G. F. (1991) Piecing together the puzzle of basal forebrain anatomy. In T. C. Napier, P. W. Kalivas, and I. Hanin (eds.), *The Basal Forebrain: Anatomy to Function*. New York: Plenum, pp. 1–42.

Hevner, R. F. and Wong-Riley, M. T. T. (1992) Entorhinal cortex of the human, monkey, and rat: metabolic map as revealed by cytochrome oxidase. *Journal of Comparative Neurology*, 326, 451–469.

Hirose, S., Ino, T., Takada, M., Kimura, J., Akiguchi, I., and Mizuno, N. (1992) Topographic projections from the subiculum to the limbic regions of the medial frontal cortex in the cat. *Neuroscience Letters*, 139, 61–64.

Hoogland, P. V. and Vermeulen-Van der Zee, E. (1988) Intrinsic and extrinsic connections of the cerebral cortex of lizards. In W. K. Schwerdtfeger and W. J. A. J. Smeets (eds.), *The Forebrain of Reptiles: Current Concepts of Structure and Function*. Basel: Karger, pp. 20–29.

Hoogland, P. V. and Vermeulen-Van der Zee, E. (1989) Efferent connections of the dorsal cortex of the lizard *Gekko gecko* studied with *Phaseolus vulgaris*-leucoagglutinin. *Journal of Comparative Neurology*, 285, 289–303.

Horel, J. A. (1988) Limbic neocortical interrelations. In H. D. Steklis and J. Erwin (eds.), *Comparative Primate Biology, Vol. 4: Neurosciences*. New York: Liss, pp. 81–97.

Jarrard, L. E. (1993) On the role of the hippocampus in learning and memory in the rat. *Behavioral and Neural Biology*, 60, 9–26.

Jones, R. S. G. (1993) Entorhinal-hippocampal connections: a speculative view of their function. *Trends in Neurosciences*, 16, 58–64.

Kawamura, K. and Norita, M. (1980) Corticoamygdaloid projections in the rhesus monkey. An HRP study. *Journal of Iwate Medical Association*, 32, 461–465.

Kevetter, G. A. and Winans, S. S. (1981) Connections of the corticomedial amygdala in the golden hamster. I. Efferents of the "vomeronasal amygdala." *Journal of Comparative Neurology*, 197, 81–98.

Kevetter, G. A. and Winans, S. S. (1981) Connections of the corticomedial amygdala in the golden hamster. II. Efferents of the "olfactory amygdala." *Journal of Comparative Neurology*, 197, 99–111.

Krayniak, P. F. and Siegel, A. (1978) Efferent connections of the septal area in the pigeon. *Brain, Behavior and Evolution*, 15, 389–404.

Krebs, J. R., Erichsen, J. T., and Bingman, V. P. (1991) The distribution of neurotransmitters and neurotransmitter-related enzymes in the dorsomedial telencephalon of the pigeon (*Columba livia*). *Journal of Comparative Neurology*, 314, 467–477.

Lehman, M. N. and Winans, S. S. (1980) Medial nucleus of the amygdala mediates chemosensory control of male hamster sexual behavior. *Science*, 210, 557–560.

Lehman, M. N. and Winans, S. S. (1983) Evidence for a ventral nonstrial pathway from the amygdala to the bed nucleus of the stria terminalis in the male golden hamster. *Brain Research*, 268, 139–146.

Leranth, C., Deller, T., and Buzséki, G. (1992) Intraseptal connections redefined: lack of a lateral septum to medial septum path. *Brain Research*, 583, 1–11.

Lewis, F. T. (1923) The significance of the term *hippocampus*. *Journal of Comparative Neurology*, 35, 213–230.

Loesche, J. and Steward, O. (1977) Behavioral correlates of denervation and reinnervation of the hippocampal formation of the rat: recovery of alternation performance following unilateral entorhinal cortex lesions. *Brain Research Bulletin*, 2, 31–39.

Lohman, A. H. M., Hoogland, P. V., and Witjes, R. J. G. M. (1988) Projections from the main and accessory olfactory bulbs to the amygdaloid complex in the lizard *Gekko gecko*. In W. K. Schwerdtfeger and W. J. A. J. Smeets (eds.), *The Forebrain of Reptiles: Current Concepts of Structure and Function*. Basel: Karger, pp. 41–49.

Lohman, A. H. M. and Mentink, G. M. (1972) Some cortical connections of the tegu lizard (*Tupinambis nigropunctatus*). *Brain Research*, 45, 325–344.

Lohman, A. H. M. and van Woerden-Verkley, I. (1978) Ascending connections to the forebrain in the tegu lizard. *Journal of Comparative Neurology*, 182, 555–594.

Martin-Elkins, C. L. and Horel, J. A. (1992) Cortical afferents to behaviorally defined regions of the inferior temporal and parahippocampal gyri as demonstrated by WGA–HRP. *Journal of Comparative Neurology*, 321, 177–192.

Martínez-García, F., Amiguet, M., Schwerdtfeger, W. K., Olucha, F. E., and Lorente, M.-J. (1990) Interhemispheric connections through the pallial commissures in the brain of *Podarcis hispanica* and *Gallotia stehlinii* (Reptilia, Lacertidae). *Journal of Morphology*, 205, 17–31.

Martínez-García, F., Desfilis, E., and Lopez-Garcia, C. (1990) Organization of the dorsomedial cortex in the lizard *Podarcis hispanica*. In W. K. Schwerdtfeger and P. Germroth (eds.), *The Forebrain in Nonmammals: New Aspects of Structure and Development*. Berlin: Springer-Verlag, pp. 77–92.

Martínez-García, F. and Olucha, F. E. (1988) Afferent projections to the Timm-positive cortical areas of the telencephalon of lizards. In W. K. Schwerdtfeger and W. J. A. J. Smeets (eds.), *The Forebrain of Reptiles: Current Concepts of Structure and Function*. Basel: Karger, pp. 30–40.

Martínez-García, F., Olucha, F. E., Teruel, V., and Lorente, M. J. (1993) Fiber connections of the amygdaloid formation of the lizard *Podarcis hispanica. Brain, Behavior and Evolution*, 41, 156–162.

Meredith, G. E., Wouterlood, F. G., and Pattiselanno, A. (1990) Hippocampal fibers make synaptic contacts with glutamate decarboxylase-immunoreactive neurons in the rat nucleus accumbens. *Brain Research*, 513, 329–334.

Murakami, T., Morita, Y., and Ito, H. (1983) Extrinsic and intrinsic fiber connections of the telencephalon in a teleost, *Sebastiscus marmoratus. Journal of Comparative Neurology*, 216, 115–131.

Neary, T. J. (1990) The pallium of anuran amphibians. In E. G. Jones and A. Peters (eds.), *Cerebral Cortex, Vol. 8A: Comparative Structure and Evolution of Cerebral Cortex, Part I.* New York: Plenum, pp. 107–138.

Nemeroff, C. B. and Bissette, G. (1988) Neuropeptides, dopamine, and schizophrenia. *Annals of the New York Academy of Sciences*, 537, 273–291.

Nieuwenhuys, R. and Meek, J. (1990a) The telencephalon of actinopterygian fishes. In E. G. Jones and A. Peters (eds.), *Cerebral Cortex, Vol. 8A: Comparative Structure and Evolution of Cerebral Cortex, Part I.* New York: Plenum, pp. 31–73.

Nieuwenhuys, R. and Meek, J. (1990b) The telencephalon of sarcopterygian fishes. In E. G. Jones and A. Peters (eds.), *Cerebral Cortex, Vol. 8A: Comparative Structure and Evolution of Cerebral Cortex, Part I.* New York: Plenum, pp. 75–106.

Northcutt, R. G. (1978) Brain organization in cartilaginous fishes. In E. S. Hodgson and R. F. Mathewson (eds.), *Sensory Biology of Sharks, Skates, and Rays.* Arlington, VA: Office of Naval Research, pp. 117–193.

Northcutt, R. G. (1981) Evolution of the telencephalon in nonmammals. *Annual Review of Neuroscience,* 4, 301–350.

Northcutt, R. G. (1991) Visual pathways in elasmobranchs: organization and phylogenetic implications. *Journal of Experimental Zoology*, Suppl. 5, 97–107.

Northcutt, R. G. and Davis, R. E. (1983) Telencephalic organization in ray-finned fishes. In R. E. Davis and R. G. Northcutt (eds.), *Fish Neurobiology, Vol. 2: Higher Brain Areas and Functions.* Ann Arbor, MI: The University of Michigan Press, pp. 203–236.

Northcutt, R. G. and Kicliter, E. (1980) Organization of the amphibian telencephalon. In S. O. E. Ebbesson (ed.), *Comparative Neurology of the Telencephalon.* New York: Plenum, pp. 203–255.

Northcutt, R. G. and Puzdrowski, R. L. (1988) Projections of the olfactory bulb and nervus terminalis in the silver lamprey. *Brain, Behavior and Evolution*, 32, 96–107.

Northcutt, R. G., Reiner, A., and Karten, H. J. (1988) Immunohistochemical study of the telencephalon of the spiny dogfish, *Squalus acanthias. Journal of Comparative Neurology*, 277, 250–267.

Nottebohm, F., Stokes, T. M., and Leonard, C. M. (1976) Central control of song in the canary, *Serinus canarius. Journal of Comparative Neurology*, 165, 457–486.

Ottersen, O. P. and Ben-Ari, Y. (1979) Afferent connections to the amygdaloid complex of the rat and cat. *Journal of Comparative Neurology*, 187, 401–424.

Pitkänen, A. and Amaral, D. G. (1993) Distribution of parvalbumin-immunoreactive cells and fibers in the monkey temporal lobe: the hippocampal formation. *Journal of Comparative Neurology*, 331, 37–74.

Phillips, R. E. (1964) "Wildness" in the mallard duck. Effects of brain lesions and stimulation on escape behaviour and reproduction. *Journal of Comparative Neurology*, 122, 139–155.

Reiner, A. and Northcutt, R. G. (1987) An immunohistochemical study of the telencephalon of the African lungfish, *Protopterus annectens. Journal of Comparative Neurology*, 256, 463–481.

Reiner, A. and Northcutt, R. G. (1992) An immunohistochemical study of the telencephalon of the Senegal bichir (*Polypterus senegalus*). *Journal of Comparative Neurology*, 319, 359–386.

Rooney, D. J. and Szabo, T. (1991) Reciprocal connections between the 'nucleus rotundus' and the dorsal lateral telencephalon in the weakly electric fish *Gnathonemus petersii. Brain Research*, 543, 153–156.

Ruit, K. G. and Neafsey, E. J. (1990) Hippocampal input to a "visceral motor" corticobulbar pathway: an anatomical and electrophysiological study in the rat. *Experimental Brain Research*, 82, 606–616.

Schwerdtfeger, W. K. and Lorente, M.-J. (1988) GABA-immunoreactive neurons in the medial and dorsomedial cortices of the lizard telencephalon. Some data on their structure, distribution and synaptic relations. In W. K. Schwerdtfeger and W. J. A. J. Smeets (eds.), *The Forebrain in Reptiles: Current Concepts of Structure and Function.* Basel: Karger, pp. 110–121.

Sherry, D. F., Jacobs, L. F., and Gaulin, S. J. C. (1992) Spatial memory and adaptive specialization of the hippocampus. *Trends in Neurosciences*, 15, 298–303.

Shibata, H. (1992) Topographic organization of subcortical projections to the anterior thalamic nuclei in the rat. *Journal of Comparative Neurology*, 323, 117–127.

Shibata, H. (1993a) Direct projections from the anterior thalamic nuclei to the retrohippocampal region in the rat. *Journal of Comparative Neurology*, 337, 431–445.

Shibata, H. (1993b) Efferent projections from the anterior thalamic nuclei to the cingulate cortex in the rat. *Journal of Comparative Neurology*, 330, 533–542.

Shinonaga, Y., Takada, M., and Mizuno, N. (1992) Direct projections from the central amygdaloid nucleus to the globus pallidus and substantia nigra in the cat. *Neuroscience*, 51, 691–703.

Shipley, M. T. (1975) The topographical and laminar organization of the presubiculum's projection to the ipsi- and contralateral entorhinal cortex in the guinea pig. *Journal of Comparative Neurology*, 160, 127–146.

Smeets, W. J. A. J. (1990) The telencephalon of cartilaginous fishes. In E. G. Jones and A. Peters (eds.), *Cerebral Cortex, Vol. 8A: Comparative Structure and Evolution of Cerebral Cortex, Part I.* New York: Plenum, pp. 3–30.

Smeets, W. J. A. J., Hoogland, P. V. and Lohman, A. H. M. (1986) A forebrain atlas of the lizard *Gekko gecko. Journal of Comparative Neurology*, 254, 1–19.

Smeets, W. J. A. J. and Northcutt, R. G. (1987) At least one thalamotelencephalic pathway in cartilaginous fishes projects to the medial pallium. *Neuroscience Letters*, 78, 277–282.

Striedter, G. F. (1991) Auditory, electrosensory, and mechanosensory lateral line pathways through the forebrain in channel catfishes. *Journal of Comparative Neurology*, 312, 311–331.

Striedter, G. F. (1992) Phylogenetic changes in the connections of the lateral preglomerular nucleus in ostariophysan teleosts: a pluralistic view of brain evolution. *Brain, Behavior and Evolution*, 39, 329–357.

Swanson, L. W. and Cowan, W. M. (1975) Hippocampal–hypothalamic connections: origin in subicular cortex. *Science*, 189, 303–304.

Swanson, L. W. and Cowan, W. M. (1977) An autoradiographic study of the organization of the efferent connections of the hippocampal formation in the rat. *Journal of Comparative Neurology*, 172, 49–84.

Swanson, L. W. and Cowan, W. M. (1979) The connections of the septal area in the rat. *Journal of Comparative Neurology*, 186, 621–656.

Thompson, S. M. and Robertson, R. T. (1987) Organization of subcortical pathways for sensory projections to the limbic cortex. II. Afferent projections to the thalamic lateral dorsal nucleus in the rat. *Journal of Comparative Neurology*, 265, 189–202.

Ulinski, P. S. (1974) Cytoarchitecture of cerebral cortex in snakes. *Journal of Comparative Neurology*, 158, 243–266.

Ulinski, P. S. (1976) Intracortical connections in the snakes *Natrix sipedon* and *Thamnophis sirtalis*. *Journal of Morphology*, 150, 463–484.

Ulinski, P. S. (1983) *Dorsal Ventricular Ridge: A Treatise on Forebrain Organization in Reptiles and Birds*. New York: John Wiley.

Ulinski, P. S. (1990) The cerebral cortex of reptiles. In E. G. Jones and A. Peters (eds.), *Cerebral Cortex, Vol. 8A: Comparative Structure and Evolution of Cerebral Cortex, Part I*. New York: Plenum, pp. 139–215.

Vogt, L. J., Vogt, B. A., and Sikes, R. W. (1992) Limbic thalamus in rabbit: architecture, projections to cingulate cortex and distribution of muscarinic acetylcholine, GABA$_A$, and opioid receptors. *Journal of Comparative Neurology*, 319, 205–217.

Voneida, T. J. and Ebbesson, S. O. E. (1969) On the origin and distribution of axons in the pallial commissures in the tegu lizard (*Tupinambis nigropunctatus*). *Brain, Behavior and Evolution*, 2, 467–481.

Wicht, H. and Northcutt, R. G. (1992) The forebrain of the Pacific hagfish: a cladistic reconstruction of the ancestral craniate forebrain. *Brain, Behavior and Evolution*, 40, 25–64.

Wilczynski, W. and Northcutt, R. G. (1983) Connections of the bullfrog striatum: afferent organization. *Journal of Comparative Neurology*, 214, 321–332.

Winans, S. S. and Scalia, F. (1970) Amygdaloid nucleus: new afferent input from the vomeronasal organ. *Science*, 170, 330–332.

Witter, M. (1993) Organization of the entorhinal-hippocampal system: a review of current anatomical data. *Hippocampus*, 3, 33–44.

Wyss, J. M., Swanson, L. W., and Cowan, W. M. (1979) A study of subcortical afferents to the hippocampal formation in the rat. *Neuroscience*, 4, 463–476.

Zeier, H. and Karten, H. J. (1971) The archistriatum of the pigeon: organization of afferent and efferent connections. *Brain Research*, 31, 313–326.

Part Six

CONCLUSION

31

Evolution of the Brain in Vertebrates

INTRODUCTION

The origin of vertebrates from ancestral chordates is one of science's persistent mysteries. The earliest vertebrates were a group of soft-bodied deuterostomes, and therefore no fossil record of them exists. Deuterostomes include three major groups, as shown in Figure 31-1—**echinoderms, hemichordates,** and **chordates.** The phylum Chordata comprises the **vertebrates** and the invertebrate chordates—**cephalochordates** (*Amphioxus,* now called *Branchiostoma*) and **urochordates** (tunicates or ascidians). Chordates and hemichordates (pterobranchs and acorn worms), diverged from each other more than 500 million years ago. Fossils of early invertebrate chordates are very scanty and have not yet provided definitive clues as to the origin of the vertebrates.

A model of the earliest vertebrates can be reconstructed from the recognized plesiomorphic features for living vertebrate groups. The earliest vertebrates would thus have been small animals of a fish-like shape with a series of gill slits in the pharynx and segmental body musculature. They would have moved through the water as the result of alternating waves of muscle contractions. These animals had a notochord, derived from the roof of the archenteron (the primitive gut), which is formed during gastrulation, and a dorsal hollow nerve cord. The nerve cord was expanded at its rostral end. Sensory receptive structures for senses including terminal and/or olfactory, visual (retinal photoreceptive), pineal-parapineal photoreceptive, taste, ocataval, and lateral line were present in the early vertebrates, but not all were evolved simultaneously. Finally, whereas in invertebrates, elaboration of the nervous system is generally accomplished by increasing the degree of complexity of individual neurons, elaboration of the nervous system in the earliest vertebrates was characterized by an increase in the number of neurons. Among the living adult invertebrate chordates and the hemichordates, the cephalochordate *Branchiostoma* most closely resembles this model.

One of the groups of deuterostome relatives of *Branchiostoma,* the echinoderms, comprise starfishes, brittle stars, sea urchins and sand dollars, sea lilies and feather stars, and sea cucumbers. In all echinoderms, the nervous system is locally organized. No central component, such as a brain or collection of central ganglia, is present.

In the worm-like hemichordates, no notochord or dorsal hollow nerve cord have been identified. A nerve net is present in the epidermis and contains sensory neurons, interneurons, and some motor neurons. The nerve net is thickened to form a dorsal, solid nerve cord in the proboscis and both dorsal and ventral nerve cords in the trunk. A **neurocord,** which does have a central lumen, is formed by invaginated nervous tissue in the intermediately located collar and connects the dorsal nerve cords of the proboscis and the trunk. The neurocord contains motor neurons and interneurons and may be homologous to the dorsal hollow nerve cord of other chordates. In the neurocord and dorsal nerve cord of the trunk is a collection of giant nerve cells, which are thought to play a role in rapid motor responses.

Among the urochordates, tunicate larvae have the best developed nervous systems. A **cerebral vesicle** is present in the rostral part of the body. It contains a structure called the **statocyst,** which is sensitive to gravity and a very simple eye called an **ocellus.** The cerebral vesicle is continuous with a neural tube that extends through the tail of the larva and lies dorsal to the notochord. The statocyst and ocellus are used to

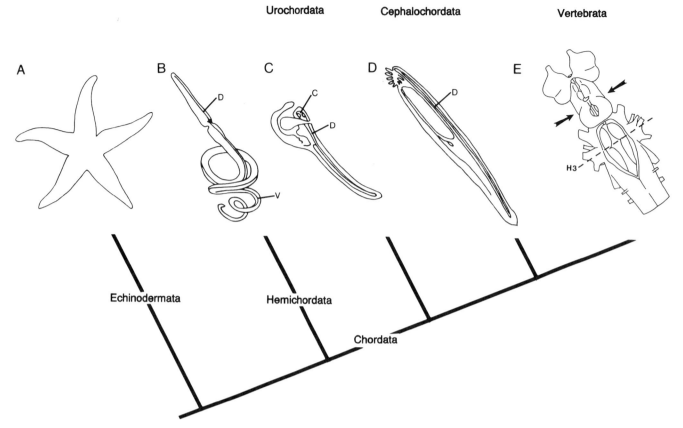

FIGURE 31-1. Dendrogram modeling central nervous system evolution in Deuterostoma. A whole starfish (echinoderm), whole acorn worm (hemichordate), sagittally sectioned tunicate larva (urochordate), sagittally sectioned *Branchiostoma* (cephalochordate), and the brain of a vertebrate (*Ichthyomyzon unicuspis*) are shown. In the latter, the rostral expression boundary of the homeobox gene *Hox*-3 is indicated by the dashed line, and the rostral extent of the midbrain is indicated by the paired arrows. Abbreviations: C, cerebral vesicle; D, dorsal nerve cord; v, ventral nerve cord.

guide the larva toward gravity and away from light in order to locate a suitable place for it to attach and begin metamorphosis to the adult form.

In adults of the cephalochordate *Branchiostoma*, a **dorsal hollow nerve cord** is present dorsal to the notochord and can be divided into a dorsolateral sensory zone and a ventral motor zone. Ocelli, which are each formed by only one ganglion cell and one pigmented cell, are distributed along the dorsal nerve cord. At the rostral end of the dorsal nerve cord is a cerebral vesicle. The dorsal roof of the posterior part of the cerebral vesicle contains cells with cilia that produce stacks of lamellae and may be photoreceptive; this set of cells is known as the **lamellar body.** At the junction between the posterior and anterior parts of the cerebral vesicle is a cluster of **infundibular cells** in the ventral part of the vesicle. A cluster of pigment-containing cells and associated flask-shaped and nerve-like cells, called the **frontal eye,** lies at the rostral end of the cerebral vesicle. A pair of **rostral nerves** arises from within the cerebral vesicle, and a segmented series of **dorsal and ventral spinal nerves** arises from the dorsal nerve cord. We will consider these and additional anatomical features of the central nervous system of developing larva of *Branchiostoma* in greater detail below.

Two basic features characterize the organization of the nervous system of chordates. The first of these features involves the regional control of cell proliferation in the radial dimension. In some nonchordate invertebrates, the nervous system is diffusely distributed over the body as a series of ganglia or an interconnected nerve net, whereas in most chordates (or their larval forms), differential regional expansion of a specific radial sector of the nervous system occurs. As a result of induction by the roof of the archenteron, the dorsal part of the nerve net in invertebrate chordates thickens to form the dorsal nerve cord or tube, as does the neural plate in vertebrates. This differential regional expansion reflects a localized and spatially limited increase in the amount of proliferation of nervous tissue.

The second of these features involves the regional control of cell proliferation in the rostrocaudal dimension. The chordate nervous system is characterized by a rostral expansion of the dorsal nervous system, that is, the brain, and this key feature is initially influenced during gastrulation by a caudorostral sequence of induction, brought about by the progressively rostral migration of the roof of the archenteron from its opening, the blastopore. Founder cell populations of neurons are established in the neural plate tissue lying dorsal to the archenteron roof that will each give rise to a specific part of the nervous system.

As discussed in Chapter 9, the process of further regional speci-fication and development has recently been found to be con-trolled by homeobox genes in chordates, as in a number of nonchordate invertebrates. The segmental, or neuromeric, or-ganization of the brain is a manifestation of the homeobox gene specification of regional development. As in the case of the radially localized formation of a dorsal nerve cord or tube, the rostrocaudal parts of the chordate nervous system reflect localized and spatially limited increases (or decreases) in the amount of proliferation of nervous tissue.

A corollary of the feature of longitudinal differentiation of the central nervous system is the occurrence of changes that affected methods of propulsion that characterized early chor-dates. Whereas some nonchordate invertebrates use cilia to move their bodies through the water, locomotion in chordate invertebrates (or their larvae) is the result of muscular contrac-tions regulated by longitudinally organized nerve fibers. In *Branchiostoma*, motor responses to stimuli are subject to con-trol by the brain due to the rostrocaudally organized, functional differentiation of the nervous system. Longitudinal transmission of information within the nervous system and the presence of rostrocaudally localized areas of integration and control are keystones of the chordate nervous system.

ANCESTRAL CHORDATES

Several attempts have been made to reconstruct a morpho-type of the brain in the earliest vertebrates, but these efforts have been hampered by a relative paucity of information on the central nervous system of the closest outgroup to verte-brates, the cephalochordate *Branchiostoma*. The marked differ-ences that exist between the brains of hagfishes and lampreys have left important questions unanswered as to the plesiomor-phic vertebrate condition for a number of traits. Recent findings on developing gastrula and larvae of *Branchiostoma* have dra-matically increased our understanding of this cephalochordate, however, and allow for new insights into how the brain of vertebrates evolved.

The adult form of *Branchiostoma* was previously used as an approximate model of the earliest vertebrates. This model implied that the central nervous system in early vertebrates consisted of a spinal cord with a slightly expanded rostral end, presumed to be a forerunner of the hindbrain. This model also implied that the vertebrate midbrain and forebrain arose evolutionarily as further rostral condensations of an epidermal nerve net.

E. Gilland and R. Baker recently proposed that the earliest vertebrates may have resembled the primary gastrula of *Branchiostoma,* as discussed in Chapter 9 (see Fig. 9-7), rather than the adult. They postulated that the entire primary gastrula of an ancestral chordate, resembling that of extant *Branchios-toma* [Fig. 31-2(A)] is homologous to the head of vertebrates. This possibility derives from the findings of P. Holland *et al.* on the location of the rostral expression border of the homeo-box gene *AmphiHox* 3 [H3, Fig. 31-2(A)] in the developing *Branchiostoma* gastrula, which appears to be a homologue of the *Hox* paralog group 3 (H3) in vertebrates. Gilland and Baker hypothesized that the first five somites (segmented mesoderm) of *Branchiostoma* are homologous with the somitomeric region

of jawed vertebrates (see Fig. 9-9) and that the head of verte-brates is thus homologous with the entire primary gastrula of the ancestral chordate stock. Developing ancestral vertebrates resembling such a gastrula would therefore have had a central nervous system that contained a midbrain, hindbrain, and spinal cord, but the prosencephalon, the paired sense organs, and the anterior part of the head would have been newly aquired with the evolution of the first vertebrates.

A more recent study by T. Lacalli, N. Holland, and J. West provides a detailed anatomical analysis of the larvae of *Branchiostoma* [Fig. 31–2(B and C)] on which out-group com-parisons with vertebrates can be based and which allows further insights of fundamental importance for our understanding of central nervous system evolution. In the larvae, four somites are present rostral to the rostral expression border of *Hox*-3 (H3) [(Fig. 31-2(B)]. The central nervous system lies medial to the somites, and several of its landmarks are projected onto the section in Figure 31-2(B) for reference. The part of the central nervous system that lies medial to most of the first somite is shown in Figure 31-2(C), expanded in scale as indicated by the thicker dashed lines.

The rostral end of the nerve cord contains an expanded portion formed by ciliated cells, the cerebral vesicle, most of which is shown in Figure 31-2(C), indicated by the shading. The cerebral vesicle contains a **central canal** that opens to the surface through a **neuropore;** it also contains an **anterior pigment spot** at its rostral end. The anterior pigment spot is formed by the pigment in the caudal ends of several pigment cells; these pigment cells and associated, putative receptor and nerve cells constitute the frontal eye. A pair of rostral nerves join the cerebral vesicle on its ventral aspect. The central canal narrows, as indicated by the arrow in the figure, and at the same rostrocaudal level, a cluster of cells that form the infundib-ulum are present in the ventral part of the cerebral vesicle. These cells secrete a noncellular strand called **Reissner's fiber,** which extends down the central canal. The narrowing of the central canal and the position of the infundibulum divide the cerebral vesicle into anterior and posterior portions. The poste-rior portion of the vesicle contains a large structure, the lamellar body, composed of the membranous lamellae in parallel stacks formed by cilia of the cells in this region.

Lacalli *et al.* discussed the putative homology of these features with structures in vertebrates. Due to multiple structural and histochemical similarities, the unpaired frontal eye of *Branchiostoma* appears to be homologous to the retinas of the paired eyes of vertebrates. The infundibular cells appear to be homologous to the infundibulum of vertebrates; infundibular cells in embryonic fish occupy the same position and also secrete a Reissner's fiber. The lamellar body appears to be the homologue of the vertebrate pineal–parapineal complex, due to positional and ultrastructural similarities. The rostral nerves may be homologues of the terminal and/or olfactory nerves, but additional study of them needs to be done.

This analysis reveals that the cerebral vesicle contains struc-tures that appear to be homologues of the vertebrate paired retinas, infundibulum, and pineal–parapineal complex, all of which lie within or are derived from the diencephalon. The rostral nerves may be homologues of rostral (olfactory and/or terminal) telencephalic nerves of vertebrates. Thus, the cerebral

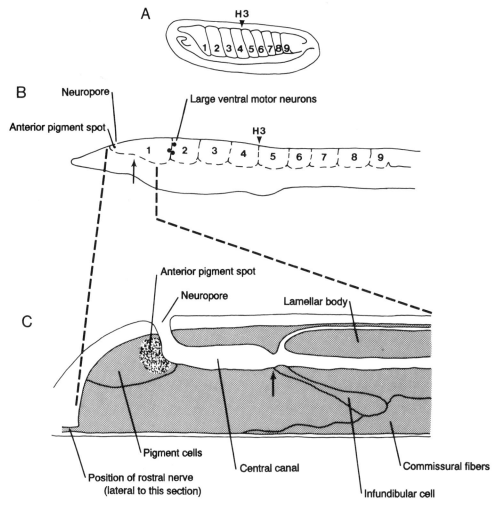

FIGURE 31-2. Drawings of semischematic parasagittal sections through an early developmental stage of *Branchiostoma* (A, also shown in Fig. 9-7) and a larva of *Branchiostoma* (B and C). Rostral is toward the left. Somites are numbered 1–9 in A and B, and the rostral expression limit of *Hox-3* is indicated by H3 in A and B. In C, an enlargement of the part of the nervous system medial to somite 1 is shown, with the cerebral vesicle indicated by shading. A adapted from Gilland and Baker (1993) and used with permission of S. Karger AG. Data for B and C from Lacalli *et al.* (1994).

vesicle appears to be homologous to the diencephalon, and possibly also the telencephalon, of vertebrates.

Immediately caudal to the cerebral vesicle are several large ventral motor neurons [Fig. 31-2(B)]. Lacalli *et al.* postulated that these may be homologous to the anterior-most of the ventral motor neurons in vertebrates, that is, the oculomotor neurons that innervate the muscles of the eye. If this comparison is correct, the region in which these ventral motor neurons lie in *Branchiostoma* would be the homologue of the vertebrate midbrain.

If we now reconsider the somites in *Branchiostoma* and their possible correspondence to somitomeres and somites in vertebrates, a new model of brain and head evolution emerges, as shown in Figure 31-3 and summarized in Table 31-1. We present this model with the strong caution that it is highly speculative and should be taken only as one of a number of possible solutions to the problem. Data yet to be obtained are needed to clarify the many complex anatomical relationships

involved here. For this new model, Figure 31-3 , which can be compared with Figure 9-9, shows the developing neural tube and rhombomeres (A), the somitomeres and somites (B), and the head segments (C) in a developing vertebrate, in comparison with the somites in a *Branchiostoma* larva (D). Table 31-1, which can be compared with Table 9-2, shows the hypothesized relationships of somites in *Branchiostoma* with somitic organization, head segmentation, and cranial nerve organization in vertebrates.

The cerebral vesicle in *Branchiostoma* is aligned with the first somite and appears to be the homologue of the vertebrate diencephalon (and perhaps telencephalon). No somitomere is present in vertebrates that would correspond to the first somite in *Branchiostoma;* whether a corresponding somitic entity was present in the common ancestral stock of cephalochordates and vertebrates is unknown. If the anterior-most ventral motor neurons in *Branchiostoma* [indicated by the large dots in Fig. 31-3(D)] are homologues of vertebrate oculomotor neurons,

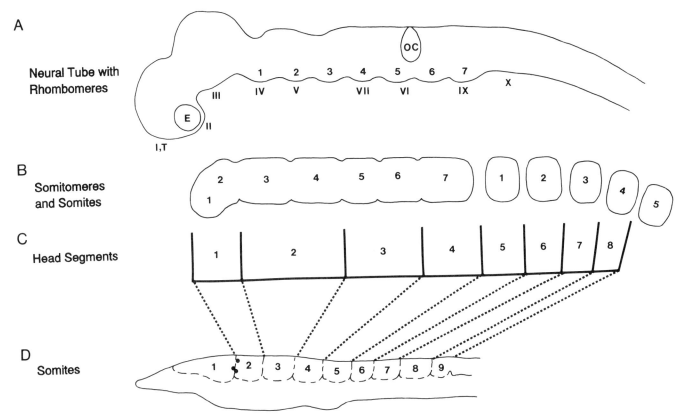

FIGURE 31-3. Speculative model of brain and head evolution. Schematic drawing of parasagittal sections in a developing vertebrate through (A) the neural tube and rhombomeres (1–7) and (B) the somitomeres (1–7) and somites (1–5), with corresponding head segments (1–8) indicated in (C). In A, the positions of some of the developing cranial nerves are indicated by Roman numerals and T (terminal nerve) and the position of the optic capsule by OC and the eye by E. A–C are similar to parts of Figure 9-9 (but with rostral to the left instead of to the right). (D) Schematic drawing of a parasagittal section through a *Branchiostoma* larva, with the somites numbered 1–9. Large dots at the border of somites 1 and 2 represent the position (more medially) of the most anterior of the large ventral motor neurons. The dotted lines indicate the correspondence of the somites in *Branchiostoma* with head segments and their respective somitomeres and somites in vertebrates, as hypothesized for this model.

TABLE 31-1. Speculative Correspondences in Chordate Brain and Head Evolution			
Branchiostoma Somite	Vertebrate Head Segment	Vertebrate Somitomere or Somite	Vertebrate Cranial Nerves
1			I, T, VN, II, and Epiphyseal
2	1	Somitomeres 1, 2	Profundus, III, LL_{AD}
3	2	3, 4	IV, V, LL_{AV}
4	3	5, 6	VI, VII_D, VIII, Otic, VII_{VL}
5	4	7	IX_D, LL_M, IX_{VL}
6	5	Somites 1	X_D, XII, LL_{ST}, LL_P, X_{VL}
7	6	2	X_D, X_{VL}, Occipital
8	7	3	X_D, X_{VL}, Occipital
9	8	4	X_D, X_{VL}, Occipital

and particularly of the neurons of cranial nerve III, then the rostral end of the midbrain in *Branchiostoma* would be aligned with the rostral end of its second somite, which would then correspond to the rostral end of the first head segment and the somitomeric region in vertebrates. The position of the rostral expression limit of *Hox-3* is at the caudal border of the fourth somite. Thus, somites 2, 3, and 4 would correspond to the somitomeric part of the head in vertebrates that comprises the first three head segments and encompasses the three oculomotor cranial nerves.

In this hypothesized model, each of the somites 2, 3, and 4 in *Branchiostoma,* would be homologous to two somitomeres in developing vertebrates: 1 and 2, 3 and 4, and 5 and 6, respectively. Somites 2–9, as a group, in *Branchiostoma* would be homologous to the somitomeric region plus the first four somites of vertebrates, and somites 5–9 in *Branchiostoma* would be individually homologous to somitomere 7 and somites 1–4 in vertebrates, respectively. These relationships are indicated with the dotted lines in Figure 31-3.

By comparing the putative homologues of parts of the brain present in *Branchiostoma* with the brain in vertebrates, some of the features of the brain in the common ancestral stock of cephalochordates and vertebrates can be identified. The brain of the common ancestor would have had the four basic divisions of the central nervous system: forebrain, midbrain, hindbrain, and spinal cord. Whether a well-developed telencephalon was present in the forebrain, other than a small number of neurons in receipt of the rostral or terminal/olfactory nerve input, cannot be determined. A diencephalon would have been present, with at least a retinal apparatus, an infundibulum, and a photoreceptive pineal–parapineal complex. The retinal apparatus was probably unpaired (as will be further discussed below). Both a midbrain and hindbrain would have been present. The rostral extent of ventral motor neurons would have marked the rostral border of the midbrain. Whether the notochord would have extended rostral to the rostral border of the brain, as it does in *Branchiostoma,* or have ended caudal to the level of the forebrain, as it does in vertebrates, cannot currently be determined.

THE EARLIEST VERTEBRATES

Vertebrates are uniquely characterized by the presence of two ectodermal tissues, neural crest and ectodermal placodes, which directly give rise to or are crucial for the development of the anterior part of the head and the special sense organs for olfaction, vision, taste, hearing, and the lateral line system. The origin of vertebrates was an event that capitalized on the evolution of neural crest and placodes in conjunction with enlargement of the brain. The same tissue that gives rise to the epidermal nervous system of extant invertebrate chordates appears to give rise to the central nervous system itself and to the two additional novel tissues in vertebrates. Of these two new tissues, the neural crest may have been the first to have been evolved, and its presence may have been the defining event of the divergence of vertebrates from their invertebrate ancestral stock.

Two Novel Tissues: Neural Crest and Placodes

As discussed in Chapter 9, the neural crest arises during gastrulation in the neural fold that borders the neural tube (see Fig. 9-5). It not only gives rise to neural tissue but also gives rise to the muscles, bone, and connective tissues of the rostral (preotic) part of the head in vertebrates, which lies anterior to the anterior end of the notochord. The new head was correlated with changes in the propulsion system and with a shift to a more actively predatory mode of life. Thus formed, it was also available to house a rostrally expanded brain and special sense organs.

Placodes are the areas of surface ectoderm that overlie migrated neural crest tissue and produce neural tissue for most of the special sense organs. Different placodes make different contributions to different sense organs. Thus, the various placodes do not appear to simply be serially homologous derivatives of an initially continuous column of placodal tissue and may not have all evolved at the same time. Placodal tissue and the sensory cells to which it gives rise are thought, however, to be homologous to the epidermal nerve net of invertebrates, as derivatives of the same ectodermal nervous tissue. The sensory neurons of the invertebrate epidermal nerve net are ciliated, as are the cells that form the cerebral vesicle in *Branchiostoma,* and the receptors of the paired sense organs that derive from placodal tissue in vertebrates are ciliated during their development.

Development of the neural plate itself is induced by the notochord that forms from the archenteron roof, along with somites and the prechordal plate (mesoderm rostral to the notochord). The induced neural plate can in turn affect other tissue in an inductive capacity. Whether neural crest induces the development of the forebrain and the development of placodes remains to be established, but circumstantial evidence favors this possibility. Neural crest tissue is present in the region of the developing telencephalon, and removal of the neural folds early in development can preclude forebrain development. Neural crest cells migrate between the surface ectoderm and the deeper paraxial mesoderm and are present beneath the parts of the surface ectoderm that form placodes before placodal differentiation and production of neural elements begins.

Thus, a mutation that resulted in the formation of the novel tissue of neural crest during embryological development—as a differentiation of the dorsolateral part (neural fold) of the existing dorsal nerve cord—could have been selected for as a result of its contribution to head formation in correlation with a more mobile and actively predatory repertoire. The development of placodes and consequently of paired special sense organs for distance reception and the development of the forebrain for analysis, integration, and retention of the sensory inputs might thus be fortuitous side effects of the presence of neural crest during development.

Sensory System Evolution

From the model of a common ancestral chordate of both cephalochordates and vertebrates, we can postulate that the earliest vertebrates would have had a pair of rostral nerves terminating in the telencephalon. The identity of these nerves as either olfactory or terminal remains to be confirmed. While paired olfactory nerves are present in all extant vertebrate groups, paired terminal nerves with LHRH-positive (luteinizing hormone releasing hormone) neurons are present only in jawed vertebrates; terminal nerve-like neurons are present in lampreys but do not contain LHRH, and a terminal nerve has not been found in hagfishes. Determination of the identity of the rostral nerves in *Branchiostoma* will thus help to clarify the evolutionary history of the telencephalic nerves of vertebrates.

A retinal structure, either single or paired, and associated with the diencephalon, would have been present in the earliest vertebrates. Lacalli *et al.* suggested that the unpaired, retinal-like, frontal eye of *Branchiostoma* may have been derived from an unpaired apical organ such as that present in an auricularia—the larval form of one of the classes of echinoderms, the sea cucumbers. In this case, an unpaired retinal-like photoreceptor would be plesiomorphic for chordates and the paired retinal photoreceptors an apomorphy in the earliest vertebrates.

Structures associated with the retina in vertebrates, such as the cornea, lens, and iris, are present in lampreys and jawed vertebrates but absent in hagfishes and in all invertebrate chordates. Thus, they may have evolved in the common ancestral stock of lampreys and jawed vertebrates rather than in that of all vertebrates. Extraocular eye muscles and their cranial nerves are also present among extant vertebrates in the same distribution; however, evidence of the presence of these muscles and nerves in fossils of osteostracans and the presence of neurons in the brain of *Branchiostoma* that are the putative homologues of vertebrate oculomotor neurons raise the possibility that muscles associated with the retinal apparatus and their innervating ventral motor neurons were present in the earliest vertebrates and have subsequently been lost in hagfishes.

The earliest vertebrates had an unpaired, median, pineal–parapineal photoreceptor apparatus in the roof of the diencephalon. A pineal–parapineal complex is plesiomorphic for jawed vertebrates. Furthermore, this structure is present in lampreys and absent in hagfishes, and the putatively homologous lamellar body is present in *Branchiostoma*. This distribution suggests that a pineal–parapineal apparatus is plesiomorphic for at least the common ancestral stock of cephalochordates and vertebrates, and its absence in hagfishes is an apomorphy.

The lateral line senses of electroreception and mechanoreception were evolved early in vertebrate evolution. No evidence for these systems have been found in invertebrate chordates (although sensory papillae that detect vibrations and/or water flow are present in the invertebrate chaetognaths). The placodal tissue for both lateral line senses may have evolved in conjunction with the early vertebrate feature of the deposition of calcium salts in the dentine, which enhance electroreception. Furthermore, depressions and canal systems that could have effectively shielded lateral line electroreceptors have been found in the earliest fishes known from the fossil record. Such shielding is essential for perceiving the direction of the stimulus source. Extant hagfishes appear to lack electroreception and may thus have lost the apparatus for it over the course of their evolution.

Otic organs for vestibular sense and hearing characterize all major groups of living vertebrates, even though not all components, such as horizontal semicircular canals, are present in all vertebrates. No trace of any comparable sense is present in invertebrate chordates. Similarly, the chemosensory system of taste and the profundal and trigeminal sensory systems are common features of vertebrates. The timing of the evolution of these various senses relative to olfactory, retinal, and pineal–parapineal senses is unknown, but all placodal sensory systems clearly evolved relatively early in the vertebrate lineage.

Organization of the Brain

At this early stage in vertebrate evolution, most of the major divisions of forebrain, midbrain, hindbrain, and spinal cord were present, as were all of the major sensory and motor systems organized for active propulsion. Most of the neuron cell bodies in the brains of these early vertebrates would have been in a periventricular position, rather than migrated laterally (centrifugally) within the various parts of the brain.

Due to the influence of placodal tissue, the earliest vertebrates would very probably have had an expanded telencephalon. Recent evidence suggests that the presence of paired olfactory organs that are connected to the forebrain is necessary for the development of normal telencephalic lobes. The neural plate cells that give rise to the definitive telencephalon thus appear to require the influence of olfactory placodal tissue to develop normally. With the development of the olfactory inputs to the telencephalon, the presence of an olfactory pallium, as is present in extant vertebrates, can be postulated. That the entire pallium was olfactory in the earliest vertebrates (as it is in hagfishes) is probable, with olfactory projections being limited to the lateral pallial region only in gnathostomes.

In all vertebrates, a diencephalon is present that contains an infundibulum and also gives rise to the retina. A diencephalon with these two structures can also be recognized in cephalochordates and is thus plesiomorphic for the common ancestor of cephalochordates and vertebrates. A pineal–parapineal apparatus is present in all vertebrates except hagfishes and is present in *Branchiostoma*. This feature of the diencephalon is thus also plesiomorphic for the common ancestor of cephalochordates and vertebrates. Other portions of the diencephalon appear to be unique to vertebrates. These include an epithalamus with paired, asymmetrical habenulae, a dorsal thalamus, probably a ventral thalamus, and ventrally, a preoptic area and hypothalamus.

In jawed vertebrates, the dorsal thalamus is composed of two divisions, the lemnothalamus and the collothalamus. A nucleus anterior that may constitute the lemnothalamus has been identified in both lampreys and hagfishes, but further study of the dorsal thalamus of lampreys and hagfishes is needed to determine whether the two divisions were plesiomorphic for all vertebrates. However this question is resolved, a lemnothalamus–medial pallium system may have been one of the earliest ascending systems established to the telencephalic pallium. A striatal area in the telencephalon has been identified in both lampreys and hagfishes, implying that the striatum is plesiomorphic for vertebrates; thus, a collothalamus relaying visual and other sensory information from the roof of the midbrain to the striatum may have been present in the earliest vertebrates. Finally, the possible presence of cell groups homologous to the preglomerular nuclear complex in the caudal diencephalon of jawless vertebrates remains to be investigated, but the presence of receptors for the senses of taste and lateral line mechanoreception (and electroreception in lampreys) in these vertebrates suggests that cell groups relaying ascending information from these systems may have been present in the earliest vertebrates.

A midbrain containing both a roof (tectum) and a tegmental region is plesiomorphic for vertebrates. The only putatively homologous part of the midbrain yet identified in *Branchiostoma* is the ventral motor cells. The midbrain sensory roof may thus be unique to vertebrates.

The hindbrain is smaller in hagfishes than in lampreys and jawed vertebrates, and fossil evidence suggests that it was small plesiomorphically for vertebrates. As the lateral line and octaval sensory receptors evolved, they can be postulated to have entered the hindbrain and synapsed in a localized region within it. This hindbrain region would probably have projected rostrally to the roof of the midbrain. Since the roof of the midbrain is devoted to a spatially organized anlaysis of sensory information in all extant vertebrates, it is highly probable that such a spatial map was present in the midbrain roof in these earliest vertebrates. The lateral line and octaval systems contribute to the localization of predators and prey, and a spatial analysis of the sensory information in the midbrain roof would have provided the basis for the profitable use of the sensory information by the animal.

Little information is available on the early evolution of motor systems and the descending pathways that form them. A striatal region of the telencephalon and both the roof and tegmentum of the midbrain may have constituted some of the earliest motor-related parts of the brain rostral to the actual motor nuclei of cranial and spinal nerves.

In summary, olfactory (and/or terminal), retinal photoreceptive, and pineal–parapineal photoreceptive senses were probably the earliest senses established and occurred in the common ancestor of cephalochordates and vertebrates, as did the major subdivisions of the brain: a forebrain with a diencephalon and possibly a telencephalon as well, a midbrain, a hindbrain, and a spinal cord. In vertebrates, electroreceptive and mechanoreceptive lateral line, taste, auditory, vestibular, trigeminal, and profundal sensory inputs were established, although not all were evolved simultaneously. Some of the major ascending sensory pathways and descending motor pathways were established. Ascending serotoninergic projections from the raphe and descending projections to the spinal cord from the reticular formation and possibly also from vestibular, trigeminal, and solitary nuclei are among the pathways that characterized the brainstem in early vertebrates.

THE ADVENT OF JAWS

The evolution of jaws that could be used for the capture and initial processing of prey into the digestive system was of seminal significance in vertebrate history. In these animals, development of the postotic skull and true, bony vertebrae were also established. The forebrain, midbrain, and hindbrain in gnathostomes are all characterized by further developments and specializations related directly or indirectly to the advantages gained by the much more mobile and actively predatory lifestyle. We will survey a number of the features of the brain that evolved within the various radiations of gnathostomes and

then consider the various mechanisms that underlie their evolution.

With increased mobility for both prey seeking and predator avoidance, changes that enhanced control of the motor system were strongly selected for. Such enhanced control was partly achieved through the development of the cerebellum in the roof, or rhombic lip, of the hindbrain. Early vertebrates had either only a small cerebellum or no cerebellum at all, with the rhombic lip poorly developed. This part of the brain may have consisted only of an emminentia granularis, a cell population associated with the cerebellum itself in gnathostomes, or of an emminentia granularis and a small cerebellum. A marked increase of cell proliferation in the rhombic lip occurs during development in gnathostomes and results in the formation of the cerebellum.

The midbrain in early vertebrates would already have been developed to the extent of relaying lateral line, octaval, and visual information to the forebrain. It would have been organized as a topographic map of external space and would also have played a role in the organization of oriented motor responses to the sensory stimuli. The midbrain roof did not undergo marked changes in organization with the development of jaws, but within different radiations of gnathostomes, this part of the brain was expanded markedly in size and was elaborated in terms of its cytoarchitectonic organization. Its usefulness for spatial analysis and orientation to stimuli continued to be selected for and thereby augmented in various groups.

Within ray-finned and cartilaginous fishes, the diencephalon is characterized more by development of the pretectum and posterior tuberculum than by development of the dorsal thalamus. The opposite condition—greater development of the dorsal thalamus than pretectal and posterior tubercular regions—characterizes amniotes, as we will consider below. The pretectum is elaborately developed particularly in some teleost fishes, in which a variety of pathways relay visual information through a varying series of pretectal nuclei to various sites in the brainstem. One pretectal pathway relays visual information to the inferior lobe of the hypothalamus, which participates in the motor control of feeding behavior.

In both cartilaginous and ray-finned fishes, the posterior tuberculum gives rise during development to a number of laterally migrated nuclei (the preglomerular nuclear complex of ray-finned fishes) that relay ascending lateral line and gustatory information to the forebrain, particularly to the (topologically) medial pallium. The dorsal thalamus in these groups relays some sensory information to the telencephalon but is more simply organized, consisting of only three periventricular nuclei.

While the dorsal thalamus is not markedly elaborated in cartilaginous and bony fishes, the telencephalon is greatly expanded in some groups within both of these radiations. The telencephalon is also large and complex in hagfishes. Within amniotes, both the dorsal thalamus and the telencephalon are expanded and complex in mammals and in some diapsid reptiles and birds. These increases in the size of the forebrain have occurred independently. Within each of these various groups of vertebrates, the increase in the size and complexity of the forebrain may be related to actively predacious lifestyles in some cases but may also be correlated with such functions as more complex analysis of sensory information, learning and memory, and social, conspecific interactive behaviors.

While the dorsal thalamus of fishes is not elaborated, it does consist of two parts, the lemnothalamus predominantly in receipt of lemniscal sensory inputs, and the collothalamus predominantly in receipt of sensory inputs relayed through the roof of the midbrain. The lemnothalamus projects to the (topological) medial pallium and to what has been identified as the dorsal pallium, while the collothalamus projects to the striatum. In some cartilaginous and ray-finned fishes, projections of both divisions of the dorsal thalamus also terminate in a pallial region that lies between the (topological) medial and lateral pallia. In sharks, this region is composed of at least two major parts, a dorsal laminar zone and a more ventrally lying zone called the central nucleus. Similarly located regions are present in the telencephalon of many teleosts, particularly in some of the euteleosts in which the telencephalon is strikingly enlarged and elaborated. That these dorsally lying pallial areas in cartilaginous and ray-finned fishes are homologous to the dorsal pallium of amniotes is dubious, however. Immunohistochemical studies in particular suggest that they have more probably evolved independently along with the independent expansion of the telencephalon.

ONTO THE LAND

The conquest of the terrestrial environment was, like the acquisition of jaws, a momentous event in vertebrate history. It allowed for a plethora of new selective pressures to operate on variations in sensory input systems, information processing areas, and motor control systems in the central nervous system and on their peripheral musculoskeletal counterparts.

In all tetrapods, a dorsal pallial region is present between the medial and lateral pallia, although the evolutionary relationship between the dorsal and lateral pallial areas in amphibians and what is recognized as the dorsal pallium in amniotes is unclear. In ancestral amniotes, two major changes occurred in the development of both the dorsal thalamus and the dorsal pallium that were strongly selected for. The dorsal thalamus underwent a dual elaboration, that is, in both the lemnothalamus and collothalamus, regionally specific cell proliferation was increased during development in correlation with the lateral (centrifugal) migration of neuron cell bodies. This change resulted in the presence of multiple, migrated nuclei within both parts of the dorsal thalamus. In ancestral amniotes, projections from the lemnothalamus to the medial pallium and to the dorsal pallium were present. Projections from the collothalamus to the striatum were also present. In addition, projections from the collothalamus to the dorsal pallium were acquired. Furthermore, ascending somatosensory projections via the dorsal column nuclei to the dorsal thalamus, both directly and relayed through the roof of the midbrain, were also either acquired or greatly enhanced at this time.

In correlation with the dual elaboration of the dorsal thalamus, a dual expansion of the dorsal pallium occurred, so that both the medial, lemnothalamic-recipient and the lateral, collothalamic-recipient divisions of the dorsal pallium were expanded by regionally specific increases in cell proliferation. In the synapsid line that led to extant mammals and in the common

ancestors of diapsids and turtles, further expansions of these two dorsal pallial divisions continued independently.

In the synapsid line, an increase in the radial organization of the neurons in both divisions of the dorsal pallium occurred, resulting in an "inside-out sequence" in the centrifugal migration of neurons during development, with the consequent formation of the six layers present in most of the dorsal pallial cortical region. Increases in cell proliferation also resulted in the presence of new cell types as defined by their immunohistochemical profiles, particularly in layers II–IV of the isocortex. In various separate lineages of mammals, other mutational events produced repeated duplications of various sensory cortical areas, resulting in multiple sensory representations.

Further mutational events in mammals resulted in a number of new organizational features and structures. The medial pallium was further expanded. A number of ascending connections from the dorsal thalamus were lost, particularly some of those from the contralateral part of the lemnothalamus. Long descending projections from the isocortex and interhemispheric connections via the corpus callosum were gained.

In the nonsynapsid line, marked increases in the area of the lateral division of dorsal pallium occurred, with the consequent formation of the anterior dorsal ventricular ridge. Expansion of the medial division was relatively minor. An increase in radial organization of the proliferated neurons did not occur in either division, and thus the dorsal ventricular ridge is predominantly characterized by a single cell lamina or by nuclear groups rather than by multiple cortical laminae.

In the diapsid reptilian stock that gave rise to birds, an independent expansion of the medial division of the dorsal pallium occurred, resulting in the formation of the Wulst. Cells with immunohistochemical phenotypes similar to some of those found in layers II–IV of mammalian isocortex were also independently evolved in the dorsal pallium. Regional increases in cell proliferation during embryological development that resulted in an enlarged medial pallium were selected for in birds as well; this phenomenon is particularly evident in birds that have sequestered stores of food for the winter to which they must repeatedly return to stock and to feed from.

The dual elaboration of both parts of the dorsal thalamus and the dual expansion of both divisions of the dorsal pallium that occurred in ancestral amniotes was strongly selected for. The massive increase in the amount of sensory information reaching the dorsal pallium was of great advantage for survival in a terrestrial environment. In correlation with the increased analysis of sensory information, the value of learning and memory is reflected in the further development of the medial pallium, particularly in mammals.

In ancestral amniotes, a new set of selective pressures also affected the evolution of somatosensory and motor control systems for survival in the terrestrial environment. These systems have been elaborated to some extent independently in the synapsid line and in birds. In mammals, sensorimotor cortex, a part of the medial (lemnopallial) division of the dorsal pallium, is well developed and distinct from other cortical areas. Similarly, in birds, the somatosensory part of the Wulst is well developed, and a more caudal, superficially lying region in the telencephalon may play a major role in motor control via influences on the striatum.

Other motor control pathways also vary among amniotes as a consequence of differential selective pressures. In mammals, selective pressures have favored motor control mediated by pathways from the striatum to the sensorimotor cortex via the dorsal thalamus and by long descending pathways to the brainstem and spinal cord from sensorimotor cortex. Pathways from the striatum to the optic tectum via the pretectum and substantia nigra are relatively minor. In nonsynapsid amniotes, selective pressures have favored striatal pathways to the optic tectum via the pretectum and/or the substantia nigra. Adaptation of the central nervous system in amniotes also encompasses examples of specialized evolution within a restricted group of species, such as has occurred in the development of complex descending motor pathways for vocal control in songbirds.

THEORIES OF VERTEBRATE BRAIN EVOLUTION

The invasion hypothesis of vertebrate brain evolution dominated the first half of the twentieth century. This hypothesis held that changes in central nervous system connections occur as a result of axon collaterals invading and forming new connections with a given group of neurons. In particular, the telencephalon was thought to have been dominated ancestrally by olfactory input and that other sensory systems later invaded the telencephalon and competed with olfactory fibers for synaptic sites, eventually establishing their own areas of pallial territory. This hypothesis may well contain at least some correct elements but for the wrong reasons. It was based on the belief that the brains of most extant anamniote vertebrates are predominantly in receipt of olfactory input and do not have the ascending sensory inputs from the nonolfactory senses that amniotes do. While olfaction was one of the earliest senses to evolve in vertebrates, the rest of the major sensory pathways to the telencephalon were also established relatively early in vertebrate history.

A more recently proposed hypothesis, Ebbesson's parcellation hypothesis, is in some respects the mirror image of the invasion hypothesis. The parcellation hypothesis holds that the brain has changed over evolution by the selective loss of connections and the concomitant segregation of neuronal populations. As opposed to new neuronal groups and connections being added in the brain over time, the parcellation hypothesis suggests that a more diffusely organized amalgam of neurons and connections goes through a process of attrition and sorting out over time, resulting in the discrete nuclei and connections present in extant vertebrate brains.

The equivalent cell hypothesis of Karten created a revolution in comparative neuroanatomy in the 1960s. This hypothesis derived from studies of ascending sensory pathways in birds and focused on the evolution of the telencephalon in amniotes. It holds that various cell populations that form nuclear groups within the dorsal ventricular ridge in birds and other nonsynapsid amniotes are comparable to groups of neurons within specific layers of mammalian isocortex that have similar respective connections. Furthermore, these cell populations in nonsynapsids and their respective comparable cell populations within isocortex in mammals were inherited from the common ancestral stock of amniotes and are thus respective homologous cell populations.

The problem of whether or not part of the dorsal pallium of ancestral amniotes was developed into a dorsal ventricular ridge or had an isocortical structure or neither has been the subject of continuing debate. Northcutt argued that expansion of the dorsal pallium occurred independently in synapsids and nonsynapsid amniotes and that parts of mammalian isocortex and the dorsal ventricular ridge are thus homoplastic.

A resolution of this question has been recently proposed by both Reiner and Butler. Each noted that, based on their distribution in extant amniotes and their absence in out-groups to amniotes, neither the dorsal ventricular ridge nor the lateral parts of isocortex were present in the common ancestral stock of amniotes. On the other hand, expansion of the dorsal pallium and the presence of certain ascending sensory pathways to specific parts of the dorsal pallium are common features of all amniotes and can thus be presumed to have characterized the common amniote stock. Thus, particular regions of the dorsal pallium in any given amniote are homologous to the respective comparable regions in other amniotes as derivatives of specific embryonic fields. The dorsal ventricular ridge and the lateral parts of isocortex are homoplastic as pallial derivatives with specific morphologies and some specific cell types.

HOW VERTEBRATE BRAINS EVOLVE

In this book, a multitude of differently evolved parts of the brain in diverse groups of vertebrates has been discussed. We can now generate at least a partial list of the variety of ways in which vertebrate brains evolve.

Induction is one of the most basic mechanisms involved in the evolution of the vertebrate central nervous system. The most salient example of this phenomenon is the origin of the vertebrate central nervous system *per se* from the nerve net-like condition of the earliest deuterostome ancestral stock. The localized condensation resulting from increased proliferation of ectodermal neural tissue along the dorsal sector of the rostrocaudal axis occurs as a result of induction by the mesoderm (the roof of the archenteron), and the caudorostral progression of the inductive process contributes to the regional specification of the nervous system. The differentiation of the neural crest, the formation of placodes, and the sensory system structures consequently induced account for some of the most momentous events in vertebrate central nervous system evolution.

A second basic mechanism of vertebrate brain evolution involves the rostrocaudal and dorsoventral specification of regions within the brain. How homeobox genes, present in the ancestral invertebrate chordate stock, shifted or extended their functional influence to specify a detailed neuromeric organization within the newly evolved central nervous system is one of the most important remaining questions in brain evolution. The neuromeric patterning established in the earliest vertebrates has been preserved in all descendant groups.

Invasion has occurred in various instances. In the earliest vertebrates, some of the new sensory systems that were developed as a result of the evolution of neural crest and placodes projected into the hindbrain and were relayed via the midbrain and diencephalon to the telencephalon. These new incoming, ascending projections and the synaptic territories that they established are perhaps the ultimate example of invasion. The evolutionary development in ancestral amniotes of new projections from the collothalamus to the dorsal pallium in addition to the existing projections to the striatum is a second example of invasion, and evidence suggests that this invasion occurred as a result of collateralization, which is one of the originally hypothesized mechanisms for invasion.

Parcellation has occurred in some instances. One of the best examples of this process may be in some of the changes associated with the elaboration of the lemnothalamus in ancestral amniotes. The single nucleus, nucleus anterior, present in anamniotes appears to be homologous as a field to multiple nuclei in the dorsal thalamus in amniotes. These multiple nuclei do not each have all of the afferent and efferent connections that the nucleus anterior has in anamniotes. The embryological development of the lemnothalamus in amniotes is such that connections and cell groups are segregated relative to the condition in anamniotes.

Loss of connections *per se,* that is, in the absence of the parcellated segregation of cell groups, has also occurred. The lemnothalamus gives rise to bilateral projections to the pallium in anamniotes. In amniotes, some of the projections to the contralateral side are maintained, but some have been lost. This loss has been most extensive in mammals.

Developmental differentiation of a specific embryonic field into multiple nuclei instead of just one nucleus has occurred in some instances. The evolution of the lemnothalamus in amniotes mentioned above is one example of this phenomenon. A second example occurs in the pretectum of acanthopterygian teleosts, where two nuclei, nuclei pretectalis superficialis pars intermedius and glomerulosus, are present instead of the single nucleus, nucleus posterior, which is present in most other groups of ray-finned fishes. Other examples of this phenomenon are the presence of new, unique nuclei within particular species or groups, such as the electromotor control systems of some electric fishes and the elaborate, descending motor pathways for vocal control in songbirds.

Duplication of neural regions has occurred in a number of instances, particularly in the evolution of isocortical sensory areas in various groups of mammals. Mutational events have occurred that have resulted in the production during development of multiple "copies" of both the lemnothalamic (VI) and collothalamic (VII) visual areas present in ancestral synapsids. Similar duplications may have occurred independently in birds, particularly in the collothalamic visual pallium.

The most frequent mechanism of central nervous system evolution may be regionally specific changes in neuron proliferation. While the specification of dorsoventral and rostrocaudal parts of the brain is regulated by homeobox gene function, differences in local neural proliferation may be under the control of other mutational events. Regional increases in proliferation usually occur in association with one of several related developmental phenomenon.

The continuation of proliferation paired with lateral migration of neurons away from the periventricular matrix accounts for the evolution of both the lemnothalamus and the collothalamus in amniotes and of the enlargement of the dorsal ventricular ridge in some diapsid reptiles and in birds. The evolutionary elaboration of the cerebellum in jawed vertebrates is a third example of this process.

The continuation of proliferation paired with migration of neurons in the presence of increased radial organization accounts for the change from an "outside-in" developmental sequence to the "inside-out" developmental sequence that produces the six isocortical layers present in most of the mammalian dorsal pallium. With a lesser degree of radial organization, nuclear groups result from neuronal migration, as is the case in the dorsal thalamus of all amniotes and in the dorsal ventricular ridge of nonsynapsid amniotes.

The continuation of proliferation paired with differentiation, as a result of mutational events, of the migrated neurons accounts for the presence of new phenotypes of neurons. An example of this phenomenon is the presence of neurons in some layers of mammalian isocortex that have neurotransmitter- and neuropeptide-specific profiles not found in the dorsal pallium of most nonsynapsid amniotes. Some neurons with mammalian-like phenotypes are present in birds, having been independently evolved as the result of similar mutational events.

Thus, evolution of the vertebrate central nervous system has occurred by the phenomena of:

- Induction.
- Homeobox gene patterning.
- Invasion.
- Parcellation.
- Loss of connections *per se.*
- Developmental differentiation of multiple nuclei
- Duplication of neural areas.
- Regionally specific changes in neuron proliferation paired with lateral or centrifugal migration, changes in the degree of radial organization, and/or differentiation of new neuronal phenotypes.

These phenomena can all be influenced by relatively simple mutational events that can thus become estabished in a population as the result of random variation. Selective pressures acting on a given population then determine whether the pheontypes produced by these random mutations increase their proportional representation within the population and eventually become established as the normal condition. The behavioral phenotypic expressions of central nervous system organization are the abilities for sensory processing, information storage, retrieval, and analysis, and motor response repertoires of the animal. The adaptive advantages conferred by an organized nervous system as opposed to a nerve net, by a variety of specific sensory input systems, by the gain of some new connections and the loss of some established connections, by the formation of multiple new nuclei through elaboration or duplication, and by regionally specific increases in cell proliferation in many different parts of the brain have determined the course of brain evolution among vertebrates.

FOR FURTHER READING

Gans, C. and Northcutt, R. G. (1983) Neural crest and the origin of vertebrates: a new head. *Science,* 220, 268–274.

Gilland, E. and Baker, R. (1993) Conservation of neuroepithelial and mesodermal segments in the embryonic vertebrate head. *Acta Anatomica*, 148, 110–123.

Kaas, J. H. (1993) Evolution of multiple areas and modules within neocortex. *Perspectives on Developmental Biology*, 1, 101–107.

Karten, H. J. (1991) Homology and evolutionary origins of the 'neocortex.' *Brain, Behavior and Evolution*, 38, 264–272.

Lacalli, T. C., Holland, N. D., and West, J. E. (1994) Landmarks in the anterior central nervous system of amphioxus larvae. *Philosophical Transactions of the Royal Society of London B*, 344, 165–185.

Northcutt, R. G. (1985) The brain and sense organs of the earliest vertebrates: reconstruction of a morphotype. In R. E. Foreman, A. Gorbman, J. M. Dodd, and R. Olsson (eds.), *Evolutionary Biology of Primitive Fishes*. New York: Plenum.

Puelles, L. and Rubenstein, J. L. R. (1993) Expression patterns of homeobox and other putative regulatory genes in the embryonic mouse forebrain suggest a neuromeric organization. *Trends in Neurosciences*, 16, 472–479.

ADDITIONAL REFERENCES

Aisemberg, G. O., Wysocka-Diller, J., Wong, V. Y., and Macagno, E. R. (1993) *Antennapedia*-class homeobox genes define diverse neuronal sets in the embryonic CNS of the leech. *Journal of Neurobiology*, 24, 1423–1432.

Allman, J. (1977) Evolution of the visual system in the early primates. In J. M. Sprague and A. N. Epstein (eds.), *Progress in Psychobiology and Physiological Psychology, Vol. VII*. New York: Academic, pp. 1–53.

Allman, J. (1990) Evolution of neocortex. In E. G. Jones and A. Peters (eds.), *Cerebral Cortex, Vol. 8A: Comparative Structure and Evolution of Cerebral Cortex, Part I*. New York: Plenum, pp. 269–283.

Barrington, E. J. W. (1979) Essential features of lower types. In M. H. Wake (ed.), *Hyman's Comparative Vertebrate Anatomy, 3rd ed.* Chicago: The University of Chicago Press, pp. 57–86.

Boncinelli, E., Gulisano, M., and Pannese, M. (1993) Conserved homeobox genes in the developing brain. *Comptes Rendus d'Acadamie des Sciences Paris*, 316, 979–984.

Bone, Q. (1960) The central nervous system in *Amphioxus. Journal of Comparative Neurology*, 115, 27–64.

Buchsbaum, R., Buchsbaum, M., Pearse, J., and Pearse, V. (1987) *Animals Without Backbones, Third Edition*. Chicago: University of Chicago Press.

Bulfone, A., Puelles, L., Porteus, M. H., Frohman, M. A., Martin, G. R., and Rubenstein, J. L. R. (1993) Spatially restricted expression of *Dix-1, Dix-2 (Tes-1), Gbx-2*, and *Wnt-3* in the embryonic day 12.5 mouse forebrain defines potential transverse and longitudinal segmental boundaries. *Journal of Neuroscience*, 13, 3155–3172.

Bullock, T. H., Bodznick, D. A., and Northcutt, R. G. (1983) The phylogenetic distribution of electroreception: evidence for convergent evolution of a primitive vertebrate sense modality. *Brain Research Reviews*, 6, 25–46.

Butler, A. B. (1994a) The evolution of the dorsal thalamus of jawed vertebrates, including mammals: cladistic analysis and a new hypothesis. *Brain Research Reviews*, 19, 29–65.

Butler, A. B. (1994b) The evolution of the dorsal pallium in the telencephalon of amniotes: cladistic analysis and a new hypothesis. *Brain Research Reviews*, 19, 66–101.

Butler, A. B., Wullimann, M. F., and Northcutt, R. G. (1991) Comparative cytoarchitectonic analysis of some visual pretectal nuclei in teleosts. *Brain, Behavior and Evolution*, 38, 92–114.

Carroll, R. L. (1988) *Vertebrate Paleontology and Evolution*. New York: Freeman.

Chalepakis, G., Stoykova, A., Wijnholds, J., Tremblay, P., and Gruss, P. (1993) Pax: gene regulators in the developing nervous system. *Journal of Neurobiology*, 24, 1367–1384.

Dowling, J. E. (1992) *Neurons and Networks*. Cambridge, MA: The Belknap Press of Harvard University Press.

Ebbesson, S. O. E. (1980) The parcellation theory and its relation to interspecific variability in brain organization, evolutionary and ontogenetic development, and neuronal plasticity. *Cell and Tissue Research*, 213, 179–212.

Ebbesson, S. O. E. (1984) Evolution and ontogeny of neural circuits. *Behavioral and Brain Sciences*, 7, 321–331.

Fritzsch, B. and Northcutt, R. G. (1993) Cranial and spinal nerve organization in amphioxus and lampreys: evidence for an ancestral craniate pattern. *Acta Anatomica*, 148, 96–109.

Gans, C. (1987) The neural crest: a spectacular invention. In P. F. A. Maderson (ed.), *Developmental and Evolutionary Aspects of the Neural Crest*. New York: Wiley, pp. 361–379.

Gans, C. and Northcutt, R. G. (1985) Neural crest: the implications for comparative anatomy. *Fortschritte der Zoologie*, 30, 507–514.

Graziadei, P. P. C. and Monti-Graziadei, A. G. (1992) The influence of the olfactory placode on the development of the telencephalon in *Xenopus laevis. Neuroscience*, 46, 617–629.

Holland, P. W., Holland, L. Z., Williams, N. A., and Holland, N. D. (1992) An amphioxus homeobox gene: sequence conservation, spatial expression during development and insights into vertebrate evolution. *Development*, 116, 653–661.

Jacobson, M. (1991) *Developmental Neurobiology*. New York: Plenum.

Kaas, J. H. (1982) The segregation of function in the nervous system: why do sensory systems have so many subdivisions? *Contributions to Sensory Physiology*, 7, 201–240.

Karten, H. J. (1969) The organization of the avian telencephalon and some speculations on the phylogeny of the amniote telencephalon. In C. Noback and J. Petras (eds.), *Comparative and Evolutionary Aspects of the Vertebrate Central Nervous System. Annals of the New York Academy of Sciences*, 167, 146–179.

Karten, H. J. and Shimizu, T. (1989) The origins of neocortex: connections and lamination as distinct events in evolution. *Journal of Cognitive Neuroscience*, 1, 291–301.

Margulis, L. and Schwartz, K. V. (1988) *Five Kingdoms*, 2nd ed. New York: Freeman.

Medina, L. and Smeets, W. J. A. J. (1991) Comparative aspects of the basal ganglia-tectal pathways in reptiles. *Journal of Comparative Neurology*, 308, 614–629.

Miller, D. M., III, Niemeyer, C. J., and Chitkara, P. (1993) Dominant *unc-37* mutations suppress the movement defect of a homeodomain mutation in *unc-4*, a neural specificity gene in *Caenorhabditis elegans. Genetics*, 135, 741–753.

Murakami, T., Morita, Y., and Ito, H. (1986) Cytoarchitecture and fiber connections of the superficial pretectum in a teleost, *Navodon modestus. Brain Research*, 373, 213–221.

Nielsen, C. (1995) *Animal Evolution: Interrelationships of the Living Phyla*. New York: Oxford University Press, Inc.

Noden, D. M. (1991) Vetebrate craniofacial development: the relation between ontogenetic process and morphological outcome. *Brain, Behavior and Evolution*, 38, 190–225.

Northcutt, R. G. (1981) Evolution of the telencephalon in nonmammals. *Annual Review of Neuroscience*, 4, 301–350.

Northcutt, R. G. (1984) Evolution of the vertebrate central nervous system: patterns and processes. *American Zoologist*, 24, 701–716.

Northcutt, R. G. (1985) Brain phylogeny: speculations on pattern and cause. In M. J. Cohen and F. Strumwasser (eds.), *Comparative Neurobiology: Modes of Communication in the Nervous System*. New York: Wiley, pp. 351–378.

Northcutt, R. G. and Bemis, W. E. (1993) Cranial nerves of the coelacanth *Latimeria chalumnae* [Osteichthyes: Sarcopterygii: Actinistia], and comparisons with other craniata. *Brain, Behavior and Evolution*, 42, S1, 1–76.

Northcutt, R. G. and Braford, M. R., Jr. (1984) Some efferent connections of the superficial pretectum in the goldfish. *Brain Research*, 296, 181–184.

Northcutt, R. G. and Gans, C. (1983) The genesis of neural crest and epidermal placodes: a reinterpretation of vertebrate origins. *The Quarterly Review of Biology*, 58, 1–28.

Northcutt, R. G., Reiner, A., and Karten, H. J. (1988) Immunohistochemical study of the telencephalon of the spiny dogfish, *Squalus acanthias*. *Journal of Comparative Neurology*, 277, 250–267.

Reiner, A. (1991) A comparison of neurotransmitter-specific and neuropeptide-specific neuronal cell types present in the dorsal cortex in turtles with those present in the isocortex in mammals: implications for the evolution of isocortex. *Brain, Behavior and Evolution*, 38, 53–91.

Reiner, A. (1993) Neurotransmitter organization and connections of turtle cortex: implications for the evolution of mammalian isocortex. *Comparative Biochemistry and Physiology*, 104A, 735–748.

Reiner, A., Brauth, S. E., and Karten, H. J. (1984) Evolution of the amniote basal ganglia. *Trends in Neurosciences*, 7, 320–325.

Reiner, A., Brauth, S. E., Kitt, C. A. and Karten, H. J. (1980) Basal ganglionic pathways to the tectum: studies in reptiles. *Journal of Comparative Neurology*, 193, 565–589.

Reiner, A. and Northcutt, R. G. (1987) An immunohistochemical study of the telencephalon of the African lungfish, *Protopterus annectens*. *Journal of Comparative Neurology*, 256, 463–481.

Reiner, A. and Northcutt, R. G. (1992) An immunohistochemical study of the telencephalon of the Senegal bichir (*Polypterus senegalus*). *Journal of Comparative Neurology*, 319, 359–386.

Ronan, M. (1989) Origins of the descending spinal projections in petromyzontid and myxinoid agnathans. *Journal of Comparative Neurology*, 281, 54–68.

Striedter, G. F. and Northcutt, R. G. (1989) Two distinct visual pathways through the superficial pretectum in a percomorph teleost. *Journal of Comparative Neurology*, 283, 342–354.

Ulinski, P. S. (1983) *Dorsal Ventricular Ridge*. New York: Wiley.

Ulinski, P. S. (1986) Neurobiology of the therapsid-mammal transition. In J. J. Roth, E. C. Roth, P. D. MacLean, and N. Hotton, III (eds.), *The Ecology and Biology of Mammal-like Reptiles*. Washington, DC: Smithsonian Institution, pp. 149–172.

Webster, K. E. (1979) Some aspects of the comparative study of the corpus striatum. In I. Divac and R. G. E. Öberg (eds.), *The Neostriatum*. New York: Pergamon, pp. 107–126.

Wicht, H. and Northcutt, R. G. (1992) The forebrain of the Pacific hagfish: a cladistic reconstruction of the ancestral craniate forebrain. *Brain, Behavior and Evolution*, 40, 1–64.

Wild, J. M. (1993) Descending projections of the songbird nucleus robustus archistriatalis. *Journal of Comparative Neurology*, 338, 225–241.

Wullimann, M. R., Meyer, D. L., and Northcutt, R. G. (1991) The visually related posterior pretectal nucleus in the non-percomorph teleost *Osteoglossum bicirrhosum* projects to the hypothalamus: a DiI study. *Journal of Comparative Neurology*, 312, 415–435.

Yoshimoto, M. and H. Ito (1993) Cytoarchitecture, fiber connections, and ultrastructure of the nucleus pretectalis superficialis pars magnocellularis (PSm) in carp. *Journal of Comparative Neurology*, 336, 433–446.

Appendix

Terms Used in Neuroanatomy

INTRODUCTION

Anatomists have devised a collection of terms that they find useful in indicating the location of structures, directions, and planes of section. These terms are nearly all in Latin or Greek, or are derived from them. Newcomers to the field often are baffled by these terms and wonder why simple words like "front," "back," "top," and "bottom" cannot be used instead. Part of the answer is that the specialized anatomical terms often provide greater precision. Moreover, the use of these terms is a centuries-old tradition that originated in the days when the language of science was Latin and when many educated persons also were familiar with classical Greek. Another reason is that anatomy is an international discipline. Anatomists publish the results of their research in many languages. By using an international nomenclature based on Latin and Greek, a number of communication barriers are lowered. This advantage applies not only to the direction and location terms but to the names of brain regions and subdivisions as well as the individual cell groups and fiber pathways that comprise the nervous system.

In this appendix we will describe and define many of the common terms that anatomists use to indicate relative and absolute direction and the location of structures. In addition, we will present a listing of the Latin and Greek derivations of many of the terms that you will encounter in this book. These include nouns and their modifiers and prefixes. Our hope is that understanding the meanings of some of these structure names will help you to remember them better because you will be able to associate a visual image with the name.

DIRECTION AND LOCATION TERMS

The most elementary of directional terms are those that indicate front and back. These are **anterior** (front) and **posterior** (back) and are illustrated in the drawing in Figure A-1(A). The drawing shows a ray-finned fish with the location of the brain and spinal cord indicated. The front and back ends of an animal can be called **rostral** (the snout end) and **caudal** (the tail end), respectively. In a legless animal, such as a fish or a snake, the terms anterior and rostral may be freely interchanged as can posterior and caudal. Another term that sometimes is encountered is **oral,** which literally means in the direction of the mouth, but which often is used interchangeably with rostral. Two other basic directional terms are shown in Figure A-1(A). These are **dorsal,** meaning in the direction of the back, and **ventral**, meaning in the direction of the abdomenal surface.

Figure A-1(B) shows the same terms applied to a quadruped; that is, a four-legged animal. Again, note the location of the brain and spinal cord. The basic directional terms apply just as well to a quadruped as to a legless animal. Some confusion occurs, however, when we attempt to transfer this simple system of terms to bipeds; that is, two-legged animals. To understand the logic of the transfer, the reader should study Figure A-2, which shows a typical biped in a rather atypical posture. When a human is seen assuming a quadrupedal posture, the basic directional terms apply rather well. This, however, is not the normal human posture. When humans are viewed in their normal, upright posture, strange things happen to their directional terms as can be seen in Figure A-3, which shows an upright human and another upright biped.

When a biped is in the upright position, dorsal and posterior become the same, as do ventral and anterior. Rostral still

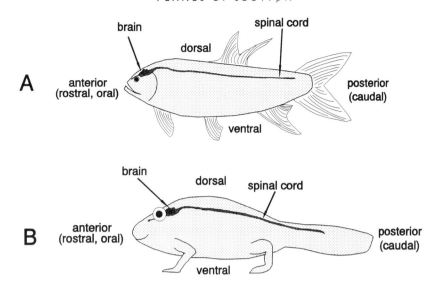

FIGURE A-1. Basic anatomical directional terms are illustrated in a drawing of a ray-finned fish (A) and a tetrapod salamander (B). The location of the brain and spinal cord are shown.

refers to the snout or head end, but that is no longer anterior. In contrast, caudal is still posterior. Notice also that the spinal cord is at right angles to the brain in the bipeds, whereas it lies on an axis parallel to that of the brain in the quadruped and legless animals. Thus, in a biped, the dorsal surface of the spinal cord is also the posterior surface.

Figure A-4 introduces several additional directional terms. **Lateral** means the side of the body, and **medial** means the midline of the body. In addition, **unilateral** means one side, **bilateral** means two sides or both sides, **contralateral** means the opposite side, and **ipsilateral** means the same side. The relative positions above and below are indicated by the terms **superior** and **inferior,** respectively. The terms superior and inferior have nothing to do with the quality or degree of complexity of the structures to which they are applied. They indicate

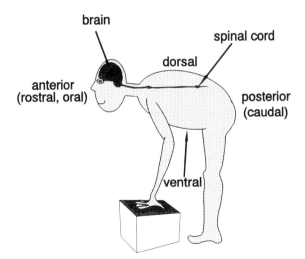

FIGURE A-2. Basic anatomical directional terms are illustrated in this drawing of a biped human who is assuming a tetrapod position. The location of the brain and spinal cord are shown.

relative position and nothing more. Other relative terms are **proximal,** meaning near, and **distal,** meaning far.

PLANES OF SECTION

Anatomists nearly always cut their specimens into thin slices called sections, which they examine with the light microscope or electron microscope in order to observe the fine details of their structure. Because a group of cells may have a very different shape when viewed from a different direction, anatomists always indicate the plane in which the specimen has been cut. Figure A-5 is a schematic representation of the most common planes of section. The planes are represented on a cylinder.

The two major planes of section are the **transverse** and the **longitudinal** planes. The most frequently used plane of section is the transverse plane. Sections cut in this plane are sometimes called **coronal** or **frontal** sections. Transverse sections are cut *across* the long axis of the cylinder; that is, parallel to the diameter of the cylinder. Longitudinal sections are cut *along* the long axis of the cylinder. The longitudinal plane that is vertically oriented and divides the cylinder into two symmetrical halves is known as the **midsagittal** plane. Sections that are cut parallel to the midsagittal plane are properly called **parasagittal** sections. In common usage, however, they usually are simply called **sagittal** sections. Sections that are cut parallel to the long axis of the cylinder and also at right angles to the midsagittal plane are known as **horizontal** sections. Both the sagittal and horizontal planes are the conventional planes for longitudinal sections.

Occasionally, an anatomist may have need to cut sections in a plane other than one that is parallel or at right angles to the main axes of the cylinder. Such a plane is known as an **oblique** plane. Sometimes anatomists deliberately use an oblique plane of section to reveal certain relationships between

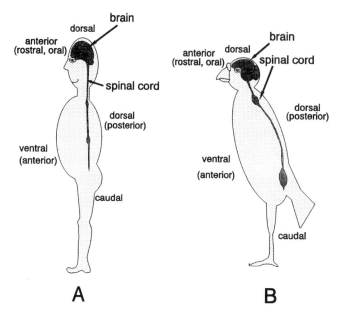

FIGURE A-3. Basic anatomical directional terms are illustrated in this drawing of two bipeds, a human (A) and a bird (B). Both animals are standing in their normal, biped posture. The locations of their brains and spinal cords are shown. Note how the terms *anterior* and *posterior* now correspond to *dorsal* and *ventral* as a consequence of the upright posture.

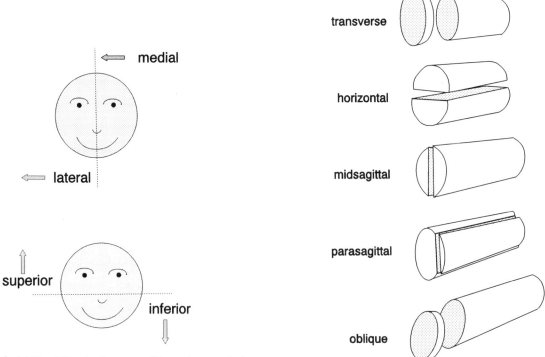

FIGURE A-4. Additional directional terms: *medial,* meaning towards the midline, *lateral,* meaning towards the side, *superior,* meaning above some reference point, and *inferior,* meaning below some reference point.

FIGURE A-5. The most common planes of section of neural tissue are illustrated as slices through a cylinder.

two or more structures that otherwise would not appear in the same section. Oblique planes of section also result unintentionally when, for some reason, the long axis of the brain or spinal cord is not aligned at an exact right angle to the plane of the blade of the cutting device. One of the infinite number of oblique planes is illustrated in Figure A-5.

NEUROANATOMICAL NAMES

Nearly all of the names given to anatomical entities, whether cell groups or gross structures, are Latin or Greek names. Not very long ago, when knowledge of Latin and Greek was common among educated persons, the terms given below would have posed no serious challenge. Today, however, when the study of these languages is a relative rarity, the terms often seem to be a hindrance to learning rather than a help. The purpose of this section is to help you to understand the meanings of many of the neuroanatomical terms that you will encounter in this book and in the scientific literature.

Two word lists are presented below. The first, which is a list of nouns and modifiers, gives the Latin or Greek anatomical name followed by an English translation. Most of the terms given are the names of parts or subdivisions of the central nervous system that can be seen with the unaided eye during a gross dissection. The majority of these terms are Latin words; the few Greek terms are indicated with a letter G. The second word list consists of prefixes that are jointed to other words or word fragments to make an anatomical name. Each prefix is followed by an example of its use. In the prefix list Greek roots are quite common.

Nouns and Modifiers

Anatomical Name	English Translation
abducens	separating
accumbens	reclining
amiculum	cloak
ampulla	flask
amygdala	almond; tonsil
annulus	ring
apex	peak; tip
arbor	tree
arbor vitae	tree of life
auriculus	little ear
bulla	bubble; blister
brachium	arm
calcar avis	bird's spur
cavum	cavity
cauda equina	horse's tail
chiasma (G)	a crossing of lines
choroid (G)	like a delicate membrane
cingulum	belt
claustrum	enclosed space
cochlea	snail
colliculus	little hill
cornu	horn
corpus	body

Anatomical Name	English Translation
corpus callosum	hard body
crista	crest
crus	leg
culmen	highest point
declive	sloping
dura	hard
falx	sickle
fasciculus	little bundle
fenestra	window
filium	thread
flocculus	little tuft
folium	leaf
fornix	arch
fossa	ditch; trench
fovea	pit
funiculus	little cord
ganglion (G)	knot
genu	knee
globus pallidus	pale globe
glomerulus	little ball of wool
gracilis	slender
habenula	little thong
hippocampus	horse-caterpillar
infundibulum	little funnel
insula	island
lacuna	gap
lamina	thin plate
lemniscus (G)	ribbon; band
limbus	border
lingula	little tongue
locus	place
locus coeruleus	sky-blue place
lunate	like the moon
macula	spot
medulla	marrow
nigra	black
nodulus	little knot
nodose	like a knot
obex	bolt
ocellus	little eye
pectin	comb
pineal	pine cone
psalterium	a stringed instrument
putamen	husk
pulvinar	cushion
pyramis	pyramid
ramus	branch
raphe (G)	seam
sacculus	little sack
septum	wall
soma (G)	body
stratum	layer
tectum	roof
tegmentum	cover
tela	web
thalamus	bed chamber
torus	bulge, mound
tuber cinereum	ashen bulge

Anatomical Name	English Translation
uncus	hook
uncinate	like a hook
utriculus	little leather bag
uvula	little grape
vagus	wanderer
vallecula	little valley
valvula	folding doors
vellum	veil
vermis	worm

Prefixes

Many anatomical names and descriptions are formed from Latin and Greek roots. The glossary below lists anatomical prefixes that are derived from Latin and Greek roots, their meanings in English, and an example of the use of the prefix in an anatomical term. The letter G or L after each example indicates whether the root is Greek or Latin.

Root	Translation	Example	Origin
acout-	to hear	acoustic	G
al-	wing	alar plate	L
alb-	white	straum album	L
all-	different	allocortex	G
apic-	peak	apical dendrite	L
arachn-	spider	arachnoid	G
arcu-	arch	arcuate	L
arch-	ancient	archistriatum	G
astr-	star	astrocyte	G
aur-	ear	auriculus	L
blast-	bud	neuroblast	G
brachi-	arm	brachium pontis	G
calcar-	spur	calcarine fissure	L
caud-	tail	caudate nucleus	L
cephal-	head	cephalic	G
cerebr-	brain	cerebrospinal	L
cervic-	neck	cervical	L
chondr-	cartilage	chondrocranium	G
chord-	cord	notochord	G
chrom-	color	chromatophore	G
cliv-	slope	declive	L
coll-	hill	colliculus	L
cort-	tree bark	cortex	L
crani-	skull	cranial	G
cruc-	cross	cruciate	L
cumb-	lying down	accumbens	L
cune-	wedge	cuneatus	L
cur-	to run	recurrens	L
cyt-	cell	cytoplasm	G
dendr-	tree	oligodendrocyte	G
derm-	skin	dermatome	G
drom-	run	orthodromic	G
embol-	wedge	emboliform	G

Root	Translation	Example	Origin
encephal-	brain	metencephalon	G
falc-	sickle	falciform	L
fastigi-	roof	fastigial	L
fer-	carry	afferent	L
for-	bore; pierce	foramen	L
fug-	flee	corticofugal	L
furc-	fork	bifurcation	L
fus-	spindle	fusiform	L
gemin-	twin	quadrigemina	L
gloss-	tongue	hypoglossal	G
junct-	join	conjunctivum	L
juxta-	close to	juxta-allocortex	L
kary-	kernel	perikaryon	G
koni-	dust	koniocortex	G
lemm-	husk; sheath	neurolemma	G
lemn-	ribbon, band	lemniscus	G
lent-	lens	lentiformis	L
lept-	thin; delicate	leptomeninges	G
ly-	dissolve	chromatolysis	G
magn-	large	magnocellular	L
met-	after	metencephalon	G
morph-	form	morphology	G
myel-	marrow	myelin	G
ocul-	eye	oculomotor	L
olig-	few	oligosynaptic	G
op-	eye; see	hemianopia	G
pale-	old	paleocortex	G
palli-	mantle	pallium	L
pappil-	nipple	papilliform	L
parv-	small	parvocellular	L
ped-	foot	peduncle	L
phy-	growth	epiphysis	G
pil-	hair	neuropil	G
plex-	braid, plait	plexiform	L
pont-	bridge	pontine	L
rect-	straight	rectus	L
resti-	rope	restiform	L
ret-	net	reticular	L
rhin-	nose	rhinencephalon	G
rhiz-	root	rhizotomy	G
rub-	red	rubrospinal	L
stri-	stripe	striatum	L
syn-	together	synapse	G
troch-	wheel; pulley	trochlear	G

FOR REFERENCE AND FURTHER READING

Ayers, D. M. (1972) *Bioscientific Terminology,* Tucson: University of Arizona Press.

Hogben, L. (1970) *The Vocabulary of Science.* New York: Stein and Day.

Lockard, I. (1992) *Desk Reference for Neuroscience.* New York: Springer.

The Oxford Dictionary for Scientific Writers and Editors. (1991) Oxford: Oxford University Press.

Glossary

Abducens Sixth cranial nerve (VI); one of three cranial nerves that innervate eye muscles.

Acousticolateralis system See Octavolateralis system.

Actinistia One of the three radiations of sarcopterygians, of which only one species, the coelacanth *Latimeria chalumnae,* is extant.

Actinopterygians The subclass of Osteichthyes, or bony fishes, that comprise the ray-finned fishes.

Adaptation General process by which a species adjusts to environmental change; a feature of an organism that is suited to the environment.

Adaptation studies Studies of similar adaptations in various taxa whether or not the adaptations have been evolved independently.

Advanced Recent, that is, more recently derived from the ancestral stock; can imply progressive "improvement" in a lineage.

Afferent An axon that projects to the central nervous system or to a specific part within it.

Agnatha Jawless fishes, also called cyclostomes: extant agnathans are the lampreys and hagfishes.

Allele Alternate form of a gene.

Allometry Study of relative or proportional size, weight, volume, or other such feature among members of a taxon, such as brain weight/body weight ratios.

Ammocoete Larval lamprey.

Amniotes Mammals, diapsid reptiles, birds, and turtles: animals characterized by the embryo having an amnion, the membrane that surrounds the embryo and the amniotic fluid.

Amphibians Group of tetrapods that generally have aquatic larvae and terrestrial adults; comprise salamanders and newts, frogs, and gymnophionans.

Amphioxus One of the common names for the cephalochordate *Branchiostoma sp.* Also known as lancelets.

Ampulla A small, flask-like organ, such as those of the semicircular canals.

Anagenesis Used variably to mean (1) progressive evolution or improvement within a lineage towards a "higher" level; see *Scala naturae;* (2) increased efficiency of design of a structure within a lineage; or (3) phenotypic or genotypic changes within a single, unbranched lineage. See also Phyletic evolution.

Analogy Similarity of function. This term is independent of considerations of phyletic continuity.

Anamniotes Include jawless vertebrates, cartilaginous fishes, ray-finned fishes, lungfishes, crossopterygians, and amphibians. An amnion is not present around the embryo during development in these vertebrates.

Anapsids Includes ancestral amniote stock from which nonanapsid amniotes are descended and modern turtles; the anapsid skull has a complete roof of bone, without fenestrae or arches, in the temporal region.

Anlage Plural: Anlagen. A German term that denotes the primordium of an organ or body part that forms during embryological development.

Anterograde The direction toward the axonal terminal area.

Anterodorsal lateral line nerve One of three preotic lateral line cranial nerves (LL_{AD}).

Anteroventral lateral line nerve One of three preotic lateral line cranial nerves (LL_{AV}).

Anurans Amphibians without tails in the adult form, the frogs and toads: one of the three extant groups of amphibians.

Apomorphy Adjective: Apomorphic. Traits that are derived specializations of a particular taxon relative to ancestral taxa in which the trait was not present, for example, mammary glands and hair are *apomorphic* for mammals in that they are not present in any of the nonmammalian tetrapod or nonmammalian sarcopterygian or nonmammalian vertebrate ancestors of mammals. See also Plesiomorphic.

Arachnoid Middle of three membranous layers of connective tissue that cover the brain and spinal cord in mammals and birds, named for its similarity in appearance to a spider web.

Archenteron Gastrocoel: embryonic gut cavity formed within the gastrula.

Autonomic nervous system Neuronal system that regulates the viscera, such as the smooth muscles of the digestive and circulatory systems, and secretion of glands. Consists of two divisions: the sympathetic division, which has its neurons in the thoracic and lumbar regions of the spinal cord, and the parasympathetic division, which has its neurons in the brainstem and sacral region of the spinal cord. Although there are exceptions, in general, the sympathetic division upregulates visceral functions and the parasympathetic system downregulates visceral functions. In addition, the two systems work in concert to regulate normal visceral functioning.

Axon The long process of a neuron that carries action potentials to the axon terminal.

Blastocoel Cavity formed within the blastula.

Blastopore Opening of the archenteron.

Blastula Early embryo stage in which the cells form a hollow sphere.

Boutons Also: Boutons termineaux. French term for the terminal synapses on an axon.

Brainstem Commonly used to refer to the hindbrain, midbrain, and diencephalon; sometimes refers to only the hindbrain (without the cerebellum) and midbrain.

Branchial Refers to the gills, as in branchial arches.

Branchial arches Part of the series of visceral arches that support the gills in fishes. Most fishes have five branchial arches that are the third through the seventh visceral arches. See also Visceral arches. The related term "branchiomeric" refers to the muscles and other structures of the gill arches.

Caecilians See Gymnophionans.

Cenozoic era Geologic time from 65 million years ago to the present, divided into the Tertiary period of 65–2 million years ago and the Quaternary period of the past 2 million years.

Central canal Longitudinal, tubular, midline cavity present in the spinal cord; contains cerebrospinal fluid.

Cephalochordates Group of chordate animals that include lancelets of the genus *Branchiostoma,* also called by the common name amphioxus.

Cerebellum The dorsal region of the metencephalon, which consists of a cortical sheet or plate, sometimes highly folded, and one or more deep nuclei. Among its many functions in various species are coordination of muscle activity and tone, balance, electroreception, and some aspects of motor learning. It is highly developed in electroreceptive aquatic nontetrapods and in birds and mammals.

Cerebral aqueduct (of Sylvius) Opening through which the third ventricle is connected with the fourth ventricle.

Cerebral hemispheres Paired structures that form the telencephalon; in mammals this term refers to the isocortical lobes (frontal, parietal, temporal, and occipital), the basal ganglia, septum, and related structures on each side.

Cerebrum Loosely refers either to most of the telencephalon or to the entire forebrain.

Cervical Refers to the neck region of the spine; may be used to indicate vertebrae, nerves or spinal cord segments.

Chemoreceptor Sensory receptor that responds to stimulation by specific classes of chemical components.

Chondrichthyes Cartilaginous fishes, comprised by Elasmobranchi (sharks, skates, and rays) and Holocephali (ratfishes).

Chondrostei The sturgeons and paddlefishes: one of the five radiations of ray-finned fishes.

Chordates Group of deuterostome animals that have a notochord, comprising urochordates (tunicates), cephalochordates (lancelets), and vertebrates.

Clade A monophyletic set of taxa of any extent.

Cladistia The reedfishes or bichirs: one of the five radiations of ray-finned fishes.

Cladistics A method of analysis for classifying animals according to their inferred phyletic relations based on sets of shared similar traits and/or for determining the polarity of a given trait based on its distribution within a set of phyletic relationships derived from traits unrelated to the trait being analyzed.

Cladogenesis See Speciation.

Cladogram A branching diagram that expresses phylogenetic relationships of a set of monophyletic taxa.

Coelacanth See Actinistia.

Colliculus Plural: Colliculi. One of four structures (arranged as two pairs: superior and inferior) that form the roof of the midbrain in mammals. Also see Tectum.

Column A longitudinally oriented group of axons, such as the dorsal column of the spinal cord; also see Fiber bundle. Also, a longitudinally oriented group of neuron cell bodies, such as the somatic efferent column.

Commissure A group of axons that crosses the midline to the contralateral side of the brain or spinal cord and generally terminates in a site or sites in the mirror-image position of the origin of the axons. A commissure may also contain decussating axons. Also see Decussation.

Contralateral On the opposite side of the brain or body.

Convergence A process that produces homoplasy in relatively distantly related taxa.

Corpus callosum Large bundle of commissural axons that interconnects the cerebral cortex in eutherian mammals.

Cortex Parts of the brain in which the neuron cell bodies are organized in a layered manner, such as in the cerebellum and the mammalian cerebral cortex.

Cranial nerves Set of nerves that emerge from the brain, in contrast to the spinal nerves that emerge from the spinal cord.

Crossopterygian Refers to two groups of sarcopterygian fishes, one that comprises the coelacanth *Latimeria chalumnae* and the other that comprises the rhipidistian fishes, which were ancestral to tetrapods. Also see Actinistia.

Cyclostomes See Agnatha.

Decussation A group of axons that crosses the midline to the contralateral side of the brain or spinal cord and generally terminates in a site or sites that are not in the mirror-image position of the origin of the axons. A decussation may also contain commissural axons. Also see Commissure.

Dendrites The multiple processes of a neuron that receive afferent synapses from the axons (or dendrites) of other neurons.

Derived character See Apomorphy.

Deuterostomes Refers to "second mouth": group of coelamate animals that comprise multiple groups including Echinodermata (which include star fishes and sea urchins), Hemichordata (acorn worms and pterobranchs), and Chordata (urochordates, cephalochordates, and vertebrates).

Diapsids Includes modern diapsid reptiles and birds; the diapsid skull has two temporal fenestrae and two arches of bone.

Diapsid reptiles Lepidosaurs (lizards, snakes, and the tuatara *Sphenodon*) and thecodont reptiles (crocodiles).

Diencephalon The caudal part of the prosencephalon, or forebrain; comprises the epithalamus, dorsal thalamus, ventral thalamus (subthalamus), hypothalamus, pretectum, and posterior tuberculum.

Dipnoi Lungfishes: one of the three extant radiations of sarcopterygians.

Dorsal facial nerve Division of the seventh cranial nerve (VII_D) that innervates facial muscles and glands.

Dorsal glossopharyngeal nerve Division of the ninth cranial nerve (IX_D) that innervates the pharynx and salivary glands.

Dorsal vagus nerve Division of the tenth cranial nerve (X_D) that innervates the viscera of the thorax and abdomen, the larynx, and the pharynx.

Dorsal ventricular ridge Part of the pallium of the telencephalon in diapsids and turtles that lies ventral to the cortical part of the pallium and comprises anterior and posterior parts; formed either by a laminar cell plate or by nuclear areas.

Dorsal root One of the two roots of spinal nerves (see also Ventral root); consists mostly of incoming sensory axons that enter the spinal cord on its dorsal side.

Dorsal root ganglion An aggregation of cell bodies of incoming sensory axons located on the dorsal root of the spinal cord.

Dura mater Means "hard mother": outer of the membranous layers of connective tissue that cover the brain and spinal cord in tetrapods.

Ectoderm The most superficial of the three germ layers formed in the gastrula stage of embryogenesis.

Efferent An axon that projects away from the central nervous system or away from a specific part within it.

Elasmobranchs Group of cartilaginous fishes that includes sharks, skates, and rays.

Electroreceptor Sensory receptor that responds to stimulation by weak electric fields.

Endoderm The deepest of the three germ layers formed in the gastrula stage of embryogenesis.

Ependyma Epithelial lining of the ventricles of the brain and the central canal of the spinal cord.

Epiphyseal nerve Cranial nerve (E) that innervates the epiphysis.

Epiphysis Part of the epithalamus; comprises pineal and/or parietal structures.

Epithalamus Dorsal-most part of the diencephalon, lying dorsal to the dorsal thalamus; comprises the habenular and related nuclei and the epiphysis.

Eutheria Placental mammals.

Evagination Process of embryological development in which the neural tube in the telencephalon (or elsewhere) expands laterally and dorsoventrally by bulging outward.

Eversion Process of embryological development in which part or all of the roof of the neural tube in the telencephalon thins, elongates, and bends outward laterally.

Evolution Change over time.

Extrasegmental Refers to axons or reflexes that involve more than one segment of the spinal cord.

Fasciculus A slender bundle of axons, such as the medial longitudinal fasciculus of the brainstem. Also see Fiber bundle.

Fenestra A small opening or window.

Fiber An axon.

Fiber bundle A group of axons; also variably called a tract, fasciculus, funiculus, or column.

Field A developmental region that gives rise to one or more structures in the adult; or a part of the brain, such as the frontal eye field or prerubral field; or a receptive area that responds to stimulation within it.

Fitness The degree to which a variant will be selected for.

Floor plate A group of cells in the midline of the neural plate induced by the underlying notochord.

Foramen of Magendie Single, median opening that, along with the paired foramina of Luschka, connect the fourth ventricle with the subarachnoid space.

Foramina of Luschka Paired set of openings that, along with the single foramen of Magendie, connect the fourth ventricle with the subarachnoid space.

Foramina of Monroe Paired openings through which the lateral ventricles of the telencephalic hemispheres are in continuity with the third ventricle of the diencephalon.

Forebrain See Prosencephalon.

Fourth ventricle Cavity present in the hindbrain and caudally continuous with the central canal of the spinal cord.

Funiculus A group of axons forming a white matter tract or column in the spinal cord. Also see Fiber bundle.

Ganglion (Plural: ganglia). A collection of neuron cell bodies. In vertebrates, most often refers to such collections outside the brain and spinal cord; sometimes refers to a collection of neuron cell bodies in the brain, such as the basal ganglia. In some invertebrates, the central nervous system is composed of a series of ganglia.

Gastrula Embryonic stage following the blastula and in which three germ layers of ectoderm, mesoderm, and endoderm are formed.

Genotype The genetic composition of an organism.

Ginglymodi Gars: one of the five radiations of ray-finned fishes.

Glia Nonneural, supporting cellular constituents of the nervous system.

Glomerulus A ball-like synaptic complex of axon terminals and dendrites, as in the cerebellar cortex and the olfactory bulb.

Gnathostomes Jawed vertebrates.

Grades An assemblage of taxa characterized by a general level of organization and ranked among other such assemblages through sequences of successive levels of organization. A grade is the unit of phyletic analysis, or anagenesis.

Gray matter Tissue within the central nervous system that is predominantly composed of neuron cell bodies and unmyelinated neuron cell processes.

Gustatory Refers to the sense of taste.

Gymnophionans Caecilians: one of the three extant groups of amphibians.

Gyrencephalic Having a telencephalon with a surface that consists of many folds or convolutions. Each fold is known as a gyrus (plural = gyri); the valley between adjacent folds is known as a sulcus (plural = sulci).

Gyrus Plural: Gyri. Ridges of cerebral cortex in mammals.

Hair cell Sensory cell of octavolateralis system that contains one or more hair-like processes.

Halecomorphi One of the five radiations of ray-finned fishes, of which only one species, the bowfin *Amia calva*, is extant.

Hemichordates Group of chordate animals that comprise acorn worms (enteropneusts) and pterobranchs.

Heterostracans One of the ostracoderm groups of extinct jawless fishes.

Higher A much misused and conceptually abused term as applied in comparative neurobiology; occasionally used to refer to the more rostral and dorsal parts of the brain, as in "higher centers"; most commonly refers to the alleged superiority of a given species as defined by human values in a *scala naturae* context. This term is best not used at all.

Hindbrain See Rhombencephalon.

Hodological Refers to axonal pathways and connections in the central nervous system; derived from the Greek word hodos, which means way or road.

Holocephali Group of cartilaginous fishes that comprises the ratfishes, or chimaeras.

Homeobox gene Also referred to as a homeobox-containing gene, or *Hox*. One of a number of genes that contain a highly conserved sequence of DNA nucleotide base pairs and that are involved in rostrocaudal patterning in a wide variety of invertebrate and vertebrate species.

Homology Relationship of traits found in members of a monophyletic taxon where the presence of that trait or its precursor can in principle be traced to a common ancestor; the specific nature of the homologous relationship must be stipulated, for example, as a discrete structure, a derivative of all or part of a developmental field, or a serially repeated, segmental structure.

Homoplasy   Adjective: Homoplastic or Homoplaseous. Similarity of traits found in a group of taxa, either monophyletic or polyphyletic, not due to inheritance from a common ancestor but to independent, similar evolutionary changes; as in the case of homology, the homoplastic relationship must be stipulated as to its nature. The process that produces homoplasy in relatively closely related species is called parallelism. The process that produces homoplasy in relatively distantly related species is called convergence.

Hypoglossal nerve   Twelfth cranial nerve (XII); innervates the muscles of the tongue and syrinx.

Hypophysis   Pituitary; comprises the adenohypophysis rostrally and the neurohypophysis caudally.

Hypothalamus   Ventral-most part of the diencephalon; set of nuclei involved in functions that include many relating to maintenance of the internal milieu of the body, emotion, motivational states, and control of the autonomic nervous system.

Interneuron   Any neuron that lies between a primary sensory afferent neuron and an effector (or motor) neuron. Sometimes used to refer to Gogli type II neurons that remain internal to a neuronal population.

Ipsilateral   On the same side of the brain or body.

Isocortex   Six-layered part of the dorsal pallial cortex in mammals; also known as neocortex.

Isthmus   Most commonly refers to the caudal part of the mesencephalon where the brainstem narrows.

Intrasegmental   Refers to axons that remain within a given segment of the spinal cord.

Lagena   A flask-like evagination of the sacculus that contains auditory receptors.

Lateral line system   An electroreceptive and/or mechanoreceptive sensory system present in many aquatic vertebrates.

Lateral ventricle   Cavity present in each of the telencephalic hemispheres.

Lemniscal   Refers to sensory pathways that ascend directly to the thalamus.

Lepidosaurs   One of the two groups of diapsids; includes the Rhynchocephalia that comprises a single species, the tuatara *Sphenodon,* and the Squamata, the lizards, amphisbaenians, and snakes.

Limbic system   Term introduced by MacLean that encompasses the limbic lobe (hippocampal formation, cingulate gyrus, parahippocampal gyrus, and related cortices) and its related subcortical structures, including the amygdala, septal nuclei, and related parts of the striatum and diencephalon. This system plays important roles in emotion and memory.

Lissencephalic   Having a telencephalon with a smooth surface that lacks sulci or gyri. See Gyrencephalic.

Lower   A much misused and conceptually abused term as applied in comparative neurobiology; occasionally used to refer to the more ventral and caudal parts of the brain, as in the "lower brainstem"; most commonly refers to relative alleged inferiority of a given species as defined by human values in a *scala naturae* context. This term is best not used at all.

Lumbar   Refers the caudal region of the spine between the thoracic and sacral levels; may be used to indicate vertebrae, nerves or spinal cord segments.

Mammals   Amniotes with mammary glands and hair; includes monotremes, marsupials, and eutherians. Extant descendants of synapsid amniotes; the synapsid skull has a single temporal fenestra, or opening.

Marsupials   Metatherian mammals: mammals with pouches in which embryological development is completed.

Medulla   Also known as medulla oblongata. Most caudal region of the brain; situated between the pons and the spinal cord. Location of many cranial nerve sensory and motor nuclei, reticular formation, and ascending and descending axonal pathways.

Meninges   Singular: Meninx. A system of membranes that surrounds the central nervous system.

Mesencephalon   Midbrain; lies caudal to the prosencephalon (forebrain) and rostral to the rhombencephalon (hindbrain). Comprises the tectum, tegmentum proper, and isthmus.

Mesoderm   The middle of the three germ layers formed in the gastrula stage of embryogenesis.

Mesomeres   Rostrocaudal segments of the mesencephalon formed during embryological development.

Mesozoic era   Geologic time from 230 to 65 million years ago, divided into the Triasic period of 230 to 181 million years ago, the Jurassic period of 181 to 135 million years ago, and the Cretaceous period of 135 to' 65 million years ago.

Metatheria   Marsupials: one of the three radiations of mammals.

Metencephalon   Rostral part of the hindbrain; comprises the cerebellum and, in mammals, the pons.

Midbrain   See Mesencephalon.

Middle lateral line nerve   One of three postotic lateral line cranial nerves (LL_M).

Monophyletic   Pertaining to taxa that have all evolved from a single ancestral taxon.

Monotremes   Prototherian mammals; egg-laying mammals.

Myelencephalon   Caudal part of hindbrain; the medulla oblongata.

Myomere   Muscle segment.

Myotome   Embryonic muscle segment.

Neocortex   See Isocortex.

Neoteny   The retention of juvenile somatic characters in a reproductively mature individual.

Neural crest   A group of cells that initially lie between the neural tube and the surface ectoderm. These cells form a variety of tissues including sensory and postganglionic neurons, the visceral skeleton, and pigment cells.

Neural plate   Thickened ectodermal tissue overlying the roof of the archenteron.

Neural tube   Structure formed by folds of the neural plate that grow dorsally to meet and then fuse into a tube of neural tissue with a central lumen during embryological development.

Neuromasts   Sensory hair cells of the lateral line system.

Neuromeres   Embryonic, rostrocaudal segments of the brain.

Neuromodulators   Chemical substances that act on a presynaptic membrane to alter the release of neurotransmitters or on postsynaptic membranes to change their sensitivity to neurotransmitters.

Neuron   A nerve cell; a cell that transmits information by changes in polarization of parts of its membrane.

Neuropil   A region with few or no neuron cell bodies in which neuron cell processes are densely intertwined.

Neurotransmitters   Chemical substances released by the presynaptic neuron at synaptic sites that diffuse to postsynaptic receptor sites and affect the degree of polarization of the membrane of the postsynaptic neuron.

Notochord   Rostrocaudally oriented structure of mesodermally derived cells ventral to the neural plate and/or neural tube; present in the adults of some invertebrate chordates and in the embryos and some adults of vertebrates.

Nucleus   In the context of neuroanatomy, a group of neuron cell bodies within the central nervous system of vertebrates;

also used in the context of cell biology to refer to an organelle within a cell body that contains the chromatin material.

Obex   Small area overlying the caudal part of the fourth ventricle at the point where the fourth ventricle is continuous with the central canal that continues through the caudal medulla into the spinal cord.

Octaval nerve   Eighth (ventibulocochlear) cranial nerve (VIII); cranial nerve that innervates the cochlear and vestibular organs.

Octavolateralis system   The auditory, vestibular, and lateral line systems of fishes and larval amphibians; also called the acousticolateralis system.

Oculomotor nerve   Third cranial nerve (III); one of three cranial nerves that innervate eye muscles.

Olfactory   Refers to the sense of smell.

Olfactory bulb   Paired structure at the rostral end of the telencephalon that receives afferent olfactory, or smell, input from receptor cells in the nasal mucosa and projects to olfactory regions of the telencephalon.

Olfactory nerve   First cranial nerve (I); cranial nerve that innervates the olfactory epithelium.

Ontogeny   Embryological development.

Optic chiasm   Point of decussation of most or all of optic nerve axons to the contralateral side of the brain.

Optic nerve   Second cranial nerve (II); cranial nerve that innervates the retina. This term is used for the portion of the optic nerve axons that lie distal to the optic chiasm.

Optic tract   Portion of the optic nerve axons that continue centripetally past the optic chiasm.

Optic tectum   Also known as the superior colliculus in mammals; part of the tectum (or roof) of the midbrain that predominantly receives visual and somatosensory information.

Otic lateral line nerve   One of three preotic lateral line cranial nerves (LL$_O$).

Organ of Jacobson   See Vomeronasal organ.

Paedomorphism   Retention of juvenile characters in the adult.

Paleozoic era   Geologic time from 405 to 230 million years ago, divided into the Devonian period of 405 to 345 million years ago, the Carboniferous period of 354 to 280 million years ago, and the Permian period of 280 to 230 million years ago.

Pallidum   Two components of the striatopallidal complexes in the striatum, or basal forebrain: In mammals, the dorsal pallidum is the globus pallidus, and the ventral pallidum consists of globus pallidus-like neurons within the substantia innominata.

Pallium   The dorsal part of the telencephalon, consisting of medial, dorsal, and lateral areas and with cytoarchitecture that is either solely cortical (see Cortex) or a combination of cortical and nuclear.

Parallelism   The process that produces homoplasy in relatively closely related taxa.

Paraphysis   Part of the epithalamus; closely associated with the epiphysis. May be glandular or photoreceptive.

Parasympathetic division of the autonomic nervous system   See Autonomic nervous system.

Paraxial mesoderm   Mesoderm that lies lateral to the neural tube during embryological development and forms somites and somitomeres in the head region.

Phenotype   The observable traits of an organism; the observable expression of the genotype.

Pheromones   Chemical compounds used as social signals, such as for attraction of mates, marking of territories, and alerting members of the same species to danger.

Photoreceptor   Sensory receptor that responds to stimulation by photons.

Phyletic evolution   Process by which a single lineage, without branching into divergent lineages, undergoes change over time. See also Anagenesis.

Phyletic studies   Studies that attempt to reconstruct the evolutionary history of a particular trait in a given lineage.

Phylogenetic scale   See *Scala naturae*.

Phylogeny   The evolutionary history of a taxon or taxa.

Phylogenetic tree   Diagram of evolutionary relationships among taxa in which the location of each taxon indicates only the relative time of its appearance in the fossil record.

Pia mater   Means "tender mother": innermost of three membranous layers of connective tissue that cover the brain and spinal cord in mammals and birds.

Placode   Area of thickened, neurogenic ectoderm in the developing head.

Plesiomorphy   Adjective: Plesiomorphic. Traits that are present in a taxon or set of taxa that are similar to the respective primitive traits present in the common ancestral taxa, for example, mammary glands and hair are *plesiomorphic* for mammals in that they were present in the common ancestral stock of all extant mammals. See also Apomorphic.

Polyphyletic   Pertaining to taxa that have evolved from more than one ancestral taxon.

Pons   Region of the brain situated between the medulla and the midbrain. Location of many cranial nerve sensory and motor nuclei, reticular formation, and ascending and descending axonal pathways.

Posterior lateral line nerve   One of three postotic lateral line cranial nerves (LL$_p$).

Posterior tuberculum   Ventral, caudal part of the diencephalon; set of nuclei involved in a variety of pathways, including some visuomotor pathways and relay pathways to the telencephalon for the gustatory and lateral line systems.

Preganglionic axon   An axon of the autonomic nervous system that terminates in a sympathetic or parasympathetic ganglion. Sympathetic preganglionic axons, which arise from neurons in the thoracic and lumbar regions of the spinal cord, terminate in sympathetic ganglia located close to the spinal cord. Parasympathetic preganglionic axons, which arise from neurons located in the brainstem and sacral region of the spinal cord, terminate in parasympathetic ganglia that are located close to the visceral organs that they control. See Autonomic nervous system and Postganglionic axon.

Postganglionic axon   An axon of the autonomic nervous system that originates in a sympathetic or parasympathetic ganglion and terminates on smooth muscles or glands. See Autonomic nervous system and Preganglionic axon.

Pretectum   Dorsal, caudal part of the diencephalon; set of nuclei involved in visuomotor and other functions.

Primitive   Early, that is, relatively early in occurrence in the phylogenetic history of a line of descendant species. Sometimes used to mean simple or rudimentary.

Primitive meninx   Single layer of membranous connective tissue that covers the brain and spinal cord in fishes.

Profundus nerve   Cranial nerve (P) that innervates the skin of the snout; called the ophthalmic branch of the trigeminal (V) cranial nerve in mammals.

Projection   Verb: Project. Refers to the group of axons that arises in a given neuron cell group and terminates in another neuron cell group, such as the projection of the retinal ganglion cells to the optic tectum or that the optic tectum projects to a nucleus in the dorsal thalamus. Also see Fiber bundle.

Proprioception   Sense of position of the body and limbs derived from sensory receptors in muscles, tendons, and joints.

Prosencephalon   The forebrain; lies rostral to the midbrain, or mesencephalon; comprises the telencephalon and diencephalon.

Prosomeres   Rostrocaudal segments of the prosencephalon formed during embryological development.

Prototheria   Monotremes—the platypus and echidnas: one of the three radiations of mammals.

Punctuated equilibrium   Evolutionary pattern of maintenance of a given form for a long period of time followed by a short period of rapid change. See also Saltatorial.

Ramus   A branch of a nerve, such as a spinal nerve.

Ray-finned fishes   See Actinopterygians.

Receptive field   The area of the body or space in which stimulation activates a sensory neuron or receptor.

Reptiles   A casual term for a set of animals—diapsid reptiles and turtles—that do not form a biologically unitary group.

Reticular formation   A coordinating system of neuronal populations located in the brainstem and spinal cord that has multiple functions, which include coordination of reflexes and the generation of motor patterns (descending reticular formation), and sleep, arousal, and attention (ascending reticular formation).

Retina   Neuronal layer of the eye containing a complex neural network that includes photoreceptors, bipolar cells, and ganglion cells.

Retinotopic organization   See Topographic organization.

Retrograde   The direction away from the axonal terminal area and toward the dendrites.

Rhinal   Refers to the nose, as in rhinal fissure or rhinencephalon.

Rhipidistians   Group of sarcopterygian fishes that were the ancestral stock of tetrapods.

Rhombencephalon   The hindbrain; lies caudal to the midbrain, or mesencephalon; comprises the metencephalon and myelencephalon.

Rhombomeres   Rostrocaudal segments of the rhombencephalon formed during embryological development.

Sacculus   Bag-like chamber of the vestibular apparatus of the inner ear.

Saltatorial   Refers to sudden or abrupt change or leap, as in a rapid evolutionary change with the abrupt establishment of new species. See also Punctuated equilibrium.

Sarcopterygians   The subclass of Osteichthyes, or bony fishes, that comprise the fleshy-finned fishes, which include coelacanths, rhipidistians, and lungfishes, and which ancestrally gave rise to tetrapods.

Sacral   Refers to the most caudal region of the spine; may be used to indicate vertebrae, nerves, or spinal cord segments.

Scala naturae   Scale of nature. Idea that animals can be ranked along a progressively ascending scale that is defined by human values and with humans ranked at the top.

Secondary meninx   Inner of two membranous layers of connective tissue that cover the brain and spinal cord in amphibians and reptiles.

Segment   One of a series of repeating, similar, rostrocaudally arrayed, morphological units.

Septum   Group of nuclei within the rostral ventromedial wall of the telencephalon.

Serial homologues   Similar, segmentally repeating structures in the body of an individual, such as vertebrae.

Somatic   Refers to structures that develop in the body wall from mesoderm, as opposed to those in the gut (endoderm) or on the body surface (ectoderm).

Somatotopic organization   See Topographic organization.

Somites   Rostrocaudal segments of the dorsal, or paraxial, mesoderm that lie lateral to the neural tube in the trunk and caudal part of the head during development.

Somitomeres   Rostrocaudal, incompletely separated segments of the dorsal, or paraxial, mesoderm that lie lateral to the neural tube in the rostral part of the head during development.

Speciation   Also called Cladogenesis. Process in which a single species gives rise to one or more new sister species or to two or more lineages of new species.

Species   A group of naturally inbreeding or potentially inbreeding individuals that do not naturally breed with individuals outside of the group.

Spinal accessory nerve   Eleventh cranial nerve (XI); innervates neck and shoulder muscles.

Straitum   General area of the basal forebrain, or, more specifically, two areas within the striatopallidal complexes in the basal forebrain: in mammals, the dorsal striatum is the caudate nucleus and putamen, and the ventral striatum is the olfactory tubercle and nucleus accumbens.

Squamates   Order of diapsid reptiles that comprises lizards, snakes, and tuatara.

Sulcus   Plural: Sulci. Valleys between the gyri (ridges) of the cerebral cortex in mammals; particularly deep sulci are referred to as fissures.

Suprasegmental   Refers to axons or reflexes that involve the brain in addition to the spinal cord.

Supratemporal lateral line nerve   One of three postotic lateral line cranial nerves (LL$_{ST}$).

Sympathetic division of the autonomic nervous system   See Autonomic nervous system.

Synapse   A specialized region of part of an axon (or dendrite) that transmits information to the receptive part of another neuron, usually by release of neuroactive substances.

Synapsids   Includes modern mammals; the synapsid skull has a single temporal fenestra, or opening.

Taxon   Plural: Taxa. A unit of classification of organisms, for example, a species, a genus, a class, a kingdom.

Tectum   Roof, or dorsal part, of the mesencephalon, or midbrain; comprises the optic tectum (superior colliculus), torus semicircularis (inferior colliculus), and, in ray-finned fishes, the torus longitudinalis.

Tegmentum   Ventral part of the mesencephalon, or midbrain; comprises the isthmus and the tegmentum proper.

Telencephalon   Rostral part of the prosencephalon, or forebrain; comprises the pallium and subpallium.

Teleostei   The largest of the five radiations of ray-finned fishes.

Terminal nerve   Also called nervus terminalis; cranial nerve (T) that innervates the nasal septum and, in ray-finned fishes, projects to the retina.

Tetrapods   Vertebrates with four feet, whether the feet are retained or have been secondarily lost or modified. Synonym of quadruped.

Thalamus, dorsal   Part of the diencephalon, lying ventral to the epithalamus and dorsal to the ventral thalamus; set of nuclei that relay information to the telencephalon. See page 322.

Thalamus, ventral   Part of the diencephalon, lying ventral to the dorsal thalamus and dorsal to the hypothalamus; set of nuclei, some of which are involved in motor and visual pathways.

Thecodonts   One of the two groups of diapsids, composed of crocodiles and birds.

Therian   Subclass of mammals that comprise metatherians (marsupials) and eutherians (placental mammals).

Third ventricle   Midline cavity present in the diencephalon; contains cerebrospinal fluid.

Thoracic   Refers to the region of the spine between the cervical and lumbar levels; may be used to indicate vertebrae, nerves, or spinal cord segments.

Tonotopic organization   See Topographic organization.

Topographic organization   The preservation of the spatial organization present on a receptor surface, such as the skin (somatotopic), the retina (retinotopic), or the cochlea (tonotopic), in neuronal populations of the central nervous system, such as the somatosensory cortex, the lateral geniculate nucleus, or the auditory cortex.

Topological   Refers to those spatial relationships among components of the nervous system that are independent of size and shape; these include sequential relationships and whether one component is contained within another.

Torus lateralis   Present only in ray-finned fishes; a lateral part of the tegmentum of the midbrain.

Torus longitudinalis   Present only in ray-finned fishes; part of the tectum (or roof) of the midbrain.

Torus semicircularis   Also known as the inferior colliculus in mammals; part of the tectum (or roof) of the midbrain that receives auditory input and, where present, lateral line input.

Tract   A group of axons. See also Fiber bundle.

Trigeminal nerve   Fifth cranial nerve (V); supplies motor innervation to jaw muscles and sensory innervation to the face, snout, and oral cavity.

Trochlear nerve   Fourth cranial nerve (IV); one of three cranial nerves that innervate eye muscles.

Tunicate   A member of the invertebrate chordate group, the urochordates, or sea squirts.

Urodeles   Newts or salamanders: one of the three extant groups of amphibians. Also called Caudata in reference to the presence of a tail.

Ventral root   One of the two roots of spinal nerves (see also Dorsal root); consists mostly of motor axons that exit the spinal cord from its ventral side.

Ventricles   Chambers or cavities: Set of cavities in the brains of vertebrates that contain cerebrospinal fluid and are continuous with the central canal of the spinal cord; also chambers within the heart.

Ventrolateral facial nerve   Division of the seventh cranial nerve (VII$_{VL}$) that innervates taste buds.

Ventrolateral glossopharyngeal nerve   Division of the ninth cranial nerve (IX$_{VL}$) that innervates taste buds.

Ventrolateral vagus nerve   Division of the tenth cranial nerve (X$_{VL}$) that innervates taste buds.

Vertebrae   Set of repeating, segmental, bony units that form the vertebral column, or backbone.

Vertebrates   Group of chordate animals that have vertebrae, that is, backbones.

Vestibulocochlear nerve   See Octaval nerve.

Visceral arches   Series of skeletal arches that support the gills in fishes; the first visceral arch is the mandibular arch, the second is the hyoid arch, and the third through seventh visceral arches are the first through the fifth branchial arches.

Vomeronasal nerve   Cranial nerve (VN) that innervates the vomeronasal organ.

Vomeronasal organ   Accessory olfactory organ typically used for detection of pheromones. Present in many tetrapods. Also called the Organ of Jacobson.

White matter   Tissue within the central nervous system that is composed of myelinated axons.

REFERENCES

Lockard, I. (1992) *Desk Reference for Neuroscience,* 2nd ed. New York: Springer-Verlag.

Margulis, L. and Schwartz, K. V. (1988) *Five Kingdoms: An Illustrated Guide to the Phyla of Life on Earth,* 2nd ed. New York: Freeman.

Purves, D. and Lichtman, J. W. (1985) *Principles of Neural Development.* Sunderland, MA.: Sinauer Associates, Inc.

Walker, W. F., Jr. and Liem, K.F. (1994) *Functional Anatomy of the Vertebrates: An Evolutionary Perspective,* 2nd ed. Fort Worth, TX.: Saunders College Publishing.

Index

Note that this index contains a mix of Latin, English, and other language terms for nuclei, tracts, and other structures. As in the text, we have presented the names of the structures here as we have variously encountered them in most of the English-based literature that we surveyed, rather than following the *Nomina Anatomica* system of Latin names. In this literature, some terms, such as *ansa lenticularis* and *substantia nigra*, most frequently appear in Latin or other non-English language, while others, such as *red nucleus* and *lateral ventricle*, most frequently appear in English. Other terms, such as *optic tectum* and *visual Wulst* are mixes of languages, and still others appear either in English or in other languages with about equal frequency, such as *brachium conjunctivum/superior cerebellar peduncle*. Just as one deals with the variation in the languages of these terms in the literature, we hope that the reader will be able to deal with the similar mix contained herein.

This index includes page listings for many of the structures as they appear in various figures as well as in the body of the text itself. The reader is also referred to a number of additional terms that are listed in the Glossary. Finally, please note that authors are cited here only for direct mention in the body of the text or in figure captions; authors cited in the references at the end of each chapter are not included in this index.